MAIZE: ORIGIN, DOMESTICATION, AND ITS ROLE IN THE DEVELOPMENT OF CULTURE

This book examines one of the thorniest problems of ancient American archaeology: the origins and domestication of maize. Using a variety of scientific techniques, Duccio Bonavia explores the development of maize, its adaptation to varying climates, and its fundamental role in ancient American cultures. An appendix (by Alexander Grobman) provides the first-ever comprehensive compilation of maize genetic data, correlating these data with the archaeological evidence presented throughout the book. This book provides a unique interpretation of questions of dating and evolution, supported by extensive data, following the spread of maize from South to North America, and eventually to Europe and beyond.

Duccio Bonavia (1935–2012) held professorships at Universidad Nacional Mayor de San Marcos, Universidad Nacional San Cristóbal de Huamanga (Ayacucho), and Universidad Peruana Cayetano Heredia (Lima), before he retired in 2005. He served as the Assistant Director of the Museo Nacional de Arqueología y Antropología de Lima and has written fourteen books, including *Perú: Hombre e Historia*, *Mural Paintings in Ancient Peru*, and *The South American Camelids.*

MAIZE: ORIGIN, DOMESTICATION, AND ITS ROLE IN THE DEVELOPMENT OF CULTURE

DUCCIO BONAVIA

Translated by JAVIER FLORES ESPINOZA
with appendix by ALEXANDER GROBMAN

CAMBRIDGE
UNIVERSITY PRESS

CAMBRIDGE UNIVERSITY PRESS
Cambridge, New York, Melbourne, Madrid, Cape Town,
Singapore, São Paulo, Delhi, Mexico City

Cambridge University Press
32 Avenue of the Americas, New York, NY 10013-2473, USA

www.cambridge.org
Information on this title: www.cambridge.org/9781107023031

First published in Spanish as *El Maíz. Su origen, su domesticación y el rol que ha cumplido en el desarrollo de la Cultura* by Universidad de San Martín de Porres 2008
First English edition 2013

Printed in the United States of America

A catalog record for this publication is available from the British Library.

Library of Congress Cataloging in Publication data
Bonavia, Duccio, 1935–
[Maíz English]
Maize: origin, domestication, and its role in the development of culture / Duccio Bonavia.
p. cm.
Includes bibliographical references and index
ISBN 978-1-107-023030-1 (hardback)
1. Corn – History. 2. Corn – America. I. Title.
SB191.M2B68413 2008
633.1′5–dc23 2012007335

ISBN 978-1-107-02303-1 Hardback

For Lucas and Stephen

Contents

Figures List

Acknowledgments from the Spanish Edition

Although it may seem an exaggeration, the acknowledgments are for me one of the things I find the hardest when writing a book. This is because no book is ever the work solely of one author, as other individuals, to whom the author is much indebted, have also taken part in one way or another. And there is always the fear that someone will slip by without being acknowledged. I therefore apologize for any involuntary omission.

Clearly this book would not have been written without the help I had from the Universidad San Martín de Porres. This institution not only took over the publication of this book but, even more importantly, allowed me to devote a whole year to the research and preparation of the manuscript. This is priceless in countries like Peru, where retired university teachers receive no support at all with which to continue practicing their profession. I would therefore like to thank the officials in the Universidad San Martín de Porres who made this project possible, particularly Ismael Pinto, Juan Carlos Paredes, and Sergio Zapata Acha, with whom I had a constant contact and who were able to understand my anxieties and needs.

Some individuals did not directly contribute to the preparation of this book, yet without them it would never have been written. Foremost among them is David H. Kelley, with whom I undertook my fieldwork as a student in the oh-so-distant year of 1958. Kelley had participated in the excavations Richard MacNeish had undertaken in Mexico, and so it was thanks to our lengthy conversations that I first became acquainted with the significance those excavations had in regard to maize. It was Kelley who realized the significance the site of Los Gavilanes – which Edward Lanning had discovered – could have. And it was Kelley who suggested to Professor Paul Mangelsdorf that I take charge of the excavations the Harvard Botanical Museum carried out in Huarmey in the early 1960s. So it was that I established contact with this master scholar, with whom I constantly corresponded almost right up to the moment he passed away. And Mangelsdorf certainly was the major influence that made me continue studying the maize problematic.

In the late 1950s I found a significant amount of maize remains dating to the Middle Horizon while excavating at Miramar (in Ancón) for my B.A. thesis. After making the due enquiries I was told that Alexander Grobman, from the Universidad Nacional Agraria de La Molina, was the person most suited for their study. I reached him, and he agreed to help me, but our contact then was very brief, for he was about to travel to the United States, where he was going to work with Mangelsdorf. We met again in 1963, and it was then that he presented me with a copy of *Races of Maize in Peru*, the book he had written with some other colleagues and Mangelsdorf himself. It was then that our friendship was formed and that we decided to work together researching preceramic maize. Any thanks will therefore always be insufficient for Grobman, because as an archaeologist, without him I would have been unable to understand the complexities that the botanical aspects of a plant enclose. This collaboration is still going on after 47 years of extended study of materials, lengthy discussions, and several publications.

In 1977 I was able to spend some time in Harvard University's Botanical Museum to prepare the final report of the Huarmey Archaeological Project. With Grobman, I took the opportunity to visit Mangelsdorf in Chapel Hill (North Carolina). He had already retired but was still active. After such a long correspondence, personally meeting him proved an unforgettable experience for me. We spent a whole morning discussing the results attained by the research carried out at Los Gavilanes.

Richard Evans Schultes was the director of the Botanical Museum while I was at Harvard. We befriended each other, and he guided me through Harvard's libraries so that I could deepen my studies and expand my knowledge. It was also Schultes who connected me with Elso Barghoorn, with whom we decided to analyze the Huarmey pollen samples, as well as with Umesh Banerjee, who worked with Barghoorn. His influence in my publications through the advice he gave me was also significant.

At the time Margaret Towle was already sick and had secluded herself in her house. I visited her several times, as I was fascinated by her knowledge of ethnobotany. She also provided advice and taught me how I should treat the assemblage of plant remains associated with maize. Her figure remains unforgettable for me.

I owe a special recognition to Richard MacNeish, who kindly sent me the manuscript of the study Walton Galinat made of the maize remains found during the work of the Ayacucho Archaeological-Botanical Project, and who allowed me to cite it. Without his generosity I would have been unable to analyze the Ayacucho samples or to draw the conclusions here presented.

Several colleagues in the United States have continuously helped me by providing me with the data and publications I required. I am particularly grateful to Ramiro Matos, Gary Urton, and Joyce Marcus. Claudia Grimaldo likewise provided me from England with several publications that I was missing.

I am particularly grateful to Pierre Drapeau, who for many years has been sending me from Canada the journals *Science* and *Nature*, thus allowing me to keep abreast with the latest advances in my field, and to amass a large part of the data used in preparing this book. This task is now being carried out by Thomas Fisher, my son-in-law.

Santiago Uceda helped me every time I needed any information on the North Coast. Joyce Marcus and Kent Flannery kindly provided me the photograph of the Guilá Naquitz Cave. The Universidad Nacional Agraria de La Molina gave me permission to reproduce on the cover of the Spanish edition of this book the beautiful charcoal drawing made by Sabino Springett, one of the most renowned Peruvian artists. Carlos M. Ochoa provided me with photographs of Cuzco maize.

One of the hardest tasks in Peru is having access to the bibliographical data required. I am thus forever grateful to Ramiro Castro de la Mata, who gave me unlimited access to his library, which clearly is one of the most complete ones in Peru on historical works. Fernando Silva Santisteban is another friend and colleague who helped me in this. Here I must also mention Ricardo Sevilla, who loaned me some specialized studies of maize that I was missing.

Some sections in the book deal with subjects of which I have an insufficient grasp, and that always leave me with lingering doubts. In these cases their revision by a specialist is of the utmost importance. I am grateful in this regard to Ramiro Castro de la Mata, Elmo León, José Iriarte, Alexander Grobman, and Uriel García Cáceres, who read some parts of the manuscript and made invaluable suggestions.

Mercedes Quispe Palomino was most helpful whenever I had doubts or problems with Quechua terms.

The bibliography clearly is the spine of a study of this type and therefore has to be as accurate as possible. I am most grateful to Juan Yataco for having helped me check the bibliographical data.

But in the modern world there is another type of support that proves essential, particularly for the elderly, that is, the intricacies of computers. The help of Ricardo Solís was invaluable in this regard.

Another essential support, without which one cannot find the peace of mind required for writing a science book, is that of the family. The permanent support I had from my children, Bruna and Aurelio; my son-in-law, Thom; and my two grandchildren, Lucas and Stephen, who have managed to be always close to me with their encouragement and affection despite the distances that separate us, are the pillars on which this book was raised. And there also is someone who is no longer here but was ever present – my late wife, Anna, who was in all of my publications a faithful partner, a meticulous secretary, and an astringent critic. Her words have lingered on and provided me the encouragement I needed whenever problems arose, the deadlines loomed closer, and it looked as if the book would never be completed.

Last of all I would like to thank all those colleagues who have harshly and unfairly criticized the position Grobman and I had, or have simply ignored us because we are not mainstream. They unwittingly provided the encouragement I needed to conclude this synthesis, which is the result of many long years devoted to the study of maize. Now the task is in the hands of young archaeologists. However, although in some pages I have been extremely harsh, this was only because my duty as a man of science demanded it, yet nothing in these harsh judgments was personal.

Postscript. Correcting a book and organizing its contents according to editorial guidelines is extremely taxing, because one must be not just extremely well prepared but also committed to the text in order to avoid changing the author's ideas. This task is even more complex in a specialized publication like the present one. The work done by Juana Iglesias in this regard is exemplary. So I must acknowledge that in all of the experience I have had with publications of this type, I have never met a person as versed in this as she is.

Duccio Bonavia

Acknowledgments to the English Edition

One of the biggest satisfactions any author can have is that of revising his work, correcting it, and particularly updating it. When Father Johan Leuridan Huys, the dean of the Facultad de Ciencias de la Comunicación, Turismo y Psicología, in the Universidad de San Martín de Porres, asked me to prepare an English edition of this book, I realized that I would be able to carry out all of these tasks. The truth is, I cannot find words to express my recognition.

I would have found it very difficult to undertake the study of maize without the help I had from Alexander Grobman, as has already been noted in the acknowledgments to the Spanish edition. His help in this new edition has once again been crucial. Even more importantly, he agreed to write the appendix on genetics, for which I am most grateful, as it is a subject I cannot dwell on.

Tom Dillehay and Jack T. Rossen kindly allowed me to cite unpublished data from their research in the Zaña Valley. Ramiro Matos, Elmo Léon, Joyce Marcus, Adolfo Gil, Rodolfo Rafino, and Britta Hoffmann provided me publications that are unavailable in Peru. Gonzalo Castro de la Mata allowed me to peruse his father's library. Any recognition in this regard is insufficient.

Last of all I would like to thank Javier Flores Espinoza for having accepted the daunting challenge of translating such a specialized work as this book is.

Duccio Bonavia

We note with great sadness the passing of Duccio Bonavia on August 4, 2012, at Magdalena de Cao, Department of La Libertad, in northern Peru. He had returned from his retirement in Canada to put some finishing touches on the archaeological project at Paredones and Huaca Prieta, which he co-directed with Tom H. Dillehay. His last stand was near the archaeological diggings in Peru, which permitted him to unlock the secrets of the past and to open a treasure chest of knowledge about the early cultures of the Peruvian coast, which were waiting to be exposed.

This book, one among many others derived from his research and writing, was, as he confessed to me, the pinnacle of his work and his greatest achievement during a lifetime devoted to archaeology and, more specifically, to the understanding of the role of maize in the evolution of cultures in the American continent.

As, due to his passing, Bonavia was unable to review the final corrections, this crucial work was done by Ms. Laura Wilmot with the assistance of Bruna Bonavia-Fisher and Alexander Grobman. They both wish to thank and commend Ms. Wilmot for her excellent contribution in editing, reviewing, and correcting the grammar, minor inconsistencies, and other details of the text of the two parts of this book and in coordinating the index.

As a result of his last work at the Paredones and Huaca Prieta sites in northern Peru, Bonavia coauthored a recent article, which was published in early 2012, while the present book was in publication. This article is of great significance and supports the main hypothesis of the evolution of maize in South America. A summary of this research is inserted at the end of the appendix.

May this book serve as a lasting milestone of the advances forwarded to science by Duccio Bonavia on the road to achieving a greater understanding of the evolution of early cultures in Peru and the Andean region.

Alexander Grobman
Lima, October 2012

1

The Maize Problematic

Maize is "... a real triumph of plant breeding (or the luckiest of accidents)...."
Harlan (1995: 187)

There can be no question that maize is one of the plants that has had a major role in the development of American cultures. Ortiz (1994: 527) correctly noted that from the Andes to Mesoamerica, and from the Caribbean to the southeast of the Woodlands, maize enabled the development of high cultures as concentrations of large populations, besides allowing them to settle down.

The subject of maize is quite complex and covers several areas that concern different disciplines, from biology to history, and although it is true that a vast number of articles and books have been written on it, it can still be said that all of these areas have not been collected into one single book. I would therefore like to begin by pointing out that this book does not pretend to be complete, nor can it be so. All it intends is to present an overview of the main issues related with this plant, at the same time paying special attention to its problematic in South America, because although much has been written on this area, as yet no attempt has been made to present a synthesis.

Throughout the text the terms "gathering," "farming/cultivation," and "domestication" will often be used. These are three words commonly used but that also often conceal some confusion. Yet all three are essential to understand the way in which a plant passed from its wild state to that of a crucial tool for mankind. "Gathering" simply means collecting and harvesting the native flora just as it appears in nature, without introducing any change to it. "Cultivation" is the act through which man manipulates the natural distribution of a plant by taking it to an environment chosen and prepared by humans so that it will reproduce, thus avoiding the competition of other species. Many plants do not change when subjected to this process, so for archaeologists it is often difficult to realize when the microenvironment has been created by humans. "Domestication" is a far more complex process wherein man handles the process of growth of a plant and plays with its genetic plasticity, so that in time a series of modifications are introduced into the plant that may lead to extreme biological changes, even some that are antinatural, and that turn the plant into an artifact. In this way mankind attains the best productive conditions for those characteristics that are of interest to it. This process was taken to extremes with maize, which as we shall see has been turned into a plant that cannot reproduce itself without human intervention. The way in which man developed this phenomenon is quite complex, but it is essentially based on the genetic plasticity of plants through two essential mechanisms: selection and hybridization. This in turn leads to some alterations in their biological mechanisms. In some cases the ability to produce seeds is lost, as has happened to the oca (*Oxalis tuberosa*) and the ulluco (*Ullucus tuberosus*); in others the ability to produce viable seeds is lost, as is the case with añu (*Tropaeolum tuberosum*), achira (*Canna* sp.), and pepino (*Solanum muricatum*), or as happened to maize, which has lost its ability to disperse its seeds. This is a long process, and it is only in some cases that this is a process in which it was not mankind but nature who intervened, as when genetic mutations take place.[1]

Mangelsdorf (1974: 9) wrote the following in this regard:

> The ear of corn enclosed in its husks has no close counterpart elsewhere in the plant kingdom either in nature or among other cultivated plants. It is superbly constructed for producing grain under man's protection, but it has a low survival value in nature, for it lacks a mechanism for dispersal of its seeds. When an ear of corn drops to the ground and finds conditions favorable for germination, scores of seedlings emerge, creating such fierce competition among themselves for moisture and soil nutrients that all usually die and none reaches the reproductive stage.

Bugé (1974: 35) believes that men in preceramic times were real biologists, in that they worked not with a static material but with a veritable process. They must have considered the production and maintenance of the different races of maize as a result of the interaction of a series of mutations, of a haphazard genetic deviation, of a natural selection, and of a hybridization. In other words, as the result of a succession of biological processes that accelerated or inhibited the attainment of certain goals that culture was establishing.

Pickersgill, one of the most renowned students of these processes, says there are four questions one has to ask when studying the origin and the evolution of cultivated plants. The first question is what plant gave rise to the modern plant; second, where it was domesticated; third, when this took place; and fourth, how these plants have changed and whether they have spread since the beginning of their cultivation (Pickersgill, 1977: 591). These are the questions I try to answer, specifically in regard to maize.

The reproductive characteristic of grasses is that they freely scatter the seeds. When man intervenes, selecting and planting, the plant depends on him, and

[1] For more information see, e.g., Harlan (1992), Helbaek (1953), Sanoja (1981: 73–74), and C. Smith (1967: 223).

the visible characteristics – the phenotypes – that man has decided to select compromise its autonomous survival. One of these characteristics is the production of more seeds through the female inflorescence. The increase in the number of seeds in the corn cob is obtained through greater condensation, that is, the number of kernels per row and the number of rows. In the case of this plant, domestication essentially consists in the elimination of the characteristic of seed dispersal through the natural separation of the rachilla in which they are inserted, and an expansion of their inclusion in the rachis in order to offer a more secure harvest for man. These characteristics are found in modern maize, and this is one of the characteristics, as we shall see, that radically distinguish it from its closest congener, teosinte, whose seeds are dispersed on reaching maturity by the fragmentation of the rachis (see Grobman, 2004: 428).

In maize, the crucial environmental phenomena are variations in temperature, moisture, the photoperiod, and the length of the day (Purseglove, 1972: 310–311). The advantage we have is that the ecological transformations that took place in the Holocene, particularly as regards temperatures and patterns of precipitation, are well documented. They may have had a role in the development of maize agriculture, and students should keep them in mind (Benz and Long, 2000: 462).

Mangelsdorf, who clearly is one of the most important and renowned students of maize, was convinced that what he called "the invention of maize culture ..." had two mothers. On the one hand, there was necessity, and it is probable that maize was originally never abundant in nature, so that it could go extinct if it was taken out of its natural habitat. And on the other hand, there were the shrewd observations made by the Indians, who noticed first that this plant had a different behavior in the fields cleared close to their encampment, and then that its planting led to a selection that enabled the preservation of the mutants chosen by man (Mangelsdorf, 1974: 167, 207–208).

The closest relative of maize is teosinte, and it will be discussed at length further on. But we should realize that the main problem when comparing these two plants is the differences in the structure of the inflorescence, that is, the ear. The major problem is that in teosinte the kernels are tightly encased inside the structures called cupulate fruitcases, whereas the kernels of maize are born uncovered on the surface of the ear. The domestication of maize brought a change in the development of the ear, for the cupules and the glumes formed the internal axis of its ear instead of casing the kernels. This is why H. Wang and colleagues (2005: 714) pointed out that "in a sense, maize domestication involved turning the teosinte ear inside out." In fact the maize cob, be it either a pod corn or a normal corn, can hardly be a functional design for seed dispersal that appeared as a result of natural selection. Its new shape does not fit in an evolutive sequence and instead represents a terminal descendant of one of the sequences. Its proliferation and the concentration of grain-bearing spikelets can be ascribed to an unconscious selection – albeit a deliberate one too – by man in his concern for finding ever more and better food (see Galinat, 1975a: 318).

Not all scholars agree on the way in which the domestication of maize took place. Some are inclined to accept the mechanisms of evolution and ecology as decisive, whereas others believe that the cause was essentially human intention that knew how to use the climatic variability. But it must be pointed out that it is likely that a chance genetic drift may have been a major factor that brought about the changes that can be seen in the development of this plant. It is very possible that the interaction of these factors combined in the process of change of the ear (Benz and Long, 2000: 460–464; Flannery, 1986a; Rindos, 1984; Tarragó, 1980: 182; Watson, 1995). Johannessen (1982: 97) accepts that there may have been in principle an unconscious selection, but it cannot have taken place in the case of the development of the large-seeded maize with the long, colored, strong ears and the many varieties that have appeared. He believes that this was only possible through a continuous and conscious selection. Iltis (1987: 208) and Grobman (2004: 428) concur, but Wilcox (2004: 145) believes that "... domestication is just one factor that could affect grain size; others include environmental conditions, genetic variability, crop processing and, for archaeological material, conditions of charring."

R.-L. Wang and colleagues (1999: 236) drew attention to the fact that domestication can strongly reduce the sequence of diversity in the genes controlling the traits that are of human interest, in that when the selection is strong, domestication has the potential to drastically reduce the genetic diversity of a plant. Doebley (1994: 106, 112) showed that when the human selection of the ear is strong, the evolutive changes will take place very fast, whereas when it is weak, they will take place quite slowly. This is why he believes that it is inappropriate to simply assume that the races of maize with similar ear morphology are phylogenetically united. This assumption is probably wrong when comparing maize from different geographical regions, different altitudinal zones, and different moments in time. Doebley points out that one must not forget that similar morphological forms may appear independently in different geographic regions.

Iltis analyzed the factors of human selection in domestication and concluded that, in maize, the major traits that appeared in domestication were the following:

1. An increase in the number of rows and kernels and in the size of the ear.
2. A hardening of the cupules and the glumes.
3. The development of tough cobs that do not disarticulate.
4. Naked, free-threshing kernels.
5. A decrease in the primary branches, that is, in the number of ears.
6. A condensation of the primary branches and the internodes of the ear.
7. An increase in leaf sheath size and number.

8. The total deletion of the peduncles of the tassels and of the space between the branches.
9. The suppression of all branches in the lateral tassels.
10. The suppression of all the lower orders of the lateral branches, including the inflorescences.
11. The synchronization of the maturing of the grains in an ear, a plant, and a field.
12. The evolution of ecogeographic and genetic-isolating mechanisms that prevent backcrossing to the ancestral teosinte[2] and thus lead to race formation. (Iltis, 1983b: 892)

Following Rindos (1984: 164–166), Benz and Long (2000: 460) suggest that the highest proportion of evolutive changes in maize took place before 5000 years BP, and they posit that the morphological modifications reflect an agriculture under domestication. In this they agree with Jaenicke-Deprés and colleagues (2003: 1208), who reached the conclusion that 4,400 years ago, early farmers already had the potential to produce a substantially homogeneous effect on the allelic diversity in three genes associated with the morphology of maize and with the biochemical properties of the cobs.

There can be no doubt, as Doebley (2006: 1318) points out, that of the achievements of the ancient farmers, the domestication of cereals is one of the major ones, that is, the triad rice-wheat-maize, which has supplied more than 50% of the calories consumed by humans. When compared with their ancestors, one finds that cereals have more grains; that these are bigger, the stalks are thicker, and the seeds are freely threshed from the chaff; and furthermore that their favor has grown. Besides, these cereals, just like other cultivated plants, have one more factor that is essential – their grains remain attached to the plants and have to be harvested by humans, instead of the seeds being scattered, as is the case in wild plants. Although it is known that these phenomena take place through a change in a small number of genes, their nature and the internal molecular variations are still not well known.

Pääbo (1999: 195) based his work on the work of R.-L. Wang and colleagues (1999), its tentativeness notwithstanding, and believes that the domestication of maize was quite rapid and that it could have taken place in a few hundred years.

Hilton and Gaut (1998) made a genealogical study of the *Zea* genus in order to contrast an artificial speciation with a natural one. There are three reasons why this work is not valid. First, for the problem raised by the antiquity of maize, they used a bibliography based on indirect data, and they did not use original sources. Second, the samples of maize they used in their experiment were not well chosen. There is no way of knowing what races they mean (see

[2] Iltis accepts that maize was generated from teosinte, a point on which not all specialists agree.

op. cit.: table 1, 864). Finally, their bibliography comprises 55 entries, but only one of them (Goloubinoff et al., 1993) is on South America.

The Geographical Distribution of Maize

Maize was under cultivation from Canada to Chile at the time the Spaniards arrived on the American continent. At present we find maize from 58° northern latitude in Canada and Russia to 40° southern latitude in the Southern Hemisphere. It grows below sea level in the plains of the Caspian Sea depression, and at altitudes over 3,600 meters above sea level (masl) in the Andes. It lives in zones that receive less than 2.5 cm of annual rainfall in the semiarid regions of the Russian plains, as well as in others with more than 1,000 cm of annual rainfall on the Pacific coastlands of Colombia. It grows in the short summers of Canada, as well as in the perennial summers of the tropical equatorial regions of Ecuador and Colombia. No other cultivated plant grows in such a large area, and only wheat takes up a larger surface area in acres. In fact, maize is maturing in some part of the world, in all longitudes, all year long (Mangelsdorf, 1974: 1–2; W. L. Brown et al., 1988: 8).

Description of the Plant

> It is an annual plant of fasciculated roots, whose stalk also has the property of forming adventitious roots. The stalk is a massive cane with a white and sugary medulla. A sheathing leaf appears on each node that is ligulate, strip-like and linear with parallel nervations. It is a monoic plant whose male flowers are born before the females in the tip of the stalks, thus forming a spike tassel. Female flowers are born in the axil of the leaves towards the mid-point of the stalk, and are grouped in rows along a thick, cylinder-like, spongy and alveolate rachis, which in some countries is called *olote* and *zuro*. The female flowers are sessile so that this inflorescence actually is a real female ament that is vulgarly known as the ear (*mazorca*, *panoja* or *choclo*); the latter is protected by large papyraceous bracts that are usually known as husk (*camisas*, *tusas* and *hojas de choclo*). Each female flower ends in a fluffy and very long (15 centimetres and more) filiform and hairy style; the styles of all the flowers come out through the end of the bracts and are first green and then reddish[3] on reaching maturity, and are known as silks (*barbas de maíz*, *pelo de choclo* and *cabello de elote*); the last name [*cabello de elote*] is because in some countries the green ears of maize are called *elote* [Mexico], and *jojoto* [Venezuela] in others,[4] which are taken as food when cooked. (Cendrero, 1943: 202)[5]

[3] This depends on dominant or recessive color genes for anthocyanin.

[4] *Choclo* is used in the central-southern Andean area.

[5] For a more detailed description, see Mangelsdorf (1974: 5–9) and Johnson (1977).

There are two positions as regards the origins of the ears of maize. One of them holds that these originated due to modifications of the pistillate inflorescence of teosinte, through a small number of key morphological changes controlled by an equally small number of major genes (Beadle, 1980; Galinat, 1983, 1985a, 1988a; Langham, 1940). The second position holds that the primary lateral inflorescence of the central spike of teosinte was transformed into the ear of maize through sexual transmutation (Iltis, 1983b).[6]

Origin of the Name

In the seventeenth century Father Bernabé Cobo, that "scientific precursor," as Porras (1986: 510) called him, wrote:

> The name of *maíz* [maize] is from the language of the Indians from the island of Hispaniola. Mexicans call it *tlaolli*, and [the Indians of] Peru *zara* in the Quechua language, and *tonco* in Aymara. The Indians of New Spain call the ears of maize *elote*, the Peruvians *choclo*, and the kernel-less heart of the ear *coronte*, which is used as fuel. The husks of the ears are very useful for the muleteers, because they fill the packsaddles with them and they remain very light. (Cobo, 1964a: 162)

Specialists agree that the word "maize" comes from Taino or Carib, where the plant was know as *mahiz*. Taino was the language spoken by an elite group of the Arawak (Beadle, 1972: 3; Ortiz, 1994: 528). Some, however, claim that the term is Arawak – *marise* – and that it became *mahiz* in the Antilles (Horkheimer, 1958: 37).

The Maya terms for maize were *Ixim*, which is a general name; *Zac ixim*, which means white maize; *Peeu ixim*, small or early maize; and *Xacin*, which are the black and white kernels (Marcus, 1982: table 1, 241). *Ixim* was also the name for the Maize god. The kernels were called *nel*, a term that means "place" in the Maya texts. The rest of the ear with the kernels removed, that is, the cob, is called *b'akal*, just like the ancient name of Palenque (Antonio Aimi, personal communication 11 October 2006). We must bear in mind that all the languages in southeastern Mesoamerica, that is, in Guatemala, Belize, and Mexico to the east of the Isthmus of Tehuantepec, including the region occupied by Mixe speakers, are all members of the Maya and Mize-Zoquean language families. During a period of 400 years, the Maya family included 29 different languages spoken in numerous communities in Mexico. It is estimated that one more became extinct since the Conquest. There are 12 Mize-Zoquean languages, one of which also disappeared after the Conquest. They are mostly found in western

[6] Readers interested in details regarding the structure, growth, and reproduction of this plant should read Kiesselbach (1949), Sass (1955), and Weatherwax (1955).

Chiapas, southwestern Oaxaca, and southeastern Veracruz. All of them have a rich terminology related with maize and its uses (Stross, 2006: 578, 581).

In Nahuatl, maize is known as *cintli* or *centli* and *teocentli*. It was the food of the gods.[7]

For the Quechua vocabulary we have the *Lexicón* Friar Domingo de Santo Tomás published 47 years before the vocabulary of Diego González Holguín (Porras, 1951: XV, XVII). Here we read, "çara, 'maize, the wheat of the Indians'" (Santo Tomás, 1951: 163, 249). González Holguín (1989: 79, 579) in turn wrote, "*çara.* Maize. *çara çara.* Maize in piles. *Viñak çaraçara*, Maize fields in canes or standing." "*Mayz. Çara*, kernel corn, *muchhascca çara, o ttiuçara.*" Interestingly enough, the words used to define maize in the Lake Titicaca basin have remained separate in Quechua and Aymara, the coexistence of these languages there since at least Inca times notwithstanding. In Aymara the term used is *tunqu*, and in Quechua *sara*. Yet the Aymara term is unknown in Cuzco, whereas in Copacabana the Quechua term is not known (Chávez, 2006: 624).[8]

Curiously enough, in the United States *Zea mays* is known as "corn," whereas in the rest of the world the terms "maize" or "Indian corn" are preferred, because in many countries the term "corn" is a synonym of "grain" (Mangelsdorf and Reeves, 1945: note 2, 235).

Taxonomy

The studies of the most distant relatives of maize are too general, and some genera have only rarely been studied scientifically (see Goodman, 1988: 203, and his bibliography). There are also some disagreements as regards the nomenclature, as Goodman (op. cit.: 204, 205, and table 1) pointed out.

In 1753 Linnaeus classified *Zea mays* in his *Species Plantarum* (Towle, 1961: 20). Maize and teosinte were long classified in two different genera, *Zea* and *Euchlaena*. It was in 1942 that Reeves and Mangelsdorf included teosinte in *Zea* (Iltis and Doebley, 1984: 591).

Zea mays L. belongs to the Maydeae tribe in the Poaceae family (Gramineae). The genus *Zea* comprises four species: *Z. diploperennis* Iltis, Doebley, and Guzmán, the perennial teosinte diploid; *Z. perennis* (Hitchcock) Reeves and Mangelsdorf, the perennial teosinte tetraploid, now extinct in nature; *Z. luxurians* (Durieu and Ascherson) Bird, the teosinte of Guatemala; and *Z. mays* or maize. This last species has been subdivided by Iltis and Doebley into *Z. mays* L. ssp. *huehuetenangensis* (Iltis and Doebley) Doebley, the teosinte from Huehuetenango; *Z. mays* L. ssp. *mexicana* (Schröder) Iltis, which corresponds

[7] For the linguistic terminology in Mesoamerica and North America, see Hill (2006).

[8] For the names of the varieties of maize in Quechua, Aymara, and A'karo, see Mejía Xesspe (1931: 13).

to the Nobogame race of the annual teosinte; *Z. mays* L. ssp. *parviglumis* (Iltis and Doebley), that is, the Balsas race of the annual teosinte; and *Z. mays* L. ssp. *mays*, common maize (Grobman, 2004: 429–430).

Wilkes (1967) proposed another classification: *Zea mays* L., maize; *Z. mexicana* (Schröder) Kuntze, that is, the annual teosinte; *Z. perennis* Reeves and Mangelsdorf, the perennial teosinte tetraploid; *Z. diploperennis* Iltis, Doebley, Guzmán, and Pazy, the perennial teosinte diploid, which he believes is the most primitive form of teosinte (Grobman, 2004: 430).

The Maydeae tribe comprises seven genera, of which only two – *Zea* and *Tripsacum* – are American. The rest are Oriental: *Coix*, *Chionachne*, *Schlerachne*, *Trilobachne*, and *Polytoca* (Galinat, 1977: 1).

One important concept that is used in this book has to be explained: race. This term, which is not much used in botany, is widely employed in the case of maize, and it often causes confusion, because it is mistakenly believed that this term is only used for humans and some other animal species.

In the case of maize, several authors have presented definitions of race. For Anderson and Cutler (1942: 71) it is "... a group of related individuals with enough characteristics in common to permit their recognition as a group.... From the standpoint of genetics, a race is a group of individuals with a significant number of genes in common, major races having a smaller number in common than do sub-races." Grobman and colleagues (1961: 51), following Mayr (1942), define it as "... an actually or potentially interbreeding population, one of the several which may form a species distinguished by having in common certain morphological and physiological traits, and, therefore, also having in common the genes which determine these traits."

This concept arose due to the many problems taxonomists had in subdividing a single species with such a vast and complex interfertilization. In 1899 Stutervant attempted a classification, and he separated pod maize from popcorn, dent corn (the ordinary maize used as fodder), flint corn, flour corn, and sweet corn. This terminology is still used in trading or by individuals who do not know botany. In 1942 Edgard Anderson and Hugh C. Cutler noted that Stutervant's classification was artificial, as it only considered the characteristics of the endosperm, whereas the full genotype had to be considered. The first complete classification of Mexican maize was made in 1943, an endeavor that reached its climax with the publication, in 1951, of Wellhausen and colleagues, *Razas de maíz en México*, which was sponsored by the Mexican Secretariat of Agriculture. This was widely applied in America, and 11 volumes were published in which 305 races of maize were defined and named (Mangelsdorf, 1974: 101–105; Sánchez Gonzales, 1994: 139). This major project was carried out by the Committee of Preservation of Indigenous Strains of Maize within the Agricultural Board of the Division of Biology and Agriculture of the National Academy of Sciences, National Research Council. The committee was headed by Ralph E. Cleland, with J. Allen Clark as executive secretary. Its members were Edgar Anderson,

William L. Brown, C. O. Erlanson, Claud L. Horu, Merle T. Jenkins, Paul C. Mangelsdorf, G. H. Stringfield, and George F. Sprague. They also had the support of the Rockefeller Foundation.

Four racial groups have been separated in Mexico and Central America. One is in western Mexico and includes the Chapalote Reventador and the Harinoso de Ocho. A second one belongs to the highlands of central and northern Mexico with the Grupo Cónico and the Sierra de Chihuahua. A third one is found in middle to low altitudes, from southern Mexico to Guatemala, and has three subgroups: (1) tropical dent corn (*dentados tropicales*), (2) a late-maturing group, and (3) short-maturity races adapted to low elevations and distributed above all on the coastal plains of the Pacific Ocean. The fourth group of mid- to high-altitude races extends from southern Mexico to Guatemala and is represented by the Serrano-Olotón type. There are more than 60 racial types in Mexico and Central America (Sánchez González, 1994: 154–155).

There are 32 races in Mexico that correspond to four major groups: Ancient Indigenous, Pre-Columbian Exotic, Prehistoric Mestizos, and Modern Incipient (Wellhausen et al., 1952: 146), whereas in Central America 25 races have been identified (Wellhausen et al., 1952). Hernández and Alanís (1970) added 5 more races for northeastern Mexico, and Benz (1986) described 5 new races (4 are the ones not defined by Wellhausen et al., 1951) and 3 new types (Sánchez González, 1994: 139, 141).

Eleven races were distinguished in southwestern North America (Adams, 1994).

The racial differentiation in the Andean region is remarkable. Goodman and Brown (1988) have pointed out that of the 252 races of maize known (here they disagree with Mangelsdorf, 1974: 103, who claims there are 305), 132 belong to the Andean region. These races have been extensively described (Grobman et al., 1961, Peru; Roberts et al., 1957, Colombia; Rodríguez et al., 1968, Bolivia; Timothy et al., 1963, Ecuador. For Brazil and other countries in eastern South America, see Brieger et al., 1958; for Venezuela, Grant et al., 1963; for Chile, Timothy et al., 1961). (See also Sevilla, 1994: 233.)[9]

Wittmack (1880–1887, 1888) was the first to present a classificatory outline of Andean maize developed from archaeological samples found at Ancón. He based his findings on morphological characteristics, on the shape of the ear, and on the characteristics of the kernels. He distinguished three groups:

1. A common maize he called *Zea Mays vulgata*, with kernels that are neither dented nor pointed and are of a somewhat irregular shape.

[9] To avoid misunderstandings, readers must bear in mind that when citing an author, the bibliography the latter used (the first set of parentheses) is often given before the actual reference for what I am citing (the second set of parentheses). For instance, if we read "(Soares de Sousa, n.d.: 1). (Goodman, 1988: 198)," it means that I am citing Goodman (1988), who in turn cites Soares de Sousa (n.d.).

2. A maize with short ears and pointed or peaked kernels that he called *Zea Mays peruviana*.
3. A variety called *Zea Mays umbilicata*, with kernels that have a groove on their outer surface.

Wittmack considered those forms transitional, with intermediate types or hybrid types among those described. The illustrations he left us of maize are of an astounding quality. Rochebrune (1879), Costantin and Bois (1910), and Harms (1922) later followed Wittmack's classification, albeit with some variations, to classify intermediate groups (Towle, 1961: 22).

Grobman and his team classified the races present in Peru in six major groups:

1. Primitive races believed to be of great antiquity due to their morphological characteristics and on the basis of archaeological specimens (5 races; see Figure 1.1, the specimen on the right).
2. Races derived in antiquity or that are primary races, that is, those that originated directly in pre-Hispanic times due to isolation, hybridization, or selection from primitive races (19 races).
3. Secondary races or races lately derived. Their origin can be outlined based on the primary races that appeared frequently after the Spanish conquest (9 races; see Figures 1.1, the specimen on the left, and 1.2).
4. Introduced races. These have been imported and retain a different morphology in the plant and the kernels despite their genetic exchange with native races (5 races).
5. Incipient races. Those that emerge in the present day as new racial entities or were well established and characterized in recent times (5 races).
6. Imperfectly defined races. These are races that have a limited geographic dispersal and some that are in an incipient state of development, and that besides do not have well-defined characteristics (6 races).

We thus have a total of 49 races. For a complete listing of the specific names of each of them and their characteristics, readers should consult Grobman and colleagues (1961: 138–336). A major detail that is worth pointing out is that the Cuzco Cristalino Amarillo race is one of the races that live in the highest-altitude environment in the world (Sevilla, 1994: 238).

I want to emphasise that popcorn – which has small and hard kernels that explode with heat, and which in Peru is known as Confite – is one of the "primitive races" defined by Grobman and colleagues. But what has to be emphasized is that the primitive races of other American countries are also popcorns, and some of them have endured from preceramic times to the present day (Grobman et al., 1961: 141). Pod corn is another peculiar type in which individual kernels are enclosed inside bracts known as glumes. This is also a primitive characteristic that is quite common and almost universal in wild grasses (Mangelsdorf, 1974: 75).

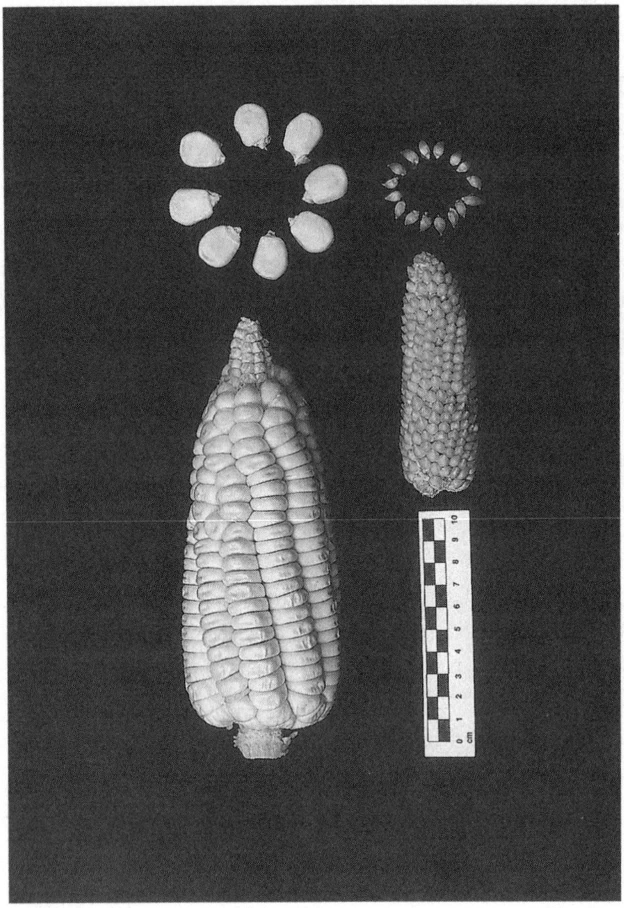

1.1. A sample of modern maize specimens. The small cob on the right is of the Confite Puntiagudo race (see Grobman et al., 1961: 149–154), and the big one on the left is of the Cuzco Gigante race (see Grobman et al., 1961: 295–299).

Pickersgill (1969: 58) has shown that there is a controversy regarding whether the pod corns found in South America are the same as the early pod-popcorns that gave rise to cultivated maize. Weatherwax (1954: 160–170) claims that the current pod corns are "monstrous" forms similar to other known maize mutants and are therefore irrelevant in a discussion of this plant's domestication. Very ancient maizes have vestigial glumes of pod corns or are "half-tunicate." Popcorns clearly are primitive types, but Pickersgill (1969) points out that it is hard to accept that Peruvian popcorns are more primitive than Mexican ones. But Pickersgill said this when Peruvian preceramic popcorns were still unknown, and she based her work on current ones.

In the 1970s Wilkes made a comparative analysis of Mexican and Andean races and came to a conclusion that is extremely interesting. Of the 32 races found in Mexico, 7 have a counterpart in Guatemala, 6 in Colombia, 5 in Peru, and 2 in Brazil. But even more important, 20 of the 32, that is, 62.5%, Mexican

1.2. Another sample of modern maize specimens. The first cob on the right is Cuzco Gigante; the two in the centre belong to the Cuzco Gigante subrace known as Saccsa (see Grobman et al., 1961: 299–300); and the one on the left is of Cuzco Gigante Amarillo, a hybrid race derived from Cuzco Gigante and Cuzco Cristalino Amarillo (see Grobman et al., 1961: 300–301). Photograph by Carlos Ochoa.

races are endemic. In the Peruvian case, he took 48 races into account, 30 of which, that is, 62.5%, are endemic (Wilkes, 1979: 5). These are the same ratios. Ten years later Wilkes made the same reasoning, but for unknown reasons – and which I was unable to determine – he ascribed 50 races to Mexico (a figure that does not agree with the data in Wellhausen and colleagues, 1951, not even if we add to them those in Hernández and Alanís, 1970, and Benz, 1986; see above), 27 of which, that is, only 54%, he claims are endemic, in comparison with 62.5% in Peruvian races (Wilkes, 1989: 445).

All of the major, contemporary commercial types – dent corn, flint corn, flour corn, popcorn, and sweet corn – existed at the time the Europeans arrived in the American continent.

The great variety in *Zea mays* L. notwithstanding, all of them hybridize, and the hybrids are all, almost without exception, fully fertile.

There is no way to establish whether maize really was a mutable species in nature. All we know is that modern *Zea mays* has mutant forms, hundreds of which have been described by geneticists. It is a fact that modern maize is a mutable species, or it at least has some highly mutable loci. Some of the mutants that appear in maize tend to make the plant more useful to man, usually at the expense of its ability to survive in nature (Mangelsdorf, 1974: 2, 133).

Galinat (1985b: 271–272) therefore says that "maize appears to be the only example of a new species or subspecies created directly by human selection."

2

Maize as Seen by Europeans

The First News

We shall see that some have accepted the possibility that maize does not have an American origin, and that it may have been known before the discovery of America. But as Mangelsdorf (1974: 1) correctly pointed out, the lack of references to this plant prior to 1492 is the best proof that this was not so.

It has been claimed that maize existed in China before the discovery of America, but this has been shown to be groundless (see Chapter 8). Some linguists have tried to prove that maize was known both in the Old World as well as in Africa before 1492, but this again proved groundless (Manlgelsdorf, 1974: 2).

It was Fernández de Oviedo y Valdéz, as Horkheimer (1958: 37) correctly pointed out, who raised doubts in this regard. Fernández de Oviedo claimed that the word *milio* from the East Indies, which Pliny, the famed Roman naturalist, mentions, could have been maize:

> As a follower of Pliny's lesson, here I will say what he points out of the millet of India. I think it is the same thing that we in our Indies call maize. Said author said these words: "Millet from India has come ten years hence, of black colour, large kernels, [and] the cane-like stalk grows seven feet ... and it is more fertile than all barleys. A grain gives sextarii. It is sown in humid places."(3) From these indications I would have it as maize, because if he says it is black, the maize in Tierra Firme is dark purple and reddish, and also white, and much of it is yellow. It may be that Pliny did not see it in all of these colours and just dark purple, which seems to be black. The stalk, which he says is like canes, is just like maize has it, and whosoever saw it in the field when it grows high, would think it is a cane field. The maize here is on the most part bigger than the seven feet he says it grows, and somewhat more, and in other places less, depending on the fertility and goodness of the soil sown. As for what he says that it is extremely fertile, I have already pointed out what I have seen, i.e. the harvesting of eighty and hundred, and a hundred and fifty *fanegas*[1] from one

1 The fanega is an ancient Spanish agrarian measure that is equal to 6,439.48 m^2.

fanega sown. He says it is sown in humid places. These Indies are a very humid land. (Fernández de Oviedo y Valdéz, 1959: 229–230)[2]

These clearly are mere speculations. In the face of this polemic regarding "Asiatic" maize, the first to claim its American origin in 1784 was Antoine Augustin Parmentier (1984), the French military agronomist and pharmacist who won acclaim by spreading the use of the potato in France.

The other hypothesis posited was that maize could have reached Europe from America before Columbus. It is known that the Vikings reached the coasts of Labrador before he did. It seems that there are old data on the Scandinavian explorations around the tenth century that mention "self-sown grainfields" and a "new-sown grain." On one occasion it was said that "an ear of grain" had been found. For Icelandic linguists, "an ear of grain" may mean the "head of wheat," but it seems instead to have been some wild grass. But even had the Vikings been able to reach as far as Cape Cod (for which there is no evidence at all), it is doubtful that maize was grown there at the time. Besides, maize is not "self-sown" (Weatherwax, 1945: 170). This clearly has no basis at all. But, interestingly enough, in the mid-twentieth century Sauer and other scholars still believed in the possibility that maize reached Europe before the discovery of America (see Goodman, 1988: 197; Jeffreys, 1971: 376; Sauer, 1969a [esp. pp. 166–167]).

In fact, the first news we find appears in the documents related to the voyage of Columbus to America. The first edition of *Le Historie della vita e dei fatti di Cristoforo Colombo* (*The History of the Life and Deeds of Christopher Columbus*), written by "D. Fernando Colombo *suo figlio*" (Fernando Columbus, his son) was published in Italian in Venice in 1571 (Colombo, 1930). A second edition appeared in Milan in 1614 (Caddeo, 1930: lix); the original manuscript was, however, lost (Caddeo, op. cit.: lx). The first Spanish edition, in 1749, reads, "[which] Alonso de Ulloa translated from Spanish to Italian, and which is now taken from the Italian version because the original Spanish [manuscript] is not found" (Caddeo, 1930: lxxiv).

Caddeo (1930: xxvi) prepared a study for the 1930 Italian edition of the works of Columbus, but here he made a mistake. Caddeo points out that the "*Historie* of D. Fernando Columbus is explicitly mentioned and widely cited" in the *History of the Indies* of F. Las Casas, which was published in Madrid in 1875. On reading the "Diario del Primer Viaje" (Journal of the First Voyage) of Columbus (1984), one finds in footnote 15 that it is "II-BN.Vitr.6–7. Copia de Fray Bartolomé de Las Casas [copy belonging to Friar Bartolomé de Las Casas]." In other words, it is an account that Las Casas wrote later. It is, however, true not only that he befriended Columbus but that he went with him on his third voyage. This is confirmed by Gil (1984: xi): "The *summary* of the

[2] Note 3 reads: "Pliny, Book xviii, Chapter vii."

Journals of the first and third voyage are preserved thanks to an autograph copy by Las Casas" (emphasis added).

Yet Gil (1984: xi–xii) also makes a mistake, for he claims that Fernando Columbus wrote the life of his father, which was published after his father's death in an Italian translation, but that this publication is "[the work of,] or was at least signed by, the adventurer Alfonso de Ulloa, which shows clear signs of interpolations (Venice, 1571)." This is not true, because the frontispiece of the first edition of the *Historie* says in large letters that the author is "D. Fernando Colombo," whereas at the bottom we read in fine type: "Nuovamente di lingua Spagnuola, tradotte nell'Italiana dal S. Alfonso Ulloa" (Once Again in the Spanish Language, Translated from the Italian by S. Alfonso Ulloa).[3]

Gil acknowledges even so that it is "an essential book" and adds that "... the Historie of D. Hernando and the History of the Indies of Las Casas thus become two essential underpinnings on which the [historical] critique of the Journal of the first voyage must rest, and whose text must always be determined in accordance with both works" (Gil, 1984: xi–xiii).

The events summarized are as follows. During his first voyage, on 2 November 1492, Columbus sent two men to visit the island of Cuba. These were Rodrigo de Xerez and Luis Torres (Colombo, 1930: 181 and note 5 on the same page).[4] They returned on 5 November accompanied by two Indians.[5] After their return, Colombo (1930: 184–185) said: "... e di un altro grano, come paniccio, da lor chiamato mahiz, di buonissimo sapore cotto, o arrostito, o pesto in polente" (... and from another grain similar to *paniccio* [foxtail millet] that they call mahiz, of very good taste stewed, roasted or ground to polenta). A similar account appears in the 1984 Spanish edition (Varela, 1984), which is a summary made by Las Casas. It actually agrees on the content, but it is not the previously mentioned account given by Columbus's son. In this text it says that on Tuesday, 6 November (1492), "[y]esterday in the evening, the Admiral says, the two men came who had been sent to see the land in the interior...." On describing it they said that "... it is very fertile and quite tilled ...," and that "*panizo* also grows there." So it is clear that Columbus saw maize and heard its aboriginal name on 5 November 1492.

It has been speculated that Columbus may have seen maize before this. Mangelsdorf claims he may have seen it before this date on the island of San Salvador in the Bahamas, on 12 October 1492. Or he could have seen it on Sunday the 14th but did not describe it in his journal, or some days later, while visiting what he called the Isla Fernandino, now known as Long Island

3 A facsimile is included in Caddeo (1930: between pp. xlviii and xlix).

4 Weatherwax (1945: 169) calls him Luis de Torres and adds that he was "a versatile linguist," according to Columbus (1492) in Varela (1984), but in Colombo (1930) his name appears without the "de." Yet Pedro Mártir de Anglería (1944) also used the "de."

5 According to Weatherwax (1945: 169), they had actually planned to stay six days on land, but they returned earlier.

(Mangelsdorf, 1974: 1). Weatherwax (1945: 169) likewise leaves open the possibility that he may have seen maize on 16 October while visiting the fields in Haiti, but he based his work on the writings of Las Casas.

Columbus gave a full account of all that had happened to the court at Barcelona in May 1493, on his return from his first voyage. Pedro Mártir de Anglería (Pietro Martire d' Anghiera) was present. In a letter to Cardinal Sforza written around mid-1493, Anglería described the plant that he had seen growing. He wrote in Latin and turned the word *panizo* into *panicum*, and he was the first to say: "... maizium, id frumenti genus appellant" (... this genus of grain they call maize) (Anglería, 1944, First Decade, book. i, chapter. iii: 8).

Anglería also tells that Antonio de Torres was one of the men on board when one of Columbus's ships returned to Spain from the second voyage in April 1494. He was an informant for Anglería. Among other things we read: "The bearer will also give you in my name certain black and white grains of the wheat with which they make bread [maize]...." (Anglería, 1944, First Decade, book. ii, chapter. vi: 25).

Besides this, one of the men brought with him a letter Guglielmo Coma sent to the king describing the new lands. This letter, along with more data from different sources, was published in 1494 or early 1495 by Nicolò Syllacio and is dated xii.1494. It reads thus: "There is here, besides, a prolific sort of grain of the size of a lupin, round like a vetch, from which when broken a very fine flour is made. It is ground like wheat. A bread of exquisite taste is made from it. Many whom are stinted in food chew the grains in their natural state" (Thacher, 1903, volume 2: 218). This fragment is quite similar to the writings of Anglería, which means they have a common origin, or it instead gives rise to two possibilities. Either the Coma-Syllacio letter is prior to that of Anglería, or the description he made was based on observations made during the second voyage and not the first one, as had been supposed. Be that as it may, the description of maize in the letter by Coma was published some 17 years before the one usually cited as the first description (Weatherwax, 1945: 172–173).

Early Data on Maize in South America

I was unable to find early descriptions of maize in South America, as I was unable to make an exhaustive search. I refer to the Andean area in another chapter. The references for Venezuela and Colombia are incomplete. In the mid-1700s Joseph Gumilla described an early and apparently distinct variety of maize that grew in the Orinoco alluvial plains (Mesa Bernal, 1957: 69).[6] In

[6] The reader must be warned that the data in Mesa Bernal have to be checked, because although this is a vast and apparently well-documented work, it does not include a single reference, so that verifying the data is difficult – I was unable to check the reference to Gumilla and Caulín.

1779 Caulín described the Cariaco race, which grew in both countries and is a race that matures quite early in Venezuela (Mesa Bernal, op. cit.: 76). There are descriptions of the Cariaco race into Yucatán that date to 1579 and 1620. It is said that this race was introduced in Yucatán in 1579, but it is not said where. The reference made to Capio and Morocho for southern Colombia, as well as for the Común race, all date to 1878. It seems that there are some earlier references to Común (Patiño, 1964: 1). There also are other descriptions from the 1700s for Colombia and Ecuador (Goodman, 1988: 198–199).

The first accounts of maize on the eastern coast of Brazil were written in the mid-1500s. They indicate that here most of the maize was white flint corn, and that cream, black or purple, and/or red varieties were less frequent. A predominantly white flour corn was also cultivated (see, e.g., Soares de Sousa, n.d.: 1) (Goodman, 1988: 198).

The Guaraní Indians of Paraguay, a part of the Tupi group of the Brazilian coast, also cultivated white flint and flour corn, just like in Brazil. They also grew an acuminate popcorn (Pisingallo) and one or several other varieties. These descriptions are from the mid- or late 1700s (de Azara, 1850: 1; Dobrizhoffer, 1822: 1) (Goodman, 1988: 198).

A History of the Name

When maize reached Europe it was called *mays* or *maizium*, but on spreading to different countries it received a name in each language. The term "maíz/maize" prevailed in some parts of Europe and particularly in Latin America (Weatherwax, 1945: 177). This process took time. *Panizo* was perhaps the oldest name used for maize in Spain.

No one used the name "maize" at first when Columbus brought it back from America. "Corn" is a general term for all types of cereals, and so it was called "Indian corn." But in England "corn" is wheat, and in Scotland, oats. In South Africa *kafir corn* is a sorghum. The word "corn" had long been used to indicate any small particle (grain) or by extension any small, round object, for example, a lump in one's foot. The plant received different names in Europe before it was classified by Linnaeus as *Zea mays*. Its dispersal was, however, rapid. According to the German botanist Leonhard Fuchs, in 1542 maize was cultivated in all gardens. Fuchs believed that it came from Greece or Asia. Others were long convinced that it had been imported from Asia. In the late sixteenth century, the English botanist John Gerard believed that maize could have originated either in the East or in the West, and he reached the conclusion that it came from America, but he called it "Turkey corn" and "Turkey wheat." Swedes adopted the latter name, *Turkiskt huete*. The Turks called it "Egyptian grain"; the Egyptians, "Syrian grain"; and the Germans, *welschkorn*, that is, "foreign grain" or *granoturco* (Kahn., 1987: 6–9).

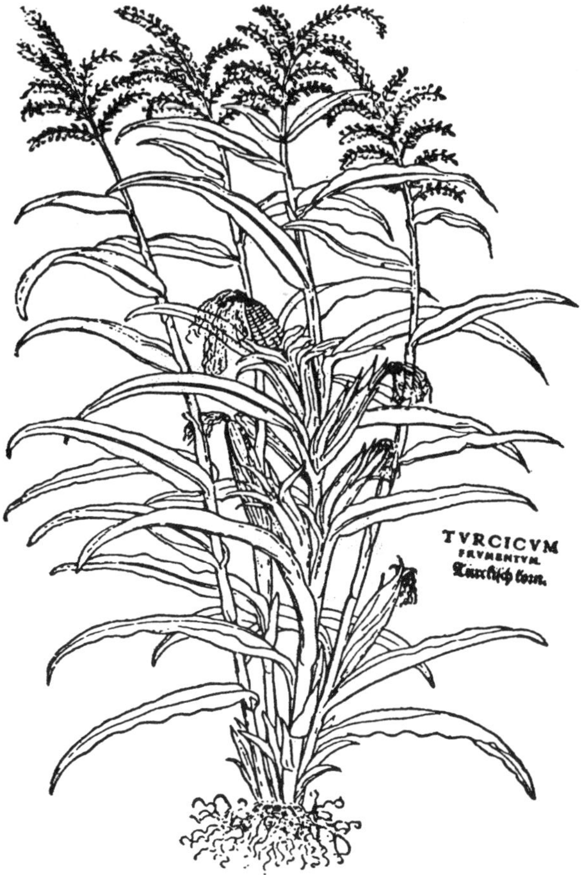

2.1. One of the earliest drawings of maize made in Europe. It was published in 1542 in *De Historia Stirpium Commentarii Insignes* by the famed herbalist Leonhart Fuchs. It was also included in the 1545 edition.

In early writings in which the authors were under the impression that maize came from western Asia, it was called "Turkish grain," *Triticum bastianum Plinii*, *Milico indico pliniano*, *Frumentum turcicum*, *Tritticum turcianum*, *Frumentum asiaticum*, and "Turkish wheat." Fuchs published one of the first illustrations of maize – a very good one at that – in his 1542 herbal (Figure 2.1). Weatherwax (1945: 175–176) credits Fuchs with being the first to depict the maize plant, but this is not true.

Brunfled's 1530 *Herbarum Vivae Icones* (Strasbourg) has what Sauer (1969b: 147) called "an emergent taxonomy." A good study of the growth of maize among sixteenth- and seventeenth-century herbalists is that of John Finan (1950). Here it is explained that it was only in 1570 that Pietro Andrea Matthioli (1501–1577), on reading the Spanish chronicles, explained the American origin of this plant that the Spaniards had imported. The herbalist Jerome Bock (Hieronymus Tragus) was the first to describe maize and to collect plants in the upper Rhine River. His book *Neu Kräuterbuch* was published in 1539 as

Welschkorn (the description is on folio 223 of the 1570 edition). He believed that it should be called *Frumentum asiaticum*, thus suggesting that he believed its origin was in Asia. For southern Germans, *welsch* means an immediate source – Italy – and *asiaticum* means that it came from Asia Minor. We owe De Candolle (1959) for the first use of *Frumentum turcicum* (1536), which he attributed to Jean Roel(ius) of Paris. By the way in which he describes the plant, it seems that he had never seen it, and that he described it because he had heard of it, and it was thus unknown in northern France.

Leonhard (Leonhart) Fuchs, who has already been mentioned, was the most renowned herbalist. His work was published in 1542 (Stuler, 1928: 231). The fine wood engraving depicted in Figure 2.1 bears the names *Turcicum Frumentum* and *Türckisch Korn*. It is explained that it was brought first from Greece and Asia, which were at the time under Turkish control. Two illustrations before that of Fuchs are known. The first of these is an Italian translation of the first book of Fernández de Oviedo, which was published in Venice in 1534. The drawing probably was of maize plants that actually grew close to the city. The second drawing may be a reduced copy of the first and was published in Seville in 1535. Another illustration appeared in the 1556 work of Ramusio, also published in Venice, which reproduces parts of Fernández de Oviedo's work. The name "Turkish grain" (*granoturco*) survived in Italy, along with several variants that will later be mentioned, but in the Friuli it was known as "Turkish sorghum" (*sorgo turco*), and as "soturco" in the Venetian dialect (Sauer, 1969b: 149–151, 159).

Much has been written on the origins of the name *granoturco*, and scholars do not always agree. It is possible that this term had its origin in Andalusia, for Arab farmers may have taken it to Turkey, where it was known as *kukuruz*. In England the name may have nothing whatsoever to do with the Turks, as it was called "wheat of turkey" because it was the grain with which the turkeys were fed.

Father Acosta, who wrote in the late sixteenth century, mentioned in his work "... the grain of maize, which in Castile is called wheat of the Indies, and in Italy Turkish grain" (Acosta, 1954: 109). Prescott (1995: note 18, 110) points out that the term *blé de Turquie* is a mistake of European origin.[7] We know that in France maize was known as *Blé turc* (Turkish wheat) for five centuries. This apparently was a transcription error made by a botanist who confused maize with buckwheat (*Fagopyrum esculentum*, of the Polygonaceae family), or perhaps the confusion arose from confusion between India and the Indies. There can be no question that the terms *blé turc* or *blé d'Inde* actually mean *blé des Indes* (Gay, 1987: 459).

Maize was given several names in Italy. The most common ones were *granoturco* or *granturco*, *granone*, *grano siciliano*, *melica* or *meliga* or *melega*,

[7] For more details, see Haudricourt and Hédin (1987: 223).

melgone, *melgotto*, *formentone*, *formentazzo*, *frumentone*, and *mais* (see, among others, Palazzi, 1940: 465, 520, 673).

In Portuguese maize is *milho*, and the same thing holds true for the western and southern coast of Africa. In South Africa it is *mielie*, whereas in eastern Africa the word comes from Asia, as it is called *hindi* in Swahili (Haudricourt and Hédin, 1987: 223). The Turkish word *kukuruz* has prevailed in Eastern Europe (Sauer 1969b: 151).

3

The Origin of Maize

... the question of the origin of maize has had a long history of controversy with vitriolic barbs still directed at researchers who challenge popular dogma. ...

Mary E. Eubanks (2001c: 92)

The origin of maize is an issue that has dragged on for more than a hundred years, as Iltis (2006: 23) correctly notes, and the debate still rages on, despite the fact that this issue has already been solved, as far as I am concerned.

In the late 1980s, when Goodman (1988: 197) discussed the origin and domestication of maize, he made some intelligent comments that are worth recalling. Goodman claimed that those who have touched on this issue did so from the standpoint of their own speciality. Thus Weatherwax was a specialist in morphology, Mangelsdorf in hybridization from a wide standpoint, Randolph in experimental taxonomy, Beadle and Kato-Yamakake in cytology, Galinat in specialized sweet corn, Wilkes in plant geography, Iltis in herbarium taxonomy, and Doebley in experimental systematics. To this list we should add the names of Pearsall and Piperno, who specialize in pollen and phytoliths, and the long list of archaeologists, headed by MacNeish, who have touched on this subject. It is true that opposite perspectives were in many cases not tolerated or were courteously ignored. But to this day as yet no serious interdisciplinary work has been undertaken by a group of specialists who gathered the data in order to prepare a genuine synthesis of this issue. The major effort in this regard was carried out by Randolph (1976), but it remained unfinished due to his death (see Anonymous, 1982). This is a real shame, for Randolph addressed this issue with great seriousness and a real knowledge of the subject.

Staller, Tykot, and Benz (2006) recently edited a book wholly devoted to the problematic of maize from the standpoint of different specialities. The expected goal was unfortunately not attained, for many of the studies are weak, and the editors were unable to direct this collection toward specific goals.

Wild Maize

As is well known, the great debate has always hinged on whether domestic maize originated from wild maize or from teosinte. The problem is that maize in the wild state has not yet been found anywhere, either in Mesoamerica or in South America. Mangelsdorf (1974: 169) was convinced that several characteristics allow wild maize to be separated from teosinte and *Tripsacum*. This hypothesis, as shall be seen later on, was based on his conviction that the archaeological maize found in the most ancient strata at Tehuacán, in Mexico, is wild.

Harlan was right when he noted that the issue of what wild maize is is independent of its time and place of domestication (Harlan 1992: 222) and is strictly related to the botanical characteristics of the samples.

There is no agreement regarding the areas where wild maize could have developed. Wilkes (1989: 443) in turn posits that wild maize was a highland plant.

Although wild maize is unknown – unless one accepts that which has been found in the deepest strata at Mexican archaeological sites – scholars have tried to reconstruct its characteristics. Wilkes (1989: 443) believes that this maize had a massive central spike with few to no branches in the tassel (in the male inflorescence of the stamens), and several small lateral ears, one in each of the upper nodes along the multiple tillers. Human selection has given rise to a plant with a simple, massive, fist-sized ear on a single stem, and a tassel with many branches. At the same time the role of the lower glume that protects the kernels has diminished, and the rachilla has been shortened and lost the abscission, so that the ear does not shatter.

Although he does not specifically mean wild maize but instead an early maize, Eubanks (1995: 179), following Mangelsdorf and colleagues (1967a), describes it in the following way: bisexual ears with male flowers subtended by female flowers on the same spike; uniformity of cob characters and size; kernel row numbers ranging from four to eight; long, soft, herbaceous glumes that partially enclose kernels; paired kernels attached to cupules that are as long as, or longer than, they are broad; a rachis composed of cupules joined together at their sides and ends; prominent rachis flaps; and cupules loosely joined that break apart rather easily in contrast to the rigid cob of modern maize.

Nowadays maize depends on man for its reproduction, but primitive maize could have dispersed itself thanks to the great ease with which the seeds could be released, due to the fragmentation of a fragile rachis and the transferral of hard and small, red- or coffee-colored seeds, which were attractive to migrating birds. Primitive maize – "possibly wild" – like that of the Tehuacán Valley, could have had 48–56 seeds, whereas a modern Peruvian hybrid maize can have between 500 and 700 seeds per ear (Grobman, 2004: 428).

Teosinte

In his 1893 study, Harshberger accurately described the area where annual teosinte grows, as well as the archaeological zones where it grew in the wild state (Wilkes, 1989: 446). It is, however, worth recalling that this plant was not studied in depth until Wilkes (1967) published his monograph; since then many scholars have taken up this subject. The study by Wilkes (op. cit.) is not only the best one ever made of teosinte; it also presents a taxonomy that differs from the previous one (as was seen in Chapter 1), and he separates maize as a species different from teosinte, whereas Iltis and Doebley considered it at the subspecie level, for they based their work only on the compatibility of crosses and not on visible characters (Grobman, 2004: 430).

Teosinte is the closest relative of maize. It has the same number of chromosomes (20), and these are similar to those of maize in their lengths and in the position of their centromeres. Teosinte hybridizes easily with maize, and first-generation hybrids are vigorous and highly fertile when self-pollinated or when they are crossed back to either parent (Mangelsdorf, 1974: 15).

The first Latin name given to teosinte was that provided by Schräder (1833), which only referred to the annual form – *Euchlaena mexicana* Schräder. In 1910 A. S. Hitchcock (1922) discovered the perennial form of teosinte, which he called *Euchlaena perennis* Hitchcock. Kuntze (1904) and Reeves and Mangelsdorf (1942) subsequently ascribed it to the *Zea* genus and renamed it *Z. mexicana* (Schräder) Kuntze and *Z. perennis* (Hitchcock) Reeves and Mangelsdorf.[1]

Randolph (1976), however, disagrees with teosinte being ascribed to the *Zea* genus. He presented three tables (Randolph, op. cit.: 1 [324], 2 [325], and 3 [326]) with 23 major differences in plant characteristics as well as in environmental responses between modern *Zea* and *Euchlaena*, which have enough taxonomical import to distinguish the two genera. In a previous study (Randolph, 1972;[2] see Randolph, 1976: 325), Randolph posited that these are two different genera. In tables IV and V, Randolph (1976: 330 and 331) shows 19 differences between ancient maize and teosinte. If we gather the results obtained in the analysis of both modern and archaeological specimens, there are 32 inheritable differences between maize and teosinte.

Randolph acknowledges the fact that more studies and experiments are required, but he goes against the hypothesis that teosinte is the progenitor of maize (Randolph, 1976: 330). In this study Randolph presents a long list based on characteristics of genetic and taxonomic significance, which can be applied to both modern as well as archaeological maize, and which are in contrast with the essential differences of both modern teosinte uncontaminated by maize and

[1] For more data regarding this point and full bibliographical data, see Mangelsdorf (1974: 19–20) and Doebley (1990: esp. p. 7).

[2] I was unable to find this.

the small number of archaeological teosinte remains available (Randolph, 1976: 344–345).[3] What Randolph emphasizes is the scarcity of archaeological teosinte in sites where maize, as well as other plants in early development stages, abound, and this would be strong evidence arguing against teosinte being considered as a desirable food plant, and, if it indeed was one, then the evidence indicates that all attempts at domestication failed.

There are six races of annual teosinte, *Z. mays* L ssp. *mexicana* (Schrader) Iltis: four in Mexico (Nobogame, Central Plateau, Chalco, and Balsas) and two in Guatemala (Huehuetenango and Guatemala). But it is also classified as two subspecies of *Z. mays* L.: ssp. *mexicana* (which includes the Chalco, Central Plateau, and Nobogame races), ssp. *parviglumis* var. *parviglumis* (which corresponds to the Balsas race), and ssp. *parviglumis* var. *huehuetenangensis* (the Huehuetenango race), as well as the *Z. luxurians* species (which corresponds to the Guatemala race; see Doebley, 1983a). Wilkes believes that the annual teosinte recently rediscovered in Oaxaca (see the following discussion), merits species status under the classification based on species and subspecies, but he prefers instead to consider it as a seventh Oaxaca race, or as part of the Balsas race instead. The latter has a limited distribution in the Jalisco area (Wilkes, 1979: 6; 1989: 447–448; for a taxonomical classification of teosinte, see also Doebley, 1990: 7–10).

Eubanks (2001b) in turn prefers to consider three species of teosinte: *Z. luxurians*, *Z. diploperennis*, and *Z. perennis*, with three subspecies *Z. m.* ssp. *mexicana*, *Z. m.* ssp. *parviglumis*, and *Z. m.* ssp. *huehuetenangensis*.

In 1979, while working in southeastern Jalisco, Iltis and his team rediscovered perennial teosinte – which was believed to be extinct since 1921 – at two sites, Población de Ciudad Guzmán and Cerro de San Miguel. In the former site they found a *Z. perennis* tetraploid, whereas the second site contained a different diploid taxa that was described for the first time – *Z. diploperennis* Iltis, Doebley, and Guzmán sp. nov. (Iltis et al., 1979: 186). Interestingly enough, Paul Mangelsdorf added a postscript to the latter (in Mangelsdorf et al., 1978, although it was actually published in 1979), in which he announced the find made by Iltis and his team. Here we find, among other things, that "... this discovery may be the key piece in the puzzle, a so-called 'missing link' in corn's genealogy. Wilkes assumes – correctly in my opinion – that hybridization between the diploid perennial teosinte and a wild annual corn could have produced all of the known annual races of teosinte" (Mangelsdorf et al., 1978: 252). Mangelsdorf liked this position for two reasons: first, because it was consistent with the archaeological evidence, and second, because it could be controlled experimentally. In a subsequent study, Mangelsdorf (1986: 80, 84–85) went over this issue once again and made a similar statement, insisting that

3 Interested readers should check the differences pointed out by Randolph in this paper, as they are far too technical to include here.

perennial teosinte is the key missing link in the genealogy of both maize (*Z. mays*) and annual teosinte (*Z. mexicana*). Modern maize and annual teosinte both descend from the hybridization of perennial teosinte with a pod-popcorn.

Galinat (1985b: 247) believes that the condensed forms of teosinte, with their triangular fruitcases and spikes borne in fascicles, may be an indirect product of human selection. Starting with such a condensed teosinte, the transformation into a botanically correct maize ear may have been simultaneously attained in several places, and it began to spread in relatively rapid fashion, perhaps in one hundred years.

There is no agreement as regards its genealogy. Doebley (1990: 13) believes that the fact that teosinte is wild, and maize fully domesticated, leads to the conclusion that the common ancestor also was a teosinte. Yet Grobman (2004: 436) believes that *Z. diploperennis* could have crossed with wild maize (an annual diploid plant) and thus given rise to a primitive annual teosinte, which would have then acquired more maize-like characteristics after an introgression with maize.

Mangelsdorf (1974: 17ff., inter alia) suggests that teosinte is more specialized than maize, at least as concerns four characteristics: adaptation to a limited range of environments; the decrease in size from a polystichous cob (i.e., one with many rows) to a dystichous one (i.e., a two-row cob); the decrease from paired kernels to just one; and the hardening of the glumes and the rachis.[4] It was for this reason that Mangelsdorf posited that maize is an ancestor and not a descendant.

Goodman (1988: 208) points out that one can object that the genera related with teosinte and maize are usually considered to be more similar to the former than to the latter. This also does not take into account the great diversity in chromosome knobs and isozyme alleles in teosinte, in comparison with Mexican maize (Doebley et al., 1984, 1987; Kato-Yamakake, 1976; J. S. C. Smith et al., 1982, 1984, 1985). Not only does teosinte have all of the knob positions known in maize; it also has an additional number of them (mostly terminal ones). Yet the origin and distribution of the chromosome knobs among maize, teosinte, and *Tripsacum*, as well as their possible relatives, cannot be explained satisfactorily except with the differentiation of populations instead of a single originator plant that acted as a progenitor. Teosinte likewise seems to have had isozymes from Mexican maize alleles, plus a few rare alleles that are restricted to the vestigial populations of teosinte. Even so, there are numerous rare isozyme alleles that appear in maize but not in teosinte (Doebley et al., 1984; J. S. C. Smith et al., 1984, 1985).

Not all scholars have the same opinion as regards the morphological differences or resemblances between maize and teosinte. Galinat (1977: 5) claims that if one observes the floral characteristics that distinguish the maize cob

[4] De Wet and Harlan (1976) instead suggest that these characteristics indicate that maize is quite specialized in both taxa.

from the female teosinte spike without taking into account the enormous power of human selection, they would have to be classified in different genera. This is so much so that Galinat insists that teosinte used to be in the genus *Euchlaena*, whereas now both are in the same genus (Iltis, 1972). Besides, a high degree of lignification takes place in the fruit-bearing cupules in all known races of teosinte. The hardening of the place where the fruit forms is adaptable to kernel production. "In contrast, all of the oldest archaeological maize cobs from Tehuacán, Mexico, and from Bat Cave, New Mexico, are soft; this does not prove that soft cobs are a trait of a so-called wild corn" (Galinat, 1975a: 318).

Harlan (1992: 224–225) believes otherwise. He claims that although the ears of maize seem to be different from the small, fragile racemes of teosinte, all parts of the flower and of the structures are present in both. The only gene from *Tripsacum dactyloides* that has been reported and described (Dewald et al., 1987) can produce all, or perhaps all, of the changes required in order to convert teosinte into a primitive corn like that from Tehuacán. Harlan wonders what attracted man to teosinte, which has a hard fruit and a relatively small output. He discards the generalized idea that maize may have been a staple crop in its condition as a harvestable cereal. Harlan tends to believe that teosinte was first used as a vegetable that was probably chewed on warm days, because its spikes are succulent, soft, juicy, sweet, and refreshing and have similar characteristics to those of immature maizes. It was perhaps for this reason that teosinte was kept in home gardens, and it was here that the critical mutation would have taken place. Harlan notes here that a gene may mutate more than once.[5]

It is worth recalling in this regard what Mangelsdorf wrote more than sixty years ago:

> There is no evidence that teosinte was ever used as a food plant by the American Indians and it is the almost universal opinion among present-day Mexican Indians familiar with the plant that it is valueless as a food plant.... Teosinte is never cultivated for food in Mexico today and, even where it is known, it is regarded as having no food value for man or beast. (Mangelsdorf, 1947: 188–189)

Mangelsdorf later returned to this issue (Mangelsdorf, 1983a: 89):

> The entire array of archaeological evidence throughout Middle America clearly suggests that teosinte itself has probably never been used either as a cultivated or gathered fruit crop. Teosinte fruit cases are distinctive, durable, highly archaeologically preservable, and yet [it is striking that they have] virtually never [been] found prior to or during incipient agriculture.

5 Interested readers can find a good description of teosinte and its relatives in Harlan (1995: 180–181).

The essential differences between teosinte and maize are as follows:

1. The ear of teosinte is fragile and breaks up at the rachis's joints. All wild cereals are fragile, and under domestication all developed non-shattering races.
2. Teosinte ears have two rows, whereas maize has four or more rows.
3. In teosinte, only one of the two female spikelets is fertile, whereas the other one is reduced. Both members of the pair are fertile in maize.
4. In teosinte the outer glumes are very hard, whereas in maize they are soft and external.
5. In teosinte the glumes cover the seeds, whereas in maize the kernels are (usually) exposed.
6. In teosinte the kernels are embedded into the deep cupules in the rachis, whereas in maize the kernels are held in place by cupules that are not too deep. This is a variable characteristic even in maize, and the cob conforms to row number and seed size as an integrated unit.
7. In teosinte the kernels are fragile, but they are not so in maize.
8. Teosinte seeds are small ones; those of maize may be small but are usually twice the size of the wild races. Increase in the size of the seed usually takes place under domestication.
9. In teosinte the primary lateral inflorescence is usually male, whereas in maize the primary lateral inflorescence is usually female.
10. In teosinte the primary lateral branches are long ones, but in maize they are short ones.

Other traits, such as the number of ears per plant or the amount of cupules per ear, are presumably secondary effects of domestication, as opposed to primary morphogenetic changes related with the transformation of teosinte into maize (Doebley et al., 1990: 9889; Harlan, 1995: 183–184).

Mangelsdorf analyzed and summarized the evidence for the flow of genes between maize and teosinte. He reached four conclusions. First, F_1 hybrids of maize and teosinte are usually vigorous, highly fertile, and easily backcrossed to either parent to produce a fertile progeny. Second, the chromosomes of both species are morphologically similar, and the synapse is more or less normal in hybrids. Third, the arrangement of the gene loci, although not identical, is similar in the two species. Finally, in both maize and teosinte, the crossing over between linked loci follows the same order as in maize, with few exceptions. Mangelsdorf therefore pointed out that a more realistic classification would have maize and teosinte represent a single dimorphic species in which one component is preserved by man, and another one by nature (Mangelsdorf, 1974: 123–124).

Clearly one of the major differences between maize and teosinte is the structure of the cupule, as has already been noted. For Galinat (1970), the cupule provides the connecting link between the maize cob and the origin of the teosinte fruitcase. In teosinte the cupules are the major component of the kernels' protective device. Cupules are obsolete in the oldest archaeological maize cobs. Here the cupules represent the remnant fingerprints from their counterpart in teosinte. A few of the oldest specimens of maize show other traits of teosinte, including two-ranking cupule interspaces and partial abscission layers (Galinat, 1975a: 317).

One problem that has yet to be solved is why teosinte pollen is smaller than that of modern maize. Mangelsdorf, Barghoorn, and Banerjee (1978) based their work on this to claim that the pollen found in Mexico City, which is better known as the Bellas Artes pollen (see the subsequent discussion), comes from wild maize (Galinat, 1985b: 273). Flannery (1973: 294), however, states that it is not true that all grains of maize pollen are larger than those from teosinte. This is true only in 4 of the 10 varieties of teosinte. There are studies showing not only that the size of pollen in many teosinte races surpasses that of the Chapalote maize race, but also that there is one teosinte – that from Jutiapa – that has grains that are significantly larger than those from the Jutiapa maize race.

For those who do not accept the origin of maize from teosinte, the transformation of teosinte kernels into those of maize seems spectacular. Galinat (1985a: 137–138) says that from the botanical standpoint, the modern kernels of maize – the result of a selection for an extremely high harvest – is a puzzle and a monstrosity. Assuming that teosinte is its wild ancestor, no other cereal underwent such a dramatic transformation during domestication. Yet Harlan (1992: 223–225; 1995: 184) does not concur. He believes there are no elements that may seem to be unique, and that the genetic base does not seem to be very complex. The case of millet is even more complex, yet it does exist.

The key to the mode of transformation may lie in *Tripsacum*. There are several species in this genus, most of them tropical ones. These species are perennial, and most of them have $Zn = 36$ diploid chromosomes, or $Zn = 72$ tetraploid chromosomes. Maize and teosinte have $Zn = 20$. In *Tripsacum* there are some triploids with 54 chromosomes, and one species – *Tripsacum andersonii* – has 64 due to an addition from the *Zea* genome. It has been clearly shown that *Tripsacum* can hybridize with maize. Hybrids are not fully sterile, and maize morphology can recover after a backcross with maize. *Tripsacum dactyloide* is the most widespread kind, and it extends as far north as Michigan and New England, and south down to South America, and it also has diploid and tetraploid races.

Dewald and colleagues (1987) found a mutant of *T. dactyloides* in two quite separate wild populations in Kansas. The genetic analysis shows that a single gene intervened here. There are major changes in a gene, and this, Harlan points out (1995: 184), makes the "mystery of maize" not much of a mystery. He believed that the position taken by Iltis (1983a) regarding sexual transmutation had to do with this. This is a condensation of the structures of the branching of the teosinte ear cluster, and the feminization of the terminal raceme of the male portion. This provides soft glumes and grains free of the fruitcase with one stroke. The *Tripsacum* gene gives all of this, but some additional modifications and development adjustments are required to have real maize kernels. It is true that this mutation has still to be found in teosinte; it may never take place, and the fact that something may happen does not mean that it will take place. Even so, Harlan concludes, the idea of feminizing the male flowers to produce ears of maize is praiseworthy (Harlan, 1995: 184–185).

Now, Mexican teosinte is sympatric with maize (Galinat, 1985b: 270). It is likewise known that maize and teosinte hybridize almost freely, and that their hybrids are fertile.

There is a large body of literature regarding the possible changes that took place in the presumed transformation of teosinte into maize, which it would be pointless to list here. Interested readers can peruse, among others, Galinat (1974b; 2001a; 2001b), Pickersgill and Heiser (1976), Flannery (1973), and Iltis (1983b). It is worth recalling that Mangelsdorf and Reeves (1939) suggested that much of the variability found in maize is due to its introgression from teosinte. Goodman (1988: 201), however, believes in this regard that although we have little direct evidence, either botanical, genetic, or agronomic, with which to support this suggestion, we do have circumstantial archaeological evidence that seems to agree, as well as considerable indirect botanical evidence. From an agronomic standpoint, the derivatives from the crossings between teosinte and maize have been frustrating (W. L. Brown and Mangelsdorf, 1951). Those who cultivate maize were unable to improve it using teosinte. Goodman (1988: 201) says that he "... knows of no currently used inbred line, hybrid, or variety that traces in any way to intentionally used teosinte germplasm." And he shows that, in the United States, most of the experiments made in this regard were unsuccessful.

Goodman points out that the botanical evidence suggests that, in Mexico, the introgression of maize and teosinte is quite limited, from both the former to the latter and vice versa. And he points out a large bibliography in this regard. He likewise indicates that the studies on chromosome knobs in maize and teosinte by Kato-Yamakake and colleagues (McClintock et al., 1981) show that the chromosomic knobs typical of maize are often found in teosinte, whereas most of the terminal knobs are found in teosinte and rarely so in maize (Goodman, 1988: 201).

Mangelsdorf pointed out, when summarizing a lifetime devoted to the study of maize, that in some fifty years spent reading and researching he had been unable to find any real evidence whatsoever of the domestication of teosinte, be it archaeological, ethnographic, linguistic, ideographic, pictoric, or historical. The only thing suggesting that it may have been the progenitor of cultivated maize is that it is its closest relative. Mangelsdorf believed that this was due to the fact that teosinte is worthless as food. This is not just because its spikes are brittle and make harvesting difficult, but also because once harvested it is of scant nutritional value. The shells enclosing the kernels, which comprise about half the weight of the harvest, are essentially formed by cellulose and lignin and have about the same composition and texture of hardwoods such as maple, walnut, and even ebony. They are so hard that squirrels and other rodents shun them, despite the eatable kernels that they hold. Once the kernels have been somehow separated from the shells, the harvest has about the same nutritional value as a mixture – in equal parts – of a whole kernel of flour corn and the sawdust of a hardwood. Beadle (1972) showed that teosinte can pop like popcorn, but no archaeological evidence has ever been found of this (Mangelsdorf, 1974: 51–52; 1983b: 235; see also Iltis, 1987: 210).

Iltis also touched on this point, but from a different angle. He agrees in that there is a complete lack of early agricultural remains of teosinte, but he draws the attention instead to what he defines as "... the astronomical rarity of any mutation affecting changes in the morphology of the teosinte CFC [cupulate fruit case] ...," which may have led to the initial domestication of maize being due not to the kernel but to other causes. Several possibilities thus arise. The first one is that the medulla of teosinte is sugary and that it may have been eaten chewed or was used to prepare a fermented beverage. A second point is that one single and very rare mutation – TGA, which corresponds to the glume of teosinte – began the domestication of the kernel; this would have required a change in the cupule, the softening of the outer glumes, and the movement outward of a kernel that is not strongly attached, thus facilitating its harvest by man. Third, there is no possibility of unconscious selection. Fourth, the domestication of maize consisted in the increase of its apical dominance. Fifth, TGA is extremely rare, which is the key to the genetic mutation (one in a million or even less), and which is lethal in nature because unprotected kernels suffer predation by vertebrates and insects. The unprotected kernels are known only in maize, but they are genetically transferred to teosinte and are even unknown in the wild state and may thus possibly lead to the conclusion that accidental discovery within a teosinte population could only have taken place under human control, for uses other than as kernels. One of the possibilities would be its use in the preparation of an alcoholic beverage. Sixth, for this same reason – that is, the extreme rarity of this major mutation and the evident fall into disuse of the CFC – the accidental discovery of this mutation took place in just one plant or its immediate descendants within a population of teosinte *parviglumis* used by hunter-gatherers. Finally, Iltis posits that the place where all this took place was the central Mexican plateau (Iltis, 2000: 29–36; 2006: 25).

Goodman (1988: 206–207) also touched on this subject, but he made some mistakes. Goodman points out that there is no evidence that teosinte was used as food before maize, even though Lorenzo and Gonzáles Quintero

(1970) claim that ancient maize is of the same age as teosinte.[6] And because the hard fruits of teosinte are better preserved than the small and fragile ones of maize, "... it is clear that teosinte was not a major dietary constituent." At Tehuacán, the earliest maize clearly is much more similar to modern maize than to modern teosinte.[7] Goodman notes that the differences between maize and teosinte are so many, and its inheritance so complex, that it is questionable that hunter-gatherers were able to produce the change in such a brief span (here he based his ranking on Mangelsdorf, 1974: 51–52, and on Randolph, 1976; the reference appears on pp. 332–333).

Iltis (2006: 29) correctly points out that the ancestry of mutant teosinte has never been seen in nature, so only an approximate reconstruction is possible. The same thing can be said for maize, because the oldest ones found at Guilá Naquitz "... are already far advanced." Here we can add that these remains are always mentioned as the oldest ones in America, but this is not true, as the maize found in the Casma Valley in Peru has slightly older dates.

Galinat (1985b: 276) was right when he claimed that "the steps described in the transformation from teosinte to maize are largely a result of imagination that is based on cytogenetic and morphological clues combined with analytical reasoning."

Tripsacum

Tripsacum has been little studied. This is a small genus, but it comprises a large variation (Goodman, 1988: 203; Harlan and De Wet, 1977: 3494; Mangelsdorf, 1961: 159–160).

The genus has 16 species, 12 of which are native to Mexico and Guatemala, 1 to Florida and Cuba, and 3 to South America. Specialists point out that there are other species in this last continent that are as yet still not described. Its probable area of origin is Mexico and Central America, and its center of variation lies in the western part of central Mexico (Wilkes, 1989: 448; see also Eubanks, 2001b: 496, who likewise includes a complete bibliography). The species most similar to maize is *T. maizar* (Mangelsdorf, 1974: 55; 1983b: 232). Galinat (1977: 34), however, believes that this genus, which is related to maize, is more varied and distant from the latter than teosinte.

Tripsacum occurs naturally from the northern United States southward to Paraguay and Bolivia and is also found in the islands in the Gulf of Mexico (De

[6] Here Goodman notes that Wilkes (1989) questions the work done by Lorenzo and Gonzáles Quintero, but this is not so – the article cited makes no reference to this. Goodman also notes that R. McK. Bird (1984) takes an intermediate position, which is also not true, as Bird rejects the work done by Lorenzo and Gonzáles Quintero. In brief, those who question the work done by Lorenzo and Gonzáles Quintero (1970) are Bird (1984: 50) and Flannery (1986b: 8). Besides, Goodman confuses Tehuacán with Zohapilco (p. 206).

[7] The same thing applies to Peru.

Wet and Harlan, 1976: 448; Harlan and De Wet, 1977: 3494; Mangelsdorf, 1961: 159).

The distribution of *Tripsacum australe* in South America is as follows. It is found in Venezuela in the savannas and on the southern slopes of the Amambay highlands; in Bolivia it occurs in the lowlands and up to 1,500 masl. In Brazil *Tripsacum* appears in areas below 800 masl; in Ecuador it is found above 1,200 masl, and in Paraguay south of latitude 26°. In Peru, the only reference available prior to the 1960s is that of Asplund, regarding the presence of *Tripsacum* close to Tingo María. Due to the attested presence of tripsacoid maize on the eastern slopes of the Andes as well as on the Amazon basin, Grobman and colleagues (1961) suggested that it was sympatric with maize. Cutler and Anderson (1941) had already verified the presence of *Tripsacum australe* in the Amazon basin. In 1963, Grobman (1967: 285–286)[8] found *Tripsacum australe* in the Huallaga River basin, on its confluence with the Mayo River, and in 1964 in Tarapoto.

"Tripsacoid" is a term much used in the literature regarding this subject that has to be explained. Anderson and Erickson (1941) were the first to use it to describe North American maize. Their definition was any combination of characteristics that may be introduced into domestic maize (*Zea mays* L.) from its closest wild relative, *Z. mays* L ssp. *mexicana* (Schrader) Iltis (i.e., teosinte), or possibly also by *Tripsacum* sp. Mangelsdorf (1961, 1968) suggested that whenever these characteristics are found in South America, where teosinte does not occur, the source of the tripsacoid traits could have been *Tripsacum*.

The genetic transfer between the *Zea* genome and *Tripsacum* is possible. According to De Wet, Harlan, Stalker, and Radrianasolo (1978), tripsacoid maize derived from the *Zea-Tripsacum* introgression resembles, in its morphological details, the teosintoid maize derived from the maize-teosinte introgression, save for its frequent perennial weakness. Yet its perennity can be introduced into maize by *Zea perennis*. For these authors, the most obvious and consistent characteristics of tripsacoid maize are, first, the increase in the induration of glumes and rachises; second, the decrease in the tissue found in the pith of the cob; third, the increase in the length of the glume; fourth, a small increase in the length of the rachis between the cupules; and finally, an increase in the length of the cupules in relation to width. These essentially are the same characteristics described by Anderson and Erickson (1941), which Wellhausen and colleagues (1952) used to determine the degree of teosinte introgression in the Mexican races, and which Roberts and colleagues (1957) used to show the exchange of genes between *Zea* and *Tripsacum* in Colombian races. De Wet and colleagues thus conclude that an "experimentally induced *Zea-Tripsacum* introgression, however, does not necessarily prove natural introgression between these two genera" (De Wet et al., 1978: 240). Because the hardening of the Mexican races

[8] The date corresponds to the edition, but the actual publication appeared years later.

is due to the introgression with teosinte, De Wet, Harlan, and Radrianasolo (1978: 741) suggest that the term "teosintoid" should be introduced in this case to distinguish it from the experimentally obtained tripsacoid maize.

There are many races in South America with a similar induration – save for their glumes and rachises – and none are morphologically similar to the North American teosintoid races. Because there is no teosinte in South America, this hardening must be due to the introgression with *Tripsacum*, as was suggested by Mangelsdorf (1961, 1968). De Wet and colleagues (1978: 741) do acknowledge that more studies are required to establish this point. Mangelsdorf (1983b: 229) insisted in this regard and defended the distinction drawn between the terms "tripsacoid" and "teosintoid."[9]

Stalker and colleagues (1977: 747–748) tried to provide a more precise definition of the tripsacoid characteristics, as opposed to teosintoid ones. They point out a series of measurable morphological traits and effects that the introgression with *Tripsacum* has on the maize genome and conclude that in this case the mutagenic effects were similar to those reported by Mangelsdorf (1958a) when maize was crossed with teosinte.

It was Eubanks who most clearly pointed out the differences present in the characteristics that distinguish tripsacoid maize from early maize. She explained that these differences "... include a strongly indurated rachis, cupules and lower glumes; two-ranked spikelets that grade into simple spikelets at the tip; lower glumes that diverge at tight angles from the cob; and elongation in the rachis tissue of the cob central axis, [thus] contributing to increased ear length" (Eubanks, 1995: 179; here she followed Mangelsdorf et al., 1967a).

The hybrids of *Tripsacum* with *Z. diploperennis* have the characteristics of both of them, that is, both of early maize and of tripsacoid maize (Eubanks, 1995: 179). It was, however, experimentally verified that populations of perennial diploid teosinte and *Tripsacum* grow in close proximity, that both can be pollenized by the wind, and that they have hybrids. When the latter crossed with their parent, the progeny reverted to the parental phenotype. But when hybrids crossed with other hybrids, some of the recombinant offspring produced ears with four to eight rows of paired kernels, in reduced cupules that exposed them and that were hence easier to remove from the cob. It was likewise verified that they are good food. Man was able to spread it over different ecologies while natural selection generated new races, and the diversified genetic pool rapidly spread among other populations with related and/or hybrid taxa. In this way maize was able to turn – in just a few millennia, through artificial selection and adaptation to a new habitat – into a genetically diverse and highly productive plant (MacNeish and Eubanks, 2000: 15, 17). Now, the critical step in the transformation from the single-rowed spike of the wild relatives of maize

[9] To avoid confusion it must be pointed out that Iltis (1969: 2) suggested that the term "tripsacoid" should be replaced with "euchlaenoid."

into multiple-row corn ears was reconstructed experimentally in the segregating progeny of hybrid plants of *Tripsacum* and *Z. diploperennis* (MacNeish and Eubanks, 2000: 13).

It is worth noting that a natural hybrid of maize and *Tripsacum* has never been found, and that the repeated attempts to experimentally produce teosinte/*Tripsacum* hybrids have all failed (Grobman et al., 1961; Mangelsdorf, 1974: 127; Randolph, 1976: 336; Roberts et al., 1957).

Goodman (1988: 212) has noted that, the speculations regarding the possibility of a maize-*Tripsacum* introgression in South America notwithstanding (e.g., Banerjee and Barghoorn, 1973a: 47; Cárdenas, 1969; Grobman et al., 1961: 55; Mangelsdorf, 1961; Roberts et al., 1957), there is as yet no publication that has described the experimental crossing of South American *Tripsacum* and maize. This is not exactly true, for as Grobman (2004: 450–451) notes, Galinat (1977: 3–4, 27–35) proved that the cross does take place and with "relative ease." Talbert and colleagues (1990) showed that *Tripsacum andersonii*, which grows in Central America as well as in northern South America, is a natural hybrid of maize and *Tripsacum*, and this shows that the introgression between *Tripsacum* and *Zea* does take place in nature. It is for this reason that Grobman (2004) accepts the possibility that there are blocks of *Tripsacum* genes incorporated into maize through natural hybridizations, even though this has as yet not taken place in nature. It has been pointed out that the difference in ploidy is a potential barrier between species. Another barrier could be the smaller size of the pollen tube in *Tripsacum*, which makes it difficult to reach the *Zea* ovules after pollination; in addition, there seem to be recent Russian studies on the significance of the ratio of maternal and paternal genomes, as well as the required induction of the maternal genome for the induction and adequate development of the endosperm in hybrid seeds.

Besides, it must not be forgotten that the study of archaeological (preceramic) pollen kernels from Peru allowed Banerjee and Barghoorn to find "convincing" evidence of the introgression of *Tripsacum* into maize (Banerjee and Barghoorn, 1973a: 48; 1973b: 34).

After analyzing tripsacoid maize, De Wet, Harlan, and Radrianasolo (1978: 744) reached the conclusion that its characteristics are exactly the same as those the Mexican or South American races would have if they were teosintoid. De Wet and colleagues conclude that, from a morphological standpoint, the "maize recovered from *Zea-Tripsacum* introgression cannot easily be distinguished from maize-teosinte hybrid derivatives on the basis of plant, tassel or ear morphology."

Wilkes (1972: 1075) in turn showed that the studies of *Tripsacum* hybrids with maize indicate that certain segments in the *Tripsacum* chromosomes can be substituted with segments belonging to maize chromosomes, with the plants continuing to be fertile and viable. Galinat (1971a) mapped more than 25 loci in the chromosomes of these genera. The data available on maize-*Tripsacum* hybrids and their derivates indicate that the respective genetic architectures of maize ($2n = 20$) and *Tripsacum* ($2n = 36$, $2n = 72$) are more similar than that what their karyotypes suggest despite their being completely different.

Galinat (1974a) pointed out that it is more difficult to transfer some *Tripsacum* chromosomes to maize than others. Eubanks (2001a) showed that a series of reciprocal *Zea diploperennis* and *Tripsacum* hybrids are very easily obtained. Besides, blocks of maize or *Tripsacum* genes can be easily transferred between them along a hybridization "bridge" by means of *Z. diploperennis*, which blocks the sterility barrier between genera. Eubanks has pointed out that intergenic hybridization was involved in the domestication of maize, because the recombining progeny of the hybridizations of *Z. diploperennis* and *Tripsacum* gives out plants that resemble the early archaeological remains of maize.

Eubanks used molecular markers in *Zea* linkage groups to show the existence of the intergenic recombination of *Zea* and *Tripsacum* genomes. Twenty percent of the *Zea* genome is compatible with *Teosinte* alone, 36% is shared with wild *Zea*, and 43% with *Tripsacum* and *Zea* (Eubanks, 2001a). Comparative studies of DNA dactyloscopy with 74 molecular markers between *T. dactiloides*, *Z. diploperennis*, three species of annual teosinte, three primitive races of maize (Nal-Tel, Chapalote, and Pollo), and one modern maize showed that maize and *Tripsacum* have in common 28% of the alleles not found in teosinte. This goes against the hypothesis of the reticulated participation of *Tripsacum* in the formation of maize (Eubanks, 1999a). The studies of genomic hybridization undertaken to see the homology of the chromosomes from *T. dactiloides* and *Z. mays* ssp. *mays* proved that the hybridization of *Zea* and *Tripsacum* is still an open question (Poggio et al., 1999) (Grobman, 2004: 450–451).

The way in which *Tripsacum* provided genes to maize and teosinte in nature is as yet unknown. The available evidence indicates that a transfer of genes may have occasionally taken place. One of the most specific possibilities that was recently studied is that *Z. diploperennis* supplied a "genetic bridge" for the transference of genes between maize and *Tripsacum*. The two species easily cross. The transfer of *Tripsacum* genes into maize was attained with the formation of *Sundance* and *Tripsacorn*, hybrids that Eubanks obtained by crossing between genes. On the other hand, the SEM (scanning electron microscope) analysis of the exine present in the archaeological pollen of Proto-Confite Morocho from the preceramic site of Los Gavilanes[10] showed either some form of very early interrelation between *Tripsacum* and *Zea* in South America or instead that early wild maize did appear in Peru. This would be a very ancient separation from its wild counterpart in Mesoamerica (Grobman, 2004: 469).

Galinat (1977: 38) believes, as regards the potential relation between Confite Morocho and *Tripsacum*, that the internodes of this Peruvian race may

[10] This is discussed in depth in Chapter 5, which deals with the archaeological evidence (see Grobman, 1982: 171, and also pp. 163, 174, 176).

come from an introgression with *Tripsacum*. This is supported by an observation made by Galinat himself and that is still unpublished, to wit, that the experimental *Tripsacum*-maize introgression elongates the internodes of the rachis and reduces the rows of the kernels, independently of any effect on induration.

De Wet and colleagues made an in-depth analysis of the issue of *Tripsacum* introgression in South American maize. Following Roberts and colleagues (1957), they point out that there are some races that are highly tripsacoid. These races, however, do not show all of the phylogenetic affinities with the Mexican tripsacoid races, and on the other hand – and as has been repeatedly stated – teosinte is not known to occur in South America. According to De Wet and colleagues, this does not rule out the possibility that teosinte was indeed present in the South American continent during the early evolutive history of maize, and that these tripsacoid races derived their teosinte traits from a local introgression. They believe it is likewise possible that South American maize races may have had their origin in highly tripsacoid races introduced from Mesoamerica (De Wet et al., 1970), or else directly introgressed with *Tripsacum* species (as was posited by Mangelsdorf, 1961, 1968). (De Wet, Harlan, Stalker, and Radrianasolo 1978: 233–234).

De Wet and colleagues note that the evidence indicating that *Tripsacum* had a role in the evolution of maize in South America is purely circumstantial, to say the least. They admit that the transference of *Tripsacum* genes into maize is possible but emphasize that distinguishing the teosintoid and the tripsacoid characteristics in maize is not easy. "However," they point out, "the probability of natural introgression in the direction of maize seems small" (De Wet, Harlan, Stalker, and Radrianasolo, 1978: 240). We have seen that this actually is not that accurate, and that there is evidence that contradicts it.

De Wet and colleagues note that Roberts and colleagues (1957) pointed out that the Colombian Chococeño race crossed with local *Tripsacum*, "... but they never experimentally demonstrated introgression ...," even though Mangelsdorf (1968) showed that several South American races are tripsacoid in the morphology of their cob. Grobman and colleagues (1961) in turn likewise pointed out that several Peruvian races have this characteristic, but it was once again not proved experimentally (De Wet, Harlan, Stalker, and Radrianasolo, 1978: 234).

For Horowitz and Marchioni (1940), the Amargo race from Argentina also has *Tripsacum* introgression (see also Mangelsdorf, 1968), and an experiment showed that certain chromosomes in the Amargo give phenotypic effects similar to those that develop in the chromosome 4 complex of teosinte when it is introduced into the genome of a standard natural maize. Mangelsdorf suggested an introgression with *Tripsacum* because teosinte is not found in South America, and because *Tripsacum* is sympatric with cultivated maize in most of this continent.

According to De Wet and colleagues, the introgression of *Tripsacum* may have had a major role in the evolution of some South American races of maize. That introgression may have taken place in areas that have preserved the traditional methods of cultivation. Yet we need not assume a *Tripsacum* introgression in order to explain the tripsacoid traits in South American maize races, for according to De Wet and colleagues, said traits may derive from teosinte. They then insist, first, that it is possible that teosinte did grow at one time in South America and, second, that the traits in question had their origin in a highly teosintoid maize that was introduced from Mesoamerica. Because no evidence of this has been found, they conclude that maize reached South America in domestic state. It was when these new tripsacoid races reached South America that the various tripsacoid races had their origin, either by hybridization or due to selective pressures. De Wet and colleagues take the argument of the teosinte-derived origin to the limit, yet they nonetheless conclude that "the possibility of introgression from *Tripsacum* can … not be completely ruled out" (De Wet, Harlan, Stalker, and Radrianasolo, 1978: 241–242).

Finally, it is worth pointing out that Stebbins (1950: 277) has suggested that the cross of maize and *Tripsacum* may have been easier in primitive maize, as well as in any species of *Tripsacum* with fewer chromosomes.

Galinat and his coauthors made a good description of the differences between maize and its relatives, teosinte and *Tripsacum*. They believe that the most conspicuous difference is the induration of the tissues, particularly those of the rachis, the cupules, and the lower glumes. Both in teosinte and in *Tripsacum*, the caryopsis is enclosed in a hard, bony case composed of an internode of the rachis, a cupule, and the lower glume. These structures become highly lignified as the fruit matures. Experiments with maize-teosinte hybrids have shown that the genes responsible for lignification appear in many of the teosinte chromosomes, if not in all of them. It is therefore hard to find individuals in subsequent generations of maize-teosinte hybrids, even in those that most resemble maize, that do not exhibit some degree of lignification of the rachis, the cupules, or the lower glumes (Galinat et al., 1956: 102–103).

The Hypotheses Regarding the Origins of Maize: Proposals and Counterproposals

In the nineteenth century several authors suggested where maize could have had its origin. In 1829 Saint-Hilaire proposed that it perhaps came from Paraguay; in 1886 De Candolle suggested New Granada, that is, modern-day Colombia; and Birket-Smith was of the same opinion in 1943 (see, Mangelsdorf, 1974: 14; Mangelsdorf et al., 1964: 438).

It is worth recalling what Vavilov believed in this regard: "Central America, including the Southern part of Mexico is … the first place as the center of origin of corn. … There is no doubt, that along with an exceptional morphological diversity of corn varieties, found nowhere else in the world, there is also concentrated in these countries the diversity of physiological and ecological types...." (Vavilov, 1931: 195).

THE POD CORN HYPOTHESIS

The pod corn hypothesis, which Grobman (2004: 431) calls a "vertical evolution," posits that cultivated maize had its origin in a wild pod corn. This is the oldest proposal ever made and was the idea of Saint-Hilaire (1829), who described *Zea mais* var. *tunicata*, a new variety of Brazilian maize that had its kernels covered by the glumes. He believed that this was the primitive state of the plant, and that its place of origin must have been in South America, probably in Paraguay.[11]

Kempton (1937) also supported this hypothesis, but it was Mangelsdorf and Reeves (1939) who were its staunchest proponents. Many specialists rejected it, arguing that it was not a legitimate race spontaneously born of normal maize crops; that it is frequently monstrous and sterile; that it essentially differs from normal maize by just one gene; and, finally, that the hypothesis that teosinte is an ancestral form is more plausible. It was then posited that pod corn is highly variable; that it is similar to other monstrosities; that it is due to the action of plant hormones; that it does not have the characteristics of a wild grass; and that it could not have existed in nature. Even so, Mangelsdorf and Reeves (1939) restated the hypothesis with some changes, and despite the objections thus raised, they concluded (Mangelsdorf and Reeves, 1959a) that it had great value and presented more supporting evidence than was expected. They also pointed out that it could be verified (for a more detailed analysis, see Mangelsdorf, 1974: 11).

It is worth noting that in the 1970s Randolph published a paper comparing the oldest-known archaeological teosinte and contemporary Tehuacán maize. Here he noted the presence of a total of 23 pairs of contrasting traits, which he listed in his tables I–III – combined with those of table IV, but not repeated in table III – with classification characteristics that grass specialists considered more taxonomically significant. The traits Collins and Kempton identified more than fifty years ago (1920) with which to separate maize from teosinte, and which can be identified in their segregating progenies as intermediate in their mode of inheritance, number fewer than 33. Randolph concluded that the archaeological and palynological data fully support the idea that maize derives from a wild maize rather than from a primitive form of teosinte (Randolph, 1976: 332, 341).

Wilkes accepted that the maize from Tehuacán is wild, with a later teosinte introgression. He indicates that the explosive evolution of maize that is clearly evident in the archaeological sequence was the result of hybridization either

[11] For more information, see Mangelsdorf (1947: 191–194).

with teosinte or with different races of maize that had teosinte germplasm from previous introgressions with teosinte (Wilkes, 1989: 446–447).

Doebley (1990: 13) questioned this hypothesis because for him it suggests that maize and teosinte come from completely separate lineages. This then requires that several teosintes are more isozymatically similar to one another than any of them are to maize. For Doebley, this hypothesis is clearly refuted by the enzymatic information and the cpDNA.

The last scholar who examined this issue was Grobman, who wrote thus:

> We are still convinced that it is plausible that modern maize derives from the domestication of an already vanished wild maize, whose characteristics were those of an annual, precocious, monoic, plant of short height, with separate female and male inflorescences, but with ears that ended in staminated spikelets, with ramified ears and which are independently covered by husks, with very small and hard kernels (Grobman, 1982). Wild maize probably was a pod corn; the pod gene has been genetically dissected and it has been shown that it is formed by two genes (Mangelsdorf and Galinat, 1964).
>
> Wild maize hybridised with *Zea diploperennis* – the form of wild perennial teosinte that is ancestral to all teosintes – may have given rise to annual teosinte through natural crossings with maize, when the first maize crops, which were not contiguous (sympatric) with the distribution of the ancestral perennial teosinte, came into contact with it due to the movement of semi-domestic maize seed by man. This is supported by the absence of teosinte in the archaeological strata of Tehuacán (Grobman, 2004: 468).... The scanning electron microscope evidence of the pollen exine from the early Proto-Confite Morocho maize from Peru [found at Los Gavilanes] would be indicating [either] that in South America there was some form of very early interrelationship between *Tripsacum* and *Zea*, or that Peru's early and wild maize had a very ancient separation from its wild Mesoamerican counterpart. (Grobman, 2004: 469; see my Figure 3.1, Hypothesis 3)

THE TEOSINTE HYPOTHESIS

The possibility that maize has its origins in teosinte dates to the late nineteenth century, when it was suggested by Ascherson (1875). It was then restated by Vavilov (1931), Beadle (1939), Miranda (1966), Harlan and de Wet (1972), Galinat (1977, 1983, 1985b), Iltis (1972, 1983a, 1983b), Doebley (1983a), and Kato-Yamakake (1984).[12]

Ascherson (1875) showed that teosinte is the closest relative of maize. In 1896 he hereupon presented two hypotheses regarding the hybrid origins of maize. One held that it had its origin in a cross of teosinte with an extinct grass closely related to maize, and the other one, that maize is a product of a cross of wild teosinte and a race of cultivated teosinte. Collins (1912) posited that

[12] See my Figure 3.1, Hypothesis 1.

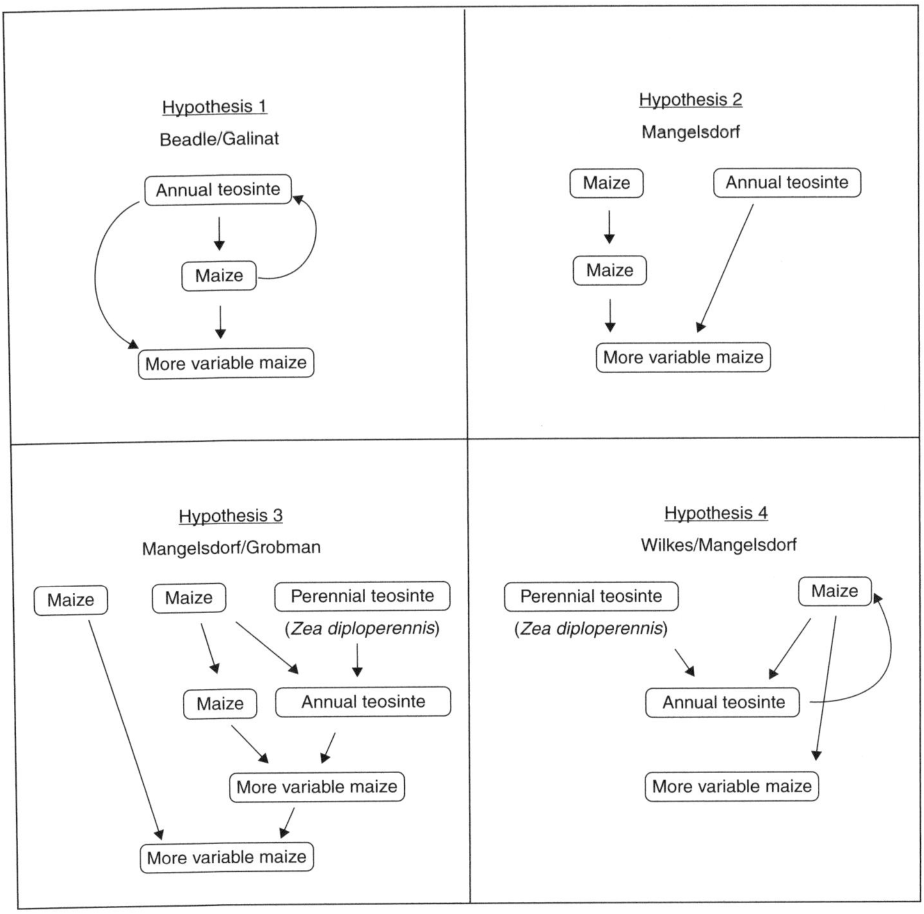

3.1. The various hypotheses regarding the origin and variability of maize according to Alexander Grobman. Drawing by Alexander Grobman.

maize is a hybrid of teosinte with an unknown grass of the Andropogoneae tribe. The studies Barghoorn and colleagues (1954) and Irwin and Barghoorn (1965) made of the Bellas Artes fossil pollen (which shall be discussed in depth subsequently), showed that the pollen from maize and from *Tripsacum* was found at a great depth, whereas that of teosinte was only found close to the surface and in the upper levels, where the abundance of maize pollen indicates the presence of agriculture, which went against the above-described position. But significantly enough, it follows from this study that teosinte is a race of maize derived from hybridization with *Tripsacum* (Irwin and Barghoorn, 1965: 43), which supports the position held by Mangelsdorf and Reeves (1939). Beadle (1939, 1977, 1980) and Galinat (1971a, 1975a, 1983) were the scholars who most strongly supported the orthodox position regarding teosinte, that is, that an ear is derived from an ear.

What is specifically being supported is that the ears of maize evolved from the female teosinte ear (female spike), which was initially considered as the earliest specimen of semidomestic teosinte (Galinat, 1971a, 1974b). The argument that claims these ancient cobs represent a semidomestic form in between teosinte and maize is based on the hypothetical transformation of the female inflorescence of teosinte into the female inflorescence of maize (Galinat, 1970, 1971a, 1974b; Iltis, 1969), which is in turn supported essentially by the common and highly suggestive observation of the series of morphologically intermediate ears of maize-teosinte hybrids (Benz and Iltis, 1990: 501; Grobman, 2004: 431; Mangelsdorf, 1974: 11–12).

Mangelsdorf (1986: 82) analyzed the position defended by Beadle, who is the staunchest defender of the thesis that claims maize had its origin in teosinte (see Beadle, 1972), and notes that what favors this position is the close genetic relation between cultivated maize and annual teosinte. But what on the other hand contradicts it is that there is no archaeological, ethnological, linguistic, ideographic, pictorial, or historical evidence showing the use of teosinte by the Indians. Besides, for Mangelsdorf the conclusive evidence is the archaeological evidence, which shows, in contrast with the thousands of maize remains found in the early Mexican levels, that only a few fragments of teosinte or teosinte hybrids are found where more remains of maize have always been found.

Iltis later on presented a new hypothesis, based on a catastrophic sexual transmutation[13] derived from an ear of maize from a teosinte tassel spike (Iltis, 1983a, 1983b; see also Doebley, 1984; Gould, 1984). Interestingly enough, Mangelsdorf long ago had made a somewhat similar proposal:

> ... It does not seem possible that maize could have been derived from teosinte during domestication by any genetic mechanism now known. If maize has originated from teosinte it represents the widest departure of a cultivated plant from its wild ancestor which still comes within man's purview. One must indeed allow a considerable period of time for its accomplishment or one must assume that cataclysmic changes, of a nature still unknown, have been involved. (Mangelsdorf, 1947: 191)

What Iltis (1983b: 888; 1987: 197; see also Iltis and Doebley, 1984: 607) posited is that the ear of maize is the transformed, feminized, and condensed central spike of the teosinte tassel, which ends in primary lateral branches. Feminization reactivated the vestigial ovary of the upper of the two florets, in each spikelet of the pair.

Iltis explains that an evolution due to an intense selection is based on the gradual accumulation of individual mutations, so that the genetic differences can be traced one by one. But a morphological evolution due to a change in function (particularly if it is due to a positional effect, as in maize) not only may be infinitely faster and more pronounced but can also initially lack discrete and identifiable genetic differences, because the switch in function may not have had

[13] This is also known as CST and sometimes even CSTT (catastrophic sexual transmutation theory).

a direct genetic cause. It may have been the subject of many changes at the same time; it may have been caused by minor multifactorial, quantitative changes; and the genetic bases of the ancestral structures that determine the new morphology may be far too removed in time (Iltis, 1983b: 893). Iltis (1987: 211) likewise believes that one of the causes of catastrophism may have been an abrupt ecological change. But it must be pointed out that he did not give any evidence of this, and that his position is purely theoretical.

Mangelsdorf (1986: 83) criticized the proposal made by Iltis, pointing out that archaeology does not show any evidence that annual teosinte existed at the time that the catastrophic change presumably took place.

Goodman (1988: 208) likewise disagrees with this "macromutation" and notes that, the complications inherent to the genetic fixation of phenotypic changes aside, there still remains a problem in that a recent genetic change or macromutation should segregate a simple 1:2:1 or 3:1 ratio in the F_2's of maize and teosinte. The genetic evidence strongly suggests that a macromutation in just one single gene has not taken place in recent evolutionary times (Galinat, 1985a). Against Iltis (1987),[14] who suggests that the genetic fixation of the hypothetical phenocopy was made through a polygenic selection instead of a single genetic mutation, Goodman (1988) instead points out that there are two problems with this proposal of a modified macromutation. One of them is that, unlike maize, teosinte rarely has a seed tassel, so that there is a high chance that polygenic selection will operate. The second problem, which suggests a very rapid evolution of maize (< 100 years) from teosinte, not only refers to the proposal made by Iltis but is also even more inconvenient for the position taken by Galinat (1988a). The rates of polygenic selection for cultivar samples rarely reach 5% for more than a year or two. They actually are typically 1–2% per character per generation, using the most advanced breeding methods and selecting just one character. For Goodman, both Galinat and Iltis require ancient farmers (or pre-farmers) to make gains of 500–1,000% or more in just a few centuries (or less), for an entire suite of characters with a taxon that has proven to be highly canalized and stable.

Wilkes made an interesting observation. He points out that maize has diverse types of endosperm, lends itself to different cooking styles and uses, and has a distinctive origin each time a landrace is formed. Heterosis between the unique gene system found in the different landraces explains in part the yield potential of this plant. The evolution of maize is clearly due more to a sequence of genetic changes throughout time than to the fixation of a particular trait. The changes that caused it to go from wild to domesticated plant were more of a process than an event (Wilkes, 1989: 441).

A large group of scholars has accepted that maize originated from teosinte. Here only some of them will be mentioned. Pearsall agrees, but what she finds

[14] The original publication mistakenly says 1988.

striking is the large variety of environments in which maize developed. This made her raise two questions: first, whether the adaptations took place at different moments in the past and, second, whether there were regions where the diffusion of maize was retarded or ground to a halt, until the ecological constraints had been overcome. For Pearsall, the archaeological evidence shows that both things took place. Pearsall based her work on Benz (1989) to explain the high adaptability of maize, which she claims is due in part to the fact that this is an organism that does not have very limited ecological preferences, and in part to the fact that it was spread by humans into habitats they maintained. She therefore does not believe that maize spread to South America on its own and considers that the geographic distribution that South American races had in the past was related, just like it is today, more with ethnic than with ecological boundaries (Pearsall, 1994a: 246). The latter point is completely lacking in support, for thus far there are no archaeological data regarding not just the distribution but even the potential presence of maize in many South American regions in pre-Hispanic times – to give just one example, there are no data for many areas in the vast Amazon forest.

Hilton and Gaut (1998: 870) also passed judgment in this regard: "Our data do not permit an explicit test of the hypothesis that parviglumis [*Zea mays* ssp. *parviglumis*] is the progenitor to maize, but the data are consistent with a domesticate/progenitor relationship for three reasons." These are as follows. First, there are no fixed differences between maize and perennial teosinte that may suggest a recent divergence between the taxa. Second, the gene trees show that maize lineages mix a subgroup of perennial teosinte lines. Finally, maize has 71% of the variation level found in both *Adh1* and *glb1* perennial teosinte. This fall in variation may be interpreted as consistent with a domestication event.

Harlan (1992: 222) interestingly enough accepts that the maize in the deepest strata at Tehuacán is domesticated yet notes that, although primitive, "… no extinct progenitor is required" because teosinte is a perfectly good wild maize.

Doebley and colleagues made several studies at the genetic level, and the conclusions they reached are that the key traits that distinguish maize from teosinte are each under multigenic control, even though for some traits – for example, the number of rows in the cupules – the data are consistent with a mode of inheritance that had to involve a single major locus plus several modifiers. For other traits, such as the presence or absence of pedicellate spikelets, the available information indicates a multigenic inheritance, with no single locus having a more dramatic effect than the others. Studies likewise indicate that, contrary to what had been held, the tunicate locus (*Tu*) did not have a significant role in the origin of maize. The major loci that affect the morphological differences between maize and teosinte are located in the first four chromosomes. Results show that the differences between maize and teosinte involve in part development modifications that enable, first of all, the primary lateral inflorescences – which are programmed in teosinte to develop into the (male) tassel – to become ears (female) in maize and, second, the expression of secondary male sexual traits on a female background in maize. Similar changes were probably involved in the origin of maize (Doebley et al. 1990: 9890–9892).

Doebley and colleagues (1997) add that the domestication of plants often entails an increase in apical dominance, that is, in the concentration of resources in the main plant stem, and the corresponding suppression of the axillary branches. One remarkable instance of this phenomenon is found in maize (*Zea mays* spp. *mays*), which shows a profound increase in apical dominance in comparison with its probable wild ancestor, teosinte (*Zea mays* spp. *parviglumis*). Previous research had identified the gene *tb1* (*teosinte branched 1*) as the major contributor to this evolutive change in maize. Doebley and colleagues cloned *tb1* by transposon tagging, and this showed that this gene encodes a protein with homology to the *cycloidea* gene of the snapdragon. The pattern of *tb1* expression and the *tb1* morphology of mutant plants suggest that *tb1* acts both to repress the growth of axillary organs as well as to enable the formation of female inflorescences. The *tb1* allele in maize is expressed at twice the level of the teosinte allele, thus suggesting that the gene regulating the changes underlies the evolutionary divergence of maize from teosinte (Doebley et al., 1997: 485). Martienssen (1997: 445) discussed this study and concluded that the conversion of teosinte into maize constitutes a change of historical proportions, and that the *tb1* gene seems to have had a crucial role in this transition.

Although supporting this position, Piperno and Flannery (2001: 2101) are cautious and warn that the issue still "... remains controversial ...," while noting that annual teosinte "is probably" the progenitor. Pickersgill (2009: 209) in turn also calls teosinte the "... presumed progenitor of maize...."

There also is an important group of specialists who have raised several objections against this position. In this case I will also cite just some of them. In the 1970s Randolph (1976) had already pointed out that the genetic stability of the main traits of maize, despite its having been domesticated and improved for more than 7,000 years, "... seems to have been ignored by the promoters of the teosinte hypothesis of the origin of domesticated maize" (Randolph, op. cit.: 344).

Iltis (1983b: 886) in turn pointed out at least five major issues that appear when trying to explain the origin of maize from teosinte. First of all, if teosinte gradually evolved, why have hybrids not been found in nature and in archaeological remains? Second, if the native population wanted its kernels, why has absolutely no evidence of this been found in archaeological remains, despite the kernals being extremely durable? Third, given the extreme hardness and concavity of the teosinte fruitcase, why is it that the glumes of the most ancient maize are soft and thin, and their cupules relatively shallow? Fourth, if teosinte ears transformed into maize ears, why is it that the ears in modern as well as in archaeological maize exhibit staminate "tails"? Fifth, if we compare the gradual evolution of all cereals, how is it that maize suddenly appeared out of ancestors that are hard to identify?

Goodman is another scholar who has been highly critical in this regard, but when he wrote in the late 1980s, the evidence available for South America was still scant, as he himself noted (Goodman, 1988: 200). He points out that a "formidable obstacle" for the hypothesis that maize was domesticated from teosinte is the fact that few indications have been found that farmers were able to turn any teosinte race into something similar to the smallest ears of primitive maize found at Tehuacán. Goodman mentions that Randolph (see Anonymous, 1982) pointed out that the essential botanical characteristics of maize have not changed since the earliest Tehuacán specimens[15] and concluded thus:

> If there was no significant change in the taxonomic status of *Zea mays* during 7,000 years of intensive selection pressure by inhabitants of a desert area successfully domesticating wild food plants, why should any biologist – knowledgeable corn geneticist least of all – continue to insist that *Euchlaeana* could have been transformed into *Zea* during a preceding time-span of a few thousand years, especially in the absence of either archaeological or anthropological evidence that teosinte ever was cultivated as a food plant or made use of at all extensively for that purpose as a wild plant? It is obviously unbelievable that any adequate appraisal of the many significant heritable characters differentiating maize and teosinte could lead to such a conclusion. Fundamental botanical characteristics of maize that separate it from teosinte include (1) paired pistillate spikelets born erect in open rachis cupules, (2) varying amounts of cupule compression effectively increasing rachis rigidity, (3) kernels borne at the surface of the cob partially enclosed by membranous glumes, (4) polystichous cobs lacking an effective dispersal mechanism, (5) a branched staminate inflorescent that included a prominent polystichous central spike, (6) an ordinarily polystichous cob with varying even numbers of kernel rows, the cob rarely distichous, (7) ears enclosed by protective husks. These chiefly polygenic differences and others of biogenetic significance in controlling growth habit, ecological and edaphic preferences must be given essentially equal consideration in any realistic appraisal of *Euchlaena* as a possible precursor of domesticated maize. (Goodman, 1988: 207)

Goodman then explains that although Randolph (1976) and Galinat (1978, 1985a) reach opposite conclusions, the genetic data are consistent. These show there are several key traits that separate maize from teosinte, and each of them is ruled by duplicate loci, as in the case of many isozyme loci. There is a minimum of six major loci that direct the following four traits:

1. Singular spikelets versus female spikelets per pair (probably loci in chromosomes 3 and 7).
2. Two rows of spikes versus many rows of spikes (probably in chromosomes 1 and 2).

[15] Goodman says the passage cited is on page 59, so it cannot be the first part published by Randolph in 1976 and must instead belong to the unpublished manuscript of the second part; see Anonymous (1982).

3. Abscission versus non-abscission of the male and female spikes.
4. Soft rachis tissue versus an indurated rachis tissue. Teosinte traits predominate in most of the genetic background; they are often simply inherited only in certain "primitive" backgrounds. Yet there are many polygenetically controlled traits that apparently completely separated teosinte from maize both 7,000 years ago and now. These differences include polygenes for condensation or shortening of the female (and male) spike; an increase in the number of seeds per spike; a shortening of the lateral branches; an elongation of the silks; the suppression of the tillers and of many low, lateral branches; and an usually earlier flowering in maize. There are indications of modifier loci of singular versus paired spikelets in chromosomes 4 and 8. For more genetic differences, readers should see Gottlieb (1984) (Goodman, 1988: 207).

Here the closing statement made by Goodman (op. cit.: 210) is worth citing in full:

> The hypothesized hybrid origin of teosinte came to be treated as an almost unquestionable fact between 1939 and the late 1960s. There seems to be an equivalent tendency today to accept the hypothesis that maize is descended directly from teosinte, probably under domestication, not as a hypothesis, but as a fact (Gould 1984) and this is based on little more evidence than that supporting the hybrid origin hypothesis. Galinat's portrait of how maize evolved from teosinte does not share the genetic weaknesses of Iltis' macromutation concept (Galinat 1975[a]). Nevertheless, it is at odds with almost all of the archaeological evidence. As archaeologists continue their efforts, we should slowly accumulate sufficient evidence if Galinat's concept is correct. Currently, however, the data at hand suggest very early (4000 to 6500 BC) use of maize in Mexico, with little or no evidence for the use of any teosinte anywhere (Callen 1965, 1967[b]). Indeed, teosinte fruitcases are no more promising a food source than are those of *Tripsacum*, and both seem inferior to *Setaria*, which was used widely at Tehuacán prior to displacement by maize (Mangelsdorf, MacNeish, and Willey 1964; Callen 1967[b]).[16]

THE COMMON ANCESTOR HYPOTHESIS

The first scholar who proposed this hypothesis was Montgomery (1906), but he did not take *Tripsacum* into account. Weatherwax (1918) posited that maize, teosinte, and *Tripsacum* have a common ancestor, and he later expanded his position (Weatherwax, 1919, 1950, 1954, 1955). Weatherwax claimed that from a morphological standpoint, all three plants show numerous rudimentary structures that are vestigial organs lost during evolution. If these rudiments could be replaced with fully developed structures, the three plants would have converged on a common form, so that all three have a single ancestral progenitor.

[16] See also Grobman (2004: 429).

Reeves and Mangelsdorf (1959) argued that by restoring the primitive organs of maize and *Tripsacum* alone, leaving teosinte aside, one would have the same type of common ancestor had the latter been included. Besides, this position does not explain all that is known and cannot be verified. Weatherwax objected that the common or immediate ancestor of maize will perhaps be found, so all that has been said is just speculation. The wild maize Weatherwax had in mind was probably a perennial plant. Judging by the characteristics given by Weatherwax, the plant would have some characteristics of the modern-day half-tunicate corn type but would not have the monstrous character of a real pod corn. Mangelsdorf (1974: 12) later indicated that the wild maize discovered in Mexico had several characteristics of the wild maize envisioned by Weatherwax, but it is not a perennial plant and does not have tillers. It is possible that other geographical races of wild maize did have tillers, but it is doubtful that any will be found that were perennial plants (Grobman, 2004: 431; Mangelsdorf, 1974: 12).

Less Important Hypotheses

There are two less well-known positions that are often forgotten but that are nonetheless worth including here.

THE PAPYRESCENT, "*SEMIVESTIDOS*" HYPOTHESIS

This is a slight modification of the pod corn hypothesis. Andres (1950) found a weak form of pod corn in Argentina, where the kernels were partially covered by soft glumes. He believed that this could have been the ancestor of maize. Galinat (1957) then called it "papyrescent," but it was shown that this was actually a defect in development.

THE CORN-GRASS HYPOTHESIS

Singleton (1951) suggested that the ancestral form of maize is a corn-grass. This is an anomalous type, the product of a single dominant gene that gives numerous tillers and small "ears," with a high proportion of single spikelets. Many of these kernels are partially enclosed in bracts, but most are not glumes but spathes. Singleton suggested that if this plant was found in nature, it would not be recognized as a maize and would be attributed to another genus. Mangelsdorf (1974: 13) admits that this may be true. If the corn-grass was the ancestral form, the mutation of just a single locus could have transformed a wild, useless plant into maize. But not all of the characteristics of the corn-grass fit this. Galinat (1954) suggests the corn-grass was a "false" progenitor of maize, which had certain traits that may have appeared in a remote ancestor of Maydeae. The archaeological evidence does not support this hypothesis. Mangelsdorf (1974: 13) believed that the possibility of this position being true was extremely remote.

THE TRIPARTITE HYPOTHESIS

In the first part of their hypothesis, Mangelsdorf and Reeves (1939) concluded that, far from being the progenitor of maize, teosinte was instead a hybrid of maize with *Tripsacum*. This does not explain the origin of maize, but other possibilities appear once teosinte has been ruled out, such as the possibility that pod corn could be the ancestor that is being sought. Second, although some scholars have said that pod corn is monstrous in several of its characteristics, Mangelsdorf and Reeves believed that its monstrosity is the result of a single relic wild gene that was superimposed over the germplasm of highly domesticated modern varieties. The third element in this hypothesis is the recognition that teosinte, if not the progenitor of maize, did at least play a major role in its evolution and domestication. Because teosinte is common in and around the maize fields in parts of Mexico, where it crosses with the latter, and given that the maize-teosinte hybrids are highly fertile and easily backcross to one or both parents, it would seem inevitable that there was, and still is, a flow of teosinte genes into maize, and hence that many modern varieties of maize can be the result of past hybridizations with teosinte. Unlike the other proposals, the tripartite hypothesis concerns only maize, pod corn, teosinte, and *Tripsacum*, all of which are alive, so that it can be experimentally tested (Grobman, 2004: 431–432; Mangelsdorf, 1974: 13; see my Figure 3.1, Hypothesis 2).

In this hypothesis, wild (pod) corn would have originated in South America. Teosinte would be a recent product of the *Zea-Tripsacum* hybridization after its introduction in Central America, and the new types are the result of mixtures with *Tripsacum*, that is, the varieties that appeared in North and Central America (Mangelsdorf and Cameron, 1942).

It is worth noting that there is biochemical evidence (Goodman and Stuber, 1980) supporting this hypothesis, which goes against maize having originated from teosinte.

Naturally there is a group of scholars who disagree with the hypothesis, but only some of them will be mentioned here. De Wet and Harlan (1972: 275–277) made a detailed and quite technical analysis of the tripartite hypothesis, and they question first of all that the "tripsacoid" South American races acquired their teosinte-like traits through direct introgression with *Tripsacum*, as was claimed by Mangelsdorf (1961). De Wet and Harlan believe that the tripsacoid traits either may be a vestigial characteristic of teosinte or may instead be derived from the teosintoid races introduced from Mesoamerica. They admit that the archaeological evidence points toward maize itself as the progenitor of maize and not teosinte. But they add that the information is incomplete. They also mention a new find of teosinte in Chalco, which has a carbon 14 date of 7040 years. We shall return to this later.

De Wet and Harlan posit three things: (1) that the progenitor of maize was by definition a wild maize, but that this was teosinte rather than an extinct race of maize – yet they admit that the maize from Tehuacán was a wild maize; (2) that cytogenetic studies indicate that maize and teosinte are conspecific, and that the chances that the female inflorescences of teosinte may have been derived from the introgression of *Tripsacum* in maize seem small; (3) that modern races can introgress with teosinte and still do so, for the two taxa are sympatric. They, however, suggest that the rapid racial differentiation in primitive maize took place in areas where teosinte was not present.

Randolph (1976: 324) in turn points out that the work done by Collins (1912) is one of the proofs used to abandon the idea of the hybrid origin of teosinte. Collins showed that *Euchlaena* is more specialized than *Zea* from an evolutive standpoint, particularly in regard to a more complete differentiation of staminated and pistillate inflorescences, and in regard to a diminution in the form and function of the outer and inner glumes of the pistillate spikelet pairs in maize.

Goodman (1988: 209) drew attention to the fact that Banerjee and Barghoorn (1972) and Banerjee (1973) present other evidence that goes against the aforementioned hypothesis, for an examination undertaken of the pollen grains with a scanning electron microscope showed that the regularity of spinule patterns in maize, teosinte, and *Tripsacum* is different. Goodman points out that it is a shame that the results of the study undertaken by Banerjee have not been published in their entirety. Goodman examined the photographs in Banerjee's dissertation (1973) and notes that there is a coincidence in the spinule pattern of teosinte and of North American maize.

Futhermore, Grobman points out,

> Some South American maize[17] has a clumped spinule pattern similar to that of *Tripsacum* introgression. Depending upon one's point of view, the latter point can be viewed as evidence that spinule patterns are not diagnostic at the generic level or that some South American maize has a history of *Tripsacum* introgression. The latter is not a new idea, but there is remarkably little evidence to support it. (Goodman, 1988: 209)

In his conclusions, Goodman (1988: 212–213) pointed out the following: (1) teosinte is not a hybrid of maize and *Tripsacum*; (2) it is believed that maize and *Tripsacum* are much more closely related now than they were in the mid-1960s; (3) of the various teosinte races, the Balsas race seems to be the one most similar to maize; and (4) the Guatemalan race is perhaps the one that most differs from all annual teosintes and is the one least similar to maize.

THE REVISED TRIPARTITE HYPOTHESIS

When Mangelsdorf and colleagues (1978) finished an article on fossil pollen and the origins of maize, which was published in 1979, Mangelsdorf added

[17] This is the archaeological maize excavated at Los Gavilanes (Grobman, 1982: 171, photograph 56).

a postscript regarding the discovery Iltis and colleagues (1979) had made of perennial diploid teosinte. Wilkes told Mangelsdorf that the hybridization between this teosinte and wild annual maize could have produced all of the races of annual teosinte, and Mangelsdorf agreed (Mangelsdorf et al., 1978: 251–252). It must be pointed out that in this same study, Mangelsdorf and colleagues (op. cit.: 249) do not deny that teosinte played a role in the evolution of maize – what they do not accept is that it was its origin.

Mangelsdorf and his team later noted that the discovery of the *Zea diploperennis* perennial teosinte (Iltis et al., 1979) dramatically changed the ways in which the issue of the origins of maize was considered and gave rise to new hypotheses. Of these, the most daring one was that of Wilkes (1967), who posited that annual teosinte was not the progenitor of maize and was instead its progeny, that is, the product of the hybridization of *Z. diploperennis* with a maize in the early stages of domestication (Wilkes, 1979). Mangelsdorf found this position correct, because unlike others it was more plausible, and verifiable and was consistent with the archaeological evidence (Mangelsdorf et al., 1981: 39; see Figure 3.1, Hypothesis 4).

The experiments crossing maize with *Z. diploperennis* (Cámara-Hernández and Mangelsdorf, 1981) are not consistent with the concept of annual teosinte as an ancestor of cultivated maize. Teosinte could instead have been a "carrier" of germplasm from the ancestral *Z. diploperennis.* The ancestor of cultivated maize is considered biparental, with *Z. diploperennis* and *Z. mays* serving as co-equal ancestors. According to this concept, *Z. mays* contributed toward the botanical characteristics of modern maize. We need not assume that numerous mutations or "catastrophic sexual transmutations" are required; ancestral teosinte instead contributed its robust root system and its resistance to many diseases. The explosive evolution of cultivated maize revealed by archaeological data may have begun when *Z. mays* and *Z. diploperennis* hybridized in Jalisco, Mexico, some 4,000 years ago (Mangelsdorf et al., 1981: 53).

When Mangelsdorf returned to this subject based on other experiments that were in turn based on Wilkes (1979), he noted that these experiments strongly supported the hypothesis that annual teosinte is a hybrid derived from perennial teosinte and maize: "In the process they confirmed my long-held belief that annual teosinte could not be corn's ancestor." Two quite different results were obtained in the hybridization experiments. First, the rhizomes of perennial teosinte, the underground stems that enable it to endure from one year to another, are accompanied by large and fleshy roots; this robust root system is in some degree transmitted to the hybrid progeny. Second, the F_2 generation included plants that had all the botanical features of modern maize. This suggests that the historical hybridization of *Z. diploperennis* with a cultivated primitive maize could have produced not just annual teosinte but also new races of maize, more vigorous and productive than any previous ones (Grobman, 2004: 469; Mangelsdorf, 1986: 85).

MacNeish and Eubanks (2000: 4) noted that the experiments undertaken to reconstruct the progenitor of maize by crossing maize and annual teosinte failed to recover the pure segregating parental phenotype. They believe that radical changes have taken place in the last 10,000 years in terms of the archaeological data, and this is too brief a span for biological modifications to have taken place only under natural conditions.

To sum up, the revised tripartite hypothesis posits the following:

1. The ancestor of cultivated maize was a form of pod corn.
2. Annual teosinte, the closest relative of maize, is not its ancestor but a derivative of the hybridization of maize and perennial teosinte (*Z. diploperennis*).
3. Many modern varieties of maize have had an introgression of teosinte, *Tripsacum*, or both (Mangelsdorf, 1983b: 233).

So restating this hypothesis 43 years after it had first been proposed apparently shows that its first and third points were valid, whereas the second one included an incorrect element.[18]

Eubanks (1995: 180) suggested another possibility, wherein the cross of *Tripsacum* and *Z. diploperennis* leads to another possible interpretation that combines both of Mangelsdorf's hypotheses. If the hybrids between the two genera simulate Mangelsdorf's extinct wild maize, domesticated maize may have appeared through human selection of natural hybrids of *Tripsacum* and perennial teosinte. Eubanks (1995, 1997a, 1997b) had already shown that the hybrids of perennial teosinte and *Tripsacum* resemble the reconstructed prototypes of primitive maize, and that the pollen of these hybrids can be distinguished from that of maize and teosinte, so his suggestion that wild maize was a natural hybrid of teosinte and *Tripsacum* is compatible with the palynological data (MacNeish and Eubanks, 2000: 14).

We must not forget that Mangelsdorf (1983b: 232, 245) had claimed that it had been proven that there is *Tripsacum* germplasm introgression in maize. It follows that maize did not have one or two ancestors but had instead at least three: *Zea mays*, *Z. diploperennis*, and *Tripsacum*. The ancestral *Zea mays*, Mangelsdorf says, was not just a single race but several. *Tripsacum* was not just a single species but was instead as many as could hybridize naturally with maize.

Grobman studied this issue and believes that the hypothesis regarding the origin of annual teosinte as a result of the hybridization of perennial diploid teosinte and wild maize, with subsequent backcrosses, was verified with the analysis of the outcome of the successful hybridizations made by Mangelsdorf and his team (Cámara-Hernández and Mangelsdorf, 1981; Mangelsdorf et al., 1981). Annual teosinte has been recovered from these crosses and backcrosses of F_1 hybrids between perennial diploid teosinte and maize, so Wilkes's hypothesis

18 For further information, see Galinat (1985b: 246), Grobman (2004: 432), and Mangelsdorf (1986: 80).

is correct. The perennial trait is ascribed to a main gene that is more recessive than dominant in manifestation. This is why Mangelsdorf (1983b) said this was the shape in which the annual teosinte presumably appeared in Jalisco, Mexico, some 4,000–5,000 years ago (Grobman, 2004: 437).

Benz (1994b: 157–158) believed that the tripartite hypothesis, both in its original and in it modified form, was not valid in regard to wild maize. He based his morphological evidence on the data in Benz and Iltis (1990) and Galinat (1983, 1985a, 1985b, 1988a). As for the antiquity of the Tehuacán archaeological maize, Benz only mentioned Long and colleagues (1989). His database clearly is restricted and therefore not valid. Yet he concludes that if the Tehuacán maize is older, a thousand years would have been required for maize to move from central Mexico to other parts of Mexico. Benz used the data in Pearsall (1994a) and Pearsall and Piperno (1990) to suggest that South American maize is older than the Mexican one, and in this case his data is once again incomplete. Even so his conclusion is interesting, because he points out that either the dates available for South America are wrong, or we instead have to find a maize in Mexico that is older than 3600 BC.

Tripsacum as a Hybrid of Maize and *Manisuris*

Walton Galinat (1964) added a fourth postulate to the tripartite hypothesis when he suggested that *Tripsacum* is a hybrid that has wild maize as one of its parents, the other being *Manisuris*, a genus of the Andropogoneae tribe that includes sorghum, millet, sugarcane, and other grasses and fodders. Mangelsdorf (1974: 13–14) believed that although this proposal was highly speculative, it relies – like the previous three postulates – on botanical entities that still exist today or are known archaeologically, as in the case of wild maize, so that experiments can somehow be carried out. Mangelsdorf added that this could be a useful addition to the hypothesis.

A Comprehensive Overview

Many scholars have tried to make a comprehensive analysis of the hypotheses presented, so a synthesis is nigh impossible. Only the two comprehensive approaches I find most significant shall be discussed here, that is, those made by Goodman and Grobman.

Goodman mentions the three major hypotheses, that is, the common origin hypothesis, the teosinte-based hypothesis, and the tripartite hypothesis. He believes that the second and third ones actually are just modifications of the first hypothesis. Both accept the idea that *Tripsacum* diverged quite early on, whereas maize and teosinte did so much later. It is quite clear from botanical and archaeological evidence, and from genetics as well as from the research undertaken by corn-breeding investigations, that whether teosinte diverged from maize or the latter from the former (due to human influence or not), the point is that both have continued their development since then. In most cases the evolution was divergent, particularly so in the case of maize. Teosinte seems to have developed little in the last 5,000 years. However, some changes in it seem to have such a high hybridization frequency with maize that they can no longer be securely used to represent teosinte in genetic and taxonomic studies (Anonymous, 1982; Randolph, 1976), and only a few races of maize show a history of teosinte introgression (Wellhausen et al., 1952) (Goodman, 1988: 204).

Goodman concludes that the isozymic and cytological results, along with the classic genetic studies, strongly suggest one or several alternative hypotheses.

1. There was just one evolutive event, involving just one pair of plants (or at most one very small population), which caused the divergence between maize and annual teosinte. This required much time long before domestication to allow the cytological and enzymatic differences – as well as the diversity present between maize and teosinte – to appear. This also entailed much subsequent parallel evolution, that is, in chromosomic knobs, isozyme alleles, and plant morphology.
2. There was one single, large, and variable population of annual teosinte that was turned into maize (or vice versa, which is an even more remote possibility). A bigger population would have been required to explain the variations present in both taxa. (It is not clear how this type of process could have taken place without ruining the progeny.)
3. Several teosinte populations gave rise independently to various maize populations (or vice versa, but this is a more remote possibility), as was posited by Kato-Yamakake (1984).
4. One of the taxa (maize or annual Mexican teosinte) gave rise to one single plant (or some plants) and acquired cytological, enzymatic, and morphological variations throughout several centuries through a combination of mutations and backcrosses to the original taxa.

Goodman, however, admits that all of these hypotheses clash both with the archaeological evidence and with the available biosystematics.

Archaeological sources suggest that both maize and teosinte were as different some 7,000 years ago as they now are (if not more so). Genetic studies suggest that maize and teosinte are basically and efficiently isolated, the occasional F_1 and backcrossing hybrids notwithstanding (Doebley, 1984; Kato-Yamakake, 1984), and both the chromosome knobs and the polymorphism of the isozyme alleles appear to be very conservatively preserved. So whatever the mode of origin of maize and annual teosinte might have been, it seems to have taken place long before domestication, and it must have involved multiple events (Goodman, 1988: 213).

Grobman in turn focuses his discussion on the two current hypotheses that are now the most likely ones. The first hypothesis is that modern maize had its origin in the domestication of several wild races of maize, an event that possibly unfolded in different places. Teosinte later interpollinated with maize, and this brought about an introgression of teosinte genes in maize and vice versa. If we compare the genome of modern teosinte with that of maize, we find that for several centuries the former coadapted and uniformized its genetic composition with maize due to a joint and reciprocal gene introgression, so that no modern-day discrimination is possible. This discrimination could have been undertaken with very ancient pre-Hispanic maize and an initial teosinte uncontaminated with maize. This hypothesis is supported by the evidence of very early remains of archaeological maize and late remains of teosinte. It should have been the other way around had teosinte been the putative father of maize. One more piece of evidence is the reinterpretation of the conclusions reached by the gene and genome studies of both species, which try to support the second position but do not do so.

The second hypothesis, the most popular one, claims that maize had its direct origin in the domestication of one or more races of annual diploid teosinte. The similarity in the number, size, and homology of the chromosomes of both taxa is presented as proof. The same thing happens with certain sequencing studies of a few defined genes that try to show the equivalency in this taxa sequence. The archaeological evidence does not support this position. Teosinte is a latecomer in regard to maize, and as yet it does not exist in South America. "It is surprising," writes Grobman, "the lack of information those who champion this theory have of the independent evolution of maize in South America, which is practically ignored in many modern publications." The assumption that wild maize could not have propagated itself in nature like teosinte "is due to ignorance" of the fact that kernels of wild maize could have become separated from their fragile rachis due to maturity and not necessarily to the fragmentation of the rachis. This is also possible in rachises as fine and thin as those of Proto-Confite Morocho, a primitive popcorn (Grobman et al., 1961, figure 49, 143) (Grobman, 2004: 465–466).

Flannery (1985: 246) correctly noted that the origins of maize remain one of the major enigmas of the main cultivated plants. The primary reason for this is that at present there is no wild maize. This means that wild maize either became extinct or descended from a different wild plant.

The Fossil Pollen from Bellas Artes (Mexico)

The discovery of the Bellas Artes pollen has been repeatedly mentioned in this chapter and must be expounded in depth, particularly because it is a most significant finding for an explanation of the origins of maize, especially bearing in mind that opinions have been given for and against it, and the existing evidence is not always explained. In the 1950s it was decided to build the first skyscraper – the Torre Latinoaméricana – in the site known as Bellas Artes, in downtown Mexico City. Several soil-sampling studies were therefore prepared by Leonardo Zeevaert, who was at the time the dean of the Faculty of Engineering in Mexico's Universidad Autónoma. Several deep cores were taken throughout these tests, which were then analyzed by Sears (1952) and by Sears and Clisby (1952).

The pollen analysis was carried out by Barghoorn, Wolfe, and Clisby, who concluded after a careful study that the large fossil grains resemble maize pollen, not just in their overall aspect but also in their size and in their pore-axis ratio whenever it could be determined. They are slightly different from modern maize kernels in that they have a slightly thicker exine and, even more significantly, a smoother contour in the folding (Barghoorn et al., 1954: 236, 238). Pollen from seven species of *Tripsacum*; from Mexican and Guatemalan teosinte; from three races of maize from the United States, from seven modern races of Mexican maize; from a modern race from Costa Rica; from three modern Peruvian races; and from two archaeological samples – an early and a late one – from Bat Cave in New Mexico were used as comparative materials for this study. The Bellas Artes fossil pollen comprised a sample of fourteen grains (Barghoorn et al., op. cit.: table 1). Figure 1 and table 2 (Barghoorn et al. 1954) exhaustively show the depths from whence came the pollen grains that had been identified. Between 74.2 m and 74.5 m there is only *Tripsacum* pollen. The grains of maize pollen appear at 70.3 m and are present up to 70.5 m. Maize and *Tripsacum* grains are common from 69.5 m to 69.7 m. Grains from both plants appear once more between 45.1 m and 45.3 m. It is only between 3.6 m and 3.8 m that teosinte pollen grains (which the authors mark with a question mark) appear alongside those of maize. Finally, teosinte is once again found at 3.3 m, always alongside maize grains.

According to a personal communication Paul B. Sears made to Barghoorn and his team, the sediments studied go back to the Wisconsin glaciation, and more specifically to the Iowa advance, that is, c. 22,500 years ago (Willey, 1966: figure 2–1, 28), a time when agriculture had obviously not yet appeared (Barghoorn et al., 1954: 239), even though the oldest samples go back much farther in time.

Kurtz, Liverman, and Tucker (1960: 92–93)[19] made an experimental analysis using modern maize pollen, replicating the environmental conditions and controlling whether the methodology Barghoorn and colleagues (1954) used as regards the pore-axis ratio is valid. Kurtz and colleagues claim that, using the methodology applied by Barghoorn and colleagues (op. cit.), one misclassifies 12% of the individual values of the samples, and more than 20% if we assume

[19] To avoid misunderstandings it must be pointed out that these same scholars published a paper this same year on this same subject, albeit with their names following a different order (Kurtz, Tucker, and Liverman). The paper here used is the second one (Kurtz, Liverman, and Tucker), which also appeared in 1960.

a normal distribution of the size of pollen traits. Of the 14 fossil grains studied, 4 (28%) are below the critical pore-axis ratio of 5.7. Of the remaining 10 grains, only 5% are sufficiently large in both the axis length and the diameter of the pore, as well as in the axis-pore ratio, to be classified as maize with a high degree of reliability. According to Kurtz and colleagues, the environment exerts a strong influence over the pollen; if the plants were under extreme climatic conditions, then the reliability of the pollen identification would be very poor. Yet they conclude by stating that "the data in the present study *do not refute* the findings of Barghoorn et al. (1954) but indicate the need for more reliable methods for the identification of corn pollen" (Kurtz et al. 1960: 94; emphasis added).

Grohne (1957) studied the pollen of "wild" and cereal-type grasses in Europe using phase-contrast microscopy and suggested that they could be separated through certain phase changes in the exine pattern. Rowley (1960) explained these phase changes by establishing that the "wild-type" grasses have three levels of phase retardation, whereas cultivated-type grasses have only two areas of phase retardation. The studies Rowley made used an electron microscope.

With these results, Irwin and Barghoorn (1965) went over the Bellas Artes pollen once more. They thus established that *Tripsacum* can be distinguished from maize and teosinte using phase optics. In *Tripsacum* the spinules can have an irregular distribution in the ektexine. On the other hand, in a large number of maize races the spinules are located very irregularly, whereas in most teosinte varieties the spacing of the spinules is less regular, and in some the spinules are rather closely aggregated and appear as clumps. Irwin and Barghoorn (op. cit.: 42) list several other changes that need not be mentioned here.

This new examination undertaken by Irwin and Barghoorn meant to study those samples in which one could not at first distinguish between maize, teosinte, or *Tripsacum* pollen grains. The conclusion they reached was that some of the samples were identified as maize and others as *Tripsacum*. None were of an intermediate or a teosinte type.

Besides the already-mentioned findings, two more significant results were also achieved by Irwin and Barghoorn (1965). First, the observation made regarding the general similitude between maize and pollen teosinte supports the idea that the latter is a maize race derived through hybridization with *Tripsacum*, as was claimed by Mangelsdorf and Reeves (1939). Second, they found that the most primitive races of maize (Puno, Chapalote, etc.) have the strongest and most regular pattern (Irwin and Barghoorn, op. cit.: 43).

Due to some doubts that had been raised, Barghoorn asked Leonardo Zeevaert whether the samples could have become contaminated. He received the following answer in a letter dated 17 October 1973:

> ... The sampling of the material was performed with a special sampler to obtain undisturbed samples of the soil, useful to determinate the natural

> compressibility and shear strength properties of the materials. Therefore, the samples taken were not disturbed or contaminated, they were "undisturbed samples" used in soil mechanics to determine "in situ" mechanical properties of the materials. *Therefore, you can be sure that the investigations made on these samples concerning the fossil maize pollen are reliable* [emphasis added].[20]

In the late 1970s, Banerjee and Barghoorn once more studied the Bellas Artes pollen and were able to recover a few more grains. They emphasized that, based on previous studies, maize had the largest pollen grains detected among grasses. Banerjee and Barghoorn therefore concluded that any grass pollen larger than 100 μ and with ektexine spinules evenly distributed was maize. They acknowledge in their study that some grass pollen with grains smaller than 100 μ and with a similar ektexine pattern could be maize. It is known that most of the living popcorn races and the pollen from some archaeological sites can reach up to 60 μ, which falls within the size range of the teosinte pollen grains. Maize-teosinte hybrids, however, exhibit a different and easily recognizable ektexine pattern when compared with the "pure races." The latter retain their ektexine patterns when crossed, even in their progeny. "If our criteria are correct," they concluded, then "the fossil pollen grains found in the deep-core either large or smaller in size, were '*pure races*' of maize" (Banerjee and Barghoorn, 1977: n.p.).

Sears (1982) sent a letter to *Science* 30 years after his original report (Sears, 1952), in which he specifically noted that the origin of the Bellas Artes pollen had been well checked, and confirmed that there had been no contamination. "The pollen was found in samples taken from the intact interior of precision cores" (Sears, op. cit.: 932). He then stated that recent studies do not show that there was a lake at the site and instead say that it was a swampy area with shallow lakes. Sears notes that Zeevaert's profile, which was based on seven cores, shows a 20 m descent in 4 km, and he thus extrapolates the possibility that the archaic and Nahua strata may have collapsed. He added, "The absence of artifacts along with the maize pollen is in no way remarkable ...," and finished by noting that "subject always to further research[,] it is my judgment that the pollen at 70 meters is an index of Archaic horticulture and not of wild Pleistocene maize" (Sears, 1982: 934). This letter, sent so many years after the original study was conducted, is striking, all the more so considering that it actually does not present any solid geological argument and is nothing more than an inference made by extrapolating data. The question that must here be asked is as follows: if there was an intrusion of archaic strata, then why is only maize pollen found, and not pollen from *Lagenaria* or any other plant that was then under cultivation?

De Wet and Harlan (1972: 273) are among the scholars who have criticized the Bellas Artes pollen study, because for them it is not wild maize. They based

[20] A copy of the letter is in my possession thanks to the courtesy of Professor Elso S. Barghoorn.

their work on a study by Galinat (1963), which shows that the size of pollen varies widely among the different races, and that its size correlates with the size of the cob. De Wet and Harlan therefore point out that it would be expected that wild maize would have small pollen, yet it falls among the upper size range of Mexican races.

Beadle (1981) likewise rejects the claims made regarding the Bellas Artes pollen, but he does not present additional evidence.

Mangelsdorf (1974) once again discussed the issue of the Bellas Artes pollen and raised the issue anew in 1978 (Mangelsdorf et al., 1978). He and his colleagues note the critiques leveled at the studies made and summarized them into three major points: first, the data are vague and ambiguous; second, the distinction drawn between teosinte and maize pollen is confusing; and finally, the samples were contaminated (Mangelsdorf et al., 1978: 238).

Mangelsdorf and colleagues give a good account of how it was that the Bellas Artes pollen was analyzed (Mangelsdorf et al., 1978: figure 1, 239–240), but it is not worth insisting on this, as it has already been explained. They point out that when Beadle's hypothesis regarding the origin of maize from teosinte was revived, those who defended this position used the study done by Kurtz, Liverman, and Tucker (1960) to rebut the Bellas Artes pollen. We saw that this study measured the axis-pore ratio of pollen. This had already been used by Barghoorn and colleagues (1954) to distinguish maize pollen from that of teosinte. We have seen that Kurtz and colleagues concluded that the axis-pore relation is not adequate for this purpose. Those who defend the teosinte position cite this study but do not make a detailed account of the data it holds, which was presented previously, particularly as regards the fact that its data "... do not refute the findings of Barghoorn et al. ..." (see Kurtz et al., op. cit.: 94). And five of the grains studied by Barghoorn and colleagues are big enough as regards the length of the axis and the diameter of the pore, and as regards the axis-pore ratio, to be reliably classified as maize (Kurtz et al., 1960: 85–93) (Mangelsdorf et al., 1978: 241–242).

Mangelsdorf and colleagues (1978) then explain that the population of the Bellas Artes fossil pollen is clearly different from any population of teosinte grains with which it has been compared. There are at least five pollen grains (Kurtz, Liverman, and Tucker, 1960), or 36% of the total, that are too big to be identified as teosinte pollen (Mangelsdorf et al., 1978: 243). For this study Mangelsdorf and his team compared the Bellas Artes pollen grains with archaeological pollen – derived from the Cueva de Coxcatlán in the Tehuacán Valley, Mexico, in two different periods – and with modern and archaeological pollen from Los Gavilanes, a Peruvian preceramic site, using a treatment known as acetolysis, which confirmed the antiquity of the various pollen grains in terms of their higher or lower capacity to respond to this chemical treatment (Mangelsdorf et al. 1978: 246–249; see also Grobman, 2004: 440).

In a later study Mangelsdorf mentioned the above-cited letter written by Sears (1982) and pointed out that it was based "… on fragile evidence …" (Mangelsdorf, 1983b: 237).

The scholars who favor the results of the Bellas Artes study include Goodman (1988: 200), who claims that had there been contamination, then there should likewise have been pollen from other plants, besides maize and/or teosinte. The contamination should have included some Old World genera, and it is hard to believe that these went unnoticed during the examinations made.[21]

[21] Interested readers will find a full summary of the data on the Bellas Artes pollen in Grobman (2004: 439–440).

4

The Domestication of Maize

> The origin of cultivated maize is a mystery. No form of wild maize is known from which cultivated maize could have arisen directly; and none of the theories proposed to account for the origin of cultivated maize has received general acceptance.
>
> Barbara McClintock (1960: 466)

If, as we have seen, the problem of the origins of maize has given rise to a large number of positions and controversies that in turn generated an abundant literature, the issue of its domestication has produced an even more abundant bibliography on which much has also been written. In both cases there have been excesses that have not brought about anything beneficial and have instead impeded the progress of knowledge. This should be avoided in science, because all that is valid in science is truth that is verifiable with evidence and concrete proof. Researchers must also be ready to accept not just opposite positions when these are presented honestly and earnestly but also the errors one may have committed in good faith when colleagues can show this with supporting evidence.

Although I have long had a very specific position as regards the domestication of maize, this chapter will try to present all of the data in the most objective way possible, without supporting any of the hypotheses or proposals that have been presented, and I will leave the discussion and my opinions for and against the various positions for the final chapter in this book.

In the case of the origin of maize there is no doubt that all specialists agree that it is Mesoamerican, and there is no way this position can be rejected with the evidence currently available. We have seen that the arguments hinge on what plant or plants may have been its ancestor. As for domestication, the controversy essentially hinges around two hypotheses: whether maize was domesticated in the Mesoamerican area and was then taken, already domesticated, to South America, or whether there were two independent centers of domestication, one in Mesoamerica and one in South America. It is true that there have been different positions that shall be mentioned throughout this chapter, but they are of little importance, because they were developed when the information available was still very poor. Some were even lacking in support.

The Hypothesis of Domestication in Mesoamerica Alone

It must be noted that this hypothesis comprises two different positions. Some believe that domestication happened in just one place in Mesoamerica, whereas for others there may have been several domestications in one general area. These two positions are discussed jointly because there are no major differences between them.

Pickersgill (1989: 433) accepts the idea that domestication took place in a single place and believes that once maize was taken to South America it had a considerable period of independent evolution in the Andean region. In addition, she believes that after the initial introduction of maize into this continent, there were no more exchanges between this area and Mesoamerica until much later.

Kato-Yamakake (1984) in turn has pronounced himself in favor of the multiple domestication of maize in the Mesoamerican area.

Galinat discussed the possibility that there have been several domestications that followed different paths. He believes that teosinte Chalco may have been domesticated by a combination of a reduction in the cupules and an elongation of the kernels, which led to such varied modern derivates as the Palomero Toluqueño, the Confite Morocho, and the Gourd Seed Dent. The majority of the maizes may predominantly come from another independent domestication, which apparently entails the tunicate locus and the Guerrero teosinte. In this case the glumes become soft and the rachilla is elongated in a way that elevates the grains almost beyond the chaff. Human selection, undertaken to attain recessive alleles to obtain a thick cob in the string cob loci, increased the vascular supply required for the more productive development of the ear. The long rachillae, plus a wider pith, enabled the attainment of the enormous cobs of contemporary maize (Galinat, 1988c: 111).

Actually, the two major hypotheses presented are known as Balsas (or River Balsas) and Tehuacán. The first one claims that mutations took place in annual teosinte (*Zea mays* spp. *parviglumis*) that led to maize in the lowlands of the Balsas River basin (Guerrero, Pacific Basin, 400–1200 masl). Benz (1999), Doebley (1990), Piperno and Pearsall (1998), and B. D. Smith (1995a) essentially support this position. They base their work on molecular data obtained with studies made in the 1980s on DNA coding for isozymes and chloroplastic DNA, which showed that teosinte is the species most closely related with maize, and which assumed a phylogenetic ascent of species with the biggest number of shared genes.

In a recent study, Piperno and colleagues (2009: 5023) explain that the archaeological evidence for maize found in the seasonal tropical forest 2,500 years before its presence in the dry highlands does not oppose and instead supports a less conflicting scenario, wherein maize was domesticated at lower and more humid altitudes in the Balsas watershed, where *Z. Mays* ssp. *parviglumis* is native. They also add that it is significant that despite the excellent preservation of macrobotanical remains in the Guilá Naquitz Cave and in the sequences from the Tehuacán Valley, as well as the most recent finding of phytoliths at Guilá Naquitz, the use of teosinte prior to the apparition of maize has not been detected in the highlands of southern and central Mexico, as one would expect if maize had originated in the highlands.

R.-L. Wang et al. (1999: 237) accept this. Piperno and Flannery (2001: 2101) point out that due to its ecological characteristics in regard to rainfall (1,200–1,600 mm) and temperature (20°–28°C), the Balsas region is a key place for the study of maize. This region is comprised of a tropical broadleaf deciduous forest. They also draw attention to the fact that although the Tehuacán zone has been well studied, the same cannot be said of the Balsas area.

The second position regarding domestication instead posits that it was in the highlands of the Tehuacán basin (state of Puebla, 1,000–1,500 masl) that the hybridization of two wild relatives of maize would have given rise to domestic maize (Mangelsdorf et al., 1967a). It was initially believed that *Tripsacum* and the extinct wild maize were involved in the hybridization (Mangelsdorf and Reeves, 1939). This was then modified, and the origin of annual teosinte was explained as the result of the hybridization of perennial teosinte diploid and maize at an early stage of domestication (Mangelsdorf, 1983b, 1986; Mangelsdorf et al., 1981; Wilkes, 1979), which evolved into the domestic maize of subsequent introgressive hybridization between maize and the newly developed annual teosinte.

MacNeish and Eubanks (2000) studied these two positions. They believe that the comparative DNA evidence from the ITS (internal transcribed spacer) indicates that maize appeared just before or at the same time as the Balsas and Chalco teosinte, as claimed by the studies carried out by Buckler and Holtsford (1996), who speculate that the domestication of maize could have taken place toward the end of the Pleistocene period or the beginning of the Holocene period, perhaps without human intervention. If this is correct, it would be unlikely that the Balsas or Chalco teosinte is the ancestor of maize, as there would not have been enough time to produce the accumulation of required mutations. In contrast, DNA studies indicate that maize, along with the Balsas and Chalco teosinte races, is derived from a natural hybridization between two of the most primitive taxa.

Other molecular data support the hybridization model. Talbert and colleagues (1990) showed that the *Tripsacum andersonii* that grows in Central America and in northern South America is a natural hybrid of maize and *Tripsacum*, thus indicating that the introgression between *Tripsacum* and *Zea* does take place in nature. There is also some molecular evidence that shows that the genes of perennial teosinte may have introgressed into South American maize in prehistoric times.

MacNeish and Eubanks concluded that at present we do not have archaeological and paleoecological evidence that support the idea of maize agriculture during the first two stages, that is, 7000–7500 years BP, according to the Balsas River model. In later studies the evidence is derived mostly from pollen, is not accompanied by solid data, and appears in many remains without contextual associations. Pollen and phytoliths refer to *Zea*, because it proves difficult – and often impossible – to distinguish pollen grains and phytoliths at the species level (M. E. Dunn, 1983; Eubanks, 1997b; Lippi et al., 1984; Piperno and Pearsall, 1993; Roosevelt, 1984; Rovner, 1999). It is also difficult to distinguish the pollen grains of *Tripsacum*-teosinte hybrids from those of maize and annual teosinte. So it is possible that wherever *Zea* pollen was found this may have been the pollen of a *Tripsacum*-teosinte hybrid (Eubanks, 1997b).

Another limitation of the Balsas River model is perhaps the reasoning that the early development in Mesoamerica parallels that of Panama. There are specific data (Ranere, 1980) that show the use of plants in Panama between 5000 and 10000 years BP, and there is no similar contemporary evidence in the lowlands of Mesoamerica. In contrast, the empirical data from the Mesoamerican lowlands, that is, the coast of Guerrero, the Pacific coast of Chiapas, Belize, and the early archaic of Veracruz,[1] indicate the exploitation of maritime and aquatic resources and little manipulation of plants: “There also is no support for the assumption that domesticated maize and other domesticated plants diffused through the tropical lowlands from Mesoamerica to Panama” (MacNeish and Eubanks, 2000: 14).

The other problem with the Balsas River model is perhaps the assumption that early maize – *Tripsacum* and teosinte – did not exist in the Mesoamerican highlands prior to 3500 BC, as held by Fritz (1994a) and Long and colleagues (1989). MacNeish and Eubanks do not accept the “contaminated” AMS (accelerator mass spectrometry) dates; they believe that the previous and traditional carbon (C14) datings are valid, and that the maize from Tehuacán was present around 7000 BP.

The Tehuacán model seems more plausible than the Balsas River model, because it is more solidly supported by archaeological and biological data. A powerful reason for the origin of maize in the highlands is that the perennial teosinte diploid is adapted to them more than to the lowlands (Iltis et al., 1979), and *Tripsacum* grows sympatrically with perennial teosinte in the same habitat.

The two models concur in that in later epochs, between 6000 and 5000 years BC, the maize found from the state of Hidalgo to the highlands of Chiapas spread to Guatemala, Honduras, and the Yucatán Peninsula. The Tehuacán model, however, emphasizes that the use of maize was limited in a large part of Central America, except for Panama, but it spread to the Pacific coast, in Venezuela and in the Andes (Pearsall, 1992b). In later times – 5500–4500

[1] Readers will find a large bibliography in the original study.

BC – the two models agree, but there still is one disagreement. The Balsas River model assumes that maize was important all over Central and South America, whereas the Tehuacán model holds that maize was of little significance south of Mesoamerica until it reached Panama, from whence it spread to northern South America (MacNeish and Eubanks, 2000: 14–17).

It is clear that there are almost no data on the process of domestication, and almost all of the inferences have been made on the basis of indirect data. Galinat (1977: 1) made a most relevant observation in this regard, that is, that the process of passing from teosinte to maize must have taken place in open fields, where cultural remains have not been preserved.

Flannery in turn developed his own proposal. He acknowledges that "… we have little data with which to pinpoint the origin of *Zea* cultivation, except the very early date of the corncobs from Tehuacán – an area where teosinte has never been collected in the wild. We know even less about why domestication began…." This is why Flannery has put forward a two-stage model (Flannery, 1973: 296). Flannery explains that the model of "density equilibrium" is not acceptable in this case, although it does apply to the Near East. He points out that there is no trace of a large population in Mesoamerica prior to 5000 BC. There is not one area where such a rapid population growth can be documented that may have affected the density equilibrium of adjacent regions. MacNeish, however, believed that this could have taken place in Puebla or Oaxaca (MacNeish, personal communication to Flannery, 1964).

There is another possibility that was posited by Ford (1968) for the southwestern United States. Here, just like in the Mesoamerican highlands, there is a great contrast in the productivity of plants between dry and wet years. Cultivation may have arisen as an attempt to overcome the differences between these two extremes by increasing the range of annual weeds. One of the major biotypes of the semiarid valleys in the central and southern Mesoamerican highlands is its tributary barrancas. A large variety of herbs and grasses that may be either uncommon or common in the valley grow on the floor of these barrancas, which are slightly more humid and are crossed by streams, which may be seasonal or permanent. Two grasses in this habitat are foxtail grass (*Setaria* sp.) and teosinte. Both were used by Indians in prehistoric times.

In the Tehuacán coprolites, Callen (1967a) found a selection of large grains and perhaps the first attempt at cultivating teosinte. The latter matures later, in autumn, and may have been gathered, but its cooked products have a bad taste. Even so, it holds a good amount of food, but it may be harder to prepare than *Setaria*. Yet Beadle experimented and showed that the daily consumption of 150 g of teosinte flour has no harmful effect at all. With the data available it seems that *Setaria* and teosinte must have composed a small part of the diet of the inhabitants. In a wet epoch, gatherers could have obtained a good amount of *Setaria* in these barrancas. This plant was obtained at a normal level in a dry period but with an increase in teosinte, which matures slightly later in the same habitat. But *Setaria* does not change its state no matter how much it is selected and planted. If Beadle is right, teosinte responded to cultivation and to the selection with a series of favorable genetic changes that went toward maize. This could have increased man's interest in the genus *Zea*.

Flannery acknowledges that in the second stage there are many variables as to why man would have wanted to increase the cultivation of *Zea*. Among these we have, first of all, the productivity of wild teosinte; second, the productivity of cultivated teosinte; third, the productivity of early maize; fourth, the productivity of the competitive vegetation that had to be removed for cultivation; and finally, the relative factor man-hours related with cleaning the land, cultivation, and so on. After experimental analyses, Flannery reached the following conclusions. He assumes that teosinte and *Setaria* were domesticated together. They drew close to the productivity of cereals in the Middle East only under the best conditions, as a second-growth pioneer in well-irrigated alluvial terrain, and besides, teosinte is harder to grow and process while it is half roughage. In many areas it was not even worth moving the mesquites to cultivate these plants. A more reasonable strategy would have been leaving the mesquites on the valley floor to grow 180 kg per hectare, and leaving teosinte in the piedmont of the barrancas, which was its home. According to the experiments made, teosinte could have yielded 150–200 kg per hectare, even reaching 600 under special conditions, but falling below 100 during droughts.

That this strategy was followed, Flannery claims, is suggested by the fact that no villages appeared on the Mesoamerican valley floors for thousands of years after *Zea* was domesticated. But the steady genetic change that brought about the growth of the ears led to a minimum productivity of 200–250 kg per hectare or more in maize. The threshold that made the Indians clean teosinte in order to expand the area under cultivation was crossed around 1500 BC. Permanent villages already existed from Puebla to Guatemala by 1300 BC. When they removed maize from the barrancas, they did the same thing with beans and cucurbits. We have a whole diet when we add the avocado (*Persea americana*), because maize gives carbohydrates, beans and squash seeds provide the plant with protein, and avocado provides fat and oil (Flannery, 1973: 296–300).[2]

The Hypothesis of Independent Domestication in the Mesoamerican and Andean Areas

Usually just a few scholars who raised this possibility are cited, but there actually are many. The most important ones are first listed, and details will follow. This is done in chronological order; the square brackets contain studies each author later published on this same subject.

[2] Matsuoka and colleagues (2002), who are cited later (see the following), posit that there was only one single domestication, but this study has serious flaws in its use of sampling data.

First in the list is Vavilov (1949/1950), with the caveat that in his first study (Vavilov, 1931) he posited the existence of just one center in Mesoamerica but then modified his point of view and accepted two centers. Then we have Randolph (1952 [1959]), Mangelsdorf and Reeves (1959b, 1959c), McClintock (1959: 456 [1960]), Grobman and colleagues (1961 [Bonavia and Grobman, 1989a; Grobman, 2004]), Mangelsdorf and Galinat (1964), Brandolini (1970), Mangelsdorf (1974 [1986]), Kato-Yamakake (1976 [1984]), R. McK. Bird (1980), and Galinat (1988a, 1988c).

Each and every position is not explained, because ultimately all of these authors hold the same position with different arguments or see these same arguments from different standpoints. The discussion is limited just to those that are most relevant.

Randolph (1952, 1959) claimed that, given the vast variability as regards cytological, morphological, and physiological characteristics, domestication necessarily had to have taken place starting from more than one race.

When Mangelsdorf and Reeves (1959b, 1959c) posited multiple origins for domestic maize, they did so based essentially on the vast South American variety and, among that, the varieties of primitive maizes. In this regard it is worth making a digression to recall that when Kuleshov (1929) studied maize from all over the world, he ascertained that the largest diversity of the *amylacea* group – that is, those with a soft endosperm – was found in Peru. He concluded that *amylacea* were precisely the groups that were most subdivided and rich in morphological and biological characters. The extreme variability is due to the ecological conditions of cultivation, mutation, strong hybridization, and a selection with established goals that has led to the existence of at least 42 races and multiple genetic variants.

Now, going back to Mangelsdorf, it is worth recalling that already in 1941 he and Reeves had commissioned Hugh C. Cutler to travel across Brazil, Paraguay, and Bolivia, with the goal of finding wild or primitive maize. Wild maize was not found, but Cutler did find a group of varieties with unusually primitive characteristics. One of these varieties is that of the Guaraní Indians of Paraguay, as well as another one that is widely spread in the Mato Grosso and other parts of the lowlands. The latter "... is the most extraordinary type of maize which we have ever studied," wrote Mangelsdorf and Reeves. This variant has the dominant genes of maize, including some that had only rarely been found before. It also has a low number of chromosomic knobs, and a few of them are small. The rachis or cob is thin and flexible, and the small spikelets can be easily removed intact. The ear has some characteristics that allow its true structures to show more clearly than had ever been possible through anatomical studies (Mangelsdorf and Reeves, 1945: 239–240).

When Mangelsdorf (1974) presented his famed synthesis years later, he quite clearly stated that Peru was an independent or a secondary center, instead of a primary center of domestication. And of the six primitive races that he proposed – Palomero Toluqueño, Chapalote or the Chapalote/Nal-Tel complex,

Pollo, Confite Morocho, Chullpi, and Kculli – the last three originated in Peru. Mangelsdorf finished by stating that wild maize evidently was a high- or intermediate-altitude plant and not a lowland one (Mangelsdorf, 1974: 113–120; see also Grobman, 2004: 438–439). It is interesting that figure 16.11 (p. 194) of his book shows three stone depictions of pre-Hispanic ears that must be Inca, one of which shows a pod corn and the other two the Chullpi race. The comment Mangelsdorf made in this regard is important, for he supported his position with Reeves (see previously): "The fact that there are no races of maize in Mexico with these characteristics is one of the reasons for concluding that there has been a separate origin of maize in Peru" (the characteristics of these ears are their ovoid form, their round kernels, and the absence of rows).

Mangelsdorf (1983b: 245) explained there are many races in South America that are different from those of Central America. He acknowledges that a problem that has still to be solved is whether these races descend from a continuity of wild geographic races, or of lineages that have their precedents in a Middle American ancestral stock. The fact that South American beans and squashes are different species from the Mesoamerican ones makes one suppose that there is an independent domestication center in the Andes. The South American races are in any case sufficiently different genetically from the Mesoamerican ones for their hybridization to have produced genetic combinations and an increase in the hybrid vigor. Another event that took place in South America is hybridization with *Tripsacum*, a distant relative. Although this is found in the "Corn Belt," in Mexico and in Guatemala, there is no evidence, other than a single case in Arkansas, that the Indians used *Tripsacum* or related it with maize. In Mesoamerica there is some indirect evidence of the hybridization of *Tripsacum* with maize. In South America the study of pollen from Los Gavilanes (on the north-central Peruvian coast) with the surface scanning electron microscope (SEM) showed there was introgression from *Tripsacum*.

McClintock (1959, 1960: 465) based her work on the evidence of chromosomic knobs and pointed out that cultivated maize may have had independent origins, starting from plants whose regions of knob formation had different capacities for the production of the substance used to form those knobs. She pointed out the existence of an "Andean complex."

It must be pointed out that, although the idea of a South American domestication of maize appears in the book Grobman published along with a group of his colleagues (Grobman et al., 1961), this hypothesis in truth was his, and he even coined the term "polyagrogenesis" for this process (see Grobman et al., 1961: 43). This position claims that because the Andean area is a place where a large number of species have been domesticated (Cook counted 70 in 1925), it is hard to accept that this happened in just one place. It is worth recalling that when Grobman and colleagues were preparing the aforementioned book, the Preceramic epoch in Peru was just beginning to be defined and the preceramic maize from the North-Central Coast had just been found and was still unpublished. But the essential argument these scholars used was the presence of several primitive races in Peru. Of Confite Morocho they said that it "... is as primitive or more so than any other known living race of maize" (Grobman et. al., 1961: 45). Besides, it was they who reinforced McClintock's proposal in regard to an "Andean chromosomic complex" (see Grobman, 2004: 437). In their study they used the concept of variability as had been put forward by Vavilov (1949/1950), even though Harlan (1956) had pointed out that the idea that centers of diversity are origin centers is debatable. Grobman and his team noted that although it can be acknowledged that the centers of diversity may appear on the periphery of the primary centers of domestication due to introgression or hybridization, the simultaneous apparition of primitive races and an extreme genetic variability in an area is best interpreted as evidence of a continuous occurrence of evolutive events, which ultimately have their origin in the domestication of these primitive races (Grobman et al., 1961: 41–47).

Let us see now how it was that Grobman and colleagues set out the evolutive history of maize. They note that the process of evolution of maize in Peru is evident. This is perceivable in the increase in the range of phenotypic and genotypic variations, the full increase in the range of adaptations of specific species to certain habitats, and the considerable increase in the potential yield in response to the improvements made in agricultural techniques (Grobman et al., 1961: 36). They likewise established several stages in this process:

1. Domestication in the mid- to low-altitude zones in the Andes
2. The formation of primitive races and the expansion of the original adaptive range of the species
3. The introgression of *Tripsacum*
4. The limited introduction of maize domesticated outside the central Andean zone
5. The interracial hybridization and formation of early hybrid races
6. The expansion of the cultivable area with improvements in agricultural methodology, and interracial hybridization and the formation of secondary hybrid races
7. The modern introduction and formation of incipient modern races (Grobman et al., 1961: 36–37)

In regard to the variability of Andean maize, they reached the following conclusions:

1. All of the range of variability in maize, in Peru and in outlying areas, is far bigger than in other primary regions in the continent.
2. This variability makes Peru an active center of evolution, both in the past and at present.
3. The presence of ancient indigenous forms and a large genetic variability make the Andean area a primary domestication center. This domestication

was followed by large-scale hybridization, and by introgression and selection. This primary center is additional to another independent primary center in Mesoamerica.

4. The biggest part of the variability can be classified into races (Grobman et al., 1961: 50–51).

We should recall here (see previously) that Mangelsdorf proposed six primitive races of popcorn (see Mangelsdorf, 1974: 113–120) from which come the current races of maize, three of which are Peruvian, through independent domestication. After discussing this issue, Grobman says that "the logical conclusion derived from the existence of these primitive races of maize would be that each of them comes from a race of wild maize through domestication and selection" (Grobman, 2004: 439). In this study he again took up the racial description he had previously made (see Grobman et al., 1961: 141 and passim) and updated it. It is as follows.

Grobman first mentions the Confite Morocho, which is probably a native of Ayacucho. This is the most secure progenitor of the eight-rowed maizes (Cuzco Harinoso de Ocho of Mexico and its derivatives). Furthermore, it is the precursor race of many others. Its antecedents have been found in the Peruvian preceramic and are known as Proto-Confite Morocho.

Kculli is the origin of the maizes that have a high concentration of anthocyanin, and it comes from the central Peruvian highlands. This race, which has its origins in preceramic times, is recognized by the intensity of its dark purple color (which is precisely the concentration of anthocyanin, both in the grains and in the cobs, due to the conjunction of dominant alleles from the *A*, *B*, and *Pr* genes, as well as the color of the Pl plants, all jointly selected). The races that may have been derived from Kculli are present not just in Peru but also in Colombia, Ecuador, Bolivia, Chile, and Argentina, with pericarp and aleurone colors, and number 29 (see Mangelsdorf, 1974: table 10.1, 115). According to Mangelsdorf (op. cit.: 114–116), Kculli is one of the most distinctive primitive races. He says that the Bolivian Kculli is slightly less pure than the Peruvian one.

The Chullpi race is perhaps native of the Apurímac-Ayacucho zone and is distinguished by its having the shape of a hand grenade. According to Mangelsdorf (1974: 109–111), this is the one that gave rise to all the types of sweet corn. Grobman believes this position "is convincing." Based on archaeological evidence, Grobman also believes that Chullpi is not the original primitive maize. Its ancestor is the Confite Chavinense, which has fasciated ears, with the kernels arranged in multiple rows in an irregular fashion, and which was originally popcorn. Many other races were derived from this primitive race; among them the sweet types for toasting, that is, the Chullpi, were selected. The Confite Chavinense does not appear much in early archaeological strata and coexists with the Proto-Confite Morocho and the Kculli.

So Confite Chavinense and not Chullpi would be the ancestor race of all the globular-shaped (hand grenade–shaped) ears such as the Andean Huayleño, Paro, Granada, Huancavelicano, and Chullpi. The latter gave origin to all of the derived races based on selection for the *su* gene of sweet corns. In Peru, Chullpi or Chispillo is preferred for toasting rather than as a sweet corn.

Grobman then discusses some characteristics of the primitive Mexican races. He mentioned first the Chapalote or Chapalote/Nal-Tel complex and Pollo.[3] Chapalote maize is native of Mexico; it gave rise to several Mexican races and is found at a very early date in archaeological sites. It is more or less contemporary with Proto-Confite Morocho, Proto-Kculli, and Confite Chavinense (Bonavia and Grobman, 1989b, 1999). The current Chapalote/Nal-Tel complex is far more evolved that Confite Morocho, its counterpart in the central Andean zone, which has retained a primitive cob structure, a very thin rachis, eight rows of kernels, and prominent navicular cupules.

As for the Palomero Toluqueño, this is a popcorn with an archaeological ancestor that is later than other primitive races from Mexico and Peru.

The Pira Naranja is a native of Colombia and the progenitor of Cateto, the Caribbean maizes, and of the Caigang groups of maizes from eastern South America (Grobman, 2004: 438–439).

Grobman prepared a summary 43 years after his original statement and stood fast by his hypothesis. He noted then that independent domestication may in fact have taken place out of races of wild maize in several places, like Mexico, the central Andes, and perhaps Colombia too (Grobman, 2004: 467).

It was in the 1960s that I began the study of preceramic maize of the central Andes (see Chapter 5) with Grobman and took the same position.[4]

When Grobman (1982: 179) analyzed the maize from Los Gavilanes, he pointed out that "until we have sources of evidence that in future show a clear relation with Mexico, and which do not appear in the Peruvian materials (for instance there is no teosinte introgression, which had already appeared in Mexico at the time of Los Gavilanes), we must sustain the hypothesis of an independent domestication for Andean maize...." In a subsequent study Bonavia and Grobman proposed that the diffusion of wild maize could have taken place in prehuman times through birds. This dispersal mechanism was posited by Pickersgill (1983) for other species, and Jaenicke-Deprés and Smith (2006: 91) proposed it for the diffusion of teosinte. Wild maize could have spread in this way from South America to Mesoamerica and then been independently domesticated by man.[5]

3 Mangelsdorf (1974: 117) doest not directly include Pollo in the complex, and in his text he associated it with a Colombian race.

4 See Bonavia (1982, 1990a, 1990b, 1991, 1991–1992, 1997, 2002b); Bonavia and Grobman (1978, 1989a, 1989b, 1998, 1999); Grobman and Bonavia (1978, 1979–1980); Grobman and colleagues (1977).

5 For more details, see Bonavia and Grobman (1989b: 462–463).

Based on the studies he has undertaken with me over a span of more than 50 years, Grobman insists that the very early evolution of different primitive races of popcorn in the central Andean area has been proven. These races are quite differentiated and have no counterpart in Mexico, but they do have a similar antiquity with the races in both centers of maize diversity. In Peru, the first primitive races appeared at quite an early date, both in the highlands and on the coast. But these primitive races "... all show a previous, primigenious Andean highland adaptation, and they only appeared on the coast in later periods." The evidence of this is the selection of high-intensity color based on anthocyanin. The amount of time its high-altitude Andean adaptation took must have been quite extended, for we must bear in mind that the preceramic maize of Los Gavilanes Epoch 2 (see Chapter 5) shows 97.1% of specimens with anthocyanic purple color selection, and 91.6% in Epoch 3 (Grobman, 1982: tables 11 and 12, 160–161). It is therefore highly unlikely that, at a time when there was a single primitive race in Mexico and three well-differentiated ones in Peru, that a single primitive race from Mexico could have reached the Andes without affecting the preexisting racial differentiation in Peru. Clear evidence of the sporadic exchange of maize between the central Andean area and Mexico appears only centuries later (Grobman, 2004: 467–468).

In the 1970s Brandolini (1970) summarized this issue and concluded that it was highly likely that the origin of cultivated maize was "polycentric."

Kato-Yamakake (1988: 109) concurs in that the domestication of maize may have taken place simultaneously in several different sites or at the same sites, but in different periods. Besides, because the primordial germplasm was taken to other lands, it found new ecological conditions in which the populations selected different materials depending on their needs and preferences. When the population turned into a farming people, this brought about a demographic expansion, and then migrations began. The different human groups established contact, and this gave rise to a kind of seed-exchange chain along specific migratory routes. The crossing of maize, with new opportunities for an additional selection of new racial varieties, took place when two or more migratory routes, each with a variety of maize with different origins, met at some point.

Brieger has exclusively analyzed living varieties of maize, because the archaeological evidence is quite scant in the area he studies, that is, the eastern part of South America, yet the conclusions he has drawn are worth summarizing. Brieger points out that there are four main types of maize that seem to correspond to successive stages of cultivation and domestication, with each stage exhibiting a different pattern of geographical distribution. These four types are as follows:

1. The most primitive type is a popcorn with quite an irregular pattern, which can be interpreted as a "relic pattern." The most promising materials are from the southern Guaraní in southern Brazil, in Paraguay, and in the Bolivian lowlands. This popcorn has a very important gene, with a supergametic

factor that impedes the growth of the pollen tubes of any other race and helps maintain these races of popcorns sexually isolated and pure.

2. The second type is a flint corn that differs genetically from the popcorn in a large number of genes with polygenic action, and that exhibits an increase in the size of the kernels. This type tends to occupy marginal areas in the northern and southern frontier latitude limits, along the Atlantic coastline and at high altitudes in the Andes.
3. The third type is a floury corn that differs from the flint type by a larger number of genes. It exhibits a slow selection and an accumulation of mutations.
4. We find dent corn at the highest level. It differs from the previous types in a number of polygenic genes that cause the presence of several cell layers of hard endosperm around the lower half of the kernel, which on drying produce a dent corn. This type evolved independently in several areas. This is the predominant type in Mexico and is the most productive type in the central Andes, as well as among the Caingang Indians in the high São Paulo Plateau. It seems to have an independent origin in at least three of the areas mentioned, but it is quite similar genetically (Brieger, 1961: 1).

Rivera (1971: 304) hypothesizes that maize appeared on the Peruvian coast in preceramic times "... from [somewhere] more to the south and to the east...." For him, this place of origin would be the Peruvian-Bolivian Altiplano. Rivera partly based his work on Rowe (1962: 51; the page he gives is wrong); however, all Rowe indicated was a possible origin in "the central sierra," whereas Rivera suggests that "... one or several races ..." were domesticated in the Peruvian-Bolivian Altiplano, from where they were taken to northwestern Argentina and the Peruvian highland. From northwestern Argentina, maize would have spread to Chile's Norte Chico "... essentially along the Copiapó, Huasco and the Elqui-Limari Complex zone, and probably slightly more to the South." From here maize was taken north, and it "... once again joined the Peruvian coastal maize route, which in turn was contacted by the Peruvian highlands." The position taken by Rivera is just a theoretical construct that does not give any solid argument or supporting evidence. It is also contrary to all the available archaeological evidence.

Rivera (1980b: 169) insisted on this point and mentioned the possibility that there was a "... probable center of maize domestication in the Southern Andean Area ..." that includes northern Chile. This hypothesis also lacks support. Yet Rivera insisted once again (Rivera, 1980c: 41), but still without adding a different idea.

In 1886 De Candolle (1959) suggested that the Chibcha area in Colombia was a center of domestication of maize; this idea was then taken up by Birket-Smith (1943) and Mesa Bernal (1955). But as Grobman and colleagues (1961: 41) pointed out, they lack a solid support base with which to prove this, and as Roberts and colleagues (1957) showed, Colombia was instead a crossroads in the diffusion of maize.

Like all hypotheses, this one has also had its share of supporters and critics. We will now go over the positions both sides hold. First we turn to those who believe this is the correct position.

Goodman (1976: 132) quite clearly noted that the greatest and sole source of variability in maize, in terms of a greater array of the kernels, cob, plant colors, and kernel size, lies in the central Andes, from midaltitude elevations to the highest ones. When Goodman coauthored a study that made a multivariate morphological analysis of 219 races of Latin American maize, the conclusion they reached was that the races found in Mexico and South America usually fit into different groups, and this supports the position that they have long developed independently (Goodman and Bird, 1977).

De Wet, Harlan, and Radrianasolo (1978) studied the morphology of teosinte and tripsacoid maize. They found morphological differences between the Mexican and the South American races that show they evolved independently under domestication. Because, as has already been pointed out, there is no evidence in South America of the presence of a wild maize or of teosinte, De Wet and his team suggested that the *Zea silvestris* Mutis (1957) found was probably a *Tripsacum* (we shall see subsequently that not all scholars agree on this). If one accepts the introgression of teosinte in Mexican maize, there is no problem in understanding teosintoid characteristics in South America. Divergent evolutive developments of indurated as well as nonindurated maize have since taken place both in Mesoamerica and in South America due to different selective pressures associated with isolation. De Wet and colleagues admit that the question that remains unanswered is whether direct introgression with *Tripsacum* has taken place in South America. They accept that this indeed happened and base their work on Horowitz and Marchioni (1940), Roberts and colleagues (1957), and Grobman and colleagues (1961).

The main argument against this, De Wet, Harlan, and Radrianasolo (1978) claim, is that no experiment has been made with this introgression. This, however, is not quite true, for "… many researchers have managed to make this crossing and managed to obtain experimental hybrids"(Grobman, 2004: 450).[6] On the other hand they point out that the hybridization of *Tripsacum-Zea* does not take place in nature. But we shall see when we turn to the issue of pollen that this also is not exactly true. All of this makes De Wet and colleagues (1978) conclude that tripsacoid maize – in the sense given to this term by Anderson and Erickson (1941) – be it either of Mexican or South American origin, seems to have been the result of the introgression of teosinte rather than of *Tripsacum*. They conclude that teosinte is ultimately a wild maize (De Wet et al., 1978: 744–746).

Although Harlan was never convinced of independent domestication, he was nonetheless one of the best examples of a scientist who managed to be objective

[6] See also Galinat (1977).

in his comments. Harlan noted that in the central Andes, maize was long different, in many respects, from the complex maizes of Mesoamerica. Bearing in mind the number of diverse domestications that have taken place in America, one more would fit in the pattern. We know, Harlan says, that teosinte had a bigger distribution than it has at present. It has been found at Taumalipas (Mexico), where it no longer exists. It has likewise been found in places where it no longer grows and only survives in herbariums (Wilkes, 1967). In 1910 a tetraploid race vanished in Ciudad Guzmán, Mexico. Harlan therefore wonders whether there was some type of teosinte in South America. The truth is that we do not know for sure. Harlan, however, cites the entry in the diary of José Celestino Bruno Mutis (Bosio) (1957) from 7 November 1777. Mutis was one of the most famed botanists of his time (Cadiz, 1732–Bogotá, 1808). He arrived at Bogotá (Santa Fe de Bogotá at the time) as the physician of the Marquis de la Vega, the viceroy of New Granada. The plants he collected have been held in Madrid's Botanical Garden since 1817. Mutis mentions a type of maize different from that which is cultivated, to which he gave the name of Maicillo Cimarrón (*Zea silvestris*). He found it in Las Minas del Sapo, close to Ibagué. Harlan says that according to the description left in the Madrid Botanical Garden, this is *Zea* and not *Tripsacum*. Harlan studied this subject and unsuccessfully traveled throughout this area trying to find some evidence, but he still insists that Mutis was the most eminent botanist in South America in his time, and that he cannot have been wrong in distinguishing wild from domestic maize (Harlan, 1992: 222–223).

It is worth noting that, despite supporting Beadle, and despite not being convinced of an independent domestication in the Andes, when he cites in his book the proposal Grobman and I made (Bonavia and Grobman, 1989b), Harlan defined the data presented as "... rather compelling evidence ..." (Harlan, 1992: 222), a statement he repeated elsewhere (Harlan, 1995: 185). Wilkes (1989: 453) also favors the proposal made by Grobman and me.

To finish this section, and before going into those who do not accept this hypothesis, it is best to cite Bruhns (1994: 95): "The age and distribution of maize in South America suggest that all of the evidence concerning the origin of this most useful plant is not in and that theories which postulate Mexico as the sole region of dispersal may well be considered reductionist and inadequate. Current theories also tend to ignore the historical framework."

The position of Earle Smith is ambiguous. He writes that there is no evidence of an independent domestication yet also points out that there is no evidence that maize was introduced into Peru from Mexico. Smith acknowledges that, based on the characteristics of modern races of maize, the "... evidence for a secondary center in the Andes is good" (Smith, 1968: 261).

Iltis on the other hand does have a very clear position. He points out that the large resemblance between certain types of Mexican-Guatemalan maizes with some Andean ones is due to the prehistoric introduction from a (secondary) center in South America, or instead to an independent and autochthonous independent origin in both regions due to an homologous variation "à la Vavilov." At present there is no way of establishing which of these possibilities happened, but the latter possibility cannot be ruled out. The fact that each major race of maize has its own popcorns suggests that homologous variations in the evolution of maize are a major issue, which has unfortunately been essentially ignored. In any case there is absolutely no evidence that the domestication of maize in South America took place independently from domestication in Mexico. Nor is the ancestor of maize the much-touted "wild" pop–pod corn, except insofar as teosinte pops quite well and has large glumes in relation to the small size of its kernels (Iltis, 1969: 3).

Pickersgill in turn also rejects independent domestication. She pointed out in this regard that several scholars (e.g., Darlington, 1963: 155) have noted that the centers of diversity may represent centers of hybridization or of ecological differentiation, and not necessarily the area where the plant has been cultivated the longest. As regards the Peruvian center of maize diversification – in connection with hybridization – it entails the presence of an indigenous form in this area prior to the introduction of any plant from Mexico. The ecological conditions in any case do not seem to be far more heterogeneous in Ancash, where this center of diversity lies, than in any other area in the Peruvian highlands, where maize is not so evidently variable. The presence of this center of diversity thus probably indicates a long history of maize cultivation in the area but is not by itself proof that it was domesticated in Peru and not introduced from Mexico at an earlier date.

For Pickersgill, the fact that the frequency of genes is different in Mexican and Peruvian maize likewise only proves that the varieties in these two areas have long been isolated. The initial introduction of a small sample of Maize from Mexico may have established – through genetic drift or the founder principle – a low frequency of genes in Peru that were not too common in Mexico. Independent domestication from a wild ancestor found in both areas is not the only possible explanation for the differences in gene frequencies (Pickersgill, 1969: 57). Yet when Pickersgill later coauthored a study on this same subject, she admitted the distinction between primitive Mexican and South American races, which allows one to conceive an independent domestication center, but she also pointed out that the two major issues are the lack of an ancestor of cultivated maize in South America and the presence of evidence equally old, or older, in Mexico (Pickersgill and Heiser, 1978: 136). Here it must be pointed out that the aforesaid evidence indeed did not exist when the article by Pickersgill and Heiser appeared, but we shall see later on (Chapter 5) that the antiquity of the early maizes of Mexico and Peru is almost the same.

A paragraph recently written by Piperno and colleagues is worth citing:

> Our evidence of maize during the early ninth millennium cal. B.P. confirms an early Holocene time frame for its domestication, as has been indicated

> by a large corpus of archaeobotanical and paleoecological data bearing on its dispersal into southern Central America and northern South America (6 [Piperno et al., 2000], 7 [Pearsall et al., 2004], 10–13 [Zarrillo et al., 2008; Piperno, 2006; Pearsall et al., 2003; Dickau et al., 2007]; 19 [Piperno and Pearsall, 1998]). *Maize and also possibly* ***C.[ucurbita] argyrosperma*** *squash join the increasing number of major and now-minor crop plants shown to have been brought under cultivation and domesticated in Mexico and South America between 10,000 and 7500 cal. B.P.*, about the same time as agriculture emerged in the Old World (6 [Piperno et al., 2000], 13 [Dickau et al., 2007], 17–20 [Piperno and Stothert, 2003; B. D. Smith, 1997b; Piperno and Pearsall, 1998; Dillehay et al., 2007]). (Piperno et al., 2009: 5023; emphasis added)

Although this is ambiguous, with this statement Piperno and colleagues apparently accept the independent domestication of maize in Mexico and South America, which they had always rejected. Even so, the article cited does not include a specific position regarding this issue, and none of the bibliography entries cited refer to this subject; besides, maize is mentioned just in a few studies. We await an explanation by the authors in this regard.

Causes That Led to Domestication

The domestication of plants throughout the world has always been attributed to man's need to become independent from nature and to improve his food base. As regards maize, there is also a different proposal that, although not important, has to be taken into account.

To the best of my knowledge, the first scholar who posited this was Iltis (2000: 23, 36), who suggested that the ancestor of maize (*Zea mays* ssp. *parviglumis*) was initially domesticated not for its kernels but for its sugary pith and other edible parts. The issue was raised anew by Smalley and Blake (2003), positing that at first the stalk was used before the kernels to get juice, which has sugar and which may produce alcohol (Smalley and Blake, op. cit.: 686). Smalley and Blake point out that Mexican caves hold many remains of chewed-up stalks – and other parts of the plant – from young maize. They also accept that there is no trace of this in Peru (Smalley and Blake, 2003: 682–683). Blake (2006: 68–69) once again returned to this issue and posited that the origins of domestication may lie iń the use of primitive maize to make alcoholic beverages.

Several objections can be levied against the article by Smalley and Blake (2003). First of all, the extensive bibliography they present – it is five pages long – lists just 14 entries for the Andean area, only one of which is related with coastal Peruvian preceramic sites. Then again one wonders how alcoholic beverages could have been prepared in the Preceramic period, when it was hard to keep a liquid boiling for a long time, as there were no vessels that could have withstood the effects of fire. The second article (Blake, 2006) does not present any actual evidence. These studies are purely theoretical and are completely unsupported.

Chávez (2003: 689) pointed out that Smalley and Blake (2003) use second-hand sources for Peru and Bolivia. No chronicler or voyager ever mentioned the use of a maize-stalk-based syrup. This is confirmed by their ethnographic experience in Peru and Bolivia. Grobman (2004: 444) also analyzed Smalley and Blake's proposal (op. cit.). He points out that in the Peruvian highland the custom of chewing the green stalks before they dry – the so-called *huiros*, or sweet corn stalks – still survives: "Yet after seeing the small size and the fragility of the first popcorns, one doubts that it may have been used to provide just a very small touch of sweetness [that was] quite different from the current evolved races."

Bonzani and Oyuela-Caycedo (2006) accept and have somewhat expanded the hypothesis of the diffusion of maize as an alcoholic beverage due to its sugar content but do not provide any solid argument.

Causes That Led to the Disappearance of Wild Maize

Although many scholars have touched on this issue, none have studied it in detail. At present not only do we still not have an answer, but this still is one of the major enigmas related with maize.

Mangelsdorf and his collaborators gave two explanations that present the two most logical and feasible scenarios for this. In one of them, the site where wild maize grew was chosen by man to sow cultivated maize. In the second scenario, wild maize grew in places that were not suitable for cultivation, but it still hybridized with cultivated maize until it lost its essential primitive characteristics and was almost unable to survive on its own. The latter of these two scenarios is the most feasible one. Maize pollinates through the wind, which can carry its pollen for many miles. It was practically impossible for any maize that grew in the wild in the valley to avoid hybridization with the cultivated maize that grew in nearby fields and that was giving out pollen profusely. The repeated contamination of wild maize by its cultivated counterpart could have eventually genetically "swamped" the former, thus leading to its disappearance (Mangelsdorf, MacNeish, and Galinat, 1964: 542).

Factors That Brought about the Major Evolutive Changes in Maize

Grobman and colleagues (1961) believe eight essential factors influenced the evolution of maize. These factors are as follows:

1. The characteristic monoecious habit of maize, and the fact that the formation of its seeds is the result of an almost complete cross-fertilization
2. Several cytological and genetic characteristics, for example, the great length of its chromosomes; the relatively high mutation rates of many genes; the high frequency of pairing of nonhomologous chromosomes; and the high

frequency of genetic recombination, which has no parallel in any of the reducing mechanisms of the crossing over that takes place in other species of plants and animals
3. A complete dependence on man for its propagation, and therefore its dependence on human preferences
4. An origin from genetically different populations, isolated by geographical barriers along large areas, and their subsequent confluence
5. A large heterotic response in the crossing of populations with a previous history of isolation
6. The genetic ability to produce a large number of seeds per plant, a characteristic sought by artificial selection in the development of some races, which is surely associated with other components with evolutive dispositions
7. Its mode of cultivation (hill planting and row spacing) and its harvesting, which facilitated the identification and selection of each single plant
8. The introgression of teosinte and *Tripsacum* species

All of these factors have been present in Peru since the cultivation of maize began. The last of the above-listed factors fulfilled a major, albeit restricted, role, when compared with the effects hybridization with teosinte had in the evolution of this plant in Mexico and Central America (Grobman et al., 1961: 36).

The Diffusion of Maize to South America

It has already been explained that Grobman and I posited that wild maize may have spread from Mesoamerica to South America (see previously). We now follow the different standpoints of those who believe that maize reached the southern part of the continent from Mexico already in a domesticated condition.

Bruhns believes that because maize agriculture had one single origin in Mesoamerica, and that because this plant then appeared virtually simultaneously in several quite separate places in South America, this all means that those who carried maize with them must have had access to a direct contact with certain populations, bypassing many nonagricultural societies on their way from Mesoamerica to the northern, central, and southern parts of the Andes, and to southwestern Brazil (Bruhns, 1994: 96). Unfortunately no more details are given in this proposal, which clearly is highly hypothetical.

Pearsall believes that the introduction of maize from southern Central America to northern South America must have taken place quite early. She suggests this would have happened around 5000 BC, that is, at a date close to its rise in eastern Panama (Piperno, 1984, 1988a; Piperno et al., 1985). Maize at the time was present both in the midaltitude areas of Colombia and on the Ecuadorean coast (Pearsall, 1994a: 251). She suggests a diffusion that took place over four stages. The process would have begun around 6700 BC, that is, before the date of introduction that had been previously proposed (Pearsall, op. cit.: 265–270). This clearly is a mostly hypothetical proposal – it even posits a maritime route without having any supporting evidence at all. The text mentions "barriers" without providing any evidence for them. And the bibliography shows that the data for the Peruvian Preceramic period are not known or were disregarded.

Stothert and colleagues (2003: 35–36) claim that primitive maize can be easily transported and that it must have been well adapted to the dry seasonal habitats of the Pacific coast of Central America, from whence it reached the Cauca and Magdalena Valleys. The populations may have taken the seeds from western Mexico and have spread them along these routes in northern South America. Once again this is a wholly hypothetical position, and no valid support is given.

Bugé also made some suggestions in this regard, but he tackled the issue from another angle. He claims that if we accept that maize originated from teosinte, it cannot then be a native of South America and must have arrived there at quite an early date, given the level of variability found in Peruvian races. Bugé believes that maize entered through the eastern Andes, from where it climbed down to the Pacific coast (he suggests Ecuador) and then climbed up once again to the highlands, from where it later descended once more to the coast. Popcorns survived in the Peruvian zone independently from Ecuador. From the eastern Andes popcorn spread to the tropical forest. Bugé based his work on native myths to establish the antiquity of maize in the jungle, and on linguistic data for its diffusion. He made a series of disquisitions on the scant use made of maize in the jungle zone, but all of these data lack support (Bugé, 1974: 53–57). Besides, Bugé is not free of bias, and he himself states at the beginning of his study that the data presented "... will be organized in terms of a model, partly derived from the data and partly following the implications of the 'Teocinte [*sic*] Origin' hypothesis proposed by Beadle (1972)" (Bugé, 1974: 29).

For Bugé, the tropical varieties of maize do not exhibit the diversity Andean races have. He accepts that popcorn must have been introduced at quite an early date in South America, around 5,000 BC or before. And he admits, based on the studies by Brieger and colleagues (1958), that although there are no absolute barriers for the diffusion of maize into South America, the major problem it had was its adaptation to altitude, and that this probably slowed its rapid diffusion. On the other hand Bugé believes that if the early varieties were taken from the Andes to the coastlands, this must have happened first in the north and then in the south. For him, the "relict distribution" of the primitive races of popcorn appears in the eastern Andes, in the *ceja de selva*, and at lower altitudes (Bugé 1974: 46). It is clear that Bugé has not read in this regard the study by Grobman and colleagues, which he, however, cites, for in this study the diffusion of the Proto-Confite Morocho and the Confite Morocho is explained, and it is clearly stated that they always had the same distribution in "middle elevations" and in the "high valleys" (Grobman et al., 1961: 147–148).

The sea is another route proposed by Bugé, from Mexico to the Guayas zone,[7] but he specifically means the Harinoso de Ocho race. His starting point is that the maize used in Valdivia was a flour or flint corn.[8] But the major entry route posited by Bugé is along the eastern side of South America, and of course from north to south, so according to him the greater development of the northern varieties is understandable. He accepts, however, the independence of the Ecuadorean and Peruvian centers due to the absence of flour and flint corn in Peru, where popcorn instead prevailed.

As Bugé himself points out, if all of this were correct, maize should be present in the cultures of the tropical forest at such an early date as 5000 BC, as is shown by the association of maize and peanuts on the coast[9] (Bugé, 1974: 29, 43, 44–47). This position could be one option in the range of possibilities, but only if one admits that maize arrived in domesticated form to South America. Bugé, however, does not present any concrete evidence; his arguments are weak, and his bibliography clearly incomplete. Besides, and as Bugé himself (Bugé, op. cit.: 29) points out, the South American tropical forest is the least-known area as regards this issue.

Lathrap's proposal is similar in some respects to that of Bugé, but the difference lies in the fact that he based his work on more concrete – albeit highly hypothetical as far as this writer is concerned – archaeological evidence. Lathrap suggests that the Nal-Tel race reached Colombia through the Panama Isthmus and from there spread to the series of Indian communities living in the tropical lowlands in the Andean piedmont, on the eastern basin of Colombia, Ecuador, and Peru. The pattern of early apparition of maize in several high Andean basins and in Peru's coastal valleys would indicate that the early races followed this path instead of the highlands or the Pacific coast. From this statement it follows that a primitive race like Nal-Tel would have reached the tropical lowlands in eastern Ecuador around 6000 BC and, once subjected to selection and cultivation, would have become the Kcello race. For this Lathrap based his work on the work done by Zevallos Menéndez (1966–1971) and Zevallos Menéndez and colleagues (1977). These studies claim that, according to the identification Earl R. Leng made, both the decoration of vessels with imprinted maize kernels and a negative cast found in a sherd show what may have been the Kcello Ecuatoriano race. Leng believes that this maize race may have been derived from Nal-Tel and would therefore be the ancestor of the Mexican Harinoso de Ocho (Lathrap, 1975: 20–21). I discuss this point when reviewing the archaeological evidence from Ecuador (see Chapter 5).

7 Bugé based his work on the research made on the Valdivia culture that shall be discussed in Chapter 5, which presents the archaeological evidence.

8 His sources for this are Wellhausen and colleagues (1952) and Brieger and colleagues (1958).

9 His only sources here are Grobman and colleagues (1961) and Pickersgill (1969).

Basing their work on the cytogenetic research by McClintock and colleagues (1981), Freitas and colleagues (2003: 901) suggest that maize was initially introduced into the central Andes. From there it spread to other lowlands and highlands in the continent without being supplemented by other types of maize until new genotypes spread southward, along the eastern Brazilian coast, but in recent times.

Now, one of the ways to approach the problematic of the diffusion of maize in South America undoubtedly is by analyzing the existing races of this plant. We must not, however, forget in this regard what Mangelsdorf (1974: 113) explained concerning the lineages. He quite clearly stated that these are not a clear cut of the taxonomic entities and may not be particularly useful for classifying races. Their usefulness is that they can show relationships. This is a very complex subject that lies beyond my scope, and it comprises a vast literature, particularly the series of publications made by the National Academy of Sciences of the United States, to which many references have been made in this book. Here, just some very general data and some specific proposals are presented.

Sanoja pointed out, in a general discussion of this subject, that the relics from early periods that were retained in Peru and Bolivia were so retained because they were races highly adapted to an extreme climate. The "presumably basic" early races – as Sanoja defines them – from Mexico-Guatemala and Peru-Bolivia may not have been related with, or derived from, one another and instead only belong to the domestication substratum (Sanoja, 1981: 86).

Mangelsdorf and his team on the other hand had a different perspective on this issue and posed a most interesting question. If the races introduced into Central America from South America were more productive than the indigenous races, and if maize had been introduced into South America, then why did it evolve more rapidly there? According to Mangelsdorf and colleagues, one of the factors behind this may have been the hybridization with *Tripsacum* (Mangelsdorf et al., 1964: 439).

Pearsall also discussed this subject and explained that when maize was isolated in South America from crossing with teosinte, the genetic suppression of the thinner wild rachis was lost. This resulted in the development, in some parts of South America, of races with thinner cobs from an ancestor with thicker cobs and developed cupules. Pearsall based these claims on the work done by Galinat (1971a: 462). So the cultural and geographic isolation may have been a good device for the development of alternative patterns, starting from the initial stock introduced. Pearsall posits that the early South American races evolved from the ancestral Nal-Tel/Chapalote (Pearsall, 1978a: 52).

It is, however, worth noting that the Group IV in the study by Goodman and Bird (1977: 208, 215), which corresponds to the popcorn from northern South America, relates Nal-Tel with Confite Morocho, with Pira, with Chirimito, and with Arguito but excludes Chapalote and Pollo and places the popcorns from the southern Andes with the South American flint corns.

Goodman and Bird (1977: 744) also say that the "Latin American" [*sic*] races can be grouped into 14 complexes, whereas the "Mexican" [*sic*] races are usually placed in mutually exclusive groups. De Wet, Harlan, and Radrianasolo (1978: 744) add that, if highly evolved races are excluded, only the Mexican Cacahuacintle, Nal-Tel, and Olotón can be grouped with the South American races.

The groups from different races with "primitive characters," as defined by Sturtevant (1899) and by Mangelsdorf and Reeves (1939), are perfectly well defined in Peru. These primitive characteristics in general are early maturity, short plants and tassels, a high index of leaf venation, small ears, long glumes, small kernels, slender cobs, a simple cob structure, large cupules, little induration of the rachis tissue, and so on. All primitive Peruvian races, like those from other countries, are popcorns, some of which still survive. There are, however, other pre-Hispanic races that still exist and that were not popcorns. Those in the latter group are not considered primitive because they seem to have derived, in pre-Hispanic times, from other ancestral races, known or unknown, whose existence can be logically inferred (Grobman et al., 1961: 141–142).

Here we shall simply list the primitive races; interested readers should see the respective literature. These races are Confite Morocho (Grobman et al., 1961: 142–149),[10] Confite Puntiagudo (Grobman et al., 1961: 149–154), Kculli (Grobman et al., 1961: 154–158), Confite Puneño (Grobman et al., 1961: 159–162), and Enano (Grobman et al., 1961: 162–164).

Grobman and colleagues, (1961: 164–165) include a series of early racial selections from populations of hybrids resulting from the intercrossing of the primitive popcorn, as well as its immediate second-step derivation. Some of the latter are the result of the hybridization of an early racial derivate with a primitive race, whereas some of the races described at the end of the list had two early derivatives as ancestors. The common feature of all of these races is their apparition in pre-Hispanic times. They are as follows (all page numbers refer to Grobman et al., 1961): Huayleño (Grobman et al., 1961: 165–169), Chullpi (170–175), Granada (175–179), Paro (1961: 180–184), Morocho (185–189), Huancavelicano (189–196), Mochero (196–201), Pagaladroga (1961: 201–205), Chaparreño (205–210), Rabo de Zorro (210–215), Piricinco (215–221), Ancashino (221–229), Shajatu (232–237), Alazán (232–237), Sabanero (237–241), Uchuquilla (241–244), Cuzco Cristalino Amarillo (244–250), Cuzco (250–253), and Piscorunto (253–256).

Pardo is one of the late races that have stirred the debate. Grobman and colleagues believe it was introduced. Its most marked characteristics are long, slender ears with large kernels and large and drooping tassels that are green-colored like the plants. Its origin is not secure. On the one hand the Tabloncillo and Harinoso de Ocho races, which according to Grobman and colleagues (1961: 305), Wellhausen

[10] For Confite Chavinense, see Grobman and colleagues (1961: 147).

and colleagues (1952) described as "exotic pre-Columbian Mexican races," are quite similar to the Pardo in their morphological characteristics, such as the type of ear and kernel, the characteristics of the tassel, the low leaf venation index, the expression of tripsacoid characters, the resistance to rust, and the adaptability to low elevations. But at the same time they share a series of major characteristics with the Cuzco race and with several related Andean races, for example, the number of rows, the general appearance of the ear, the type of cupules, the internode pattern of the stem, the low number and the typically Andean position of the chromosome knobs, and the adaptation to the mean low temperatures during their growing season. Besides, Pardo grows well at intermediate altitudes.

This race may derive from the introduction of Tabloncillo into the coast, as it resembles that race in several morphological characteristics. Although it is not fully established, it is possible that this introduction took place at the end of the Late Horizon, or more likely after the Conquest, given that there is no archaeological evidence of it. It was probably brought in the galleons that sailed to Peru from the western coast of Mexico. We find in the chronicles that maize was being toasted aboard the ships slightly after the Spanish conquest. Wellhausen and colleagues (1952), however, note that there is an inconsistency, to wit that Tabloncillo has a range of seven to nine chromosomic knobs, whereas the range in Pardo is zero to two. The reduction in the number of knobs, its flour-like characteristic, and its overall resemblance to Cuzco maize can be explained through the hybridization of Tabloncillo and Cuzco on the coast, and the subsequent selection of the current characteristics of Pardo (Grobman et al., 1961: 304–306).

Several of the scholars who have already been cited have pointed out that the most remarkable characteristic of Peruvian maize is its great variability, which is greater than that found in any other part of the world. This variability shows in the array of adaptations to a vast range of ecological conditions, as well as in its morphological, cytological, and genetic characteristics. Based on all that has been thus far pointed out, Grobman and colleagues state that the main point is that this is an agricultural product that has been used since preceramic times, thus enabling its selection for specialized uses as food. Although maize is quite varied in nature, this has led in many races to the development of large floury kernels. Except for a few late introductions, all of these races are native to the Peruvian or Bolivian area, and they had their origin in a small number of precursor races. Two of the latter races were popcorns: Confite Chavinense, a fasciated-spherical eared race, and Proto-Confite Morocho, which has a slender ear with 8–10 rows of imbricated kernels. Proto-Kculli, the third precursor, may have been a popcorn that was domesticated independently or that was derivative of Proto-Confite Morocho; it is the forerunner of some of the highly anthocyanin-pigmented races of the Andes. The popcorns in Peru and neighboring countries, both archaeological and living ones, are derived from these races; such is the case of Confite Puntiagudo, Pisankalla, Confite Iqueño, Confite Puneño, Enano, Proto-Pichinga, and Polulo.

There also is strong evidence that maize was domesticated in the central Andes independently of Mesoamerica, as has already been explained (see previously). This hypothesis gains support with the discussion of the racial problem because, although wild maize has never been found in the Andes, we can infer from the distribution of archaeological maize and the derived living races that the zones where this took place probably lie between 2,000 and 2,800 masl. This maize was taken from the highlands to the coast. As corn cultivation expanded in area covered, given enough time, small isolated populations or ecological races acquired a genetic diversity that allowed them to turn into individual races. Some of them hybridized among themselves or with precursor popcorns, which gave rise to an additional number of races: the Anciently Derived races, which, as we have seen, number 19. Thirteen of these developed in the Andes at mid- and high altitudes, four on the coast, and two on the eastern slopes of the Andes, at mid- or low altitudes.

Cultivated maize established contact with *Tripsacum australe* in the periphery of the central Andes, probably in the Bolivian lowlands, and an introgressive hybridization of these two species could then take place. At least three tripsacoid – Confite Puntiagudo, Enano, and the Uchuquilla – with a most ancient origin are known. The introduction of *Tripsacum* genes passed from there to other derived races. There were at least 24 races of maize in Peru at the time of the Spanish conquest, and this is based on archaeological data. A major hybridization then took place with some introduced races, and 10–20 races known as Lately Derived races and Incipient New races were created. Interestingly enough, most of these late groups are cultivated at lower altitudes than the earliest races. We have seen that only the Pardo maize may have come from Mexico in late Inca times or during the early Spanish conquest. All other introductions are no more than 100 years old.

The evidence of the influence of Mesoamerican maize on its Andean counterpart was in truth quite limited until very recent times. In the central Andes, the movement of maize was essentially centrifugal. Some coastal maizes, however, show a strong Colombian influence, particularly from the Chococeño race (itself a derivative of a Peruvian popcorn, the Confite Iqueño with *Tripsacum* introgression), as well as from races from the Bolivian lowlands and highlands. The very productive coastal and highland races gave rise to the most complex hybrids. From all this it is clear that the evolution of maize in the central Andean area, as well as in other areas, seems to have been the result of a continuous process of accumulation of genetic diversity and residual heterosis, brought about by interracial hybridization and subsequent selection. Peruvian maize still retains a considerable genetic variability (Grobman et al., 1961: 337–343).

A significant fact that has to be emphasized is that the distributional range of several Peruvian races extends far outside the frontiers of the central Andes, that is, essentially to present-day Peru and Bolivia, for we find them in Ecuador and Colombia to the north, and in Chile, Argentina, and neighboring zones to the south. Although some additional peripheral variations appear in these countries, it is clear that the central Andean region is the area that currently exhibits the highest variability of maize in regard to the whole South American continent. Much of it is relatively recent and is due to the hybridization of Andean and exotic races – with a subsequent selection and recent introductions – for the process continues in a relatively large aggregate of microcenters all over Peru. This variability forms part of the concept of "genes-center" or centers of reproduction, with a diversity that Harlan posited in 1951 and that continues to grow (Grobman et al., 1961: 50). Just one example in this regard is given here. The Guaraní race is a popcorn with two kinds of kernels, some with pointed kernels and one with round ones. According to specialists, it is related with the early Andean popcorn (Bugé, 1974: 51).

It is, however, worth noting that things are quite different and complicated in the large area east of the Andes. To the southwest of the Guaraní soft corn area there is a wide belt of dent corn races. The Calchaquí race in northern Chile and northwestern Argentina is a flint corn. To the east and the southwest of the Guaraní area there is a race of flint corn known as Orange Hard Flints (some call it Cateto). Among the Guaraní, soft corn and Cateto are completely different races.

There are special races associated with particular Indian tribes, and there are others that occupied, and still do occupy, vast areas of distribution that go back to the time of the European discovery of America. We thus find a very peculiar maize race – Interlocked (Entretrabado) – that extends from Iquitos (Peru) to the northern lowlands of Bolivia, and eastward up to the Ilha do Bananal (in central Brazil). The soft corn extends toward the south, from the lowlands of Bolivia across Paraguay and up to northern Argentina, southern Brazil, and Uruguay. The Guaraní have been moving a lot since antiquity. Maize from this tribe has been found in the state of Bahia. There also are two racial groups – Interlocked and Guaraní – which are a type of soft corn with an intense yellow to orange color. This particular color is typical of central South America and is found in Colombia (in high-altitude racial types and in the northern lowlands), but it does not exist in the Andes of Peru and Bolivia or in Mesoamerica.

Brieger concluded that there are at least five completely independent and very large areas with a typical maize: two that are related with the largest development of flour maize, two with flint corn, and one with dent corn. Three races seem to be limited to definite indigenous tribes, that is, the Guaraní, Calchaquí, and Caingang races. The Interlocked race is in turn associated with a larger number of tribes. And there is data that indicates the presence of many other races in connection with small tribal groups. It is for this reason that Brieger concludes that agriculture in the central South American area attained a considerable level of development, which has neither been considered nor studied in depth (Brieger, 1968: 553–555).

There are several controversial hypotheses regarding the origin of the Maíz de Ocho race. Upham and colleagues (1987: 410) believe these can be divided into two groups. One of them vouches for its similitude with the Cabuya maize from Colombia – a derivate from the Peruvian Confite Morocho – a position held by Grobman and colleagues (1961), among others, who thus uphold their position of a South American origin. The other group follows the position of W. L. Brown (1974) and W. L. Brown and Anderson (1947), among others, who suggest a Guatemalan origin.

This is a variety that adapted itself to arid conditions; has large, floury kernels; is more productive; and is easily ground, all of which makes it important for the human diet.

When discussing the problematic of the diffusion of South American races to Mexico, Mangelsdorf (1974: 188) used the name "Pre-Columbian Exotic," which Wellhausen and colleagues (1952) gave to these varieties. Mangelsdorf included among them the Harinoso de Ocho, that is, he accepted its South American origin. He likewise noted that the only primitive race that exhibits most of the characteristics of this maize is the Peruvian Confite Morocho popcorn, and he thus concurs with Grobman and colleagues (1961) in that all of the corn known as Maíz de Ocho had its origin in this primitive race (Mangelsdorf, 1974: 113–114).

Upham and colleagues (1987: 417), however, reject a South American origin. They point out that the oldest date for this type of maize is 1225 years BC (according to the radiocarbon dating), for what Galinat called the Proto-Maíz de Ocho in southwestern New Mexico, as well as its hybrids with Chapalote and the recombinations that led to the modern Maíz de Ocho, visible in archaeological studies; these studies evince that this maize developed in the dry southwestern deserts (see especially Upham et al., op. cit.: 413–417). The dates for other zones where this type of maize is found are the late second millennium and the early third millennium AD.

Galinat (1988b: 682) disagrees and believes that the Maíz de Ocho had an independent origin more than once. Peru's Cuzco Gigante is the most evolved form in terms of the size of the kernels.

As was already pointed out when discussing Mangelsdorf's position (see previously), Grobman and colleagues (1961: 337) have noted that Peru is unquestionably the center of diversity of flour maize. Pickersgill (1972: 99) initially agreed with the South American origin, from whence it spread to Mexico via Colombia, and from thence northward. Bugé (1974: 34) also supports this hypothesis. Yet Pickersgill, basing her work on a study by Whitt and colleagues (2002), later pointed out that the "sweet" mutation may have taken place independently at least twice, and that the sweet corns from North America and Mexico cannot be the result of a South American influence (Pickersgill, 2007: 936). She therefore asks: "The mutation to sweet corn evidently occurred twice in North America, but what about South America?" She claims that the proposal

Mangelsdorf (1974: 111) made of a common origin in South America must be revised, and that it will be interesting to establish whether the Peruvian sweet corn has the same mutations as all of the North American sweet types or not (Pickersgill, 2009: 206).

Genetic Information

Although it is true that genetic information is not a subject per se and that it should be included among the remaining data mentioned here, it was decided to discuss it on its own because in the case of maize it developed a lot in the late twentieth century. I do not intend to discuss this information, because it lies beyond my specialty; all that is sought here is to present evidence so as to be able to establish in the general discussion (Chapter 10) whether it fits with the remaining data, and which position(s) it supports. The following discussion is limited to the *Zea* and *Tripsacum* genera.[11]

First let us consider some general concepts regarding this subject. Selection for domestication is the degree in which the desired allele becomes more abundant than those that are not wanted in successive generations. The more intensive the selection in domestication, the more the desired alleles increase in regard to the less-wanted ones (Benz et al., 2006: 74). The studies undertaken by Harlan and De Wet (1971; 1972: 274) showed that the primary genetic pool of all cereals lies both in spontaneous and in cultivated plants. It follows that the primary genetic pool of maize includes all cultivated races as well as those of annual teosinte. The secondary genetic pool of maize probably includes all species of *Tripsacum*.

The process of domestication is still a mystery, but what is most known is that many domestic plants have fewer genetic variation than their wild ancestors. This reduction in genetic variations probably is the result of a small population of plants, which is associated with an intense agronomical selection of certain traits. This is what is defined as a population "bottleneck," and what Eyre-Walker and colleagues (1998) called "domestication bottlenecks." The effects of this phenomenon are important both because they limit the genetic variation of cultivated plants and because the lack of genetic variation in modern plant diversities is a topic related with their growth.

[11] Given that there has been an explosion of research activity in the areas of classical and evolutionary genetics, archaeobotany, cytogenetics, and the molecular biology of maize in recent years, I have felt it is very important to add an appendix to my book dealing with such advances and their interpretation. I am very fortunate that a specialist in these fields, Alexander Grobman, my closest collaborator and partner in the maize ethnobotanical aspects of my archaeological research, has accepted this task, for which I thank him. He has wide expertise and knowledge of maize evolutionary genetics, racial classifications, molecular biology, and breeding and can present a scientific survey and balanced interpretation of such voluminous information. His contribution, specifically prepared for this English edition of my book, is included as an appendix at its end.

To study the domestication bottleneck of maize, Eyre-Walker and colleagues (1998: 4441) analyzed the genetic differences between two wild relatives *Zea mays* ssp. *parviglumis* and *Z. luxurians*. They point out that the domestication of maize is intriguing for two reasons. First, maize is morphologically different from its wild relatives. The authors claim that the wild progenitor of maize has been "unambiguously" identified as an annual member of the genus *Zea* "... only through the application of molecular markers," according to the studies made by Doebley and colleagues (1984; Doebley, Ranfroe, and Blanton 1987). Second, according to the studies undertaken by McClintock (1978; see subsequently), maize is genetically diverse, as evidenced by, among other characteristics, the chromosome knobs.

For Eyre-Walker and colleagues, these two characteristics of maize are somewhat contradictory, because on the one hand its high genetic diversity implies that it had a historically big population size, and on the other hand the high degree of morphological divergence between maize and its wild ancestors suggests that maize underwent a selection for its morphological traits. In other words, the morphological difference between maize and its wild ancestors suggests that maize experienced a domestication bottleneck (Eyre-Walker et al., 1998: 1441).

We know that maize is genetically diverse, and many measures bear witness to its high genetic variation. This implies that maize historically had a large-sized population, as was already noted. A computer simulation was run to analyze the domestication bottlenecks of maize. This showed that the sequence diversity found in the locus *Adh1* of maize can be explained with a founding population of very few *parviglumis* individuals. Eyre-Walker and colleagues (1998) claim that we can think of populations with 10–20 individuals in 10 generations. What this study shows is that the sequence of diversity in maize is consistent with a small initial population of only a handful of individuals representing a quite diverse progenitor. The simulation shows that the size of the founding population depends on the duration of the domestication event.

Because nothing is known of maize, we can take einkorn wheat as an example. Einkorn wheat requires at least some centuries, and if we assume maize proceeds in like fashion (e.g., 300 years), then the bottleneck in a population of 586 individuals of *Z. mays* ssp. *parviglumis* suffices to explain the diversity of the sequence found in the *Adh1* of maize. If we base our thinking on the existing archaeological data,[12] it has been estimated that domestication was attained some 7,500 years ago, and that maize was introduced into other regions in Mexico around 4,700 years ago. If we substract 4,700 from 7,500, we get 2,800 years. The authors believe this estimate is far too big for two reasons: First, the archaeological finds are scattered, and it is therefore hard to know which is the earliest area of maize distribution. Second, some believe that maize

[12] Eyre-Walker and colleagues published their work in 1998.

was domesticated at a later date. In any case, based on population models, the domestication bottleneck of 2,800 years corresponds to a population of about 5,600 individuals. The authors conclude that the length of the domestication bottleneck of maize cannot be established. Their studies, however, show that a few to a few hundred individuals of *Z. parviglumis* sufficed to capture the amount of genetic diversity found in the locus *Adh1* of maize. They calculated that the separation of *Z. luxurians* and *Z. mays* ssp. *parviglumis* must have taken place about 1.02 million years ago.

Eyre-Walker and colleagues conclude that the exploration undertaken shows that maize, its high genetic diversity notwithstanding, could at first have had a small population of quite diverse progenitors. Genetic studies have shown that the morphological differences between maize and its wild relatives can be attributed to just five loci, as was shown by Dorweiler and colleagues (1993). It is possible that the domestication of maize was based on the crossing of individuals with the appropriate alleles from those five loci with a small number of additional wild individuals, and with a continuous selection for morphological traits. Eyre-Walker and colleagues accept that their study cannot reject more complex scenarios of hybridizations, introgression, and large population sizes, but it suffices to explain both the morphological divergence of maize and the extent of its diversity (Eyre-Walker et al., 1998: 4445–4446).

Wright and colleagues (2005: 1310–1314) likewise touched on the issue of the domestication of maize. They believe that this resulted in a highly modified inflorescence of the plant's architecture. The enhancements after domestication also brought about remarkable changes in yield, in the habits of the plant, in its biochemical composition, and in other traits. At the genetic level, these phenotypic variations were the result of a strong (artificial) directional selection over certain genes. Wright and colleagues admit that most plants and animals have had a domestication bottleneck that reduced their genetic diversity in regard to their living ancestors. The selection is similar to the more severe bottleneck that removes most (or all) of the variants from a given locus.

The data on polymorphism are generally consistent with the population bottleneck during the domestication of maize. It has to be pointed out that these authors accept the work done by Matsuoka and colleagues (2002) (see subsequently) – which, as has already been pointed out here, has serious problems – in which a single domestication of maize in Mexico is posited. Wright and colleagues estimate that 2–4% of maize genes were selected during domestication and subsequent enhancements. If we tentatively accept that the maize genome has 59,000 genes, 1,200 of these were then chosen during domestication. They do, however, admit that some of these "candidate genes" could be false ones (Wright et al., 2005: 1310, 1313–1314).

Jaenicke-Després and Smith (2006: 84) also posit that genetic changes reflect an unconscious selection and adaptive responses to new selective pressures, which are associated with human planting, harvesting, storage, and methodical selection. The major changes in maize are the modifications in the vegetative structure (viz., the reduction of the branches), in seed head morphology (viz., number, placement, size, shape, and number of rows of cobs), in the characteristics of the seeds (viz., kernel shape, hardness, and color), and in several properties of starches and proteins.

Teosinte has a 90% genetic identity with maize, and 10% of its very important genes allow it to survive in the wild state. Maize does not have these genes, and they characterize annual teosinte as a different taxon (Wilkes, 1979: 12). Many of the traits that distinguish maize from teosinte are controlled by just one or two genes (Beadle, 1972; Galinat, 1971a; Langham, 1940) and, given the occurrence of appropriate mutants, they can be quite rapidly fixed during domestication (Pickersgill and Heiser, 1978: 136).

Mangelsdorf pointed out that a series of studies of DNA organelles, along with other data, lead to the conclusion that annual teosintes are derived from a mixture of *Zea diploperennis* with maize (Mangelsdorf, 1983b: 241–243).

Doebley and colleagues (1990: 9888) in turn analyzed the segregation of both molecular marker loci (MMLs) and the key of morphological traits that distinguish maize from teosinte in an F_2 maize-teosinte population. This enabled them, first, to make a more accurate estimate of the number of genes controlling the traits that distinguish maize from teosinte. Second, it allowed them to characterize the chromosome situation of the genes, and finally to establish the relative contributions made by the different chromosomic regions of the key traits. The biosystematic evidence indicates that the annual Mexican teosintes *Z. mays* ssp. *mexicana* and ssp. *parviglumis* show closer genetic relations with maize than with other teosinte species, and it suggests that the latter probably is the progenitor of maize (and this supports Beadle 1980). Doebley and colleagues admit that although this is accepted by the majority, there is still no consensus as regards the genetic and morphological steps in this transformation. The main problem is that maize and teosinte differ dramatically in their morphological characteristics, and that the alternative perspectives on the transformation require different and major morphological changes.

Goloubinoff and colleagues (1993) studied the changes with a significant sample. They worked with two modern specimens from the United States; two specimens from the Peruvian coast; one from central Mexico; two ancient Peruvian specimens, one from the coast and one from the highlands; an ancient sample from Chile; four samples of modern teosinte (*Z. mays mexicana*, *Z. mays parviglumis*, and *Z. diploperennis* from the Mexican lowlands and highlands, and *Z. luxurians* from Guatemala); and one sample of modern Mexican *Tripsacum* (*T. pilosum*) (Goloubinoff et al., op. cit.: table 1, 1998).

Goloubinoff and colleagues explain that, although a correlation between morphological and genetic diversity still has not been seen in other organisms, the dramatic morphological and genetic diversity of maize has made some geneticists contemplate the possibility that the molecular evolution in this plant took place at a faster rate from the moment it was domesticated. The extent of the variation in the sequence among the ancient alleles is similar to that which exists among contemporary alleles. Even the two 4,700-year-old alleles are more closely related with some of the modern ones (1%) than with each other (3.1%). The extent of the difference in the sequence remains constant instead of diminishing, from the present day up to the time period that is approximately halfway through to domestication. Goloubinoff and colleagues claim that if maize originated from one domestication event and subsequently developed at an accelerated rate, they would have predicted less diversity among the ancient alleles than among the modern ones.

The range of divergence of the sequence in the grass family has been estimated at 1.6% per million years in sites with similar characteristics (synonymous) and 0.05% for those with different characteristics (nonsynonymous). Among the *Adhz* maize alleles, one same sequence shows percentages of 2.5% ± 0.9 (4% for untranslated regions and 1.3% for translated ones). The genetic pool of maize may therefore be old by at least several million years, and it may widely outstrip the era of domestication. One possible explanation for the existence of a deep gene pool in maize is that a constant flow of alleles, from teosinte to domesticated maize, took place due to cross-pollination. But teosinte does not grow naturally in the Andean area, from whence the ancient samples come. The introgression of the teosinte alleles therefore took place before the introduction of these races of maize into South America, which in the case of the Peruvian samples was before 4,700 years ago. The majority of the most significant contributions made by teosinte thus took place early in the history of maize.

Goloubinoff and colleagues prepared "parsimony trees" relating the *Adh2* teosinte alleles to those of ancient and modern maize (for two maize and two teosinte alleles). Here we see that neither the alleles of maize nor those of teosinte form monophyletic groups. Many maize alleles are instead more closely related to teosinte alleles than to other maize alleles, and vice versa. This applies not just to the alleles in the teosinte line that are known to be related with maize – for example, *Z. mays parviglumis* and *Z. mays mexicana* – but also to more distant taxa like *Z. luxurians* and *Z. diploperennis*, which do not commonly cross-pollinate with maize. A phylogenetic analysis therefore does not have evidence with which to support the notion that the modern races of maize emerged from one single common ancestor, as a specific line from *Z. mays parviglumis* or *Z. mays mexicana*, or as hypothetical lines of "wild maize." Despite a spectacular display of morphological variability, the remains of domestic maize are genetically undistinguishable – from the standpoint of the *Adh2* gene – from morphologically more uniform species of teosinte.

The same tree shows that the rates of evolution of maize and teosinte alleles are similar in relation with another group, like *Tripsacum*. Here we likewise find that the alleles of modern maize have not developed more extensively than those of ancient maize in comparison with *Tripsacum*. The relative rate controls thus do not show any indication of a particular acceleration in the evolution of maize. Here we see that several ancient alleles are more closely related with modern ones. This close association between ancient and modern alleles is incompatible with the notion that there was an acceleration in the base substitution rate in maize, given that it would be expected that even as low an acceleration as a tenfold increase would produce detectable differences in the periods examined.

The demonstration that there was no acceleration in the evolutive sequence of maize DNA, and that the genetic pool predates domestication, leads to three possibilities that are not mutually exclusive for maize domestication. One of them is that maize was domesticated from one single wild ancestor that was subsequently introgressed by wild teosinte before its export to South America. A second possibility is that maize was domesticated from a population of wild ancestors that initially had, and later perpetuated, a high degree of allelic polymorphism. The third possibility is that maize was domesticated independently from several different wild ancestors that have been subsequently interbred among themselves and with wild teosinte (Goloubinoff et al., 1993: 1997, 2000–2001).

Goloubinoff and colleagues (1994) then add that if the nuclear evolution of maize was ruled by the same rules that apply to wild species (Li and Graur, 1991: 86–88), less than one chance substitution of nucleotides per 15,000 pairs of DNA bases would have taken place since the domestication of maize began. The question Goloubinoff and colleagues raise is whether all of the morphological and physiological variations that are now observable between the modern races of maize and teosinte can be explained within the limits of this genetic variation, or whether some unknown mechanism of "accelerated evolution" intervened as a result of the process of domestication.

Several factors may have contributed to the increase in the fixation of the mutations that took place in the maize genome throughout its several millennia under cultivation. The increase in the population of this plant may have enlarged the total number of spontaneous mutants that appear in each generation. A system of constant selection of advantageous mutations for agriculture, as well as the practice of a hybrid vigor crossing, could have enriched the "mutator" elements present in maize populations. The practice of monoculture may have stimulated the horizontal transfer of transposable DNA elements. The rapid adaptability of maize to different altitudes, latitudes, and humidity regimes may have provided natural barriers along with geographical isolation, thus avoiding the dilution of new mutated alleles that takes place in a large gene pool (Goloubinoff et al., 1994: 115; see also J. B. Jones and Brown, 2000: 771–772).

Other scholars made more detailed studies. Dorweiler and colleagues (1993: 233–234) thus established that the *tga1* (teosinte glume architecture 1) genetic locus that controls a key difference in the development of teosinte and maize ears alters the development of the cupulate fruitcase of teosinte, so that the kernel is exposed on the ear for an easy harvest. According to Dorweiler and colleagues this study shows what Beadle (1939, 1972, 1980) claimed, that is, that a small number of changes in a single gene may bring about the transformation of teosinte into maize (H. Wang et al., 2005, concur). The cupulate fruitcase of teosinte is composed of a rachis internode (or rachid) and the attached spikelet. In teosinte ears, the rachids are deeply invaginated so that when the spikelet is mature (the kernel included), it fits within this invagination or cupule. The spikelet comprises a female flower and a series of bracts that subtend it. The lowest of these bracts is in the outer glume, which seals the opening to the cupule so that the kernel is not visible and is protected from plagues and from granivores. When mature, the rachid and the outer glume become extremely hard (indurated) and give the cupulate fruitcase the aspect of a polished pebble. The rachid and the glume are present in maize, but they do not form a casing around the kernel, which is instead exposed to facilitate its harvesting.

The discovery of *tga1* showed that the kernels of maize were derived from those of teosinte by modifying one or two genes. For Dorweiler and colleagues this denies the alternative proposal made by Iltis (1983b), for whom maize ears are derived from the feminization of a central tassel spike, whereas soft glumes were the automatic result of feminization, because the tassel spikelets have soft glumes. The effects of *tga1* on the maize background are quite severe and reduce the usefulness of maize as a cultivated plant. If teosinte is the ancestor of maize, then *tga1* may be the most significant step in the evolution of maize, as without it maize could not be a usable cultivable plant.

Fedoroff (2003: 1158) attempted to outline the differences between the genomic regions of maize and teosinte. She believes that this is possible in up to five regions. The differences in two of them have been attributed to alternative alleles of one single gene – *tga1* (*teosinte glume architecture*) and *tb1* (*teosinte branched*) – that affects the structure of the kernel and the plant's architecture. Gene *tga1* controls the hardness of the glume, its size, and its curvature. The kernels of teosinte are surrounded by a stonelike fruitcase, thus ensuring it will pass unscathed through the digestive tract of animals, which was essential for the dispersal of its seeds. Reproductive success, is however, the nutritional failure of its consumers. One of the major differences between the kernels of maize and teosinte therefore lies in their structure, that is, the cupule and the outer glume that encloses the kernel. Maize kernels do not develop a fruitcase, because its glume is thinner and shorter, and the cupule is collapsed. The hardness of teosinte kernels is due to the silica deposited in the epidermal cells of the glumes, and to the impregnation of glume cells with the polymer lignin. The *tga1* allele of maize gives a slower growth of the glumes and less deposition of silica and lignification, which is what the *tga1* allele of teosinte does.

We thus see that the locus *tb1* is amply responsible for the difference in architecture in these two plants. Teosinte gives out many long, lateral branches at most nodes on its main stem. Each of these branches is tipped by male inflorescences (tassels), whereas its slender female inflorescences (ears) are given out by secondary branches that grow in the axils of the leaves of primary branches. In contrast, modern maize has a main branch with a tassel at its end. Its branches have lateral branches only on two or three nodes on the main stem, which are short and are tipped by ears. Its lateral branches are short and support the big ears. Many of the differences are attributable to the gene *tb1*, which was originally identified as one of the mutants similar to teosinte. The mutations often annul the function of genes, indicating that the maize alleles act to suppress the development of lateral stalks, thus turning herbaceous teosinte into modern maize, with a single stalk and with male structures turned into female ones (Doebley et al., 1997).

Furthermore, a later study analyzed this problematic and concluded that much remains to be studied. That is, we still do not know which specific molecular events changed the expression or function of the genes of teosinte if it were assumed that they gave rise to the maize phenotypes, and which transposable elements were involved. The maize genes involved in the differences with teosinte were found located near transposable element insertions; these elements can be neutral or not in regard to their effect on gene regulation. In conclusion, the study posited that

> this question will be answered by comparative analyses of multiple alleles of individual genes to determine whether these insertions have any functional consequences. In this regard, it will be particularly interesting to define the mutations responsible for changes in the regulation of function of genes, such as *tga1* and *tb1*, and so learn what molecular magic caught the eye of ancient teosinte farmers some 7000 years ago. (White and Doebley, 1998: 331–332)

R.-L. Wang and colleagues (1999: 238) also studied the significance of *tb1*, and they emphasized that their analysis indicates that the ancient farmers exerted an energetic selection over *tb1*, which drastically reduced polymorphism in its regulatory but not in its encoding regions, so that alterations in the regulation of *tb1* led to the change in the plant's architecture, from teosinte to maize. It must, however, be pointed out that the sample used is not significant for maize, as can be seen in their table 1 (p. 236), where it is compared with *Z. mays* ssp. *parviglumis* and ssp. *mexicana*. For maize 13 samples were used, 23% of which were South American, 8% Central American, 46% Mexican, and 23% from the United States. The South American samples comprised a sample from Bolivia, one from Ecuador, and one from Venezuela. We thus see that the sample clearly is not representative either in number or as regards the zones chosen, for the Andean zone is poorly represented. Pääbo (1999: 195), however, believes that the work done by Wang and colleagues (op. cit.) is important.

It must be pointed out that Jaenicke-Deprés and Smith (2006) studied the effects of the following genes: *tb1* (*teosinte branched 1*) on the overall structure of the plant; *pbf* (*prolamin box-binding factor*), which regulates the storage of protein in the kernels; and *su1* (*sugary-1*), which is involved in the biosynthesis of starch. Jaenicke-Deprés and Smith worked with archaeological samples of maize from Mexico and the southwestern United States. The results indicate that the typical alleles of contemporary maize were already present in Mexican maize some 4,400 years ago (the samples used came from the caves of Ocampo and Tularosa). Yet the allelic selection of one of the genes had not been completed at such recent a date as 2,000 years ago. I find this study is hampered by three major flaws. First, they consider that the maize from Guilá Naquitz is the most ancient one and ignore the Peruvian finds. Second, the study was made with Mexican and North American samples alone, without including South American samples. Finally, they blindly accept the work done by Matsuoka and colleagues (2002), which also has serious flaws, as has already been pointed out, and to which we shall later on return.

Several scholars emphatically claim that the molecular evidence does not support a multiple domestication of maize. These scholars are Doebley and Matsuoka and their collaborators. Doebley (1990: 15–16) claims that the maize from all regions in Mexico comprises one single group, which is closely related with the ssp. *parviglumis*. The data from the isozymes do not support the hypothesis that some forms of maize were domesticated from ssp. *mexicana* or any of the teosintes section of Luxuriantes. If maize were domesticated several times, then the ssp. *parviglumis* was in each case the ancestral teosinte. There is no reason, Doebley claims, to argue that maize was independently domesticated several times from ssp. *parviglumis*. Because any transformation of teosinte into maize may have involved a series of unlikely mutations, it would be far more plausible to present the hypothesis that this transformation took place only one time. In a later study, Doebley (2004: 49) reached the conclusion that the teosinte population has a genetic pool with cryptic genetic variations, over which selection may have acted during the domestication of maize, as was proposed by Iltis (1983b). But the first study does not even mention South American evidence, whereas the second one is based on the work done by Matsuoka and colleagues (2002) for the molecular tests of the origin of maize.

Matsuoka and colleagues (2002: 6082) studied the phylogeny based on microsatellites with samples of maize and teosinte. They claim to have presented a monophyletic lineage derived from ssp. *parviglumis*, and this would therefore prove there was only a single domestication of maize. They use the same arguments to indicate a single origin. The *mexicana* subspecies is removed from all samples of maize, whereas the specimens of ssp. *parviglumis* overlap those of maize, which would thus document the close relation of ssp. *parviglumis* and maize and would support the phylogenetic results insofar as this species would be the sole progenitor of corn. Two points have to be made here that invalidate this study. First of all, there is an evident confusion between origin and domestication, which clearly shows on page 6082. Second, and more importantly, Matsuoka and colleagues worked with modern races alone and did not take into account archaeological ones. Significantly enough, they themselves acknowledge this, for they point out that the DNA analysis of archaeological materials "... could place archaeological specimens in phylogenetic trees such as those presented ..." (Matsuoka et al., op. cit.: 6084).[13]

Freitas and colleagues (2003) is another study to which attention must be drawn. Based on DNA analysis, they managed to separate three groups of alleles that have a different distribution in South America. Their data support the model wherein two lowland and highland agricultural systems generated separate expanses of maize crops in South America. One of these focused on a highland culture that would have spread from Central America to the Andean regions along the highlands of Panama, on the western side of South America, whereas the second one focused on a lowland culture that expanded along northeastern South America and entered this part of the continent following the river systems.

It has to be pointed out that this study is completely unsupported. First, it is clear that Freitas and colleagues are not familiar with the literature on the area they are discussing, nor with its geography and ecology. They go as far as to claim that the archaeological evidence suggests there were no contacts between lowlands and highlands in early epochs, due to the barriers the mountains and the forest presented. The only source Freitas and colleagues have is a study by Bennett that was published in the *Handbook of South American Indians* in 1946; they used the 1963 reprint without realizing this, that is, that it was the reproduction of an old text without any changes whatsoever. They used 21 samples, 11 of which are modern races, comprising 52% of the samples. Of these, 90% are Brazilian, and 10% are from Paraguay. Of the 10 archaeological samples (48%), 7 (70%) are Brazilian, 2 (20%) Peruvian, and 1 (10%) Chilean. The sample clearly is not significant. Besides, the origin or provenance of the Peruvian specimens is not indicated, and all we find is "Peru highlands 440 ± 40" and "Coastal Peru 4500 ± 500" (table 1, 902); the same holds true for the Chilean data. This study is scientifically worthless.

Grobman in turn believes that when the article by Freitas and colleagues (2003) is stripped bare, one reaches the conclusion that Andean maizes are different from Brazilian ones in regard to the alleles in the DNA site known as *Adh2*, which for Freitas and colleagues (2003: 904) would show that "... the two Central American agricultural systems – highland and lowland – generated separate expansions of maize cultivation into South America." This latter claim is extrapolated from the results of a study that says nothing in this regard, and that besides does not include any Central American maizes for comparison. This is a capricious repetition based on a dogma without the support of experimental data (Alexander Grobman, personal communication, 18 August 2003).

[13] Grobman also discussed this study, but his comments appear further on to avoid decontextualizing them.

Another group of scholars has a different position. Pickersgill (1989: 431), for instance, shows that the study of DNA from almost all of the Mesoamerican races – including three of the four ancient Mexican indigenous races – confirmed they were introduced into 3 of the 18 known racial groups, whereas the maize from the Andean complex belonged to two different groups. For Pickersgill, the analysis of mitochondrial DNA fits in with the morphological studies as well as with the available information on chromosome nodes, showing differences between Mesoamerican and Andean maizes. But the affinities between the ancient indigenous maizes and those from the Andes, which are apparent in the data on chromosome nodes, do not show affinity in mitochondrial DNA.

Grobman (2004) made an in-depth study of these issues. He believes that the domestication of maize from annual teosinte – possibly *Zea mays* ssp. *parviglumis* – is supported by several scholars and recent molecular studies (Gaut and Doebley, 1997; White and Doebley, 1998). Alleles that control the same function, albeit at different levels in maize and teosinte, have been described (Gaut and Doebley, 1997). For instance, *tb1* (*teosinte branched 1*) appears at higher levels in teosinte than in maize. Multiple axillary branches form in teosinte, from which come subbranches where multiple female inflorescences appear, whereas in maize there are few condensed branches shaped as a peduncle where only one ear is inserted per axil. In Peruvian archaeological maize we find a plant with branching of several ears per axil (see Grobman, 1982: 166, 168). Doebley and colleagues (1997) suggest these changes took place during the process of domestication. As for the other allele – *tga1* (*teosinte glume architecture 1*) – Grobman says that Dorweiler and Doebley (1997) state that it is a piece of evidence that claims that the evolutive road led from teosinte to maize through stages. Dorweiler and Doebley consider that the differences of expression in teosinte and maize gave rise to different paths in the development of the plant throughout its evolutive process. This would hold if annual teosinte originated from perennial diploid teosinte crossed with maize, or if the latter came directly from teosinte (Grobman, 2004: 449–450)

As for the maize genome, Grobman wrote thus:

> It is estimated that the maize genome has between 2,000 and 3,000 million base pairs (about 2,500 Mb) in its composition. This makes it some 20 times larger than the genome of *Arabidopsis thaliana*, a plant that was adopted, because of its precocity and the smaller size of its genome, as a plant of a superior model for basic studies of plant genomes, and which has the first fully-sequenced genome of higher plants. Maize however probably has just twice the number of genes that *Arabidopsis* has. The remaining DNA in maize comprises repetitive elements whose exact function is not that of active genes. It is estimated that 80% of the maize genome comprises retrotransposons. (Grobman, 2004: 461; for more details, see 461–465)

To understand the genetic side of maize, it is worth recalling a point that has already been emphasized but that is of crucial significance – the astounding genetic variability of maize in the central Andes when compared with the quite limited variability of the species in pre-Hispanic times; the enormous increase in the size of cobs and kernels that took place in this period, which denotes a most accelerated evolutive process, at a speed almost without parallel; and the magnitude of change when compared with other cultivated plants (Grobman et al., 1961: 35).

It is also worth looking at the relations that exist between maize and its closest relatives at the molecular level. The taxonomic studies J. S. C. Smith and Lester (1980) made with biochemical techniques and antibodies show electrophoretic bands that do not distinguish maize from the Mexican annual teosinte or the tetraploid perennial teosinte. They do distinguish teosinte from southern Guatemala and Honduras, as well as *Tripsacum dactyloides*, *Coix*, and *Manisuris*, and other Andropogoneae. The data have been used to explain the origin of maize from teosinte. However, the teosinte from southern Guatemala lacks the bands that do appear in maize and in Mexican teosinte.

Goodman and Stuber (1980) employed electrophoresis of isoenzymes to identify maize lines used in hybrids. Of the 22 isoenzyme-producing alleles tested, only 11 were found in annual teosinte, whereas 21 were in maize. The data confirmed the origin of annual teosinte from maize × perennial diploid teosinte, but not of maize from teosinte. Sederof and colleagues (1981) analyzed mitochondria genomes with molecular hybridizations between annual teosinte from Guatemala, annual teosinte from the central Mexican plateau, perennial tetraploid teosinte, and maize, and they reached the same conclusions as J. S. C. Smith and Lester (1980).

Changes in the position of homologous sequences in different taxa were analyzed using DNA transference techniques and cloned fragments of mitochondria from maize DNA. A third of the cloned fragments showed the sequences preserved in homology and the position of fragments of the *BamHl* restriction. The patterns shown in other fragments indicate that there were extensive rearrangements in the DNA sequence. This type of modification of the genome in mtDNA may be important and different from the simple DNA-based mutations.

R.-L. Wang and colleagues (1999) "speculate," as Grobman pointed out, that domestication diminishes the diversity of sequences in genes controlling traits that are of human interest. We have seen they tried to prove their hypothesis by sampling a 2.0 kilobase (kb) region in the transcription unit (TU) where we find the *tb1* (*teosinte branched*) gene, which is the main gene responsible for the expression in long branches in teosinte, with tassels at their ends. Maize has short branches that end in ears, so the gene acts like a repressor of the elongation of the branch. Samples of genetic diversity in the TU show that maize has 79% of the diversity found in teosinte, and just 3% of the NTR (nontranscription region). Wang and colleagues were surprised when they found that maize remained polymorphic throughout time for this gene. For Wang and colleagues, the ancient farmers would have exerted a very strong selective pressure on *tb1*. But, Grobman (2004: 453) points out, if "... Wilkes' theory were the correct one, since it likewise explains this situation, [it would do so] in a better way."[14]

Matsuoka and colleagues (2002; this study has already been mentioned here, and some objections have already been raised; see previously) posit the domestication of maize as a single event that took place in the highlands of southern Mexico. Two hundred and sixty-four plants representing variations of maize from Canada to Chile, and of annual teosinte from Mexico and Guatemala, were genotyped by means of 99 microsatellites (i.e., DNA fragments). The microsatellites determined patterns of mutations that are verified with a clusters tree. For Matsuoka and colleagues, the teosinte subspecies that gave rise to maize is *Z. parviglumis*. The comments Grobman made (2004: 453–456) on the work of Matsuoka and colleagues are very important:

1. The study did not include *Zea diploprennis*, under the excuse that it is a different species that is not involved in the origin of maize. To this end Matsuoka and colleagues cite Doebley, "whose position is known," and who besides is one of the coauthors in the study by Matsuoka and colleagues. This gives "... a circular and biased argument, for they want to show the phylogeny of maize and this is invalidated with premises that are the conclusion itself." It is amazing, as Grobman points out, that this incorrect position and the biased analysis made in this study have gone unnoticed by those who reviewed the paper.
2. When analyzing the phylogeny of the races of maize in the central Andes, Matsuoka and colleagues present these races as being the most distant ones from the Mexican maizes, which is correct. In fact, the peculiar development of the races of Andean maize with fasciated ears shaped like a hand grenade and multiple rows of kernals – whose ancestor is the Peruvian Confite Chavinense race, which is not duplicated in Mesoamerica or in Mexico – cannot be explained with migrations of maize from Mexico. The primitive races from Mexico and Peru are practically contemporary. On the other hand, the presence and dispersal of the primitive Confite Chavinense and Proto-Kculli races were a local, independent, and quite early development in the central Andes and "cannot be ignored."
3. The third error is giving these races an origin in the South American lowlands, from whence they would have moved to the highlands. The work done at Los Gavilanes (north-central Peruvian coast, see Grobman, 1982) shows the opposite. Matsuoka and colleagues seem not to know this study, "... like others who do not cite the literature [that is] in Spanish."

[14] Readers should recall that I also disagree with the database used by Wang and colleagues; see previously.

4. Matsuoka and colleagues accept the lack of agreement with the archaeological data, and that there are other distributional paradoxes in the current populations of the *parviglumis* subspecies that require explanation.

Now, Matsuoka and colleagues accept that the divergence of maize with teosinte in Mexico is of 9188 years BP (with the confidence limits range of 5683–13093 years BP), which would be the upper limit of domestication and which agrees with the data derived from the pollen grains from Guilá Naquitz,[15] but they do not consider the pollen from Bellas Artes in Mexico City (which was extensively discussed in Chapter 2).

Based on molecular data, Matsuoka and colleagues, on the other hand, posit 6250 years BP as the date maize and teosinte separated, which is consistent with the pollen data from Mexico – if we reject that of Bellas Artes. But what they have not taken into account are the dates from the Casma Valley, Peru (which shall be explained in depth in Chapter 5), which have about the same age.

If we admit the previous existence of wild maize, the alternative hypothesis of the rise of annual teosinte from the crossing of *Z. diploperennis* "... could explain the data with a different interpretation" (Grobman 1982: 454). One of the major differences between maize and teosinte is that the latter has large axillary branches that end in tassels, whereas maize has short branches that end in ears. This difference is mostly controlled by the *tb1* gene, which acts by accumulating its mRNA in the organs where the gene intervenes, thus limiting the growth by elongation of the organ. In the case of maize ears, the primordial cells that accumulate the mRNA produced by the allele of the gene in maize do not cause elongation or the growth of a long-branch-shaped organ, as in teosinte. This has been interpreted as a change brought about by the selection of the *tb1* alleles in maize, under the presumed domestication of maize from teosinte (R.-L. Wang et al., 1999). Wang and colleagues (op. cit.) posit that because of domestication, the diversity of the gene sequences controlling the traits that are of interest to man must be reduced. They found, using the analysis of diversity through nucleotides, that, in the TU, maize only has 39% of the variability of teosinte, which is not different from a neutral gene like *Adh1*.

Although R.-L. Wang and colleagues only compare maize hybrids from Mexico, and a few from the United States and Venezuela – tripsacoids in fact – without including any from the Andean area,[16] they conclude nonetheless that maize is derived from teosinte. Wang and colleagues (1999) have as their starting point a hypothesis that may be wrong, and the arguments are adapted to it. Had they began with the hypothesis that annual teosinte was derived from the hybridization of perennial teosinte and maize, they would have had all the more

[15] Grobman made a lapsus calami when he mentioned the pollen from Gatún, Panama, for Matsuoka and colleagues do not mention it; see Matsuoka and colleagues (2002: 6083).

[16] Here Grobman makes an omission, for Wang and colleagues include one single sample from Bolivia, which obviously does not change his position; see previously.

reason to find more variation in the resultant hybrid – annual teosinte, that is, the progenitor of maize. This shows that the same result can be used to prove an alternative hypothesis.

This type of study, Grobman says, was expanded by Clark and colleagues (2004) to determine the presumed domestication of maize from teosinte – the de facto working hypothesis, which would have the selected allele *tb1* of annual teosinte exerting an influence at a certain distance from the neighboring genes or gene sequences. The result would have been negative, as the selection of *tb1* did not affect the genomic diversity of the other genes. Clark and colleagues add that the limited impact *tb1* selection had on the genomic diversity is significant, for *tb1* was under a strong selection during domestication. In contrast, the regions selected in other species had multiple genes.

"It is clear," Grobman writes,

> that had the proponents simply wanted to change hypothesis and accept that it is not maize which comes from teosinte, but teosinte from a crossing of maize × *Zea diploperennis*, the results would have perfectly fitted the alternative hypothesis as there would have been no need for such selective pressure in order to create maize, which was pre-existing, and the *tb1* allele of annual teosinte would have come from perennial wild teosinte; hence the results obtained would fit the logic of the alternative hypothesis. (1982: 455)

The explanation sought by Clark and colleagues (2004) in regard to the use of large populations during the process of domestication, which would have affected the result, is unacceptable when explaining the minimum genomic effect brought about by the presumed selection. There is no evidence of large populations in emerging cultivation during the process of maize domestication – quite the contrary. This is not even visible in many maize archaeological samples many generations afterward (see Bonavia and Grobman, 1979 [and Bonavia, 1982]), which evince small areas of maize population. Besides, Clark and colleagues contradict themselves, for they use a small number of generations as the reason for the bottleneck with which to explain the lack of effect of genomic variability, but on the other hand they claim that domestication must have necessitated many generations. The maize alleles were found scattered inside clades that included teosinte, but not in homology with the latter. Clark and colleagues tried to show statistical differences to sustain the position of a teosinte-based domestication through an analysis of the *tb1* locus. For Grobman, all of the statistics used cannot clinch the case, and a whole range of possible interpretations can be proven with them.

Interestingly enough, Pickersgill also touched on this issue and made some remarks regarding the work by Clark and colleagues (2004). Pickersgill clearly states that the effect of the maize allele of *tb1* on the genetic background of teosinte "*is not clearly established*." Besides, "the mutation rate may therefore have been less of a constraint on domestication than the ability of early farmers to detect and fix favourable phenotypes" (Pickersgill, 2007: 933; emphasis added).

It may be true that the *tb1* effect was maintained in the alleles from certain races of wild maize that generated ramified ears, alongside the gene *branched* with multiple-ear peduncles (Grobman, 1982: photograph 62 [lapsus calami: it actually is 52] and drawing 60). It is likewise possible that it appeared in primitive Mexican races like the Palomero Toluqueño, and in intermediate form between modern maize and teosinte, and that it has survived to the present day (Mangesldorf, 1983b: photograph 2C, 229).

The study by Goloubinoff and colleagues (1993, already widely mentioned; see previously) – which as we recall included a very representative sample of specimens, both ancient and modern, from the United States, Mexico, Peru, and Chile as well as four specimens of modern teosinte and one of *Tripsacum* – showed that the divergence in DNA sequences in members of the Poaceae family for *Adh1* is 1.6% per million years. The observations made for the variation of sequences in gene *Adh2* in ancient maize is a mean of 2.8% with a maximum of 3.7%; in modern maize the mean is 2.2%, with a maximum of 3.7%. The degree of difference in the sequences remains constant throughout time. It was found that the variation in sequences of ancient alleles is of the same order as in modern ones; even the alleles of two[17] archaeological samples from Los Gavilanes are more related to modern alleles than to each other, at the same age level. Had maize originated in one single domestication event, less variation of the gene under study would have been predicted among ancient alleles than among modern ones. Goloubinoff and colleagues (1993) therefore believe that the genetic pool of maize must be very ancient and must have preceded the process of domestication by several million years. If we want teosinte to intervene in the temporal equation, one possible explanation would be that there was a continuous flow of teosinte genes into maize.

The problem raised by this explanation is that in the Andean area there is no evidence of teosinte in its natural form. The similitude of certain alleles is greater between teosinte and maize than between races of maize, even with the *Z. luxurians* and *Z. diploperennis* taxa, for which frequent crossings with maize are not known. This means that the evidence in Goloubinoff and colleagues (1993) does not support the hypothesis that maize is the outcome either of a single domestication event from a hypothetical line of "wild maize" or from a specific line of *Z. parviglumis* or *Z. mays mexicana*, as held by Doebley and Galinat. Nor should a faster evolutive rate be distinguished in *Tripsacum* than in maize, in teosinte, or vice versa.

Given the extraordinary variability of maize, Grobman (2004: 457–458) points out, only one of the following four hypotheses can be accepted:

[17] Grobman here made a slip of the pen, for it is actually just one sample.

1. Modern maize is the domestication of an ancestral wild maize that then underwent the introgression of teosinte before its export to South America. Yet the evidence from archaeological maizes does not show the introgression of teosinte, nor does the distribution of chromosomic knobs support this thesis.
2. Maize was domesticated from a wild ancestral population with a high level of polymorphism, which was then perpetuated.
3. Maize was domesticated independently from various ancestors of wild maize that intercrossed (Bonavia and Grobman, 1989b; Grobman et al., 1961; Mangesldorf, 1974).[18] But it subsequently crossed with teosinte too.
4. Maize was domesticated independently in multiple occasions, in different places, and from different races of teosinte. Galinat (1988a) proposed two different domestication processes, and Kato-Yamakake (1984) suggested several in different Mexican regions based on the diversity of chromosome knobs. This is the hypothesis that Matsuoka and colleagues (2002) tried to prove.

For Grobman, the classical bottleneck effect in domestication, due to the limited initial population that causes a diminution in the variability of the domesticated species in regard to the wild one, does not apply to maize. The latter has an extraordinary morphological variability, quite higher than that of teosinte, and besides, it is morphologically quite different.

The greater variability of maize was measured through the analysis of isozymes; chromosome knobs; the different racial frequencies of heterochromatic zones marked in the chromosomes (particularly in chromosome 10) and in supernumerary chromosomes; RFLP (chloroplast restriction fragment length polymorphism) data; internal transcription-spaced nuclear sequences; and nuclear sequences of simple copies discussed by Eyre-Walker and colleagues (1998; see previously). They claim, contrariwise, that there is a great variability in the *Adh1* gene, which forms the basis of their studies of DNA sequence diversity in *Z. mays* ssp. *parviglumis*, and that a founding population of 20 individuals in 10 generations would suffice to explain the domestication of maize from *Z. mays* ssp. *parviglumis*. This species is the one that was proposed by Doebley and others as the ancestor of maize. However, notes Grobman, Eyre-Walker and colleagues admit that the model may be correct for the gene *Adh1* but is perhaps not applicable to other genes, like *cl*, in which both maize and *Z. luxurians* (teosinte from Guatemala or Florida) have identical sequences. *Z. luxurians* and *Z. mays* ssp. *parviglumis* have such different sequences in *Adh1* that it is estimated that these two species separated 1.02 million years ago. The problem is that *Z. luxurians* has less base variation than maize or *Z. mays* ssp. *parviglumis*, and no crosses of maize with *Z. luxurians* are known.

[18] This has been posited several times by different authors.

The discovery of a new teosinte – *Z. nicaraguensis* – and the study of the development of its inflorescence made by Orr and Sundberg (2001), who compared patterns of development in the inflorescence of this species with other Poaceae like maize, teosinte, and *Tripsacum*, led Orr and Sundberg to conclude that the mechanisms of development and of male and female flowering are the same in *Zea* and in *Tripsacum*. Orr and Sundberg suggest a novel system of polystichy that is not frequent in teosinte *Z. nicaraguensis*. This led to an intriguing question, that is, whether polystichy is an inherent characteristic of all the grass families in the Andropogoneae group, or whether polystichy in maize could have had its origin in this already preexisting genetic potential. It is striking that this new species grew in flooded soils and formed stable inflorescences, whereas under these same conditions maize suffers significant changes in gene expression.

The studies of the hybridization of maize and teosinte that Lauter and Doebley (2002) carried out to establish the effect of quantitative cryptic genes – some of which could have epistatic effects – would prove the presence of this type of gene, whose aggregate action would explain discrete differential morphology effects between maize and teosinte. These studies changed the previous position, according to which there were only four or five differential genes between maize and teosinte (Grobman, 2004: 452–459).

The maize genome has been sequenced with 95% certainty for a maize line B73. As a careful analysis of Peruvian maize, of Mexico's annual teosinte, and of perennial diploid teosinte and *Tripsacum* is being completed, variations in genome size and in the location of genes in the genome, and other significant genomic characteristics, are being established. Of special interest is that some 75% of the maize genome is formed by transposons and repeat sequences of genes. Therefore, when the comparative study of the genomes of early archaeological specimens from Peru and Mexico, annual teosinte, perennial diploid teosinte, and *Tripsacum* are completed, we will have much more evidence than that which is based just on a few genes of neutral action, like *Adh1* and *Adh2*, as well as the more recent data on *tb1*, *d8*, *ts2*, and *zagl1*,

> ... which have been the basis of speculations regarding the changes in genetic diversity of the presumed domestication of maize through teosinte. This form of the origin of maize supposedly should be – in theory – accompanied by a reduction of diversity, when it necessarily passes through the bottleneck of small original populations in the process of domestication, a reduction that has not taken place [and thus] removed one of the most solid bases of the hypothesis of the domestication of maize from teosinte. (Grobman, 2004: 469)

Finally, it must be pointed out that Eubanks (2001b: 509) undertook a RFLP genotyping, that is, a form of DNA fingerprinting that is used in the maize-seed industry to verify the parentage and pedigree purity. Eubanks reached the conclusion that domesticated maize is clearly a composite of the genomes of teosinte and *Tripsacum*. This agrees with the results obtained in experimental crosses of *Tripsacum* and perennial diploid teosinte as well as with the archaeological data. Thus this supports the hypothesis that maize had its origin in a natural recombination of the genes of teosinte and *Tripsacum*. For Eubanks, the rapid apparition of maize in some millennia, as happens archaeologically, can be explained through human selection in mutations due to genomic responses in answer to the shock brought about by the intrusion of foreign DNA in the *Zea* genome. This change could have taken place in a few generations, as was suggested by Gould (1984). This is consistent with the archaeological data, insofar as the most ancient remains have all of the characteristics that distinguish maize from its relatives.

MacNeish and Eubanks (2000: 4) were therefore right when they noted that although geneticists have identified and mapped the five major genes responsible for the transformation of maize (Doebley et al., 1990; Galinat, 1977, 1985b; Langham, 1940; Mangelsdorf, 1947; Rogers, 1950), how this anomalous form originated is still a scientific enigma.

Chromosome Knobs

Ting (1964) showed that the chromosomes of maize and teosinte have the same size, appear in the homologues of both species, form 10 bivalents at diakinesis, and produce bridges that indicate the execution of the crossing over and the exchange of chromosome sectors. The relatively high frequency of homozygote inversions in teosinte chromosomes in Mexico is a potential mechanism of autocompatibility in crosses within homozygote populations for similar inversions, but it is also a signal of the incompatibility of the heterozygote state in crosses with maize or with other individuals from teosinte populations that did not have such inversions.

We have known since the work done by Longley (1937) that chromosome 10 of maize appears abnormally in certain maize groups or plants, and it differs from normal chromosomes in its greater length and in the difference in the chromomere patterns. Grobman and colleagues (1961) detected this abnormal chromosome in several Peruvian races. Its origin and significance are still unclear.

The supernumerary chromosomes, which are smaller than the 10 normal pairs, may appear in certain races of maize more frequently than in others. They are known as chromosome B. These appear in quite variable number in several races of maize in Peru and North America, as well as in teosinte. Several researchers have accepted the possibility that the B chromosomes are fragmentary residues of strange chromosomes that entered via hybridization between species. In light of recent research on the genomic hybridization between maize with chromosome B and *Tripsacum dactyloides*, these chromosomes could be evidence of past hybridizations between these two species (Grobman, 2004: 459–461).

Knobs are conspicuous protuberances, deeply colored and thicker than chromomeres, that appear in certain positions in the chromosomes of some species of *Tripsacum*, teosinte, and certain maize races that presumably had an introgression from the former species (Grobman et al., 1961: 45–46).[19]

Chromosomic knobs are useful when classifying races of maize, and when trying to establish the relations among them. They have been used as part of racial descriptions in most of the studies on this subject undertaken in both hemispheres (Mangelsdorf, 1974: 118–119). For McClintock (1959), these knobs are regulated by their geographical situation, as well as by their racial classification.[20]

It is worth noting that the structures of the chromosomes in *Zea* have remained relatively conserved in comparison with the reorganized chromosomes of *Tripsacum* and other more distantly related grasses (Galinat, 1975a: 317).

Galinat (1977: 25–27) points out that although maize and teosinte are polymorphic due to the number and size of their chromosome knobs, the variation in number of knob-forming positions and their size are greater in teosinte. Galinat then points out a large number of differences between teosinte and maize in this regard.

For Iltis (1969: 2), the characteristics of Euchlaenoids[21] are not only the induration of the lemmas and the cupules but also many chromosome knobs, all of which reflect a recurring injection of the germplasm of wild teosinte in maize. This had a great influence in the evolution of this plant's races. Outside Mexico and Guatemala, and beyond the genetic control of teosinte (in Peru, for instance), the human selection of maize would have been able to produce a greater and surprising morphological diversity, superior in some regards to that of Mexico. Here, however, there are races of maize that are more basic and more clearly distinguishable than in any other country. Yet not all scholars agree.

Reeves (1944) has shown a statistically significant relationship between the number of chromosome knobs and the distance from Central America. He believes that maize was distributed over parts of north-central and South America before the arrival of teosinte (see also Mangelsdorf and Reeves, 1945: 241).

For Mangesldorf, the modern varieties of Latin America and the United States contain chromosomes or segments of chromosomes that produce tripsacoid effects that originated with teosinte or *Tripsacum*, or with both. Those in Mexico and Central America are due to teosinte, whereas it is most unlikely that the South American varieties have the same origin, because teosinte is not known there. The tripsacoid characteristics of South America that have no counterpart in Central America are explainable with the tripsacoid effects derived

19 For a definition of chromosome nodes, see also McClintock (1960: 462).

20 Interested readers should see Grobman and colleagues (1961: 45–46, 49) and Moreno and colleagues (1959).

21 This is a term coined by Iltis as a replacement for "tripsacoid."

from *Tripsacum*. The fact that they have some of the same effects as the teosinte chromosomes is consistent with the idea that teosinte itself is a hybrid of maize with *Tripsacum*. This is at the same time consistent with the hypothesis that *Tripsacum* is a hybrid of teosinte with *Manisuris*, as was claimed by Galinat (1970). Teosinte and *Tripsacum* would in both cases have had common genes (Mangelsdorf, 1974: 130–131).

Teosinte in fact has a large number of chromosome knobs (Galinat, 1985b: 276). In teosinte *Z. luxurians*, the chromosomes are cytologically different from those of maize (Beadle, 1932). In annual teosinte, *Z. mays* ssp. *mexicana*, the chromosomes are cytologically similar to those of maize (Beadle, 1932). In annual teosinte, the length of the arms, the position of the centromere, and the size and the position of the knobs are identical to those of maize (Longley, 1941). Maize and Mexican annual teosinte have chromosomic knobs in the same positions and similar frequencies (Doebley, 2004: 40–41).

De Wet and colleagues (1971: 261) have pointed out that the chromosomes of *Zea* and *Tripsacum* show the presence of "pycnotic knobs," which are variously distributed among the chromosomes. The Mexican species of *Tripsacum* are all characterized by the presence of knobs (Prywer, 1960), whereas the chromosomes of *T. australe* seem to lack them. They are likewise present in the chromosomes of all races of *Z. mays* ssp. *mexicana* and in many of those of *Z. mays*. Mangelsdorf and Cameron (1942) posited that the presence of knobs in maize indicated an introgression from *Tripsacum*. Yet W. L. Brown (1949) noted that northern flint corn – presumably tripsacoid – also lacks knobs. The high number of these knobs in races of maize is probably due to hybridization with *Z. mays* ssp. *mexicana*. Wilkes (1967) indicates that the races of maize with the highest number of knobs in Mexico and Guatemala have parallels with the distribution of teosinte. The natural introgression between maize and teosinte is usually found in this region of sympatric distribution (Wilkes, 1970).

The presence or absence of chromosome knobs has been used to accept or reject the tripartite hypothesis. *Tripsacum* in general has terminal knobs, but many seem to be absent in *T. australe*, as was already noted; *Z. mays* has intercalary knobs, but their number varies enormously. *T. perennis* essentially lacks knobs for Reeves and Mangelsdorf (1959), whereas *Z. mexicana* occupies an intermediate position between *Z. mays* and *Tripsacum* as regards their number and position in the chromosomes. Reeves and Mangelsdorf assumed that the knobs have been transferred from *Tripsacum* to maize through teosinte. That maize may have received its knobs from teosinte seems to be a well-established fact. The races with a higher number of knobs are essentially sympatric with *Z. mexicana* in Mexico and Central America (Longley and Kato-Yamakake, 1965; Wilkes, 1967). It was then established (Wilkes, 1967) that the number of knobs in maize is essentially the same as in the teosinte of this same region. However, the high number of knobs is not always associated with tripsacoid characteristics. The early Mexican races of Nal-Tel and Chapalote maize have a high number of knobs, but it is assumed they are non-tripsacoid in terms of their small spikelets and pollen structures. The tripsacoid maize from South America and the North American Corn Belt may similarly have a low number of knobs in their chromosomes. The fact that the number of knobs and their position show an almost complete homology between the chromosomes of *Z. mays* ssp. *mexicana* and *Z. mays*, in which these taxa are sympatric, seems to provide a poor support for the hypothesis that teosinte originated as a hybrid derivative of maize and *Tripsacum* (De Wet et al., 1971: 263).

Mangelsdorf and Reeves (1939) long ago posited that chromosomic knobs may trace their origin to hybridization with *Tripsacum*, and then to the repeated hybridization of maize with teosinte. Because it is assumed that the latter did not exist in the Andes, there should be no chromosomic knobs there. The confirmation of this latter point is strong evidence for Peru as a primary center of maize domestication, and that the knobs are a proof of hybridization with teosinte and finally with *Tripsacum*. There is good evidence that some species of *Tripsacum* do not have chromosomic knobs, as we have just seen, so their absence in certain races of maize is no longer secure evidence to claim that the hybridization of teosinte and *Tripsacum* did not take place (Mangelsdorf, 1974: 118–119). Mangelsdorf (op. cit.: 124) insists on the fact that some forms of *T. australe*, as was already noted, do not have chromosome knobs (Graner and Addison, 1944), so their absence in a variety of maize cannot be considered as evidence that there was no mix with *Tripsacum*. This does not invalidate the assumption that the presence of knobs indicates teosinte introgression – rather, it reinforces it.

Grobman and colleagues (1961) likewise analyzed this problem in regard to what happened in the Andean area. They indicate that there is a low number of chromosome knobs in the early popcorns, as well as in their early hybrid races from the Peruvian highlands and coastlands. The ample occurrence of maize races with a limited number of knobs in Peru does not fully agree with what is known for Guatemala (Mangelsdorf and Cameron, 1942) and Mexico (Wellhausen and Prywer, 1954; Wellhausen et al., 1952) in regard to the frequency of chromosome knobs and their distribution at various altitudinal levels. Although it is true that in Peru there is a gradual increase in the frequency of the knobs as the altitude diminishes, just like in Guatemala and in Mexico, the localized distribution of the small number of chromosome knobs in certain coastal races is, however, in stark contrast with the overall pattern (Grobman et al., 1961: 45–46).

Galinat (1977: 24) also passed judgment in this regard. He believes that the difference in the number of positions of the active chromosome knobs is frequently taken to be the result of natural selection. Maize, when adapted to high latitudes or altitudes, tends to have a low number of knobs, whereas the races found at low latitudes or altitudes have a high number of them (W. L. Brown, 1949; Longley, 1938; Mangelsdorf and Cameron, 1942). The constitution of the knobs and the distribution pattern on maize may change during the process of adaptation of a race to a new area. Galinat acknowledges that little is known regarding the origin of the position of the knobs, or of the process whereby they become manifest in the form of a visible knob, or in the degree of stability of the knob in various genotypes and cytoplasms; there is evidence that shows that the knobs can be reduced in size or increased in size through duplication (Longley and Kato-Yamakake, 1965). The origin of polymorphism for the knobs of American Maydeae and for the *Coix* in the eastern Maydeae is still a mystery.

Pickersgill (1969: 57–58) also broached this subject and wrote that, based on the study of the number of chromosome knobs, the cytological evidence can allow several interpretations, as we saw was pointed out by Grobman and colleagues (1961: 46). There is no consensus in that the knobs found in Mexican maize are derived from their hybridization with related wild grasses, that is, teosinte, which does not exist in Peru. If we accept this premise, then the low number of knobs in Peruvian maize can likewise be explained, positing an independent domestication of maize in Peru, or the early introduction of Mexican maize prior to intense hybridization with teosinte. This position of Pickersgill's is interesting, insofar as it entails the presence of a wild maize.

Iltis (1969: 3) has another outlook on this issue. He considers that the well-known and intriguing reduction of chromosome knobs in maize, particularly in those races far removed from Mexico, can be equally related with a genetic "relaxation" of teosinte under an extreme selection in isolation.

Now we know that annual Mexican teosinte and the maize races in the lowlands of Central America and the Caribbean have a high number of chromosome knobs and similar nuclear and cytoplasmatic DNA (Kato-Yamakake, 1976; Timothy et al., 1979).

It is, however, worth noting that the Palomero Toluqueño race does not have chromosome knobs (Longley and Kato-Yamakake, 1965; Wellhausen et al., 1951), whereas the statuses of Pira Naranja and Chapalote/Nal-Tel are not clear. Roberts and colleagues (1957) have noted an average of 7.0 chromosome knobs, but they worked with just two plants of Pira Naranja. For Chapalote and Nal-Tel, Wellhausen and colleagues (1951) pointed out an average of 6.0 and 5.5 with knobs, but in a small sample. For this same race Longley and Kato-Yamakake (1965) gave an average of 11.7 and 10, respectively. It was on this basis that Mangelsdorf (1974: 119) said that "these high knob numbers might appear to support the suggestion that there may have been two kinds of primitive races of maize, one with virtually knobless chromosomes, the other with numerous chromosome knobs." Besides, Chapalote and Nal-Tel in no way represent pure races. McClintock (1960: 470) has shown that in the latter, the chromosomic constitution of most of the plants studied reflects complexes with a prevalence of knobs found in plants of other races that grow in the same region (see Mangelsdorf, 1974: 113–120).

McClintock (1960: 467–468) herself has shown that in Mesoamerica there are three maize complexes. The first of these, in central Mexico, is a type with several distinctive knob complexes. Among them there was one without knobs (the "no-knob" complex). The second complex produces large knobs in most of the knob-forming chromosomal regions, as well as another original knobs complex, which contributed to the origin of the types of maize that grow in the central-western and northwestern parts of Mexico. The third complex is restricted to the central highlands of Guatemala. It has relatively small knobs in many of the knob-forming regions and none in others (the "small-knob" complex).

On the other hand, the work done by Reeves (1944) verified that all of the races in the eastern South American area, as well as those in Venezuela and Dutch Guyana, have chromosome knobs. Yet most of the Andean maize and those from the southern Amazon have just two knobs (R. McK. Bird, 1984: 50).

The Andean complex was defined thanks to the work undertaken by McClintock (1959). This complex is characterized by the fact that most of the races have a mid- or small-sized chromosomal knob in the intercalary position in the long arm of chromosome 7, and – less frequently – a medium to small knob in the intercalary position on the long arm of chromosome 6. This happens in 30 high-altitude races from Ecuador, Bolivia, and Chile, with the sole exception of two races. Among the 30 races McClintock included Confite Puneño, Chuspillu, and Kculli, which Mangelsdorf (1974: 116) says "form a lineage of ... the ancestral form." The same thing happens with Confite Morocho, which Grobman and colleagues (1961: 142–143) have pointed out has a small knob subterminal on the long arm of chromosome 7, and a small knob subterminal in the long arm of chromosome 6. This pattern is stable and unique, as well as different from that of other regions, and it evinces a great antiquity (Grobman, 2004: 437).

McClintock (1960: 466) wrote: "The constancy of the knob-forming capacities of particular knob-forming regions was spectacularly revealed in the preliminary study of races of maize of western South America." In the high-altitude Andean valleys of Ecuador, Bolivia, and Chile that were previously under the sway of the Inca, "... one particular knob complex was present in plants of nearly all races examined": "the Inca-Andean complex."[22] A quite different knob complex was found in the races from the high-altitude Andean valleys of Venezuela – which were not under Inca control – that have been studied. Yet within the land of the "Inca-Andean complex" there are a few exceptional races with chromosome types that have various knobs that do not belong to this complex. Scholars concluded, on the basis of morphological traits, that these exceptional races have a "foreign germplasm," and that in two of these it was probably introduced from Mexico. The types of knobs support this deduction. It is surmised, based on the evidence of the studies undertaken with Mexican

[22] We have seen that most scholars call it the Andean complex.

and Central American maize, that one of the two races is the Bolivian Pisinkalla, whose germplasm derives from maize from central Mexico. The other race is Canguil from Ecuador, whose knob types suggest that its strange germplasm comes in part from the southern region of the state of Chiapas, in Mexico. But the germplasm was in both cases diluted with Inca-Andean germplasm. The study of the makeup of the knobs in the lowlands of these countries likewise showed a contribution from Inca-Andean maize.

There is evidence along the eastern and western Andean basins of an extensive mix of the germplasm of Inca-Andean races with the native ones, or with the maize introduced in lower-altitude lands. The Inca-Andean germplasm predominates into some low-lying areas, like northern Bolivia, whereas in the lowlands of southeastern Bolivia it is far more diluted with that from other sources, one of which seems to come from maize that currently grows in the Antilles. In the northeastern region of coastal Ecuador and in the adjacent lands inland, Inca-Andean germplasm seems to be mixed with what seems to be the same germplasm from the maize that grows in some parts of Central America. Germplasm has mixed further south with that of Inca-Andean origin.

In Chile, some germplasm apparently comes from Central America and Mexico, but some unknown source also contributed to the mixed germplasm of the races that grow on the eastern and coastal areas of Chile (McClintock, 1960: 466–467).

Interestingly enough, the racial variants of Paraguay also have a low number of chromosome knobs (Mangelsdorf and Reeves, 1945: 241). On the other hand, the Confite Puntiagudo and Pisankalla races from Bolivia and Argentina lack the typical pattern of the Andean chromosomal knob. According to Sevilla (1994: 235), some knobs from these races are typical of the early races from northwestern Mexico, as well as from some small-ear races that are related with Mexican dent corn. This analogy must have a long history due to its great variability in Argentina, Chile, Paraguay, Uruguay, Brazil, and Bolivia.

When discussing this Andean complex, Mangelsdorf (1983b) cited McClintock:[23] "The extent of distribution of maize with this one chromosome constitution is truly extraordinary. It demands explanation as it contrasts so greatly with the many different combinations of chromosomal components that are found elsewhere in the Americas." The concentration, in the highlands of Guatemala, of races that have small knobs, as well as others in which chromosomes have no knobs, may indicate that this is the origin of the Andean complex. Yet there are reasons to believe that Guatemala is a secondary center. When Mangelsdorf and Cameron (1942) became aware of the low number of chromosome knobs in Guatemalan varieties growing at high altitudes, and after recalling that Reeves (Mangelsdorf and Reeves, 1939) had discovered Peruvian races

[23] Mangelsdorf did not provide the reference; the citation probably comes from McClintock and colleagues (1981).

without chromosome knobs, Mangelsdorf and Cameron (1942) concluded that they had been introduced from South America and called them Andean varieties. Wellhausen and colleagues (1957) reached the same conclusion (Mangelsdorf, 1983b: 240–241). McClintock (1978), however, believes that Andean maize comes from Guatemala, and Kato-Yamakake (1984) concurs. This hypothesis has also been supported by Bretting and Goodman (1989).

Finally, Matsuoka and colleagues (2002: 6083) claim that chromosomal knobs have never been the subject of formal phylogenetic analyses to explain the origins of maize. We have seen that this is groundless, and that Matsuoka and colleagues also show they do not know the literature. They likewise believe that the data in the chromosome knobs may not be appropriate for phylogenetic studies, because the frequency in many chromosome knobs can change in a concerted and nonneutral fashion due to a meiotic drive.

Pollen

Pollen grains are the male element in flowering plants and are produced in great number in the stamen of the flowers (Piperno, 1995: 131).

Maize is a naturally cross-pollinated plant that produces pollen grains profusely, so that the latter are easily carried by the wind; it is thus inevitable that two races growing close by will have some degree of interracial hybridization. Man, on the other hand, carried the plants when moving, and the plants themselves moved due to exchanges of items, thus giving rise to crossings and intercrossings that gave rise to racial differences (Mangelsdorf, 1974: 121).

Banerjee and Barghoorn (1972) showed that maize and teosinte have similar but not identical pollen and have different "clumped" spirals than *Tripsacum*. As Randolph (1976: 323) points out, "... in several of Galinat's maize-teosinte derivatives even one tripsacum [*sic*] chromosome produced indications of clumping in the spinule pattern."

Because Eubanks (1995, 1997a) showed that the perennial teosinte × *Tripsacum* hybrids resemble the reconstructed prototypes of primitive maize, and because the pollen derived from the hybrids cannot be distinguished from either maize or teosinte, then her suggestion that "wild" maize is a natural hybrid of teosinte and *Tripsacum* is compatible with the palynological data (MacNeish and Eubanks, 2000: 14).

One of the questions that have been raised is whether the pollen of the ancestral teosinte can be distinguished from that of early maize. Iltis (1987: 212) believes that this possibility not only does not exist but never will exist, yet Sluyter and Domínguez (2006: 1151) disagree and claim that it is possible. They even believe that it can be distinguished from the pollen of the teosinte-*Tripsacum* hybrid.

Some scholars even doubt that an identification of maize based on the size of the pollen grains is conclusive (Schoenwetter, 1974: 300). As for fossil remains,

Kurtz and colleagues have pointed out that "... because of the great variability of size and the lack of distinguishing characters of corn pollen, it has been recognized that it is difficult to determine with any degree of reliability that a fossil pollen grain is corn" (Kurtz, Liverman, and Tucker, 1960: 85). An explanation here is in order. The term "fossil" is often used for archaeological maize remains, but this is wrong. A fossil is "a substance of organic origin that is more or less petrified due to natural causes, and is found in the layers of the earth" (Real Academia Española, 2001: 732), which clearly is not the case of archaeological remains. A good example of fossil pollen is that of Bellas Artes in Mexico, which has already been discussed in depth.

Barghoorn and his team have made many studies of pollen. In the study of the Bellas Artes pollen, they worked with some 1,000 grains and established consistent differences in regard to individual and average grains, as well as the pore-long axis ratio, which always are valid means with which to distinguish maize pollen from that of *Tripsacum*, and in some cases from teosinte. The latter, which Mangelsdorf and Reeves (1939) posit as a hybrid of maize and *Tripsacum*, shows an intermediate value in overall size and is perhaps more significant in its pore ratio. The intermediate value agrees with the idea that teosinte had its origin as a hybrid (Barghoorn et al., 1954: 232).

Irwin and Barghoorn (1965: 40, 42) studied the ektexine spinule of pollen and concluded that in *Tripsacum* the spinules are irregularly distributed in the ektexine, whereas in maize they are very regularly located. In teosinte they are spaced in a less regular fashion and in some are rather closely aggregated, appearing as clumps. When different races of maize and teosinte hybridize, the pollen grains in the derived hybrids have a pattern wherein some ektexine spinules are occasionally missing and thus give rise to empty spaces (Banerjee and Barghoorn, 1972). This happens in many Mexican popcorn races, and in the Peruvian Confite Morocho. This would indicate the presence of teosinte germplasm. An explanation is here in order, which shall be repeated in Chapter 5, when discussing the archaeological remains. When Banerjee and Barghoorn studied the preceramic sample of a Proto-Confite Morocho from Los Gavilanes, in Peru, they wrote: "The pollen grains from this site show a distinct spinule clumping and demonstrate *the oldest convincing archaeo-palynological evidence of introgression of* ***Tripsacum*** *with maize*" (Banerjee and Barghoorn, 1973a: 47–48; emphasis added; see also Mangelsdorf, 1974: 184; and my Figure 5.13).

Some problems appear in regard to the study of pollen. First of all, usually, when working with these grains, the differences at the species level cannot be distinguished (M. E. Dunn, 1983; Eubanks, 1997b; Lippi et al.: 1984; MacNeish and Eubanks, 2000: 14; Piperno and Pearsall, 1993; Roosevelt, 1984; Rovner, 1999). Distinguishing *Tripsacum*-teosinte hybrids from those of maize and annual teosinte is also difficult. So it is possible that, wherever *Zea* pollen was found, it may have been a hybrid of *Tripsacum* and teosinte (Eubanks, 1997b).

On the other hand, Kurtz, Liverman, and Tucker (1960) pointed out there are variations in the pollen, even in plants that are in the same environment, thus suggesting that genetic and environmental modifications may exert an influence.

Phytoliths

Phytoliths have recently been the subject of much work to solve the problems maize raises at the archaeological level. Because this issue is discussed in the chapter on pre-Hispanic remains, all that is in order here is to present some general ideas.

Piperno notes that the term "phytolith" literally means "plant stones." These actually are secretions formed both by the opaline silicate and by the calcium that usually develops in the cells of livings plants in nonflowering organs; these secretions are later released into the environment, where plants die or decay. They probably are the most durable known remains of plants (Piperno, 1995: 131, 135). These phytoliths characterize particular taxa at various levels, genera, tribes, subfamilies, and families, for there is a close correlation between the types of phytoliths and the taxonomic affinities of the plants that have them (Piperno, 1988a). Phytoliths at first could not be used to identify genera or species (Piperno, 1984: 362), but this later became possible (Piperno, 1995: 136; for more information see Piperno, 1985a; 1988a; 1994a; 2009: 147).

Nowadays maize phytoliths can be separated from those of other wild grasses. We have seen that there is a high degree of variability within the *Zea* genus. The data from the phytoliths indicate that the combination of their size and tridimensional form may be used to separate many races of maize from teosinte (Pearsall, 1994b: 116; Pearsall et al., 2003: 612; Piperno, 1984: 370; 1988a; 1991; Piperno and Pearsall, 1993). The characteristics of maize phytoliths are based on their cross-shaped form, which is found mostly in leaves and rondels, which are mostly found in glumes and in the cupules of the cobs (Piperno, 2009: 149).

Piperno developed a new method with which to establish the three-dimensional structure of cross-shaped phytoliths, which according to her allows one to distinguish cultivated maize from wild hybrids (Piperno, 1988b; it is, however, worth noting that Piperno worked exclusively with Mexican races; see table 10.3, 210).

When I asked José Iriarte what were the difficulties present in distinguishing the type of cross-shaped phytolith produced by maize leaves from panicoid grasses, he explained that this is not a one-on-one correlation. One should instead analyze a collection of phytoliths and apply the multivariate statistical form known as discriminant analysis, in regard to the three variables that are best distinguished among the groups of maize crossings and the forms composed by the crossing of composite panicoid grasses and other grass subfamilies

(Poaceae). These variables are determined by the three-dimensional form and by the size of the cross-shaped phytoliths (for a detailed explanation of the technique, see Piperno, 2006: 52–65). Some races of maize give crosses of variety 1 that are bigger than 21 microns, and that have yet to be recorded in any panicoid grass. At present the varieties or races of maize cannot be distinguished through the study of phytoliths, be they from the leaf or from the ear. They can, however, be perfectly distinguished from wild grasses, teosinte included (José Iriarte, letter to the author, 2 May 2003; see also Pearsall 2000 and Piperno 2006).

The analyses Iriarte made for southeastern Uruguay confirm that the determination of the cross-shaped phytoliths and the morphological characteristics of neotropical grasses pointed out by Pearsall and Piperno are indeed valid for the identification of maize in archaeological contexts (Iriarte, 2003: 1092). This same study specifies that the cross-shaped technique is a conservative method that probably does not allow one to identify all races of maize (Piperno, 1988a: 173–174), but neither does it result in an erroneous identification of wild grasses as maize (Iriarte, 2003: 1086).

Staller and Thompson (2002; see also Staller, 2003: 373–374) have questioned the work done by Pearsall and Piperno as regards their technique for the identification of cross-shaped phytoliths, but as Iriarte correctly points out, they simply based their work on previous critiques (e.g., Doolittle and Frederick, 1991) without adding their own arguments; they furthermore used erroneous and misleading concepts in regard to the techniques used (Iriarte, 2003: 1086; see also Piperno, 1998: 427–434).

To finish this argument, I would like to bring up a datum on which specialists must pass judgment. When the study of the archaeological wheat found in the Middle Eastern site of Çatalhöyük began, it was established that when this plant grew in irrigated fields with clay-rich alluvial soils, its more prolonged exposition to the silicate present in the water would lead to a wider formation of phytoliths, and that often large groupings of silicified cells form. Yet when wheat is cultivated under dry conditions, the phytoliths usually consist of single cells or small groupings (Balter, 2001: 2279).

Rovner (1999) made a similar claim, in that soil conditions and, particularly, the available moisture may cause substantial changes in the mean and the size range of the values for the size of the populations of phytoliths, derived from members of the same species from one year to another, or from one place to another.

To clarify what Balter (2001) and Rovner (1999) claimed, I passed the question on to José Iriarte. He explained the following:

> Just like in the macro-botanical remains (e.g. squash seeds), there also is a range of variation in the size of the sets of phytoliths of each species. Yet the intra-specific variation of maize phytoliths does not impede our distinguishing

it from wild grasses. But two points have to be explained. First, the diagnostic phytoliths of maize ears known as "wavy top rondels" are given out only by maize and are morphologically unique, no matter under what conditions the plant grows. Second, the phytoliths of maize leaves, which are known as "cross-shaped bodies," are determined not only by their size but also by their three-dimensional form through multivariate analysis. (José Iriarte, letter to the author, 13 November 2006)[24]

[24] A detailed review and description of the technique used to identify maize from the phytoliths appears in Piperno (2006: 52–65).

5

The Archaeological Evidence

Nothing proves more damaging for a theory than some facts.

Anonymous

This chapter does not intend to present an exhaustive survey of the archaeological evidence, for that would mean writing a book on its own. Besides, on the one hand this is not the aim of this study, and on the other I actually did not have either the time or the means with which to gather all of the information available on this subject. But because acquiring an overall picture requires having at least some general evidence, I limit myself here to a succinct summary of the available data. I emphasize, through a more in-depth analysis, those areas that are at present believed to be crucial to at least have an understanding of the issue at hand, if not to actually solve it. By this I essentially mean the Mexican area, the Ecuadorean area, and what Bennett (1948) defined as the central Andean area, which mainly comprises what now are Peru and Bolivia.

The only thing this chapter aims to do is to present the dates in which maize appeared or was used in different parts of the continent without entering into the contexts and the associations, for which points I refer the reader to the respective sources in the bibliography. Furthermore, all references made are in general to radiocarbon dates without any kind of calibration, except in exceptional cases that are indicated.

To avoid any misunderstanding it is also worth noting that the information different authors have presented on this subject is used, but without a major critical assessment and relying on the aforesaid database. The one area for which an in-depth examination of this kind is made is the Andean area, my branch of knowledge – and besides one on which there are major controversies.

The sources available to me are discussed from north to south, and only the data regarding the most ancient finds of maize is presented.

Canada

The earliest evidence available for Canada is the Princess Point culture, on the Grand Banks site at Ontario, which has an antiquity that ranges between AD 400 and 600 (Crawford et al., 2006: 549).

United States

According to Schwarcz (2006: 318), in North America maize is a cultigen that was introduced at a late date, starting at dates that range between c. 600 and 2000 BP, although as we shall see there are some earlier dates. The most ancient remains are in the Southwest, close to the area where maize had its origins. The latest apparition of maize in the human diet took place in the northeastern United States and was dated to just a few hundred years prior to the arrival of the Europeans.

Until 1900 only one race – Maíz de Ocho – predominated over a vast expanse that extended from the Dakotas to New England, and as far northward as the Gaspé Peninsula in Canada.

The limited variability of early maize in this area was the result of a time lag of several thousand years in its dissemination northwards since its original arrival from the short-day zone in Mexico and Central America. The first domestic plants must have been ones that flower during short days, like teosinte and most of the tropical types of maize. Besides the photoperiod factor, an area with arid conditions and poor soils like the southern and the southeastern United States, as well as the thin strip of vegetation in the prairies in what now is the Corn Belt in the United States, all require the development both of the adaptation of maize and of an agricultural technology (Galinat, 1985b: 245).

In general we can see that maize spread gradually south of Ontario (in Canada) and in the northern United States, and in a more abrupt fashion in the south (Pearsall, 1996: 3).

The earliest date available for the eastern United States, in the Tennessee and Ohio areas, is 1800 BP (Fritz, 1990: 397; Wagner, 1994: 342), but it seems to have become a major staple only around 1200 BP (Van der Merwe and Vogel, 1978) (Van der Merwe and Tschauner, 1999: 526).

According to Conard and colleagues (1984: 444–446), in Illinois the AMS (accelerator mass spectrometry) dates for the maize found in archaic periods are not valid. For them, the introduction of maize in mid-Woodland times, which has been dated at c. 2000 BP, is incorrect and should be 1550 BP.

B. D. Smith (2006: 12228) in general believes that maize reached the eastern United States c. 2200 BP. For the New York zone, the date is c. AD 1000. We must, however, bear in mind that this date corresponds to macrobotanical remains. But there is a phytolith-based radiocarbon dating of 2270 BP (Hart et al., 2007: 564), which may be the date Smith means. But in the zone of Alabama, a pollen grain from the sediments of Lake Shelby, on the coastal area, was dated to 3500 BP. This would be the earliest date available for maize in the United States (Fearn and Liu, 1995).

Maize appears in the Mississippi basin between c. 170 BC and AD 60 (Fearn and Liu, 1995; Riley et al., 1994). Maize has in fact been dated at 100 BC for the Middle Woodland Holding site, in the mid-Mississippi Valley (Fritz, 1993).

Maize appeared in the central plains of the United States in AD 250–400 (Adair, 1994: 332), whereas the eastern Basket Maker group in Durango (Colorado) depended on maize toward the mid-first millennium BC and reached the peak of its reliance on maize in AD 500–1000 (Coltrain et al., 2006: 285).

For the Midwest, Berry (1985: 304) gives dates between 500 and 200 BC, whereas the dates Wills (1988: 37) gives for maize in the southeastern United States lie between 2000 and 3000 BP.

In the 1990s, dates of slightly more than 3000 BP were used for maize in the southwestern United States (Adams, 1994: 293; B. D. Smith, 1994–1995b: 179). More recently there are dates of 2017 BC (c. 4000 BP) (Smith, 1997a: 379; Wills, 1995) and 4300 BP for McEuen Cave, and 3620 and 3680 BP for Old Corn Site (Huckell, 2006: table 7–1, 100). Also in the Southwest, Simmons (1986: 85) has reported pollen remains dated to 2000 BC, whereas the remains of maize found in the LA 18091 site in the Chaco shelters have a date of 1000 BC. Upham and colleagues (1987: 410) concur, for they write that the maize from the Chapalote series actually occurred slightly earlier, in the second millennium BC. Doolittle and Mabry (2006: 117) likewise believe that the maize found in this part of the United States is the same one that was domesticated in southern Mexico, even though it is somewhat different from the Mesoamerican maize at the race or variety level.

For Vierra and Ford (2006: 507), maize was introduced into the region north of the Rio Grande, in New Mexico, before 1000 BC. Cave Cebollita (Cebolleta Mesa) was in fact the first stratified site where a large collection of maize was removed and where statistics could be applied to the results derived from the analyses. This is a maize that is related with Chapalote in its deepest levels and was then replaced by a tripsacoid maize. Its date has been given as ranging between AD 1050 and 1200 (Mangelsdorf, 1974: 161–162).

A famed and much-cited cave in New Mexico is Bat Cave, which was excavated by Herbert Dick first in 1948 and then in 1950. He found 766 shelled cobs, 125 loose kernels, 8 pieces of husks, 10 leaf sheaths, and 5 tassel fragments – remains that were considered the oldest in the United States, while the specimens in the deepest levels were believed to be the most primitive ones. In addition, this is the largest sample of maize that has been excavated, on which basis it was possible to develop a sequence. This was also the site where it was first possible to take radiocarbon dates. We must, however, point out that on both occasions the excavations were carried out using artificial strata. The date of 5605 years BP (c. 3600 BC), established using carbon, was rejected, and it was instead believed that the correct date was the indirect date of 2300–1500 BC (Galinat, 1985b: 247; Galinat et al., 1956: 101; Mangelsdorf, 1974: 147–152; 1983b: 229). Berry (1985: 281–284) made a critical analysis of Bat Cave and showed that the oldest date is not valid. Simmons (1986: 73) concurs and points out that the date of 3600 BC is unacceptable and that the correct date must be 1600 BC, but even this date has also been questioned.

The most ancient maize in this cave is a form of pod corn, and it probably is also a kind of popcorn. It is related with the Chapalote with a brown pericarp. The most modern sample presents evidence of the introgression of teosinte (Mangelsdorf, 1974: 147–152). One of the husks, which was open and did not stick to the ear like current ones, and which covered ramified groups of ears, is quite similar to those found at the Peruvian site of Los Gavilanes (Grobman, 2004: 440–441).

New World Indians used the slash-and-burn technique, which is inadequate for areas with grasses. The large plains, which are now the most highly productive agricultural areas in the world, were not of great significance for agriculture until the adoption of the steel plough in 1800 (Weatherwax, 1954: 1). Very little or no maize at all was cultivated in the western United States, except in the areas occupied by the Hopi and other southwestern tribes (Goodman, 1988: 197).

Schoeninger recently attempted an overview of the evidence available on the adoption of maize agriculture in North America, based on the data from stable isotopes. Based on a selection of the available data from human bone carbon and stable nitrogen isotopes, she concluded that these are consistent when compared with other archaeological data. Everything seems to indicate that the adoption of maize agriculture all along North America was not a single event. Along the northern tier, the evidence available for Ontario indicates that since its introduction, most of the population obtained most of its energy from maize. The same thing can be said of the Southwest Grasshopper Pueblo. The major difference between these regions is the source of protein in the diet of their inhabitants – in Ontario they ate fish and waterfowl, whereas in the Southwest the major source of protein was turkey or another protein source with a C_4 signal. A marked diet range along the Mississippi and Illinois Rivers indicates different subsistence strategies between foragers and farmers, and even between hunters and fishing people. Maize was used in a limited way in the region now known as Georgia during the period when it was beginning to become significant in other North American regions. The coastal populations of Georgia did not rely heavily on maize before the coming of the Europeans, thus suggesting that this was a forced change in their subsistence strategy rather than a freely chosen one (Schoeninger, 2009: 637–638). Grobman disagrees with this position and believes instead that it is strange that in the curve for the presence of effects in bone collagen due to feeding on maize, the indications of its use increase only in such late periods as 1000 BP (see Schoeninger, 2009: Fig. 1, 634). This would question the credibility of the technique, as there are older dates in the South Western United States. It would be far more logical to interpret the curve, not as if there was no feeding on maize prior to the date specified, but as if it increased violently for some reason that has yet to be studied. (Alexander Grobman, letter to the author, 15 September 2009).

Mexico

The Cueva de La Perra in Taumalipas (northwestern Mexico) was studied by MacNeish in 1949. Remains of maize related with the Nal-Tel race were found in its deepest part, with ears and grains larger than those from Bat Cave, and without any evidence of contamination with teosinte. The latter does appear in other races in later samples. The most ancient remains were dated to 4445 radiocarbon years (Benz, 1994b: 169; Grobman, 2004: 441; Mangelsdorf, 1974: 152–154; Mangelsdorf, MacNeish, and Galinat, 1964).

Remains of Chapalote were found in the Cueva de La Golondrina, in northwestern Mexico, as well as evidence of the introduction, in prehistoric times, of the eight-row Harinoso de Ocho maize from South America (Wellhausen et al., 1951). However, there is also evidence of the introgression of teosinte after the Christian era (Grobman, 2004: 442).

In northwestern Mexico there also is a series of caves that have been studied and are known under the names of Swallow, Slab, Tan, Olla, and Dark. The best samples come from Swallow Cave, which was studied by Robert H. Lister. Most of the maize found here was related with Chapalote. The earliest maize is the precursor of this modern race, but it is more primitive. There is also evidence of the introduction of eight-row flour corn, as well as of hybridization with teosinte. The emergence of modern races is visible in the upper levels. The maize excavated here was classified as Pre-Chapalote, Early Chapalote, tripsacoid maize, and Harinoso de Ocho. There are no radiocarbon dates for this site (Mangelsdorf, 1974: 157–160; Mangelsdorf, MacNeish, and Galinat, 1964).

There is a group of caves usually known as the Cuevas de Ocampo. The best-known of them are the Romero, Valenzuela, and Ojo de Agua Caves. They are found in Infiernillo Canyon, north of Ocampo in the eastern Sierra Madre of the Tamaulipas Mountains. More than 12,000 specimens of maize were found in these caves, but only Romero Cave was studied in depth. However, the remains from Valenzuela Cave were also included when the maize from Romero Cave was analyzed, as we shall soon see. Nine races or subraces were identified here. By far most of them are Chapalote. The oldest samples were called Pre-Chapalote and must have been cultivated in 2350–1850 BC. Tripsacoid Chapalote appears in 1500–1200 BC. Teosinte and its hybrids have been identified. These are the first remains of teosinte found, and they correspond to the Guerra phase, c. 1850–1200 BC. A few remains of *Tripsacum* and of maize contaminated with teosinte were also found (Mangelsdorf, 1974: 154–157; Mangelsdorf et al., 1967b; B. D. Smith, 1997a: 351).

In Romero Cave 3,472 intact or semi-intact cobs were found, along with 8,099 tassels or tassel branches. Nine remains are of teosinte, five are of *Tripsacum*, and four are hybrids of maize with teosinte, whereas all the rest are maize. The oldest cobs are not tripsacoid (in this they are similar to the remains from Tehuacán) but predecessors of Chapalote; as was pointed out previously, this is a Pre- or Proto-Chapalote. Later maize was tripsacoid and exhibited teosinte-Chapalote hybridization. There are nine specimens of teosinte and three maize-teosinte hybrids. At the time the study was carried out, these were the only archaeological remains of teosinte (see previously), but now there apparently are more (see subsequently). Interestingly enough, however, the fruits of teosinte were found in the feces, without their hard pericarp having been altered by their passage through the digestive system. It must be emphasized that nowadays there is no teosinte in Tamaulipas. Five specimens of *Tripsacum* were also found (Grobman, 2004: 441–442; Mangelsdorf et al., 1967b).[1]

Maize appears for the first time in Romero Cave in Level 7. Of the nine specimens in this occupation, five come from a disturbed context. Two AMS dates of 2560 and 3930 years BP were obtained from the two specimens found in another "relatively undisturbed" context (B. D. Smith, 1997a: 364–365).

In Valenzuela Cave, maize first appeared in its sixth occupation. Of the three units excavated, one held remains of leaves, another one fragments of stalks, and the third some cobs. These materials were within secure stratigraphic contexts, and their antiquity according to the AMS method is 3890 BP (B. D. Smith, 1997a: 370).

However, it must be pointed out as regards the study made by Smith that when he analyzed the remaining dates and included his data in two tables, the information that resulted is incoherent, incomplete, and in some parts unintelligible (B. D. Smith, 1997a: 372, 375, table 10 [373] and table 11 [374]). He admits that his datings show a very close agreement with other occupational episodes in other sites that were dated 40 years ago, with both the AMS and the traditional methods (Smith, 1997a: 371).

Benz (1994b: 169) has criticized the classification and pointed out that there are no specific data with which to distinguish Pre-Chapalote from Early Chapalote or Chapalote, but his arguments are not convincing.

De Wet and Harlan report very well-preserved fruits of teosinte found in Chalco, which were dated to 7040 BP. No more details are given, beyond that the information was a personal communication from Lauro González Quintero of Mexico's Instituto Nacional de Antropología e Historia (De Wet and Harlan,

1 Grobman mistakenly wrote that there were 3,015 intact or semi-intact cobs.

1972: 276). These are the remains that were published by José Luis Lorenzo and Gonzáles Quintero (1970) along with Chalco-type teosinte fruitcases in the Playa Level of Zohapilco, which were found in Zohapilco, Tlapacoya, in the Mexico basin, and which were dated to 5090 BC (see also Galinat, 1977: 6; Niederberger, 1979).[2] As regards this point it must be noted that Mangelsdorf (1983b: 238–239) convincingly questioned it and considered that the specimens are probably intrusive. He cited Wilkes, who analyzed the samples, in regard to the fact that these "… may be less than a hundred years old." Iltis (1983b: 886) believes they were "probably not used by man." While discussing the work of Mangesldorf (1983a: 89), who claims that teosinte was never collected nor cultivated by man, Flannery (1986b: 8), however, criticizes the fact that Mangesldorf did not consider the discovery of the Chalco teosinte and argues that it cannot be ignored even though it precedes the radiocarbon date of Coxcatlán, in the Tehuacán Valley, by just 40 years.

Whether or not the findings of Chalco teosinte are valid, it is worth recalling that hybrids of maize and teosinte are particularly common in central Mexico, and that in 1943 Mangelsdorf was able to verify that hybridization took place close to the hamlet of Chalco (Mangelsdorf, 1974: 123).

Zea pollen dated to 4100 BC has been found in Lake Cotrina, close to Veracruz (on the Gulf Coast), as well as in Lake Pompal in Los Tuxtla, there with the date 2900 BC (Benz, 2006: 16–17).

Pope and colleagues (2001: 1372–1373) reported the finding of *Zea* pollen in San Andrés, Tabasco (15 km south of the Gulf of Mexico). According to these authors, the first grains, dating to 5100 BC (6208 C14 years), must be exotic, for there is no cultivated maize on the coast close to the Tabasco zone. And given the small size of the sample, they assume it was due to the forest being cleared. The biggest grains probably are domestic maize, and they appear 100 years later, that is, in 4800 BC. It is worth noting that a pollen grain of *Manihot* sp. – which to judge by its characteristics is domestic manioc – was found in one of the parts dated to 4600 BC (5805 C14 years). It must be explained that the dates obtained from pollen were not direct ones and were instead obtained by association. Table 1 (Pope et al., op. cit.: 1372) points out that wood was dated; this same table shows the oldest date obtained from a cob, which is 2,565 radiocarbon years.

When the San Andrés finds were later discussed in regard to the pollen extracted from sediments dating to 6200 [6208] years BP, Pohl and colleagues said, "We deduced that the *Zea* … was cultivated because it was outside its natural habitat and appeared abruptly with other indicators of land clearance" (Pohl et al., 2007: 6870). When discussing the 6200 and 6300 BP maize phytoliths we read that "… data demonstrate that the introduced cultigen was maize rather than teosinte" (Pohl et al., op. cit.: 6872). Finally, a conclusion is drawn that clearly is an error, for it is stated that the maize pollen and phytoliths "…

2 There is a confusing datum in Beadle (1980: 99), which seems to refer to this same point.

are 5,800 years older than the earliest maize macrofossils ..." (Pohl et al., op. cit.: 6874). We have seen that the pollen and the phytoliths have an age of 6200 years BP, and that the oldest date obtained from a cob is 2565 years BP, so the difference can in no way be the one indicated. It must likewise be pointed out that the term "fossil," used for these remains, is incorrect.[3]

A famed series of caves was discovered in the Tehuacán Valley, south of Puebla; these are known as the Coxcatlán, Purrón, San Marcos, Tecorral, and El Riego Caves. The most important studies on the maize problematic were carried out in these caves by a team headed by MacNeish, with Mangelsdorf in charge of analyzing maize. A total of 24,860 specimens of this plant were found, 12,875 of which were complete cobs. Of the fragments, 3,941 have been identified, whereas 3,878 could not be classified. There were also roots, stalks, remains of leaves, tassels, petioles, and fragments of tassels as well as 600 grains. Quids of stalks and husks were also found (Mangelsdorf, MacNeish, and Galinat, 1964: 541).

A large amount of material was recovered from San Marcos Cave, which was initially dated to between 5200 BC and AD 300. Mangelsdorf reconstructed what he defined as wild maize based on the materials recovered here. These are small and uniform cobs (19–25 mm in length), usually with eight rows; an arrangement of small staminated spikelets on the upper part of the ears, followed by the small pistillate spikes that form the grains; a polystichous arrangement of the grains; small paired spikelets; fragile rachises; and long and very soft glumes that would have covered the grains, partially or in full. The ears had about 50 kernels enclosed when young in a husk system that opened up at maturity, thus allowing the seeds to disperse. The kernels were round, brown, and partially enclosed by their glumes (Mangelsdorf, 1974: 166 and passim, figure 15.24, 180). It was thus a type of wild or newly domesticated maize that did not even remotely exhibit, in the oldest strata, any external signs of having been subject to a flow of teosinte genes (Grobman, 2004: 433–434). An AMS dating was later taken, and a date of 4700 BP was obtained (Long et al., 1989; Piperno and Flannery, 2001: 2101).

It is worth recalling the comment Mangelsdorf (1974: 168) made regarding an ear derived from the oldest phase (Coxcatlán) in this cave, based on the report he and his team made (Mangelsdorf et al., 1967a). He pointed out that prehistoric wild maize resembled the genetically reconstructed ancestral form – both have small male and female spikelets in the same inflorescence (Mangelsdorf, 1958a) – as well as certain races of modern maize that are considered primitive – Mexico's Nal-Tel and Chapalote, Colombia's Pollo, and Peru's Confite Morocho – and *Tripsacum*, a relative of maize that has regular pistillate spikelets below and staminated small spikelets on top on its lateral spike. Mangelsdorf considers the earliest maize in this cave and in that of Coxcatlán to be wild.

[3] See the explanation in Chapter 4 where pollen is analyzed.

In fact, it was in Coxcatlán Cave that Mangelsdorf found what he defined as the first remains of wild maize. Here there were 28 occupation strata and maize was dated to between 1050 and 7050 BP (Mangelsdorf, 1974: 177–178; Mangelsdorf, MacNeish, and Galinat, 1964). A subsequent AMS dating gave the dates "4040 and 4090 BP" (Buckler et al., 1998: table 5, 160; B. D. Smith, 2005: table 1, 9441).[4] Smith wrote regarding these new results, which were based on six cobs of maize from preceramic contexts, that they were "... considerably younger than the temporal span of the cultural zones and phases from which they were drawn, further increasing the percentage of anomalous preceramic dates and providing further evidence for postdepositional disturbance" (Smith, 2005: 9443, based on Long et al., 1989, which I mention later). As we shall see, there are major discrepancies in this regard.

Before continuing with this review, an explanation is in order, because otherwise confusions and mistakes may arise. In some cases the Chapalote and Nal-Tel races are mentioned, whereas in others the terms "Chapalote/Nal-Tel" or "Nal-Tel/Chapalote complex" are preferred. Wellhausen and colleagues (1952) classified both races as "ancient indigenous." Mangelsdorf (1974: 173) believed that Chapalote is one of the most distinctive Mexican races, particularly in the northwest. Nal-Tel is closely related with it and essentially differs in the orange pericarp. It likewise tends to have shorter ears with slightly more rows than Chapalote. The two races are quite similar as regards the other characteristics, and usually it is not possible to distinguish their cobs. It was for this reason that the term "Nal-Tel/Chapalote complex" was coined, to avoid confusion.

The work undertaken in the El Riego Cave only corroborated the conclusions reached with the research done in San Marcos Cave (Mangelsdorf, 1974: 177). There is, however, one point that should be clarified. A recent study by Benz and colleagues (2006) accepts MacNeish and García Cook's (1972) stratigraphy, but it is qualified as of "apparent stratigraphic integrity" (Benz et al., 2006: 73–75). Benz and colleagues then add: "The excavator's field catalog indicates excavation proceeded in *arbitrary levels*, whereas published accounts indicate natural stratigraphy was followed during excavation" (Benz et al., op. cit.: 75; emphasis added). This is not true. Kent Flannery, one of MacNeish's closest associates, notes that "the first 1 m by 1 m test pit was made by arbitrary 10-cm levels. Once the natural stratigraphic levels could be seen in the profile of this first test, MacNeish assigned the letters A, B, C, etc. to each level, *and the excavation proceeded totally by natural levels.* This was MacNeish's standard way of working" (Kent Flannery, letter to the author, 5 September 2006; emphasis added).

Purrón Cave, close to Coxcatlán, had 25 strata dated between 7000 BC and AD 500, but these held very few plant remains (Mangelsdorf, MacNeish, and Galinat, 1964: 540).

[4] We shall see subsequently that these dates are incorrect.

For Mangelsdorf, the materials from the San Marcos and Tecorral Caves are the most interesting ones in the Tehuacán Valley, for several reasons. First, this is where the oldest remains were found. Second, the oldest ears are of wild maize. Third, this maize was the progenitor of Chapalote and Nal-Tel, two indigenous Mexican races. Besides, the specimens of all parts of the plants are well preserved, and this supports the fossil pollen evidence that the ancestors of cultivated maize are maize. Finally, the collection presents a well-defined evolutive sequence covering a period of c. 6,500 years (Mangelsdorf, 1974: 167).

The characteristics of the wild maize from the San Marcos and Coxcatlán Caves – still according to Mangelsdorf – are remarkably uniform in size and other characteristics, which is a peculiarity of wild species. The ears have fragile rachises just like many wild grasses; this provides them with a means of dispersal that modern maize lacks. The glumes are relatively long vis-à-vis other structures, and they may have partially enclosed the grains, just like in other wild plants. There are some places in the valley, below San Marcos Cave, that are well adapted to annual grasses that probably included wild maize. There is no good evidence available, in terms of the other species found, that agriculture began at that time. The predominant maize in the following epoch (called Abejas phase, 3400–2300 BC), in which there definitely was an established agriculture, was larger and more variable (Mangelsdorf, 1974: 169; see also Mangelsdorf, MacNeish, and Galinat, 1964: 541; 1967a: 180; and see my Figure 5.1).

Wilkes (1972: 1076) fully accepts the results and interpretations of Mangelsdorf regarding Tehuacán.

Galinat has made several observations regarding the Tehuacán materials, namely, that the cobs do not exhibit induration, and that they may be indicating only that the introgression of teosinte does not produce a significant induration in the context of primitive maize. A situation like this can be expected if the primitive maize has intermediate tunicate alleles like those studied by Mangelsdorf and Galinat (1964), and like those present in the primitive race of Chapalote, and perhaps in others too (Galinat, 1977: 7). Galinat likewise points out that one can see that the oldest cobs in Tehuacán have a higher level of condensation than those of teosinte, and that they have a reduced teosinte fruitcase like that which can be produced by the tunicate trait that is also visible in these cobs; this is a major piece of evidence supporting the teosinte origin of maize (Galinat, 1985b: 253).

After studying the rachises in the Tehuacán maizes, Benz and Iltis (1990: 507, 508) concluded that they belonged to a cultigen, and then Benz and Long (2000: 460) claimed that the morphological changes showed that significant efforts had taken place in the earliest maizes to produce genetic variations in the ear, thus suggesting human dependence for this plant at an earlier date than had been thought.

De Wet and Harlan (1976: 452) in turn believe that the so-called Tehuacán wild maize (Mangelsdorf et al.: 1967a) lacked the capacity of natural seed

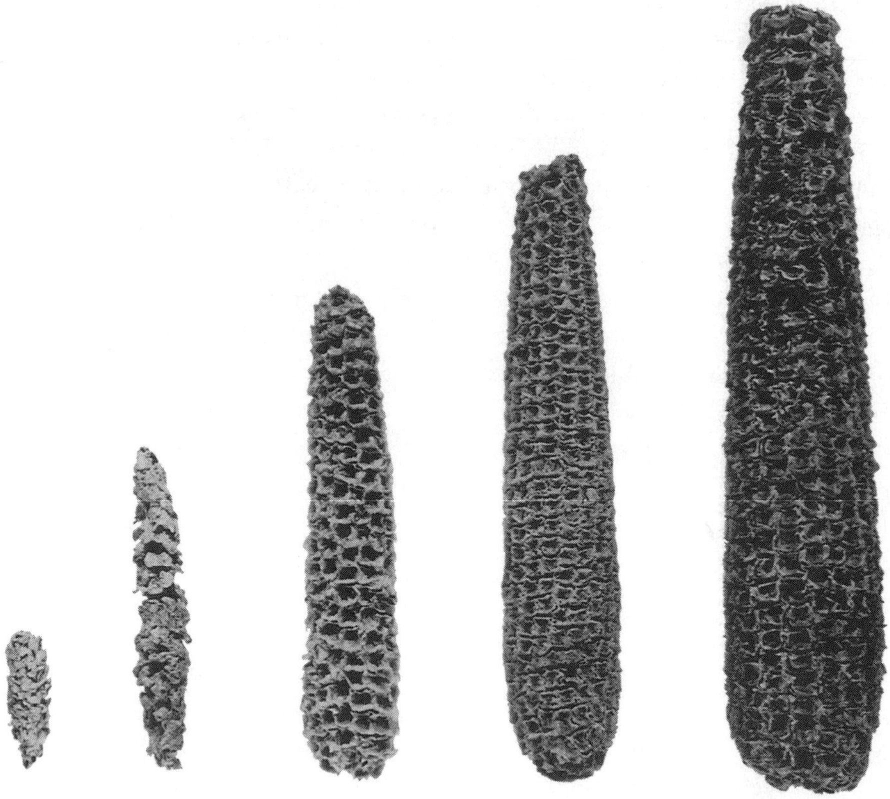

5.1. Cobs from the Tehuacán Valley, Mexico, showing the full evolutive sequence of domestication from c. 5000 BC (the small cob, on the left) to AD 1500 (the largest cob, on the right). From left to right the images are as follows: wild maize according to Paul Mangelsdorf, Coxcatlán phase, Marcos Cave; early cultivated maize, Abejas phase, San Marcos Cave; Chapalote cob, Palo Blanco phase, San Marcos Cave; Chapalote cob, Venta Salada phase, Coxcatlán Cave; Cónico cob, Venta Salada phase. Readers should bear in mind that the smallest cob is 23 mm long. Figure 122 in *Prehistory of Tehuacan Valley*, Vol. 1, 1967. Robert S. Peabody Museum of Archaeology, Phillips Academy, Andover, Massachusetts.

dispersal. The rachises are intact even though the grains were removed from the cobs, with no indication whatsoever of articulations between the cupules. This strongly suggests that the races of maize in the Coxcatlán phase of Teotihuacán had already been domesticated to the point that they depended on man for the dispersal of the seeds through planting and the harvest, as is common among cultivated cereals.

Benz has criticized the work done by Mangelsdorf with the Tehuacán maizes (Mangelsdorf, 1974; Mangelsdorf, MacNeish, and Galinat, 1964, 1967a, 1967b) because he believes there was a considerable confusion in the analysis made of the cobs; Benz particularly wonders whether or not these were adequately described and correctly identified. He does, however, admit that no one has questioned the accuracy of the racial characteristics of the maizes in the Post-Coxcatlán phase. Benz insists there are no descriptive data, and that the original description of this assemblage is fully lacking the data that characterize the races and superraces identified, other than the number of rows and spikes per row of the "wild maize" (Benz, 1994b: 172–173, 177). Benz wrote the following when he returned to this issue shortly afterward: "Archaeological context provided the bias for a long held misconception about the age of domesticated maize in the Tehuacán Valley because its age was inferred from associated radiocarbon determinations whose reliability for estimating corn cob age was low" (Benz, 2006: 10). Three things have to be noted regarding this point. First, the critique here made of Mangelsdorf is unacceptable. All who knew him can vouch not only for his great experience but also for the high reliability of his work. Questioning this is not honest. Second, the critiques leveled against the datings obtained for Tehuacán, which shall be mentioned in the following pages, were countered by MacNeish and Flannery. Third, the fact is that rejecting the validity of the associations goes against one of the main tenets of the science of archaeology – a point I return to in depth when reviewing the Peruvian archaeological materials, in this chapter and in the final one (Chapter 10).

Benz (2006: 16) insists that, based on the analysis of the group of morphological or genetic changes throughout time ("evolutionary rates") of the Tehuacán maize, one reaches the conclusion that it was of limited importance in the human diet, despite the fact that full domestication and intensive selection were present in the San Marcos Cave.

As was already mentioned, an attempt was made in the late 1980s to redesign the Tehuacán chronology based on the AMS method. We saw that the Coxcatlán phase was dated between 5350 BP and 7000 BP using the traditional carbon 14 method. When the maize (materials from San Marcos Cave) was used – long after the work at Tehuacán had been done – to get dates using the AMS method, the oldest date obtained was 4680 BP (which once calibrated becomes 3500 BP). This means that on average there is a difference of at least 1,500 years between the initial datings and the new one (Long et al., 1989: table 1, 1037, 1039). Fritz (1994a) later defended the AMS method datings pretending that agriculture in America was a far later occurrence than had been believed. MacNeish (1997: 666–670) defended his position and showed that, during storage of the samples used for the AMS analysis, they were an "uncurated and contaminated corn," and he furthermore specified that "in the late 1960s the corn specimens in Mexico were sprayed or soaked with a preservative called metacreal. Of the 12 dates done by the University of Arizona on the contaminated corncobs, only one is acceptable (Fritz 1994[a])" (MacNeish, op. cit.: 668; see also MacNeish and Eubanks, 2000: 15). Flannery (1997: 662) wrote in regard to Fritz (op. cit.) that it was "... an unfortunate overreaction to AMS dates on Tehuacán maize (Long et al. 1989)." Long and Fritz (2001) rejected the objections and considered the dates as valid. It must, however, be pointed out that in this text the elementary principles of archaeology are ignored. MacNeish (2001) gave a solid reply and clearly showed the inconsistencies and problems raised by the AMS datings done for Tehuacán. He showed that the AMS dates were rejected because they did not agree with the stratigraphic sequences and the materials the strata held. MacNeish also showed that the new "unacceptable" datings are due not so much to a possible contamination with Bedacryl (the substance used to preserve the maizes), as to their processing in the Arizona laboratory.

Going back over this issue, MacNeish and Eubanks (2000: 15) clearly pointed out that the problem that arose between the traditional carbon 14 datings and the AMS ones as regards Tehuacán were significant but not crucial, for there is pollen from *Tripsacum* and *Zea* that was identified in Oaxaca in the 10000 years BP range, and maize from Oaxaca (from the Guilá Naquitz site, to which I shall shortly refer) from 6500 BP. Eubanks went over the Tehuacán AMS issue once more and quite clearly stated that with the available information on the dates for the remains of maize from Guilá Naquitz (Piperno and Flannery, 2001), and the *Zea* pollen from San Andrés, in Tabasco (Pope et al., 2001), which "securely" place early maize in the 4000–5000 calendar years BC range – and which fall nicely in line with the original C14 dates – it proves likely that the original dates for the first appearance of maize in the Tehuacán Valley some 7,000 years ago are closer to the other finds than had been previously believed (Eubanks, 2001b: 499). The doubts raised are thus removed. Eubanks was furthermore quite clear when she stated that "*the Tehuacán Valley macrofossils represent the full spectrum of maize evolution*" (Eubanks, op. cit.: 499; emphasis added).

Pearsall (1994b: 120) likewise considers that even with the modification of the date in Long and colleagues (1989), the Tehuacán maize "... is still the oldest available collection." B. D. Smith (2005), however, went over this issue once more some years later and wrote thus:

> In contrast to these studies, however, the first set of direct AMS radiocarbon dates from Coxcatlan Cave, obtained on six corn cobs selected from secure and well dated preceramic contexts, all produced dates considerably younger than the temporal span of the cultural zones and phases from which they were drawn, further increasing the percentage of anomalous preceramic dates and providing further evidence for postdepositional disturbance. (Smith, op. cit.: 9443)

Smith himself notes that instead of obtaining direct AMS dates from all the cobs included in the study measuring the rate of early maize evolution, based on the analysis of a series of 26 temporally organized maize cobs from San Marcos and Coxcatlán Caves in Tehuacán (Benz and Long, 2000), most samples were simply dated by association. During the analysis three AMS dates were obtained from maize cobs found in the western part of Coxcatlán Cave (B. D. Smith, 2005: table 1; AMS dates 1860, 1900, 4090), which were then used to establish the age of 10 additional contemporary specimens derived from different units in the caves. The excavation units that yielded the dates 1860, 1900, and

4090 showed considerable evidence of disturbance, thus rendering all dating by association problematic. For instance, the date 4090 (2600 BC) comes from a square that yielded a much younger sample (470 [AD 1435]) based on the following lower zone, immediately adjacent to the squares that yielded older dates from zones that are superimposed (the dates 5560 [4360 BC] and 6925 [5780 BC]).

Older dates than those of the upper zones (4040 [2570 BC] and 5240 [4040 BC]) were likewise obtained in the grid immediately adjacent to those that yielded the dates 1860 (AD 130) and 1900 (AD 100). Given these evident indications of a vertical displacement, the apparent projection of the contemporaneity of the three AMS dates of the dated cobs to include additional, nonidentified maizes from the grids in the Coxcatlán and San Marcos Caves is unjustified, so other detailed and time-sensitive analyses of the rates of evolutionary changes are necessary (B. D. Smith, 2005: 9443).

The methodology expounded by Smith clearly has no scientific value whatsoever and is, besides, contradictory. It uses the "associations" of some samples in order to date others, which is precisely what is rejected by those who claim that the traditional C14 datings were used by associating them with other specimens, specifically maize in this case. Smith also shows a lack of understanding of the basic tenets of archaeology.

An interesting datum that specialists should bear in mind is given by Farnsworth and colleagues (1985: 114). They made a reappraisal of the isotopic and archaeological reconstructions of the diet in the Tehuacán Valley. Farnsworth and colleagues concluded that the analysis of carbon and nitrogen isotopes in bone collagen suggests that cultivation, and even domestication, were practiced at an earlier date and in a far wider way than had been previously believed.

Eubanks (2001b: 499) believes, in regard to the sequence of the Tehuacán maizes, that the first cobs were transformed in the intermediate stages of early cultivation and were followed by the first tripsacoids, after which came the Nal-Tel/Chapalote complex, a late tripsacoid, and a faint popcorn that still existed at the time of the Spanish contact.

As for the Tehuacán zone as a primary domestication center, there are some specialists who believe it cannot have been a primary center because the AMS dates are more recent than some of the South American dates (Tschauner, 1998: 321; Van der Merwe and Tschauner, 1999: 526).

It is worth recalling that teosinte was not found in Tehuacán, but it does appear in Mitla (700–500 BC), in the Oaxaca Valley, and in Romero Cave (900–400 BC), and it coincides with the later hardening of the cob in Tehuacán (Wilkes, 1979: 15).

Several sites have been studied in the Iguala Valley, in the central Balsas watershed, where the work done dealt essentially with the remains of pollen and of phytoliths. One of these sites is Laguna Ixtacyola. Interestingly enough, the authors of this study stated that "we emphasize at the outset that we do not identify any *Zea* pollen as maize because neither grain size nor morphology allow teosinte and maize to be distinguished" (Piperno et al., 2007: 11876). *Zea* pollen was found close to the bottom of Core 1, and there is one date that is 20 cm above the deepest zone with *Zea* pollen, that is, 6290 BP. The authors state that "it is clear that *Zea* has been a continuous part of the vegetation of the Iguala region probably since the end of the last ice age" (approximately 10000–12000 years BP; Piperno et al., op. cit.: 11876). They conclude that the *Zea* pollen associated with vegetational disturbance is quite probably maize, but the presence of this plant cannot be confirmed because no phytoliths were found (Piperno et al., 2007: 11877).

Ixtapa is another site in this same zone. Piperno and colleagues point out that the rich phytolith record notwithstanding, "... no teosinte, maize ... were found." They however add a contradictory phrase, for they claim that "a fragmented pollen grain from *Zea* (... associated with a date of 10,850 B.P.) is the only evidence of this taxon with the exception of grains found in surface sediment. Concordant with this picture is the absence of human artifacts" (Piperno et al., 2007: 11878). The status of this site is therefore not clear.

Laguna Tuxpan is another one of the sites studied. *Zea* pollen was found there in many levels. So "maize cob and leaf phytoliths occur at the bottom of the sequence ... There is no sign of teosinte phytoliths" (Piperno et al., 2007: 11879). *Zea* grains, as well as cob phytoliths and perhaps leaf phytoliths, were found despite the poor preservation of the pollen. It is, however, explained that the "phytoliths commonly found in the fruitcases and leaves of teosinte are absent" (Piperno et al., 2007: 11880). The authors conclude that *Cucurbita* and maize seem to have been planted in fertile lands close to the lakeshore in the period approximately 5000 years BP–10000 years BP. Given the C14 datings and the fact that the dated collections of phytoliths show many indications of human disturbance, it is possible that maize and squash were deposited at some moment during the first half of that time span, along with the disturbance taxa (Piperno et al., 2007: 11880). As we see, these are mere elucubrations.

When discussing these finds the authors claim that the pollen record in one of the oldest sites studied, that of Ixtacyola, "... indicate[s] initial *Zea* presence immediately above a sediment level dated to 22,110 B.P." They add that if their interpretation is correct, in that the hard-water error in these sediments formed on a dry lake bed is either not a factor or is limited to 1,000 years, then the pollen would probably be from teosinte, as the domestication of maize before the end of the Pleistocene period is not to be expected. An already-mentioned *Zea* pollen grain recovered from Ixtapa, with an age of 10850 BP, supports this position, particularly because *Zea* is later missing, and no other evidence of agriculture or human activities in general was found at the site. The authors then speculate that the downslope movement of vegetation "... may well have involved various types of teosinte, including the race Chalco ... ," and that the teosinte Balsas, which is nowadays absent below 400–500 m, "... could have descended into lower-lying tropical areas...." When discussing the Tuxpan maize phytoliths alongside *Zea* pollen in sediments dated to 5000–10000 BP, they say these "... indicate that the cultivars were probably deposited earlier rather than later in that interval" (Piperno et al., 2007: 11880).

We see that for all of these sites there are no actual data but only mere assumptions, something the authors confirm when they write: "Because pollen from maize and teosinte cannot be distinguished, the pollen *record is equivocal* as to which taxa contributed the *Zea* grains, which are continuously present in Holocene pollen records at most of the sites" (Piperno et al., 2007: 11880–11881; emphasis added).

The Xihuatoxtla Shelter is another site studied in the Guerrero zone of the central Balsas River. Here starch grains and phytolith remains have been studied. The analysis of the former showed they present the characteristics of the grains found in modern samples of teosinte (the "transverse" fissure). But without giving any real explanation, the authors add that "the archaeological frequencies are characteristic of Mexican popcorn." They insist in that "grain morphology suggests the presence of popcorns or other hard-endosperm maize types" (Piperno et al., 2009: 5021). The study of the phytoliths points out that "short-cell[s] ... diagnostic of the glumes and cupules of maize cobs, called wavy-top rondels and ruffle-top rondels, are present in a number of different contexts ... ," that is, in remains extracted from preceramic and ceramic grinding stones, in sediments associated with these artifacts, and in column samples from the area under study. Then they point out that "the types of long-cell phytoliths that always occur in high numbers in (and are diagnostic of) teosinte fruitcases ... are not present in any samples." Other rondel phytoliths, designated as "maize type" in their table 1, "... cannot be unequivocally assigned to maize" because they appear in some non-*Zea* grasses, and yet they are of the commonest rondel type found in maize cobs, and because they co-occur with ruffle-top and wavy-top forms, they probably are from maize. Besides, these phytoliths lack edge ornamentation, and therefore are not the rondels characteristic of teosinte. The authors therefore conclude that the latter was not used in Xihuatoxtla, and that the "... *Zea* remains are exclusively from maize." According to these authors, all of the data indicate that maize was cultivated during the early ninth millennia cal. BP (Piperno et al., 2009: 5022).

And then, despite repeating what had been stated by Piperno and colleagues (2007: 11876) – and which has already been pointed out – claiming that "... pollen from teosinte and maize cannot be reliably distinguished," they insist that the pollen recovered from Lake Ictaxyola (which is 20 km to the west of Xihuatoxtla) "may suggest" there was teosinte in the region in the Late Pleistocene period, and they conclude that in Xihuatoxtla, the phytoliths and the starch remains indicate that teosinte was not used either as a grain or for its stalks, which is incongruent.

5.2. The Guilá Naquitz Cave, 5 km to the northwest of Mitla (Mexico). Photograph courtesy of Kent Flannery and Joyce Marcus.

The study ends with a statement I believe is quite right, that is, that "… our data shift the focus of investigation back to lower elevations" (Piperno et al., 2009: 5024). But this is nothing new – Sauer said so in 1952 (Sauer, 1969a: 40 and passim), and Grobman and I have repeated it more than once (e.g., Bonavia and Grobman, 1978: 87).

Hastorf has commented on the research undertaken in the Balsas River basin. She notes that although the evidence locates the process of domestication further back in time, it is "… still *without concrete support* for the mechanism that triggered these results, and more importantly without information of timing." She ends by stating that although the process of domestication of maize has been pushed back in time, we really do not have the first evidence of the use of teosinte by man in the Balsas River area (Hastorf, 2009: 4958; emphasis added).

A major site for the problematic under discussion is the already-mentioned site of Guilá Naquitz. It lies 5 km away from the town of Mitla, to the east of the Oaxaca Valley (Figure 5.2). Here "… four small primitive-looking maize cobs …" were found, "but their provenience was such that little can be concluded from them." The cobs were found in small ash lenses that lay stratigraphically above Zone B1, which corresponds to the oldest preceramic level, and below scattered ceramic sherds that preceded the deposition in Zone A: "Since these lenses of ash had no artifacts, all we can do is guess that they date somewhere between 6700 BC and the period of the earliest Formative sherd scatters, 1000 BC (?)." Richard I. Ford and George Beadle studied the maizes, and both agreed that interpretation hinges on from what side one looks at them in terms of the maize-teosinte argument. They can thus be considered maize-teosinte hybrids or an early maize that exhibits a strong influence of teosinte in its ancestors. The report clearly states: "In view of the specimens' undated stratigraphic position, there is little more to say at this time" (Flannery, 1986b: 8).

Buckler and colleagues (1998) and Grobman (2004) are the only ones who have cast some doubt on the remains of maize from Guilá Naquitz. Buckler and colleagues (op. cit.: 159) note that "the small number of remains makes stratigraphic intrusion a strong possibility." Grobman in turn notes, while discussing Piperno and Flannery (2001), that although Eubanks – who is mentioned – was among the scholars who believed that maize was derived from annual teosinte, she has since changed her mind. Grobman then points out that the only thing the four ear fragments found in Guilá Naquitz indicate are "occasional visits" and an "ephemeral occupation." He then adds: "This is not the best context wherein to pass judgement, since no stratigraphy is presented that can allow us to define maize in secure archaeological contexts." And, he continues: "Benz and Piperno however, passed judgement independently as regards the evolution of maize from teosinte, without acknowledging that there is another possible explanation, or that the evidence of the maize pollen in Guilá Naquitz is older than the other fragments of maize." Grobman then pointed out the greater age of the remains from Casma, Peru, and noted that "like all other archaeological finds of early preceramic maize, that of Casma does not exhibit the influence of teosinte" (Grobman, 2004: 443–444).

When the cobs in question, which had been studied in 1970, were reanalyzed by Piperno and Flannery (2001: 2102), they opted for the second possibility suggested by Ford and Beadle, in that it is a primitive maize with a strong teosinte influence (and they relied on Benz, 2001).

Eubanks also commented the Guilá Naquitz cobs, but she claimed that they resemble the experimental segregating populations of *Tripsacum*-teosinte. She also pointed out that because the pollen found in the stratigraphic layers below the cobs has the morphological attributes of modern specimens of *Tripsacum* and maize (Schoenwetter and Smith, 1986), it can be inferred that it represents a segregating population of *Tripsacum*-teosinte recombinants that were collected in the wild or at the beginning of domestication. It is, however, striking that in her paper Eubanks claims to agree with Beadle and Ford, for they do not mention *Tripsacum* at all (Eubanks, 2001b: 500–501).

Benz also discussed the maize remains from Guilá Naquitz, which he indicates have rachises that do not disarticulate; he also points out that there are ruptures close to the apical inflorescence through the internodes instead of the nodes, which is an index of domestication given that the dispersal of the kernels depends on man. The three inflorescences are distichous (i.e., they have two rows), and a fragment has four rows. Like teosinte, the two-row specimens are distichous and have small, single-grain-bearing spikelets per segment of the rachis. The four-row specimen is like maize. Benz, searching for two rows, suggests these are domestic plants subject to human selection. The two-row cobs from Guilá Naquitz are similar to those of teosinte thanks to several other morphological characteristics. They differ from the latter in that the two-rowed inflorescence fragments are perpendicular to the rachis, and that the cupules are very short and shallow, with rachises that do not disarticulate. He admits that the four-row specimen is identical to the specimens from San Marcos Cave in Tehuacán (Benz, 2001: 2105).

Benz later discussed the maizes from Guilá Naquitz once more. He pointed out that the rachises of the "three" cobs have a single spikelet that is distichous, indurate and mottled, and smooth and that has a shiny surface that is very similar to the sheaths of teosinte.[5] What Benz actually studied were the remains of two ears, a complete one and another one that is broken in two. The four original specimens were stored in the Prehistory Department in Mexico City. When Dolores Piperno wanted to study them many years later she only found two of them, which are the ones Benz analyzed (Joyce Marcus, letter to the author, 25 January 2006).

We have seen that two of the specimens have two rows of grains just like the teosinte's inflorescence, and the third has four rows of grains. This subtle difference is one of the vital characteristics distinguishing maize from teosinte. Benz then compared the size of the Guilá Naquitz grains with the most ancient ones from San Marcos cave and then claimed that there were no significant differences, even though the inflorescences of the specimens from San Marcos do not have the same shape in the cupule, or the mottled rachises in the Guilá Naquitz specimens. Besides, the San Marcos samples are in general more recent.[6] He also states that the early specimens from San Marcos are typical of maize with one significant exception; a large number (c. > 5%) are different and have four rows, like the Guilá Naquitz specimen (Benz, 2006: 15–16).

Grobman (2004: 442–443) analyzed the comments Benz made (he used the first paper: Benz 2001) and believes that what Benz claims are evidence of domestication actually are not so. Benz (2001: 2104) wrote: "... efforts to domesticate teosinte were successful at least 700 years before the earliest maize cobs were incorporated into the preceramic refuse of San Marcos Cave in the Tehuacán Valley." Grobman (op. cit.: 442–443) believes that this claim by Benz is "... definitely biased because his evidence is non-conclusive as he claims – when arguing in favour of the theory of the domestication of maize from annual teosinte – since the data he presents are questionable...."

Benz (2001) believes the three maize cobs that do not disarticulate are proof that maize descends from teosinte, but "... this argument only shows his

5 We have seen that Flannery actually excavated four samples (see previously).

6 Here Benz accepts the AMS dates.

ignorance of how the real dispersal of the seeds in primitive maize must have operated, through the fragility of the rachilla and not of the rachis" (Grobman, 2004: 443). Grobman adds that the new and equally valid theory of Wilkes, of a cross of wild maize × *Zea diploperennis* and a backcross to maize, would also explain these ears as an alternative interpretation verified by recent research. These ears could likewise be used to show that what Benz posits as maize in an early stage of domestication of annual teosinte is simply a hybrid of feral maize × perennial teosinte, given that similar materials are obtained by segregation in backcrosses of maize × perennial teosinte. Grobman draws attention to the fact that there are photographs of many ears that coincide with those in Benz (2001: figure 1, 2104), and that come from F_2 populations and BC1 backcrosses between maize and *Zea diploperennis* carried out by Cámara-Hernández and Mangelsdorf (1981: plates IV, V, VI, X, and XI).

It so happens that two samples of Guilá Naquitz cobs were used for dating with the AMS method; the dates obtained are 5410 and 5420 BP (i.e., c. 6250 calendar years), and so they are about 700 years older than those from Tehuacán (Piperno and Flannery, 2001: 2102). It must, however, be noted that the title of the paper reporting these new dates reads: "The Earliest Archaeological Maize ..." (Piperno and Flannery, op. cit.), which is actually incorrect, as Peru has maize specimens from the coastal Casma Valley that are older, whereas those from Rosamachay, in the highlands, have the same dates. These are discussed later.

Pollen remains have also been found at Guilá Naquitz. A grain found in Zone B1 is "unhesitatingly" *Zea*. Four others have been classified as teosinte/*Zea*, as well as other "graminoid" grains defined as *Tripsacum*. This pollen was subjected to an exam using the Tsukada method (Tsukada and Rowley, 1964), and it resembles the modern teosinte. Schoenwetter believes that a plant similar to maize was cultivated in the preceramic phase of Guilá Naquitz or thereabouts (Schoenwetter, 1974: 301–302). Piperno and Flannery (2001: 2103) later actually wrote that the analysis of the oldest phytoliths from this site did not identify maize.

Mangelsdorf (1983b: 237–238) analyzed this pollen and concluded that "as a proponent of the theory that one of the ancestors of cultivated corn was a wild corn, I see features of resemblance between the Guilá Naquitz and the Bellas Artes fossil pollen. Both are pre-agricultural and both are associated with grasses including *Tripsacum* and with species of Compositae and Chenopodinae." Flannery (1986b: 8), on the other hand, reports a small *Zea* grain that Schoenwetter and Smith (1986) believe is from teosinte and says that it comes from the Zone C that has been dated to 7450–7280 BC.

In a later study Piperno (2003b: 834) notes, in regard to Guilá Naquitz Cave, that it "... was not a center of maize production when those four cobs were deposited, but the possibility is real that even earlier maize-growing than presently evidenced took place in that part of Oaxaca."

Schoenwetter and Smith (1986: 216) reported finding pollen from cf. *Zea mexicana* in Abrigo Martínez, a shelter close to that of Guilá Naquitz, whose age is estimated between 9500 and 10000 BP. Pollen from cf. *Tripsacum* was found in a nearby settlement known as Cueva Blanca, which has an estimated antiquity of 5000–6500 BP. Schoenwetter and Smith point out that the reliability of the Maydeae tribe (e.g., maize, *Tripsacum*, or teosinte) is quite restricted. There may be different species contributing to this pollen, or it may all derive from one single species. The plants that produced the pollen may have been either wild or cultivated but were genetically unaffected by human selection, or domesticated, or some combination thereof.

Flannery discussed the discovery of teosinte in Guilá Naquitz and believes that the sole presence of the pollen is not proof that it was eaten. But its "indisputable" use came in the time of the Oaxaca Formative (1500–500 BC), where it appears combined with grains of maize in Tomaltepec and San José de Mogote. Ford found maize cobs with teosinte introgression at Fábrica de San José. Flannery believes that "... Mangelsdorf may have underestimated the use of teosinte in early Mesoamerica, especially since we have few excavated sites from the most relevant periods." He, however, acknowledges that Mangelsdorf is right in that there is no excavated site that documents this gradual genetic change from teosinte to maize. There are thus two possibilities: either this change did not take place or we have very few sites in too few areas, depending on what hypotheses one wants to accept. Iltis posited his (already mentioned) theory of "sexual catastrophism" when *Zea diploperennis* was discovered, that is, that a mutation turns the inflorescences of teosinte into the female ear of maize. This modified Beadle's hypothesis and accommodated some of Mangelsdorf's critiques, as Iltis himself acknowledges. If this were correct, Flannery explains, the transition from teosinte to maize would have been so rapid that it would be hard to detect it archaeologically (Flannery 1986b: 8).

Guatemala

There are some data related to pollen grains extracted from a sediment sample of a series of cores from Lake Petenxil, which "... could represent possibly wild maize types...." It was given a date of 3950 radiocarbon years (Irwin and Barghoorn, 1965: 43; see also Bartlett et al., 1969: 389). Randolph (1976: 341), however, notes that pollen from maize and teosinte appear at this site alongside agricultural activities, and this questions the assertion made by Galinat (1973a), for whom Guatemala's modern teosinte is the primitive maize of the modern races of teosinte.[7]

[7] The second part of Randolph's paper never appeared (see Randolph 1976 in the bibliography), so we do not know what bibliography he used, nor do we know which of the studies by Galinat he was citing.

Pollen was apparently found on the Pacific Ocean coastline of Guatemala with an apparent antiquity of 4600 BP, but I was unable to find more data in this regard (see Neff et al., 2002).

Belize

Pohl and colleagues (1996: 368) report the finding of maize pollen remains dating to 3400 BP on the Caribbean coast, but they do not indicate the exact provenance. Piperno (1995: 134) mentions the Cobweb Swamp site, which dates to c. 4700 BP, and the Cob III site, dated c. 4600 BP, published by J. G. Jones (1991).

According to Pohl and colleagues (1996: 368), the first domestic plants that appeared in northern Belize were "perhaps" cassava (manioc) and maize in 3400 BC.

Honduras

Maize pollen has been found in a sediment-sample core from Lake Yojoa, which has been ascribed to 4770 years BP (Pearsall, 1996: table 1; Rue, 1989: 178), and pollen with about the same age has also been found in the sediments from Pantano Petapilla (Benz, 2006: 17; Webster et al., 2005: 107).

El Salvador

Dull (2006: 363) notes that there is secure evidence of the cultivation of maize in western El Salvador c. 3700 BP (calibrated). He likewise mentions the discovery of *Zea* pollen in the sediments of Laguna Verde (in the Sierra de Apaneca, Llamatepec) that would be the earliest evidence, with c. 4440 years BP (calibrated), but he also emphasizes that there is no association with archaeological sites. Pollen remains have likewise been found in Chalchuapa, and they would indicate the cultivation of maize some centuries prior to c. 3710 BP (calibrated). There is also *Zea* pollen in the sediments from Lago Llano, which was dated to 2910 BP.

Dull also mentions that a cob of maize was found in the coastal site of El Carmen, which was dated to 3430 BP (Dull, 2006: 360–361).

Costa Rica

Horn (2006: 372–375) reports the oldest dates for which evidence of maize is available in the central subarea of the Atlantic basin to be in the central highlands of Costa Rica. He mentions the sites of Laguna Bonillita, Machita Swamp, and Laguna María Aguilar, all of which have similar dates, that is, c. 2500 BP (see also Northrop and Horn, 1996). But the oldest dating of maize cultivation is in the Arenal-Tilarán subarea. This is a charred kernel dated to 4450 BP. There is another place on the Pacific Ocean drainage, in the Guanacaste-Nicoya region, where maize pollen has been dated to 4760 BP (Arford and Horn, 2004: 112; Horn, op. cit.: 375–376). Bradley and Vieja (1994) and Sheets (1994) apparently report the indirect dating of maize with an age of 4450 years BP, but I was unable to consult the data. This may be the aforementioned charred kernel. Dickau and colleagues (2007: 3652) note that in the southern Pacific drainage, "... a lake core from Laguna Zoncho revealed earlier maize pollen (3317–2952 ca. BP) ...," but I was unable to find more data in this regard.

Panama

Dickau and colleagues have reported three rock shelters in the Chiriqui Province: Casita de Piedra, Trapiche, and Hornito. The analyses made were based on grains of starch. The radiocarbon dates for Casita de Piedra range between 2890 and 6560 years BP (Dickau et al., 2007: table 1, 3653), and for Trapiche between 2300 and 5850 BP (Dickau et al., 2007: table 1, 3653). It must, however, be pointed out that at this second site, "... preceramic strata were capped by a 15 cm level that contained a small number of ... ceramics ..." (Dickau et al., 2007: 3651), so their preceramic status is not all too clear. The radiocarbon dates for Hornito range between 5880 and 6270 BP (Dickau et al., 2007: table 1, 3653). The authors subdivided the preceramic occupation into two phases: Talamanca, which they place between 5200 and 8000 cal. years BP, and Boquete, placed between 2100 and 5200 cal. years BP (Dickau et al., 2007: 3651). It is claimed that in all three sites "... maize (*Zea mays*) was processed alongside these root crops [they mean arrowroot, *Maranta arundinacea*, and manioc] in both preceramic phases ...," and they insist that "we recovered maize starch at all three preceramic sites." Besides, the starch from these three sites is the earliest available evidence found of the presence of maize in this region (Dickau et al., 2007: 3652).

In Lake La Yeguada, maize pollen and phytoliths were extracted from a core sediment sample that has a date slightly less than 7000 BP (Piperno and Holst, 1998: 769).[8]

Piperno and Holst (1998: 773) mention the Sitio Sierra and state that it is the earliest evidence of maize "macro-fossils"; they also point out that the analysis of the bone isotopes indicates that this plant was consumed. The dates they give range between 2200 BC and AD 50. There is a radiocarbon dating of 3000 BP for maize pollen from this same site (Piperno, 1995: 139).

Maize pollen and phytoliths have been found in the Cueva de los Ladrones, on the southern slopes of the Cerro Guacamayo, at 300 masl and 25 km away from the Pacific Ocean. The phytoliths were dated to 4910–1820 BC (Piperno, 1984:

[8] Previous publications had given a younger date; for example, Piperno (1995: 149; based on Piperno et al. 1990) gives 5700 years BP, and Pearsall (1996: table 1) gives 4200 BP.

381; 1985b; 1988a; Piperno and Clary, 1984; Piperno et al., 1985: 874, 876). However, it was later pointed out that the pollen associated with carbon was dated to 7000 BP (Piperno, 1995: 139). Dickau and colleagues confirm the radiocarbon date of 6860 for pollen and add that phytoliths were also found in the same preceramic levels. They likewise indicate that "new starch data from our analysis provide additional evidence that the first preceramic occupants of Ladrones were using maize by 7800 cal BP [6860 radiocarbon years]." Because of the characteristics of these starch granules, they must be "... hard endosperm varieties of maize (e.g. popcorn, not flour corns)" (Dickau et al., 2007: 3654). Piperno (2009: 155) adds that recent studies she made indicate that "... cob phytoliths are present in the same sediments that contained the earliest maize leaf phytoliths and pollen ...." Fritz (1994b: 641) questioned the work done in this cave.

A proportion of the isotopes detected in bone collagen at the site of Cerro Mangote indicate a moderate use of maize between 5000 and 7000 BP. Phytoliths from this plant were also found (Norr, 1995; Piperno, 2003a: 695; Piperno et al. 2000, 2001).

Another site studied is Cueva de los Vampiros, just a few kilometers away from the Pacific coast. Here, phytoliths extracted from preceramic deposits dated around 8600 BP were classified as wild grasses on account of their characteristics (Piperno, 2006: 145). However, it was later claimed that maize-leaf phytoliths were found in a context dating to c. 7000 years BP (Piperno, 2009: 156, table 3).

The Aguadulce rocky shelter is a major site on the coastlands of central Panama. The discovery of maize phytoliths with an antiquity of 4500 BP was initially reported (Piperno, 1985b; 1988a; 1995: table 6.1). The presence of phytoliths from maize glumes was later reported on top of the preceramic stratum, with an antiquity of c. 7000 BP (Piperno and Holst, 1998: 770, 773; Piperno and Pearsall, 1998). There are in fact two AMS datings of 6207 and 6910 BP for starch grains that have been identified as coming from maize, and it is definitely stated that they are "not wild Poaceae" (Piperno et al., 2000: 896). It is also worth emphasizing they were associated with *Manihot esculenta*, that is, manioc, which definitely is not wild. We thus see a well-developed agricultural system that comprised a series of plants in 6000–7000 BP. Starch from arrowroot (*Maranta arundinacea*) was also identified (Piperno et al., 2000: 895–896). The presence of manioc is significant, as it is a South American plant, a point that is discussed in depth in the final section of this book.

When mentioning a "... 5 cm-thick level from a column sample dating to shortly before 7000 BP ...," in a recent study, Piperno (2009: 156) explains that "... there is a unique maize cob phytoliths assemblage." This assemblage includes the oldest maize from the site, and the morphology of the phytoliths comes close to that of the modern teosinte fruitcase rather than to modern maize. Piperno then points out that the cob phytoliths are more similar to those from modern maize after c. 7000 BC (Piperno, op. cit.: 157).

In the Gatun basin there is a lake with the same name, from which a core sample of sediment was extracted. Maize appeared in it after 7000 BP. The deepest remains of this plant are not associated with any sign of agriculture. The pollen grains have the exine with a very regular spinule pattern, without traces of spinule clumping or of a *Tripsacum*-type incised reticulum, and without teosinte's more easily deformed arrangement of irregular and thinner spinules. The differences between the oldest and the most recent grains have been established at the exine level. The dates for the oldest maize are 6230 and 7300, and it has been classified as wild maize. Manioc (*Manihot esculenta*) appears at the 1800 years BP level (Bartlett et al., 1969: 389–390). However, in later publications maize was given a date of 4200 BP (Bartlett and Barghoorn, 1973: 247; Pearsall, 1996: table 1; Piperno, 1985b). When Pickersgill and Heiser (1978: 136–137) discuss this issue, they refer to Bartlett and colleagues (op. cit.) when stating there was contamination with recent carbon but without evidence of a nonindigenous pollen, and they raise some doubts, albeit without pointing them out explicitly. In this same year Ranere and Hansell (1978: 55) stated that the earliest date for cultivated-maize pollen in the Gatun sediments was 1200 BC.

Piperno (1994b) reports the presence of maize phytoliths in the sediments from Lake Wodehouse, with an age of 3900 years BP; they were also found in Monte Oscuro around 7500 BP (Piperno and Jones, 2003: 81). Maize phytoliths were at the bottom of the preceramic deposit of a site known as SE-189, and they correspond to the "seventh millennium BP" (Piperno, 1995: 141). Piperno (1994a: 638) cites the work of Norr (1991, 1995)[9] when discussing Monagrillo, in regard to the analyses that have been made of isotopes in human skeletons, which indicate the consumption of maize between 5000 and 7000 BP, which agrees, as Piperno points out, with the data obtained from pollen and phytoliths.

When discussing the finds made in Panama, Piperno says that an interesting characteristic is that maize phytoliths are not uniformly present in the preceramic phase in all of the sites studied. She ascribes the differences to the seasonality of the occupations, to functional variability, and to the ecological conditions of the sites studied. "They also tell us that it might be a mistake to assume that all residential groups in Panama between 5000 BP and 7000 BP were cultivating crops, the same crops, or the same crop mixture" (Piperno, 1995: 141). I find Piperno's argument logical, but it is telling that the same reasoning has not been applied to the finds made in the Andean area, as is discussed at the end of this book.

Dominican Republic

For the Dominican Republic we have references to the El Cerro site, where maize pollen has been dated to 1450 BP; remains with the same date were also

9 The article gives Norr no date because it was in press and was published in 1995.

found at Puerto Alejandro (Y. R. Ortega and Guerrero, 1981: 48, 86; Sanoja, 1989: 532).

Laguna Castilla is another site on the Caribbean side of the Cordillera Central. Archaeological studies have been made in this zone, but the only evidence found of prehistoric human occupation in the lake's basin are remains of maize pollen extracted from sediment cores. Maize pollen has also been found in the nearby Laguna de Salvador (Lane et al., 2008: 2121). The first appearance of maize in another lake, known as Castilla, was between approximately 815 and 900 cal. years BP, and the authors "... suggest that the majority of sedimentary carbon produced by C_4 plants and entering Laguna Castilla originated either from maize itself, or from C_4 weeds associated with maize agriculture" (Lane et al, 2008: 2128).

Puerto Rico

"Beginning in levels dated by extrapolation ..." there is maize pollen in the Maisabel site towards the second millennium BC, and in the Maruca and Puerto Ferro sites "c. third to first millennia BC" (Newsom, 2006: 331; Newsom and Pearsall, 2003).

Venezuela

According to Pearsall (1994a: 250–251) there is no early maize in Venezuela, and B. D. Smith (1994–1995b: 179) agrees.

In the Orinoco zone, at the Rancho Peludo site of the Dabajuro tradition, there "... perhaps [is] evidence of corn agriculture" around 1800 BC (Bruhns, 1994: 153). Van der Merwe and colleagues (1993: 66) point out in this regard that in the lower Orinoco tropical zone, both the archaeological and the isotopic evidence concur in that maize agriculture was missing around 2800 BP (Roosevelt, 1980; Van der Merwe and Medina, 1991; Van der Merwe et al., 1981). On the other hand, the identification of maize phytoliths goes back 6,000 years in the Ecuadorean Amazon (Bush et al., 1989), thus suggesting that it may have been viable a long time before. "The simplest explanation," Van der Merwe and colleagues claim, "for this contradiction is that the presence of phytoliths identified as maize does not imply the presence of maize agriculture of sufficient importance to be visible in other forms of evidence." Good evidence for maize in the Orinoco zone goes back only up to 800 BC (Van der Merwe et al., 1981).

The site of Parmana is located on the southeastern part of the state of Guárico, on the left bank of the Orinoco River, some 600 km away from its delta. We know that here maize first appeared in the earliest Corozal phases, "... but is very scarce...." The beginning of the Corozal phases is c. 400 years BC (Roosevelt, 1980: 195, 235).

Segovia and colleagues (1999) have pointed out that Nicolás de Federman (1579), as well as Fajardo and Losada (which is cited in Oviedos 1824), mentions maizes with three and four forms and colors that were used by the ethnic groups along the Barquisimeto River. Gumilla (1791) mentions a unique species planted by the Otomacas, the Guanco, and the Paos, which was called *onona* or two-month maize. But these are data that correspond to the sixteenth to nineteenth centuries.

Sanoja says that Colombia's Pollo race is the only variety that was cultivated in southwestern Venezuela. It has been found in the temperate valleys of the Andes and in the savannas around the eastern piedmont since the first centuries of the Christian era. Sanoja gives three possible explanations: either the race was introduced from Colombia, it grew in wild form in southwestern Venezuela, or it was introduced into both regions from Central America (Sanoja, 1981: 112).

Colombia

Oyuela-Caycedo (1996: 73) claims that maize was introduced into northern Colombia from Mesoamerica some 3,000 years ago, but this statement is unsupported.

In one of her studies, Pearsall (2008: 110) mentions the site of Páramo de Peña Negra I in the Bogotá plains, where maize pollen dating to 6320–3210 BC was presumably found. No reference is, however, made, so I was unable to check the data or expand them.

Pearsall (1996: table 1) points out that the oldest date for the pollen from the Hacienda Lusitania, in the Calima Valley, is 5150 BP.

A pollen core from the Hacienda El Dorado, also in the Calima region, has been analyzed. Maize pollen was identified with an age of 6680 years BP (Bray et al., 1987; Monsalve, 1985; Pearsall, 1994a: 250; 1995b: 127; Piperno, 1995: 134).

It must be noted that Fritz (1994b: 640) questions the finding of pollen in the Calima sites.

According to the report presented by Correal Urrego and Pinto Nolla (1983: 180–181; see also Bruhns, 1994: 68), maize pollen and rachis, with an age of 3270 years BP, were found in the deepest layer of the rocky shelter of Zipacón, in the municipality of the same name in the Cundinamarca zone.

The data regarding the Abeja site (in the Araracuara region of the Caquetá River, in the Colombian Amazon forest) are contradictory. For Bonzani and Oyuela-Caycedo (2006: 349–350), the pollen from maize dates to between 2745 and 2380 years BC, whereas for Benz (2006: 18) the dates range between 3450 and 3000 years BC. The present author was unable to check the original data (e.g., Herrera et al., 1992; Mora, 2003).

In regard to Colombia's primitive maize, Sanoja writes that Pollo, a popcorn with a distribution limited to the eastern drainages of the Cordillera Oriental, is one of the oldest races of maize, and that it seems to derive from the Confite Morocho or from an archaic, locally domesticated form (Roberts et al., 1957; Sanoja, 1981: 111).[10]

Ecuador

Lippi and colleagues (1984: 118) say there is "... a preceramic site on the coast of Ecuador ..." with maize phytoliths dating to 6000 years BP. Their source is a personal communication by Karen Stothert. It is hard to establish exactly what site they mean.

Staller and Thompson (2001: 127), on the other hand, point out that both the archaeological as well as the ethnobotanic and isotopic evidence indicate that "... the initial introduction of maize on the coast of Ecuador took place between c. 2200 to 1950 BC." We shall see that this does not agree with other available data.

There is an abundant bibliography available for this area, for example, inter alia, Pearsall (1987, 1992a, 1992b, 1993), Pearsall and Piperno (1990, 1993b), Pearsall and colleagues (2004), Piperno (1984, 1985a, 1990, 1991, 1994b), and Piperno and colleagues (1985).

The only information on maize available for the province of Manabí corresponds to the Chorrera period, 2500–3500 BP (Pearsall, 1994b: 129).

Pearsall and colleagues (2004: 424) made an overview of the findings of maize on the Ecuadorean coast. They point out that the evidence of macro-remains is more recent than that of micro-remains, phytoliths in this case. The macro-remains appear in the Machalilla phase of the Middle Formative at La Ponga (1200–800 years BC), and in the Chorrerra phase of the Late Formative.[11] Micro-remains, on the other hand, are present in preceramic times at Las Vegas (5000–4700 BC) and persist in the Valdivia tradition at the Real Alto site (4400–1800 BC).[12]

But let us see specific data for the different sites. There is no question that the Valdivia culture, in the Guayas zone, is one of the most significant cultures for the issue at hand. According to specialists, some of the early charred maize remains from early Valdivia contexts are the only available direct evidence of food plants (Pearsall, 1988b). Remains of phytoliths in early (calibrated) Valdivia contexts gave dates of 7000–5500 years BC (Pearsall, 1979; Piperno, 1988b; Stothert, 1985;) (Van der Merwe et al., 1993: 65).

[10] Ficcarelli and colleagues (2003: 842) made a very general statement indicating that "... the cultivation of *Zea mays* is found as early as 8000 yr BP in the Colombian Cordillera...." They give Kuhry (1988) as their source, a dissertation I was unable to peruse.

[11] See in this regard Pearsall (1988a) and Zevallos Menéndez and colleagues (1977), among others.

[12] See, for example, Pearsall (1988b: 118–133), Pearsall and Piperno (1990: 325–330), and Piperno (1991: 164; 1995: 136).

Regarding the phytoliths from Valdivia phases 1 and 2 (which, according to Zeidler, 1991, and Pearsall, 2003b: 224, can be dated to 4500–2900 BC), Rovner (1999: 488–489) states that "... control studies [show] that soil conditions, especially available moisture, can cause substantial variation in the mean and range of size values in phytolith populations derived from members of the same species from one year of one place to the next. On the other hand, shape remains stable even in the presence of significant size modulation." Rovner believes that the classificatory methods developed by Pearsall and Piperno to identify remarkably early maize, in both South and Central America, "... emphasize the use of size parameters in highly questionable ways." The increase in the size values of the phytoliths distinguishing them from wild grasses once again "... [is] supposed ..." (Pearsall, 1989; Piperno, 1988a). However, Rovner says, the size values in the sets of archaeological phytoliths presented as evidence of domestic maize in the earliest Valdivia 1 and 2 contexts are larger than the size values presented for each and every modern race of maize that has been examined (Pearsall, 1989: 331, table 5.2).

I do not intend to enter into a debate that lies beyond my field. This issue must be solved by specialists. I do, however, go back to this issue and make a general comment at the end of this book, emphasizing the different points of view.

It is interesting that Pearsall (1978a: 54) claims that the smallest race from Valdivia is quite similar to Confite Morocho both in size and in shape. This assertion, however, lacks proper support.

Lathrap (1975: 19–21, 64) drew attention to the fact that the rims of Phase 3 Valdivia (c. 2900–2600 BC) vessels were decorated with impressions of maize kernels. In fact, according to the report by Pearsall (2002), there are new data on the phytoliths from this Valdivia phase, even though Staller (2003: 376) claims that this identification is "highly suspect."

In the 1970s, Zevallos and his team started a debate when they published evidence of the trace of a grain of maize, which had been imprinted on the inside of a sherd. The evidence they claimed to be presenting composed "... some imprints ... of maize seeds ... as well as [a print of] another grain of maize; the latter was present in a small fragment of what had been a plate ..." (Zevallos Menéndez, 1966–1971: 17–18). According to Zevallos, the second print was left by a grain of maize that was in the process of germination and that fell into the potter's fresh clay; germination continued while the clay was in this condition and so the small root and the stalk sprouted. Once the vessel was fired, the grain was burnt and its print remained on the clay (Zevallos Menéndez, op. cit.: 19). The other two pieces of evidence Zevallos presents are the presence of metates and the presence of ceramic motifs in pottery showing ears of maize (Zevallos Menéndez, op. cit.: 21–22). The fragment in question corresponds to the Valdivia Phase 5 or 6 (to which a date ranging between 2600 and 2100 BC is assigned; see Zevallos Menéndez et al., 1977).

It was based on this information that Zevallos concluded that this Valdivia maize "corresponds to the flour corn or sweet corn," which is different from the pod corn MacNeish found in Mexico and other northern sites and which is also found in South America, for instance the Peruvian Confite Morocho popcorn or the Pollo. Zevallos posited an independent domestication in South America, because flour corn or sweet corn "... appear at an early date in South America and not in Mesoamerica" (Zevallos Menéndez, 1966–1971: 25–26). He added that "... the imprint of a grain of maize in a sherd from the Valdivia culture, recovered in a midden that corresponds in its entirety to this culture, and through *rigorously controlled excavations, with artificial levels of 0.10 cm* ... plus the modelled and incised depiction of the ears of maize in Valdivia ceramic decoration ..." were proof that it had been "... conclusively shown ..." that the cultivation of plants was known in the Ecuadorean Formative (Zevallos Menéndez, 1966–1971: 27; emphasis added). But Zevallos flagrantly contradicts himself, for he states in a later study, coauthored with other colleagues, that the print corresponds to the Kcello Ecuatoriano race, which is a variety of maize with a low number of rows, as well as to three other possible popcorn varieties (Zevallos Menéndez et al., 1977: 388). It should be noted that when archaeology is done in earnest, no excavation uses artificial strata, nor must we forget that Kcello is a form of the Cuzco Cristalino Amarillo race (which is known in Cuzco as Ckello-Sara) that was taken to Ecuador in Inca times (see Bonavia, 1982: 383; Grobman et al., 1961: 250).

It must be pointed out in this regard that Paul C. Mangelsdorf (1977) sent a letter to the editors of *Science* immediately after the publication of this study (Zevallos Menéndez et al., 1977) that was never published. I have a copy of this letter in my files. Here Mangelsdorf made the following comments. First of all he accepted that the charred remains and print found in the ceramic vessel did belong to a maize kernel, and then he also acknowledged that Galinat's opinion was correct as regards its characteristics.[13] Mangelsdorf also believed that the idea that this grain was germinating was questionable, and the experiments made showed and yielded different measurements. He also doubted that it could be the race mentioned (Kcello Ecuatoriano), because it grows at high altitudes (2,000–2,600 masl) and would hardly have survived in the humid lowlands of the Ecuadorean coast. The identification of the vertical appliqué strips in Valdivia ceramics is also questionable.

Zevallos Menéndez claims the prints show anatomic characteristics that are unique to maize, but these are not mentioned, and they cannot be distinguished in the photographs (Zevallos Menéndez, 1977: figure 2, 387). Everything seems to indicate that the print in the vessel is from a Chococeño, which is not an eight-row productive maize but a primitive popcorn of early agriculture.

[13] Galinat apparently examined the sample, but I was unable to find anything in this regard. Perhaps Mangelsdorf meant the report by Earl Leng mentioned subsequently.

Assuming that the people of Valdivia cultivated a lowland race instead of a highland race, we would then have to relate it with the Mochero maize from northern Peru (see Grobman et al., 1961: 196–201). Based on several reasons, Mangelsdorf believed that the maize in question was used to make chicha, which would explain the metates, particularly bearing in mind that there is no tortilla-making tradition either in Ecuador or in Peru. Mangelsdorf ended by stating that, far from being efficient, the early agriculture of Ecuador was instead primitive and that nutrition would have been based on popcorn and on chicha.

Lippi and colleagues (1984: 119) have also criticized the study by Zevallos Menéndez and colleagues (1977).

Staller and Thompson (2001: 150) and Tykot and Staller (2002: 669) pointed out the presence of ceramic sherds with imprints of maize kernels belonging to Phases 7 and 8 (c. 2200 BC) of Valdivia, as well as dental calculus related with this plant (see also Staller and Thompson, 2002; Thompson and Staller, 2000).

Pearsall (1994b: 129) reported the presence of maize in the terminal phase of Valdivia (c. 3600 years BP), at a site called C69 in "... contexts at San Isidro [Manabí Province] and in Mafa [in the province of Esmeraldas] (c. 3000–2500 BP)...." It would be the most ancient maize in these regions. It is associated with pottery in both cases.

Turner (1978: 694–696) studied the caries in the skeletons from the Valdivia and Machalilla cultures. He reached the conclusion that in Valdivia there was no evidence of caries, whereas in Machalilla, the culture that came immediately after it (c. 1500–1100 BC), there were "two possible carious lesions...." Turner admits that the sample examined is very small, but the data would indicate that maize agriculture was not then intensive, as it has been shown that the physical and chemical properties of this plant, in combination with the common methods used to prepare food, have a strong cariogenic potential in many environments. On the other hand, the studies Van der Merwe and Tschauner (1999: 526) made analyzing isotopes from human skeletons do not support a maize-based diet in the early Valdivia period either and point instead to the use of forest and river resources (Van der Merwe et al., 1993). Van der Merwe and Tschauner state that both the isotopic data and the macrobotanic remains (Lippi et al., 1984) "... provide more solid evidence ..." of maize on the Ecuadorian coast in Middle Formative times, c. 3500 years BP.

According to Pearsall (2003b: 223–224), maize grains were recovered from six different contexts through flotation at the Loma Alta site (in the Valdivia Valley, some 12 km from the sea), but the AMS dates do not support their belonging to the Early Formative. Phytoliths from this plant were found in two samples. All of these specimens correspond to the Valdivia Phases 1 and 2. But when Van de Merwe and colleagues (1993: 81) made their study of the human collagen values at this same site, the results indicated that during these

Valdivia phases neither maize nor marine remains played a major role in the human diet.

Zarrillo and colleagues made a more detailed report of Loma Alta. They confirm that the site corresponds to the two earliest phases of the Valdivia culture. They analyzed the starch found in ceramic remains and grinding stones associated with Early Valdivia. Most of the 116 starch granules recovered in the cooking pots were identified as maize, "such as flint or pop and 15% soft endosperm (flour) maize" (Zarrillo et al., 2008: 5007). Twenty-one samples from the grinding stones, which represent 60% of the starch granules, are consistent with "... a soft endosperm maize variety, whereas nine (40%) were characteristic of a hard endosperm variety" (Zarrillo et al., 2008: 5008). But they state that domestic maize was commonly used "... by at least 5300–4960 ca. BP...." They then add that "... although maize is securely present at Loma Alta by 5300 ca. BP based on direct date ... maize may well be present as early as 6250 cal. BP." And it is most interesting that they acknowledge this happens "altough diagnostic *maize phytoliths are not well represented in the cooking pots or grinding stones tested* ..." (Zarrillo et al., 2008: 5009; emphasis added). The authors then explain that

> with respect to the maize starches (soft versus hard endosperm) recovered from the cooking-pot residues, if ground soft endosperm (flour) maize and hard endosperm (flint/pop) kernel maize were both cooked in the pots, the indurate aleurone of the flint/pop maize kernels may have provided protection to the endosperm starch, delaying or eschewing gelatinization of flint/pop maize starches resulting in a higher recovery rate from the pot residues compared with flour maize. (Zarrillo et al., 2008: 5009)

They then add that "... the identification of starch from both flint/pop and flour races indicates that more than one maize variety was being cultivated ..." (Zarrillo et al., 2008: 5010).

It is worth recalling that phytoliths from manioc (*Manihot esculenta*), arrowroot (*M. arundinacea*), chili peppers (*Capsicum* spp.), and *Canavalia* have been found in the ceramic sherds alongside maize starch.

This study has some problems that are worth pointing out. First of all, "an accelerator mass spectrometry (AMS) date on rare carbonized maize kernels from the lower levels ... was far too young (2730–2350 cal. B.P., at two-sigmas; Beta-103315) to be associated and demonstrated that *there was some mixture* of small remains between the occupation layers of the site" (Zarrillo et al., 2008: 5007; emphasis added).

In addition, it is striking that the excavation used arbitrary levels instead of natural ones, as is the case in modern archaeology. This is clear in the following statement: "The dates reported ... for the grinding stone tools are based on charcoal found in the same *10-cm arbitrary level* as the artifacts" (Zarrillo et al., 2008: 5007; emphasis added). It also is a shame that no stratigraphic cut is presented and that "maize kernels" are mentioned without specifying the number or location of these remains, or that all of the information is based on general statements in which specific data are missing.

It is also worth recalling that the study clearly points out that the contexts of the microbotanic remains "... are usually dated by association ..." (Zarrillo et al., 2008: 5006). Not only is this a standard practice in archaeological excavations, it is in fact a law in our science. This point is here emphasized because North American colleagues who have criticized our work in Peru deny our work by arguing that only direct dates are valid, and yet when they themselves do use indirect dates, these are valid.

Alexander Grobman analyzed the paper by Zarrillo and colleagues, and made some comments that are important and have to be included here. Grobman wrote thus:

> Whereas in the early epochs we only find porcorns in Peru, just like in Mexico, the presence of starch found in the grinding tools – which the authors point out corresponds to a large extent to flour corn – would lead to the conclusion that the types of more floury corn would have been developed and existed in those times in the tropical zones of South Western Ecuador, something that does not agree with the more secure data [available] on the exclusive presence in this period in Mexico of races of maize that were only popcorns.
>
> The presence of flour corn, which according to the authors reaches 60%, would lead to the conclusion that these maizes are of Andean highland origin, and that they would not have arrived as rapidly from Mexico as the authors assume, because for that period only the Chapalote and Nal Tel races are known, both [of which] are variants of one single racial type [present] at that time in Mexico. On the other hand it is inconsistent that hard popcorn and flour corn were both cooked in the same vessel, as both types have a different use. It would be an excessive use of energy to cook maize when it can be used popped. We should remember that the sites where these maize residues were found lie to the north of the Guayas-Babahoyo [River] mouth in a rainy zone – not in the part of Ecuador [that is] more to the south and is drier. It is therefore difficult for races of flour corn, which Zarrillo et al. find in great abundance in the area under study in past times, to have lasted for long. The most likely [answer] is that if the identification of flour corn versus popcorn is correct, [then] the flour corn would have an adjacent, high-altitude Andean origin, as is posited for maize in Peru.
>
> The presence of manioc and *Canavalia* in the same residues of flours points towards an agriculture with more complex origins than the mere transfer of maize from Mexico to Ecuador in very early epochs, to join previously established crops in this part of Ecuador. Manioc has Amazonian origins and its early presence in the zone indicates an already-established agriculture, that it is not easy to assume was superimposed with the presence of maize from Mexico at such an early date and in such an evolved state, and with the widespread presence of flour races – as is posited by Zarrillo et al. – that did not exist in Mexico

> at this time. The position taken by the authors [Zarrillo and colleagues] that maize joined an already-established agriculture with other crops, at a late phase of presence in said agricultural complex, cannot therefore be accepted.
>
> The arguments regarding the joint cooking of various foods including maize, raises a series of questions due to the speculations made by the authors, and leaves much space for research before we can accept their arguments.
>
> Zarrillo et al. accept that maize was already an integral part of the diet of the peoples of South Western Ecuador by the time they entered the stage of the development of settlements. This confirms – so they claim – that maize arrived at a very early date and before the Formative Period of this region. This agrees with our position. Where we part ways is on that maize arrived to the coasts of Peru and Ecuador directly from Mexico, as they state in the first paragraphs of their study. This section would be invalidated by the very information provided by these authors, which gives more validity to the hypothesis of a direct Andean origin of their maize. The flour corns developed in the highest cold zones, because in the tropics they would have selective disadvantages in the face of insect attacks. We have detected the presence of *Heliothis zea* on the Peruvian coast in very early epochs, and it is very likely that in Ecuador, like in Peru, the first types of maize were popcorns with hard and small grains, to which were added, at such early stages, races of flour corn that had developed quite early and which had developed previously, and which came from the Andean highlands.... (Alexander Grobman, letter to the author, 2 April 2008)

Another major Ecuadorian site is Real Alto, in the southwestern Guayas Province, close to the port of Chanduy. This site has a Valdivia occupation that comprises the seven first phases of this culture (c. 4500–2000 BC). The first studies reported that

> ... large numbers of maize prints have been found [that were] used as a decorative element on the rims of the vessels from the Valdivia 3 Period, i.e. at the end of the A Phase, with an absolute dating of 5500 years BP. Studying the accidental print found by Zevallos, as well as the different prints found ... at the Real Alto Site, Doctor Earl Leng identified the maize the Valdivianos used as a flint type with eight straight rows of very large and regularly spaced grains; this Kcello Ecuatoriano race is still sown in large number in the Cuenca Valley. (Lathrap and Marcos, 1975: 44)[14]

Based on the above-cited arguments and others – the presence of grinding stones; the remarkable wear on the teeth of adults and of adult dogs; the large number of *Cerithidia pulchra* in the refuse; and finally, the significant proportion of deer remains in the Valdivia levels – it is likewise claimed that there was an "... intensive cultivation of maize ..." (Lathrap and Marcos, 1975: 58).

[14] The Cuenca Valley lies at 2,500 masl.

Seventeen stone artifacts (manos and metates) were found in Structure 20 of this site, which has no absolute date but which was assigned to Valdivia Phase 3 on the basis of its ceramic associations, and because it is identical to Structure 1, which has dates of 3845 and 4050 BP. Grains of starch and phytoliths of maize cobs were in all of these pieces. This would confirm the previous studies made by Pearsall (1978b; 1979; 2002: 53; and Pearsall et al., 2004: 425–426, 429, 438). According to Pearsall (2003b: 225–226, 235) and Pearsall and colleagues (2004: 438), maize phytoliths are present in the Phases 1 and 2 of Valdivia, and by Valdivia 3 maize abounds – there even are house floor deposits that store this plant. Burleigh and Brothwell (1978: 360) studied the remains of a dog from Real Alto and came to the conclusion that its diet was essentially maize. Roosevelt (1984: 10), however, questions the presence of maize at Real Alto in the Valdivia phases.

I believe the evidence is not as solid as is claimed. We can on principle accept that the prints found are of maize. What is not clear is whether the stone artifacts contain just remains of maize or also remains from other plants. Then it is noted that the dental wear in adults and dogs is remarkable, and to explain this they start with ethnographic data from populations that prepare maize tortillas, which include particles from the grinding stones used, thus bringing about dental wear in those who eat them. In this ethnographic case the argument clearly works, but it is naïve to extend it to the case of Valdivia, because other examples could be given, some of them for and others against. Here just two examples are cited. There is a marked dental wear in the people of the Peruvian highlands because they continuously make a great use of toasted maize – the so-called *cancha*, which is chewed slowly. But a remarkable dental wear is also found in early preceramic populations, prior to the introduction of maize. See, for instance, photograph 62b, which shows the dentition of a child from pre-maize times found at Los Gavilanes (Bonavia, 1982: 207), or the teeth of the Lauricocha skeleton (Cardich, 1964: figure 90, 104). We must not forget that although it is true that dental attrition is associated with the ingestion of foods that contain abrasive materials, it is also related with other, nonmasticatory functions (Wing and Brown, 1979: 91). We can therefore wonder how valid the argument of Lathrap and Marcos is.

Lathrap and Marcos then bring up the large amount of *Cerithidia pulchra* that has been found, which as they note is a small beach snail. The ethnographic evidence shows that the wet pastry from which maize tortillas are made is prepared by boiling the maize with lime, which is obtained from calcining snails. In this case I also acknowledge that this may have taken place at Valdivia, but the authors do not provide any actual proof that this was actually so. The final argument, that is, the presence of a large number of deer remains, is as speculative as the former. After camelids, cervids are among the animals most frequently found in archaeological deposits right from the first human occupations in the Andes. Ethnographic sources showing how, in some specific cases, deer are associated with the cultivation of maize for quite particular reasons may become evidence alongside other archaeological proofs, but they are not proofs in themselves.[15]

Much has also been written regarding the site of Las Vegas, in the Santa Elena Peninsula.[16] Piperno points out the presence of two contexts with maize phytoliths in Las Vegas (site OGSE-80). Late Las Vegas was dated with traditional radiocarbon to 8170 BP, but the upper level has two more recent dates of 7150 and 7440 years BP. For Piperno, "... we can conclude with certainty that cultivated maize appeared between 8000 and 7000 years BP" (Piperno, 1988b: 211). Pearsall (1996: 6) later confirmed there are phytolith remains in the oldest strata of the Las Vegas phase, which corresponds to the Ecuadorian Late Preceramic, and which were dated to 4600 BC. There is no evidence of maize in the strata belonging to early Las Vegas. But when Piperno again discussed this point, she explained that "an age of 7170 ± 60 BP was obtained on the earliest maize-bearing assemblage" (Piperno, 2003b: 834).

It is striking that in regard to the phytoliths from OGSE-80, Stothert calls them "primitive maize" and then claims they "could be primitive maize" (Stothert, 1985: 621). Races cannot be identified using phytoliths to the best of my knowledge, so this claim, besides being worthless, is also groundless. Stothert clearly based her work on the work done by Piperno (1988b: 214), who posits groundlessly that the phytolith evidence from the Late Las Vegas context "... supports the hypothesis that there was an early dispersal of some domesticated form of teosinte/maize, from its home in Mexico to South America." Stothert insisted on this later, when she claimed that "... the inhabitants of Site 80 began cultivating a primitive variety [of maize] shortly before 6600 BP" (Stothert et al., 2003: 35). It must be noted that two dates are given for the phytoliths in this same study (Stothert et al., op. cit.) that do not agree with the previous ones (see previously): 5780 and 7170 years BP.

La Emerenciana, in the El Oro Province, on the Ecuadorean coast, is a very debatable site. Staller and Thompson worked here with opal phytoliths, that is, microsamples formed by amorphous silicates exuded by plants that have been extracted from ceramic sherds. Staller and Thompson insist that these phytoliths are reliable, whereas the others that have been removed from sediments and so on may be intrusive. They removed these samples from three Valdivia pottery sherds that correspond to Phases 7 and 8, and that have an age of 3860 BP. They also extracted remains of dental calculus with remains of maize phytoliths (Staller and Thompson, 2002: 33–35, 39, 42, 44). Another paper avers that this plant was at first used as the main ingredient for the consumption of "fermented intoxicants" rather than as solid food (Tykot and Staller, 2002: 674–675). No evidence was, however, presented that supports this claim.

15 For the arguments advanced by the authors, see Lathrap and Marcos (1975: 59).

16 See, among others, Pearsall (1988a; 1994a: 250; 1994b: 121; 1995b: 127; 1996: table 1; 2002: 54; 2003b); Pearsall and Piperno (1990); Piperno (1988a, 1988b); Stothert (1985, 1988); and Stothert and colleagues (2003: 37–38).

The most serious flaw of the work done at La Emerenciana is that the authors have not shown the site's stratigraphy, just an "... idealised profile of the natural stratigraphy ..." (Staller and Thompson, 2001: figure 6, 141). This "idealized" stratigraphy was presented once again in another paper (Staller and Thompson, 2002: figure 6, 40) with horizontal levels all of the same width, even though the text claimed the excavation followed the natural strata. When Staller (2003: 373) later returned to this argument, he again insisted that his data from La Emerenciana were "... internally consistent with related lines of evidence from other regions of the Andes," which is not really true. It is therefore hard to accept the results of this work.[17]

The site of La Chimba, in the northern Andean zone of Ecuador, is one of the sites with the largest amount of botanic samples; maize appears in all strata, with the oldest specimens dating to 2640 BP (Pearsall, 2003b: 234). Maize pollen, which has been dated to 4000 years BP, has been extracted from the sediments of Lago San Pablo, also in the northern highlands (Athens, 1990, 1991; Pearsall, 1994b: 122; 1996: table 1).

Maize pollen and phytoliths have also been extracted from the sediments of Lago Ayauch, at the foot of the Andes, in the southeastern Ecuadorian Amazon (Bush et al., 1989: 304; Pearsall, 1994a: 250; 1994b: 122; 1996: table 1, 2003b: 232–233; Piperno, 1990; Tschauner, 1998: 322).

An explanation is in order here. Bush and colleagues (1989: 304) are quite clear, for they state that "the first occurrence of *Z. mays* pollen and phytoliths is at 2.4 m in sediments bounded by uncorrected radiocarbon dates of 4,570 ± 70 years BP ... and 7,010 ± 130 years BP...." They comment that the presence of maize 6000 years BP in the Amazon basin matches the preceramic data from coastal Ecuador, which is in turn consistent with the hypothesis of the dispersal of maize from Mexico to South America (Bush and colleagues, op. cit.: 305). Pearsall has discussed the work done by Bush and colleagues on several occasions (and also discussed Piperno, 1990: 667) and has pointed out that there are several things that indicate the beginning of cultivation in this zone (Pearsall, 1995b: 128). Bruhns (2003: 159) is more cautious. Although she does not discuss the presence of maize, she does point out that "... there are no cultural

[17] See also Piperno 2003b: 832–834. Staller recently published a book that includes a long list of the work undertaken at La Emerenciana (Staller, 2010: 205–219), but no new data were added. Although it includes many illustrations, not one of them shows the stratigraphy. This book supposedly is a "history of *Zea mays* L.," but it actually just presents the biased outlook of its author, who only includes the data that suits his own point of view and ignores all evidence to the contrary. The data available for South America is ignored – yet the bibliography does list some references for this continent – the sole exception being La Emerenciana, which in itself says much.

This study also evinces that the references cited have often not been read, particularly because they do not deal with the issues discussed herein, and in some cases even contradict them. Most of the references given in the book are not precise, and the bibliography has many mistakes, and many of the studies cited are not mentioned at all in the book. This study therefore does not add anything to the study of maize.

associations for this material, and all of the known archaeological sites are some millennia later." Piperno (1995: 134) is, however, categorical in claiming that is "... the earliest evidence for maize in the Amazon Basin...."

Some have questioned these finds. Staller (2003: 377) is one of them, but his sole argument is that this is a single sample from a "questionable context," and he did not even take the trouble of looking up the original work and instead based his thinking on the paper by Pearsall (2002: 54). Bonzani and Oyuela-Caycedo (2006: 350) are other critics, and they believe that "... the human socioeconomic contexts of these finds are unclear." One can disagree with the ideas Bush and colleagues have, but we cannot ignore the fact that deducing data on the "socioeconomic contexts" from sediment samples is very difficult, to say the least. All this shows is that they have not read the work done by Bush and colleagues.

On the basis of their research on stable isotopes, Tykot and his team concluded that maize was not significant in the diet of the Ecuadorian coastal populations until the end of the Valdivia culture, and that maize was used in the Initial period, but not as was done in later times. They, however, believe that in those times maize was used more in Ecuador than in Peru, and it was used in the highlands before it was on the coast (Tykot, 2004: 439; Tykot et al., 2006: 196).

Smith has raised some doubts regarding the discovery of phytoliths from 7000 and 8000 years BP, because "this is more than 3,000 years before maize appears in the archeological record in Mexico, where it was domesticated." But it must be noted that his sole source is a paper by Pearsall and Piperno (1990). Smith acknowledges that these dates were more plausible when the traditional C14 datings were accepted, and he finishes by asking the following question: "If maize had already been domesticated in Mexico and had reached South America by 8,000 B.P., why is it that no cobs or kernels of that antiquity have been recovered anywhere in the Americas?" Smith is convinced that we must accept the AMS datings, and that maize reached South America only by 3000–3200 years BP (B. D. Smith, 1994–1995b: 179–180). He would probably have reached different conclusions had the data he used been complete. Smith fully ignores the literature on South America and shows he does not know at all the finds made and datings obtained in the central Andes, so his position is hollow.

It must likewise be pointed out for the benefit of the reader that Staller and Thompson (2001: 133; 2002: 45–46) have also questioned the work done by Pearsall and Piperno. Again, I do not intend to enter into a technical debate that does not fall to me. Even so, none of the arguments are convincing. But what can be stated is that the literature these authors exclusively based their work on is Bruhns (1994), Fritz (1994a, 1994b), Rovner (1999), and B. D. Smith (1998). Of these, only Bruhns uses a good bibliography, so the data in Staller and Thompson are not valid.

Peru

There is a vast literature on the issue of maize in pre-Hispanic Peru that has never been compiled. I do not intend to do so here and only refer to the major publications. As for the rest, interested readers are referred to the original sources, where they will be able to find at least part of it.

The handbook written by Towle (1961) is beyond doubt a classic work of ethnobotany that is still the only valid one on this subject, the many years gone by since its publication notwithstanding.[18] We must bear in mind that the Preceramic period was just being discovered at the time that Towle wrote her book, so the usable bibliography in her text goes from the Initial period to the Late Horizon. Of the studies she lists (Towle, op. cit.: 22–24), the most useful ones for this subject are Towle (1952a, 1952b, 1954), Yacovleff and Herrera (1934), Cutler (1946), and Grobman and colleagues (1956). The book by Grobman and colleagues (1961: 75–92) includes archaeological data that may be of use, but we must not forget the year it was written. This book includes references to the literature on maize that may have been valid at the time, even if they are not now; besides they were not meant solely for archaeologists (Grobman et al., op. cit.: 72–75).

Although it applies to the Preceramic *sensu lato*, the bibliography in Bonavia and colleagues (2001) lists a sizable portion of the data regarding maize in this period. We will have a very complete idea of what has been published on this subject if we add to it the bibliographies in Bonavia and Kaplan (1990), Bonavia (1982: 449–490), Bonavia and Grobman (1989b: 467–470), and Bonavia and Grobman (1999: 257–261). On preceramic maize, see also Grobman (2004: 446–448).[19]

Perhaps the first mention of archaeological maize in modern times was made by Darwin. In chapter XVIII of the journal of his travels, in the entry for 19 July 1835, he noted the following while on the island of San Lorenzo: "I was much interested by finding embedded, together with pieces of sea-weed in the mass of shells, in the eighty-five foot bed, a bit of cotton-thread plaited rush, and the head of *a stalk of Indian Corn*" (Darwin, 1839: 451–452; emphasis added).[20] I return to this later.

[18] The study Donald Ugent and Carlos M. Ochoa published in 2006 (*La Etnobotánica del Perú, Desde la Prehistoria al Presente*, Lima, Consejo Nacional de Ciencia, Tecnología e Innovación Tecnológica-CONCYTEC) has not been considered here because it has far too many mistakes and inaccuracies, and its bibliography is incomplete.

[19] In this publication the bibliography was left out due to the neglect shown by the editors; it was later added as a "corrigendum," albeit in a most disorganized, hard-to-use fashion and still with errors.

[20] For reasons I do not know, the following was added after the first edition: "I compared these relics with similar ones taken out of the Huacas, or old Peruvian tombs, and found them identical on appearance." This likewise appears in the Spanish translation (1921: 165), which even

The first thing that has to be emphasized, and for nonspecialists in particular as it can lead to serious doubts and confusion, is that two major aspects must be borne in mind to properly understand the problems raised by the study of archaeological maize, especially as regards preceramic times. First, the study of ethnobotany in Peru is relatively recent. Although there have been some isolated efforts, these actually began only in the 1940s with the research carried out at Huaca Prieta by Junius Bird, who was the first to draw attention to the significance of cultivated plants. Margaret A. Towle was a great specialist who dedicated herself to the study of botanical remains from Peruvian sites. For a long time there were no specialized researchers in Peru, and Peruvian archaeologists showed no interest in this subject. The truth is that after Huaca Prieta, the first study organized in the 1960s with the study of plants as its main goal was undertaken by the present author in Huarmey, and it reached its climax with the Huarmey Archaeological Project. Although nowadays some researchers study this subject, Peruvian archaeologists practically have no interest at all in this type of study. The only ones who have dedicated themselves to the analysis of maize are Alexander Grobman, an agronomist who has devoted a lifetime to the study of this plant *largo sensu* – he certainly is one of the major specialists on this subject in the world – and I. We must also bear in mind that one needs a special training for this type of study, which as yet is missing in Peruvian universities.

The second point is that the research actually undertaken has in fact been most limited precisely due to this lack of interest among archaeologists. Even in the coastal area, where research faces no major problem, studies have been limited to some sites, but of these only Los Gavilanes and Tuquillo, in Huarmey, have been systematically excavated. For the vast expanse of the highlands, studies are only available – and in quite limited fashion – for the Callejón de Huaylas and the Ayacucho zone; all the rest is terra ignota as regards preceramic ethnobotany. And bearing in mind that specialists agree that the midaltitude Andean valleys were zones of crucial significance for plant domestication (see inter alia Bonavia 1991: 128 and Grobman et al. 1961: 36), we see that we know practically nothing in this regard. On the other hand, all of the eastern Andean slopes and the tropical lowlands are unknown areas from an archaeological standpoint. These factors must be taken into account when drawing conclusions. We must realize that we are dealing with a very limited, and definitely nonrepresentative, sample. A new find may change the present picture at any moment.

Here a detailed account of the archaeological contexts where maize has been found cannot be presented, nor is it the goal of this book to do so. First of all, because the point here is to shed light on the origins of this plant, this review limits itself to the finds made for preceramic times, as has already been done for

makes a mistake, as it mentions "una mazorca [ear] de maíz" where the original text read "a stalk of . . . corn."

the rest of the American continents. The general characteristics of each site is then presented, listing the publication(s) that include(s) the specialized bibliography to which interested readers can turn to expand the information. Specific data thus do not appear in my bibliography. The only exception here made is in the case of a few sites for which the available data were not published, and to which I had access. In these cases it is essential to give a detailed account and make the data public. The reliability of the data is emphasized in all of the sites discussed herein, and attention is drawn to those I believe are doubtful.

The sources on preceramic maize can be grouped into two large categories. One is those sites that are questionable, or that have not received sufficient attention, and must therefore not be taken into account until they have been properly researched. The second category collects all those sites for which we have reliable data. The analysis in both cases goes from north to south and touches on first the coast and then the highlands.[21]

We begin with the first category. The northernmost site is La Cocina, in the Lacramarca Valley (south of the Santa River), which is also known as Besique A. Although there is one radiocarbon date that matches the antiquity of maize on the Peruvian coast, there is no data on this site that can be considered reliable (Bonavia, 1982: 363).

As for Río Seco del León – which actually is not in the Chancay Valley, as has always been said, but in the desert to the north of this valley – an explanation is in order. When discussing this region in the 1980s, I based my work on the work done by Collier (1961: 103), who in turn mentions two sources: Lanning (1959: 48) and a personal communication from this same archaeologist. At the time I did not have access to Lanning's work (op. cit.), but now I can confirm that this was a mistake Collier made, for this site is not even mentioned on page 48, nor in any other page. I talked with Lanning many times, and he never said anything in this regard. Matos Mendieta (1966: 513), on the other hand, is one of the scholars who mention this site. He explained that he took the data from the notes he took in Lanning's classes, but he later realized his mistake and vaguely recalls that Lanning actually meant the site of Bandurria (Ramiro Matos, letter to the author, 15 September 2006). But we also know that maize has thus far not been found at Bandurria either (Fung Pineda, 2004: 325–334). Work was done later at Río Seco del León, and no mention was ever made of maize having been found. All references to this site must therefore be considered as errors.[22]

Maize was at one time reported as being present in the area that lies between Chilca and Ancón on the Central Coast. This datum has figured in the literature since the 1970s, but no actual evidence has ever been presented (Bonavia, 1982: 358). The only possible instance, which cannot ever be verified, is Ancón:

[21] For an overview of this subject, see Bonavia (1982: 346–356).

[22] For more information, see Bonavia (1982: 359).

Lanning told me in 1980 that he had managed to detect the remains of maize in a coprolite from the early Final Preceramic, but this was never published (see Bonavia, 1982: 359).

In 1978 the discovery of preceramic maize at the site of El Paraíso was reported, but this was a mistake, as the author was able to prove (Bonavia, 1982: 359).

Some excavations were undertaken in the island of San Lorenzo in the early 1970s. Maize was purportedly found in a pottery-less stratum. The work done by Rosello and his group did not follow any scientific methodology at all and is worthless. Even so, it is possible that some preceramic site may have held maize, because when Darwin visited the site (see previously) and described the shell mounds with the remains of this plant, he did not point out the presence of pottery; this is interesting because his observations were quite detailed (Bonavia, 1982: 358–359).

Establishing the truth in the case of Chira-Villa, now inside the southern urban area of the city of Lima, is not easy. The report of the work done here raises some doubts, but it does not conclusively rule out the possibility that maize was indeed found in a preceramic stratum. Even so, this can no longer be proven, for the site has been destroyed (Bonavia, 1982: 357).

Tablada de Lurin is another site inside the urban area of Lima. It was studied by a team from the Pontificia Universidad Católica del Perú and has the code name of PV48-II. Preceramic maize has also supposedly been found here, but unfortunately the study is scientifically worthless (Bonavia, 1982: 357).

As regards the site of Chilca, in the homonymous zone, I wrote the following: "... this is a site that should be excavated once more in earnest, not only because it holds significant data, but because there is a remote possibility that preceramic maize is present here. This may be assumed after interpreting with great difficulty the almost unintelligible data presented by Frédéric Engel" (Bonavia, 1982: 356).[23] I do not know whether this site still exists or whether it has been destroyed. Chilca clearly is one of the many sites that could have yielded crucial data and that were not used, due to carelessness.

Finally, there is a site close to the mouth of the Ica River, of which it has also been said that it held preceramic maize. In this case it was a mistaken citation and definitively must not be taken into account (Bonavia, 1982: 356).

We turn now to the sites where preceramic maize has conclusively been found.

In the department of Lambayeque (province of Chiclayo), there is a site known as Cerro Guitarra that lies on the right bank of the Zaña River, 14 km east of the coast. Jack Rossen studied this site, where "plant remains were recovered from a few houses with midden deposits.... Maize (*Zea mays*) is present in several site contexts ... probably representing its first appearance in the valley" (Dillehay et al., 2011: 156).

[23] For more details, see Bonavia (1982: 356–357).

Rossen informed me that

> a total of seven cobs was found in House 42, 45 and 73 at Cerro Guitarra, with the best preserved (and photographed) specimens located in House 45. These cobs are in association with varying radiocarbon dates (3867–3481 cal. BP – GX 25033-LX [3630 ± 80] from House 73; 2919–2721 cal. BP CAMS46613 [2730 ± 50] and 1610–1410 cal. BP CAMS46614 [1350 ± 50] from House 45). We reject the last date, from *Donax* shell, as too recent [this was done as an experiment to compare the dating of the shell versus the carbonized wood]. We view the first two dates as a reasonable bracket for the site (3835–2867 BP), placing it as Terminal Preceramic or aceramic, in that perhaps a preceramic lifestyle in the lowest Zaña valley after ceramics had appeared in the upper valley. (Jack Rossen, letter to the author, 18 September 2011)

Rossen was generous to send us the photographs of the two cobs coming from House 45; these have been examined by Alexander Grobman. Because this description is unpublished, I present the following report by Grobman:

> Cob 1. Race Proto-Confite Morocho; large cob fragment with 8 rows of kernels with a potential of 12–14 kernels per row, although there is some loss of grain formation that may be due to lack of adequate pollination or lack of water; apparent spiral arrangement of cupules; length of cob fragment nearly complete 4.5 cm; cob diameter 1.1 cm in the center; prominent outer glumes apparently coriaceous; internal glumes long and soft appearing morphologically as semi-tunicate; the color of the bottom of the cupules seems to be purple. The cob has lost a segment in its base [Figure 5.3]. Cob 2. Race Proto-Confite Morocho; cobs with very slender rachis with 8 rows of kernels; potentially 10 kernels per row; cob length 3.5 cm; cob diameter 0.7 cm from tip to tip of lower outer glumes; lower outer glume length very small; internal glumes soft and prominent, hairiness in cupules; semi-lanceolate cupules with purple bottom; cob appears nearly complete [Figure 5.4]. Neither one of the cobs exhibits morphological features associated with the introgression of teosinte. (Alexander Grobman, letter to the author, 8 September 2011)

In the book published by Dillehay (2011), Piperno has reviewed the maize found in Zaña and compares it only to the ones found in Mexico, Ecuador, and Colombia but does not mention any other finding of Peruvian maize, and even the date attributed to the maize found in Cerro Guitarra is mistaken. Her conclusion that "the South American record is fully in accord with recent archaeobotanical evidence from the Central Balsas region of Mexico ..." (Piperno, 2011: 279) has no support and demonstrates that she has not analyzed the specimens of Cerro Guitarra's maize.[24]

[24] While this book was in publication, results from the Huaca Prieta Project, conducted by a team led by Tom Dillehay and Duccio Bonavia, were published (Grobman et al., 2012). This work took place at the sites of Huaca Prieta and Paredones, which are located in the department of La Libertad in the coastal part of the Chicama Valley. Huaca Prieta was first excavated in the 1940s by Bird (Bird et al., 1985). At that time, no maize was found in preceramic strata but

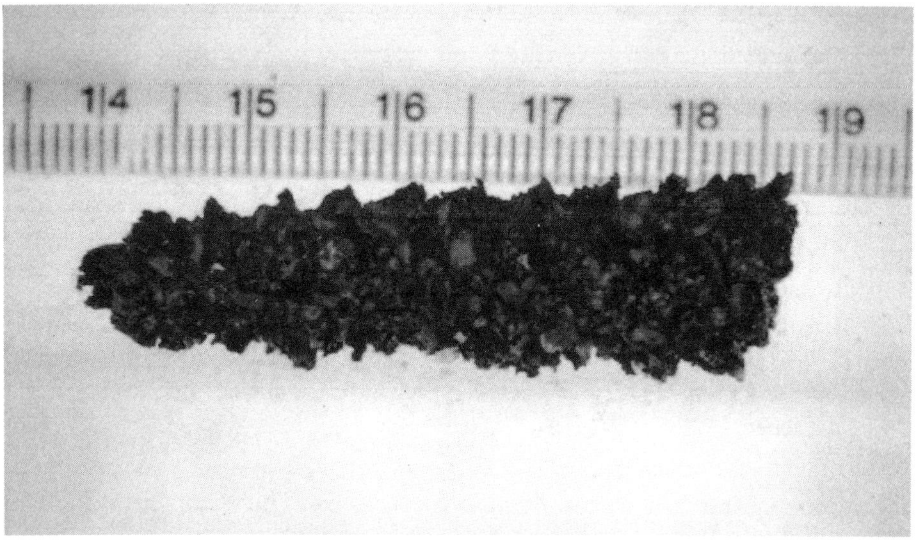

5.3. A Proto-Confite Morocho cob fragment with eight rows of kernels and prominent outer glumes, apparently coriaceous. The internal glumes are long and soft, of a semi-tunicate type. Provenance: Cerro Guitarra. (This determination was made from this photograph of the specimen.) Photograph courtesy of Jack Rossen.

5.4. A Proto-Confite Morocho cob with eight rows of kernels. The internal glumes are soft and prominent semi-navicular cupules with their floor exhibiting purple coloring. Provenance: Cerro Guitarra. (This determination was made from this photograph of the specimen.) Photograph courtesy of Jack Rossen.

only in those associated with the Cupisnique and Gallinazo cultures. With the work of this new project, maize has been found in strata of the Middle and Final Preceramic. In total, 288 maize cobs, 1 husk sheath and husk fragment, 2 pieces of stem, a sheath part, and a maize kernel were found. Of these, 234 specimens are associated with the Middle to Late Preceramic period. In addition to the macro-remains, corn starch and phytoliths were also found. Dates were obtained using the AMS method on the maize cobs, husk, and shank range from 3839– 4149

There are two sites in the Casma Valley known as Cerro El Calvario and Cerro Julia. In these two cases an exception is made and the sites are discussed at greater length, given their significance and the fact that unpublished data is available.

Both Cerro El Calvario (PV32–1) and Cerro Julia (PV32–2) were excavated by Santiago Uceda. In the former case a test pit was dug using the natural stratigraphy, with a very clear sequence (Uceda Castillo, 1986: 229–261; figure 91 shows the stratigraphy in detail). A maize cob was found in Level 5 (Uceda Castillo, op. cit.: 259). A radiocarbon date of 6070 ± 70 (Gif-6773) years BP was obtained for this level, and it is clearly stated that it was "... in association with preceramic maize" (Uceda Castillo, letter to the author, 2 September 1985; Uceda Castillo, 1986: 261).

Grobman analyzed the sample and believes that it is a very ancient maize related to that of Los Gavilanes and Guitarrero Cave. The specimen was so frail that part of it broke up into bits while it was being taken to Lima, so the length of the cob could not be established. This is a hybrid of Confite Chavinense and Proto-Confite Morocho, but it is closer to the former. The cob is fasciated, that is, it has two axes of uneven diameter. This is a very typical characteristic of the Confite Chavinense. The cupules are interlocked and the glumes have a purple coloration. The latter is a characteristic of the preceramic maize from Huarmey and shows a connection with highland races, as shall be seen later on (taken

(Beta 278050) to 6504–6775 (OS 86020) cal. years BP to 2Σ and are therefore roughly contemporary in age to the Mexican samples at Guilá Naquitz, Oaxaca, whose AMS dates cal. to 2Σ are 6170–6290 and 6015–6305 BP (Piperno and Flannery, 2001: table 1, 2102).

On the basis of cob morphology, number of rows of kernels, and size and shape of their cupules, three races of maize were identified from the earlier period: Proto-Confite Morocho, Confite Chavinenese, and Proto-Kculli. The race Proto-Alazán appears in Late Preceramic times and is the first identification of this race in this epoch. None of the cobs showed any morphological evidence of having descended from or having introgression from teosinte, as is the case in the cobs at Gilá Naquitz, the earliest found in Mexico. The cobs from Paredones and Huaca Prieta are all polystichous and do not exhibit the typical induration of the lower glume associated with teosinte introgression, but they show a surprisingly more advanced stage of selection than similar period cobs in Mexico due to high frequency of fasciation of the cobs of Confite Chavinense, allowing for a larger number of kernel rows. The cobs exhibit, with a high frequency, purple coloring in the cupules, a trait that results from the interaction of alleles of four genes in four different chromosomes, and that is associated with highland maize in Peru. These suggest that the maize at these sites was immediately derived from highland Andean maize.

The three early maize races that were identified at these sites are the same as those that have been found at other preceramic sites, such as Cerro Julia, Cerro El Calvario, Los Gavilanes, and Áspero on the coast, and Guitarrero Cave and Rosamachay Cave in the highlands of Peru (see information in the present book).

The significance of these early finds is that they are different from the Mexican races and that they are different at such an early period, as is reinforced by the chromosome knob data discussed in the main text; this suggests an early appearance of a maize racial complex in the central Andes region independent of the evolution of Mexican maize races and without signs of participation of teosinte in their formation. For further information and details, see Grobman and colleagues (2012) and the forthcoming report by Bonavia and Grobman (2012).

from a letter from Bonavia to Uceda, 21 September 1985, a copy of which is in the possession of the author; Uceda Castillo 1986: 279; see also Bonavia and Grobman, 1999: 241; Grobman, 2004: 447).

In Cerro Julia, a test pit was dug using the same methodology. Here too, maize leaves and fragments of stalks were found in the level 3, which has been dated to 6050 ± 70 (Gif-6772) radiocarbon years BP (Uceda Castillo, letter to the author, 2 September 1985; Uceda Castillo, 1986: 91). Uceda clearly says that the maize was found in the two sites mentioned, and within a context that he defines as "recent Preceramic" (Uceda Castillo, op. cit.: 225); he insists that "... two of the three radiocarbon dates obtained come from sites associated with the late preceramic with maize ..." (Uceda Castillo, 1986: 279). Uceda is quite clear in his conclusions: "The late Preceramic occupation is of specific interest due to the presence of maize dated in the sixth millennium. This means first of all that it is the earliest preceramic maize found thus far on the Peruvian coast.... *It is therefore a supplementary evidence of the presence of preceramic maize, and of its antiquity in the Central Andes*" (Uceda Castillo, 1986: 288; emphasis added). Uceda later mentioned these finds in several publications, and I along with Grobman also mentioned them (Bonavia and Grobman, 1989a: 839; 1989b: 459; Uceda Castillo, 1987: 23; 1992: 49). The dates of the Casma maize correspond to Lanning's Preceramic IV (1967: 25).

It is worth recalling what Burger said of the Casma finds. Burger pointed out that Uceda

> ... recovered maize in *unambiguous* Preceramic contexts in two excavations. At the site of El Calvario, located on the south bank near the mouth of the Casma [River], a maize cob was recovered along with fish bone, cotton, and squash from the fifth of six strata; a radiocarbon measurement from this layer produced a date of 6070 ± 70 B.P. (GIF-6772). At Cerro Negro (or Cerro Julia) near the Panamerican Highway on the north side of the valley, Uceda encountered maize stalks and husks in the third of five strata. Material associated with maize yielded a 14C date of 6050 ± 70 B.P. (GIF-6773). *These radiocarbon measurements suggest that maize has greater antiquity on the Peruvian coast than realized previously.* Alexander Grobman examined the maize cob and concluded that it was intermediate between Confite Chavinense and Proto-Confite Morocho. (Burger, 1989: 190; emphasis added)

In regard to this maize, it is worth recalling that Grobman (2004: 444) later wrote that "like all the other archaeological findings of preceramic maize ... that of Casma does not show the influence of teosinte."

Sevilla (1994: 226) accepted the evidence from Casma. Robert McKelvy Bird (1990: 831) is the only one who has questioned it, with a reasoning that proves untenable due to a lack of arguments. For more information on this issue, the reader is referred to the work by Bonavia and Grobman (1999: 241–242).

In the case of Las Aldas (south of the Casma Valley; some prefer the alternative spelling Las Haldas), we have the following proof to show the presence of maize in preceramic strata. Lanning (1960: 587), the first to study the site, certified he found maize and was categorical in claiming that it appeared in the upper preceramic levels (Lanning, 1967: 67). Lanning also provided this information to Rowe (1963: 5), who accepted it. David H. Kelley mentioned the preceramic maize Lanning had found at Culebras in a letter he sent to Mangelsdorf on 20 March 1970.[25] He then added the following: "He makes the same statement with respect to Las Haldas. I believe him...." I personally went over this point with Lanning on 7 June 1980, some time before his death. We discussed his work at Las Aldas, and Lanning certified that there was no question regarding the presence of maize in the upper preceramic levels. These findings, to the best of my knowledge, have not been discussed and have instead been accepted by most specialists (e.g., Bruhns, 1994: 106; Cohen, 1978: 259). Willey (1971: note 61, 186) initially had some reservations but then fully accepted them, just like Moseley (Moseley and Willey, 1973: 458). I was a firsthand witness of how earnest Lanning was in his work and cannot therefore question his statements.

Here it must be pointed out that no maize was found in the preceramic strata when Fung later excavated Las Aldas. But it is to be noted that neither was it found in the Initial Period strata. Fung was therefore cautious and wrote thus: "... Our excavations did not record maize in the preceramic and pre-Chavín ceramic strata ... *perhaps by chance*" (Fung Pineda, 1969: 188; emphasis added).[26]

Culebras is the next major site on the North-Central Coast where preceramic maize has been found, but little has been published on it. It may seem strange that Lanning, who excavated it in 1958, did not do so. It is important that the truth be known in this regard. For those who do not know it, we must point out that Lanning was working at the time with Frédéric Engel. Lanning himself told me just before his death that the contract he had with Engel did not allow him to disclose the data of the work he was carrying out. Because enough time had gone by, Lanning felt relieved of this bond and was ready to publish the data on the excavation of Culebras (Edward Lanning, personal communication to the author, 7 June 1980). He unfortunately passed away before he could do so, and I have thus far been unable to find where his field notebooks are.

There is one paper that was distributed in mimeographed format and that had a restricted distribution at a meeting held at the University of California at Berkeley, which I was able to find some years ago thanks to the kindness of John H. Rowe. Here Lanning (1959: 48) quite clearly says that maize and plain weavings appear in the upper preceramic levels. This study essentially refers to the pottery of the Initial period and the Early Horizon and was used by some scholars at the time, like Collier (1961: 103), who mentioned it several times.

The significance of Culebras is reflected in the observations found in Lanning's dissertation (1960: 476–482, 589): "Maize is usually absent from

[25] A copy of this letter is in my possession.

[26] For more information in this regard, see Bonavia (1982: 362–363).

preceramic refuse, but occurs in the uppermost levels of Culebras culture refuse *at several sites*" (Lanning, 1960: 40; emphasis added). One could object that it is not clear whether he means the site of Culebras or instead what he called the "Culebras Complex," which extends over several valleys in the North-Central Coast, which Lanning combined into one entity because they have common cultural characteristics (see Lanning, 1967: 66–68). Either way, the site of Culebras was involved, for the definition of the complex was derived from the work carried out at the homonymous site. Lanning was furthermore categorical when he discussed the Initial period occupation: "... small maize cobs occur with slightly greater frequency than in the uppermost preceramic levels ..." (Lanning, 1960: 484). He later confirmed this quite clearly: "... this vital grain also appears in the uppermost preceramic levels at Culebras 1 ..." (Lanning, 1967: 67).

The testimony given by several scholars can be used to validate this information. First, I visited the site along with Jorge C. Muelle and Ernesto Tabío when Lanning was excavating it. On that occasion he showed us three or four cobs and clearly explained they came from the upper part of the preceramic level (for more details, see Bonavia, 1982: 361). Then we have the fact that Lanning gave Collier these same data in 1959 (Collier, 1961: 103). Besides, I discussed the significance of these finds with David H. Kelley, who was also with Lanning during the excavations at Culebras. Kelley said he himself had personally verified the presence of maize in the preceramic strata (David H. Kelley, personal communication, 18 January 1960). Kelley told Paul Mangelsdorf this several years later (letter, 20 March 1970; a copy is held by Duccio Bonavia). Finally, I had a long conversation with Lanning on this subject before his death, as has already been pointed out, and he ratified it (personal communication, 7 June 1980).[27]

The preceramic maize from Culebras has been accepted by most archaeologists (e.g., Cohen, 1978: 227, 259; Moseley and Willey, 1973: 458; Pearsall, 1992b: 191; Willey, 1971: 96). Bird (1990: 829) simply omits the discussion and qualifies it as "superficial CP," that is, "Preceramic with Superficial Cotton." This is false, for we know there was a "... deep preceramic deposit ..." in Culebras (Lanning, 1960: 476, 477). This is corroborated by Tabío (1977: 90–93) and Engel (1958: 10).

Research from Tuquillo (PV35–7), a site close to the homonymous seaside resort, just to the north of Huarmey, is never mentioned in the literature, despite having been published. A large part of this site was destroyed when materials were removed to build the highway that goes from the Pan-American Highway to the resort. I was able to intervene and save the evidence from a small part of the site – an edge – that had remained intact. Here there was a very clear stratigraphy. Six maize cobs were found in the preceramic stratum, 3 of which were almost complete, as well as 19 fragments and several stalks. This maize is part of a racial complex derived from the Proto-Confite Morocho. The preceramic

[27] For additional references, see Bonavia (1982: 359–362).

strata correspond to the final phase of Los Gavilanes, that is, what was called Epoch 3 (see subsequently). For the full final report on this site, see Bonavia (1982: 233–236).

Clearly Los Gavilanes (PV35–1) is the major site for the maize problematic. It lies on the right desert bank of the Huarmey Valley and was discovered by Edward Lanning. To avoid mistakes, we must bear in mind that it was at first known as Huarmey Norte 1. This is the only site that has been systematically excavated and studied, from 1957 to 1979, in order to study maize. Although David H. Kelley initially took part, it was the present writer who carried out all of the research, which reached its climax with the Proyecto Arqueológico Huarmey (Huarmey Archaeological Project, 1976–1977).

Los Gavilanes is the only Peruvian preceramic site with a complete monographic report (Bonavia, 1982) in which 12 noted specialists participated. Besides this monograph, several studies have also been published that are here listed in chronological order: Grobman and colleagues (*1961: 74), the first report on the site while the materials from the first campaign were still under study, followed by Kelley and Bonavia (*1963), Mangelsdorf and Cámara-Hernández (*1967), Banerjee and Barghoorn (*1972), Banerjee (*1973: 63–71), Banerjee and Barghoorn (*1973a, *1973b), Grobman and colleagues (*1977), Grobman and Bonavia (*1978), Bonavia and Grobman (*1978, *1979), Grobman and Bonavia (1979–1980), Castro de la Mata and Bonavia (1980), Bonavia (*1981), Patrucco and colleagues (1983), Weir and Bonavia (*1985), Bonavia and Grobman (*1989a, *1989b), J. G. Jones and Bonavia (*1992), Goloubinoff and colleagues (*1993), Bonavia (*1996b), Bonavia and Grobman (*1998, *1999), Bonavia (*2000, *2002b), Y. Ortega and Bonavia (2003), and Bonavia and colleagues (2004). The publications marked with an asterisk refer to maize.

With the help of some specialists, namely, geologists, we covered everything in regard to the geology, geography, geomorphology, and climate of the zone studied. I also studied the lithic, textile, and wood materials; some of the zoological aspects; and animal paleoscatology (i.e., the study of animal coprolites). Botany, palynology (from different angles), zoology, physical anthropology, and human coprolites were also studied by other specialists, while the study of animal coprolites was expanded.

Because interested readers can find all the information they need in the aforementioned studies, here only some very general characteristics of the site will be mentioned. This is quite a unique site, as it is located in a sandy area with a settlement with stone architecture of which not much is known – in order to study it I would have had to remove all the materials belonging to the following occupation, and this would have entailed more work than could be done. A system of storage pits was built on top to store maize. These buildings were in the shape of an inverted frustum, with an approximately circular opening. The storage pits excavated in the sand had walls made of stone. Ears of maize were stored in them covered with sand. This is a very ancient method that was essentially used in the

5.5. A reconstruction of Los Gavilanes (Huarmey) by Félix Caycho Quispe, following guidelines laid down by Duccio Bonavia. For reasons of space the size of the figure is out of proportion. The scene shows the storage pits in use as they looked during the final occupation of the site. In the foreground is a herd of llamas arriving, loaded with maize plants, from a neighboring valley. At left, a group of people are separating the ears from the plants. Some men are filling the storage pits in the center using netting, while two others are covering with sand one of the already full bins. At top are storage pits that have already been filled in. At right a man goes towards the public building to feed the fire that had a religious purpose. Drawing by Félix Caycho Quispe, the following instructions given by Duccio Bonavia. (After Bonavia, 1982, drawing 64: 272–273.)

North and North-Central Coast, which I was able to verify had been in use from preceramic times to the present day (Bonavia, 2002b). In this way maize can be stored in a perfect state of preservation, as was verified in the excavations made. In these holes, as I have called them, I found not only cobs – some of them still with grains – but also the remains of different parts of the plants (Figure 5.5).

The site extends over an area of about 1.7 hectares. Here 47 of these storage pits had been built, and the diameter of their mouth ranges between 2 m and 20 m, whereas their depth ranges between 0.48 cm and 1.75 m. The estimates made calculated that the 47 holes have a volume of 1,590 cubic m. A special study was made to establish exactly what amount of maize could be stored, using the average size of the preceramic ears here excavated. Two estimates were obtained, a low one and a high one. The low estimate gives 461,128 kg, and the high one 712,364 kg, assuming in both cases that the holes were filled to the brim and were tightly packed. This is obviously debatable, but the amount is still considerable even if they were only partially filled.[28]

[28] For more details, see Bonavia (1982: table 1, 67; 2002b) and Bonavia and Grobman (1979).

The occupation of the site was subdivided into three epochs, which were labeled Los Gavilanes 1, 2, and 3, with the former being the oldest one and the latter the most recent one. Maize appeared in Epoch 2. We have two dates for it, a radiocarbon date of 4140 ± 160 (GX-5076) years BP, and another one obtained by thermoluminescence of 4800 ± 500 (BOR 20) years BP. For Epoch 3 there are four radiocarbon dates, which respectively are 3750 ± 110 (GIF-3564), 3755 ± 155 (GX-5078), 3595 ± 140 (GX-5079), and 3250 ± 155 (GX-5082) years BP (see Bonavia, 1982: 73–75, 276–277). In other words, they fall within the chronology Lanning (1967) proposed for preceramic times, with Epoch 2 corresponding to the Preceramic V, and 3 to Preceramic VI.

There are two more dates, one in radiocarbon years of 2080 ± 130 (GX-5077) and another using the AMS method of 1610 ± 40 (B-18297). Neither of them is valid, because both samples were contaminated. The first case has been analyzed in depth (see Bonavia, 1982: 176–277), and it was shown that the charcoal used for the dating was probably wetted by urine or seawater. This caused the absorption of salts, acids present in the soil, and so on, and gave rise to a distorted date. The second dating with the AMS method was based on a fragment of a maize cob from Los Gavilanes, Epoch 2. The cob was handled to study it and take photographs, because it was not intended to be used for dating. Besides, it was stored for more than 25 years in a storeroom that did not have adequate preservation conditions. Pohl and colleagues (2007: 6873) have noted in this regard that "vigilance is needed to guard against [the] growth of fungi and bacteria that can cause modern contamination of organic material that results in artificially young ages." They likewise insist on the "... potential hazards of contamination in long-term storage ..." (Pohl et al., op. cit.). I believe that in this case the date is also distorted, for various reasons. First, for reasons that are still not clear, and that specialists must explain, with the AMS method some dates come out wrong (e.g., Pearsall, 2003b: 223–224; Rossen et al., 1996). Second, there is some problem – which specialists likewise have not solved – with dates based on maize cobs (e.g., Fernández Distel, 1980: 90, 96; Rivera, 2006: 404, 409), which give more recent dates than they should.

There is an additional point that has to be explained, as it could give rise to doubts. There are two dates that were obtained from materials excavated in 1960, during the first excavations made at the site. The first botanical report in fact said the following:

> When the site was excavated in the year 1960 carbon samples were gathered in order to be utilized for dating through C14. Due, however, to an error, samples of corn were sent to laboratories for determination. The results obtained were totally inconsistent since the date fluctuations were between 200 and 800 years before the present era, with a margin of error varying between 70 and 95 years. (Grobman et al., 1977: 224)

Although this same report explained the problems present at the time with maize-based datings, the possibility (later discarded) that there was humidity at the site due to the fossil lagoon adjacent to it, the handling of the samples, and the addition of new dates obtained through thermoluminescence and C14 (Grobman et al., 1977: 225), this argument was used to criticize the work done at Los Gavilanes and to question its real age (e.g., R. McK. Bird, 1990: 829; Pearsall, 1992b: 184, table 9.6).

After the promising preliminary tests conducted by David Kelley in 1957 and 1958, in 1960 I was charged with excavating the site to recover more botanical materials. Once these were unearthed, I sent them to the Botanical Museum at Harvard University. Among these materials were samples of charcoal for radiocarbon dating. Since then, I have not been involved in the analyses conducted by Paul Mangelsdorf and his collaborators at Harvard. More than a decade later, as has already been noted (Grobman et al., 1977: 224), "due ... to an error, samples of corn were sent to laboratories for determination." In fact, I was not informed of the destination of the samples or even to which laboratory the samples were sent. As a result, the report accounts for the samples were written in an informal manner, as has been pointed out above. We assumed that the erroneous dates could not be significant because the new and coherent dates were included in this publication, as has been indicated. I subsequently obtained a letter written by David Kelley and submitted to Paul Mangelsdorf (dated 20 March 1970; a copy is in the possession of the author). Here Kelley writes:

> ... Bonavia did collect specimens of charcoal for dating, carefully in accordance with the best available techniques. However, he probably did not expect that corn specimens would be used in this way and they may not have been handled properly. Also, I looked at some of the specimens and handled them. Hence, there may be some possibility of contamination over and above what Vescelius suggested (which is probably most basic).

It was also Kelley himself who told me (in a letter, 19 November 1970) that "... without consulting me [they] sent off some of the corncobs instead of the samples you have collected," and that the samples were submitted for radiocarbon dating after the cobs were analyzed (and handled at Harvard). Therefore, one can conclude that the possibility of contamination was high. Another secondary issue is the accuracy in dating corn at that time (probably in the early sixties). In the same letter that Kelley sent to Mangelsdorf (see previously), he explains the following: "With respect to the dating I was talking with Gary Vescelius ... and he told me that corncobs were particularly bad to use for C14 dating, because growing corn had a most unusual rate of absorption of C14, giving a high content and indicating spuriously late dates." Maize is a plant that consumes more C14 during photosynthesis than other woody plants, and the result therefore must be normalized (Creel and Long, 1986: 827). In fact, the techniques for improving the C14 measurements of maize through isotopic fractionation are relatively recent (e.g., Damon et al., 1978; Taylor, 1987: 48). Nevertheless, the

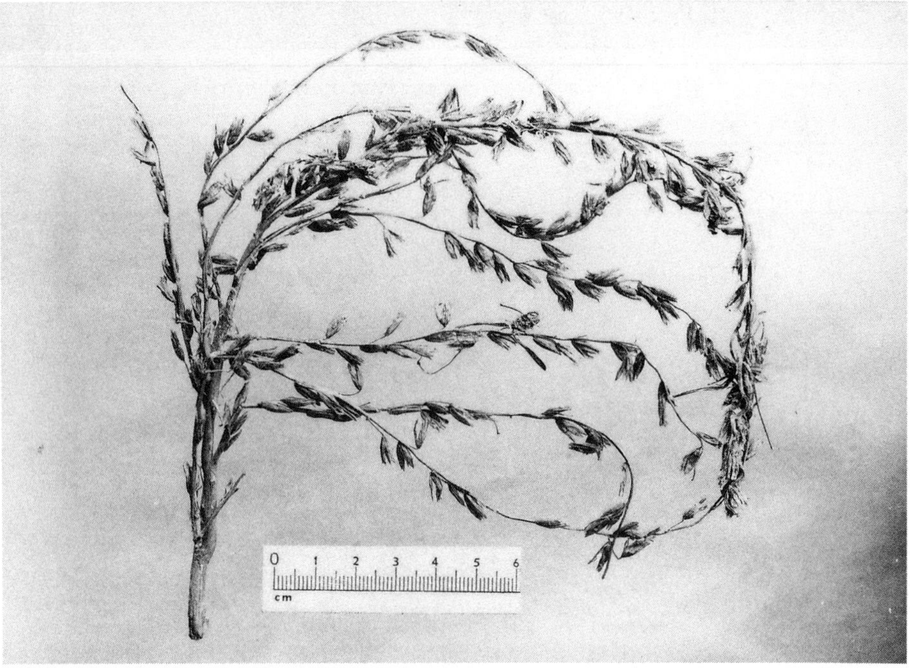

5.6. A tassel with primary branches distributed along a central branch that ends in a formation of virtually polystichous spikelets from Los Gavilanes. Photograph by Duccio Bonavia.

fact that the samples were handled is believed to be the principal reason for the anomalous dates.

In any case I trust the chronology of Los Gavilanes, because even bearing in mind the two datings discussed previously, only 4 of 10 radiometric dates conflict with our sequence. Experienced fieldworkers demonstrate that it often occurs. Kent Flannery clearly states: "No matter how carefully you select your radiocarbon samples, in any group of 10 there are almost sure to be some that come out looking aberrant" (Flannery, 1986a: 175). In his fieldwork at Guilá Naquitz, he obtained "six of these dates [that] look to be internally consistent ...," and he therefore explicitly argued that "... the other four dates are too young" (Flannery, op. cit.: 175). In the case of Los Gavilanes we can also argue that we obtained six dates that seem to be internally consistent and four that we consider too young.

Overall we recovered 2 complete plants, 20 stalks, 202 complete cobs, 109 incomplete cobs (i.e., a total of 311 cobs), 19 tassels (Figure 5.6), 1 leaf sheath, 37 loose kernels (although there is also a large number of kernels attached to the cobs), kernel pericarps, silks or stigmas, fragments of leaves, anthers, and pollen grains in indefinite number. The samples studied by Grobman were classified as belonging to the two major racial types, Proto-Confite Morocho (Figure 5.7) and Confite Chavinense (Figures 5.8 and 5.9), and to the racial type Proto-Kculli (Figure 5.10). Besides, morphologically intermediate cobs

5.7. A Proto-Confite Morocho cob showing soft and extended membranous glumes of a semi-tunicate type. Provenance: Los Gavilanes. Photograph by Duccio Bonavia.

5.8. A typical Confite Chavinense cob with 16 irregular rows, fasciated, with large glumes and slightly tripsacoid. Provenance: Los Gavilanes. Photograph by Duccio Bonavia.

were distinguished among the preceding ones that were defined as segregating, derived from hybridizations between the race types. These segregating types were defined as Proto-Confite Morocho/Confite Chavinense, Proto-Confite Morocho/Confite Chavinense/Proto-Kculli, or Confite Chavinense/Proto-Kculli, depending on their greater or smaller phenotypic proximity to the proposed progenitor types. For the details, the reader is referred to the study by Grobman (1982). It is worth recalling that, basing his work on some specimens,

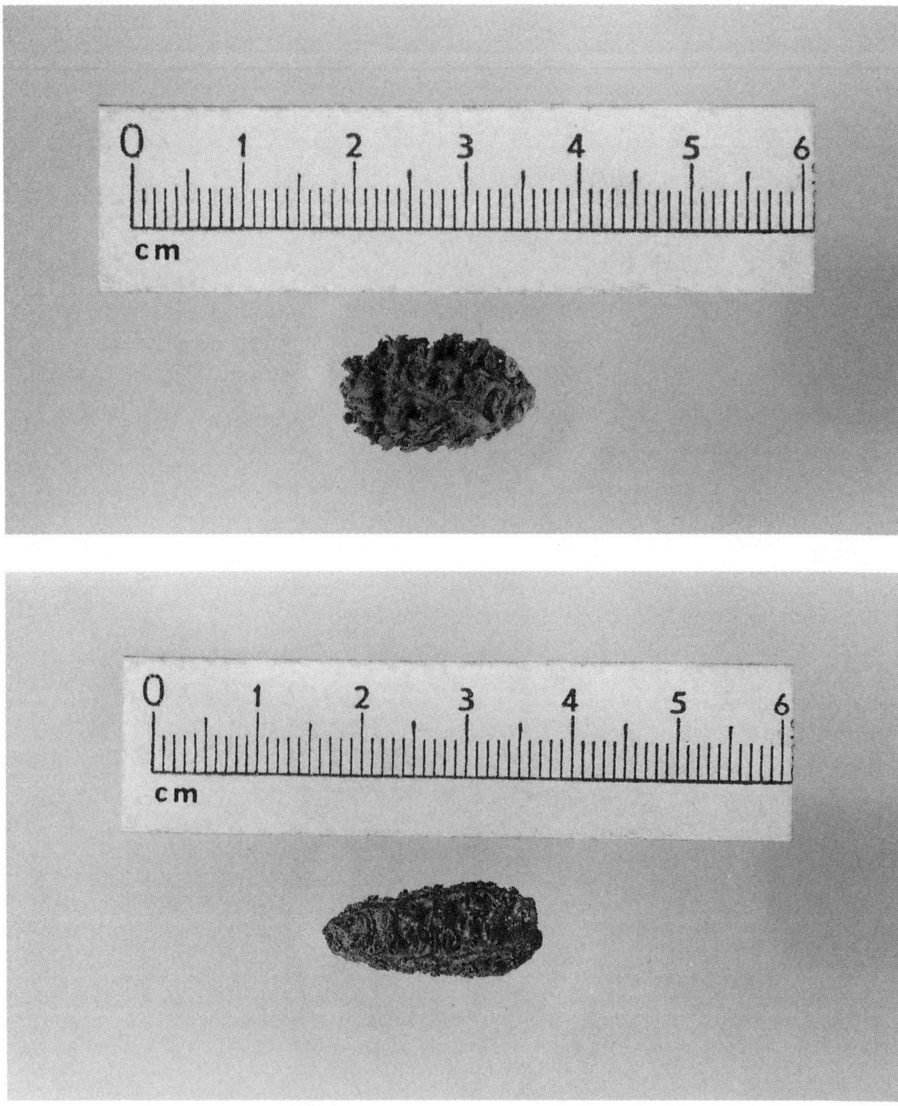

5.9. Three Confite Chavinense cobs that correspond to semispherical, short-length ears, similar in length to the remains found in the most ancient strata at Tehuacán (Mexico). Provenance: Los Gavilanes. Photograph by Duccio Bonavia.

Grobman managed to reconstruct an ideotype of ramified ears that may have been an original form of the wild maizes, with axilar hermaphroditic inflorescences. Some of the phenotypic evidence of the genetic system of ramified ears was found in the maize cultivated in the Preceramic period and the Middle Horizon of Huarmey (Grobman, 1982: drawing 60, 167; see my Figure 5.11). Another important detail is that the pollen grains extracted from the tassels found were analyzed by Umesh Banerjee in the laboratories of Harvard University with a scanning electron microscope, the results of which have been

5.9 *(cont.)*

5.10. A typical Proto-Kculli cob from Los Gavilanes. Photograph by Duccio Bonavia.

mentioned innumerable times in the specialized literature (e.g., Banerjee, 1973: 63–71; Grobman, 1982: 171, inter alia).

Because Banerjee's work is little known, it is worth giving some details of his study here. Banerjee explains that with electron microscopy one can distinguish the differences in the characters of the various genera by examining the pattern of the pollen's ektexine grains. This pattern is quite similar in "pure races" of maize and teosinte and is represented by uniformly distributed spinules, with the exception of most popcorn races. Pure-race ektexine has been found

5.11. An ideotype of ramified ears that may have been the original form of wild maize, with axillary hermaphroditic inflorescences. Some phenotypic evidence of the genetic system of ramified ears has been found in the cultivated maize from the Preceramic period and the Middle Horizon. Drawing by Félix Caycho Quispe, following instructions given by Alexander Grobman. (After Bonavia, 1982, drawing 60: 167.)

in Tehuacán (Mexico) and in the more recent remains from Los Gavilanes (Figure 5.12).

When different races of maize hybridize with teosinte, in nature or artificially, the pollen grains from the derived hybrids show a new phenotypic pattern wherein some ektexine spinules are occasionally lost, thus leading to a new pattern due to spinule loss. The presence of this pattern represents the introgression of teosinte into maize or vice versa, that of maize with teosinte. The ektexine pattern of the pollen grains from *Tripsacum* (both diploid and tetraploid spp.) shows a "spinule clumping" that is quite different from that called "negatively reticuloid." *Tripsacum* retains its spinule clumping when it hybridizes both with maize and with teosinte. This means that the phenotypic pattern of ektexine in *Tripsacum* is dominant over maize and teosinte, and that the grains of pollen from these derivative hybrids somehow retain the grouping of the spinules (Banerjee, 1973: 70–71). The study made of the pollen from Los Gavilanes shows this introgression of *Tripsacum* (Banerjee, op. cit.: 69–70).

The changes Banerjee spotted when analyzing the maize pollen from the different levels of Los Gavilanes are interesting. The ektexine pattern in the oldest levels was "entirely new" (Banerjee, 1973: 64). This pattern can only

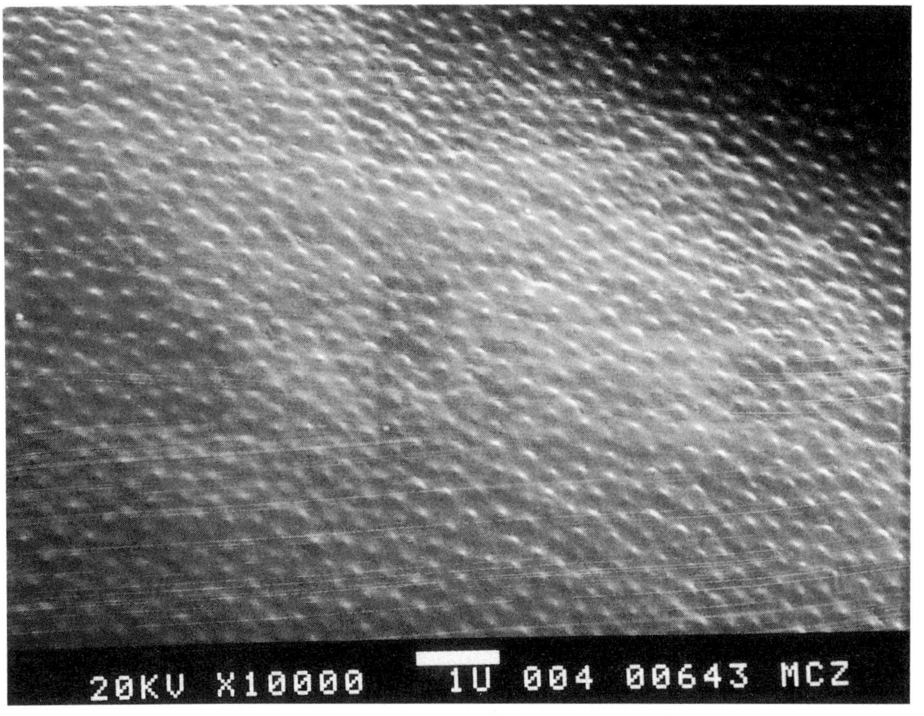

5.12. Pollen grain, Los Gavilanes, Epoch 2, with its surface enlarged by 10,000 and showing the uniform distribution of the spinules. This corresponds to a pure maize. Photograph by Umesh Chandra Banerjee. Photograph courtesy of Umesh Chandra Banerjee.

be derived through a direct introgression from *Tripsacum*, which has an ektexine pattern that dominates over maize and teosinte (Banerjee and Barghoorn, 1972, 1973a, 1973b). Maize retains the spinule clumping or spinule block in the immediately higher level, but there is also an occasional lack of ektexine spinules. This suggests either that there was a constant introduction of new races of maize, which may have had teosinte germplasm, or instead that its original maize stock began to hybridize with already existing but varied races, through which means the teosinte germplasm could have been introduced into the Huarmey stock.

In the following level we have a quite complex palynological picture, for we are before the presence of a "polymorphic" ektexine patterning. In other words, this is a pollen with very large grains, implying hybridization, that show an outgrowth of exine. This may mean a maximal hybridization, or perhaps it means that new races were introduced at this time from another place. In the last level, that is, the upper one, we notice the establishment of a "pure-race type of ektexine pattern," which might mean a continuous inbreeding in a small plant population, or – and this is less likely – that this race was intentionally selected by the people of Los Gavilanes (Banerjee, 1973: 63–65). It was for these reasons that Mangelsdorf (1983b: 231) classified the pollen from this site as tripsacoid.

To avoid mistakes it is worth going over how the position of the scholars who studied this pollen has changed. At first Banerjee and Barghoorn (1972: 226) wrote as follows:

> When we examined the pollen grains of "Confite Morocho" [this actually is Proto-Confite Morocho – D.B.] maize, an ancient and primitive domestic race of maize from Peru ... we were surprised to find [an] ektexine pattern very similar to [the] pattern shown in Figure 2,[29] which may be interpreted as a sign of teosinte contaminated maize. Since teosinte only occurs in the wild in Mexico and Central America, we would suspect that this race of maize "Confite Morocho" may not have originated in Peru.

However, a comparison of the plates IV A and B published by Irwin and Barghoorn (1965: 50–51) with photograph 56d in Grobman (1982: 171) shows a remarkable resemblance. Plate IV A corresponds to a (pollen) grain of maize from the Chapalote race, IV B is a (Proto) Confite Morocho, and the picture in Grobman is of Proto-Confite Morocho. When comparing A with B, Irwin and Barghoorn wrote that both show a relatively pure pattern. If we compare these photographs with those of teosinte races, we can see a second type of maize-teosinte exine. The spinules in the kernels of maize are darker and are therefore more distinctive. This suggests that the spinules of the maize-type pollen exines are larger, thus giving rise to a retardation phase that is longer than that of teosinte. A practical result is that the larger areas of maize exine may be brought into sharp focus (with grains of the same size).

When Banerjee and Barghoorn (1973b: 34) later returned to this subject they categorically stated that the pollen grains from Los Gavilanes in the Epoch 2 "... show pollen grain ektexine with distinct spinule-clumping and indicate *the oldest convincing evidence of the introgression of Tripsacum with maize*" (emphasis added; see my Figure 5.13). In another study presented this same year, they said exactly the same thing and added the following: "Moreover, we found that a collection of pollen grains of the extant race of Cuzco maize (*Zea mays* L.) also shows a distinct spinule clumping, and we may assume perhaps that this race of maize has likewise been derived through natural introgression with *Tripsacum*" (Banerjee and Barghoorn, 1973a: 48). Galinat accepted this (1977: 37).

To avoid any misunderstanding it must be pointed out that photograph 58d in Grobman (1982: 171) reads "possible relation with teosinte" but should actually say "*Tripsacum.*" This was explained by Grobman (2004: 451–452):

> This at the time was established as a clear evidence of introgression from *Tripsacum*, but in his note Banerjee erroneously indicated that it was evidence of introgression from teosinte, which we copied in the caption to the

[29] These photographs show a pattern of F3 ektexine pollen grains artificially hybridized with teosinte/Chalco and Chapalote, which Banerjee and Barghoorn consider a sign of contamination with teosinte on maize.

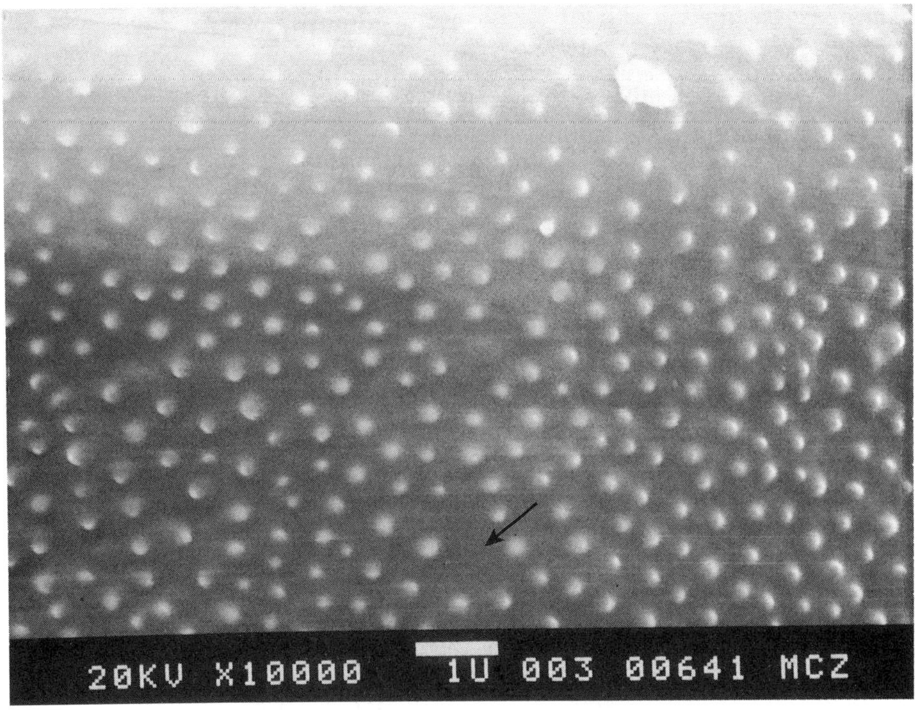

5.13. Pollen grain, Los Gavilanes, Epoch 2, with its surface enlarged by 10,000 and showing the loss of the spinules (see arrow), which according to Umesh Banerjee indicates a relation with *Tripsacum*. Photograph by Umesh Chandra Banerjee. Photograph courtesy of Umesh Chandra Banerjee. (After Bonavia, 1982, photograph 56d: 171.)

> photograph indicated. The issue of this mistake was later cleared up and it was determined that it is an introgression from *Tripsacum*. The dispersal pattern of the pollen exine spinules in teosinte and in maize is similar, and both differ from that of *Tripsacum*. (Mangelsdorf, 1983 [184])

When MacNeish and Eubanks (2000: 14) refer in passing to the maize from Los Gavilanes, they mention the molecular evidence that indicates that the genes from perennial teosinte may have entered the South American maize at a very early age. They also recall that Goloubinoff and colleagues (1993) pointed out that the maize from Los Gavilanes has an *adh2* allele that is identical to that of perennial teosinte. Grobman, however, says that the plants from Los Gavilanes have few axilar ears, but many of those found on each axil are ramified, which

> ... indicates that the mode of growth is quite determinate and very different from the indeterminate mode of teosinte. With subsequent selection the number of ears was reduced, probably to avoid internal competition over the resources translocated from the leaves, in order to ensure a stable production within an ever more determinate mode of growth. (Grobman, 2004: 429)

Besides the aforementioned studies, human coproliths found at Los Gavilanes have also been studied (Weir and Bonavia, 1985). The presence of *Zea mays* pollen was established both for Epochs 2 and 3. In all, 44 samples were studied, 22 for Epoch 2 and 22 for Epoch 3. Maize was detected in 6 of the samples for Epoch 2 and 10 in Epoch 3. In all, these samples respresent 27% and 45% of the samples from Epochs 2 and 3, respectively. (Weir and Bonavia, 1985: table 3, 130–131; graph 1, 136; graph 7, 139). Llama (*Lama glama*) coprolites were also studied, and it was found that three of the four samples had maize pollen. This represents 50% in Epoch 2 and 27% and 6% in Epoch 3 (J. G. Jones and Bonavia, 1992: 839–840).

The scholars who have accepted the work done at Los Gavilanes are here listed in chronological order, so that readers can judge on their own: Rowe (1960: 141), Willey (1971: note 61, 186), Flannery (1973: 304, table 3), Moseley and Willey (1973: 458), Moseley (1975), Lynch (1978: 525), Pearsall (1978a, appendix 2, 68), Wilson (1981: 103), Vescelius (1981b: 10), Hastorf (1985), Lathrap (1987: 351), Burger and Van der Merwe, (1990: 91), Harlan (1992: 222), Quilter (1992: 114), Sevilla Panizo (1994: 232), Harlan (1995: 185), and Van der Merwe and Tschauner (1999: 526, 528). Those who reject it are as follows: R. McK. Bird (1978: 92; 1984: 49; 1987: 298), R. McK. Bird and J. B. Bird (1980: 330), R. McK. Bird (1990), Moseley (1992: 19), Feldman (1992), Pearsall (1992b: 184, table 9.6), Hastorf and Johannessen (1994: 429), Pearsall (1994a: 225, 258, table 15.2), Hastorf (1999: 55), Tykot and Staller (2002: 667), and Pearsall (2003b: table 3, 238).

It must be pointed out that since the report on Los Gavilanes was published more than 28 years ago, not a single critique of the methodology used, nor of the stratigraphy, has ever been made. Not even the findings of any of the other plants associated with maize have been criticized.[30] The objections leveled at it were solely aimed at the remains found of maize but without bearing in mind their context, using arguments that I believe are not valid, and using the data in a biased and frivolous fashion to boot (see Bonavia and Grobman, 1989a, 1999). We shall return to this issue in the discussion in the final part of the book (Chapter 10).

Excavations have been made at the Cerro Lampay site, in the lower Fortaleza Valley. This is a ceremonial center that dates to the Final Preceramic period and in which evidence of ritual practices has been found. A large fireplace was found on the upper part of the central platform of the building, close to the stairway in the central corridor. A set of organic remains labeled TD3 was found in this corridor, on top of the mat and the stairway steps (Vega Centeno Sara-Lafosse, 2007: 161). The refuse here held a cob of maize (Vega Centeno Sara-Lafosse, 2007: table 2, 163). The report does not give additional data on the botanical remains. Although figure 4 (Vega Centeno Sara-Lafosse, 2007: 158) states that 25 AMS dates have been obtained, no indication is given that can allow us to

[30] There was only one objection in regard to the finding of chirimoya (*Annona cherimolia*; see Pozorski and Pozorski, 1997), but this reservation was easily dispelled (Bonavia et al., 2004).

accurately date TD3. The only thing we can conclude is that the site was used in 3734–3984 BP (2400–2200 cal. BC), that is, that this occupation dates to the Late Preceramic.

The next site to be discussed is Áspero, in the mouth of the Supe Valley. This was the first preceramic site excavated in Peru, at a time when this epoch was still unknown, to the point that no importance was ascribed to it. Willey and Corbett (1954) are quite clear in their report, even though there seem to be some inconsistencies that have not been pointed out, and that will now be explained.

There can be no question regarding the preceramic epoch of this site. Willey and Corbett are categorical in this regard, when they note that it held no pottery (Willey and Corbett, 1954: 151). The report specifies that four maize cobs were found in "Room 2" below the floor (Willey and Corbett, op. cit.: 27). Another cob was found while excavating "Room 4," in the fill of the room (Willey and Corbett, op. cit.: 28). No other finding of maize is reported. There is one detail in the study included in the report that has gone unnoticed not just by Willey and Corbett but also by the team who subsequently worked at Áspero under the direction of Moseley, as well as by the archaeologists who used this information. Here 49 maize cobs are recorded (Towle, 1954: table 14) that came from "… one location" (Towle, op. cit.: 131). Towle later explains this: "It was beneath this floor that a cache of maize containing forty-nine whole and broken cobs was discovered" (Towle, 1961: 119), and she meant the floor of Platform 1. In these reports Towle did not mention the 5 cobs found in Rooms 2 and 4, whereas Willey and Corbett (1954) did not point out the discovery of the 49 cobs.

I was able to clear this up by asking Gordon Willey, who answered thus:

> That Corbett and I failed to mention the cache of many maize cobs which came from underneath Platform 1 in Room 4 was an oversight which I regret and for which I accept responsibility. The "4 maize cobs from Room 2" and the "1 maize cob from Room 4" – found in refuse beneath the respective floors of these rooms – were quite separate from the cache of cobs under Platform 1. I am not sure if Towle included the "4 maize cobs from Room 2" and the "1 maize cob from Room 4" in her cache count of 49. I would doubt that she did. We apparently did not make a field count of the cobs in the Platform 1 cache, for which I, again, must accept responsibility. All I remember of the circumstances of the [discovery of the] cache beneath Platform 1 in Room 4 was that there were a great many cobs, or cob fragments, in it, and that these were in a little pile in the loose dirt that underlay the hard fired clay surfacing of the platform. (Gordon R. Willey, letter to Duccio Bonavia, 29 February 1996)

So the 1941 campaign definitely found 54 maize cobs at Áspero.

As for the context of this find, it is clearly stated that "there was no evidence which would indicate that the structure represented more than one building period." It is likewise specified that this "… was built after the site has been occupied by peoples who were familiar with maize horticulture" (Willey and

Corbett, 1954: 29). Besides, the authors insistently repeated that "... no ... ceramics ... were recovered ... ," and they explained that "Aspero differs from these other two middens in having no pottery" (Willey and Corbett, op. cit.: 25). Willey and Corbett finished thus: "*No potsherds ... were encountered either above or below the floor*," and "*no pottery was found in any of the rooms or in the excavations outside the building*" (Willey and Corbett, 1954: 25, 28; emphasis added).

Willey and Corbett then said: "The Aspero Temple belongs to an agricultural period ... as corncobs were found in the rubbish beneath the temple, sealed in by the prepared clay floor. *There is no possibility that these finds are intrusive....* At the same time, the midden, which is extensive and fairly deep, is without pottery" (Willey and Corbett, 1954: 151–152; emphasis added).

The comments made by Willey and Corbett are interesting to read nowadays because they were made, as was already noted, at a time when the Preceramic period was not yet known, although they foresaw it. Willey and Corbett stated the problems they had in explaining maize agriculture before pottery in terms of the knowledge then available. They noted that the hypothesis of contemporaneity with pottery would mean that the people of Áspero "... for some strange reason ..." kept the mound scrupulously clean of pottery and at the same time placed the vessels and sherds on top of the area close to the village. "We do not lean toward this explanation.... *Tentatively, we incline toward the interpretation that places it* [*the site*] *as antecedent to the pottery-bearing sites*" (Willey and Corbett, 1954: 152; emphasis added).

Towle (1954: 131–134) studied 49 cobs, 36 of which were whole and the rest broken. None of them had kernels. Towle did not attempt any racial classification. She later reaffirmed (Towle, 1961: 119) what Willey and Corbett (1954) had stated: "No pottery was found at either the midden or in the 'Temple' structure...." Now we know that the site belongs to the Late Preceramic.

Many years later Moseley and Willey (1973: 455) certified that the maize cobs from Áspero[31] "... were definitively not intrusive" (see also Moseley, 1975: 80; Moseley and Willey, op. cit.: 458). They furthermore reported that "in the restudy of the site 1 cob was found in the canal bank profile on the eastern margin of the site"; they immediately commented on the preceramic maize found on the Peruvian coast and finished by stating that "Áspero is the southernmost of these settlements ..." (Moseley and Willey, 1973: 458). They also explained that maritime subsistence was successfully practiced at Áspero, albeit with the adoption of a new mode, that is, maize agriculture (Moseley and Willey, op. cit.: 466).

Feldman later excavated at Áspero as part of his doctoral dissertation. Before submitting it he told me that he had found preceramic maize in three components – As1D-1 = 2, As1V-3 = 3, and As1V-4 = 5 – and that they all belonged to

[31] They mean the 49 cobs, without noticing the presence of the other 5 noted previously.

the Preceramic VI (in Lanning's terminology 1967), that is, the Late Preceramic (Robert Feldman, letter to the author, 21 November 1978). More details of the finds appear in the dissertation, where it is indicated that maize was found both in the refuse and among the architectural remains. In all, maize was found in four middens and in three associations with architecture. In all there were 12 cobs, but it is immediately added that "many more were found in the single excavated pit in Li-31" (Feldman, 1980: 182–183). According to table V (Feldman, op. cit.: 178), 96 cobs were found that have never been studied and whose chronological position was never explained. Furthermore, at present the location of these materials is unknown. The only comment Feldman made in regard to this find was the following: "In sharp contrast to Áspero, the Li-31 test pit (As2A) produced an abundance of maize cobs from every level except the lowest, which was the interface between the midden and the sterile sands underneath" (Feldman, 1980: 185).

Feldman is quite clear in regard to other finds. He explains that one cob (As1V-5) was found in the southern component of Áspero with pottery, and three (As1N-3 = 5; 4 = 3) to the east of the base of the Huaca de los Ídolos, in a disturbed test pit with pottery. Feldman then says the following: "Eliminating these samples leaves only 3 maize cobs from the midden *that do not have obvious questions about their preceramic associations*: As1V-4 = 5 and As1D-1 = 2" (Feldman, 1980; emphasis added). Here Feldman obviously makes a lapsus calami, for he initially says "3 maize cobs" and then only mentions two at the end of the sentence. The missing cob is in fact As1V-3 = 3. This is recorded in the list of maize remains that Feldman sent me, which I have in my files. Feldman then explains at length the provenance of each of these samples and adds that the group of cobs associated with architecture (five out of seven) comes from the outer, northeastern part of the walls in the Huaca de los Ídolos, where "intrusive" pottery was found. The sixth fragment lay 23 cm below the surface, beside a ruined wall and to the east of the place where Willey and Corbett found the maize (see previuosly) in the "Aspero Temple." In this case the maize "... was superficial and late." The final sample was found on the outer part of a small hole with stone-lined walls (Feldman, 1980: 183–185).[32] On 4 September 1975, Feldman handed the samples of maize to the Laboratorio de Prehistoria of the Universidad Peruana Cayetano Heredia in Lima, which was under the direction of the present author.

Given the significance of these three preceramic samples, and considering that the full data were never published, I believe that these must be made public now. The information is included under the heading "Áspero maize" in an undated document already mentioned, which Feldman later gave me. Here the provenances of the maize in question are listed under the subheading "Midden Cuts" and read as follows: "As1D-1 = 2. From the top level (0–25 cm) of test

[32] See also Bonavia (1982: 359–360).

pit D, northern part of the site"; "As1V-3 = 3. Against a wall 23 cm below surface in a ruined room complex just west of the structure excavated in 1941 by Willey and Corbett"; and "As1V-4=5. From a test pit placed in the depression between the northern two pits of the 1941 excavation,[33] northern part of the site. Found about 10 cm below the surface." I here certify that in all three cases, Feldman himself wrote on the margin "Basural precerámico" (Preceramic midden) and "BC 2400–2000."

These preceramic samples were analyzed by Alexander Grobman. The report that he sent to Robert Feldman in August 1980 is cited here to the letter. It reads thus:

> An examination was made of three cobs from pre-ceramic levels, according to the information left with us at the time the corn material was deposited for study. All other cobs were left unstudied in detail since they are supposedly from ceramic horizons.
>
> The sample As1D-1 = 2 is a fragment from a cob that measures 4.2 cm long, is 1.2 × 1.1 cm wide, the width of the rachis is 0.4 × 0.6 cm. Its cupules have a length of 2.5 mm, a width of 5 mm and very few silks. The sample is cylindrical, 8-rowed, probably purple cob, and glume colored, and bears the morphological characteristics of an evolved form of Proto-Confite Morocho. A very slight fasciation would indicate racial introgression, and backcross to an 8-rowed form [Figure 5.14]. The other two cobs (As1V-4 = 5 and As1V-3 = 3) are larger, appear more evolved, and could be a transition to more advanced races. They exhibit ten rows of kernels, fasciation and cupules compressed along the longer axis of the cob [Figures 5.15 and 5.16].
>
> As1V-4 = 5 is 6.3 cm long, the cob has a width of 2.1 × 1.9 cm, the rachis is 12 × 0.8 cm wide, 10-rowed in irregular spirals, probably purple cob, overall cylindrical form. Its cupules have a length of 3 mm, a width of 7 mm, [and] very few silks. Marked fasciation, the glumes are hard and evolved. Sample As1V-3 = 3 is fragmented, its width is 1.9 × 1.6 cm, the width of its rachis is 0.9 × 0.8 cm, 10-rowed in spiral cylindrical form. The characteristics of its cupules could not be analysed due to its poor condition. It has soft and evolved glumes.
>
> The maize cobs as a group are no different from an average sample of cobs found in the Los Gavilanes (Huarmey) site in morphological characteristics. The reduced size of sample, corresponding to a pre-ceramic context, however, does not give good information on the possible existence at Áspero of earlier morphologically evolved forms, which do exist in the Los Gavilanes site.
>
> The other cob sample appears much more evolved morphologically. Their length and width is greater, they are definitely tripsacoid in terms of glume consistency. There is a high frequency of purple pericarp of kernels, and purple glume and cob color, which coincided with the observation we made at Huarmey. The extent of fasciation frequency is also high. There is considerable

[33] This obviously means the work of Willey and Corbett.

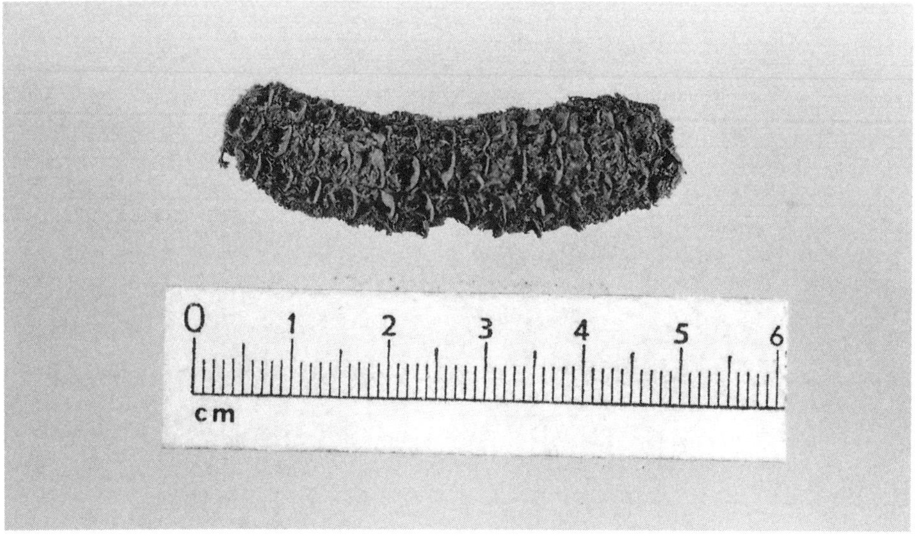

5.14. A fragment of a preceramic cob (AS1D-1 = 2) from Áspero. It is an evolved form of Proto-Confite Morocho. Photograph by Duccio Bonavia.

5.15. A fragment of a preceramic cob (AS1V-3 = 3) from Áspero. It is a more evolved race than Proto-Confite Morocho and may mark the transition toward more advanced races. Photograph by Duccio Bonavia.

heterogeneity among the cobs in the sample. In all, it could be said that these cobs represent a sample of a more evolved population, and definitely later than the sample of three cobs described above. The racial composition of this population is undefined, and they could well be precursors of the modern races of maize found in the area at present. (Grobman, Ms. 1980; see also Grobman, 1982: 176)

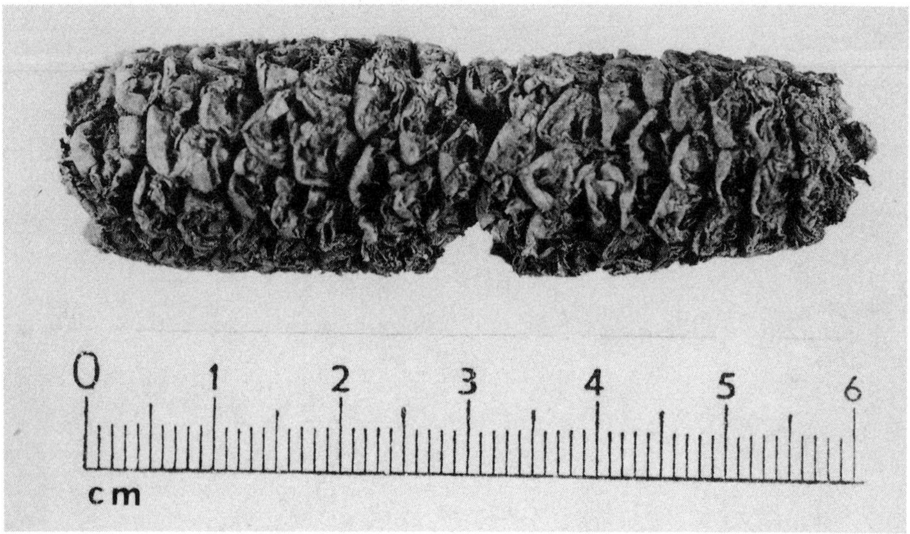

5.16. A preceramic cob from Áspero (ASIV-4 = 5) with 10 spiraling irregular rows, possibly of a purple cob color and an overall cylindrical form, with horny glumes. Like the specimen in Figure 5.15, this is a race that is more evolved than Proto-Confite Morocho and may mark the transition toward more advanced races. Photograph by Duccio Bonavia.

These materials were stored in the Laboratorio de Prehistoria of the Universidad Peruana Cayetano Heredia from 1975 to 1981, and on the orders of Feldman they were then handed over to Robert McKelvy Bird. I do not know their current whereabouts.

There is one significant fact that has gone unnoticed. In 1959, when Junius Bird wrote the prologue for the second edition of the famed handbook he coauthored with Bennett, he mentioned: "… the pre-ceramic maize growers of Áspero…." He then clearly made an observation that was ahead of its time: that pottery and maize "… *are not coeval throughout the Andean Area*" (J. B. Bird, 1960: 5; emphasis added). Since then almost all archaeologists have accepted the maize from Áspero (see, e.g., inter alia J. B. Bird, 1960: 5; Cohen, 1978: 259; Lanning, 1967: 68; Osborn, 1977: 182–183; Pearsall, 1978c: 395; Quilter, 1992: 114; Willey, 1971: note 61, 186).

The present author would like to emphasize the fact that Moseley insistently pointed out the preceramic maize from Áspero (Moseley, 1975: 80, 82, 89–90; 1978: 10) and then, for reasons that cannot be fathomed, changed his mind and questioned its preceramic status (Moseley, 1992: 21). But what cannot be explained is that, years later, Feldman would make the following statement: the "… excavated cobs … come from mixed or surficial [*sic*] contexts and cannot definitely be associated with the preceramic occupation" (Feldman, 1992: 72). With this, Feldman is trying to prove one of two things: either his work at Áspero was poorly done and he has misinterpreted the data – which if true would be quite serious, as it would question a dissertation defended at Harvard

University – or instead that he now has to adopt another position, for reasons of which I am not aware.

There is no point in wasting time discussing the position taken by Robert McKelvy Bird, which is weak, and lacks argument and is besides contradictory, for in an initial study Bird (1970: 48) did accept the preceramic maize from Áspero and then later rejected it (Bird, 1984: 43; Bird and J. B. Bird, 1980: 330). This last bibliographical reference requires an explanation. There can be no doubt that either Junius Bird was led into coauthoring this paper by his son or else he did not read it, because having met him and discussed with him the work done at Áspero on more than one occasion, I know that he did not doubt its preceramic status which, as we have seen, he recorded at an earlier date (J. B. Bird, 1960: 5; see previously).[34] I can certify that McKelvy Bird was well apprised of the preceramic context of the samples, for I have a copy of a letter dated 13 November 1980 that Feldman sent to McKelvy Bird, which among other things says the following: "These cobs came from 4 midden contexts and 3 architectural contexts, dating to 3,000–2,500 BC (corrected C-14 dates) period, [that is,] the *late Cotton Preceramic Period*. Three of these cobs were measured by Alexander Grobman" (emphasis added). Lathrap staunchly defended the work at Áspero precisely in regard to the work of Bird. Among other points he made the following one: "... it is difficult to imagine 'intrusion' as a likely explanation for an occurrence of maize that does not fit one's model" (Lathrap, 1987: 352).

The site now known as Caral, in the mid-Supe Valley, which had previously been known as Chupacigarro since the 1940s, when it was discovered by Paul Kosok (1965: 219, 220–223), has caused a most serious problem. This is one site that has been much talked about in recent years and over which legends have been spun, but the fact of the matter is that as yet not a single scientific report has appeared. A book was published on occasion of an exhibition displaying the findings made at Caral, which compiled the work Shady and her team had published in different journals (Shady Solis and Leyva, 2003). On examining these writings, the interested reader will realize that the claims made therein are completely unsupported and that, furthermore, the authors ignore the basic literature published on this subject. To give but one example: a study that pretends to discuss the origins of civilization in "the North-Central area and in the Supe Valley" ignores the "Culebras complex" that Lanning developed and that the present author has mentioned. Besides, the research done by Kosok, Willey and Corbett, Frédéric Engel, Williams, and Elzbieta M. Zechenter is not even mentioned (Shady Solis, 2003: 51–91). The book published in 2004 is meant for the public at large but has no scientific value whatsoever (Shady Solis, 2004).

[34] Interested readers who want more details should read Bonavia and Grobman (1999: 246–247).

And as far as maize is concerned, the situation is not just confusing: once again no support is offered to draw conclusions. Let us go over the facts. Two reports published in 2000 read thus: "The presence of maize is rare. Only two specimens were found, associated with the late occupation phases of Caral" (Shady Solis, 2003: 117). "A maize cob was recovered, which is in process of being identified" (Shady Solis and Machacuay, 2003: 185). The second case seems to be a mistake, because there were two cobs, but what matters is that none of the two texts present any information on the exact location of the find, the stratigraphy, or its associations. When Shady later published a study on Caral-Supe, she wrote the following: "The presence of maize is rare; only two specimens were found associated with the late occupation phases of Caral" (Shady Solis, 2001: 70). Yet when another paper was published this same year dating the site, it categorically claimed that "corn (*Zea mays*) is absent" (Shady Solis et al., 2001: 725).

This last claim not only contradicts the previous claims but it also refutes the evidence Shady gave to Grobman and me. In 2000, we were given to study first one cob and then a second one. We were told that these were in a "preceramic context" but were not given any additional information. The first specimen is just a fragment of the Confite Chavinense race, whereas the second is a whole popcorn cob of the Proto-Confite Morocho race, with eight-row boat-shaped cupules. Grobman jotted in his notes that it was not a very primitive specimen. This is certified by him (Grobman 2004: 467), who said the following of Caral: "Ruth Shady showed the author and Duccio Bonavia a fragment of a cob and a whole maize cob." I cannot tell whether these two specimens were included with those that were later published, and that shall be immediately mentioned. What is clear is that Shady had these data, for it was given to her on 28 October 2000, that is, before she published the 2001 paper in which she denies the presence of maize at Caral.

Yet a book on maize has recently come out (Staller et al., 2006) that includes an article by Shady that pretends to be "The History of Maize in the Land Where Civilization Came into Being" (Shady Solis, 2006: 381). The title clearly is far too pretentious, because the article does not outline the "history" of maize in Peru's North-Central Coast but limits itself just to the remains of this plant found at Caral and in a subsector of this site known as Miraya, which lies on the left bank of the Supe River, in the lower mid-valley to the south of Caral (see Shady Solis et al., 2003: figure 1, 55, figure 2, 56).

According to what is stated in this article, 15 cobs, 2 husks, and 1 tassel have been found at Caral and Miraya, and this is repeated three times (Shady Solis, 2006: 381, 387, 401). But it turns out that only 14 cobs from Caral and 2 from Miraya are described. There clearly is a mistake in the numbers. Once again there is no way of knowing whether these specimens included the two previously mentioned, whereas the articles cited previously in which the presence of maize is first stated and then is denied do not even appear in the bibliography of this new publication.

In the entire article we find just six sketches of the stratigraphy but no detailed and complete drawing, so the location of the specimens cannot be checked. Let us see now what evidence we are presented with.

"The oldest specimen of maize ..." comes from the Residential Sector A, Subsector A1, and "... was found in the construction *fill*...." This is a cob of the Proto-Confite Morocho race. "The second and third maize cob samples were associated with the *fill* that was sustained by the third south perimeter wall" (Shady Solis, 2006: 389; emphasis added). The "second specimen" is then described as a "race *Derived* From Confite Chavinense [*sic*]," whereas the third cob is from the race "*afin* a Proto Alazán" (Shady Solis, 2006: 389; emphasis added). As for their chronological status, it is stated that the specimens were in the "... strata of the Middle and Late periods of occupation of this subsector. The first sample is from a late phase of the Middle period (2300–2200 BC), and the second and third are from a late phase of the Final Late period," which according to table 28.1 corresponds to 2100–1800 years BC (Shady Solis, 2006: 389, 391, table 28.1, 382).

Shady then turns to Subsector A5 of Caral, where a "... maize cob, identified as belonging to a variety of race of maize that was a predecessor of the *Brown or Cusco race*, was in the *fill* ..." belonging to the last phase of the Middle period (2300–2200 years BC) (Shady Solis, 2006: 391; emphasis added). While describing the "Stratigraphic Interpretation of Sector A" we read the following: "The oldest maize, found in the two residential complexes of Sector A, shows a similar stratigraphic location; this indicates the same position in time because they are associated with the late phase of the Middle period (2300–2300 BC)" (Shady Solis, op. cit.: 391).

Later we are presented with Sector 12, which corresponds to the "Residential Units." Here it is explained that "one maize cob was found *in the fill*, which covered the floor of the sunken patio...." This, from a racial standpoint, is "... more developed Confite Morocho...." As for the chronology, it corresponds to the late phase of the final Late Period (2100–1800 years BC) (Shady Solis, 2006: 391; emphasis added).

A husk of maize was found in Sector H1, "The Gallery Pyramids," "... in a Late stratigraphic location, where some solid stone platforms were buried under several meters of *shicra*....[35] A specimen of maize was placed in the *fill* material ..." when these areas were covered over to build a platform. But a contradiction appears when the "context in which maize was found" is described, for then we read: "A maize husk was found with a small ear of maize, of an *undetermined race*, deposited in the *shicra fill* which completely covered Room 2 and served as a base for the construction of a platform, in association with the Gallery...."

[35] *Shicra*, also known as containment bags, retaining bags, or bagged fill, are woven bags made out of vegetable fibers. The bags are filled with stones and placed inside the buildings as filling. This construction technique appeared in the Preceramic period, was used in the Initial period and the Early Horizon, and was then discarded.

This corresponds to the final Late Period phase (2100–1800 years BC) (Shady Solis, 2006: 392–393; emphasis added).

Sector C, Subsector C2, east of the Central Pyramid, is then described. Here "the maize cob [*sic*] was found among the materials *of a fill* contained by the wall built of large stones. . . ." This is then contradicted when Shady explains that the specimen is ". . . a fragment of cob . . ." that belongs to the "afín a Confite Chavinense (related to Confite Chavinense [*sic*]). . . ." And when Shady establishes the "Chronological Correlation" we read thus: "The maize was found *among the fill materials* from a late phase of the final Late Period (2100–1800 BC) . . ." (Shady Solis, 2006: 393, 395; emphasis added).

Reference is then made to the Residential Sector NN2, where ". . . seven specimens of maize . . ." were found: "1) Three cobs. One of a race that shows affinity with the Confite Chavinense, another of the race that preceded the Brown (*Pardo*) and one more of an undetermined race were found *in the fill material* covering the floors of the outside patio of Housing Unit 3." Then ". . . 2) One cob was found *in the fill* covered by floor 3 . . . and is of an undetermined race. 3) Another cob, also unidentified, was found *in the fill* of the platform. . . . 4) Two cobs, of the *more developed* Confite Morocho race, were found in a hole, associated with *the fill* of the platform annexed. . . ." The chronological relation is with the phases of the final Late Period (2100–1800 years BC; Shady Solis, 2006: 396; emphasis added).

As for the settlement of Miraya, in Subsector C4 there were "three specimens of maize . . . two cobs and a husk. They were found *in the shicra fill* that covered Room 4. . . ." Yet it is immediately added that "[t]hey were *underneath a shicra bag* filled with stones . . ." (Shady Solis, 2006: 397; emphasis added). This is another flagrant contradiction.

The first cob is of the Confite Chavinense race whereas the second one, which is fragmented, was identified ". . . as an intermediate race *More Developed Confite* [*sic*]. . . . It has been suggested that it might possibly also belong to the race *Derived* from Confite Chavinense. . . ." The context corresponds to the final Late Period (2100–1800 years BC) (Shady Solis, 2006: 397–398; emphasis added).

Finally, in Sector C5 of Miraya a tassel of "undetermined race" [*sic*] was found on the floor of a small room. It corresponds to the final Late Period (2100–1800 years BC) (Shady Solis, 2006: 398–399).

We now have to discuss some of the statements made by Shady. She writes that "the oldest cobs from the settlement of Caral and Miraya have been recognized as belonging to the racial types Proto-Confite Morocho and Confite Chavinense." Shady then acknowledges that there are specimens of races that are more developed than the previous ones, ". . . such as the Proto Alazán, derived from Confite Chavinense, more developed Confite Morocho and another race that was the predecessor of the Brown [Pardo] or Cusco" (Shady Solis, 2006: 399).

Shady insists that "the specimens come from the layers *of construction fill* of both residential and public buildings" (Shady Solis, 2006: 399; emphasis added).

Then it is claimed that maize appears late in Caral, with a "... small representativity in comparison with other cultivated plants, [which] indicate[s] that maize production was not destined for the daily diet of the people, nor did it have a relevant role in the formation of civilization." The same thing is repeated with almost the same words in a subsequent paragraph (Shady Solis, 2006: 401). It is then added that as from the late Middle Period phase (2300–2200 years BC), "... maize continued to be consumed until the last phase of occupation, but only in small quantities for ritual purposes" (Shady Solis, op. cit.: 391). These claims cannot be verified because the list of plants presented by Shady (2006: table 28–4, 387) is just that, with no quantitative indications, and so is of little value.

Before beginning a discussion of what has thus far been presented, and to avoid misunderstandings, it is worth explaining that the specimens of maize were shown to Alexander Grobman, who made a very quick and preliminary identification. He was never sent a draft of the manuscript to check it, nor did he receive the final published text. Shady and her team thus worked with observations that required a subsequent, in-depth analysis, for example, the study of the cupules, and the indices of fasciation, a description of the glumes, and so forth (Alexander Grobman, personal communication, 26 August 2006). Even so, Grobman stands by the racial classification he made, but he does not assume any responsibility for the arbitrary interpretation Shady and her team made through ignorance (Alexander Grobman, personal communication, 1 May 2007).

Some comments are in order as regards this article by Shady. First, there is no description of the contexts or of the associations, nor has any clear stratigraphy been published that may allow the reader to understand the provenance of the specimens. Without this the study is worthless. Second, there is not a single absolute date that would allow one to have an idea whether or not the relative chronology presented is valid. Besides, it is almost impossible to relate the contexts discussed in this article with the dating that Shady and colleagues (2001) had previously presented. Finally, and worst of all, the samples come from fills; this shows the site had several reoccupations that the excavators apparently have not understood, in which materials from several epochs are jumbled. Remains from fills are always questionable. This is corroborated by the data available here for maize. The lack of any grounding in ethnobotanical studies, specifically regarding maize, in the knowledge base of the team that worked with Shady and in Shady's own knowledge base is clear, and this prevented them from seeing the incongruities found in her article.

I discussed the comments made here on maize with Alexander Grobman, so these are ideas we both share. If an analysis is made of the races present in Peruvian preceramic sites, they turn out to be clearly limited to just three races: Proto-Confite Morocho, Confite Chavinense, and Proto-Kculli, as well as their hybrids. In the list that Shady presents, there are just two samples that can be defined as preceramic; one comes from Sector A, Subsector A1, and is a Proto-Confite Morocho (Shady Solis, 2006: 389), and the other is from Miraya, Subsector C4, and is a Confite Chavinense (Shady Solis, op. cit.: 397). All the rest are definitively later. Let us look at the facts. A maize "Derived From Confite Chavinense" was found in Subsector A1 of Caral (Shady Solis, 2006: 389). In technical terms, "derived" means that a race had its origin in another, and this takes time. Then the presence of an "afín a Proto-Alazán" is mentioned in the same context (Shady Solis, op. cit.: 389). This means that it is a more evolved Proto-Alazán. There are remains of a Proto-Alazán race mixed with Pagaladroga in a Middle Horizon site (PV35–4) in Huarmey (Bonavia et al., 2009: 256–257). To the best of my knowledge, it was only in Puémape (which corresponds to the Initial period) that a single specimen of a Confite Chavinense hybrid "derived from the Alazán race" was found (report of Alexander Grobman to Franco Léon del Val, 1 August 1992, copy of which is in the possession of the present author). In other words, this race was not yet defined in the Initial period. The Proto-Alazán is a race that coexisted alongside the ancestral forms Mochero and Pagaladroga and was typical of the Early Intermediate period on the north Peruvian coast (Grobman et al., 1961: 236).

It is then claimed when discussing Subsector A5 of Caral that there was a "Brown or Cuzco" cob (Shady Solis, 2006: 391). This actually is a mistake, for the reference is to the Pardo race (a generic term used by Grobman [Grobman et al., 1961: 302–306] that does not mean the color and cannot therefore be translated). First, this is a race that did not exist on the coast in preceramic times. If it is a predecessor of the Pardo, it is very late, probably Late Horizon. Even this is not fully certain, and it is possible that it was introduced into the early colonial period (Grobman et al., 1961: 306). And this is confirmed because the cob has "kidney-shaped cupules" (Shady Solis, 2006: 391).

A "more developed" Confite Morocho was found in Sector 12 (Shady Solis, 2006: 391). This clearly is a descendant of the Proto-Confite Morocho. It certainly is a more evolved race than that which was found in the Preceramic period. In Sector C, Subsector C2, a cob was classified as "afín a Confite Chavinense" (Shady Solis, 2006: 393). This means that it is also a more evolved Confite Chavinense. Similar evidence appears in the Residential Sector NN2, which comprises a cob that has "affinity with the Confite Chavinense," another one that is "Brown (Pardo)," and finally a third one that is the "more developed Confite Morocho race" (Shady Solis, op. cit.: 396). In all three cases we are before specimens whose preceramic status is quite doubtful.

In Subsector C4 at Miraya, a cob was found that is "an intermediate race *More Developed Confite*," and "it has been suggested that it might possibly also belong to the race Derived From Confite Chavinense," that is, that it is more evolved than Confite Chavinense (Shady Solis, 2006: 397–398).

Shady's ignorance is evinced even more when she states that "a panicle of maize of an undetermined race ..." was found on the floor of a small room (Shady Solis, 2006: 398), for it is well known that panicles that are not found in association with other remains of maize cannot be classified at the racial level.

All of the discussion by Shady (2006: 399–401) regarding the interpretation of the maize remains is based on five references, with no page numbers given. She shows she has not read the original literature and has instead based her work on the writings of others, from where she took the references. For instance, Shady ascribes the discovery of maize in "Las Haldas in Casma" to Lanning (1963). First, Las Aldas is not in Casma, and second, although it is true that Lanning did find maize in this settlement, the article cited analyzes the preagricultural occupations in the Central Coast, so Las Aldas is not even mentioned. Shady then writes: "The same author [Lanning] mentioned maize cobs were found in Pre-ceramic Period strata at the site of Culebras in Huarmey [1]" (Shady Solis, 2006: 399). Here Shady says that Lanning mentioned the findings made at Culebras, but her note [1] cites Bonavia; moreover, Culebras lies in the homonymous valley north and not in the Huarmey Valley. Her ignorance is certified by figure 28.1 (Shady Solis, op. cit.: 383), in which the location of the sites of Culebras and Los Gavilanes (in Huarmey) is mistaken.

When discussing the finds made at Cerro (El) Calvario and Cerro Julia without giving any bibliographical reference, Shady says these were "large excavations," which is not true (see previously). She also claims that "... according to Grobman, [the samples found there] are of the *same family* [*sic*] as the maize from Huarmey and the Callejón de Huaylas" (Shady Solis, 2006: 399; emphasis added). No more comments are needed, other than to note this shows ignorance of the basic principles of botany. Besides, this is something Grobman never said and that is nonetheless attributed to him. Shady mentions other sites (e.g., Guitarrero Cave) in the same way, without giving any bibliographical reference.

Until a full report of the work made at Caral is published, with good pictures of the stratigraphy and specific data on the finds, with their respective dates, all the information regarding the maize from this site cannot and must not be taken into consideration.

To finish with the coastal evidence, I would like to present some information I received directly from Lanning (personal communication, 7 June 1980). During the research he made at Ancón in the 1960s, Lanning found some coprolites in the strata corresponding to what he called Playa Hermosa (i.e., the early Final Preceramic). Maize pollen was detected when analyzing one of them. Lanning explained to the author that he believed this was maize cultivated in the adjacent valley or somewhere else, and that it was probably transformed into flour, which was then transported. For him this was one of the best ways of transporting pollen. One can accept or reject Lanning's explanation, which without supporting evidence stands as mere speculation, but what is important here is that maize was found in association with a preceramic context. Lanning unfortunately never published the evidence. Patterson (1971) did not mention maize at all in an overall presentation he made of the Central Coast, but he did include it in a table showing the main preceramic plants. Because Patterson worked with Lanning, it can be assumed he meant the aforesaid evidence (see Bonavia, 1982: 359).

Let us turn now to the evidence available for the Peruvian highlands. In a preliminary report, Burger and Salazar-Burger (1980) indicated they had analyzed carbon isotopes in human bone, and that the results showed that the diet in the preceramic Callejón de Huaylas, and specifically at the site of Huaricoto, included maize. Burger and Van der Merwe then wrote that the 13C value obtained in samples corresponding to the late Preceramic (Chaukayán phase) was 18.9%, thus "... suggest[ing] that the diet of the late Preceramic occupants of Huaricoto ..." included about the same amount of maize consumed by those who lived in Chavín de Huántar 1,500 years later. They add: "It also provides independent evidence that the people responsible for the earliest shrines at Huaricoto were probably farmers, and that maize was among the crops consumed." Burger and Van der Merwe finished by stating that "the isotopic analysis of the Chaukayán-phase sample from Huaricoto confirms that maize was already being cultivated in highland Peru during the late Preceramic" (Burger and Van der Merwe, 1990: 91).

The case of Guitarrero Cave poses serious problems in its stratigraphy. Lynch himself, who excavated it, quite honestly pointed this out in his report (Lynch, 1980a). The reader is referred to Bonavia (1982: 366–367) so that he or she is not tired and can find an overall initial presentation of this issue. I later went over this problem with Grobman (Bonavia and Grobman, 1999), and given the controversies Guitarrero gave rise to, it is worth discussing this subject. Of essential interest here is all that is related with Complex III, from whence the maize that has been called into question comes. Lynch (1980b: 40) acknowledges that it "... is thoroughly enigmatic ... ," but claims that "it is essentially preceramic in content and stratigraphically superposed to Complex II.... Nevertheless there were hints of disturbance and possible contamination and redeposition." Lynch presents two possible interpretations: either Complex III is essentially a Complex II that was excavated in antiquity and is "minimally contaminated" with more recent materials or Complex III is a fully preceramic component that followed Complex II with fewer signs of contamination (Lynch, 1980b: 41). Lynch later concludes that "whatever the beginning date for Complex III, the corn recovered from that stratum must belong with the preceramic materials" (Lynch, 1980c: 305). Lynch's reasoning is that, besides the morphological characteristics pointed out by C. Smith (1980b),

> ... it may be more significant that the cobs from excavation units 35, 36, and 37 of grid square B2 display hints of a morphological progression that

> corresponds to the internal stratigraphy of Complex III. If the cobs were intrusive from Complex IV, this would be a most improbable outcome. It is also significant that the slim cobs of Complex III show no clear relationships with more modern races of Peruvian corn, as would be expected in the case of modern mixture and intrusion. Similarly, Kautz [1980: 49–51] notes that the pollen evidence from Complex III integrates exceedingly well with the pollen record from Complex II below it. This would be unlikely if there had been substantial mixture and intrusion of plant remains. (Lynch, 1980c: 305)

On the basis of the botanical data, Lynch believes that Complex III must be considered as a unit, and that in chronological terms it must be placed at the end of Complex II and at the beginning of the early material from Complex IV. He finishes by stating that "... we may assume that Complex III is basically a primary deposit, to which all or most of the corn belongs, but that it is minimally contaminated ..." by two fragments of textiles that correspond to burials from ceramic times (Lynch, 1980c: 306).

Smith, who studied the botanical materials, admitted that dating Complex III is difficult and discussed the proposal made by Lynch, in that Complex III is part of Complex II mixed with later materials. But he added, "However no ceramic sherds were found in Complex III fill ..." and pointed out that the place where the maize was found "... did not seem to be disturbed ..." (C. Smith, 1980b: 122, 138). But his botanical argument is significant: "Inasmuch as the cobs make a morphologically earlier series than the cobs from Complex IV, they may represent a late preceramic occupation.... In view of the difficulty in dating material from Complex III, the morphology of the cobs must stand as a firm indication of the antiquity of Complex III maize over Complex IV maize" (Smith, 1980b: 122, 138). Grobman agrees with this statement.

Kautz analyzed the pollen. When he mentions the results obtained from the sample that corresponded to what he called the "Pollen Zone 3," which included Complexes II and III, he quite clearly says that "... with the exception of only one category (Alstroemeriaceae), the pollen evidence from the stratum [Complex III] integrates exceedingly well with the pollen evidence from immediately below it [Complex II] ..." (Kautz, 1980: 49).

The only one who has criticized the work done by Lynch with firsthand experience is Vescelius (1981a, 1981b). However, those who have used his argument to question the evidence from Guitarrero Cave have not read his papers carefully and are not aware that Vescelius also made a slip. He wrote: "Under any circumstance, Complex III itself is likely to be an aggregate of early preceramic and Early Horizon or post-Early Horizon materials, so that we have only two options: either the corn cobs from that unit date from the seventh millennium BC, or they date from the first millennium BC or thereafter" (Vescelius, 1981b: 11). In other words, all that Vescelius did was repeat what Lynch had pointed out with great honesty. Yet on taking this position, he simply rejected the first option without presenting any argument, which is not scientific – it is just an opinion. But as regards the maize from Complex III, Vescelius does admit that "perhaps they are a bit more primitive in character than the cobs from Complex IV" (Vescelius, 1981b: 11).

It is interesting that the complex that Vescelius least discussed is Complex III, of which he claimed there "... is no good reason to suppose that it is anything more than a mixture of Complex II and Complex IV soil and refuse ..." (Vescelius, 1981b: 9). But despite having carefully read the articles by Vescelius, I do not find solid arguments with which to refute Lynch, as regards his position on Complex III. His claim that "... I am inclined to doubt that it [the maize from Complex III] dates from anytime prior to the middle or the first millennium BC" (Vescelius, 1981b: 13) is an honest opinion based on a logical reasoning, but Vescelius does not have enough arguments.

Following the critique leveled by Vescelius, Lynch and colleagues (1985) reanalyzed the chronology of Guitarrero Cave with the AMS method. The result was that "the accelerator dates support the antiquity of the Guitarrero artifacts" (Lynch et al., 1985: 864). This removed the objections raised by Vescelius (1981b) as regards Complexes I and II. As for Complex III, Lynch and colleagues concluded that it "... consists of restratified material from Complex II that has been *minimally* contaminated by the modern remains from badly mixed Complex IV." And when discussing the upper part of Complex IIe, they again state that it "... might be reassigned to *minimally* mixed Complex III" (Lynch et al., 1985: 865; emphasis added). In other words, they restate what Lynch stated in his final report (see previously).

What I find striking is that at the end, Lynch and colleagues (1985: 866) conclude that "maize, which was found only in Complexes III and IV, may be less than 2,000 or 3,000 years old...." I have the impression that this statement was hurriedly made under the pressure of the critiques by Vescelius (1981b), R. McK. Bird (1987, 1990), and all who have followed them. No one can doubt there are problems in Complex III, and I said this right from the beginning (Bonavia, 1982: 366–367). But I would like to draw the attention of my colleagues to one specific fact. Lynch excavated the cave and admitted the presence of a mix in this complex with the intrusion of more recent materials from Complex IV, yet he has repeatedly insisted that this context was "minimally contaminated" (Lynch, 1980b: 41; 1980c: 306; Lynch et al., 1985: 865). At the same time it is clear that no pottery was found in Complex III (Lynch, 1980b: 40–42; Lynch et al., 1985: 866; C. Smith, 1980b: 122).

Now, if they claim that in this "minimally" contaminated context there are later remains from the upper stratum, then all the maize is intrusive (i.e., the 26 or 27 cobs).[36] Here one thing the critics have missed has to be pointed out. The maize remains from Complex III were found in three excavated units labeled

[36] In table 6.1 C. Smith (1980b) indicates 26 specimens, which he repeats on page 125, but on page 138 we find 27; it can be assumed this is a lapsus calami.

"Samples 35, 36, 37." Most of the cobs come from Sample 35, which stratigraphically is the uppermost, but two come from 36 and one from 37, which are lower. It is worth recalling here that according to Smith, the cob from sample 37 is the most primitive one (C. Smith, 1980b: 125). Had these specimens all been together, it would perhaps be possible that they somehow came from the upper strata. But it is hard to accept a selective intrusion of isolated samples of maize, and all the more so if we analyze the botanic aspect, which openly contradicts this possibility (Smith, 1980b: 112). I wonder: why did the cobs not get also mixed up with pot sherds? I discussed this with Lynch, and he answered thus:

> It is merely a reasonable, but untested, hypothesis that the cultigens in [Complex] III came from Complex IV. With all the radiocarbon tests in hand now, it is difficult to argue that Complex III has much integrity, but it could be a combination of remains of various ages. Kaplan's dates on the beans ... show that they are not all of the same age. Similarly, the maize cobs could be from [contexts from] two or more ages. C. Earle Smith argued that, morphologically, the cobs from Complex III were significantly different, as a group, from the much larger sample collected from Complex IV. *And it has always troubled me that no potsherds were found in Complex III*; 26 or 27 cobs might be expected to have brought along a pretty good sample of pottery as well. One might easily argue *that 26 cobs is not "minimal contamination" – or at least not so minimal that a good sample of potsherds would not also be present if the source were of ceramic age....* Gary's [Vescelius's] argument that there was only a single, relatively short occupation during Guitarrero II makes sense with the new dates, *but Complex III might still be something on its own, rather than a simple mechanical mixture of II and IV.* (Lynch, letter to the author, 7 March 1996; emphasis added)

I tend to believe that there may be a mix in Complex III, which includes preceramic maize from this complex, and others that may come from Complex IV. I thus repeat the position I have always held with Grobman, which we summarized in one of our papers (Bonavia and Grobman, 1989a: 839) and on which we later insisted (Bonavia and Grobman, 1999: 248–250). In this regard we concur with Aikens (1981: 225), for whom "... there seems little doubt that the earlier specimens [of maize] are preceramic."

For the benefit of the reader, the dates obtained with the traditional C14 method are as follows: for Complex III, 7730 years BP and, for Complex IV, 2315 and 8225 years BP. Complex IIe was dated to 7575 and 8175 years BP. The date obtained with the AMS method for Complex IIe was 9600 years BP (Lynch, 1980b: table 2.1, 32; Lynch et al., 1985: table 1, 865).

As regards the botanical aspects, the reader is referred to the study by C. Smith (1981b) and to the comment made by Grobman (1982: 176); they do not agree on all points, but there clearly is a racial relation with the maize from Los Gavilanes. In his most recent paper, Grobman (2004: 445) said of the maize in Complex III that "its racial context seems to be identical to that of the coast at Los Gavilanes...."[37]

I have already discussed elsewhere the problems regarding the finds made at Ayacucho (see Bonavia, 1982: 363–366), but I did not then have the information now available. Explaining the situation here is crucial, for several reasons. First of all, these finds have been cited innumerable times without any real understanding, and this has given rise to many mistakes. Second, the volume on botany that was going to be published in the series in which all the other reports of the Ayacucho Archaeological-Botanical Project appeared, and to which I will refer, will not come out, and so much data has remained unpublished. Third and last, Richard MacNeish, the author of these reports and the man who headed the project, has passed away. Part of these data should have been included in the section discussing the sites to which the discovery of maize has erroneously been ascribed, but it was instead decided to keep all of the data regarding the Ayacucho sites together to preserve the context, and because there often is a relation between them.

To tell the truth, it was Alexander Grobman (1974: 3) who suggested to MacNeish that he study the Ayacucho zone.

The case of this project is problematic because only three volumes of the final report have been published, and the one on botany has not appeared, as has already been noted (MacNeish, Nelken-Terner, and Vierra, 1980; MacNeish et al., 1981, 1983). Some data have already been presented in the preliminary reports (García Cook, 1974: 21, 24; MacNeish, 1969; MacNeish et al., 1970), but they are not considered – except for the work of García Cook – because the final reports are a more extensive source.[38] The data included in the three volumes of the final report are, however, chaotic and contradictory and have been presented in a most confusing way.

Walton Galinat was able to examine the Ayacucho maize, and he had the advantage of using the data of the provenance, association, and stratigraphy that were given him by MacNeish and his team. What proves striking here, as we shall now see, is the discrepancy between these data and those that figure in the final reports. Galinat gave a copy of this manuscript study to Grobman in 1973, when the latter went over the Ayacucho maize. We were unable to use it for many years, because we did not know the codes used to define the sites and their stratigraphy. With the help of the final reports, Grobman and the present author managed to reconstruct the data in the 1990s, and it is used here with the permission Galinat gave Grobman (letter to Grobman, 6 February 1996). Additional help came from a manuscript by Walton Galinat, prepared for volume 1 of the final reports, which was never published and which Richard

[37] Shady (2006: 399) tried to refute the findings made at Guitarrero Cave with the isotopic analyses from Chaukayán and Huaricoto but without any real knowledge and without even giving a single bibliographical reference.

[38] The final reports were discussed in Bonavia (1982: 363–366).

MacNeish sent me in 1997 (letter to the author, 24 November 1997), giving me permission to use it.

Galinat's manuscript report indicates the provenance of the samples by site and by levels and groups them into "good," "medium," and "poor" contexts. In some cases the corresponding phases are indicated, and the maize is grouped by race according to Galinat's classification. In this specific case, because we are essentially interested in establishing whether or not there is a secure preceramic maize in Ayacucho, the racial aspects will not be considered, even though these will be mentioned further on. It must likewise be explained that in this manuscript the cobs are grouped in one table and the stalks, husks, and tassels in another table.

The first site in question is Pikimachay (Ac 100). Here in Zone F (Cachi phase; MacNeish, 1981a: 53), according to MacNeish (1981b: 203), a cob was found in "the final preceramic ... occupation ..." (Occupation 27; MacNeish, 1981a: 55). The data coincides with that of Galinat, who also points out the presence of a stalk in a context that he calls "good." Zone G also corresponds to the Cachi phase, Occupation 26 (MacNeish, 1981a: 55; MacNeish and Vierra, 1983: 163). Here the presence of a husk and a cob is indicated, but intrusions caused by rodents are acknowledged (MacNeish and Vierra, 1983: 163). And here we have a problem. Galinat's notes mention a "Zone g [*sic*]." It turns out that this corresponds to the southern part of the cave and has a date of 9000–7000 years BC. This must be a mistake made by Galinat, and it probably was Zone G, which is also a disturbed zone but corresponds to the Cachi phase (MacNeish, 1981a: figures 2–10, 30). In this context Galinat indicates the presence of a stalk found in a "good" context, and 119 cobs in a "bad" context. Yet for Zone G, the manuscript MacNeish later sent me mentions 5 cobs in a "bad" context and 1 in a "medium" one. At present the truth cannot be established.

Zone VI also corresponds to the Cachi phase, Occupation 25 (MacNeish, 1981a: 55, 56; MacNeish and Vierra, 1983: 160). Here MacNeish and Vierra (1983:160) say "maize" when mentioning plants and feces, whereas MacNeish (1981a: 38) says "vegetal materials." For Galinat, on the other hand, there are two cobs in this zone, two husks and a tassel in a "good" context, and two stalks and a tassel in a "bad" context. This is confirmed in Galinat's final manuscript. But in this case we find an additional problem. In the report, MacNeish (1981c: table 6-9) mentions the presence of "maize" that corresponds to the Cachi phase in a zone he calls "V1." This presumed zone does not exist in the reports, so it must be a typo for VI, in which case the data in Galinat is confirmed. Zone H also corresponds to the Cachi phase, Occupation 24 (MacNeish, 1981a: 53, 55, 56; 1981c: figures 6–9; MacNeish and Vierra, 1983: 160). According to MacNeish and Vierra (1983: 160) "... a tassel of corn, corn leaves, [and] two corn cobs ..." were found here. MacNeish (1981c: table 6-9) once again says "maize." Yet we find a contradiction with the data in Galinat, because his table (Galinat's) shows the presence of a cob and two husks in a "good" context.

Zone VII belongs to the Chihua phase, Occupation 23 (MacNeish, 1981a: 53, 55, 56; MacNeish and Vierra, 1983: 160;), and it is clearly stated that it essentially is a rockfall with "... a considerable number of crevices and hollows between the rocks and obvious intrusions by rodents ..." (MacNeish, 1981a: 38). Here in two feces there were "... possible corn fragments ..." (MacNeish and Vierra, 1983: 160), "maize" (MacNeish, 1981c: table 6-8), and "corn fragments" (MacNeish, 1981c: 163). The data in Galinat disagree, for his notes indicate the presence of two cobs and a husk in a "good" context, of three cobs in a "medium" context, and of a cob, four stalks, and two husks in a "poor" context. Yet the later manuscript gives two cobs in a "mediocre" context and one in a "good" context. In this case the doubt persists.

Finally, on reading the report, it follows that Zone VIII, which also belongs to the Chihua phase, Occupation 22 (MacNeish, 1981a: 55, 56), is a zone with intrusions from later zones (MacNeish, 1981a: 38–39, 40). It is said that there was maize here (MacNeish, 1981c: table 6-8). We once again run into a contradiction with the data Galinat had in his notes, for he reports the presence of three cobs and a husk in a "good" context, a husk in a "medium" context, and three stalks, two husks, and two tassels in a "poor" context. The manuscript, however, mentions three cobs in a "poor" context, as well as four kernels whose context is not qualified. Here the doubts also linger.

In his manuscript Galinat also points out the presence of a cob found in Zone Xd in a "poor" context that I was unable to locate, and 143 cobs whose provenance is not mentioned, and which were in a "disturbed" context.

Another site with serious inconsistencies is Big Tambillo (Ac 244). Four zones are indicated here. In the first three – A, C, and D – there was maize (but there are discrepancies between the data in Vierra 1981: 134–136, 138, and Galinat), but it will not be discussed, as it corresponds to ceramic contexts. The only preceramic zone is E. Vierra (1981: 136) is quite clear in not pointing out botanic remains, and just pointing out scant cultural ones. He assigns them to the Cachi phase. Galinat, however, indicates the presence of stalks and two husks in a "good" context.

The case of the Cueva Tambillo Boulder (Ac 240) is pathetic. Excavations were begun, but to avoid carrying the materials, the bags with all of the botanic materials recovered were left behind, hidden. The peasants who arrived later not only stole them but also destroyed the site (MacNeish and Wiersum, 1981: 128). No comment is needed. MacNeish and Wiersum wrote the following in this regard: "... zone H, containing a considerable number of corn cobs and other plant remains associated with possible late preceramic Cachi remains. Further, there was a chance of earlier preserved plant remains under Zone H ..." (MacNeish and Wiersum, 1981: 128). They then add that on their return, the rubble from the excavation made by the peasants had many plant remains, including cotton, maize cobs, and squash (MacNeish and Wiersum, 1981: 129). In fact, when summarizing the "occupations," they said "corn cobs"

(MacNeish and Wiersum, 1981: 128). This is confirmed when MacNeish reiterates the presence of "... very late-type corn cobs ..." (MacNeish, 1981b: 203). Yet when mentioning the site, MacNeish and Vierra (1983: 182) assign Zone H to the Cachi phase but only mention "... a possible corn cob ... ," whereas MacNeish (1981c: table 6-9) merely writes "corn." Galinat only saw the materials from the upper stratum G with pottery and does not mention stratum H.

Another serious problem is that raised by the Puente site (Ac 158). MacNeish (1981b: 203) indicates the presence of "... very late-type corn cobs ..." in the IIc context that García Cook and MacNeish (1981: 107) date to 4610 years BC, or between 4725 and 4325 years BC (García Cook and MacNeish, 1981: figures 4–10). There is another ambiguous phrase of MacNeish's (1981b: 203) in this regard: "Only zone H of Ac 240, with very late-type corn cobs like zone F of Ac 100 and zone IIc of Ac 158...." It is known that Zone IIc is a thin stratum 10 cm thick and only 4 m² where some artifacts were found, and which has been considered as perhaps a one-man occupation (corresponding to Occupation 20) (García Cook and MacNeish, 1981: 99, 109). On the other hand, García Cook and MacNeish (1981) never mention finding maize. The site is not recorded in the notes taken by Galinat.

Finally there is an additional site that I believe is important: Rosamachay (Ac 117), specifically Zone D, which corresponds to the Chihua phase, and for which context "the corn cob ..." is twice mentioned (MacNeish and Vierra, 1983: 179). MacNeish and García Cook (1981: 123–124) wrote the following in this regard:

> Because [of the presence] of a corn cob and late preceramic tools, a sample of charcoal from zone D was sent for radiocarbon determination. The date was 3300 ± 105 BC radiocarbon years (I 5688) [5250 ± 105 BP; Ziólkowski et al., 1994: 332]. Since this seemed too early for corn in Peru, another sample was sent that included a piece of corn leaf. This sample was dated at 3520 ± 110 B.C. radiocarbon years (I 5685) [5470 ± 110 BP; Ziólkowski et al., op. cit.: 332]. Thus, we changed our minds about the antiquity of corn in Peru and shifted the date for the appearance of corn back from the end of the Chihua phase to about 3100 BC.

The finding of a cob is then repeated (MacNeish and García Cook, 1981: 124). Yet on one occasion MacNeish (1981c: 163) uses the plural: "corn cobs." He ratified this and wrote (MacNeish, 1981b: 213):

> Since the artifacts from zone D of Ac 117 were not numerous, it was difficult to put the zone in its correct chronological position. Because it contained some of Peru's earliest corn, we sent in [to the laboratory] a piece of carbon from square N1E4 on the top of the zone. The date came back as ranging from 3405 B.C. to 3195 B.C. (I 5688) [5250 ± 105 years BP; Ziólkowski et al. 1994: 332]. Since we did not believe it because it was so old, we sent

> on I 5685, carbon from square N2E4 of the same level, which dated from 3630 B.C. to 3410 B.C. [5470 ± 110 BP; Ziółkowski et al., op. cit.: 332], *confirming the previous date* as well as putting zone D in its correct alignment. (emphasis added)

Yet in his notes, Galinat confirms there was only one cob in Zone D from a "good" context. And in the (unpublished) final report he wrote: "This specimen is not only the best dated but also in good cultural contexts" (p. 8 of the manuscript in Bonavia's files). So this definitely was only one cob, and MacNeish got mixed up.

The general reference made to the Chihua phase insists on the presence of maize. For instance, when mentioning fecal analyses, we read that "... at the very end of the phase [there was a] primitive Ayacucho-type corn" (MacNeish, Nelken-Terner, and Vierra, 1980: 10); "... a number of corn cobs and leaves ... during late Chihua times ..." are also mentioned (MacNeish and Vierra, 1983: 158). Finally, MacNeish and Nelken-Terner (1983: 10) conclude that "... our plant remains are few, and only a dozen or so feces have been analyzed,[39] but there does seem to be some evidence that, in addition to the gourd, squash, and quinoa used in the previous phase, the occupant had now acquired common beans, achiote, tree gourd [*sic*], lucuma, coca, perhaps potatoes, and, at the very end of the phase, primitive Ayacucho-type corn." Now, the Chihua phase is given a date that falls between 4400 and 3100 years BC. In the case of Pikimachay, Zone VII has a date of 3350 years BC, and Zone VIII, 3600 years BC (MacNeish, 1981b: table 8-10). We saw in Rosamachay that Zone D has a date of 3300 years BC, and another of 3520 years BC (MacNeish and García Cook, 1981: 123–124).

In regard to the Cachi phase it is clearly stated that in the case of Pikimachay, "foodstuffs and feces ... included corn ..." (MacNeish, Nelken-Terner, and Vierra, 1980: 11). MacNeish insists on a "... horticultural subsistence pattern using corn ..." (MacNeish, 1981b: 222), and then with Nelken-Terner he writes that "foodstuffs and feces ... include corn ..." (MacNeish and Nelken-Terner, 1983: 11). The date assigned to the Cachi phase ranges between 3100 and 1750 years BC, and in the case of Pikimachay, Zone F is given a date of 1900 years BC, G of 2200 years BC, VI of 2250 years BC, and H of 2300 years BC. As for Cueva Big Tambillo, Zone E has a date of 2300 years BC, and Zone H, in Cueva Tambillo Boulder, is dated to 2800 years BC (MacNeish, 1981b: table 8-11). And when discussing the Chihua phase, MacNeish and Vierra (1983: 185) point out the finding of maize in Zones K and X, which certainly is a mistake, as this does not appear in the report (MacNeish, 1981a: 34, 43, 48, 55).

I therefore conclude that of all the sites excavated by the Ayacucho Archaeological-Botanical Project, the only secure find of maize corresponds to the site of Rosamachay, and its date places it in Lanning's (1967: 25) Period IV. The

[39] Nothing has been published on them.

problems raised by Tambillo Boulder cannot be solved because the only evidence available is the word of the excavators.

As regards maize, Galinat (1972: 108) wrote that the archaeological data from Ayacucho are confusing, a point that confirms all that has just been stated here. When discussing maize, besides the ancient known races, Galinat proposes a new one he calls Ayacucho, which is "… the more primitive and ancestral to several indigenous races in Peru" (Galinat, 1972: 108). Flannery, based on a study by Galinat and MacNeish that was apparently never published, mentions the maize from Rosamachay and defines it as a "… teosinte-influenced race, most closely related to Mexico's Nal-Tel and presumably introduced from that region" (Flannery, 1973: 302). At about the same time García Cook stated, when discussing in general the Ayacucho maize from the Chihua phase, that these were "… primitive types of maize, perhaps the ancestors of one of the most ancient species [*sic*] of the current ones in Peru – Confite Morocho – that according to Dr Galinat (MacNeish et al., 1970: 38) could be pointing towards an independent domestication of maize in the Andean highlands" (García Cook, 1974: 21). The truth is that the above-cited report by MacNeish and colleagues has nothing of what García Cook mentions; Galinat does not even appear, and maize is only vaguely mentioned on pages 42 and 44. Yet in this same study García Cook (1974: 24) claims that "… maize agriculture was practiced" in the Cachi phase.

Alexander Grobman examined the Ayacucho maize in 1973. With the permission of Galinat and in his presence (see Grobman, 1974: 3; 2004: 446), Grobman reclassified it into two races, Proto-Confite Morocho and Confite Chavinense. He made the following comments:

> In the case of Ayacucho, the specimens were initially assigned to the Proto-Confite Morocho race [lapsus calami: Confite Morocho in the original manuscript] (MacNeish et al., 1970, pp. 38); Galinat then referred to Confite Puneño and Confite Morocho, also establishing the presence of some hybrids from these two races, and some intermediate specimens he called Ayacucho, a now-extinct race (Galinat, 1972, pp. 107–108). Yet Galinat himself later changed his point of view and posited that the cobs from Ayacucho were of Confite Morocho and Pollo (Galinat, 1977, pp. 38). Then, in a personal communication to Pickersgill and Heiser (1978, pp. 137), Galinat claimed that the Ayacucho race was related with the Nal-Tel of México. Our revision of the materials however made us conclude that all the specimens of Ayacucho can easily be grouped as intermediate types of all the possible range between the two Andean races Proto-Confite Morocho and Confite Chavinense, thus discarding the presence of the "Ayacucho race" posited by Galinat (Grobman, 1974). We thus consider unfounded the disquisitions Galinat (1977, pp. 38) made trying to show a contradiction between his data and those published by myself (Grobman et al., 1961) and Mangelsdorf (1974). (Grobman, 1982: 176–177)

MacNeish and colleagues (1975: 32) insist that the Ayacucho maize "perhaps" moved from the north: "… the evidence of teocinte intergression [*sic*] in these cobs suggests its ultimate origin was Mesoamerica." Grobman observed in this regard that the opinion of MacNeish – that the Ayacucho corn is tripsacoid and came from the north – "... has no scientific grounds nor is it demonstrable, it is based only on suppositions." Galinat concurred when he reviewed the Ayacucho maize with Grobman, who showed him the "Ayacucho race" was an error. All of the specimens were segregants of the interracial cross of Proto-Confite Morocho and Confite Chavinense (Grobman, 2004: 446).

Grobman is right when he points out that "it is a pity that the Ayacucho archaeological process did not follow a more stringent method, because the opportunity of assigning exact archaeological contexts to the residues of maize found in several caves was lost due to endogenous and exogenous factors" (Grobman, 2004: 446).

Another site that must be discussed is Waynuna, in the Cotahuasi Valley, department of Arequipa (3,625 masl). Here the remains of a circular house 2.10 m in diameter were found, of which "... only about a quarter ..." was excavated" (Perry et al., 2006: 76). This means that the excavation had less than one square meter, yet they pretend to have been able to carry out stratigraphic work here (Perry et al., op. cit.: 76).

"Unscreened" [*sic*] samples of soil from all preceramic strata, from Level 3b up to the bottom, "... including the preoccupation surface (level 7), and three unwashed fragments of grinding stones from levels 3a and 3b ... ," were analyzed for plant microfossil remains by the L.P. (Linda Perry) and D.R.P. (Dolores R. Piperno) standard methods; 1,077 granules of starch were recovered. "Maize remains ... were the most prevalent: we found 970 definitively identified starch granules and 49 *probable* maize granules" (Perry et al., 2006: 76; emphasis added). Five phytoliths of leaves and cobs were recovered in Level 5. Other samples have phytoliths of cobs, but no secure identification was possible because they were smaller than the maize phytoliths (Perry et al., op. cit.: 77)

Starch granules were recovered in this excavation. The maize granules were subdivided into categories based on their morphological characteristics and on the resulting damage from the grinding and pounding. The starch granules from the flour corns differ from the harder endosperm of maize, such as popcorn or dent corn, and this distinction has been noted in starch residues from Ecuador, Panama, and Venezuela (Pearsall et al., 2004; Perry, 2004; Piperno et al., 2000). Although many modern Peruvian maizes from the comparative collections the authors have assembled have both morphologies in the same kernel, the presence of two distinct assemblages in the lithic artifacts, each dominated by the starch-type endosperm – either floury or hard – indicates that at least two races of maize were used in Waynuna. Due to the characteristics of the starch granules, Perry and colleagues (2006: 78) concluded they were ground. They finished by stating that

> the starch and phytolith assemblages from Waynuna testify to the use of maize by 4,000 cal yr BP, whereas damage to starch granules extracted from the worn surface of the grindstone tool fragments confirms on-site processing of maize for food use. The presence of both leaf and cob phytoliths in the same context strongly suggests that maize was cultivated and processed on site. (Perry et al., 2006: 78)

In their discussion, Perry and colleagues (2006) show they are unaware of the existing literature on this subject (in regard not just to maize but also other plants, but we will not go into this). They pretend that "... the earliest conclusive evidence for maize previously reported from the Central Andean highlands date to ~2,500 cal yr BP ... ," which Burger and Van der Merwe (1990) obtained from carbon isotopes. First of all, the age of the Chaukayán phase ranges between c. 2200–1800 years BC (Burger and Van der Merwe, op. cit.: table 1, 90). Second, Perry and colleagues (2006) ignore the sample from Rosamachay, in Ayacucho, which as we have seen has two dates: 5250 and 5470 BP.

From the data presented the presence of "two races" turns out to be highly doubtful. It is likewise inadmissible that one should try to draw conclusions based on the "stratigraphic" work done in an excavation that is smaller than one square meter. This goes against all principles not just of archaeological methodology but also of professional ethics. If the work proceeds and the results are confirmed, they will certainly be significant, but until then they simply cannot be taken into account.

Finally we have a study Bush and colleagues made in the tropical forests of Peru close to Puerto Maldonado, in the department of Madre de Dios, in the province of Tambopata. Bush and colleagues analyzed the sediments from four lakes but found remains of maize in only one of them, Lake Gentry. The lakes selected for the study were those that are not directly influenced by rivers, and that remain in terra firme forest, and where replication could be achieved within a small geographical area (Bush et al., 2007: 210). Close to the lake, about 1 km away, is a modern farmhouse. The local people say that the knoll holds much pottery and stone artifacts, and they showed Bush and colleagues a collection of axe heads and other stone tools they found close to their homes. No archaeological study of the zone has been made (Bush et al., op. cit.: 211).

Sediment cores were taken from Lake Gentry, along with the remains of pollen of *Zea* found that fell within the 500–3700 years BP range, and *Manihot* from c. 2400 years BP, which evinces agriculture (Bush et. al., 2007: 211). Bush and colleagues insist that "only Gentry provided direct evidence of cultivation with pollen of *Zea* being found regularly between 3700 and 500 yr. BP ..." (Bush et al., op. cit.: 215).

They explain that "there is no clear spike in charcoal associated with the onset of maize cultivation; therefore, the field systems were already cleared or the cultivation took place on exposed mud when lake levels were low" (Bush et al., 2007: 215).

According to Bush and colleagues, the beginning of maize cultivation coincides with a period of slow sedimentation in the lake, which is consistent with the intermittent accumulation of sediments and the low levels of the lake, which presented the chance to cultivate exposed wetlands. The finding of maize pollen in these times cannot be a coincidence, as the granules of pollen (larger than 90 μm) are poorly dispersed, and cultivation close to the point from where the sample was taken improves the odds of collecting fossil granules.

Bush and colleagues conclude that if we consider together the remains of carbon and of the crop, it could be inferred that the Amazon region was occupied by man for more than 8,000 years, and that the landscape was altered using fire at least on a local level. From the data obtained it cannot be determined whether the occupation was seasonal or permanent, but the abundance of pottery close to the lake suggests some degree of permanence (Bush et al., 2007: 215).

It is a pity that the paper under consideration only includes general data, and that specific information that would enable a better understanding of the finds is missing. We have seen that it is twice stated that the *Zea* pollen was found between "3700 and 500 yr BP" (Bush et al., 2007: 211, 215), yet in table 1 (Bush et al., op. cit.: 213) five C14 AMS dates are shown, without considering that one of these is a modern date. The dates are: 940, 2250, 2610, 4070, and 5440. There is no way of knowing which of these corresponds to the finding of *Zea*. It is to be hoped that Bush and colleagues clear this up and present a fuller study of the remains, which are clearly important.

To finish this section on Peru, some general data will be given so that the reader is aware of what position various authors have on this subject. Pickersgill accepted that the most ancient Peruvian maize comes from the coastal Late Preceramic (it mistakenly reads Period IV instead of VI) and points out that

> ... the majority of cobs found in these sites are unlike races of maize of the same age found in Mexico, or indeed any races yet found in Mexico (Mangelsdorf and Cámara-Hernández 1967). There may have been some introduction of maize from Mexico in the late Preceramic, [Pickersgill points out,] to account for the minority of Peruvian cobs which do resemble Mexican maize, but the data available suggest that maize was probably present in Peru considerably earlier than the late Preceramic. Peruvian maize then evolved independently along slightly different lines from Mexican maize to produce the racial differences which were already established by the late Preceramic. Although maize occurs first in the late Preceramic in Peru, it was not necessarily first introduced from Mexico during that period. (Pickersgill, 1972: 99)

Major Goodman has some important ideas, even though he wrote when preceramic maize had just begun to be studied; there is, however, no way of knowing what his sources were, as his bibliography does not cite any specific work on Peru. Among other things he wrote the following:

> Complete ears dated at about 500 B.C. are clearly similar to Andean races still found in Peru and Bolivia and are quite distinct from current or archaeological Mexican maize. At the earliest levels of domestication, it appears that kernel size was small; thus it is believed that the earliest maize was a popcorn. Later, larger-kernelled types of maize appear. On the basis of the variability of currently grown maize races, these appear to differ from the popcorns in the constitution of the endosperm of the grain.

Goodman then added: "While conclusive archaeological evidence for this hypothesis has not appeared, the earliest known South American maize, dated at about 1000 B.C., differs substantially from any recovered at Tehuacán" (Goodman, 1976: 131–132). These two positions will be discussed in the final chapter.

A group of students do not accept the presence of preceramic maize. One of those who most denies it is Robert McKelvy Bird. Here only a small reference shall be made to one of his studies to show his inconsistency; interested readers can expand the data by reading Bonavia and Grobman (1999: 244 and passim). Bird says: "Coastal maize purported to predate 1500 B.C. is much more recent in appearance or gives late radiocarbon dates or comes from disturbed contexts ..." (Bird, 1984: 49). His references here are Towle (1954), Grobman and colleagues (1977), and Feldman (1980). The report by Towle mentions the maize Willey and Corbett found at Áspero, which we saw is definitely preceramic; the study by Grobman and colleagues is one of the first reports on Los Gavilanes, in which two dates – radiocarbon and thermoluminescence – are published that show it is a context from preceramic times; and the study by Feldman is his dissertation, which as we have seen clearly states that his findings were preceramic maize. In this same study Bird presents two tables (1984: 1a and 1b, 58–59) that summarize the maize found in South America. For Peru he exclusively based his work on samples from the Moche, Cupisnique, and Gallinazo cultures, but he does not provide one single datum for the Preceramic period, despite once again citing Grobman and colleagues (1977), and despite the fact that the final report on Los Gavilanes had already appeared. All of the studies done by Bird regarding preceramic maize are like this and do not deserve further comment.

Pearsall (1978a: 53) at first accepted the preceramic maize from the North-Central Coast, but then she changed her mind and systematically rejected it. Again, all of her studies are not summarized here, and instead just a few examples are given. In one of her studies Pearsall (1994a) wrote that "the earliest remains of maize on the Peruvian coast ... dates [*sic*] to the Cotton Preceramic, 2700–2200 B.C., at the Los Gavilanes site (Bonavia 1982; Bonavia and Grobman 1979, 1989a, 1989b)." However, she immediately added that no maize had been found at the site of Paloma, and that the finds of Los Gavilanes are controversial because R. McK. Bird (1987, 1990) questions them. In the conclusions she adds her opinion vis-à-vis "... a number of preceramic sites with later occupations or intrusions (pits, burials, and the like), such as Asia, Áspero,

El Paraíso, and La Galgada where maize occurred only in Initial Period [first ceramic period, 1800/1500 B.C. to 750 B.C.] ... or later deposits at those sites, and not in preceramic strata. There is some additional support for Cotton Preceramic maize from the older excavations at Áspero, however, as well as from Culebras I, Río Seco, and Los Cerrillos (Pearsall 1992b). Even given this, most Cotton Preceramic sites ... lack maize" (Pearsall, 1994a: 255). Readers who have followed the description made of maize preceramic sites will realize that not only does the argument presented not make sense, it also shows misinformation. Instead of making a direct critical analysis to refute the validity of the work done at Los Gavilanes, Pearsall based her work solely on the biased opinion of Bird. It has been shown here that Río Seco never had maize in its context, and that there is no preceramic occupation at Los Cerrillos. Later, questioning preceramic maize, Pearsall pointed out that the Pozorskis did not find it in the preceramic strata of their excavations in Moche and Casma (Pearsall, 1994a: 260), yet she does not say that Uceda did find it in two sites in Casma, as has already been shown here. In a more recent piece, Pearsall (2003b: table 3, 238) presents a site-by-site synthesis of botanical data for the Peruvian coast in preceramic times, and, according to her, none have maize. Yet her list includes Los Gavilanes, Áspero, and Las Aldas, all sites that we have seen had maize. And in note 7 (Pearsall, 2003b: 241) we read: "As discussed in Pearsall (1992[b]), evidence for late Preceramic maize on the Peruvian coast is clouded by problems in establishing clear Preceramic context of remains in some cases, and contradictions in dating in others." In other words, in each and every case discussed, all we have are words but no solid argument.

The attention of nonspecialists has to be drawn to the fact that much in the literature shows an ignorance that is hard to believe, which is often repeated and piles error upon error. Such is the case of Cutler and Blake (1971: 369), who not only are completely ignorant of preceramic maize but even dare claim that it "... appears after ceramics," or Staller and Thompson (2001: 132–133), who not only show they are not well acquainted with the bibliography but even make a poor use of it. To show that maize was late on the Peruvian coast, Staller and Thompson (2002: 47) base themselves on R. McK. Bird (1978, 1979a, 1984, 1990) and R. McK. Bird and J. B. Bird (1980). In other words, not only do they show they are not acquainted at all with the specialized literature, they also pretend that no study was published between 1990 and 2002. Such is also the case of Benz (2006: 18), who says the following after mentioning the data on pollen in the Colombian Amazon and in the Orinoco basin: "If one accepts the aforementioned pollen and phytoliths evidence, then Coastal Peru obtained maize relatively late in comparison with coastal Ecuador and Colombia. Not until 2500 BC was maize present in 50% of sites." The bibliography he used is as follows: Dillehay and colleagues (1989, 1997) and Piperno and Pearsall (1998). First of all, the data for Dillehay and colleagues (1989) in Benz's bibliography are wrong. Second, Dillehay and colleagues (1989, 1997) do not refer to the problem of maize, nor do they discuss it. Something is mentioned superficially in Piperno and Pearsall (1998). And Benz shows he has not read Burger (1989).

Several scholars have tried to explain from whence it was that maize reached the coast. Rowe (1962: 51) suggested, at a time when there was almost no evidence, that it may have come from the central Sierra. Although now the central highlands are a real possibility, as is discussed in the final chapter, the important thing here is that he focused on the highlands.[40] Yet Van der Merwe and Tschauner (1999: 526), following Sánchez Gonzáles (1994), posit that maize reached the Peruvian coast from Ecuador. But there still is no evidence in this regard.

A recent current tries to show that in the end, maize did not have much use in ancient Peru. We cannot go into this subject at length, but it has to be mentioned. Tykot (2004: 439) has claimed, based on an analysis of stable isotopes, that maize was not significant in the diet of the early populations up to the Initial period (c. 1800/1500–900 years BC). Van der Merwe and Tschauner (1999: 529) go even further, for they pretend that maize had a secondary role in the diet between 1400 and 2200 BP, and that it was essentially a marine diet. Their source is Gumerman (1994), who studied the Moche culture. Yet although Van der Merwe and Tschauner do not state it openly, they do imply that maize was a "prestige and state crop" even in Inca times, and for this their only source is Murra (1973).

Blake (2006: 65–68) tries to do the same thing. He claims that, except for Ecuador, in South America maize was not a major element in the diet until after 2000 years BP. Here it suffices to show how it was that Blake handled the data. In his table 2, Blake (op. cit.: 61) summarizes all of the radiocarbon dates that may indirectly date maize. For Peru he mentions the following sites: Chavín de Huántar, La Galgada, Casma PV32–1, Cardal, Caral, and Los Gavilanes. Let us focus only on the way he handled three of these sites: La Galgada, Casma PV32–1, and Los Gavilanes. For La Galgada, Blake cites Grieder (1988a), "p. 69," and C. Smith (1988), "p. 126," and the date number is "TX 4446." On the aforementioned page of Grieder's text we find table 2 and the date "TX 4446," but on reading Smith in the same book and page indicated by Blake we find that the cob was found in "D-11:AD-4/7, Floor 3" (Smith, 1988: 126), whereas the already-cited table 2 points out that the date TX 4446 comes from "G-12: h4, Floor 8, firepit" (Grieder, 1988a: 69). In table 2 there is no date associated with maize. This simply does not make sense.

For site PV32–1 in Casma, that is, Cerro El Calvario, Blake refers not to the original source in Uceda (1986, 1987, 1992) but to Bonavia and Grobman

[40] Some studies, like that of Shady (2006), are not even worth bearing in mind. She claims, without a single argument, first that maize arrived to Caral from the highlands and then that it came "... through long-distance exchange networks" (Shady Solis, 2006: 381, 401).

(1989a: 839), and he mentions just one site, although we have seen that there actually are two sites in this valley, both of them with maize. As for Los Gavilanes, Blake first says it is in Supe (*sic*). He only gives Bonavia and Grobman (1989a: 838) as reference; only gives one radiocarbon date – he chose the oldest one but ignored the thermoluminescence date, which is even older; and claims that the "dated material" is "unknown," which is not true. Had Blake gone to the original source (Bonavia, 1982: 74), he would have found all of these details. But he also ignores all of the sizable literature published after 1989.

As for Shady (2006: 401), she has also joined this current but without providing a single argument with which to support her position. We have already seen that the evidence from the findings made at Caral does not support this. Shady wrote:

> Maize was an object for special use only, and most probably, not used for dietary sustenance by the majority of the population. Its significance appears to be framed in religious beliefs and, perhaps we could trace back to these times the importance that was given to maize in later periods, when it was grown in the lands of the Sun, of the *huacas* or of the Inca.

No comment is needed.

Pearsall raises two issues that have to be explained. She wonders where on the coast maize could have been grown in preceramic times. She acknowledges that it was cultivated in the valleys but believes that this required water control; she then adds that there were no irrigation canals (Pearsall, 2003b: 243–244). First, we now know that irrigation canals were already in use in preceramic times (see Dillehay et al., 2005), but it is true that we as yet do not know how extended their use was. This will not be easy to determine, for in the valleys the evidence was destroyed, first by later pre-Hispanic agriculture and then by modern farming. But what Pearsall has not considered is the *décrue* technique, that is, using the waters that flow out of the riverbed. This is an old custom used by farmers on the Peruvian coast at least until the early twentieth century. I made the corresponding calculations for a small valley like Huarmey and showed that 273 hectares could have been cultivated if half the land available by flooding was used – a significant amount (see Bonavia, 1982: 275–258; 1998: 49–50).

When Pearsall (2003b: 236) presents an analysis of Peruvian preceramic sites, she points out that she used two criteria to select among those that have been researched: sites where the main goal was to study subsistence, and sites where a systematic collection of botanical data was made using "quarter-inch excavation screens," fine sieving, flotation, and coprolite, phytolith, and pollen analysis. Pearsall enumerates a long list of sites, and had she studied them carefully, she would have realized that most of them did not follow the rules she had laid out. One of the few sites studied that followed exactly these procedures is

Los Gavilanes. On reading the report, one will find that "quarter-inch excavation screens" were used, as well as 1.5 by 1.5 mm screens (Bonavia, 1982: 34). The reasons why flotation was not used were clearly explained (Bonavia, op. cit.: 147). Besides, Pearsall does not point out that coprolites and pollen were analyzed at Los Gavilanes (Banerjee, 1973; J. G. Jones and Bonavia, 1992; Weir and Bonavia, 1985). Phytoliths were not analyzed, due to the abundance of maize remains found, and to the significant amount of pollen that was recovered. Yet Pearsall precisely ignores this study and denies the presence of preceramic maize at Los Gavilanes.

The goal of this study, as was pointed out at the beginning, is to analyze the problems regarding the coming of maize and its early development as a cultivated plant. So the period that extends from the Initial period (1800/1500 BC) to the Late Horizon (AD 1534) shall not be discussed. Instead here some titles are listed that interested readers can look up as regards maize in later periods. There are data in Grobman and colleagues (1961: 92–111), particularly in regard to the Moche and Chimú cultures. Vargas (1962: 109, figures 4, 5, 6, 108) has good ceramic depictions of maize. Yacovleff and Herrera (1934: 258–259) have an excellent depiction of an ear eaten by mice and macaws, as well as a vessel depicting maize that was taken from a mold using the cob itself. Eubanks (1979) analyzed 35 vessels with depictions of maize in Mochica ceramics and managed to distinguish 19 races (she notes the same potential holds true for the pottery of the Zapotec culture in the Oaxaca Valley, in southern Mexico). In her conclusions, Eubanks identifies 9 Peruvian races and 10 from Chile, Colombia, Ecuador, and Venezuela. For her this suggests that the pre-Hispanic distribution of the races was bigger than it now is. Eubanks also establishes the presence of possible contacts between the Chilean coast and the northern Peruvian coast, as well as with Colombia. She also admits the possibility of relations with Central America, because for her the Pollo race (from Venezuela and Colombia) is related with the Mexican Nal-Tel and Chapalote. This supports Mangelsdorf (1974: 117–118). Eubanks acknowledges that more studies are required, but she seems to suggest that they are one same lineage. She likewise acknowledges the presence of Bolivian races, which showed an exchange between the coast and the highlands (Eubanks, 1979: 769).[41] Not all specialists agree with this study (e.g., Benz, 2000).

Finally, a study by Yacovleff and Herrera (1934) includes an anthropomorphic Nazca figure with one or several maize plants, complete with leaves, roots, ears, and floral tassels (figures 2a, j, h, 255). Also shown are a heavily stylized plant (figures 4t, u, v, 258), or a complete one that is also stylized (figure 8a, 264; see also Yacovleff and Herrera, op. cit.: 259).

41 For more information, see Eubanks (1999b), where she distinguishes 14 different races of maize that now extend from Mexico to Brazil, and 8 found nowadays from Brazil to western Mexico.

Chile

Latcham (1936) made a description of Chilean plants in pre-Hispanic times, but this does not provide any useful data here.

The first Chilean site in our discussion is Quiani. Here Junius Bird (1943; 1946: 590) found maize in a preceramic context (see also Núñez and Moragas, 1978: 65). This was confirmed by Willey (1971: 206), who gave a date of 3660 BC (he means Quiani II). The site was later mentioned by Rivera (1980c: 42), who gives two dates – 5630 and 6170 years BP – and somewhat doubtfully states that the maize could be of the Coroico race. But in another study that same year, Rivera (1980b: 159) gives the same dates and claims "... they could be related with maize," which he insists is a modern Coroico popcorn. Years later, when mentioning Quiani, Rivera (1991: 10) said that the "average" date of 6500 years BP was not secure.

Yet Bruhns (1994: 74–75) clearly specifies that Quiani I appeared c. 5000 years BC with maize, and that this plant is also present in Quiani II, c. 3500 BC, and she emphasizes that these are domestic plants and an "early maize agriculture." Rivera, however, discussed this issue once more and claims, when reviewing the work done in 1941 and 1970 (J. B. Bird, 1943, 1970; J. B. Bird and Rivera, 1988), that Bird found a cob. Nickerson (1953) follows Wissler (1945) and suggests this is a variety of popcorn because it resembles the modern Coroico popcorn. On this occasion Rivera accepted two radiocarbon dates for Quiani – 5630 and 6170 years BP – which he had previously questioned (see previously). Rivera also claims that Junius Bird tried to verify this find and unsuccessfully excavated the site in 1970 in search of more maize (Rivera, 2006: 406–407). Grobman (2004: 449) does not question the chronological [temporal] status of these remains but does point out that the maize of Coroico race "does not make sense."

Camarones 14, in the ravine of the same name, close to Tiliviche, is a questionable site. Rivera (1980c: 42) initially indicated the presence of maize but noted that its variety could not be established, and he gives three dates: 6615, 6659, and 7420 years BP. For Robert McKelvy Bird, the maize from Camarones 14 is similar to that of San Pedro Viejo de Pichasca.[42] However, in that same year Rivera (1980b: 159) claimed the association with maize was not clear, a point he repeated later (Rivera, 1991: 10). Núñez (1986: 40) is likewise cautious, for he wrote that "at Camarones 14, extreme caution was had when considering the record of maize (*Zea mays*) ... four pieces of evidence (two of them intrusive) were verified and the rest as having been relocated while cleaning the profile. If this did not happen, maize is supported by a date of 5470 BC." In his last study, Rivera (2006: 404, 406) explains this by indicating that the maize

[42] There are notes in Rivera's manuscript (1978a) that were not included in the publication. This corresponds to note 5.

appeared in ceramic strata. Its association is questionable in three preceramic contexts (Schiapacasse and Niemeyer, 1984: 81–82). In fact, we know from the data in Schiapacasse (1988: 7) that the remains of maize are not secure due to their problematical stratigraphic status.

True and colleagues (1970: 179) reported the site of Tarapacá (Grupo 5, Tarapacá 14A), which is a habitation site. They note that "some maize was recovered from the house fill...." The stone tools are similar to those from Conanoxa (close to Quebrada Camarones, see Núñez, 1965: 107–109). It is suggested that Tarapacá has a date ranging between 3500 and 4000 years BP, but it is cautiously noted that it "is tentative." Núñez and Moragas (1978: 62) accept the "... evidence of the maize" and assign it a date of 4480 years BC. Rivera (1980b: table 1, 122; see also Rivera, 1980c: table N° 1) later said that maize was found in a coprolite, and also reported the finding of macro-remains of maize that were studied by Williams (1980), which were dated to between 4780 and 6830 years BP. Núñez (1986: 36, 40) contradicted himself in this regard, because he initially mentioned "sites with doubtful indicia" that included Tarapacá-1, yet when he later mentioned Tarapacá (Tr-12 and Tr-14A), he admitted the "... presumption [of the presence of] maize ..." based on the report by True and colleagues (1970).

Tiliviche is a major site in the arid Chilean north, in the hinterland of Pisagua, in the province of Tarapacá, some 40 km from the coast, in the Atacama desert. Here six encampments and a cemetery were found. Til-1B is the largest type, from whence most of the data comes. Site 1B did not have any disturbance, and it was here that evidence of "...possible initial harvests of maize" was found in the three stratigraphic divisions (Núñez and Moragas, 1978: 53, 55, 58). Then it was confirmed that "the presence of maize is continuous but in very low frequency," although it does tend to increase in later strata (Núñez and Moragas, op. cit.: 58). There are ears, kernels, and leaves found in the three stratigraphic zones. The early evidence is associated with the stratum dated to 5900 years BC. Grinding tools were found with these remains. Galinat studied these corns (Ms. 1975b) and assigned them to the Piricinco Coroico racial group. The kernels of maize appear in the "first well-stratified refuse, dated to 5900 years BC (7850 years BP)." Núñez and Moragas suggest that the evidence "... connects it [Piricinco Coroico] with an eastern descent through the highlands" (Núñez and Moragas, 1978: 59, 62). This was reaffirmed by Núñez Enríquez and Zlatar Montan (1978a: 739–740). And yet these same scholars believe that maize "... did not manage to determine a new mode of life that would bring about radical transformations in the social organisation ..." (Núñez Enríquez and Zlatar Montan, 1978b: 744, 754).

Rivera (1978a: table 2, 164; 1978b: annex 27) also provides important data on Tiliviche 1-B, for he specifies that here a whole cob and a broken one were found, and he also gives three radiocarbon dates: 7850, 6950, and 6060. He assigns these corns to the Piricinco Coroico race and adds that "according to Robert Bird (personal communication), it is best related with the maize from Early Horizon Peru and with the modern Chutucuno Chico from the Loa River than with other races or archaeological material. However, the relation with Cupisnique maize is not stronger than that between Cupisnique and Chutucuno Chico corn." Rivera (1980b: note 4, 124, see also table 2, 123; 1980c: 42; 1991: 10) simply repeats this. He, however, later made a long comment (Rivera, 2006) based on the work done by Núñez and Moragas (1976), Staal (1974), and Núñez Enríquez and Zlatar Montan (1978a, 1978b). Rivera points out that according to Staal (1974: 21), "leaves and cobs" were found in Tiliviche 1-B. Rivera agrees with the date and comments that the maize could actually be varieties of Chutucuno Chico and Capio Chico Chileno, which are varieties of popcorn associated with Confite Puneño.

According to Núñez (1986: 34–35), 18 samples were removed from Unit 1 and 24 from Unit 2. The presence of maize in Unit 1 is denied for the early stratigraphic zone, and only 1 sample from the intermediate stratigraphic zone and 6 from the late stratigraphic zone are accepted. There are 10 samples from the small mound in Unit 1, all from the late stratigraphic zone. There are 2 samples from the early stratigraphic zone in Unit 2 and 10 from the intermediate stratigraphic zone, 9 good samples out of 13 in the Intermediate reoccupation stratigraphic zone, and 3 more from the "Late Reoccupation Stratigraphic Zone." Núñez (1986: 34–35) wrote thus: "The association between maize and typical preceramic traits is eloquent.... It is posited that in Unit 1 (bigger mound), which has a coherent radiocarbon sequence ranging between 7810 and 4100 BC, maize was introduced in the first refuse placed on the upper floor, sometime after 5900 years BC." For Núñez, these maize traits indicate they are not intrusive, but neither are they exclusive of a stratigraphic zone or associated with archaic characteristics.

The first findings of maize are on the floors dated to 5900 and 4955 years BC. The best evidence comes from between 4955 and 4110 years BC, particularly after 4850 BC. There is no maize in Unit 2 in the early levels; maize only begins to appear on the floor dated between 5255 and 4760 years BC. There are good samples in two levels dated between 4760 and 3235 years BC, as well as in the levels from 3235 and 2720 BC and later (Núñez, 1986: 40, 43). Núñez (1986: 35) wrote the following in regard to the remains of maize:

> The studies of R. Bird (Ms. [1979b]) based on Unit-1; 2 ears, 12 kernels and leaves, have established that the husks[43] recall the *Midwest dent* and *Amazonian interlocked flour* specimens. In general, the Tiliviche specimens do not have the extreme kernel interlocking (the interdigitation of adjacent rows) typical of the western and southern band of the Amazon Basin, such as the races of Piricinco Coroico Amarillo. This identification rectifies or competes with the eastern taxon *Piricinco coroico*, proposed by Galinat (Núñez and Moragas, 1978). In fact, the [corns] from Tiliviche in general have cupules and interallicoid

[43] In Spanish the term *envoltorio* is incorrectly used.

> longitudinal spaces [that are] too short to [be] similar to the races from the eastern lowlands.... Cobs and kernels show in this regard a homogenous race that exhibits a close relation with *Chutucuno Chico* (from Northern Chile) and *Capio chico chileno*. It [the homogeneous race] also joins the *Altiplano Small flour* pattern (*Altiplano Huayleño*, *Negrito chileno* and *Confite puneño*). This is a race of popcorn with a closer proximity to *Chutucuno Chico* and *Confite puneño*. Bird (op. cit.) finds similitudes with the archaeological remains of Huaca Prieta (HP5-House 2), dated from 1300 BC, and the Cupisnique phase, [dated to] around 800 BC.

Only "... remains of maize leaves ..." were found in the cemetery, but these are late, for they date to 1830 BC (Núñez, 1986: 43). In this same study Núñez shows with solid arguments that maize is not intrusive, nor is it exclusive to one stratigraphic zone. He believes that in Tiliviche there was a "... scant selective, and perhaps ritual, consumption of maize ..." (Núñez, op. cit.: 40) and concluded that

> ... maize and *cavia* are slightly recorded from the first refuse strata [found] over the early habitational floors. But they are more common in later strata that comprise a range between 4760 and 2720 BC. Of the twelve datings, ten fall within an archaic sequence order for the site between 7810 and 2720 years BC, and one for 1830 BC that corresponds to the cemetery Til-2, that is correlated with the local habitat. It is established that at least guinea pig and maize formed part of the archaic contexts of Tiliviche since at least 4760 to 1830 BC. (Núñez, 1986: 40)

A maize kernel and cob from Level IV were recently dated with the AMS method, which gave the dates 850 years BP and 920 years BP (Rivera, 2006: 404). This issue is discussed in the final chapter.

There clearly is much confusion as regards the racial classification of the Tiliviche maize and its origin, as can be perceived in the passage previously cited by Núñez and Moragas (1978: 62). It is extremely unlikely that the maize in question corresponds to the Piricinco race. The latter, as Grobman and colleagues (1961: 218) noted, "... is perhaps the most widely distributed corn race with a single continuous geographical range." It extends from the eastern ranges of the Peruvian Andes to the south, to the department of Pando and the Bolivian lowlands, and to a large area of Brazil (Brieger et al., 1958). But Piricinco is primarily cultivated by the Indians of the Amazon basin (Grobman et al., 1961: 215–221). It is thus a tropical race. It is worth noting that the only evidence available of the "Piricincoid" traits on the coast is that which appears in some late Moche vessels from the north Peruvian coast (e.g., Grobman et al., 1961: figure 33, 103). It is therefore extremely unlikely that this race managed to reach Chile, particularly at such an early date.

To avoid confusion it must also be explained that the name "Piricinco Coroico" is wrong. This maize race was originally described by Cutler (1946), who called it "Coroico." Grobman and colleagues (1961: 221) opted for "Piricinco" because this is the most common name among the Indian population.

From the site of Tulán, to the southeast of San Pedro de Atacama, only one datum was found in Rivera (2006: 408), who says that here there is maize that has been dated to 2500–3000 years BP.

San Pedro Viejo de Pichasca, in the province of Coquimbo, in the department of Ovalle, 900 km to the south of Tulán, is another site with remains of maize. It was initially said that there were "still tentative" samples of maize dated to 4700 years BP. It was hypothesized that this was a "... center of agricultural experimentation thus far unknown" (Rivera, 1971: 304). It was then verified that there indeed was "... maize ... ," which for Level III was dated to 9920 years BP, and in Level II to 7050 BP, with a mean level of 4700 years BP. In Level II the "... maize [was] in the upper part" (Ampuero Brito and Rivera Díaz, 1971: 65–66; Lynch, 1983: 128). Galinat (1971b: 306) studied the remains and noted that there are "three different specialised types": Capio Chico Chileno, a native of Chile; Curagua, a popcorn resembling the Colombian Pira; and the Pororo from Bolivia. Rivera discussed the report presented by Galinat and stated that "it follows ... that the relation between Curagua and certain popcorn races could yield new landmarks in the process of maize domestication" (Rivera, 1978a: 160). Besides, the Negrito Chileno, which is red and specialized, resembles the types from Bolivia and Peru.

According to Robert McKelvy Bird, the maize from Camarones 14 is similar to that of San Pedro Viejo de Pichasca.[44] The data on maize were later repeated, but it was then noted that the dates for the specimens are 1285, 2375, and 4700 years BP (Rivera, 1980b: 110; 1980c: 42; 1991: 10). Yet Rivera himself recently (2006: 409) stated that the lower levels had been dated anew with the AMS method. The date for a kernel of maize had been AD 925 and that of a cob AD 850.

R. McK. Bird (1984: 46) does not accept the early Chilean maizes and argues that their form "... is far too close to modern maize ... and [the samples] appear fresh and not deteriorated...." When Pearsall (1994a: 263) reviewed this issue, she concluded that doubts linger because the maize had not been directly dated,[45] and because many sites have more recent strata on top, as is the case of San Pedro Viejo de Pichasca and Tarapacá. There was also the possibility that rodent and human activity may have disturbed the contexts. Pearsall was struck by the fact that the Chilean dates seemed to be older than the Peruvian ones, yet she said: "*It is also difficult to obtain the original reports of these finds; I have relied on two recent overviews for most of the information discussed here*" (Pearsall, 1994a: 263; emphasis added). One wonders how such

[44] Once again, the original study by Rivera (1978a) has notes that were not included when it was published, in this case note 5.

[45] The dating of Tiliviche and San Pedro Viejo de Pichasca with the AMS method had not yet been done when she was writing.

a critique can be leveled using just secondhand sources. This unfortunately has become a habit in many North American colleagues dedicated to analyzing the problematic of maize.

Núñez is quite clear in his observations: "The record of sites with questionable evidence (Quiani, Camarones 14, Tarapacá-1, etc.) shows the controversial fate of preceramic maizes, and the need to intensify some internal critique of their associations" (Núñez, 1986: 36). But he agrees with Lynch (1983) as regards the presence of preceramic maize: "*Its presence in preceramic context*[*s*] *is out of the question*" (Núñez, op. cit.: 35; emphasis added).

In a recent publication Rivera made a major point – that the study of this subject in Chile has not moved forward in the last fifteen years, and that it is now time to reassess the results with new methodologies. As for the serious divergences between traditional radiocarbon datings and the new AMS datings, he does not believe this is a significant issue, and claims it is just "a cautionary note" (Rivera, 2006: 403, 411). All that I would add here is that Rivera has wholly ignored the literature on the central Andean preceramic maize, and that his reading of this subject is far too biased toward the Chilean issue, which cannot be studied in isolation from neighboring areas.[46]

Brazil

Unfortunately very little information was found on Brazil. Lathrap (1987: 359) is vague. All he says is that on the eastern side of Brazil there are dry caves where "string" cobs have been found as early as 3000 BC (Brochado, 1984). Lathrap claims that these resemble the Coxcatlán maizes and the Confite Morocho and adds that "by 3000 B.C., maize had saturated the tropical lowland alluvian network and was penetrating beyond its limits."

Bush and colleagues (2007: 212) reported core sediments they analyzed from Lake Geral, close to Prainha, on the northern zone of the state of Pará. Maize pollen and phytoliths were "found consistently" after c. 850 BP.

Bruhns (1994: 77, 95) discussed the finds made at Santana de Riacho, a rocky shelter in southeastern Brazil. She points out although there are no other remains of cultivated plants, maize does appear around 3000 years BC. For Bruhns, this "... is a rather aberrant appearance of corn and no other sites have, as yet, yielded similar remains."

Uruguay

A site known as Los Ajos has been studied in the wetlands of southeastern Uruguay, in the La Plata basin. It was occupied by man in the Middle Holocene. Los Ajos is one of the largest sites in the area and covers 12 hectares.

[46] Interested readers should see Bonavia (1982: 378–380).

This is a circular village where domestic and public areas can be distinguished. It has been established that the Preceramic period here went from c. 3000 to c. 4190 radiocarbon years BP. The data obtained in the excavations made show that throughout the preceramic occupation, the population had a mixed economy in which they incorporated domestic plants.

No botanical macro-remains were recovered, despite the flotation method having been used. Phytoliths from maize cobs were, however, found 15 cm on top of the oldest context in the Preceramic Mound Component, dated to "... about 4,190 14C yr BP...." The phytoliths are found throughout the upper sequence of the site. Starch grains diagnostic of maize kernels were also recovered from a subspherical mano in the excavation, 5 cm above the remains dated to "... about 3,460 14C yr BP ..." (Iriarte et al., 2004: 614–615).

This same report indicates that "... starch grains from maize kernels were documented in contexts dating to 3,600 14C yr BP at Isla Larga and 2,800 14C yr BP at Los Indios ... ," which are other mounds found in the study area (Iriarte et al., 2004: 615). We are also told that phytoliths of maize kernels and starch grains were found in milling stones from later, ceramic times (Iriarte et al., op. cit.: 616).

Iriarte and colleagues conclude that these finds provide "... the first evidence of permanent village living in southeastern South America by people who subsisted on mixed economies and adopted major crop plants such as maize (*Zea mays* L.) and squash (*Cucurbita* spp.) long before previously thought" (Iriarte et al., 2004: 617; for more data, see Iriarte, 2006).

Given the lack of information for the Brazilian area and the vague and contradictory sources available for Argentina, the data for Uruguay prove most significant, not just because of the antiquity but also – and this is essential – because they derive from methodical excavations with secure datings. The only problem here raised, which has been repeated throughout this book, is that as long as no methodology that allows the racial identification of maize micro-remains is available, it will prove very difficult to relate the Uruguayan finds with those from other areas of the South American continent.

But this study by Iriarte and colleagues shows that the significance maize had for the early South American populations is becoming ever clearer, and that its antiquity is moving backward in time. This opens a new window for the problematic of maize in an area where, as Iriarte and colleagues (2004) point out, no one ever expected to find evidence of this kind.

Argentina

Parodi (1935, 1966) described the aboriginal Argentinean agriculture, but there is no data on maize prior to the Conquest.

Gil recently noted that although cultigens – maize included – have an antiquity of 2,000 years in Argentina, the most ancient ones come from a funerary context. The remaining maize has been dated to about 1,000 or more years later. The isotopic data obtained from human skeletons indicate that maize probably was not a major element in the diet. Despite pre-Hispanic maize being 2,000 years old, it has been claimed, Gil says, that it reached the southernmost reaches of the frontier between central-western Argentina and northern Patagonia 1,000 years later (Gil, 2003: 299). Gil and colleagues (2006: 201) point out that there are at present two positions regarding central-western Argentina. One of them holds that maize arrived c. 4000 years BP (Bárcena, 2001; Roig et al., 1985), and the other that it did so c. 2000 years BP (García, 1992; Lagiglia, 1980, 2001). The direct radiocarbon data agree with the second position, but very few dates are available (Gil, 2003).

The site known as León Huasi I is in the province of Jujuy. The corresponding report is confusing and does not have concrete data. It says that Level B2 (the next-to-last one, as B3 is the deepest one) has a date of 10,559 ± 300 years BP. According to the report, maize was present in all the levels, mostly as kernels. But no more details are given (Fernández Distel, 1989; the data on maize are on pp. 7–8). Maize exhibits a considerable degree of variation according to Cámara-Hernández (1989: 21–22), who studied it. He notes that "... despite the presence of small kernels, their size is not as small as that of the more primitive archaeological maizes from Peru, and its texture does not correspond to that of a popcorn like that of the former." In his comparative survey Cámara-Hernández did not reach clear correlations with modern races. These remains at most resemble Pisingallo and Bola.

Fernández Distel (1974: 118, 122) reports the finding of maize in Huachichocana Cave, in the department of Tumbaya, in the province of Jujuy, to which he ascribed a tentative antiquity of 3000 years BC. Three dates were later obtained from this same level ("Layer E") – 9620, 8670, and 8930 years BP. In the table "Economic Vegetables," the entry "cultivated vegetables with a certain highly advanced degree of domestication, applied to feeding ... ," reads thus: "Maize (*Zea mays* L.), with primitive characteristics and maize with a certain highly advanced degree of variation" (Fernández Distel, 1975: 13). Fernández Distel then refers to an analysis made by Julián Cámara-Hernández and describes this "maize with primitive characteristics" as having "... cobs with a narrow diameter, long and soft glumes, small and ... flint kernels that pop, possibly of a reddish colour and acuminate. A variety currently widespread in the area of the Quebrada Humahuaca, and which bears these primitive characters, is the one known as 'pisincho' (Oryzaea)" (Fernádez Distel, 1975: 16). In a subsequent publication (Aguerre et al., 1975) it is once again stated that "there are three probable cultivated vegetables in Ch. III, Layer E3: ají (*Capsicum baccatum* or *Capsicum chacoense*), common bean (*Phaseolus vulgaris*) and maize (*Zea mays*). The first two, as was specifically explained by the botanists who analysed them, may be wild specimens for they thrive naturally in the area." This same article cites a report by Cámara-Hernández (Ms., n.d.), who said the following in regard to maize: “The cob fragment lately described and the part of a cane may belong to a maize with primitive characteristics, small plants, thin canes, with small, probably hard, kernels and thin cobs. The remaining cob fragments would indicate the presence of variations in maize. The scant amount of materials does not allow other observations [to be made].” The authors then add: “In another part of his report, the above-cited scholar [i.e., Cámara-Hernández] establishes a relationship between this maize that denotes variations and the current races of maize from the Quebrada de Humahuaca” (Aguerre et al., 1975: 212). Aguerre and colleagues in fact repeat the presence of maize in Layer E3 of Huachichocana and insist on the date of 8930 years BP for this stratum (Aguerre et al., op. cit.: 211–212).

Fernández Distel later commented on sample P2608, which was used for radiocarbon dating: “Maize was recovered in the 1972 excavations that was only dated in 1978. In the meantime it was deposited in the Botanical Chair in the Agronomy Faculty of the University of Buenos Aires, where it was cleaned, measured and photographed.” And she adds:

> The date is as given, i.e. it did not respond to the expectations of antiquity that the Layer (E3) in which it was found seemed to indicate. We asked the laboratory [what] the possible reasons of contamination of the sample [were], for it was not preserved in a sterilised environment and was in contact with cotton. The laboratory did not confirm the contamination, but left this possibility open. With the analysis in question, the sole testimony of maize (stalks and [kernel-less] cobs) that we had from Layer E3 was destroyed. *The date in the 3rd century AD made us place it in Layer E1*, for which we have another date (from charcoal) in the 5th century. We do not think we are mistaken in this regard because the layer also has maize. (Fernández Distel, 1980: 90; emphasis added)[47]

In a more recent study, Fernández Distel (1986: 416–417) again insists that maize “... appeared in all of the layers ... with remarkable rarity in Layer E1 (inclusive), [and] was extremely popular in the hearths and refuse[48] of Layer C ....”

The evidence presented is neither complete nor clear, and its stratigraphical position is not convincing. It is even clear that Fernández Distel is not sure of the context of her materials, for as we have seen, the radiocarbon dating is far more recent than expected, so she accepts that the remains she presumed came from layer E3 actually corresponded to Layer E1, which is associated with pottery. A more detailed report was never published. Cohen believes that “... the dates on this assemblage are not firm and it is unclear that the preceramic levels

47 The date for P2608 is AD 340 ± 190 (Fernández Distel, 1980: 96).

48 The text says “fogones basural,” but this almost certainly is a mistake and should read “fogones y basural.”

at Huachichocana predate the early dated ceramic assemblages discussed above" (Cohen, 1978: 262). I concur.

Pearsall's position is inconsistent. In an early study she claimed that "... [the remains] cannot be evaluated until ... [they] are more fully reported" (Pearsall, 1978c: 399), yet she later accepted in a critical fashion the finds of Huachichocana (Pearsall, 1994a: 260), but basing her thinking on the work of Tarragó (1980) and showing (see subsequently) that she had not read it. When discussing the research at Huachichocana, Tarragó said that "the most ancient evidence of maize in Huachichocana III, which is associated with Level E3 (7000 years BC) does not suffice to draw conclusions. It would be necessary to reinforce these data with new evidence" (Tarragó, 1980: 208). From the work of Tarragó it also follows that only "four cobs" were found at Huachichocana, and that "... only one of them and part of a cane, could belong to a maize with primitive characteristics ..." (Tarragó, 1980: 193). Her study later concluded that the evidence "... stills does not suffice" (Tarragó, 1980: 208).

In this case Lynch (1983: 129) also has a mistaken position, for he notes that "... corn seems to have been abandoned after the introduction of pottery."

Lagiglia (2001: 55) is, however, categorical: "... We believe that the cultural contents from the upper layers are consistent with the presence of an incipient agriculture, so an initial agriculturalisation around 7970 years BC, as Fernández Distel and his team believe, must be discarded."

Based on the report by Cámara-Hernández (Fernández Distel, 1975: 16), Grobman points out that "... the description comes close to that of the maizes from the Central Andean Zone. The reports however are not clear and there are serious doubts regarding the context of the finds" (Grobman, 2004: 449).

In Antofagasta de la Sierra, Catamarca, Rodríguez and Aschero mention the sites of Punta de la Peña 4 and Punta de la Peña 9, but the maize found here is late (c. 500–1979 years BP; Rodríguez and Aschero, 2007: 259). This study also mentions the site of Quebrada Seca 3, also in the Catamarca region, for which the presence of "... starch grains of *Zea mays* ..." dated to 4510 radiocarbon BP is indicated (Rodríguez and Aschero, 2007: 261). Yet on another page we read the following: "... The first micro-remains of maize were found at Quebrada Seca 3 (c. 4700 BP) ... (Rodríguez and Aschero, 2007: 268)," which clearly is a contradiction. The presence of maize in Peñas Chicas 1.1 with a date of 3590 years BP is also indicated. For this information they based their work on two studies by Babot (2004 and 2005) that I was unable to find. This is a very superficial study that shows the authors do not know the literature on this issue. This text raises serious reservations.

The final piece of information I found on Argentina is in the province of Mendoza. This concerns the Gruta del Indio, where "cultigens" have been dated to 1900 and 2200 years BP with the AMS method. They are part of the Contexto Atuel, but we must bear in mind that the remains come from a burial context (Lagiglia, 1980). It is not clear whether maize is present, and I was unable to obtain the direct source (Gil, 2003: 296). But a later study (see subsequently) sheds some light on this issue.

Gil and colleagues (2006: 212) have reported that in south-central Mendoza, "the oldest corn is recorded at the highland site of El Indígeno." This is the only mention made of this site, and no bibliographical reference is given. These same authors point out that in the province of Mendoza there are 17 sites with cultigens, 16 of which have maize. It appears more frequently in the mid-Atuel Valley. It begins to appear c. 2200 radiocarbon years, but there is only one direct date in the Gruta del Indio to 2065 years BP (Gil et al., 2006: 202, table 15–3, 205). Gil and colleagues believe that south-central Mendoza marks the southernmost limit of the expansion of maize in pre-Hispanic South America (Gil et al., op. cit.: 211).

Gil and colleagues later studied several sites in the western zone of central Argentina based on human stable isotope data. They concluded that "… corn was clearly significant in the human diet only after ca. 1250 years BP but [was] probably used in low levels from 2000 years ago …" (Gil et al., 2009: 229). And when analyzing the samples from San Juan and northern Mendoza, they concluded that "in all of these cases only for the last part of the Late Holocene was maize, on average, significant in the human diet" (Gil et al., op. cit.: 231). This study makes generalizations that are based on far too little data, and all of the sites studied are late.

6

The Role of Maize in Andean Culture

Maize has had a quite specific role in almost all American populations. For instance, in Guatemala anyone can plant whenever and whatever plants he or she feels like, without this being a matter of concern for the community, and the planting does not usually involve any rituals, or, if it does, there are only a few of them. Not so with maize. This plant undergoes a series of ceremonies when it is planted (Johannessen, 1982: 92, 93–96). Furthermore, it is significant that the term "teosinte" comes from the Aztec word *teocentli*, which means "the ear of God," and that the conquistadors called it "mother of maize" (Kahn, 1987: 22). For North American Indians, maize is, metaphorically speaking, their mother, an enabling being, a transformer, and a healing being (Ortiz, 1994: 527). There still are legends concerning this plant. In the Pueblo culture maize is the mother. It is far more than just a plant or a food, even though it is eaten daily. Maize, as a material object and as an idea, permeates every object in the life of the Pueblo, from birth until death, from the present to the future (Ford, 1994: 525).

For some Mesoamerican groups maize was a creation of lightning or its progeny. For instance, among the Zapotec of Loxicha, in southern Oaxaca, maize is taken to be the progenitor of lightning (the father) and the earth (the mother). In a Zapotec account collected in Mitla, lightning is considered the creator of the multi-colored maize (Marcus, 2006: 231).

Among Mesoamericans maize is far more than just a grain. It is a powerful cultural icon that keeps societies united around a group of beliefs, economies, annual ritual cycles, and work patterns. These shared beliefs and activities were at the same time adapted throughout time, from the transhumant living conditions of the hunters and gatherers, to the Spanish conquest, and to the modern era of globalization (Alcorn et al., 2006: 599).

In Andean society we need only recall the dynastic origin myth of the Inca to see the significance of maize. The wife of Manco Capac taught the people how to plant maize. And all throughout the Late Horizon the annual cultivation cycle of this plant was inaugurated by the Inca (Murra, 1975: 54).

Maize has been used as food by Andean society since preceramic times, but it also had other, quite complex roles of which little is known prior to the Late Horizon.[1] Finucane and colleagues (2006) recently showed, based on the study of collagen in human and animal skeletons, that maize was the subsistence basis of the Huari Empire in Middle Horizon times (c. AD 550–1000), and not just in humans but also in animals. Finucane (2009: 542) reaffirmed his position in a later study – "... maize was the staple that supported both the urban and the rural population ..." – adding that it was the major staple in the Ayacucho area since at least approximately 800 years BC (i.e., since the Initial period). The time period for which we do have much information is the Late Horizon. It is well known that the consumption of certain foods – maize was one of them – and beverages was used as an explicit and implicit form of state control of local populations. The output of maize and chicha had, as we shall see, a major role in the political banquets, and this reached its climax in the Inca Empire (Dillehay, 2003: 356–357). In other words, there was an intimate relationship between maize, social relations, and power. And we must not forget, as Finucane (2007: 2122) points out, that in Inca times, in many regions in the central Andes, "rather than serving merely as a ceremonial cereal or one component of a diversified economy, maize was the staff of life."

On reading the Spanish chroniclers, one has the impression that in the highlands, maize was a much-desired and festive food in contrast with, for instance, potatoes and chuño (freeze-dried potatoes). During the harvest, maize was carried to the house amid great celebration, with men and women singing and praying to the plant so that it would last for a long time. They would drink, eat, and sing for three full nights while they watched over the *Mama Zara* (Mother of Maize). The best ears were wrapped with the finest blankets the family had. Maize was integrated into the life cycle in the villages, even if it was not locally grown. And the Incan state devoted a considerable technological and magical effort to ensure the propagation and harvesting of this plant (Murra, 1975: 53–54; see my Figure 6.1).

We must bear in mind the double role that maize fulfilled. If on the one hand it was cultivated to prepare the chicha used in ceremonies and as part of Andean hospitality, it was also at the same time not just a major staple but even the favorite food. For instance, we know from the chroniclers that the Inca army preferred maize to any other rations (see Murra, 1975: 53–56). Besides, this plant had been important since pre-Inca times. For example, in Mochica society, maize was part of the oblations made in tombs, along with other specific

[1] A study made in Virú (on the northern coast of Peru), based on the study of coprolites and a stable isotope analysis of human bones, estimated a 40–60% dependence on maize. If we combine the analyses of coprolite remains with faunal, floral, and stable carbon and nitrogen isotope data from the Early to the Middle Horizon, we find that "... there is a continuous and gradual increase in the use of maize through time" (Ericson et al., 1989: 94).

6.1. A drawing by Felipe Guaman Poma de Ayala (1936: 1047 [1157]) showing the harvest of maize in May, "... when they have to pile up the maize, peel it and shell it, removing the seeds and having the best maize placed aside to eat, and setting out the worst to make chicha...." Drawing by Felipe Guaman Poma de Ayala. After the 1936 facsimile edition.

types of food. Interestingly enough, agricultural produce predominates among the offerings – in the case of maize, the ears with the highest number of rows (Gumerman, 1994: 410).

Some ceremonies still endure. In the Cuzco zone, planting maize is almost a religious task. The seeds, which are mixed, are carefully selected. The mix is not just tolerated; it is fostered by the peasants. Many small farmers intentionally mix the seeds to increase the possibility of hybridizing varieties, with the subsequent genetic differentiation (Sevilla Panizo, 1994: 224).

It has already been pointed out (see Chapter 5) that a group of scholars has recently appeared who raised the hypothesis that maize was exclusively used as a sacred plant. It has even been claimed that its "consumption ... as a ritual intoxicant may help explain why it spread so rapidly all over the Andean world" (Staller and Thompson, 2001: 150). Staller and Thompson are the scholars who have most frequently insisted on this point (Staller, 2003: 377; Staller and Thompson, 2002: 34–44). But this has no support at all, nor do they have enough arguments to buttress their position. Besides, they have essentially worked with Ecuadorean data and do not know the Andean reality. They have even contradicted themselves. One of their studies is based on an analysis of phytoliths, yet they themselves have said that "while phytolith assemblages can reflect the presence or absence of maize in a food residue sample, phytoliths alone do not reflect the percentage of the food residue[s] which represent maize" (Staller and Thompson, 2002: 34, 44, 46).

On the other hand, it is often forgotten, as Morris noted, that alcoholic beverages played a major role in all societies in the world, so it should come as no surprise that the same thing happened with chicha in Inca times (Morris, 1979: 21). We must also remember that the production of chicha is quite ancient in the Andes; the Inca only regulated and expanded it to further the goals of the state (Dillehay, 2003: 362).

To understand the role chicha played in the Andean world, we must understand that this was not a market economy. Products were exchanged based on reciprocity and redistribution. This tradition has some resemblance to the value gifts have in other societies. This naturally stands in stark contrast with the modern market economy. The value of the gift lies not in its intrinsic value but in who gives it. In the case of reciprocity, the exchange is usually not completed in just one act. The gift establishes an obligation that may be returned at some other moment by whoever has received it. The gift not only may be returned but must be returned, and this ensures that the relation between the participants will endure. The difference from contemporary society, as Murra (1960) explained, is that the gifts were exchanged between members with a different sociopolitical status, usually between a chief and his people. The chief acted as a centralizing entity and was given gifts by different individuals in reciprocity for gifts he received from various specialists, which came from different ecologies. So the gifts were permanently redistributed, thus allowing products from different regions and specialists to be in reach of all of society without a market. This is why chicha had a major role in this context (Morris, 1979: 25–26).

The association of chicha with political and religious ceremonies was essential for the maintenance of the political and economic system of the Inca Empire. It was not just that millions of gallons of chicha were annually prepared and consumed; we must likewise consider the way in which these were redistributed, and how this was essential for the lords to retain their authority. The skill shown by the state in expanding the output of chicha proved crucial for its political and economic expansion (Morris, 1979: 32).[2]

The chicha made out of fermented maize was the essence of hospitality and was a kind of common denominator of ceremonial and ritual relations. This was

2 Readers interested in the political significance of chicha on the North Coast of Peru should read Rostworowski (1977: 240–244).

the beverage that generous leaders had to provide for their people, as part of the duties of leadership. In the Inca state, chicha had a formal political role in the maintenance of political authority through redistributive hospitality and tribute. But its production and consumption at the same time also had an economic role in the mobilization of the corvée labor known as the mita (the compulsory, labor-intensive work performed by men of 25–50 years of age), via the attempt to institutionalize the political banquet (see Dillehay, 2003: 360–361; Morris, 1993: 43; Valdizán, 1990: 139).

The political significance of chicha as an element that could be converted into labor was used not only in Inca times but also in pre-Inca polities. What is now known as a "reciprocal hospitality" is well documented on the northern Peruvian coast, where personal work in public projects was fêted with chicha and food supplied by the curaca (headman; Netherly, 1977: 214; Rostworowski, 1977: 241–242). The importance chicha had in these encounters involving reciprocal obligations was shown in 1566, when Gonzáles de Cuenca, the *visitador* (inspector) appointed for the North Coast, prohibited the distribution of chicha and other alcoholic beverages to the Indians (Netherly, 1977: 216–217). The curacas were thus unable to command the workers and therefore could not pay the tribute meted out by Spanish officials unless chicha was distributed. The ban had to be lifted (J. D. Moore, 1989: 685).

In the Inca state the form, use, and significance of chicha were culturally well defined. As Dillehay points out, chicha is like a liquid form of material culture that also has some specific properties. In most Andean societies, beverages cannot be stored for long, which is why they must be consumed. This means their significance is immediately evident; their ingredients therefore acquire value in their culinary transformation and in the process of consumption as well as in the ritual and social contexts themselves, but not in their accumulation. Chicha is the means that allowed – through a mechanism, to wit, the banquet – the agricultural surplus to turn into prestige, political power, or "... perhaps into objects with a lasting value that could be used to convert economic capital into symbolic capital in a multi-ethnic economy" (Dietler, 1996; cited by Dillehay 2003: 358).[3] Chicha, as a food product with specific psychoactive properties that are due to specific preparation techniques, represents a peculiar type of material culture that is frequently turned into a particularly significant ritual and social artifact. The same relevance applies to all of the technical paraphernalia related with the consumption and preparation of chicha, such as the vessels in which this task was carried out and the labor drafts required for its production (Dillehay, 2003: 358).

In Inca times the priests had many maize-related duties. Every year they had to ask the gods whether or not they could plant. They had to follow the

[3] This citation has been retranslated into English.

movements of the shadows in the *Intihuatana*[4] in order to regulate the periods when the land would lie fallow, when it was to be irrigated, and when it was harvested. They kept a quipu system[5] with which to check the sequence of dry and rainy years. They fasted from planting season until the sprouts were the size of a finger. They organized processions with war drums, in which the participants shouted to scare away the drought and the frosts. They sacrificed llamas in gratitude and to request a good harvest. The Intiwasi, the famed Temple of the Sun, was decorated with symbolic gold plants to stimulate the harvest (Murra, 1975: 54–55).

The production of chicha, furthermore, had a quite well-defined social context. In Inca times chicha was made by the chosen women. Guaman Poma de Ayala left an invaluable testimony in this regard. He mentioned the Virgins of the Sun and said that they made "... nice chicha, which was so good that it matured in a month called Yamor Toctoy Asua ..." (Guaman Poma de Ayala, 1936: 300 [302]). We shall see that on the Peruvian coast there was another context wherein it was men who made the chicha. But there also was a type of home-brewed chicha (J. D. Moore, 1989: 688–689: Rostworowski, 1977: 241).

Rostworowski has shown the differences that existed between specialized trades on the coast and in the highlands. As regards the preparation of chicha, in the highlands the women dedicated themselves to this task and made chicha for domestic use, but this task was taken over by the *mamacuna* (chosen women) when large amounts of chicha were needed for the cult or for the Inca. On the coast it was instead a male occupation. For instance, we have the deposition made by Don Pedro Payampoyfel: "... We do not have any other occupation [other] than making chicha ..." (AGI [Archivo General de Indias], Justicia 458, fol. 2090v). The significance of chicha comes through in the *ordenanzas* (regulations) Dr. Cuenca gave in 1566 for the North Coast, during his first inspection in Trujillo, as well as those issued by Juan de Oces in another inspection he

4 "Archaeological literature has used intihuatana to designate certain outcrops of bed rock carved so as to leave an irregular vertical protuberance in the middle which is assumed to have been some sort of a sundial for calendrical observations.... The word is good *Quechua*, and means 'hitching-post of the sun'" (Rowe 1946: note 39, 328).

5 Rowe explains that the *quipeu* "... (*khipo*, 'knot') ... consisted of a main cord from which hung smaller strings with groups of simple knots on them at intervals. Frequently, subsidiary strings are attached to the main pendant strings, and often the strings are distinguished by color or method of twisting.... A quipu represented a series of numbers which could, perhaps, be read by any trained *Inca* accountant, but, in order that anyone but the original maker might understand what the numbers referred to, the quipu had to be explained.... The quipu is excellently adapted for recording numbers, but would be an exceedingly clumsy instrument with which to calculate.... In addition to recording numbers, the quipu was used as a memory aid in reciting genealogies, liturgical material, and narrative verse, so that some chroniclers (e.g. Valera and Morúa) speak of *Inca* history as based on the quipus in such a way that they might appear to have been a form of writing, which they certainly were not" (Rowe 1946: note 39, 328).

made of this city in 1574. To understand these laws, which proved insufficient in the highlands, we have to analyze the Yunga (coastal) customs. Here the caciques and *principales* (chiefs and headmen) had public places where they drank chicha, and this caused "... the drunkenness of the Indians, and in this they use many Indian men and women in making the chicha ..." (AGI, Patronato 189, Ramo II).

From all of the documents it follows that part of the prestige of a coastal lord lay in giving his subjects drink, and in having a large number of hammock bearers. The more prestigious a lord was, the more magnificent his public houses had to be. When a lord went out, he would take with him an entourage of bearers carrying jugs with chicha that was offered to passersby to refresh themselves as the litter passed by. The caciques protested when Cuenca drastically banned this custom. In Chicama and San Pedro de Lloc they requested that they at least be allowed to drink while carrying out their agricultural labors. And Cristóbal Payco, from Jequetepeque, clearly explained that the Indians obeyed their caciques because they gave them drink, and that they would not work their land if they did not receive it. Chicha provided the lords with "... the complicated web of reciprocities that could not be suppressed without it raising serious problems." Years later, the *visitador* Juan de Oces had to lay down detailed regulations on the process of preparation and exchange of chicha. The presence of some taverns was accepted, and "... all the Indians who make chicha should be in there, and there they must do it" (AGI, Lima 28-A). Here chicha was exchanged for maize. In return for their work the inspectors (*veedores*), marshals, and measurers received eleven measures (*medidas*) of chicha and kept one for themselves, and at the end of the week the pile of maize obtained was distributed by the priest. A part of it was distributed to the poor and to the other specialists in equal parts. A daily arroba was given to the cacique and the *segunda persona* (second person) of the *señorío* (chiefdom), half to the *principales*, and an *azumbre*[6] to the commoners (Figure 6.2).

The ordinances issued by the *visitador* Juan de Oces forbade the use of any beverage, be it from the yucca, carob tree, or *jora*,[7] subject to punishment in the public square and having one's hair shorn. No one could make chicha at home, not even the lord of the chiefdom. The chicha-making specialists were freed of any other occupations and could not be forced to participate in the corvée for the encomendero, the cacique, or the *principales.* The only thing that could be asked of them was that they participate in repairing the main irrigation channel of the repartimiento.[8] When the cacique or any other *principal* went from one repartimiento to another, the lord of wherever they

[6] An *azumbre* is an ancient Spanish measurement equal to 2.017 liters (Llerena Landa, 1957: 24).

[7] Jora is germinated maize flour used to make chicha.

[8] The encomienda was a system in which allotments of men were made by the king on the conquest of America in return for the services rendered by individual conquistadors.

6.2. A drawing by Felipe Guaman Poma de Ayala (1936: 776 [792]) showing a cacique *principal* (Indian chieftain). Elsewhere the Indian chronicler explains that these officials drink chicha and wine, chew coca and gamble, and are always drunk. On the bottom right we see an aryballos (a type of pre-Hispanic earthen vessel) with "fresh chicha," and on the left another vessel with "vintage wine." Drawing by Felipe Guaman Poma de Ayala. After the 1936 facsimile edition.

were going to was forced to provide the daily *arroba* of chicha to which the visiting lord was entitled, so that he could have his people drink without having to take the chicha bearers with him. It was also established that the new ordinances would be published in the Yunga language (Rostworowski, 1977: 240–244).

In the Inca Empire, the consumption of chicha in the context of state banquets probably had three major social functions. First of all, it facilitated social integration and channeled the flow of social relations. It was a crucial part of the protocol of hospitality. The association with hospitality at the moment of making a toast took on a very powerful social value, which was transformed into a nexus that established relations of reciprocal obligation between the host and his guests. The act of toasting promoted both solidarity and social inequality through social and religious rituals, which were established and "paid" for by the state. Second, the institutionalized status and the distinctions of roles by age, gender, and class were often established symbolically through patterns that were implicit in the act of drinking chicha: the specific order in which the people were seated, their location at the moment the beverage was consumed, the order in which it was taken, the different types of vessels used to drink it, and even one's required behavior. Finally, the production of chicha had an economic role in the mobilization of the mita groups, which aimed at institutionalizing the political banquet (Dillehay, 2003: 360–361).

Data found not only in the chroniclers but also by archaeologists show that in pre-Hispanic times chicha was produced in very large quantities, and that its consumption formed part of religious and political ceremonies, as has already been noted. It was a significant and crucial element in the process of reciprocity, the structure of which was broken when the Spaniards limited its use because they interpreted it as drunkenness and disorder and associated it with paganism (Morris, 1979: 26; see also Rostworowski, 1977).

Now we know that chicha was the everyday beverage taken by the Andean peoples, but it was also at the same time a crucial element in all ceremonies, in large quantities, and was accompanied by ritual dances. In these ceremonies people would drink until they fell asleep. For the Incas, intoxication was a religious act and not an individual vice. The Indians did not drink in excess, except in the prescribed ceremonies (Rowe, 1946: 292). Alcoholism was actually rare in most Andean indigenous societies. But the toasts with chicha and ritual intoxication would take place in all major ceremonies. Protocol dictated toasting and drinking. And those who served the chicha had to serve it to all. As Dillehay put it, "liquor accompanied the toasts, while the declarations and commemorative orations were the essence of the Andean ceremonies and rituals."

Participants were seated and served following a formal order and in terms of their status. Alcohol was not taken in large amounts by all; rather, it was taken for several days, depending on the rite or ceremony. Status also determined the type of vessel used to drink. Common people used gourds, those higher up used elegant vessels, and the upper tiers of the social hierarchy used gold or silver cups. There are precedents for this in Moche and Chimú with stirrup spout vessels, or among the Inca with the *quero*, which in turn has its antecedents in Tiahuanaco. Ritual intoxication is an ancient tradition in the Andean area, for broken vessels used for this purpose have been found in tombs (e.g., among the Mochica and the Huari; Dillehay, 2003: 356–357).

The Spanish chroniclers have information on this subject. For instance, Father Acosta mentioned the drunkenness of the Indians in these terms: "... Our Indians take the must from their chewed maize, which they then mix with water and boil; others use rotting maize and from thence call it *sora*, which [beverage] is stronger than any grape wine " (Acosta, 1954: 493). Interestingly enough, Acosta showed he knew nothing of the real intent of this drunkenness, despite having dedicated his chapters XX and XXI to this subject. He wrote thus: "It is shameful for Christians that an Inca, a barbarous and idolatrous king, would rein his subjects in their drunkenness, while our people, instead of correcting their customs, have allowed it to grow so much" (Acosta, 1954: 495). In chapter XXII, Acosta tried to suggest "in what ways the Indians can be dissuaded from drunkenness." He concluded that the danger lay not in individual but in collective drunkenness, still without understanding its cause. He ended by admitting that the Spaniards allowed it to take place because "they benefit with the work of the Indians," whereas others did so with the idea of profit (Acosta, 1954: 500). He, however, also acknowledged that "... no one can deny that this beverage ... the Indians make out of maize... strengthens and is healthy and of good taste for those who are used to it. And it is from all standpoints inhuman to pretend to deprive this lineage of poor and helpless people, who have no other pleasure, of this sole means of relief and recreation" (Acosta, 1954: 498).

The Spaniards, in fact, planned a campaign to eliminate all that they considered as drunkenness in the "pagan" rituals (Morris, 1982: 166; Rostworowski, 1977: 241). Many Spanish priests even railed against the drinking orgies because they considered them pagan rituals, and their elimination was part and parcel of the campaigns launched to extirpate the idolatries. Some of this still endures in the Peruvian highlands (Rowe, 1946: 292–293).

It must, however, be pointed out that Polo de Ondegardo, who "... was the first to methodically study the institutions of the conquered peoples ... their juridical, civil, and penal organisation ..." (Porras, 1986: 335), made an interesting assessment around 1561. For Polo the Indian leaders

> ... made sure that the common people did not get drunk, and set penalties for those who got drunk, except in cases when it could be done, such as wakes, weddings, and at the time that the fields of the Inca or the Sun were being worked, when each of these tasks was finished and in all the things the community as a body took part, but in the latter case drinking was only allowed when the community was making a general sacrifice. Beyond this no one got drunk, nor was any more punishment required than the ban that forbade all of this. (Polo de Ondegardo, 1940: 193)

It is worth recalling in this regard the ordinances issued by Francisco de Toledo, who was the "Viceroy and Captain-General" of Peru in 1569–1582. Toledo reported that "... chicha taverns have recently appeared among the freed negro and mulatto women, as well as other people who have this business ..." (Toledo, 1867: 90). Among other things, the ordinance said the following:

> ... I order and command that no Spaniard, Negro, Mulatto or Indian can make chicha to sell, nor have a tavern where it is sold in his home, nor should they consent that their Negroes, Indians or Mulattoes do so, on penalty that if it were a Spaniard, he will pay fifty pesos the first time, and the second time he will pay the same amount and will be banished from this city and its jurisdiction for five full years. And if said Indian wine or chicha is made in the house

of any Spaniard he shall pay the same fine even if it is not of his interest, and the same thing holds if the drunken orgy takes place in his house, and all the *botijas* [jars] shall be smashed. And if it was a Black man or woman, Mulatto or Indian, they will be fined twelve pesos and will be publicly given a hundred lashes. And if the blacks or mulattoes are *horros* [those who had been slaves and attained freedom] the penalty shall double and they will be banished from this city and its jurisdiction for five full years. And since this is such a great harm it is well that the remedy be universal, as well as the duty of enforcing it, [and so] I order and command that if Indians are found drinking in the house of a Spaniard of any state and condition, they will be fined fifty pesos meted out as noted. And the fine will be twice the amount should the Spaniard forbid the entrance of the marshals, or presented any resistance, plus his banishment from this city for ten full years, all of which fines be divided by three as is said, between the chamber, the denouncer and the judge. (Toledo, 1867: 90)

Title XXI of the ordinances, which is on "the drunkenness and the beer taverns the Indians have," says that "... on Sundays and on feast days, and sometimes on any day, the Indians do" indulge in this, which is "... a harmful vice for health...." It is then stated that "... all of the idolatries they have are drunken orgies, and that none of these take place without superstitions and witchcraft ..." (Toledo, 1867: 89–90). Here we clearly see that the Spaniards were completely unaware of the role these "*borracheras*" had in the Andean culture.

Valdizán (1990) studied these drunken orgies and their consequences to see if they were harmful.[9] He noted in this regard that chicha "... is a harmless beverage in its early stages of fermentation ..." but not so in the final ones, when the alcoholic degree increases and "a highly toxic ptomaine is even formed ..." that was discovered in Colombia by Dr. Zerda (c. 1898).[10] He then partially cited a study by Torres Umaña (1917) that explains that ptomaine, just like phosphorus, carbon oxide, arsenic, morphine, and so on, causes immediate variations in the normal chemical functioning of the organism vis-à-vis the nutritional processes in general, and urinary excretion in particular.

Valdizán tended to believe that highly fermented chicha proved harmful when taken in large amounts and with the addition of some substances to make it "stronger." But he added that this was not so in the case of the "sweet and soft chicha we drink in our time and whose low alcoholic content is traditional amongst us." Valdizán studied what he called "*chichismo*" among the inmates of the "Asilo Colonia de La Magdalena" (a lunatic asylum in Lima). He reached the conclusion that "... even after this selection, it is not possible to present a single case of toxicophrenia due to the drink of the ancient Peruvians." Valdizán concluded thus: "Either chicha is completely harmless, as is pretended by those who disdain its role in the process of degeneration of the Indian race, or chicha,

[9] We must bear in mind that he wrote in the early twentieth century.

[10] Valdizán did not provide more information in this regard and admitted he only had bibliographical data.

while producing the nutritional disorders mentioned by Doctor Torres Umaña, acts in a completely different way from that in which habitual agents of the alcoholic toxicophrenias act" (Valdizán, 1990: 149–153).

There is a good archaeological example of chicha brewing in the Chimú site of Manchán, in Casma. Here we see that it was prepared in a nonspecialized social context, similar to the domestic production the chroniclers mention. The methods and equipment were quite spread out over the various precincts. The way chicha was prepared at Manchán suggests that the Chimú state was able to obtain specific and significant types of resources even without economic specialization, or without it entailing an extensive output (Moore, 1989: 686–687, 692).

The city of Huánuco Pampa, in the highlands some 150 km from the city of Huánuco, is an Inca site where the manufacture of this beverage was well studied. Here large vessels have been found that recall those used nowadays when brewing chicha. There is evidence of large-scale production that was probably carried out by people who dedicated themselves to this full time. The area devoted to this task provided – through the thousands of vessels found there – one of the major investments made by the Inca.

Chicha was consumed in the settlement and was not stored, and it is even unlikely that it was distributed outside it. Here archaeologists found large numbers of *quero*-type cups, and mortars with their pestles and a large ceremonial area around the *ushnu*, in the central plaza, which is where the banquets and feasts were held (Morris, 1982: 166; Morris and Thompson, 1985: 90–91). It is worth recalling that Inca cities did not have a concentrated population like Western European cities did, and that they served as the site where the peasant population went to deliver their products (Bonavia, 1972). So in terms of the relations between the state and the local population, the manufacture of chicha at Huánuco Pampa had a crucial role in the activities that supported the state. We know that similar cases exist in Tiahuanaco (Bolivia), Argentina, and Chile (Dillehay, 2003: 359).

The ceremonial use of chicha has not been forgotten by contemporary communities. The present work is no place to explain and describe this. Only two examples are given here. When describing the ancient customs, Valcárcel compared them with modern usages: "Beverages like chicha were poured on the tomb or on the altar, and all the libations made were preceded – as is still the custom nowadays among the Indians – by the type of offering known as *tinca*. Three fingers are placed in the glass from which one is drinking and then two of the fingers are moved against the other one and the drops are directed towards the mountains, or wherever it is believed that the propitious spirit lives" (Valcárcel, 1959: 160). And Gillin, who studied the contemporary community of Moche in the 1940s, explained that a great insult amongst the Mocheros is to reject chicha when it is offered (Gillin, 1947: 48).

The use of maize as food has been pointed out, and it was used in this way in various forms and as a raw material for the brewing of chicha. But maize also had many other uses in the Indian world. Here I mention just some of them. The stalk and its leaves were used as fodder for the camelids, and after the Spanish conquest for the imported animals. Their bracts, which cover their female flowers, were used to make *humintas* (*humitas*) and tamales. The stigmas were used in medicine as diuretics. The cobs (*marlos ckoronta*)[11] were used as fuel, and the term *ccorunttani* even means "to rummage among the ccorunttas for fuel" (González Holguín, 1989: 69) (Yacovleff and Herrera, 1934: 256).

According to Garcilaso de la Vega (1959: 130; 1966, volume 1, book VIII, chapter IX: 499), maize was also used to prepare "... a very good vinegar. An excellent honey is made from the unripe cane, which is very sweet." Besides, "... the leaves from the ear of maize and the stalks are used by those who make statues who thus avoid weight."

Finally, I would like to insist that maize had a very significant role in Andean beliefs – a role that was reflected in its depiction in art by the majority of the pre-Inca cultures and by the Inca themselves. And it is significant, as Mangelsdorf (1974: 187) pointed out, that whereas in ancient Peru this plant was depicted in stone, ceramics, textiles, gold, and silver, in ancient Mexico there are images of it only in stone and pottery.

[11] González Holguín (1989: 69) writes "*ccoruntta*."

7

Maize as Seen by the First Europeans

When the Carmelite Vázquez de Espinosa mentions the island of Hispaniola, "... that the Indians call Haiti ... ," which he must have heard of in 1612–1613, he reports that "there also was maize in abundance – this is the wheat of the Indies – of yucca and maize they made their wine to drink, and at present the Indians do so ..." (Vázquez de Espinosa, 1948: [99] 36–37).

Fernández de Oviedo made an excellent description of this plant in the sixteenth century: "Maize is born from some canes that give out some spikes or ears the length of a *jeme*,[1] and smaller or bigger, and thick as an arm wrist or less, and full of thick grains like chickpeas (but not fully rounded)" (Fernández de Oviedo y Valdéz, 1959: 226). He then added:

> This bread has the cane and pole in which it is born as thick as a spear ... some are as thick as a thumb and others more or less so, depending on the goodness of the land where they are planted. And they usually grow far more than the height of a man. The leaf is like that of the common cane of Castile, and is far longer and much wider, and greener, and the leaf is more tameable or flexible, and not as rough. Each stalk gives at least one ear, and some give out two or three, and in each ear there are two hundred and three hundred kernels, and even four hundred more or less, and [there even are] some with five hundred, depending on the size of the ear. And each spike or ear of these is wrapped around by three or four leaves or skin, wrapped tightly over the kernels, one over the other, and they are somewhat coarse, and almost of the complexion and type of the cane leaves in which they are born. The kernels are so tightly held by this bark or peel that not even the sun nor the air harm them, and they ripen inside. (Fernández de Oviedo y Valdéz, 1959: 228)

The observations Cabello Valboa made in 1586 are also of great interest:

> God gave these aborigines a seed on their entrance to this rustic and peasant New World. And they, with their industry, made it domestic and so useful and

[1] A *jeme* is the distance from the tip of the thumb to the tip of the forefinger, with both fingers extended.

> superior that it excels over those that men use. And what I will say of it is that in no other part of the universe (not in the Islands, nor on its Tierra Firme) has a seed like this one been found except just in this strip of the world we are discussing, as well as in the islands contiguous and close to it. But on paying close attention to the manner in which it [this plant] is born ... as well as to the fit of its grains and the organisation of its spikes and the making of its leaves, it [seems to be] of the species of that seed so used by the Andalusian *Moriscos*, which they call Panizo. The exception is that the grain in the Indian seed is incomparably bigger than the *Morisco* one. In the New World this seed has been and is given many different names, but the one which is quite predominant and has been taken in our country is Maiz, because this was the name given to it by those in whose power the plant was first seen by our Spaniards in the Island Hispaniola, where the natives call it Maiz.... It would be a long account if I were to write down the good lands in which this extremely excellent grain has been found. All that I will say is that if our Spain introduced its use it would not be so plagued by famine and extreme need as it is, particularly in those provinces that lie on the Mediterranean Sea.... The natives of this New World have used this seed both for delicacies and for brews, and since it is excellent for both one and the other, they have always esteemed it [maize] much. (Cabello Valboa, 1951: 181–182)

Father Acosta was very careful in the descriptions he made, but interestingly this was not so in the case of maize:

> I believe that the grain of maize is not inferior to that of wheat in virtue and sustenance. It is thicker and warmer, and engenders blood, so that if those who eat it again do so in abundance, they often suffer from swellings and the mange. It is born on a cane and each carries on it one or two ears to which the grains are affixed, and although these are big kernels, they have many – in some we counted seven hundred grains. It is planted by hand and is not scattered. It requires warm and humid land. It appears in many parts of the Indies in great abundance. Harvesting three hundred *hanegas*[2] from the land sown is not unusual. (Acosta, 1954: 109)

It is striking that not even the mestizo chronicler Garcilaso de la Vega left a good description of maize:

> Of the fruits that grew above ground the most important was the grain the Mexicans and the inhabitants of the Windward Islands call maize and the Peruvians *sara*, for it is their bread. It is of two kinds, one hard kind called *murchu* [*sic*; i.e., *muruchu*], the other soft and very tasty, called *capia*. They eat it instead of bread, roasted or boiled in plain water.... In some provinces it is softer and tenderer than in others, especially in the province called Rucana. (Garcilaso de la Vega, 1966, volume 1, book VIII, chapter IX: 499)

2 A fanega or (hanega) is an old Spanish unit that is equal to 6,439.48 m^2 (Llerena Landa, 1957: 80).

Garcilaso is far more specific as regards the flour of maize:

> I have seen all this with my own eyes,[3] and until I was nine or ten years old I was brought up on *sara* or maize, the bread of which is known by three names – *çancu*, bread for sacrifices; *huminta*, special bread for celebrations; and *tanta* ... common bread. Roast *sara* is called *camcha*,[4] "toasted maize." ... Cooked *sara* is called *muti* (by the Spaniards *mote*): the word includes both the noun maize and the adjective cooked. With maize flour the Spaniards make little biscuits, fritters, and other dainties.... The same flour is mixed with plain water to brew their beverage, which can be soured in the Indian fashion to make a very good vinegar. An excellent honey is made from the unripe cane, which is very sweet. The dried canes and their leaves are of great value, and cattle are very fond of them. The leaves from the ear of maize and the stalks are used by those who make statues who thus avoid weight. (Garcilaso de la Vega, 1959c: 130; 1966, volume 1, book VIII, chapter IX: 499)

Garcilaso also emphasized its healing aspects when expounding "the medicinal herbs they used":

> ... Later the Spaniards experimented with many medicinal products, especially maize, which the Indians call *sara*. This was partly due to the information the Indians gave of the little they knew in medicine, and partly because the Spaniards philosophized about what they found and discovered that maize, as well as being such a substantial foodstuff, is of great benefit in diseases of the kidneys, pains in the side, stone, stoppage of the urine, and pains in the bladder and colon. They realized this because very few or no Indians have those diseases, and attributed the fact to the habit of commonly drinking a brew of maize. Many Spaniards who suffer from such diseases therefore drink it. The Indians also use it as a plaster for many diseases. (Garcilaso de la Vega, 1959a: 199; 1966, volume 1, book II, chapter XXV: 123)

Garcilaso also attributed curative properties to maize flour: "As a remedy in all sorts of treatment experienced doctors have rejected wheat flour in favor of maize flour.... Thus the advantages I have mentioned are all derived from the various parts of the *sora* [*sic*], and there are many other medical derivatives, both beverages and plasters, as we shall have occasion to mention later" (Garcilaso de la Vega, 1959c: 130; 1966, volume 1, book VIII, chapter IX: 499).

But there can be no question that we owe the masterpiece in descriptions to Father Bernabé Cobo. "His botany," as Porras pointed out, "is deeply poetic despite its scientific anticipation. It connects us directly with the flowers, the plants, the fruits and the trees of America sans the torment of nomenclature" (Porras Barrenechea, 1986: 510). The notes Cobo made on maize are extensive but well worth citing in full: "... Maize is as common throughout all of America

[3] Garcilaso means here the way the flour of maize was prepared.

[4] This is now known as *cancha*.

as wheat is in Europe, both in Tierra Firme as well as in the islands adjacent to it" (Cobo, 1964a: 159). He continues:

> ... Its leaves are quite similar to those of canes except that they are wider and not so rough. The stalk or cane of maize usually reaches a height of an *estado*,[5] and it gets to be more or less as thick as a thumb. It has nodes at equal intervals, just like the common cane: it is tender, slim and easily broken. It gives at its end a spike or plumage that is between white and red in colour, and several small shoots. This plant produces its fruit not at its tip, like other legumes, but around the cane, and from one up to four *choclos* (this is what they call the spikes or ears of maize in Peru) on each sprig or cane. On being peeled, each choclo is almost as thick as the wrist, and some are a *tercia* in length, and the usual [thing is for them to be] a jeme and smaller. The choclo is covered with some thin, rough and rubbery tunics or cups, and between them and the kernels there are many silks the colour of maize, which surpass the length of the choclo come out from its tip in a small bunch as thick as a finger. The kernels of maize are the size of not fully rounded chickpeas; they are placed lengthwise in the choclo in rows and in great order, like the seeds in the pomegranate, and are so tightly packed that on removing them from the choclo the difficult part is removing one, for on doing so the rest follow.... Maize is such a common seed that it grows not just in temperate lands, but also in many others with various climates, as in cold and warm lands, dry and wet, in mountains and in plains, in winter and summer lands, in irrigable and rain-fed land. There is difference between maize and wheat: all of the lands that carry wheat can also take maize, and those which do not grow wheat because they are too cold, also do not grow maize. But here wheat has the advantage over maize, as it stands more cold than maize, because in temperate lands that tend more to cold than to warmth, they plant the wheat in the upper parts and on the slopes, so as to leave the plains and the warmer land for maize. And in a year with ice, if it is not too strong, the land sown with maize is usually lost while wheat survives, even though one and the other are in the same lands, as is commonly experienced in the region of the city of Cuzco and throughout all of Peru. But it does not happen so but quite the contrary, because maize is abundantly collected in all temperate *yunca*[6] lands but no wheat grows [there] because it is too humid and warm, so that wheat, although it is born, it does not give fruit and instead everything goes in lushness.... Maize does not grow everywhere in the same size or in equal abundance. In warm lands it grows so luxuriant and lush that some maize fields hide a man on horseback. And from here it diminishes as the land goes colder, until it reaches or does not rise over the earth for more than a *codo*.[7] In fertile and thick land it usually gives two hundred per hanega, and sometimes four and five hundred, but in thin and ordinary land it often gives a hundred and less, up to ten.

5 An *estado* is a unit that is equal to 1.67 m or 6 Spanish feet (Llerena Landa, 1957: 79).

6 A *yunca* is a warm valley.

7 A *codo* is a unit that is equal to 1.4149 m (Llerena Landa, 1957: 41).

Cobo then distinguishes the varieties of maize:

> There are many differences in maize, because first of all it comes in all colours: white, black and yellow, purple, light and dark red, and mixed in many colours. Besides this it is also distinguished by the size of the kernels. The biggest ones found are slightly smaller than broad beans. There is a very tender maize, with a very white and soft flour, and another very hard one that the Indians call *murucho* and the Spaniards *morocho*, which is that which mounts eat, and to all these differences the Indians have given proper names.... [T]his seed is so useful that besides being sustenance for men, it also is for animals, because it is given to mounts instead of barley; it is the grain that poultry – hens, turkeys, pigeons and ducks – eat and fatten with it, and the fattened animals gain weight with it better than with acorns. Nor is even its cane worthless, because the Indians suck it when green as if it was sugarcane, and in some parts they make juice, honey and wines [out of it]. Its leaf, green and dry, is a wonderful fodder for mounts. And in New Spain they make nice images in statues that come out very light, even when they are very large.

Cobo finishes by mentioning the healing aspects of maize:

> *Maize* is very medicinal because the juice of its green leaves heals fresh wounds, and heals the pain of flatulence and removes the exposure to cold when the grain, toasted and sprinkled with wine, is applied hot in a small sack. It eliminates bruises when its flour is mixed with radish-leaf juice. Finally, the *pleada* or *atole*[8] made of it with sugar is a most delicious and easy-to-digest food, that is given both to the wounded as well as to those sick with fevers. (Cobo, 1964a: 159–160, 162)

It is quite interesting that the Spaniards not only accepted maize immediately as food but even made it a staple in their voyages. This follows quite clearly from reading the accounts they left. Xerez (Jerez) tells that when Pizarro left Panama for the first time and his men landed at Puerto del Hambre, the ship was sent back for food. At its return, "... the provisions the ship brought were maize and pork, [so] the people who were still alive recovered ..." (Jerez, 1968: 196).

When recounting the first voyage of Pizarro, Cieza de León also mentions the hunger the expedition endured. Pizarro sent Montenegro to the Isla de Las Perlas in search of food, and there "they placed much maize in the ship" (Cieza de León, 1987: 15). This was repeated in several passages. Cieza recounts the second expedition led by Pizarro and notes that on reaching the town of Tacámez (Atacames),

> the Spaniards, happy with the much maize they had found, ate pleasantly because on having need [i.e., being hungry], the men do not feel it if they have maize, because with it a very good honey is made, as all who have made it know, and as thick as they want it to be. I have made some of it in this life ... (Cieza de León, op. cit.: 37)

[8] Atole is a warm beverage made out of maize flour dissolved in water.

Diego Silva y Guzmán also gave an account of this voyage. His rhymed chronicle, written in 1538–1539, has an accuracy as regards the dates and places that "... is far superior to that of the other chroniclers...." This is the most detailed account of the maritime voyages of Pizarro (Porras Barrenechea, 1986: 56–57). The poet recounts a moment in the second expedition when they were unable to land and lost the "maize and food" (Silva y Guzmán, 1968: CXLIX, 70). Then, before Almagro reached the Isla de Gallo, they "found the maize harvested ..." (Silva y Guzmán, op. cit.: CLX, 74). And when Tafur left, leaving Pizarro and his men behind on the Isla de Gallo, Pizarro headed north to an island called Gorgona, and "to Tierra Firme a journey they made. And of maize they brought it [the brig] loaded" (Silva y Guzmán, op. cit.: CLXXII, 78).

Cieza later says that when the Spaniards reached Tumbes, the Indians sent raftsmen to the ship with some produce. Among them was "chicha" (Cieza de León, 1987: 53). And when Pedro de Candia returned after Pizarro had sent him to find out whether what Alonso de Molina had said was true, the "lord" [of Tumbes] ordered that "... many rafts [loaded] with much maize" should go with him (Cieza de León, op. cit.: 58).

Ruiz de Arce tells that when he traveled from Nicaragua to catch up with Pizarro, he stopped in the Bay of St. Matthew. Here he landed "in search of maize for my horse ..." (Ruiz de Arce, 1968: 414). Ruiz says that on reaching Tangaraya (i.e., San Miguel de Tangarará, in the Chira River valley), he saw that here they "do not eat bread; they eat the maize toasted and cooked, and this they use as bread. They make wine out of this maize, in large amount" (Ruiz de Arce, op. cit.: 420).

Mena in turn says that when Hernando de Soto went to the "town of Caxas" before the showdown at Cajamarca, he found there "... very large houses, [where] they found much maize ... ," and there were "... over five hundred women who did nothing but clothes and maize wine for the men of war; there was much wine in those houses" (Mena, 1968: 137). Mena himself mentions the gifts the Inca sent Pizarro "... one day before we reached the encampment of Atabalipa...." These included "... many cooked sheep and maize bread and jugs with chicha" (Mena, op. cit.: 140–141). Xerez (i.e., Jerez) tells that while on his way to Cajamarca, Pizarro received the messengers sent by Atahualpa a second time: "This ambassador brought with him the retinue of a lord, and five or six cups of fine gold with which he drank, and with them he gave the Spaniards to drink the chicha he was bringing ..." (Jerez, 1968: 220).

Mena in turn tells that the houses of Atahualpa in Cajamarca "... were full of women who made chicha for the encampment of Atabalipa" (Mena, 1968: 141). Xerez explains that when Pizarro reached Cajamarca he sent messengers to Atahualpa. When they reached the chambers of the Inca, "... women came [to them] with golden cups in which they had maize chicha. When Atabalipa saw them, he raised his eyes towards them without saying a word [and] they left rapidly and came back with other bigger golden cups, and made them drink with them" (Jerez, 1968: 225).

Betanzos also described the departure of the messengers Pizarro sent to the Inca at the Baños and says the king offered them "... maize tortillas ..." and "... told Unanchullo to bring out *chicha* in tumblers of fine gold ..." because he wanted to verify whether Hernando de Soto and his horsemen would stay with them, but everything was returned (Betanzos, 1987: 270; 1996: 254).

Estete left an account of the journey Hernando Pizarro and a group of Spaniards made to Pachacamac and to Jauja, where he mentions the presence of maize fields in the Callejón de Huaylas. Estete insists that "all of that land abounds in ... maize ..." On climbing up the highlands on the return journey from the coastlands, they also saw "... much ... maize" (Estete, 1968: 245–246). And when Hernando Pizarro wrote a letter to the Real Audiencia of Santo Domingo giving an account of his journey to Pachacamac, he wrote that in the highlands they "... make chicha to pour on the ground." He added that in each town there were "houses of women" that among other things "... have the task of making chicha for when the men of war pass by." Hernando also noted how the Inca used the *quipus* to count (he did not call them by name), but he did specify that every time the Indians brought the Spaniards "... maize or chicha they removed [the amount handed to the foreigners] from the knots ... so that in this all they have very good accounts and method [*muy gran cuenta y razón*] (H. Pizarro, 1968: 126).

Pedro Pizarro also provides very interesting data. He tells how, on leaving from Jauja to Cuzco, he was "wandering in search of maize or other things to eat ..." (Pizarro, 1968: 491) – in other words, maize already was a major staple for the Spaniards. Pedro Pizarro also notes that on returning to Cuzco from the Altiplano, "... food was scarce ... in just a few days ... ," so they sent men to Xaquixaguana, where "... there was much maize ..." (Pizarro, op. cit.: 526). Contemporary chronicles explicitly point out that the Spaniards sought maize to feed their horses. I will just mention Pedro Pizarro (op. cit.: 582 [also on p. 583]), who points out they sent out for maize "... for the horses, who were exhausted. ..."

There is one document that bears the date 1534 – this date itself is in question – and that was inspired by several letters. One of these letters was sent to the king in 1533, and it was originally sent by Pizarro to the Cabildo of San Miguel. Here we read that before Pizarro left Cajamarca and headed south, he received many golden objects from a cacique. These included "... deux lits de maïz, dont chacun supporte deux mazorcas d'or ..." ("... two bases of maize, each of which holds up two golden ears ..."; Anonymous, 1992: XXXVII). Hélène Cazes, who transcribed the document from old French to modern French, says that "mazorcas" – as it appeared in the original text – is a "Castilian measure of variable size" (note 25, XLII). Although it is true that this word can mean a "portion of linen or wool already spun in the spindle" (Real Academia Española,

2001: 998), this clearly is a misinterpretation. The correct meaning here is a *mazorca* (ear), that is, the fruit of maize.

When the Spanish chroniclers described the coast, they said there was "a great abundance of maize, with which they make bread and pies and great beverages, like the beer they drink ..." (Estete [Anonymous?], 1968: 396). When describing in general the "people who live below the line of the Equinox," Zárate in turn noted that "the Indian women plant, knead and grind the bread that is eaten in all of that province, that in the language of the islands is called maize, and in Peru zara" (Zárate, 1968: 119). When describing the North Coast, Zárate specified that it is "a most fertile land, and maize ... is planted and harvested all year long, without having to wait for a specific period" (Zárate, op. cit.: 125). As regards the Trujillo (Moche) Valley, Zárate said that it "... abounds plentifully ... in maize" (Zárate, op. cit.: 127).

When describing the department of Tumbes, Cieza de León (1984: 186, 202) noted that "maize comes twice a year ... ," a point he repeated in his general description of the coast. Molina "El Chileno" also described the coast and explained that "... each town of these had a large number of storehouses where they collected the maize and all of the provisions they gave the Inca in tribute ..." (Molina, 1968: 316). Vázquez de Espinosa (1948: [1221] 398) discussed the North-Central Coast, of which he said that "... a large amount of maize is harvested...." Vázquez then describes the Locumba Valley on the South Coast, "... formed by two rivers that come down from the highlands ...," which were "... sown ... with maize ... for everything [gave fruit] in great abundance because the land is quite fertile ..." (Vázquez de Espinosa, op. cit.: [1411] 477).

Molina (El Chileno) mentions the way maize was sown:

> ... in some parts of this coast there are other ways, never heard of, in which they plant their seeds and maize. Because where they do not have water nor does it rain, they fish a small, anchovy-like sardine. When they till the land they place two or three grains of maize on each anchovy they bury in the fields, and a very nice maize is born. They plant many good fields, three or four times a year...." (Molina, 1968: 313–314)

This also caught the attention of Cieza de León, but he specifically referred to the Chilca Valley, which was well known for this:

> ... [The Chilca Valley is] full of maize fields.... It is a remarkable thing to hear what is done in this valley. In order to have the required humidity [for the plants], the Indians make some wide and very deep pits in which they plant and put what I have already mentioned. And with dew and humidity God allows the maize to grow. But maize could not grow, nor the grains germinate [?: *mortificarse*] were it not that one or two heads of the sardines they catch at sea with their netting are placed with each [kernel]. And so, on planting they place them together with maize in the same hole they make to cast the grains

> in, and in this way it is born and grows abundantly. It certainly is a remarkable and never-seen thing that people may live as they will in a land where it does not rain, nor does anything fall other than a little dew. (Cieza de León, 1984: 215–216)

This also surprised Vázquez de Espinosa: "And with the freshness of the sand they sow in it maize in sardine heads – which are there called anchovies – and in the heads of other fishes that abound in that sea . . . thus the yield is abundant . . ." (Vázquez de Espinosa, 1948: [1333] 440).

A group of chroniclers wrote on the highlands. Sancho de la Hoz was one of these. He noted that "... little maize is harvested because it [the land] is too cold, and this [plant] does not exist except in the same type of land" (Sancho de la Hoz, 1968: 327). When describing the Collao, Sancho pointed out the problems living there raised, but noted that maize was one of the things that supported the people (Sancho de la Hoz, op. cit.: 331). This was mentioned by Zárate (1968: 131) too, who likewise mentioned that "the food the Indians in that land [the Altiplano] eat is stewed and toasted maize instead of bread. . . ."

Cieza de León (1984: 228) discussed Cajamarca and emphatically stated that here "... there is abundant maize ... ," and he pointed out the same for Huánuco (Cieza de León, op. cit.: 233). As for the department of Junín, he noted that "little maize is found here as the land is so cold ..." (Cieza de León, 1984: 240).

The zone of Jauja was described by Ruiz de Arce (1968: 427), who claimed that "in this land only sheep [i.e., Andean camelids] and maize grow ... ," whereas in the Cuzco area "nothing is grown but maize, and it is harvested only once a year." He then added that "the bread they eat is as follows: toasted or stewed maize" (Ruiz de Arce, op. cit.: 433). Pedro Pizarro also described Cuzco and pointed out that the Incas "sowed maize in all" of the *andenes* (agricultural terraces) (P. Pizarro, 1968: 514). As for the Altiplano, he pointed out that "... maize does not grow [there] ..." (Pizarro, op. cit.: 506). Zárate (1968: 197) also mentioned this region: "... No maize is grown in it because [the land] is very cold. . . ." And when Vázquez de Espinosa (1948: [1690] 599) described Santa Cruz, in Bolivia, he explained that they "... make a very good bread out of maize; wheat is not grown. . . ."

Pedro Pizarro mentioned the *mamaconas* and noted that "... the women busy themselves making chicha, which was a kind of brew they made out of maize, which they drank just like us wine ..." (P. Pizarro, 1968: 497). And while describing the Indian women, Pizarro tells how they walked behind the soldiers carrying "... the chicha," among other items, "... which was a certain brew they make out of maize – like wine. With this maize they made bread and chicha and vinegar and honey, and it is used as barley for the horses" (Pizarro, op. cit.: 578). Pizarro also explained that "[m]aize was the food of the poor Indians . . . ," and pointed out the significance of chicha had for the armies of the

Inca: "... Wherever they [the army] arrived they had large amounts of chicha that the mamaconas gave them ..." (Pizarro, op. cit.: 578, 555).

Several chroniclers mention the storehouses the Incas had, which so caught the attention of the Spaniards. Borregán thus wrote: "... they place ... maize in those storerooms ..." (Borregán, 1968: 464). In a 1534 manuscript, Sancho de la Hoz tells how, while taking a new group of Spaniards to people Jauja, they had a major clash with the Indians, who burned the city; here there was "... a large building that was in the plaza ... with much clothes and maize." It was burned so that the Spaniards would not seize it (Sancho de la Hoz, 1968: 291). One of these chroniclers discusses the storehouses of Cuzco and says these housed "... the provisions [for the] ... men of war," that is, "maize and wine of the type they usually do ..." (Estete [Anonymous?], 1968: 393).

Much has been written on Cuzco, and it is impossible to gather all the existing data in a study like the present one. Here only the more significant data are mentioned. For instance, Estete (Anonymous?) mentioned the Temple of the Sun when describing Cuzco, specifying that it had "... eight silver bins in which they had ... maize for the temple ..." (Estete [Anonymous?], 1968: 392). Cieza de León (1967: 93) described the Coricancha: "Around this temple there were many small dwellings.... They had a garden in which the clods were pieces of fine gold, and it was skilfully sown with maize plants, which were of gold, the canes as well as the leaves and ears. And they were so well planted that they were not pulled out even if strong winds blew." Molina "El Chileno" (1968: 329) likewise mentions this. He says that in the first patio there was "... a very big and well-crafted stone fountain where they offered chicha, which [is] a beverage made out of maize – like beer. They said the Sun came down to drink here. [The temple] had a golden maize field, with its canes and ears, before they entered [i.e., before the entrance to the enclosure] where the image of the Sun was...."

Interestingly enough, Garcilaso de la Vega (1959a: book 3, chapters XX–XXII, 268–275) describes the Temple of the Sun but does not mention the garden with the golden plants. All he says in chapter XXIV (op. cit.: 278–279; 1966: 187 [English citation]) is as follows: "That garden, which now serves to supply the monastery with vegetables, was in Inca times a garden of gold and silver ...," and he adds that "there was also a great maize field...." The chronicler who did leave a description of the Temple of the Sun is Father Cobo (1964b: 169). He mentions a "small patio" where the "statue of the sun was placed during the day," whereas at night it was kept "in its chapel," where it slept with "many *mamaconas*," that is, "women of the Sun." Cobo then adds that "in front of this chapel they had an orchard where, on days when they held feasts for the Sun, they drove canes of *maize* [into the ground] with their leaves and ears made of very fine gold, [and] which they kept for this purpose." Betanzos (1968: 247; 1996: 46) was able to collect accounts of the ceremonies Inca Yupanqui held on finishing the erection of the Temple of the Sun. He says the Inca ordered his people to "... have ready ... [many] provisions of maize ...," and that he then "... ordered that a big fire be built ... ," where camelids that had been beheaded, clothes "... and maize [were cast into] as a sacrifice to the Sun." Finally he decreed that his people should fast "... and eat only raw maize and drink chicha...."

When recounting the Andean customs in Inca times, Santillán notes that "... in places where no maize grew, the Inca gave out of his own" (Santillán, 1968: 399). And Betanzos collected an account of the legend of Manco Capac, where it was said that he "... and his companion, with the four women, planted some land with maize [at a given time]. It is said that they took the maize from the cave ... named Pacarictambo ..." (Betanzos, 1968: 214; 1996: 16).

The significance of chicha as a crucial element in reciprocity has already been discussed, and we shall return to this issue when it is discussed once more. Interested readers can find data in Vázquez de Espinosa (1948: [1219] 398), who even wrote: "... *Mingando* all relatives and friends, that is the same as to inviting them to work and to a feast. And so one and the other are both done with a solemn dance, feast and drunkenness."

Garcilaso de la Vega (1959b: 16–17; 1966, book IV, chapter III: 199) left an account of the duties of the "chosen [women]" or Virgins of the Sun, which among other things included "also brew[ing] the drink the Inca and his kinsfolk drank on the festivals, [which they] called in their language *aca*...."

Betanzos was quite meticulous when describing the rules Inca Yupanqui laid down for those who were to become *orejones* ("big ears," i.e., noblemen). Here it is worth going into detail. This was a very complex ceremony. At the beginning a festival was held in which a group of women "... would make four jugs of *chicha*. These jugs of chicha would be ready from the time they were made in this fiesta until the end of the fiesta of the Sun. And the jugs should always be well covered, and each jug should contain five arrobas [i.e., about 322.66 liters]...." The participants had to follow several rules, among them not "eat[ing] anything other than raw maize." A "young maiden" who fulfilled certain conditions was asked to "... make a certain jug of *chicha* called *caliz* ... ," and she should "... always walk along with the young man and serve him [the chicha she had prepared]...." Later on, and after fulfilling several requirements, the candidate went before a huaca "... with the maiden carrying that little caliz jug[, and] she will fill two small tumblers of *chicha* and give them to the neophyte. He will drink one of the tumblers and give the other to the idol by pouring the *chicha* out in front of it...." The neophyte then returned to the city and went to a huaca, where he made a sacrifice by "... offering a certain *chicha* and making a fire before it. In the fire they will make [some] offerings of maize ..." and other items. This burning of maize was repeated later (Betanzos, 1968: 263–265; 1996: 60–61). When the ceremony was over, the neophytes returned to their homes and "... will get out those four jugs of *chicha* that they made at the beginning of the fiesta. They will drink from the jugs. And they will get the neophyte so drunk on the chicha that he will pass out...." It was then that they would "... pierce his ears" (Betanzos, 1968: 267; 1996: 63 [English citation]).

Betanzos also left testimony of the time when Inca Yupanqui assembled the lords of Cuzco to organize a meeting: "The lords told him that it was a good thing and well conceived. They decided to give the order to make a large amount of chicha." The day of the festival "... many large jugs of chicha were brought out on the square." Betanzos likewise describes how in some ceremonies "... they started drinking the *chicha* that they had there. According to what they say, they had an immense quantity of it there." The Inca then asked that they fill the storehouses with many items, and the lords sent messengers to have this done (Betanzos, 1968: 257–258; 1996: 56).

Betanzos wrote a magnificent account of the way in which Pachacuti Inca Yupanqui gave orders to his lords before beginning his campaigns to conquer new lands. First a ritual had to take place.[9] So he "... ordered that in their land they should leave a large garrison of *principales* and majordomos. Each of the Cuzco *orejones* should pour into the river certain tumblers with chicha, and they should likewise take other tumblers with chicha, pretending to drink with the waters." The instructions the lords received were as follows:

> If a lord or lady goes to the house of another one to visit or see him, he should take with him a jug of chicha, if a lady. And on getting to where the lord or lady he is visiting is, he shall have his chicha poured into two tumblers; from one will drink the lord that he is visiting, and from the other the lord who gives the *chicha*, and so the two drink. And the host does the same thing. He has two tumblers with chicha brought, and he gives one to the visitor and drinks himself from the other one. This is done between those who are lords, and this is the highest honour amongst them. If this is not done when they visit each other, the visitor feels dishonoured because he was not given to drink, so he will excuse himself of going to see him again. He who gives another to drink likewise feels dishonoured if the drink is not received. So when they make the sacrifice to the waters mentioned above, they say they drink with them [and] pour a tumbler of chicha into the river, and as this is done they drink the other one. (Betanzos, 1968: 288–289)

Betanzos also tells how on reaching the province of Soras, Inca Yupanqui, to subdue the "lords" of the land, "... he ordered them to ... splash a certain amount of *chicha* over themselves ..." (Betanzos, 1987: 93; 1996: 87). And when this same Inca returned to "... the city of Cuzco [and was] enjoying himself ..." he "made a maize god whom he called çaramama, saying that was the mother of maize, [and] another chicha idol ..." (Betanzos, 1987: 99; 1996: 92). And when Topa Inca Yupanqui, the son of Pachacuti, left to conquer

[9] The passages cited here and in the following paragraph are missing in the manuscript published in 1987 by María del Carmen Martín Rubio, which was translated into English by Ronald Hamilton and Dana Buchanan (1996).

the "Andesuyo," he found that the Indians there "... cultivated some fields of maize ..." (Betanzos, 1987: 134; 1996: 125).

Cieza de León (1985: 48) likewise mentioned the lands occupied by the Incas and explained that they ensured that those lands without maize were supplied in such a way that they eventually had it in abundance.

There is a vast amount of information in the chronicles on the rituals and sacrifices made in Inca times. I mention here just some wherein maize is somehow involved. For instance, Garcilaso de la Vega says that "for their solemn sacrifices they used ... a maize loaf called *çancu*, and they made the same bread to eat as an occasional delicacy: they called it *huminta*. The two names were applied, not because the bread was any different, but because one kind was used for sacrifices and the other simply for eating...." He also added that "they also made porridge, which they call *api*, and ate it with great relish, because it was only consumed on rare occasions" (Garcilaso de la Vega, 1959c: 129; 1966, book VIII, chapter IX: 498).[10]

Father Cobo describes what was offered in sacrifice to the gods, and he explains that with the offerings "... they poured chicha ... ," and he repeats that "... the foods were burned and the *chicha* was poured on the ground" (Cobo, 1964b: 203).

Molina "El Chileno" (1968: 340) recounted the festival held in Cuzco in April, "... when the maize and the fields were harvested in the valley of Cuzco, in which harvest the lords of Cuzco used every year to make a great sacrifice to the Sun and to all of their *huacas* and shrines in Cuzco, here and in all the provinces and kingdoms...." He then added that

> More than 200 young women left Cuzco at eight in the morning, each with a big new jar of slightly over an arroba and a half [i.e., slightly more than 24 liters] of *aca* [chicha] that were sealed with a mud lid. All of the jugs were new, and with the same new lids and the same sealing they came in groups of five in great order and disposition. From time to time they waited and offered it to the Sun...." (Molina, 1968: 341–342)

Acosta in turn describes the festival held in the month of May:

> The feast called *Aymoray*, much used among the Indians nowadays, was held in this moon and month, which is when maize is brought home. This festival is held coming home from the fields. They chant some songs in which they beg maize to last long. They call maize *Mamacora* and take from their fields part of the maize that most stands out in [the] amount [given out], and place it in a small bin they call *pirua* along with certain ceremonies. They stay awake watching over the maize for three nights and put it in the richest blankets they have. When the maize is covered and dressed-up, they adorn this *pirua* and hold it in great veneration, and say it is the mother of the maize that is in their fields, and that with this maize [it] grows and is preserved. Around this month

10 *Api*, according to González Holguín (1989: 31), is "maçamorra."

> they make a specific sacrifice and the sorcerers ask the *pirua* if it has enough strength for the coming year. If it says no, it is taken to the fields to burn it with as much solemnity as possible, and they make another *pirua* with these same ceremonies, saying they are renewing [it] so that the maize seeds will not die. If it answers that it has strength to endure more they let it be until the following year. This impertinence has lasted to the present day, and nowadays it is quite usual for the Indians to have these *piruas* and to hold the festival of the *Aymoray*. (Acosta, 1954: 175)

Acosta (op. cit.: 160) also tells us that "they have a particular reverence and veneration for the confluence of two rivers. Here they wash to heal themselves, covering themselves first with maize flour or with other things, and adding different ceremonies …"

When describing the sacrifices made in Inca times, Cobo points out that "… they cast flour of white maize to the sea as an offering"; he also explains that for the sacrifices they prepared a mixture of "… maize flour …" with other substances (Cobo, 1964b: 203). When describing the activities of the Indians, Gutiérrez de Santa Clara (1963: 231) specifies that they "also made offerings of much fruit, bread, [and] wine of the land to the Sun and Moon. …"

There is also abundant information on the ceremonies held in homage to the sun god. When recalling the "things they sacrificed to the Sun," Garcilaso de la Vega says they "… offered as a sacrifice much of the brew they drank, made of water and maize …" (Garcilaso de la Vega, 1959a: 153; 1966, book II, chapter VIII: 86). And when recounting the festivals held for the sun, Garcilaso notes that "that night the women of the Sun busied themselves with the preparation of enormous quantities of a maizen dough called *çancu*, of which they made little round loaves the size of an ordinary apple …" (Garcilaso de la Vega, 1959b: 200–201; 1966, book VI, chapter XX: 357). And in chapter XXI Garcilaso mentions this brew as "the beverage they drink"; interestingly enough, he does not use the term chicha, nor does he say it was made out of maize (1959b: 202–203; 1966: 356). In chapter XXIII Garcilaso details the rules followed when "they drank to one another, and in what order," that is, the Inca, his captains, and the curacas (1959b: 207–209; 1966: 363).

Another curious testimony is that given by Andagoya, who according to Porras Barrenechea (1986: 70) wrote in 1541–1542:

> The ceremonies and rites they have in this land have the Sun as a divine thing, to whom they make sacrifices and offerings. The order they have in this is that when the sun comes out they take out to the plaza many jugs of *chicha* – the wine they make – and other provisions, which they place in the plaza for the Sun. Here they pour the wine in certain ceremonies, and they worship [*haciendo la mocha*][11] the Sun. … (Andagoya, 1954: 247)

[11] The term is *mochar* in Quechua, from *muchay*, which means to venerate, to worship, or to idolize.

Cieza de León in turn left a description of the Hatun Raimi festival. After the sacrifice was made, the high priest went to the Temple of the Sun along with other priests,

> ... and after saying their cursed psalms they made the Virgin mamaconas come out, richly dressed and with much *chicha* of that which they had already made. And between all those in the great city of Cuzco they ate the livestock and birds they had killed for this vain sacrifice and they drank that chicha, which they held as sacred. They gave it out to drink in large golden cups, and it was in large silver jars, of the many that were kept in the temple. (Cieza de León, 1967: 104)

Molina[12] mentions the "toasted maize" and says that "... maize was burned." Of the Capac Raymi, Molina reports that "... in Cuzco they made a large amount of *chicha* ...," and he explains that on orders of the "sorcerer," the sick gave the dead to eat by "... pouring *chicha* on them" (Molina, 1916: 28, 34, 59, 102).

There are also interesting data on the funerary ceremonies. For instance, Sancho de la Hoz (1968: 330) tells that the "maize fields" of the dead lords "... were sown for them, and a bit is placed on their tombs." Betanzos (1987: 142; 1996: 131) found out that when Inca Yupanqui felt death drawing near, he himself stipulated "the pagan rites" that should be followed. Among other things, he ordered that the people of his lineage "... should drink much *chicha* and get intoxicated, and while they were drunk they would be strangled ... [and] buried." He likewise specified that besides jewels, the women had to take with them "... small jars full of *chicha* ... and pots full of toasted and cooked maize...." Inca Yupanqui also ordered that in the provinces, "... a distribution should be made of all the maize ... in the storehouses of each province ...," and that "much *chicha* would be made and given to these [the local population] ..." (Betanzos, 1987: 142; 1996: 132, 133).

Cieza de León (1984: 275) tells how the funerals were carried out among the Colla in the Altiplano: "In the days when they cry their dead before burying them, they made a lot of their wine or beverage to drink with the maize of the deceased, or that which the relatives had offered. And if there was a large amount of this wine they have the deceased for more honoured than if little was spent."

Finally, there is an interesting and truly impressive account of what happened after Cuzco was seized by the Spaniards. Quizquiz, one of the generals of Atahualpa, was mustering his forces outside the capital. Pizarro then sent Almagro with a detachment of Spaniards and a large number of Indians. At their return a festival was held in Cuzco, and the mummies of the Incas were brought out. Estete (or the anonymous chronicler) describes the drunken orgy that followed:

[12] Molina was a Spanish parish priest in the city of Cuzco.

They drank ... the wine, because although that which they drank was made out of roots and maize – like beer – it sufficed to make them drunk. There was so much people and they defecated such good excrement – both men and women – and so much was placed in those skins because all that they do is drink and not eat, that it is unquestioningly true that all day long the two canals of more than half a *vara*[13] each, that empty below the slabs into the river from which they drank, and which had been made to clean and as an outlet for the rains that fell on the plaza or by chance [*o por ventura*], were full of the urine which they peed into them, in such abundance that it seemed there were fountains there that gave out urine. The people who drank it are not to be wondered at, although seeing it is a marvel and something never before seen ... [These celebrations lasted for] over XXX unbroken days, and so much wine of that kind was used up that all of the gold and silver seized would not be enough to buy it. (Estete [Anonymous?], 1968: 400–401)

[13] A vara is an ancient Spanish unit that is equal to 0.8359 m (Llerena Landa, 1957: 205).

8

The Dispersal of Maize around the World

Columbus introduced maize into Europe in May 1493, after his first voyage, when he gave an account of his journey to the court in Barcelona. Pedro Mártir de Anglería was there, as was pointed out in Chapter 2. And it was he who, in a letter to Cardinal Sforza in mid-November 1493, wrote the following of the New World:

> Bread they also make – with a slight difference – from a certain floury wheat of which the people of Insubria and the Spaniards of Granada have an abundance. The ear has more than a *palmo*[1] in length, tends to form a point and is almost as wide as an arm. The kernels are admirably arranged by nature: in form and size they resemble the *legumbre alverjón*;[2] when green they are white, when they mature they become black, [and] when ground they are whiter than snow. This kind of wheat is called maize. (Anglería, 1944, First Decade, book I, chapter III: 8)

Mártir himself evidently saw the plant grow, and although he wrote in Latin and turned "panizo" into "panicum," he was the first to describe it as "... maizium, id frumenti genus appelant" (this genus of grain they call maize) (Anglería, op cit.; see also Sauer, 1969b: 153–154).[3]

The testimony given by Fernández de Oviedo – who wrote in the first half of the sixteenth century – is interesting:

> ... I say that when Her Majesty the Empress was in Ávila at the time that the Emperor, our lord, was in Germany, I saw inside a house in that city, one of the coldest in Spain, a good maize field with the canes ten *palmos* high, more or less, and as thick and green and beautiful as can be seen in these parts

1 *Palmo* is a unit of measure with two meanings. It can either be the distance between the thumb and the little finger, when the hand is fully extended (Real Academia Española, 2001: 125), or an old Spanish unit that is equal to 0.2089 m (Llerena Landa, 1957: 151).

2 This may refer to the plant known as the hairy tare (*Vicia hirsuta*).

3 The data Mangelsdorf (1974: 206) gives on this point, based on Weatherwax (1954), is wrong.

> [America] where it best grows, and there it had a waterwheel with which it was watered every day (Fernández de Oviedo y Valdéz, 1959: 230).

Also interesting is the testimony of Garcilaso de la Vega, who was already familiar with this plant, and who on reaching Spain in 1560 noticed that "the seed of the hard maize [*muruchu*] is the kind that has been introduced into Spain: the soft sort has not been brought here" (Garcilaso de la Vega, 1959c: 128; 1966, book VIII, chapter IX: 498). But it is not clear whether or not this is correct, for the first varieties to reach Europe were the West Indian ones that gave rise to the current varieties, and Garcilaso in all probability was not familiar with them (Haudricourt and Hédin, 1987: 223). In fact, the "Early Caribbean" race from Haiti, which is considered an ancient race, was one of the earliest ones introduced into Europe (W. L. Brown, 1953), and there are archaeological remains of it (Newsom, 2006: 330).

The first document believed to deal with the cultivation of maize in Europe is a reference that appears in the *corps de ville* of the city of Bayonne in 1570. Another document was then found in the municipal archives of Bayonne with an older date, which would be "… the first secure trace of maize in Europe. …" This is an order given by Odet de Foix, viscount and marshal of Lautrec and seneschal de Guyenne to the magistrates and "sergents de Labour," dated in Bayonne, on 14 May 1523 (Goyheneche, 1966: 114; see also a copy of this document on pp. 118–120). The term *arthomayro* or *arthomayre*, which must be translated as "maize," is used in this document. Nowadays maize is *d'arthoa* in Basque, a word that once was used for millet. An analysis of the order given by Odet de Foix shows that the description of the plant coincides with that of maize; besides, the document gives very interesting details on the cultivation and use of the plant. One interesting detail is that maize "… could be used only to feed the pigs" (Goyheneche, 1966: 115; see also p. 118 of the document, and also Lefebvre, 1933). Goyheneche says that "… maize was known and farmed a lot since 1523, i.e. barely thirty one years after the discovery of America … ," and he believes that it came from what is now the Dominican Republic (Goyheneche, 1966: 115–116). Goyheneche concludes that "… the Basques certainly were amongst the first to farm maize in Europe and this lets one presume, as Larramendi claimed, that they introduced it into the Old World" (Goyheneche, 1966: 116). Larramendi ([1882] 1969: 66–67) said of maize that "… it is more useful in Guipúzcoa. It was brought from the Indies first to this province, and it was brought by Gonzalo Percaistegui, a native of Hernani, and it was then passed on to other provinces." No information on Percaistegui has apparently been found, but there is a document supporting this statement, because according to Berraondo (1927: 305), a "Historia del Emperador Carlos V" (A History of the Emperor Charles V), written by Friar Prudencio de Sandoval (1560–1612), claims there already was maize in San Sebastián in 1521.

When Father Cobo wrote between 1613 and 1653, he noted that "the maize plant is already well known in Spain as the *wheat of the Indies* . . ." (Cobo, 1964a: 159).

Yet after its introduction in Spain, maize was cultivated only in some areas, like Galicia and nearby Portugal, where it became a staple. Even so, the history of maize in Spain has yet to be written, as was pointed out by Sauer (1969b: 152–153). It is known that maize was cultivated in Castile in 1498, but it was only grown in Andalusia in 1826 (Soukup, n.d. [1987]: 434). It is interesting that there is little evidence that maize went east from Spain. Its use was more significant in the eastern Mediterranean and in Italy than in Spain (Sauer, 1969b: 151).

After its introduction by the Spaniards or by other explorers, maize spread rapidly in all the countries where it could be grown. It was considered a botanical curiosity, and its agricultural potential began to be explored. In Western Europe information on maize was first sought in the classical authors Theophrastus, Dioscorides, and Pliny, but nothing was found in them on the American plants (Weatherwax, 1945: 173–174).

We must bear in mind that maize is a plant that had lived in tropical regions with short days. It is likely that on reaching Europe, the first plants had a lush vegetation. They had to adapt, and it was only after several centuries that the European varieties appeared (Gay, 1987: 460). The introduction of this new plant in the Old World pushed back the cultivation of millet. In Europe maize took over the zones where foxtail millet (*panizo*, *Setaria italica*) was grown (Haudricourt and Hédin, 1987: 223). The advantage maize has over its plant counterparts in the Old World is that it can thrive in areas that are far too dry for rice, and far too humid for wheat. In geographical terms, maize is clearly anchored between these two plants. Its most invaluable characteristic is its high output per unit of land, which has a world average of twice that of wheat. There are very few other plants that give such a large amount of carbohydrates, sugar, and fat with such a short growing season (Crosby, 1975: 171).

It was for these reasons that maize spread all over the world in less than 300 years and became a major staple in many of them (Dowswell et al., 1996; Paliwal, 2006: 6).

This American plant was used at first in Europe as an ornamental plant and was grown here and there in the sixteenth and seventeenth centuries, but it was not significant as a major crop in large areas until the late seventeenth century (Crosby, 1975: 176; Trucco, 1935: 973). Europe in fact had still not understood the real value of maize 250 years after the discovery of America. The study of this plant began first in the New World, by learning from the natives, and this knowledge was then taken to Europe (Weatherwax, 1945: 177–178). White men at first accepted as is the Indian races of maize, and it was only in the late eighteenth century that an increasing effort was made to try and improve them (Mangelsdorf, 1974: 209).

Arthur Young (1793, volume 1: 643, 645, 647, 650; volume 2: 353), an expert farmer and journalist, saw a large number of maize fields in northern Spain in the late eighteenth century, and travelers who journeyed to Portugal at this same time noted that this plant was the major staple of the peasants. It is significant that the Spanish population diminished in the seventeenth century and grew in the eighteenth (Crosby, 1975: 179).

John Locke noticed that around 1670 there was much maize in many parts of southern France, where it was known as *blé d'Espagne* and was used to feed the poor. It continued spreading in the eighteenth century and became a key diet component in southern France, and it may have been behind the rise in population after the population decrease in the early seventeenth century. The name given to it implies that maize had been imported from Spain (Crosby, 1975: 179). Thanks to Arthur Young, we know that it was a major product in southern France by 1780: "Where there is no maize, there are fallows: and where there are fallows, the people starve for want. For the inhabitants of a country to live depending upon that plant, which is the preparation for wheat, and at the same time to keep their cattle fat upon the leaves of it, is to possess a treasure" (Young, 1793, volume 2: 241).

It seems, according to Dodoens (1578), that maize was taken from France to England in the late sixteenth century.[4]

In Italy, thanks to the 1656 "Relazione Grimani" (Grimani Account), we know that maize was a part of the diet of rural populations in the Friuli area after 1630. And according to a 1656 inquiry, the cultivation of maize was competing with that of other cereals. In the piedmont, its cultivation was limited to the areas where sorghum was grown. In Lombardy, maize did not alter the traditional cultivation systems, and the same thing happened in the Veneto and in the Romagna. In Friuli, in 1782, 1795, and 1804, the output of maize could be 10 to 15 times higher than that of all other grains. Maize was well established in many Italian communities by 1656, yet in others it was not unknown but was almost ignored (Fornasin, 1999).

Maize was cultivated from quite early on in the Po Valley. When Goethe (1962: 20) made his famed journey to Italy in 1780, he discovered that polenta was an essential component of the diet in this region. Maize had some kind of role, at least in northern Italy, in the recovery of its population in the second half of the seventeenth century (Crosby, 1975: 179).

Curatola says of this plant that

> ... just like in Veneto and Lombardy ... where it began to be grown for human consumption, maize was long considered a secondary product. It was only since the eighteenth century, with the development of a large-scale market economy, that maize gradually became the predominant, if not exclusive, crop

4 Several authors, for example, De Wet and Harlan (1972: 271), cite this work under the name of Lyte, who was not its author but its translator.

> in some regions of Southern Europe. Its high output, far beyond that of any other cereal, and hence its enormous profitability in economic terms, made the big landowners of Spain and Northern Italy adopt it as a monoculture wherever conditions were more favourable. Maize ... thus became the major, or even better, the sole source of sustenance of whole rural populations, ever more oppressed and impoverished, [who were] forbidden even of growing their traditional and most needed food. (Curatola, 1985: 22; 1990: 130–131)

Curatola then says that "from the fertile Po Valley ... where it had shown all of its potential, maize spread to Southern France and then along the Danube River basin, and thence passed over the sea to Egypt, South Africa, India, the United States and other parts of the ecumene with a warm and temperate climate" (Curatola, 1985: 11; 1990: 131; he based his work on Messedaglia, 1927; Roe, 1976: 1–2, 20–29; and Rousell, 1845: 345–376). This is not wholly correct, for we have seen that maize passed from Spain to France, and as we shall later on see it reached Africa following other routes. As for the United States, we saw in Chapter 5 that maize has an ancient history since pre-Hispanic times.

When discussing the early maize found in the Balkans and in Turkey, Cutler and Blake (1971: 374) recall the routes the Spanish used after its discovery, which explain the similarity in the crops found here. Besides, flint was traded for firearms between the Balkans and Spain, and there was a long list of exchanges between Italians and Greeks. In 1800 maize was known and was widely used in the Balkans. In the nineteenth century the population increased, and one of the factors behind this was the cultivation of American crops. Many peoples followed the example set by Hungary and moved from pastoralism to farming, with maize as their major crop. The Serbs are the best example (Crosby, 1975: 180). Maize is nowadays more important in southeastern Europe than in the southwest (Crosby, op. cit.: 179).

The lands of Romania and the former Yugoslavia are among the biggest producers of maize in the world. But the significance of this plant in the Balkans and its vicinity seems not to predate the early seventeenth century. Geographers and travelers who wrote on this region rarely or never mention maize. This and other American plants, like potato and squash, began to expand when the population pressure increased in the eighteenth and nineteenth centuries (Crosby, 1975: 179–180).

Romania is a classic example of the use of maize in the Old World. Maize was not introduced there before the eighteenth century, or at least it was of no significance before then. Yet in the late nineteenth century, the Romanians used maize and depended on it as much as did the Mexicans. They cultivated wheat and maize, the former for export and the latter for food. Maize adapts well to rotation farming, and this enabled the Romanians to become one of the major European producers. *Mamaliga* is a dish prepared with maize that has become the major, if not the only, dish in the food of the Moldavian peasants. And whenever there is a celebration, they drink spirits made from maize, just like the mountain people of Tennessee. No other country in the Balkans adopted maize so strongly as the Romanians, but it extended to other neighboring areas from 1900 onward. In the late nineteenth century, an expert on the Balkans, when describing a typical Macedonian town, said that its houses were surrounded by fields of maize (Eliot, 1965: 328). Halpern (1958: 75–58) noted of a Serbian village that the poorest peasants of Orašac still ate maize instead of wheat bread and grew at the time more hectares of maize than wheat, due to its far superior yield (Crosby, 1975: 180–181).

Hungary is another good example. There was a massive immigration when the Turks abandoned it, and it shifted from being a stockbreeding society to one of farmers. In the late eighteenth century, maize was the main product of eastern Hungary. And it was mostly due to Hungary that the Hapsburg Empire was the biggest European producer of maize in the nineteenth century (Crosby, 1975: 180).

Maize reached the Caucasus only in the seventeenth century and was known as "the food of the Narts," who are very ancient mythological heroes in the northern Caucasus (Haudricourt and Hédin, 1987: 113).

In Africa, the slave trade carried out in Columbus's time ensured that the transfer of flora would take place before it did in Europe (Crosby, 1975: 196). It is believed that Portuguese sailors introduced maize into Africa in the early sixteenth century, precisely within the context of the slave trade. Maize was imported in various places at the same time (Miracle, 1966).

In western Africa, maize was under cultivation at least by the second half of the sixteenth century, and perhaps even before. It spread rapidly in rainy forest areas. In the seventeenth century there was abundant maize in the Gold Coast, as well as on the coasts of the Congo and Angola. It spread inland at around the same time. In 1900 maize spread all over Africa, except for Uganda. The Boers found the Bantu growing maize when they moved northward from the Cape Colony in the early nineteenth century. *Mealies*, as this maize is now known, is the main staple of the Bantu diet. Nowadays South Africa is one of the largest producers of maize in the world – 70% of its surface dedicated to agriculture is devoted to this plant. Maize continued its expansion in the twentieth century and became for the first time the mainstay of eastern and central-tropical Africa (Crosby, 1975: 186–187). Maize has displaced sorghum in southern and eastern Africa, whereas in western Africa it is instead a garden plant (Harlan, 1995: 186).

Yet on analyzing the words for this plant in the African languages and dialects, one reaches the conclusion that maize in many cases arrived not through the Atlantic Ocean but from Egypt, through the Lake Chad region, and from Arabia via Zanzibar, Madagascar, and Mozambique. When Napoleon was in Egypt in the late eighteenth century, the Egyptians called maize "Turkish wheat" or "Syrian wheat." When Leonhard Rauwolf, a "proto-botanist," was in the Middle East in 1570, he mentioned "Indian millet," that is, millet, along the

Euphrates River and in the fields around Aleppo and Jerusalem. A maize specimen he collected in the Euphrates Valley in 1577 is in the Leyden herbarium (Dannenfeldt, 1968: 97, 254).

Maize apparently reached the Middle East in the sixteenth century. There is little documentary or archaeological evidence, but the linguistic data abound. The early names that appeared in Europe, some of which are still used – as was seen in Chapter 2 – were *granoturco*, *blé de Turquie*, *Türkisher Korn*, *Turkie wheat*, and *trigo de Turquía*.

Nowadays maize has a secondary role in the Middle East, except in Egypt, where the *fellaheen* depend on it. Maize reached Egypt quite early, in the sixteenth century, but it did not become important until the eighteenth century. Egyptian soils receive much sun, have abundant water, and are of high quality, and they are thus ideal for maize (Crosby, 1975: 188–190).

There is not much early information on India. It seems that maize spread there in the early nineteenth century; however, there are some regions in which it was somewhat important. Here it is known as *Mecca*, *Makka*, *Makkaim*, *makai*, and *mungar*, which indicates that it is a food from "Mecca," which means "God," or that it reached India from the Islamic area – a more believable explanation. In any case it spread rapidly and displaced millet. In the late eighteenth century it spread all over India, where the people of the mountains are highly dependent on it, and it is the major staple in the north, in the Punjab, in the northwestern provinces, and in the Oudh. George Watt, a botanist who studied India toward the end of the nineteenth century, noted that *Makkal*, that is, maize, was extremely widespread and was essential for the common people (Crosby, 1975: 192; Watt, 1888–1899).

Maize probably reached southern Asia in the early sixteenth century (Brandolini, 1970) from Zanzibar through the Portuguese and Arab merchants. It is likely that it was introduced into the northeastern Himalaya region by merchants along the Silk Road, from whence it spread to neighboring regions (Dowswell et al., 1966) (Paliwal, 2006: 6). We know that by 1659 it was a major crop in Indonesia, the Philippines, and Thailand (Paliwal, op. cit.: 6). In 1699 it was the most important product in Timor. Maize, however, was of little significance in the East Indies in the seventeenth century, but by 1800 it had become the second major crop, at least in Java. By the mid-twentieth century maize had a secondary role among the cereals in Indonesia, and a major role in the Celebes, Timor, Lombok, eastern Java, and the island of Madura (Crosby, 1975: 195).

Maize entered China in the early sixteenth century, following maritime and land routes (Ho, 1956). It spread into southern China in the provinces of Fujian, Hunan, and Sichuan around 1750 (Paliwal, 2006: 6). The first Chinese drawing of maize was published by Li Che-Tchen in Pen-ts'ao kang-mou in 1590, and it is a pod corn (Haudricourt and Hédin, 1987: 223). No human group in the Old World adopted the American food plants as rapidly as the Chinese. Peanuts were already growing close to Shanghai while those who seized Tenochtitlán with Cortés were still living, and it, along with the sweet potato, were turning into a major staple for the poor people of Fujian. By the late eighteenth and early nineteenth century, maize had become a major plant in vast areas of the highlands of southwestern China. The northern farmers took a long time in accepting maize and did not cultivate it in significant amounts until the nineteenth century. Nowadays a large part of the food in northern China is based on maize. In China maize feeds men and not animals, just like in Egypt, India, and Indonesia, and definitely not like in the United States (Crosby, 1975: 199).

Maize was introduced in Japan c. 1580 by Portuguese sailors (Paliwal, 2006: 6; Suto and Yoshida, 1956), but it never gained significance there (Crosby, 1975: 197).

I was essentially unable to find data on the dispersal of teosinte over the world. According to Iltis and Doebley (1984: 590), seeds of this plant were taken to Göttingen around 1832, where they grew as an unknown maize-like grass. It was called *Euchlaena mexicana* and was classified in the Olyrineae family, but its similarity to maize was never mentioned. Teosinte appeared once again in Europe in 1849, but with the generic name of reana (from *Reana luxurians*, a synonym of teosinte), and it was correctly placed beside *Zea*.[5]

[5] Readers interested in the dispersal of maize throughout the world should read Weatherwax (1954).

9

Chicha

This chapter does not intend to present an exhaustive description of chicha from an ethnographic standpoint, that is, the different ways it can be prepared, the various customs associated with it, or the role it has in Andean society. All it intends is to show, by comparing current-day data with historical evidence, that part of the ancestral knowledge that existed in the Andes regarding this beverage is now lost.

"Chicha" is such a common name in Peru and Bolivia that many think it is a native term. Interestingly enough, the original name has been forgotten. The word "chicha" was first collected by Giovanni Battista Ramusio around 1521. It later on spread to Hispanic America as a label for alcoholic beverages made out of grains and fruits. It is possible that it had its origin among the Cuna aborigines of Panama, the Arawak in the West Indies, or the Aztecs in Mexico (Chávez, 2006: 627–628; Corominas and Pascual, 1989, volume II: 354).

Pedro Pizarro, who took part in the conquest of Peru right from its very first stages, used the word "chicha" in his chronicle but did not explain it (P. Pizarro, 1968: 474). A long time afterward Father Cobo clearly said that "the name *chicha* is not of this kingdom; I believe the Spaniards took it from the language of the Island Hispaniola. In the Quechua language of Peru it is *aca* [azua (?)], and *cusa* in Aymara" (Cobo, 1964a: 163).

For Nicholson, the origin of the word "chicha" is not clear, but it does seem to be a Carib (Arawak) term derived from *chichal* or *chichiatl*. *Chichilia* means "to ferment," and *atl* means "water." But there may also be another explanation. *Chi* means "with," and *chal* means "saliva"; together they may mean "spitting" or "to spit." This term describes the main way in which chicha was prepared in past times, using saliva to convert the starches into sugars, as is explained later on, and so to facilitate fermentation and increase the content of alcohol (Figure 9.1). The standard practice was to chew maize flour, a task usually carried out by women. The term in Quechua is *akka* or *acca*, and *kufa* in Aymara. The chewed-up flour is *muko* in Quechua, and those who chewed it were called *muccupuccuk*. The practice of chewing maize flour is nowadays quite rare in the highlands and is practically missing on the coast (Nicholson, 1960: 291). Sauer

9.1. A drawing from Girolamo Benzoni's *Historia del Mondo Nuovo di M. Girolamo Benzoni Milanese. La qual tratta dell'Isole, & Mari nouamente ritrovati, & delle nuove Città da lui propio vedute per acqua & per terra, in quattordici anni. Con Priuilegio della Illustrissima Signoria di Venetia. Per anni XX. In Venetia. A presso Francesco Rampazetto.* It shows Indians preparing chicha in the Antilles. In the middle ground two Indians are straining the bolus while another one is stirring the mixture in a vessel over a fire. In the foreground an Indian spits into a vessel the maize kernels he has chewed. Engraving from Benzoni (1565). Carlo Radicati di Primeglio kindly gave this picture to Bonavia while he (Radicati) was preparing the Spanish editon of Benzoni.

(1950: 494) believed that "chicha" was an Arawak word. The *Diccionario de la Real Academia Española* (2001: 355) says the following of "chicha": "From the aboriginal Panamanian word *chichab*, maize."

Valdizán, on the other hand, based his work on a study by Barberena (1894; nota bene: Valdizán mistakenly gives 1895 as its date of publication, and the reference he gives is incomplete) and ascribed it to Nahuatl, because for Barberena "chicha" means spitting. *Chi* would be "with," and *chal*, "gob of phlegm" or "with a gob of phlegm," that is, a beverage "prepared with saliva or by the action of the saliva." The word *azua* in turn would come from *at* or *ah*, which mean "water," and *zu*, "to suck," that is, it would mean "water or liquid that is sucked" (Valdizán, 1990: 137). Staller reports a personal communication he had with Brian Stross, who suggested to him that the term is not Arawak but Nahuatl, which would support Barberena (see previously). In Nahuatl the term would be *chichiya*, which according to Molina, the chronicler, means "to drink bitter bitter," which apparently refers to the process of fermentation. Stross says the term is still in use in Guatemala (Staller, 2006: 449–450). Staller unfortunately does not give the reference to Molina, so the information cannot be checked.

Chávez (2006: 628) says that the equivalent word for chicha in Aymara nowadays is *k'usa*, just like it appeared in the vocabulary published by Bertonio (1956, part I: 160; part II: 66), where we read "*kusa*." In contemporary Quechua the word used is *aqha*, which is sometimes pronounced and written as *aha*. In his *Vocabulary*, Gonzáles Holguín wrote: "Aka The açua or chicha," whereas under the word "chicha" we find the following: "Aka strong chicha viñapu aka" (González Holguín, 1989: 18, 470). Chávez suggests that it must have been pronounced as *aqha* and not as *aka* or *aca*, which means human feces (Chávez, op. cit.: 628). According to Gonzáles Holguín, *aca* in fact means "all dung from man or an animal that is not small" (González Holguín, op. cit.: 11). Yet *aka* only means "chicha," as has already been pointed out. No other meaning was found for it. The interesting thing here is that no one has pointed out the fact that this word does exist in Quechua but with a completely different meaning, for under "*chicha*," González Holguín (op. cit.: 107) gives, "ossota or shoe with two or three soles," and under *chichani* we read, "to sew the shoes or ossotas that have two soles."

Santo Tomás (1951: 229) gives *azua* or *açua*; the latter term could be derived from some extinct native language (Chávez, 2006: 628).

Morris (1979: 22) calls the chicha *aqa* and explains that it can be made out of several products – peanuts and yucca, as is used in the tropical forest. He points out that chicha can be prepared in various forms; these forms may be regional variants, but at the same time they can also be related with the personal specialties of those who prepare them.

Gillin discussed this subject and said that chicha is a general term for a beverage that is common to all of Hispanic America. Arona (1938: 165–166) defined it as an old Spanish term but later on says (Arona, op. cit.: 177) that it could be from the Caribbean. Based on Valdizán and Maldonado (1922, volume 2: 6), he then gave the Quechua terms *acca*, *aka*, *asuha*, and *khusa* for chicha and pointed out that Farfán (1941: 234) also uses *aqha* (Gillin, 1947: 46).

The testimony of Tschudi is significant in this regard:

> Among the Khetsuas this beverage was known as *Akha* or *Asiva*; among the Tsintsaysuyus it was usually known as *Asiva*; [and] among the Kol'a's (mistakenly known as Aymara) as *Khusa*. The Spaniards gave the name of chicha (tsitsa) to this maize beer, a word the Conquistadors heard for the first time in the West Indies to designate similar beverages and which they later spread all over Central and South America, to the confines of the Spanish language. (Tschudi, 1918: 39)

Tschudi then added:

> Of the multitude of expressions for the different classes of maize beer that abound in both Khetsua and Aymará, I here list only the main ones. In Khetsua, *kul' akha*, dark coloured; *gelu akha*, yellow; *tsumpi akha*, reddish; *tsuya akha*, clear and settled; *puxtsko akha*, acid; *l'oxl'a akha*, oily [aceitosa]; *kaymaska akha* "sehalee," chicha; *akha pl'oxl'a*, the foam of the chicha, and so on. Among the Aymara, *piske kusa*, white; *tsuri kusa*, yellow; *yuu kusa*, reddish or dark yellow, *vila kusa*, *Kami kusa*, reddish; *kul'ku kusa*, a strong red colour; *kol'yu kusa*, dark reddish chicha. Besides these colour labels there are many technical terms for the preparation, *resistencia* [?], etc. of chicha.... The men who prepared chicha or sold it were called *akha asiwax* or *akha kamayox*.... (Tschudi, op. cit.: 42–43)

The definition of chicha given by the Real Academia Española (2001: 355–356) is the following: "Alcoholic beverage that results from the fermentation of maize in sugary water, used in some countries in America."

Father Cobo also discussed this:

> This name of *chicha* includes all of the beverages the natives of this New World used instead of wine, and with which they frequently got drunk.... *Chicha* is made out of many things, each nation adapting those seeds and fruits that their land most abundantly gives to make chicha out of them.... But the best *chicha* of them all and which is usually drunk in this land, which has the pre-eminence amongst all the other Indian beverages, is that which is made out of maize. (Cobo, 1964a: 162)

What Cobo says is somewhat confirmed by Cooper, who explained that of all the alcoholic beverages of aboriginal South America, those made with maize are the ones most widely used by the largest number of people, and that have the biggest geographical distribution. They are widely and continuously spread over the vast expanse that extends from Honduras to the isthmus in Central America; along the Andean cultural strip, from the isthmus and the northern and western parts of Colombia, through Ecuador and Peru, to western Bolivia, Atacama, the land of the Diaguita, Middle Chile, and Chiloé; and – as a late introduction – from Chiloé to the southern confines of the growth of maize in the Guaitecas archipelago. They also appear in the West Indies, Venezuela, and the Guianas; in the Tupí-Guaraní region of the Amazon forest; in the zone of Paraguay-Paraná; and on the Brazilian coast and its hinterland. They are also common over a large part of the *montaña*, in the Orinoco River valley, in the upper and middle Amazon River and its tributaries, in the eastern zone of Bolivia and the Chaco, in the Mato Grosso, and in the non-Tupi eastern Brazil (Cooper, 1949: 539–541).

Rowe discussed the chicha of the ancient Peruvians. He explains that in Inca times *aqha* was made out of maize, quinoa, oca, and molle but was not a distilled liquor. This beverage was prepared by women who chewed the pulp of the fruit and would spit the mass into jars of hot water. This was then left to ferment as desired (Rowe, 1946: 292). Guaman Poma de Ayala says that the Inca gave him with his food a "... very soft chicha that matured for a month, which they called *yamor aca* ..." (Guaman Poma de Ayala, 1936: 332 [334]).

We must distinguish the types of alcoholic chicha from the nonalcoholic ones, as Cutler and Cárdenas (1947: 33) point out. Sweet corn is in fact distinguished by having a large amount of sugar with a high content of alcohol (Mangelsdorf, 1974: 108). It is possible that the selection of different varieties of maize, particularly when preparing chicha, may in fact have been one of the criteria used of old in producing some of the many varieties that now exist in the Andean area (Nicholson, 1960: 291). Mangelsdorf personally experimented with the purple Kculli race that is used not just to make nonfermented chicha and *mazamorra* but also as a dye. This race retains its color when dried under the sun and no difference is perceivable. When observing the piles of kernels left by the Indians, we clearly see that the selection was quite stringent. This variety is also used to make fermented chicha, as it contains a large amount of sugar. The same can be said of the Giant Cuzco race (Mangelsdorf, 1974: 114, 208).

Nowadays one of the most well-known ways to make chicha is to do so straight from the germinated maize flour known as *guiñapo* or *jora*. Strangely enough, this is also known as "false chicha [*chicha postiza*]" (Sevilla Panizo, 1994: 223). But this is not the only way of preparing this beverage; the other kind is known as "chewed chicha." A review of the testimonies left by the Spanish chroniclers in this regard is interesting. But before proceeding we must explain that the ptyalin in saliva is an enzyme known as alpha-amylase that turns starches into fermentable sugars. The germination of maize frees diastase, which is far more effective than that obtained by salivation.

Zárate described the chicha that was prepared in the Andean highlands and explains that the Indians

> ... drink a beverage instead of maize, that they prepare placing maize with water in some jars they keep below ground, and here this is boiled. Besides the raw maize they place in each jar a given amount of chewed maize, for which there are men and women for hire, and this acts as yeast. They believe that [the chicha] prepared with dammed and not running water is better and stronger. This brew is commonly called chicha in the language of the islands, because in the Peruvian languages it is known as azúa. It is white or red, like the colour of the maize used, and it gets one drunk far easier than the wine of Castile – but if the Indians could have it according to their liking for it, they would abandon that of their land. (Zárate, 1968: 132)

Father Acosta also gave a detailed account of chicha:

> Maize is used by the Indians not just as bread but also as wine, because they prepare their beverages with it, with which they get drunk far quicker than with grape wine. Maize wine, which in Peru is known as *azúa*, and *chicha* in

> the word common in the Indies, is prepared in various ways. It is stronger, like beer, [and it is prepared] first wetting the kernel of maize until it begins to sprout, and then cooked in a given order it comes out so strong that it knocks you down in just a few bouts. This they call *sora* in Peru, and it is banned due to the serious harm that getting seriously drunk entails.... Another way of preparing azúa or chicha is by chewing the maize and making a yeast, and what is thus chewed is then cooked. The Indians even believe that in order to make a good yeast it must be chewed by corrupt and old Indian women, that just hearing this is revolting, and they do not drink this wine. The cleanest and healthiest way, and that which least makes one lose his head is toasted maize.... (Acosta, 1954: 110)

It is striking that Garcilaso de la Vega once again left a quite superficial description of chicha, and that he does not mention the chewed type at all. He wrote thus:

> Some Indians who are more intent on getting drunk than the rest, place the *sara* in steep and keep it there until it begins to sprout. They then grind and boil it in the same water as other things. Once this is strained it is kept until it ferments. A very strong drink, which intoxicates immediately, is thus produced. It is called *viñapu*, or in another language *sora*. The Incas forbade its use since it at once produces drunkenness, but I am told that it has recently been revived by some vicious peoples. (Garcilaso de la Vega, 1959c: 130; 1966, book VIII, chapter IX: 499)

In his dictionary, González Holguín (1989: 470) in fact calls the "strong chicha" *viñapu aka*.

Father Cobo left a very detailed description:

> ... It is prepared in many ways. What distinguishes one from the other is that some *chichas* are stronger than others and of different colours, because they are made of red, white, yellow, ashy and other colours. A very strong one is called *sora*, that is made with a maize that first spends some days buried until it sprouts; another one is made out of toasted maize; another one with chewed maize; and in other ways too. The most common one that the Indians of Peru drink is made out of chewed *maize*. For this we see – not just in their towns, but also in many towns of Spaniards where many Indians are, like Potosí, Oruro and others – rings of old Indian women and young Indian boys seated chewing maize, just the sight of which is most revolting for the Spaniards but not for the Indians, as well as [the idea of] drinking a brew made in such a filthy way. They do not chew all of the maize with which chicha is made, just part of it that acts as yeast when mixed with the rest. The Indians believe this is needed to make the chicha be just right, that when the maize is ground to this purpose in our watermills they chew the flour until they wet it with the mouth and make a bolus. And those who have this occupation of chewing maize or flour take their pay, besides what they get swallowing what they want to kill their hunger. (Cobo, 1964a: 162–163)

Vázquez de Espinosa is another author who left a good description of chicha:

> In the Indies, and particularly in the Kingdom of Peru, the Indians make their beverage with maize in many [different] ways.... That which is made out of maize, which is the wheat of the Indies, is made in various ways. The ordinary one is called *jura* or *asua*; this gets the Indians very drunk and is a not-all-too clean beverage. To make it they soak the maize and then cover it with a mat or something else, and leave it alone for some days until it has all sprouted. Then they grind it well and they strain this mass with boiling water and pour it into their jars and jugs until it has boiled for two days like wine. After it has boiled it has a spice to it and they drink it and use it in their drunken orgies. They build their houses and till their fields preparing a lot [of chicha], and they *mingar* all of their relatives and friends, which is tantamount to inviting them to work and to a feast. And so one and the other are done with a solemn dance and drunkenness.
>
> Others make the maize with old Indian women and boys, and whosoever they find for it chewing it, which is quite a revolting beverage. [They do this] so that it is prepared more quickly and has the same effect as the others. Others make it toasted, which is more pleasant, healthier and fresh, and almost has the taste of a good *aloja*....[1] (Vázquez de Espinosa, 1948: 1219 and 1220 [397–398])

Once again it is Tschudi who made an exhaustive study of this subject, and although long is well worth citing, as he knew whereof he spoke:

> The primitive preparation of akha is quite simple. Boiling water was poured over the maize that was more or less grounded, and was immediately mixed, after a given time, with a certain amount of water [that was] established through practice and cooked. It was stirred after cooling with the stick and allowed to ferment. Once this was almost over, the beverage was consumed because it sours easily or evaporates. This akha that is being made at present, frequently in the same primitive way, is a yellowish, more or less acid beverage, somewhat similar to a light, sour beer.... The Indians probably discovered by accident that they could prepare a better product with fermented maize, so they placed it in a container with water so that it would soak for several days until it sprouted (until some roots and leaves were given out). They immediately squeezed it, ground it and followed the standard procedure to make akha. But they found that the mix came out much better if they only placed a part of this liquid in their common chicha, and slightly afterwards all of the product was prepared with sprouted maize, and they turned the plain beverage into another spicy, strong and intoxicating one by adding some plants. The akha of germinated maize they call Wiñapu (wiña, sprout, grow) or Sora.... By instinct Peruvians got to chew (mutki) maize or sprouted maize instead of squeezing it, putting the gob mixed with saliva in a vessel and then proceeding as with the common akha. This procedure, which the Indians somehow

[1] *Aloja* is a beverage made with water, honey, and spices.

> discovered, that is unknown to us, is in all respects physiologically correct as it relies upon a process of transformation of which they could have no inkling, but which they instinctively reached. In fact, the saliva abundantly segregated during mastication and mixed with the sprouted and well shredded maize that the khetsuas call *muku*, contains ptyalin, which transforms the glucose in maize, just as the diastase[2] in the malt acts in the same way in the preparation of beer. Since ptyaline also acted as a ferment, the Indians attained a more complete fermentation of their mix by adding a bit of lees, than was the case with other types of chicha. The akha of chewed maize as is prepared in the highlands for some festivals, is called texte, it had the consistency of a mazamorra and was extremely capricious. In the time of the Inkas, those who were in charge of chewing the maize were women and girls. They were forced to fast throughout all of the procedure, which sometimes lasted for several days in a row, i.e. they were not to eat salt and ají (utsu, capsia, sepec), and the married women had to abstain from the marital bed. The akha meant to be consumed by the Inka and the royal family was prepared by chosen virgins. (Tschudi, 1918: 40–42)

In our time there are several good descriptions of chicha. Morris (1979) made one of them. He points out there are two major variants that are based on a source of diastase used to expand the content of alcohol and improve the taste. Saliva is a common source of diastase in a large part of the Americas. Dried ground maize is placed in the mouth in the form of a slightly moist ball, which is moved with the tongue until it absorbs the saliva. This salivated maize flour is then dried and is the raw material for chicha. Nowadays this is made with a malted maize called *jora*, Morris explains. It is prepared by soaking the maize overnight in water in a ceramic vessel. The following day the grains are placed over a thin layer of leaves or straw and left to germinate until the sprouts are about the size of the kernels. These sprouts are dried under the sun. The resulting *jora* is then ground or, more correctly, cracked to prepare the base of the beverage.

Chicha nowadays is made by filling a vessel with a large mouth up to about a third of its capacity with the malted maize or *jora*. Hot but not boiling water is added until it is just below the rim. The mixture is carefully stirred and allowed to cool for about an hour. The liquid on top is separated in another big vessel. A semicongealed layer that remains atop the sediment in the jar is also removed and then condensed by boiling it over a low flame to turn it into a sweet, caramel-like paste. The liquid is left unstirred for about two days and is then strained, boiled for about three hours, and immediately left to cool in a jar with a wide mouth. Once it is cold, they add some of the products that were previously separated from it. Once fermentation begins, the liquid is transferred to another jar with a smaller mouth, where fermentation proceeds, and the chicha is served from this

2 The Spanish edition of Tschudi says "diástasis," but this clearly is a mistake and should be "diastase."

vessel. The speed of fermentation varies depending on the temperature in different regions; it also depends on the preference as to the degree of fermentation. Some prefer chicha that has just begun to ferment, others prefer it with five or six days of fermentation (Morris, 1979: 22–25).

Muelle (1945) studied the chicha made in Cuzco, in the precinct of San Sebastián. He explains that here they do not prepare the *sutay-akha*, that is, the chicha that is buried for several days and that is prepared elsewhere. Muelle described how this beverage was prepared, the Quechua terminology used, and the customs associated with it. The present author presents just some details that are relevant. Interested readers should peruse this work for more information.

Muelle explains that the word *jora* is a synonym of *guiñapo*. The first meaning is erudite but is already of popular use, whereas *guiñapo* comes from *wiñay*, which means to grow, that is, it refers to the sprouting of maize during germination. *Tekhte* is a chicha that looks like milk and is prepared with white maize, to which a bit of sugar and quinoa are added to create (Muelle, 1945: 147). In San Sebastián, Muelle says, chicha "... cannot be left over for another day ... as it sours" (Muelle, op. cit.: 146).

Gillin explains that in the mid-twentieth century *chancaca* (brown-cake sugar) was added to chicha on the North Coast to increase its alcoholic content (Gillin, 1947: 47).

Hocquenghem and Monzón described the chicha made in Piura, on the North Coast, and provided recipes for *chicha de jora*, *chicha casera* [domestic], *serrana* [of the highlands], and that which is known as *loja* or *aloja* in Catacaos. They explain that bran is chewed and cast into jars only in the case of the *chicha de jora*; this procedure is called "*ensalivamiento* or *muqueado*" (Hocquenghem and Monzón, 1995: 112–113). Interestingly enough, in the Bolivian highlands and valleys, the chicha made out of quinoa is known as *loja* or *aloja* (Cutler and Cárdenas, 1947: 34), whereas that of Catacaos is made with maize.

One of the best studies of chicha was made by Cutler and Cárdenas. They point out that a simple alcoholic chicha is made by mixing a substance that has starch or sugar with water, and letting the liquid ferment. Few varieties of chicha are prepared like this. Most use methods that enhance the alcoholic degree and the taste. The increase in alcohol is attained by turning some starches into sugars that are more usable in fermentation. An enzyme – diastase – causes this change, and in South America its most common source for chicha is saliva. The custom of chewing roots, fruits, and grains when preparing beverages is quite widespread. In the case of maize, we find this custom from coastal and central Brazil to the Peruvian highlands. Malting is another way of introducing diastase, that is, by soaking the grains in water to make them sprout. We have seen that this method was described by the chroniclers. Cutler and Cárdenas, however, claim the chroniclers do not provide the full information on how to prepare chicha. Yet they contradict themselves, because they do say that "it is probable, however, that malting is a pre-Columbian development" (Cutler and

Cárdenas, 1947: 34–35), when just a few lines before they had pointed out that Garcilaso de la Vega, Acosta, and Cobo describe this process, which we have seen is correct.

Cutler and Cárdenas indicate that the process of malting is common in Bolivia and Peru, particularly in the highlands. They include two photographs for interested readers that show the *muko*, that is, the salivated maize flour, and Bolivian women *mukeando*. Cutler and Cárdenas (1947: 39–41) give a good description of how maize is collected for chicha and the variety used, but they discuss the Bolivian zone of Cochabamba.

It is important to note the distinction Cutler and Cárdenas draw between salivation and mastication. In addition, this is the best description found of these two processes. For them, salivation is when "the flour is moistened very slightly with water, rolled into a ball of convenient size and popped into the mouth. It is thoroughly worked with the tongue until well mixed with saliva, after which it is pressed against the roof of the mouth to form a single mass, then shoved forward with the tongue and removed with the fingers. The teeth play very little part in the process." Mastication, on the other hand, is when one needs to crush certain hard materials with the teeth, such as the carob or sweet potatoes (Cutler and Cárdenas, 1947: 41).

In the paper by Cutler and Cárdenas, interested readers will find all the preliminary steps taken when preparing chicha and its day-to-day processing, along with the native names for each phase and product (Cutler and Cárdenas, 1947: e.g., 41, 45–52). Their claim that the process of making chicha is far simpler in the eastern Bolivian lowlands is also interesting (Cutler and Cárdenas, op. cit.: 53–59).[3]

Cámara-Hernández and Arancibia de Cabezas discuss the chicha prepared in the Quebrada de Humahuaca, Argentina. This is the fermentation of a watery extract of pre-germinated kernels. The different types of beverage depend on the variety of maize used; the most common type is the Morocho, which gives a condensed, strong chicha, whereas other varieties are used to make a clear (*clara*) chicha. Cámara-Hernández and Arancibia de Cabezas note that the term *mukeado* is also used in the Quebrada de Humahuaca. They explain that the unsalivated chicha is called *chicha postiza* or *chicha falsa* (phoney chicha). (Significantly enough, we saw [previously] that it is given the same name in Peru.) The fermented flour is boiled with water to prepare a dish known as *apuña*, which gets sweeter the more it is boiled. Several chichas are made this way. After decanting, the different levels are used in various ways. The first level has an aromatic purpose and is used to make bread, the second to make chicha for Carnival, and so on (Cámara-Hernández and Arancibia de Cabezas, 1976: 223).

3 Interested readers should also see Nicholson (1960: esp. 291, 294–298, 290).

When studying the Bolivian chicha, Nicholson pointed out that chewed *chicha de jora* (once again known as *muko*) is also made in this country, but it is a method that is more common in the Puno region and on the Peruvian side of Lake Titicaca than in Cuzco. We saw in Cutler and Cárdenas (1947) (see also Ramírez et al., 1960) that the use of salivated maize to make chicha is quite common in Bolivia. The method used in Cuzco and Puno is essentially similar to that of Cochabamba, but the "salivating" is traditionally carried out by young girls who have never chewed coca (Nicholson, 1960: 298). Yet Muelle gives different information. He noted that "*muqu*," the stronger chicha prepared in Puno by having girls who have not tried coca chew the *jora*, is not made in Cuzco (Muelle, 1945: 152; Kahn, 1987: 40, also has data on chicha).

Here it is worth going over the terms González Holguín included on this subject in his 1608 dictionary (González Holguín, 1989). He noted that "chewing maize to make chicha" is "*muccuni*" (p. 583); "to dissolve flour in the mouth, that is making maize mucu with saliva in order to make chicha" is "*Muc chhini*"; and "making the mass of mucu sparse with water in the mouth, dissolving it," is "*Muccucta mucchhicuni*" (p. 245). "To make chewed muccu of toasted maize" is "*Muccuni*," whereas "*Muccu* ... is the chewed mass out of which the yeast for the azna is made." "To chew maize to make chicha for pay" is "*Muccupucuni*," and "those who hire themselves to chew" are "*Muccupuccuk*." "Making chicha yeast" is "*Mucucta* or *muccuscacta puchcuchini*" (p. 248), whereas "[to] Smell of chicha" is "*Aka akam aznani*" (p. 18). Under "chicha," González Holguín also included "*Aka*, strong chicha [is] *viñapu aka*," and "chicha, to make, ... *Akuni*" (p. 470).

I would like to insist a bit on the preparation of chicha through salivation, because as we shall see in my conclusions, this is a modality that is now almost lost. For instance, when Gutiérrez de Santa Clara mentions the offerings made to the gods, he includes the "... wine of the land made out of chewed maize ...," and when describing the North Coast he says there was "... another kind of wine they call chicha, which is made with maize chewed in the mouth ..." (Gutiérrez de Santa Clara, 1963: 231, 241). When describing the customs of the Indians, Guaman Poma de Ayala says that Indians

> should not drink chicha chewed with the mouth that they call *moco* [maize chewed for chicha], *acto* [flour chewed for chicha], *haca* [chicha], *mocchi* [chewed for chicha], [and] *pururo* [?] because it is a dirty and disgusting thing. They should instead drink a chicha of germinated maize they call *sura asua* [chicha of germinated maize], so that Christians drink it and benefit from it.... (Guaman Poma de Ayala, 1936: 881 [897])

As for the "caciques principales," Guaman Poma pointed out that

> neither said caciques principales nor the other *curacas*, *mandoncillos* [minor officials], mayors and aldermen, or the common Indians, or anyone in this kingdom, are to make the Indian women, [be they] single, widows or married,

or girls and boys chew – what they call moco [chewed maize for chicha], acto, mocchi [flours chewed to make chicha], pururo [?], haca [chicha] – in order to get drink by making them prepare hurcan [chicha]. . . ." (Guaman Poma de Ayala, op. cit.: 788–789 [804–805]).

Tschudi, who was in Peru in 1838–1842, also described this way of making chicha:

> In some parts of the Peruvian highlands there still survives to this day the custom of preparing this chewed chicha. It usually is old women who devote themselves to chew germinated maize; I have seen some who have this occupation, whose teeth have worn down almost to the gums. The procedure is often also done in family or with some guests; the fermentation and decanting over, a piece of boneless, fatless and muscle-less flesh in each jar, and after sealing them hermetically they are buried in a convenient location. The vessels are taken out only when it is the birthday [*santo*] of a child. By then the vessels hold an exquisite and strong, dark yellow beverage that resembles the Spanish wines. Nothing remains of the flesh for all of it has dissolved and the insoluble remains lie on the bottom. The Aymara called this chicha that was long stored, *l'utapu* or *yanuna kusa*. (Tschudi, 1918: 42)

Valdizán, writing in 1918, left the following testimony:

> This way of making chicha still exists nowadays. The number of regions that preserve this Inca tradition is very limited; in some populations in South and Central Peru it still is possible to drink the *Akha* that so pleases the primitive inhabitants of Peru. One can still find individuals who actually practice the chewing of the maize and who dedicate themselves to this task with some pecuniary advantage. In Huarochirí, according to the references given [me] by my friend Doctor Tello, the primitive way of preparing chicha is still preserved. The vessel where the boiling takes place is called *pampana*, and that in which fermentation takes place is called *macma*. My uncle, engineer Darío Valdizán, who has travelled … almost the whole length of our territory, provides me with identical information regarding some localities in the department of Ayacucho … [i]f some settlements in Central Peru still have the custom of chewing maize in identical fashion to the way the primitive Peruvians did, [on the other hand] most of the settlements [have] adopted germinated maize, reduced to a fine pap and which is sold under the name of jora and as a dark-coloured dusty mass, ready for the preparation of the fermented beverage. (Valdizán, 1990: 147–149)

Polo de Ondegardo collected data on the care that was taken in Inca times regarding the use of chicha:

> There likewise was great vigilance that no chicha was made out of new maize, and much less was eaten on the cob before it being dry; nor were the *viras* – the maize canes – to be eaten when they are green. This all is the most pernicious and harmful thing for the Indians of all they do. There was much vigilance

> in each town regarding each and every part of this.... (Polo de Ondegardo, 1940: 193)

And when discussing the chicha called *sora*, Tschudi tells us that "the Inkas completely forbade the people to prepare this *sora* for their use, because it drove the Indians to [commit] greater excesses and [even] more uncontrolled orgies. Only the Inkas and the aristocracy could freely indulge themselves in enjoying this capricious beverage" (Tschudi, 1918: 41).

Father Cobo gave a detailed account of the medicinal qualities of chicha:

> All kinds of maize *chicha*, when taken, are useful against urine stoppage, [and] against sand and stones in the kidneys and the bladder, for which reason Indians, both old and young, never have these ailments, because they drink *chicha*. Taking half a *cuartillo*[4] that has been soaked overnight with the segments of half a white onion and a bit of sugar over an empty stomach stops the evacuation of the bowels, or at least calms it down so that it does not skin and cause ulcers in the urinary tract. And if half a warm *cuartillo* of this beverage is taken when it is not too sour or mature over an empty stomach, it helps against diarrhoea and all urine stoppages and the *mal de ijada*.[5] The lees or sediment of the mass *chicha* is made with are also useful, because when applied to feet with the gout it eliminates the burning sensation and mitigates the pain. (Cobo, 1964a: 163)

Acosta noted in this regard that "... the most polished Indians and some Spaniards use this as a medicine, as they find that it actually is a very healthy beverage for the kidneys and the urine; hence you rarely find such an ailment in the Indians because of their custom of drinking their chicha" (Acosta, 1954: 110).

It is interesting that the Spaniards rapidly adopted the custom of preparing chicha. Father Cobo says that in the early seventeenth century they already made it, "... but cleaner and with more curiosity than the Indians ... ," and they even modified the Indian formula (Cobo, 1964a: 163).

As was pointed out at the beginning of this chapter, the author does not intend to discuss the details of the preparation of chicha, but it is worth adding here an overview that will help readers who are not acquainted with this subject. The vessels used for this vary a lot. The first step in making chicha is choosing the maize, removing the kernels from the cob, and then soaking them to make them germinate. They then have to be dried and ground or crushed. This is what is known as *jora*. The cooking of the chicha varies a lot, from about 12 hours up to one or two days. The separation of the liquids from the residues is then done by filtration or by sedimentation (Moore, 1989: 686).

What Valdizán points out about the chicha consumed in Piura is invaluable. He noted that here there was "... an extraordinary consumption of the jora

[4] A *cuartillo* is a measurement that is equal to a quarter of an *azumbre*, that is, 504 ml (Real Academia Española, 2001: 471).

[5] *Mal de ijada* is a pain felt in the cavities between the false ribs and the hip bone.

chicha called *claro* just because it has been carefully filtered or decanted, and because it has a lighter look than the common chicha" (Valdizán, 1990: 149). We will return to this point in the discussion, because the *claro* or *clarito* chicha is also characteristic of the zone of Guadalupe, in the province of Pacasmayo.

To finish this chapter we shall go over the case of Huarmey, in the province of the same name, on the north-central Peruvian coast. Here the *jora* maize is traditionally used to prepare a chicha that has a long history in the valley and is quite typical, as it involved a very special procedure, different from that used in all other coastal valleys. Raimondi had already noted this down in his travel notebooks when he passed through the valley in 1859:

> Huarmey does not have much trade, but it instead has one specialty that has made it famous: the preparation of its chicha, which is highly esteemed and is even often sent to the capital as a gift. Sometimes they let the good chicha settle, and when it is quite clear they bottle it and so preserve it for a long time, serving it afterwards as wine. It has quite a high alcohol content, so that it has the effect of a very strong wine even when taken in small amounts. Those who take this chicha sometimes have a very strong headache. To recover they eat an egg with a lot of ají (*Capsicum*), and they say they can then continue drinking without suffering any harm. (Raimondi, 1942: 170)

Middendorf made a similar observation about at the same time: "Huarmey is famed in the coast for its chicha. Clay vessels or bottles are filled with it and are buried. In this way the chicha is stored for years. The chicha from Huarmey is dark but not thick, tastes like wine and it intoxicates rapidly" (Middendorf, 1973, volume II: 209). This is, in fact, a most pure chicha to which no other ingredient is added other than maize. The present author can bear witness to the fact that according to the elders of Huarmey, there was until the 1970s a lady there who was the only one who retained the ancient way of making chicha. This author was unable to find what procedures she employed, but from the descriptions left by Raimondi and Middendorf it seems to have been the same, because the one the author drank had the same characteristics. The *chicha del año*[6] is still consumed in La Libertad, but this author does not know how it is made.

Interested readers who want to expand their knowledge of this subject should read, among others, the studies by Camino (1987), Cavero Carrasco (1986), Gómez Huamán (1966), and Vásquez (1967).

[6] This means the chicha that is stored underground for a year.

10

Discussion and Conclusions

> It is certainly desirable that all of the facts adduced in a work of history have been carefully verified, if only to deprive pedants of a weapon they insidiously employ – and not without success – to discredit strong and genuine historical writings; but [they should also be verified] because accuracy is in any case a moral duty.
>
> Croce (1960:8)

One of the most serious issues that bear on the scientific discussion, and specifically on the maize problematic, is the misuse of the sources, or their partial use. This is harmful because not all those who have to use the data are specialists with a command of the subject matter, and they rely on the data others have presented, which they believe are complete and correct. When this is not so, a chain begins of mistakes that are repeated by other scholars until at some point or another the wrong or incomplete account acquires the status of an established truth. It is for this reason that some specific examples from texts perused while writing this book are now given, for although the errors have been pointed out, it is worth insisting in this regard to raise an awareness regarding the serious risk this entails.

Perhaps the best example in this regard is given by B. D. Smith (1994–1995b), which is a synthesis that tries to present a sweeping overview of the origins of agriculture in the Americas. Yet a careful examination of the bibliography shows that out of the 42 entries, not one was authored by a South American scholar.

Van der Merwe and Tschauner (1999: 532–534) provide another good example. They discuss the adoption of maize as the basis for the rise of social inequality. In this case the authors use the same measuring stick for all societies, from the United States to the Andes, an approach that makes no sense at all, and that shows an absolute lack of familiarity with the Andean area to boot. On reading this study one gets the impression that only the Incas imposed maize in the central Andean area, which is clearly wrong.

Doebley (2004) is a serious case in point. When discussing the 6,000-year-old maize from San Marcos Cave and from Guilá Naquitz, Doebley claims that it included "... an ear of only 6 cm in length with as few as 28 kernels." The references given here (Doebley, 2004: 55) are Benz and Iltis (1990) and Benz (2001). First of all, it must be pointed out that neither of these two studies mentions the number of kernels, nor is it clear whether Doebley means total kernels or just their number per row. In the case of Guilá Naquitz the study involves fragmented cobs, so it is not easy to infer their length (Flannery, 1986b: 8). In San Marcos, the length of the cobs ranged between 1.9 and 2.5 cm and had on average 55 kernels (Mangelsdorf, 1974: 168). Beadle (1972: 2) concurs, for he wrote that the Tehuacán cobs have 50–60 kernels (to be precise: 36–72 kernels; Mangelsdorf, 1974: 168).[1] So the maize Doebley mentions, with a 6-cm-long cob, definitely cannot have had 28 kernels. It is likewise evident that when Doebley (2004: 39) describes the kernels, he was thinking of large ones and not the small, acuminate, and red ones.

Jaenicke-Deprés and Smith (2006: 90) studied ancient DNA and tried to "integrate" all genetic data into their work. And yet they only examined those from the southwestern United States and Mexico and complained that no research had been undertaken in the last 40 years. For them the Andean area simply does not exist, and they show they are not acquainted with it because they generalize and claim that the kernels have "only occasionally [been] preserved" (Jaenicke-Deprés and Smith, 2006: 88), whereas the exceptional preservation conditions of the Peruvian coast are well known and are only equaled by the Egyptian zone, as well as by some dry highland caves.

Another study that is scientifically worthless is C. H. Brown (2006), which presents a glottochronogical linguistic analysis that purportedly tries to establish a chronology of maize in the Americas. Brown has a poor grasp of the archaeological data, and he does not even mention the central Andes, as his study only extends as far south as Ecuador (Brown, 2006: 655–656). Brown studied 591 languages but did not consider either Quechua or Aymara (see Brown, op. cit.: table 47–4, 658–659), to mention just two major examples.

Finally we have the work done by Benz and Staller (2006), which is a clear example of a misinformed study. They claim that just like MacNeish and Flannery were influential in the problematic of the origin and dispersal of early maize, so "... Lathrap [11–13] was to the theoretical development regarding domestication and the spread and role of *Zea* to Andean prehistory" (Benz and Staller, 2006: 665). Notes 11 and 12 refer to Lathrap (1968a) and (1968b) respectively, neither of which broach this subject. Note 13 cites Lathrap (1970), which likewise does not make a direct reference to this issue and only mentions it in passing on pages 59 and 67. And yet no reference is made to Lathrap (1987), where this subject is indeed discussed. On the other hand, all of the

[1] For comparative purposes, the reader should bear in mind that the maize from Epoch 2 at Los Gavilanes, Peru, has cobs that are on average 4.9 cm long, and that have 179 kernels on average (Grobman, 1982: table 11, 160).

conclusions drawn by Benz and Staller (2006) are distortions, for they claim to study the "Antiquity, Biogeography, and Culture History of Maize in the Americas" based solely on the papers included in the volume they themselves edited (Staller et al., 2006), without comparing and contrasting this database with the quite vast existing information, which was not taken into consideration by the scholars included in the aforementioned book. This is borne out by the fact that out of the 33 entries in their bibliography, not one refers to Peru or was authored by a Peruvian scholar (Benz and Staller, 2006: 673).

Another serious problem to which attention must be drawn is the way in which the different specialists who intervene in the analysis of archaeological materials study the issues solely from the standpoint of their specialty, ignoring those of other disciplines and distorting the truth. This clearly is one of the most difficult issues to solve, for if a specialist turns to an issue that deals with archaeological specimens, or analyzes the latter without having at least an overall understanding of this discipline, he or she will be unable to understand the true value of the data he or she is using. A general example, which may seem trivial, though it actually is not, will suffice here before we move on to the study of some specifics. When mentioning archaeological specimens, botanists generally call them "fossils," which is clearly wrong. Paleontological remains are fossils, not so archaeological remains. (To prove this we need not turn to specialized sources – the *Diccionario de la Real Academia Española* will suffice, 2001: 141, 1122.) MacNeish (2001: 104) also drew attention to this point.

Bennetzen and colleagues (2001) are one specific case; they reject the data in Eubanks (1995, 1997a) and MacNeish and Eubanks (2000) for the simple reason that they take the position of Beadle (1939, 1972), albeit without any argument. They acknowledge, to boot, that there is a rift between archaeologists and botanists (Bennetzen et al., 2001: 85).

The geographers Johannessen and colleagues (1970) are another clear case in which the truth was distorted: they claim that archaeology shows "... only the times and places of changes, without illuminating the human actions that may have caused these changes" (Johannessen et al., op. cit.: 394). This shows ignorance, for although it is true that archaeology has its limitations, it does reconstruct the context and the historical process of the sites it studies.

Some scholars have actually realized this and have drawn attention to this problem. For instance, when pointing out that botanical collection can only be studied by specialists, Cutler and Blake (1971) underlined the fact that few botanists take the time to do so "... largely because they do not understand that archaeological sites may provide usable records of the past history of plants and that these materials can often be dated or placed in order by archaeologists" (Cutler and Blake, op. cit.: 367).

Bugé (1974: 34) likewise emphasized the problems we have when trying to understand the so-called primitive races of maize, when considered from different standpoints. So whereas botanists see them from the standpoint of natural factors, anthropologists do so from a cultural perspective.

It must be clearly pointed out here that archaeologists also make this mistake. Many of them have made serious mistakes when touching on aspects that concern other specialties, particularly botany and zoology (e.g., Shady Solis, 2006). There is a reason for this in the specific case of Peru, for archaeologists study in humanities or social sciences faculties and do not follow courses in the natural sciences. And it is almost impossible to discuss subjects such as plant domestication when one does not have at least a general command of biology as well as some other fields, along with the respective technical language that gives access to the data. We must not forget – as Eubanks (1995: 172) correctly notes – that nowadays the study of maize requires the help of several disciplines or of specific aspects of some of them, such as systematics, morphology, cytogenetics, molecular biology, the experimental cultivation of plants, and archaeology.[2]

There also is another problem that can be as critical as the previous ones – the way a scholar may approach an area from the standpoint of another area, something that may be the correct approach in some cases but not in others. Bruhns (1994) drew attention to this and quite clearly noted that the research on early agriculture in Mesoamerica "... has likewise influenced studies in South America, as investigators have tried to *impose* Mesoamerican-based models of agricultural development *upon an entirely different continent*" (Bruhns, op. cit.: 89; emphasis added).

It is clear that at present, the origin and the domestication of maize are the two issues in this problematic that are most hotly debated, and over which disagreements are strongest. We must, however, acknowledge that although one can take a position regarding these issues, there is as yet no way of knowing what the truth is, as we still lack the information required for this.

I believe that the position that holds that maize originated from a wild maize is the most acceptable one and fully concur with Grobman's position (2004), which was expounded at length in Chapter 3. In brief, Grobman posited that wild maize has disappeared, and that it must have been an annual, precocious, and short monoecious plant, with separate female and male inflorescences, but with cobs that end in staminated spikelets, with branched ears and an independent husk covering, and with very small and hard kernels. Wild maize probably was a pod corn; the pod gene has been genetically dissected, and the studies made by Mangelsdorf and Galinat (1964) showed that it comprises two genes.

Grobman (2004) likewise notes that wild maize probably hybridized with the form of wild perennial teosinte (*Zea diploperennis*), which is the ancestor of all teosintes. This may have given rise to annual teosinte through natural crosses with maize, when the latter began to be cultivated in areas where it

[2] See also Wilkes (2004: 7), who reminds us that this was also posited by Alphonse de Candolle (1959).

was not sympatric with the ancestral perennial teosinte. These plants may have established contact thanks to man, who carried the seeds of a semiwild maize. This is clearly borne out by the absence of teosinte in the archaeological strata at Tehuacán. Another piece of evidence that we have gone over is the pollen exine in the Proto-Confite Morocho, the primitive Peruvian maize, which clearly shows there was an interrelation between *Tripsacum* and *Zea* in South America, or at least that the early and wild Peruvian maize had long become separated from its wild Mesoamerican counterpart (Grobman, 2004: 468–469).

It is worth recalling in this regard some of the ideas held by Mangelsdorf, who showed in his work that at present there is no teosinte or *Tripsacum* in the Tehuacán Valley, and that the remains of these plants have not been found among the archaeological specimens (Mangelsdorf, 1974: 171). On the other hand he showed that the wild and domestic maizes of Tehuacán were essentially identical as regards their botanical characteristics, the former being simply smaller in its parts. Wild maize, with its small cobs, must at first have been of little use as a food plant, and besides, its aspect probably was not too promising. Yet it responded favorably to the environment and improved thanks to man, and it grew noticeably in size. Its subsequent hybridization with its relatives – *Tripsacum* and teosinte – led to an explosive evolution whose result was a tremendous variability and increase in size. Even so, it is worth recalling that maize has not undergone substantial changes in 7,000 years (Mangelsdorf, 1974: 180). In the end, size is the only major difference between the Tehuacán corn and modern maize (Mangelsdorf, 1974: 169; 1986: 82).

Grobman (2004: 434) explains that the corn that is believed to have been wild or at a primary stage of domestication in Tehuacán probably scattered its seeds through the disarticulation of a small and brittle rachillae, not through the disarticulation of the rachis, as happens in teosinte. These different characteristics of both species are typical of a wild plant.

Besides, the most conclusive evidence of this position is the discovery of the Bellas Artes pollen, which was discussed in depth in Chapter 3. The sequence of the apparition of the pollen of the different plants is clear and speaks by itself, and the doubts that have been raised are groundless. Here we see that *Tripsacum* appears in the deepest strata, and then we find maize. This is still present, accompanied by *Tripsacum*, whereas teosinte appears only in the upper part of the drill core and is associated with maize. In other words, the presence of maize long before teosinte is clearly evident. This concurs with the data provided by archaeology, for in the Ocampo Caves, teosinte only appears in 1850–1200 BC (Mangelsdorf, 1974: 154–157; B. D. Smith, 1997a: 351).

We must recall in this regard that an earnest scholar like Wilkes (1979: 13) accepted that the Tehuacán corn was wild, whereas Randolph (1976: 344) had an intermediate position as regards "the unexplained absence of Teosinte from Tehuacán valley during early stages of plant domestication." Randolph pointed out that those who posit that maize originated from teosinte have not borne in mind that the early remains are not of wild maize but of semidomestic teosinte, which was taken to Tehuacán from somewhere else, as there is no teosinte in the archaeological sites or in the valley's flora. And yet there are remains of other food plants that were brought from outside this area, some of them from southeastern Mexico, where teosinte did exist. Randolph then recalls that the fruit of teosinte is very hard and should therefore have been preserved, as was the case of *Setaria* (in Coxcatlán, 6500 BC).

Interestingly enough, Beadle, the foremost opponent of Mangelsdorf vis-à-vis the origin of maize, made this statement, which is quite significant: "While it is logically impossible to prove that such corn never existed as a wild plant, I see no compelling evidence whatever to indicate that it did" (Beadle, 1972: 9). This actually is a vacuous argument that does not solve or refute anything.

We should recall here that Beadle presented four arguments to claim that the most ancient corn found at Tehuacán was not wild maize, but rather a transitional stage between teosinte and maize brought about by humans (see Chapter 3). First of all, he argued that early cobs are morphologically and genetically closer to teosinte than to modern corn. Second, these are brittle cobs, thus indicating that the ancestor was teosinte. Third, some of the earliest cobs are two-ranked, which is a teosintoid trait. Fourth, the earliest corn cobs had soft glumes (Beadle, 1972: 9; 1980: 116).

The reply Mangelsdorf gave suffices to show that Beadle was mistaken. Mangelsdorf showed that there are major differences between the earliest Tehuacán breeds of corn and teosinte. First of all, the archaeological spikelets are paired and are not solitary. Second, most of the spikes are many-ranked but not two-ranked, and only a few have just two of them. The kernels are not sessile but are borne on rachillae. They are round and not pointed. The axes of the spikelets have an angle to the right of the axis of the rachis, and not parallel to it, and finally the leaf sheaths are glabrous and not pilose, as in the Mexico Valley teosinte (Mangelsdorf, 1974: 180–181).

Mangelsdorf furthermore made other significant comments that his colleagues have not taken into account. If teosinte had had a major role for man, it would be very strange that it did not in turn leave the traces that maize has indeed left behind. It is not just a case of a lack of archaeological specimens of teosinte, which are scant, are late, and when they do appear are associated with maize, but there are no linguistic, ethnographic, ideographic, pictorial, or historical data regarding this plant (see Mangelsdorf, 1986: 82).

Significantly enough – and this is something that will have to be researched more, as was pointed out by Harlan (1992: 222–223) – teosinte had a bigger area of diffusion, which then grew smaller. And if the data provided by Celestino Mutis are correct (see Chapter 4), teosinte may have existed in South America.

It is worth recalling here what Mangelsdorf wrote after the discovery of *Zea diploperennis.* He stated that maize and teosinte must have diverged from a common ancestor long before they reached the Tehuacán Valley. As for this ancestor, all that can be deduced is that it must have belonged to the Andropogoneae; everything else is mere speculation. A wild relative of maize was found with the discovery of *Z. diploperennis*. For Mangelsdorf, the debate regarding whether the ancestor was a cultivated corn-teosinte or a wild maize is irrelevant, for it was both things. It was for this reason that he believed, shortly before he died, that the mystery of maize had essentially been solved (Mangelsdorf, 1986: 86).

The five arguments that Iltis (1983b) presented against Beadle's hypothesis have already been noted. Flannery was closer here to Beadle than to Mangelsdorf, and yet he acknowledged that there was not one single excavated site that documented the gradual genetic change from teosinte into corn. He likewise noted that if Iltis's hypothesis of a "catastrophic sexual transmutation" is correct, it could never be detected archaeologically (Flannery, 1986b: 8).

Grobman in turn presented solid arguments against Beadle's position, but these have not been taken into account by the specialists. The first question that Grobman raises is whether maize comes from teosinte, and why the early archaeological corns from Tehuacán have none of the characteristics that would be typical if teosinte were their ancestor. These instead begin to appear in more modern phases many centuries later, as a result of the introgression of teosinte in maize. On the other hand, if maize were derived from teosinte, then all of the changes in the traits that separate both species should appear suddenly and with no transition whatsoever, which seems unlikely given the study of the evolution of cultivated plants. It was precisely for this reason that Iltis presented his hypothesis of a catastrophic sexual transmutation theory (CSTT), as this allowed him to evade this objection.

Corn and teosinte definitely were very different in the past. Maize and teosinte fields lie side by side, one as a crop and the other as weeds, and the latter has grown so close to the former in its plant characteristics that it is very hard to distinguish them. "Maize was not derived from the phenotypic traits of teosinte, and was instead the other way round," says Grobman. Strangely enough – and no one has tried to explain this – maize and teosinte still have quite different characteristics in their inflorescences and seeds, the reciprocal flow of genes throughout centuries of interpollination notwithstanding, just as may have been the case at the beginning of their association, despite the fact that it has been pointed out that few genes separate them (Grobman, 2004: 434, 436).

Grobman furthermore comments on the graph Gloria Cadell prepared for Mangelsdorf (1983b: figure 4, 237), which shows the apparition in chronological order of maize and annual teosinte in archaeological sites from Panama to New Mexico.

> The dates range from 5000 years BC (Tehuacán) to 1000 CE [Common Era; a synonym of AD]. *We clearly see that in all archaeological sites, the evidence of the presence of maize precedes teosinte and the introgression of the latter into maize* (Guilá Naquitz, Tehuacán, Cañón del Infiernillo, Cueva de la Perra,

> Cueva de las Golondrinas y Cueva del Murciélago, as well as Gatún), *in some cases by thousands of years*. This is *a powerful argument with which to question the origin of maize from annual teosinte*, but does favour instead Wilke's hypothesis of an inverted origin. (Grobman, 2004: 444; emphasis added)

We must not forget that Mangelsdorf (1983b: 237–238) found similarities between the pollen from Guilá Naquitz and that of Bellas Artes (see Chapter 5).

Eubanks reexamined this issue with new biological data and explained that there is no archaeological evidence showing a gradual evolution in which the mutations transformed teosinte into maize. Segregating experiments of crossings of teosinte and *Tripsacum* show that the transition from teosinte into maize could have been rapid and may have required just a few generations of intercrossing (Eubanks, 2001b: 498). If this is so, it will likewise be difficult to find archaeological evidence of this.

Goodman has also made some significant observations. He pointed out that for those who defend the descent from teosinte thesis, the difference between the latter and maize is essentially of an agronomic nature and is in effect based on the structure of the female inflorescence of maize, as well as on the changes that have taken place in it (e.g., Iltis, 1969, 1985). The resulting cultigen was easily cultivated and was reproduced in an abundant amount. Besides, Beadle (1939, 1972) claimed that teosinte could be eaten both popped and without popping.

It is undeniable that there is genetic and cytological evidence that shows that maize and teosinte are quite closely related. The data of the isozymes agree with the racial classification presented by Wilkes (1967). The enzymatic and cytological similitude between maize and the Balsas teosinte can be used to suggest a direct lineal relationship, which supports Wilkes's position, as well as the other hypotheses regarding the origin of maize in the Balsas River basin (Iltis, 1987; Iltis and Doebley, 1984). Goodman, however, claims there is another possible interpretation of the data. Those populations that have continuously had the largest size were at least affected by genetic drift, with or without an endogenous mixture, and therefore are even more similar today. Both the origin effect and the endogenous mixture may have severe influences on the multivariate dimensions of genetic similitudes; most of the teosinte races, with the possible exception of the Balsas one, may have suffered an endogenous mixture due to the small size of the past and present population (Goodman, 1988: 205–206).

Buckler and colleagues (1998) restated this issue, but only from a geographical standpoint. It is, however, worth noting that their starting point was that the AMS dates for Tehuacán are correct, a point that we saw was quite questionable (see Chapter 5). It is for this reason that they accept that in the Tehuacán case we are before two possibilities: either this is a most recent domestication, or instead these semiarid regions adopted domestic maize at a very late date. Buckler and colleagues believe that maize (*Zea mays* ssp. *mays*) was domesticated from annual teosinte populations from central Mexico, in accordance with the molecular data in Doebley (1990) and Buckler and Holtsford (1996), and that the "wild maize" theories held by Mangelsdorf (1986) and MacNeish (1992) have been refuted. For Buckler and colleagues the dates of domestication have not yet been established, and they essentially defend three arguments that suggest domestication may have taken place long before its appearance at Tehuacán. First of all there is the high level of molecular differentiation, which suggests an early divergence (their information comes from Buckler and Holtsford, 1996; Doebley, 1990). Second, the study of molecular evolution reveals that, according to the data in Gaut and Clegg (1993), the original population of maize was a big one, and this must have extended the time required for domestication by perhaps several thousand years. Finally, there is good evidence of phytoliths from Panama and Ecuador with an age of 7000 years BC (Pearsall and Piperno, 1990; Piperno et al., 1985).

For Buckler and colleagues (1998), this all suggests that the domestication of maize began in central Mexico (probably in the Guerrero zone) prior to 9000 years BP. Given the colder temperatures present in Mexico before 10000 BP, it is posited that teosintes would have had their ranges depressed some 1,000 m, thus implying that maize's teosinte ancestor was in the lowlands of Guerrero: "We speculate that domestication may have begun among coastal people cultivating teosinte to maintain populations near the coast. This theory suggests that the semiarid highland regions of Mexico were late in the adoption of maize agriculture" (Buckler et al., op. cit.: 159–160). This study is essentially unsupported, because its starting point is the reliability of the AMS dates, a point that has been proven to be quite debatable, to say the least, whereas their other arguments are, as they themselves acknowledge, simple speculations with no solid grounding.

Wilkes (1989: 449) concluded that all the theories regarding the origins of maize fit into one of the three postulates of evolutionary patterns. The first one holds that there was a direct evolution through the domestication of a wild ancestor, be it teosinte, wild maize, or a wild grass. The second postulate is based on a hybrid origin from two dissimilar relatives, and the third one has as its starting point the hypothesis that maize had its origin in a wild ancestor and with repeated hybridization with teosinte, its closest wild relative. Yet Wilkes also made a statement (2004: 22 and 23) that is worth repeating: "The evidence of teosinte introgression into corn in the archaeological record remains circumstantial at best because teosinte and hybrids have been recovered only at Romero's Cave and at Guila Naquitz, yet the dates fit Tehuacán ..."

After reviewing all of these hypotheses and their respective arguments, I insist that the wild maize hypothesis is still the most plausible one.

Domestication is the second point of contention in the maize problematic, in terms of whether it happened only once in Mesoamerica or whether it happened two or more times, in different parts of the continent, particularly in

South America. But besides this, there is also another point that must be borne into account, and that is usually forgotten – the time period from the moment that the plant passed from wild to domestic state, that is, the time domestication took up. This is certainly important in trying to understand the former aspect and may well be a strong argument for one or another position.

We must unfortunately acknowledge once again that we do not have enough data to show – using actual data – that one position or another is correct. At present all we can do is just give an opinion. We saw in Chapter 1 that Pääbo (1999) pointed out the possibility that domestication was a rapid process. He, however, essentially based his work on genetic information and did not take into account geographic or anthropological factors. It is true – and many scholars agree – that some changes may take place in a plant in a very brief span, but I believe that it is very hard for this to have taken place in the set of processes that domestication signifies or represents.

Nowadays no one questions the fact that maize had its origin in Mesoamerica, and that it spread north and southward from there. Interpreting the data becomes even harder given the early dates available for South America, which were presented in Chapter 5 and which shall be discussed subsequently. But there is one specific fact that cannot be left aside – geography, that is, the tremendous physiographic differences between Mesoamerica and South America. We must not forget that the Andean area is far more complex than Mesoamerica from an ecological standpoint. We are often unaware that of the 103 life zones present in the world – which were established crossing data on latitude, altitude, humidity, temperature, and evapotranspiration – the Andean area has 84, 17 of which are of a transitional nature (Holdridge, 1967; ONERN, 1976; Tosi, 1960). It therefore does not matter whether maize reached South America in wild, domestic, or semidomestic state, because in either of these cases it was faced with having to adapt to a harsh and rugged land. And once there, maize had to endure many other changes when it was taken by man from the high Andean altitudes down to sea level or to the humid Amazon basin, passing through all of the intermediate ecosystems. Of course, for this we have to accept that its arrival took place along the Andean mountains. This new environment certainly constituted an unequaled experimental field.

But if, as some posit, maize arrived first to the great Amazon area and from thence to the Andean zone, the changes it underwent would essentially have been the same. To give just one extreme example, in Bolivia there are at least two varieties of maize on the slopes and mountains around Lake Titicaca, from its lowest level at 3,810 masl up to 4,100 masl. This is the highest-altitude agriculture practiced in the Andes, and perhaps in the whole world (Chávez, 2006: 623). It is therefore unlikely that domestication took place within a short span, and this is why I have long defended the position that domestication is not an event but a process (see, e.g., Bonavia, 1997: 82). This by no means is an attempt to be original, for this idea has already been posited by other scholars (e.g., Wilkes, 1989). Brieger (1968: 548) pointed out that with the technology available, a modern maize farmer would require at least ten years to produce a new population or a new hybrid. And the Corn Belt's famed Dent Maize, which was attained in the United States around 1800, required at least three generations before the goal was reached.

The opinions of various scholars as regards the two major positions on domestication were presented in Chapter 4. From this it follows that as regards domestication there are disagreements even in Mesoamerica, concerning whether it happened in just one site or in several of them. Here only the two most plausible positions are mentioned, which are known as the Balsas and Tehuacán hypotheses. I do not have a sufficient command of Mesoamerican archaeology to pass judgment, but I find that the Tehuacán hypothesis has a stronger archaeological support, whereas the Balsas hypothesis has a stronger theoretical underpinning.

Here it is worth insisting on a significant fact. Although most scholars lean toward a Mesoamerican domestication and emphatically reject the possibility of there having been an independent domestication in South America, none of them has made a full, detailed analysis using the existing evidence. One other point that has passed by unnoticed is that whereas independent domestication in different geographical areas is accepted without any discussion whatsoever for other plants – squash (*Cucurbita*), to give just one example – when we come to maize this is a taboo that must not be discussed at all (see Balter, 2007: 1833). Not only has independent domestication as a cultural phenomenon been widely proven, its foci continually increase. Whereas, at the turn of the twenty-first century, 7 of these foci had been accepted (K. Brown 2001: 633), nowadays there are at least 10, one of which is the Andean area (Balter, 2007: 1831). The study of rice in China is significant in this regard, for there are indications that there was a multiple domestication of this plant (M. Jones and Brown, 2000: 773).

Pickersgill (1989) is one of the few scholars who has tried to explain this problem, and who has an adequate grasp of the subject. We have seen that for her maize reached South America in the domestic state, and that it then long stayed isolated, evolving independently, until in late times there was once more contact between Mesoamerica and South America. Pickersgill (1972) later accepted the independent evolution of Mexican and Andean maize.[3] We saw in Chapter 5 that the differences between them were also pointed out by Goodman (1976).

It is true that since the late 1940s, with the work done by Vavilov, several scholars posited an independent domestication of maize in South America, yet it is clear that no one has defended this position more strongly or cogently than Alexander Grobman, who marshaled a solid scientific argument in support of polyagrogenesis, as he called it. Here we need not expound this point at length,

[3] Yet in a recent paper (Pickersgill 2007: 929), her position is not altogether clear.

as this has already been done in Chapter 4. It was in the late 1950s that I became interested in the maize problematic under the influence of David Kelley, who had excavated in Mexico with MacNeish. Subsequent contact with Paul Mangelsdorf allowed me to become acquainted with this subject. It was also at the same time that my association with Grobman began, seeing as we both had similar points of view, and since then we have collaborated, because to have an overall approach it was essential that the archaeological data be combined with the botanical information. We also often had the help of other specialists.

There are five major arguments, supported with concrete evidence, that promote the independent domestication hypothesis. The first argument is the presence of three races of maize in the preceramic Andean area, that is, Proto-Confite Morocho, Proto-Kculli, and Confite Chavinense, along with their hybrids, whereas in Mexico there was practically only one race – Chapalote/Nal-Tel – which Mangelsdorf initially separated into two but then considered one single complex.

The second argument concerns the antiquity of the samples. We have seen that in Mexico, the oldest dates are for Igualá Valley, that is, for Laguna Ixtacyola, Ixtapa, Laguna Tuxpan, and the Xihuatoxtla Schelter, where dates of up to 10000 years BP are given, based on phytoliths and pollen remains. Yet on analyzing these finds, the data is always ambiguous and raises serious doubts. We do not find a solid body of evidence that supports them. The same thing holds for San Andrés (Tabasco). The only solid evidence comes from Guilá Naquitz. Now, if we do not consider the Peruvian finds made at Guitarrero Cave due to the problems they raise – which have dates that are earlier than those from Guilá Naquitz – we still have the reliable dates of Cerro Julia and Cerro El Calvario, which are on average c. 650 years older than those for Guilá Naquitz. Given the relativity with which we must approach radiocarbon dates – a point that is discussed later – we can say that in terms of the information as yet available, domestication took place simultaneously in Mexico and the Andean area or happened at a slightly earlier date in the latter.

The third argument is the essential difference found in the composition of the chromosal knobs in early Mexican and Andean maize. This was discussed in depth in Chapter 4, and we will return to it later on. The fourth argument is that the Casma maize shows no evidence at all of introgression with teosinte. Finally, the last argument is the high variability of Andean corn, a point to which we shall return subsequently.

I fully agree with Grobman in that maize is quite ancient as a species, in both its domestic and its wild form. The oldest Peruvian specimens are primitive popcorn races. It has been shown that in Mesoamerica, maize preceded annual teosinte as a wild plant species. Corn must probably have left Mexico in the wild state and without human intervention, possibly carried by birds (Bonavia and Grobman, 1989b: 462), and it penetrated South America through the Panamanian isthmus. It is possible that it was domesticated in several places, that is, Mexico, the central Andes, and perhaps Colombia.

It is a fact that the races that developed in these two centers did so independently. The study by Goodman and Bird (1977) showed that the Mexican and South American races fall into mutually exclusive groups and agrees with the idea of an independent development in both areas. This somewhat agrees with one of the ideas set forward by Goloubinoff and colleagues (1993: 1997–2001), claiming that maize was independently domesticated from various different wild ancestors, which then intercrossed among themselves and crossed with wild teosinte.

Goloubinoff wrote the following in this regard, in a letter sent to Grobman and me:

> My findings are that there is a high sequence variability in maize, however it is probably not due to an acceleration of the molecular clock but rather to the presence of a very ancient gene pool in maize. There is no evidence that a single "bottleneck" ever existed in the history of maize neither that *Z. diploperennis* was one of the progenitors of all races of maize. However, the data from South American ancient specimens help me to minimize the role of teosinte introgression in the diversified pattern of the maize alleles. My data suggest – although do not prove yet – that maize have [*sic*] been domesticated on several independent instances, from various ancestors. Thus, your position of an independent domestication of maize in the Andes is strengthened by this work. (Pierre Goloubinoff, letter to Grobman and this author, 8 March 1991; the original letter is in the possession of Bonavia)

Grobman in turn correctly noted the following:

> ... the domestication of maize is a process that is still not well-defined. Maize on the one hand has its closest relative in Mexico – teosinte. On the other hand, the corns in the most ancient contexts in Peru have characteristics that are more similar to a species other than teosinte. The drawing of the primitive Los Gavilanes maize, made on the basis of the specimens found there, is an attempt [at reconstruction] [Grobman, 1982: drawing 60, 167; see my Figure 5.11]. But having said this, the comparative evidence shows us that it may have been a domestication from a wild progenitor that was transported to Peru from Mesoamerica by human or animal means, or a differentiation from a semi-wild progenitor under selection in different ecological areas in Peru, [which proceeded] independently from a very primitive *popcorn* maize and formed more primitive and quite differentiated races, even before Mexico's Nal-Tel. (Alexander Grobman, letter to Bonavia, 5 May 2003)

Raymond and De Boer (2006: 340–341) made the interesting observation that maize requires a very simple technology for its development, and that of all cultivated plants it is the one that most easily adapts to mobile societies.[4] To

[4] Maize has a relatively short growing season, requires little technology, is easily transported, and can be stored in such a way that it does not cause problems for hunter-gatherers in their seasonal economic activities.

cultivate maize there is no need to think of sedentism, and these scholars give its use in the Amazon forest as an example. This falls under what the present writer would prefer to call horticulture (Bonavia, 1991: 121 and passim).

Another significant fact that Grobman (2004: 434) underlines is that "... the maize that is believed to have been wild or in a primary stage of domestication, probably scattered its seeds through the disarticulation of a small and brittle rachillae, not through the disarticulation of the rachis, as happens in teosinte. These different characteristics of both species are typical of a wild plant." A great number of variations later appeared in Tehuacán, which may be interpreted as due to the presence or introgression of teosinte in maize, and it happened 3,500 years after the first appearance of wild maize (Wilkes, 1989). Yet by then there was in Peru a non-tripsacoid maize, that is, one that did not have the characteristics of teosinte.

It is also worth recalling that the earliest maize from Mexico's San Marcos Cave are cobs 1.9 to 2.5 cm long, with an average of 55 kernels (Mangelsdorf, 1974: 168), whereas the maize cobs from Epoch 2 at Los Gavilanes – the most ancient ones – are on average 4.9 cm long and have an average of 179 kernels (Grobman, 1982: table 11, 160). This indicates a different evolutive direction.

In her study, Eubanks (2001c: 96) admits that teosinte had a key role in the ancestors of maize but shows at the same time that there are no archaeological data that show that it slowly turned into maize through a steady accumulation of the mutations that set these two plants apart. Based on the experiments that reconstruct the archaeological evidence, we reach the conclusion that these same differences between maize and teosinte may have originated in almost sudden fashion and due to human selection and cultivation of recombinants, derived from the introgression of *Tripsacum* and a primitive teosinte. But as Doebley (1990: 16) correctly noted, there are problems when it comes to documenting the introgression. And it is even harder to try to establish its direction. The question is as follows: did the genes flow from the cultivated plant to the wild one, was it the other way round, or did they instead flow in both directions?

In his most recent study, Grobman (2004: 466) discussed the possibility that annual teosinte was formed by the crossing of *Z. diploperennis* and wild maize, something that was experimentally proven. For him this would supplement the alternative hypothesis of a modern maize derived from wild maize, with the participation of teosinte happening only subsequently. The study of Jaenicke-Deprés and colleagues (2003), which showed that primitive maizes have the same alleles that modern maizes have in Mexico, whereas teosinte is missing some of them, is compatible with this hypothesis. According to the interpretation made by Jaenicke-Deprés and colleagues (2003), the early selection of alleles would not be necessary if modern maize was a direct descendant of wild maize. Their data therefore fit this hypothesis.

Peru is noted for its extreme diversity of maize races (Harlan, 1995: 143). Mangelsdorf (1974: 105) wrote in this regard: "Of the 32 Mexican races, 20, almost two-thirds, are 'endemics,' races not found in other countries. This suggests that Mexico is one of the principal centers of the origin, evolution, and diversification of maize. This is also true of Peru; some 30 of its 48 races occur only in that country." If we express this statement as percentages, it turns out that the 20 Mexican races come up to 62.5%, and the 30 Peruvian races likewise come up to 62.5%. No comment is needed here.

If we study the evolution of maize in the Andean area, we reach the conclusion that many of the adaptive mechanisms that allowed this plant to grow in various harsh environments became fixed quite early on in the evolutive process (Sevilla Panizo, 1994: 243).

We saw in Chapter 4 the major differences between Andean and Mexican maizes, as regards the structure of their chromosomal knobs. McClintock explains that the analysis of the chromosomal makeup of the different races and their purported relatives could confirm their hybrid origins. It has been shown that this is correct. It has now also been proven that the makeup of the chromosomal knobs can show how an alien maize, when introduced into a given area, may contribute to the origin of new races, and it is sometimes even possible to infer the source of the maize so introduced. Given that the whole chromosomal organization is quite similar in *Euchlaena* and in maize, and that the two genera can easily cross, it follows that an exchange of chromosome segments clearly took place between them, including in the regions where the knobs are formed (McClintock, 1960: 462, 466–467).

McClintock continues, "Because of the types of knobs observed, and their individual distributions in the examined races in both North and South America, I am led to consider the possibility that *cultivated maize my have had several independent origins*, from plants whose knob-forming regions had distinctly different capacities for producing knob substance." She then added that a cultivated type may have had its origin in plants wherein all of the knob-forming regions were of such limited capacity, in which case the derived maize has no detectable knobs, or just a small one in one or several of the knob-forming regions. Another type may have had its origin in plants whose knob-forming regions were less limited in capacity, so the chromosomes of the cultivated plants had either small or medium-sized knobs. However, another type could have originated from plants whose knob-forming regions were able to produce a large amount of knob substance; large knobs will be present in these cultivated types (McClintock 1960: 465; emphasis added).

We can thus conclude that much of the maize that is cultivated at present in western Mexico, on the coastal areas of Central America, and in northern Venezuela derives from original types wherein most knob-forming regions have a well-developed capacity to produce knob substance, and that on the contrary, the maize growing in west-central Guatemala derives from an original type whose knob-forming regions were limited in capacity (McClintock, 1960: 465–466).

We saw that Pickersgill (1969) and Grobman and colleagues (1961) pointed out that the chromosomal knobs can be interpreted in various ways. This means that no agreement has been reached on whether the chromosomal knobs in Mexican maize are the result of hybridization with teosinte. If we accept this position, then the low number of knobs in Peruvian maize can be explained through an independent domestication of this plant in the Andean zone, or through the early introduction of Mexican maize, prior to an intensive hybridization with teosinte. But as has already been noted, this means that Pickersgill and Grobman accept the existence of a wild maize.

According to Rivera (1980b: 108), after studying the maize M. West excavated in the Puerto Morin site, in the Virú Valley, Robert McKelvy Bird also raised the possibility – in a personal communication – of an independent domestication.

We saw in Chapter 4 that in a recent study by Piperno and colleagues (2009: 5023) we find a statement that, although it is not clear, does seem to suggest the possibility that an independent domestication of maize took place in Mexico and South America. This is striking, for the main author of this study has systematically rejected this. A change in attitude would be quite well received, but this article unfortunately does not make a complete or specific presentation of this issue, and the references given are not specifically concerned with this issue.[5] A clarification by these scholars is in order.

One question many scholars have posed is why maize was domesticated, that is, what made hunter-gatherers initiate the process that led to the domestication of this plant. This is clearly a difficult question to answer, for no reliable traces remain in archaeological sites, and the same thing holds true for most of the plants that underwent this process. We saw in Chapter 4 that Iltis (2000) was the first to suggest that what led to domestication was not the fruits of maize but the juicy and chewable pulp of the stalk. This certainly may have happened, but I believe that the main reason for domestication was that man realized that the kernels could be used as food. We must not forget that the first maizes were popcorns, so the idea may have risen on observing that the kernels pop and are nice to eat when heated. This may have happened accidentally, when the maize plant was placed over the fire to use it as fuel.

On the other hand, and as was noted by Tykot (2003: 695), there is no evidence of "quids" in the earliest maize epochs, either in Tamaulipas or Guitarrero Cave, or in the Ayacucho settlements. I would like to add that the quids found in Mesoamerica are all late ones (Mangelsdorf, 1974: 177). When discussing the possibility that maize was first used for its pulp, Grobman noted that the amount of juice that could have been extracted from small stalks of maize like those found at Los Gavilanes is minimal, and so this does not make sense at all (Grobman, personal communication, 28 February 2004).

5 Interested readers should see the detailed comments made in Chapter 4.

Smalley and Blake (2003) later suggested that the reason maize was domesticated was to use the plant's juice to prepare an alcoholic beverage. We have seen that this idea has been recently followed by other scholars. On one hand, it has been shown here that they have not presented any concrete evidence in this regard. On the other hand, it is hard to imagine the possibility of attaining an alcoholic beverage in preceramic times that requires reheating and a long boiling period. In Peru, at least, it was established – and this is well known – that water was heated and perhaps even boiled for a brief span, by placing warm stones inside containers made out of gourds (*Lagenaria* sp.). An experiment would be interesting, but it is doubtful that an alcoholic beverage can actually be prepared in this way. Moreover, I was unable to find any ethnological evidence that indicates the use of maize plant juice in this regard. We have seen – and it is quite well known – that kernels were and are used for chicha.

When Piperno and colleagues (2009) published their research on the central Balsas River valley, they mentioned the position held by Smalley and Blake (2003) and pointed out that "this hypothesis is testable using the idiosyncratic short-cell phytoliths that maize and teosinte stalks produce in significant quantities and that would be expected to occur in the phytoliths record." Besides, they point out that they looked for phytoliths with these characteristics, which "... were not seen in any context at Xihutoxtla [*sic*; i.e., Xihuatoxtla]." Piperno and colleagues (2009: 5022) therefore conclude that "... the major focus of maize utilization was directed toward the cob of the plant." They also insist in that "... maize kernels were commonly processed and consumed, indicating that early domesticated maize was a more significant grain crop that some investigators have supposed." Finally, they categorically conclude that "... the theory that the use of stalk sugar for the production of alcoholic beverages or other purposes was the primary motive for the early cultivation and diffusion of maize (16 [the reference here is to Smalley and Blake, 2003]) is not supported by current data" (Piperno et al., 2009: 5023).

Some clarifications are due in regard to the term "tripsacoid," for its inaccurate use or understanding may lead to mistakes or misinterpretations. When these characteristics are mentioned, it means that there is a greater hardening in the horny part of the glumes, particularly in the lower ones (see Grobman, 1982: 174; see also Mangelsdorf, 1974: 125–126). And yet, as Grobman showed,

> the horny glumes can be assimilated or not to the so-called tripsacoid characteristics. Their appearance in such early levels on the coast of Peru, without their predominance in the population and without evidence of the presence of teosinte or *Tripsacum* in nearby areas, suggests the presence of genetic mechanisms for the increase of fibres (sclerenchyma cells) in the glumes, independently of the contribution made by teosinte or *Tripsacum* (Grobman, 1982: 163)

On the other hand we must not forget that not only has the presence of *Tripsacum* been verified in South America, but also its hybridization is not only possible but even takes place in nature, as was discussed in Chapter 3.

Another point that has been much discussed and on which there is no agreement concerns the movement of plants in terms of whether their direction was from north to south or vice versa. We must once again admit that we lack evidence in this regard. In the specific case of maize there is a serious problem, which I have addressed on several occasions (e.g., Bonavia and Grobman, 1989b: 461), and on which I will insist once more when discussing pollen and phytoliths. Although it is true that northern South America, and specifically the Colombian and Ecuadorian zones plus their border zones, has yielded very early dates for maize pollen and phytoliths, all that this indicates is the presence of this plant. It does not at all help us to answer the question of whether maize moved from Mesoamerica to South America, or whether the movement went in the opposite direction. The reason for this is quite simple, for as yet it cannot be established what variety or races these remains belong to using pollen or phytoliths.

Bugé (1974: 33) also passed judgment in this regard. He noted that because identification below the family level is impossible, it cannot provide us evidence at the racial level. I would therefore insist that all that this can tell us is the presence or absence of this plant in a given site. And unfortunately in the northern South American area, preservation is very poor due to climatic conditions, and it is almost impossible to find macro-remains, that is, cobs, that might help us solve the issue. If the remains of maize in this area are related with the Chapalote/Nal-Tel complex, there would be no question that the movement was from north to south. But if the relation was with any of the three races that appear at an early date in the central Andean area, that is, Proto-Confite Morocho, Proto-Kculli, or Confite Chavinense, the movement would instead have been northward. The debate will continue as long as this is not somehow solved, but it will be forever based on assumptions and not on concrete facts. MacNeish and Eubanks (2000: 14) were quite clear in this regard when they noted that there are no data with which to retain the assumption that maize and other domestic plants spread from Mesoamerica to Panama through the tropical lowlands.

On the other hand, those who have handled phytoliths and pollen data have thus far been unable to present solid arguments in this regard. Pearsall is a good example. Her starting point was the hypothesis that maize came to South America from the north, and this is why it was a "foreign" plant. Pearsall believes that its initial use was "… of low utilization as a vegetable or curiosity. …" She admits that the type of maize cannot be directly reconstructed with data from phytoliths and pollen but proposes but that "… it was probably similar to the early maize of central Mexico; small, fragile cobs with few rows of small kernels. Low productivity makes it unlikely that maize would immediately supplant other, more productive resources" (Pearsall, 1995b: 128). So we see that we are facing not just mere assumptions but even a flagrant contradiction, because on the one hand the limitations of pollen and phytoliths evidence is acknowledged, but on the other hand the characteristics of this maize are explained without any basis whatsoever. Pearsall also incurs another contradiction, for it is hard to estimate the relative abundance of the use of plants with data derived from phytoliths (Pearsall, 1995b: 128).

There also are some finds that I believe are most important, which although they have been published, have not been awarded their due value, that is, the early evidence of yucca (*Manihot esculenta*). These findings were discussed in Chapter 5, so here are just mentioned. The first specimen comes from the San Andrés site in Mexico, where *Zea* pollen was found that was believed to be domestic and that dates to 5100 BC. In the 4600 BC level there was a grain of *Manihot* sp. whose characteristics resemble those of the domesticated yucca "*Manihot esculentum* [*sic*; i.e., *Manihot esculenta*]," but it is later said that "... the species cannot be positively identified from the pollen." Yet it is immediately stated that "the occurrence of *Manihot* sp. at San Andrés indicates indirect contact with farmers in the Amazon basin, where DNA evidence suggests that manioc was domesticated (Olsen and Schaal, 1999: 5586)" (Pope et al., 2001: 1372–1373).

Another find corresponds to the Aguadulce site in Panama, where *Manihot esculenta* was found in the levels dating to 6000–7000 BP (c. 4000–5000 BC) alongside remains of maize (Piperno and Pearsall, 1998; Piperno et al., 2000: 896). Arrowroot (*Maranta arundinacea*) remains were also found. Yucca remains, albeit somewhat later ones, were also found in Belize that dated to around 3400 BC (Pohl et al., 1996: 368).

Here it is worth recalling what Piperno and colleagues (2000: 896; emphasis added) stated in this regard: "Manioc ... was previously thought to have been domesticated in both Mesoamerica and South America, but recent botanical and molecular studies *indicate a South American origin* (Piperno and Pearsall 1998), particularly a region of southwestern Brazil."[6] This was confirmed by the work done by Rival and McKey (2008: 1120), who noted that the domestication of yucca took place in the southern part of Brazil-Rondônia and Acre, even though they also based their work on Olsen and Schaal (1999, 2001). On the other hand, as regards arrowroot, although its origins are not clear, it is believed to come from northern South America (Piperno and Pearsall, 1998).

This is clear evidence that plants were moving from South America to Mesoamerica between the fourth and the fifth millennium before the Christian era, but the reverse movement has not been proven. Even more important is that the evidence shows that maize was already being cultivated in Mesoamerica and Panama alongside South American plants.

[6] The reference cited here is Olsen and Schaal (1999).

It was pointed out in Chapter 5 that there are some differences among specialists in the study of pollen. For instance, we saw that Schoenwetter (1974) raised serious questions regarding the analyses made using archaeological pollen. Dull (2006: 359) also noted that from the published bibliography one comes to the conclusion that there is not one single standard with which to identify maize pollen that is used or accepted by all scholars. We have a most confusing database. Besides, many researchers have unconsciously accepted ancient dates based on pollen studies without realizing that these samples have been called into question.

We are before a very similar problem as far as phytoliths are concerned. Rovner (1996: 431) criticized the work done by Pearsall and Piperno (1993a): "No mention is made of recent control studies, which show that phytolith size is modulated by soil and moisture conditions imposing uncontrolled variables on Pearsall's rigid size-based typology." What followed was a serious critique of Piperno (1988a). Rovner (1999) later returned to this issue. I myself pointed out, when touching on this possibility, that a similar observation had been made in studies prepared in the Middle East (Balter, 2001).

When discussing this with me, Grobman gave an idealized example that is significant. If photo-period-sensitive Ecuadorean maize was planted in Illinois, these plants would grow, changing their vegetative apparatus. The deposits of silica that form the phytoliths would surely vary. The same thing holds for Peruvian and Ecuadorean maize sown in Missouri; their behavior will not be the same as in their normal habitats (Grobman, personal communication, 3 May 2003).

I do not intend to join the debate or support any of the parties concerned, because this subject is not my field. Even so, it would be best if the specialists themselves clear this up, for in its current state some serious doubts are raised.

Pearsall, who as we have seen is highly critical of the position Grobman and I have, once acknowledged that the characteristics of maize phytoliths correspond to domestic maize, which was used to establish these definitions. And Pearsall herself notes that it will not be possible to check "the model expounded by Bonavia and Grobman" until some criteria can be established for the identification of wild maize, as these are different from those that characterize teosinte or maize (Pearsall, 1994a: 248). I understand that this has already been done with the Bellas Artes pollen, which for some reason Pearsall does not consider.

There is one point that has not been made as regards both pollen and phytoliths, which must be explained. There clearly are no problems when these are found among archaeological remains. However, problems do arise when these samples are found in lake or swamp sediments, because in this case there is no possibility of clearly establishing their association with possible archaeological remains, even when they lie close by. But what is even more serious is that one can actually wonder whether or not these really are cultivated remains. Pearsall (1994a: 248) herself noted that wild and domestic maize cannot as yet be distinguished through pollen alone. I insist that a comparison should be made with the Bellas Artes pollen, which according to the studies made by Barghoorn and colleagues (1954) has a strong similarity with that of modern maize.

Although the problem of how plants spread has been somewhat addressed, we must now return to it, because this is a major component of the issue we are trying to elucidate. We shall not consider the proposal made by Lathrap (1987: 355), who suggested an early large area (7000 years BC) of "sedentary fishing" (*sic*), which comprised the land extending from Tamaulipas to the estuary of La Plata. These people would have taken the "proto-crops" cultivated from north to south and vice versa. Primitive maize would have come from the north. This proposal is completely unsupported, and it is striking that in 1987, the sole supports a serious researcher like Lathrap had for his hypothesis were Sauer (1952) and Tello (1943), which by the way make no contribution to this issue. We shall also not consider Roosevelt (1980: 62–67), because when discussing the diffusion of maize in South America, she says that this was a late event, thus ignoring all of the existing literature at the time of her writing.

Pearsall proposed that there were three stages in the dispersal of maize in South America. The initial stage saw an exchange of products on the eastern side of the Andes, as was suggested by Bugé (1974), or along the low intermontane Andean valleys. The second stage was when the plant adapted to the new conditions and was moved to higher altitudes, and finally the third stage was when maize was used as food in long-distance journeys (Pearsall, 1978a: 44–45). We see that this is a purely theoretical position that is completely unsupported by archaeological evidence, which is furthermore almost nonexistent for the eastern side of the Andes.

Matsuoka and colleagues also broached this issue and pointed out that there was an initial diversification early on in Mexican soil. Maize would have spread from thence along western and northern Mexico to the southwestern and eastern United States and toward Canada. The other route would have run from the Mexican highlands to the west and south; to the Mexican lowlands, Guatemala, and the Caribbean Islands; and from there to the Andes (Matsuoka et al., 2002: 6084). This also is a hypothesis that has no archaeological support. For instance, to the best of my knowledge there is no early maize in the Caribbean Islands.

On the basis of the differences found between Peruvian and Mexican maize, Pickersgill and Heiser (1978: 137) suggested that it was possible that a race similar to Nal-Tel spread from Mexico to Peru between 5000 and 7000 years BP. After this, only minor exchanges took place, except for the late movement northward of flour and sweet corn, perhaps around 1450 BP (Pickersgill, 1972). The great racial differentiation that distinguishes the Andean maize would have taken place throughout 3,000 years of evolution in isolation. This, of all of the positions here reviewed, is clearly the most reasonable one, but it unfortunately also is not supported with concrete evidence.

Grobman and I posited that a wild maize may have been carried to South America by migratory birds, as has happened and has been shown for other grasses prior to the arrival of man, with humans later on accelerating the process. Wild maize, as Mangelsdorf (1974: 178, 180) explained, must have had male and female flowers in the same structure. The seeds were partially covered and protected by soft glumes. Additional protection was later needed with the adaptation of leaves and shortened internodes of the flower branches to cover the kernels, as they form the husks. Grobman (1982: drawing 60, 167) reconstructed an ideotype of a probable wild maize inflorescence based on the finds made at Los Gavilanes (see Grobman, 1982: photograph 52, 167; my Figure 5.11). This type of inflorescence allows for the spillage and dispersal of the seeds when a brittle rachilla that supports the seed on the cob breaks, or due to a violent separation caused by birds. The small, flinty seeds, of red, brown, or purple color (the latter case being the Proto-Kculli race) would have easily been individuated and eaten by birds. After a few hours in their digestive tracts, the seeds would have been deposited in their excreta, in places far removed from where the kernels were originally seized.

The dickcissel (known as "rice bird," "pájaro arrocero," "arrocero Americano," or *Spiza americana*) is one of the birds that may have been quite effective in dispersing the seeds of wild maize. This migratory species invariably travels every year from the Northern to the Southern Hemisphere and passes through Mesoamerica on its way to South America, eating and damaging rice, wheat, sorghum, and other crops. Grobman observed the dispersal effect these birds have on sorghum seeds on the northern coastal region of Peru, in the same way as has been posited for maize. This can explain the evidence that all samples of very early maize seeds in South America are of popcorn types with small kernels, formed in a very dense protein matrix in the seed's endosperm, with pointed or acuminate forms as a possible protection against birds.

This seed dispersal mechanism, which Pickersgill (1983) documented for other species, may explain the diffusion of maize in the preagricultural times in which the evidence currently available suggests this dispersal took place. Wild maize may have dispersed from Mesoamerica to South America and the intermediate areas, in a way similar to that which has been accepted for other species, and it may have been domesticated independently in both Mesoamerica and the central Andes. This hypothesis may also explain the greater racial variation – in comparison with Mesoamerica – in the types of maize that developed immediately afterward in the central Andes under the impact of domestication (Bonavia and Grobman, 1989b: 462–463).

The truth be told, Grobman and I acknowledge that our position also lacks the evidence required to show that it is correct, but we believe that even so it is based on some concrete facts, which is what the other hypotheses lack.

Some scholars have posited that both the north–south movement of maize and its opposite took place by sea or along a coastal route (e.g., Benz and Staller,

2006: 669; Pearsall, 1978a: 54, 58, 61, 63, 65; Tykot and Staller, 2002: 670). They thus show they are not familiar with the evidence of the early Peruvian maize, which as has already been shown has a high percentage of purple color (82% in all cobs from Epoch 2 at Los Gavilanes, and 91.6% in Epoch 3; Grobman, 1982: 161); it has been proven that this is a result of the genetic fixation of anthocyanin at high altitudes (Grobman et al., 1961: 48–49; Greenblat, 1968). The early maize developed in midaltitude valleys in the Andes and was taken from there to the coast.

Piperno claims that many swamp sequences, from Belize to Ecuador, show the presence of maize in 4700–7000 BP, that is, that maize was prior to, or contemporary with, the Tehuacán remains (Piperno, 1994a: 638). She insisted on this anew and noted that skeptics will argue that a direct date from a cob is more reliable than that obtained from a geological sediment; "however, the paleoecological age determinations were carried out, respectively, on polliniferous sandy peats, gyttyas,[7] and peats. All of these types of sediments are reliable indicators of age, and it is unlikely that the sediments are much older than the pollen grains they contain." Piperno added that the Mexican maize is not the most ancient type, and that all of the botanical information available would indicate that the Tehuacán maize was domesticated elsewhere (e.g., Benz and Iltis, 1990; Doebley, 1990) (Piperno, 1995: 135).

Three comments are in order here. First of all, Piperno does not state how the sediments may be related with cultural remains. Second, as we have already seen, those who posit that maize was domesticated in a low-altitude area and was then taken to Tehuacán have not presented any solid evidence. But the third point is a major one, and it must be emphasized. It so happens that in this specific case, Piperno accepts the association – this being one of the major tenets of archaeology – of pollen remains with the sediments they were extracted from. Yet in another context, when criticizing the dates of Los Gavilanes and other Peruvian sites, the association of a group of remains with those of maize have been rejected, under the pretense that only direct dates are valid. And it so happens that one of those who most strongly took this stance is Piperno. This is a flagrant inconsistency.

There is another major point that has only been touched tangentially. If we analyze the dates available for South America (see Chapter 5), particularly those for the Ecuadorean zone, we will find that there is a marked difference between the antiquity of pollinic data and that of macro-remains. Benz and Staller (2006: 667) pointed this out, but they made a mistake, for when discussing the pollen found in lacustrine deposits and on the Ecuadorean littoral, they defined it as "maize pollen dated by association." Once more, the point here is that in the sediments there is no association with other material remains, the only possible association being in regard to geological strata. We can therefore ask whether all

[7] Gyttya is the acid substratum of peat.

of these dates are valid, and what they actually represent. Besides, many of these dates from Ecuador really are older than those available for the Mexican specimens, and this, as Staller and Thompson (2002: 46) point out, would indicate an independent domestication in South America.[8]

Other scholars have taken a stand on this issue, but not all of them with solid arguments. Such was the case of Benz (1994b: 157–158), who claims that either the dates for South American maize are wrong, or earlier dates have yet to be found in Mexico. The arguments Benz presents here are not valid, even though the question is correct. Blake (2006: 65) points out that if we compare the map of pollen distribution with that of the distribution of his phytoliths (in his figures 4–2 and 4–3), we find a similar pattern. Both have early dates. Several samples from Mexico, Panama, Colombia, and Ecuador are in fact older than the Guilá Naquitz maize: "This is hard to accept because the Guilá Naquitz maize is primitive – any more primitive and it would still have been close in morphology to teosinte." The maps likewise indicate that, for Blake, there are discontinuities in the distribution of the samples. If maize spread southward at the time suggested by the pollen and the phytoliths, the population should have passed around certain zones. They returned back north many centuries later, passing through areas they had not gone through before. For Blake, another possible interpretation would be that maize passed southward either eventually or continuously, and that we have not yet found the evidence for this. Blake's position is clearly biased, for he interprets the data starting from a significant but unproven hypotheses, that is, that teosinte was the origin of maize.

Let us see now what the facts say. It so happens that if we analyze the dates presented in Chapter 5, the picture is not exactly as that which the aforementioned authors describe; even worse, there is no consistency that can allow us to establish a logical sequence of the available data. To avoid any misunderstanding, we must once again bear in mind that I have considered the dates just as the authors presented them, and without any correction. An in-depth archaeological study would require a more critical analysis, and the dates would have to be calibrated. But this is not required for this present effort. So let us see what the situation before us is. The following analysis will be based on BP dates, and only the most ancient dates shall be taken into account. The traditional carbon 14 method is used here unless otherwise stated, and whenever the AMS system has been used it shall be noted.

For the United States we have dates ranging between the fifth and the fourth millennia, whereas the pollen-based dates are concentrated in the third millennium.

The dates for Mexico range between the seventh and the fourth millennia, and with AMS dates they fall between the fourth and the third millennia. The pollinic data in turn range between the tenth and the fourth millennia.

[8] Staller insisted on this point (2003: 376–377).

For Belize we have little data, but what is available falls within the fourth millennium, whereas the pollinic data falls within the third millennium.

For Honduras we only have pollinic data for the fourth millennium. In El Salvador maize was present in the third millennium, whereas pollen data is available for the fourth millennium.

Costa Rica is somewhat homogeneous, for both the macro-remains and the pollinic remains fall within the fourth millennium.

The picture for Panama is likewise interesting in its consistency. Here maize ranges between the seventh and the second millennia. The pollinic data range instead between the seventh and the first millennia, and those from phytoliths, between the seventh and the third millennia.

For Puerto Rico there only are pollinic data that range between the third and the second millennia.

There are scant data available for Venezuela, and what little there is falls within the third millennium.

In Colombia, although maize has an antiquity that falls within the third millennium, the pollinic data are far older, for they range between the sixth and the second millennia.[9]

In Ecuador the dates for maize range between the fifth and the first millennia, but the pollen data fall between the seventh and the fourth millennia, and the same thing holds for phytoliths.

In Peru the data range between the sixth and the third millennia, and the single phytolith datum available dates to the fourth millennium.

For Chile we only have dates for macro-remains, ranging between the ninth and the fourth millennia.

As for Brazil there is almost no data available, and what little there is falls within the fourth millennium.

In Uruguay the phytoliths of maize cobs and starch grains from this plant's kernels have been dated to between the fourth and the late second millennia BP.

For Argentina, for which we likewise do not have good data, those that are available range between the tenth and the eighth millennia.

The analysis of these results shows that things stand thus. Let us turn first to the dates of the macro-remains. The most ancient date available is for Argentina. The difference between the dates for Peru and Mexico show that if we uphold the results initially obtained for Tehuacán with the traditional C14 method, the Mexican results are older by some thousands of years. The Peruvian dates are older if we take into account AMS dates, but if we compare the dates obtained with AMS dates for Guilá Naquitz and compare them with their Peruvian C14 counterparts, we find the latter are more ancient.

9 The data mentioned by Ficcarelli and colleagues (2003), which we reviewed in Chapter 5, are not considered here because no details are given, and because the original source is a dissertation (Kuhry, 1988) that I was unable to look up.

The comparison of the results obtained between Mexico and Panama evidently lacks consistency, for Panama falls within the Mexican range. There also is no consistency between the Panamanian and the Peruvian dates.

So it turns out that the oldest South American dates are for Chile (9900 years BP) and Argentina (10559 years BP). In other words, the oldest dates are for marginal areas where it is hard to envision the domestication of plants taking place, at least with the evidence thus far available. As for Brazil, we simply do not have data. And it is precisely here that more research is needed, as this is a promising terrain.

The scant data available for Venezuela and Colombia are consistent (within the third millennium BP), and the difference between the Peruvian and Ecuadorean remains is not that big, considering the flexibility required when going over radiocarbon results.

If we bear in mind the dates obtained from pollen grains, we find that the result is different. The oldest date available is from Mexico – 10,000 years. We must not, however, forget that here the Bellas Artes samples have not been included. The date is consistent if we are considering wild maize, but it clearly is not if we are dealing with domesticated maize. Further south there is some consistency. The pollen found from Guatemala to Costa Rica gives dates in the fifth millennium BP. Panama has earlier dates that fall within the seventh millennium BP. There is scant data available for the Dominican Republic, Puerto Rico, and Venezuela. As for Colombia, the dates fall within the sixth millennium. We thus see that with the exception of Panama, the data do not disagree. In general, the potential arrival of pollen from South America to Mexico seems to be far more consistent when we examine the results derived from the macro-remains.

There is scant information as regards phytoliths. There is some consistency from Panama to Peru, but we have similar datings, so there is no way that a possible vertical movement can be argued from this data in any sense at all.

When assembling all of these data, we find that the oldest dates for macro-remains, as well as for pollen that fall within similar ranges, all come from Mexico and Argentina. When we consider only the data obtained from pollen and phytoliths, we have consistency only from Costa Rica to Panama, but it turns out that the dates for Mexico, Colombia, and Ecuador are the most ancient ones. There actually are very few places where the dates for pollen and phytoliths are consistent.

In sum, and bearing in mind all of the data, it follows that the results are not exactly what the scholars who presented them pointed out (see previously), and that there actually is much confusion. Establishing a consistent route for the diffusion of maize with this information is almost impossible. To clear this up, it is of the utmost importance that we have more studies, but ones that try as far as possible to obtain samples of macro-remains, phytoliths, and pollen from the same assemblages, so as to first establish an internal consistency for the site before comparing it with other sites.

It is on the other hand clear that pollen and phytolith analyses have focused on some areas while ignoring others. There are practically no studies of this type south of Ecuador or for all of the vast expanses of the eastern South American areas. And it is essential to finally establish whether there actually is some difference between the pollen of cultivated maize and the possible "wild" maize, and for this the study of Bellas Artes is essential, and it could even be repeated. This is crucial because the pollen extracted from lake or swamp deposits often leaves a lingering doubt concerning whether these actually are archaeological or natural remains. It is likewise vital that specialists establish whether or not environmental changes can induce variations in the phytoliths, given that the differences in South America are quite marked, even in regions that are quite close to one other.

Although several factors are suggesting a north–south movement, it cannot be denied that several others, some of them early ones, suggest the opposite, as is the case of the aforementioned manioc samples. But other evidence is also available. For instance, McClintock (1960: 469) has pointed out that the Inca-Andean complex is present in Guatemala; besides, not only was this complex found, it also happens that its components are mostly concentrated in the maize from the west-central highlands.

Finally, it must be pointed out here that Benz (2006: 18) claims there is a difference between the genetic information in maize and the archaeological information. This issue is not clear, and so I prefer to leave it open.

As for the central Andean area, few scholars have tried to explain the diffusion of maize, and besides, paleobotanical studies are relatively recent in Peru. Collier (1961: 108, note 3) clearly shows that almost nothing was known in this regard in the 1960s. Rowe tried to present a hypothesis in this regard, and it is worth going over his position, as it has not been taken into account by the studies made in this area. We must, however, bear in mind that the remains of preceramic maize were just being discovered at the time when Rowe made this proposal, in the mid-1960s.

Rowe posited the presence of three agricultural traditions in the central Andes, one in Lake Titicaca, another one in the Marañón basin, and a third one on the coast, each of which had marked differences vis-à-vis the others. The overriding common characteristic between them was maize cultivation. This was done in all sites where the ecological conditions allowed it, and with fewer technological variations than those required, for instance, by the potato.

Preceramic maize was known at the time only for the Huarmey-Supe area, that is, on the North-Central Coast, and Rowe realized that because these were cultivate remains, they had to have antecedents elsewhere. He was well aware that maize could not have had a costal origin and pointed out that it must have originated somewhere in the highlands immediately behind the Huarmey-Supe zone, namely, inside what he called the Marañón agricultural zone. Rowe, however, did admit that the evidence available did not allow one to rule out the alternative possibility – that cultivated maize had been brought over from southern Mexico to the Ancash and Huánuco zones, where it developed before its diffusion. Rowe pointed out that an argument could be made for both these positions. He believed that the Huánuco-Ancash zone had been a "... vigorous focus of experimentation in farming at all altitudes." Regardless of whether maize reached this favorable context or instead developed locally, in any case a variety with "primitive" characteristics developed here before spreading to the coast and the highlands. The areas lying close to the Marañón tradition were clearly the first to be affected (Rowe, 1965a: 4, 8–10, 13).

The hypothesis presented by Rowe is important because his ideas were ahead of his time, they showed great intuition, and they have been somewhat supported by subsequent research. We must not forget, as has already been pointed out here, that before beginning our collaboration, Grobman had already established that the purple color is a genetic fixation for altitude (Grobman et al., 1961: 48–49), which was subsequently verified by Greenblat (1968). This fixation is codified by genes located in four different chromosomes (Grobman, 2004: 467). Hydrosoluble pigments known as anthocyanins are responsible for the color of flowers and fruits and sometimes of leaves too. These pigments are always found in the heterosid form (Casteñeda Casteñeda et al.. 2005: 107). Grobman later found a high percentage of purple-colored maize at Los Gavilanes (Grobman, 1982: 174).

I presented some ideas regarding this aspect of the maize problematic in the early 1980s that are still valid. I wrote at the time that it seemed that there was one maize domestication center somewhere in the Peruvian departments of Ancash and Huánuco; this plant later spread throughout the highlands and from there to the coast through some valleys, such as the Huarmey Valley, and finally all over the coastlands. The high variability of the Huarmey maize indicates this. C. Smith (1980a: 111) likewise points out that the only way of explaining the variation found in the maize from Guitarrero Cave is by accepting the presence of maize populations under selective pressures in other parts of Peru that had access to the Callejón de Huaylas. At present we cannot explain the Ayacucho maize, which may have been part of the same complex or instead a separate domestication area, but to judge by the available botanical evidence this does not seem possible (Bonavia, 1982: 371–372).

Grobman and I have long held the great antiquity of maize in Peru, based on solid and tangible evidence derived from the study of a large number of cobs, husks, kernels, and parts of maize plants recovered from secure archaeological contexts (Bonavia and Grobman, 1989a; 1999). Robert McKelvy Bird disagrees. It has already been pointed out that his arguments, which are now reviewed, are mostly of a botanical nature. His opposition to the hypothesis presented by Grobman and me has convinced other scholars, who are not necessarily familiar with botany and genetics, and made them question the position held by Grobman and Bonavia. Although pointing this out may seem redundant, the disagreements here not only concern the thesis but also focus on the fact that the arguments Bird raised against Grobman and me use incomplete information; that Bird and the others have not taken into account new data that have superseded past information; that they use irrelevant data; and that they use information selectively to bias it against one side of the argument, which distorts the facts.

One of the major arguments R. McK. Bird (1990) raised against the antiquity of maize in Peru, and therefore against its presence in the Preceramic period (Bird, 1990: 833), is that he believes the cobs and kernels of maize that date to this period are too big to be ancient maize; they should therefore be dated in more recent times (AD). Bird claims that many kernel specimens were heavier or larger than maize kernels documented as being from the first millennium BC or older. He also adds that the maize samples are not uniform (see also Bird, 1978: 92).

Let us see the evidence. We start with the size of the kernels. The three maize races present at Los Gavilanes (Confite Chavinense, Proto-Confite Morocho, and Proto-Kculli) are all primitive popcorns, are characteristic of the central Andes, and, as was already seen, have also been found in Guitarrero Cave and at other sites. They have no counterpart in the Mesoamerican area as regards morphology and the external phenotypic aspect and the color of the cob (see the discussion in Bonavia and Grobman, 1989b; Grobman, 1982; Mangelsdorf, 1974: 194). If we divide the mean length of the cobs in the 85 ears from Los Gavilanes Epochs 2 and 3 by the mean number of kernels per row, we get a width of only 2.8 mm for the Confite Chavinense ears and 3.9 mm for those of Proto-Confite Morocho. The kernels from Los Cerrillos (Wallace, 1962), which are at least 1,600 years younger, have on average a width of 4.5 mm.

We must now compare these measurements with those available for the maize from Cueva de San Marcos, one of the two caves where the oldest Mexican maize was found, and where we only have one primitive race – a popcorn that we can call the Nal-Tel/Chapalote complex, a conclusion reached by Mangelsdorf (1974: 174). Using this information and analyzing the data presented for the maize in each zone in the Cueva de San Marcos, which totals 171 intact cobs, Grobman calculated the following width for the kernels per zone: B, 3.7 mm; C1, 3.7 mm; C, 3.8 mm; D, 3.16 mm; and E-F, 2.9 mm. A comparison of these figures with those for the Los Gavilanes maize shows kernels whose width is bigger in the primitive Mexican maize than in the 85 intact ears classified as pure Confite Chavinense and Proto-Confite Morocho racial types (no intermediate hybrids). Furthermore, none of the measurements obtained in Cueva de San Marcos were as low as the average width that the Confite Chavinense kernels from Los Gavilanes must have had. The kernels of the latter race are isodiametric, that is, they are almost the same size in all three dimensions.

R. McK. Bird (1990) particularly questions the smaller size of some of the maize kernels found at Los Gavilanes. Grobman (1982: 164–166) pointed out that out of the few complete kernels that were found and measured, some were popcorns (n = 6 and 7) that had the size typical of this group (mean length: 5.16 mm; width: 4.42 mm). Out of the whole sample of 35 kernels, only a few (M = 6) have a mean length of 9.66 mm and a mean width of 8.36 mm, and only one is 10.0 mm long and 8.0 mm wide (see Grobman, 1982: table 15, 165). Grobman assigned these few bigger kernels to the emerging types of the Huayleño race, a maize used roasted (not a popcorn) that has a higher floury endosperm content typical of the resulting heterotic effects of interracial cross-breeding. It is very likely that the introduction of maize from the Callejón de Huaylas – whence the Los Gavilanes maize clearly comes – most probably already included hybrids with a slightly bigger kernel size, but clearly in not-too-large number in comparison with later periods. Despite the scant evidence of heterotic effects found at Los Gavilanes, it does point us toward the expected greater length of the cobs characterized as intermediate forms between Confite Chavinense and Proto-Confite Morocho (see Grobman, 1982: table 11, 160). The situation as regards heterosis for kernel size is therefore not expected to be substantially different.

In view of this, the evidence from Los Gavilanes shows a very high percentage of small-sized kernels in the ears, but these vary in size, and the change is not unlike what has been found in the southern United States and Mexico during the earliest stages in the evolution of maize. The corn found in Bat Cave (New Mexico) actually also has a range of variation in the size of the kernels that is almost exactly in the same order of magnitude as that found at Los Gavilanes (5–9 mm long, and 4–8 mm thick; see Mangelsdorf, 1974: figures 14.1 and 14.3, 150–151).

It was noted, in the discussion made in previous publications (Bonavia, 1982: 366–367; Bonavia and Grobman, 1989a, 1989b; Grobman, 1982), that the maize from Complex III at Guitarrero Cave, which was studied by C. Smith (1980b), has quite strong similarities – and no differences – with the corn from Los Gavilanes as regards racial characteristics, including the size of the ears. This maize is closer to Confite Chavinense but also shows the presence of Proto-Confite Morocho, which may have come from another geographical area – probably from Ayacucho, which is nowadays where Confite Morocho, the race derived from it, is currently distributed (Grobman, 1982: 176). It is therefore clear that Complex III of Guitarrero Cave held preceramic maize.

The data provided by Towle (1954) on the excavations Willey and Corbett (1954) made in Áspero cannot unfortunately be verified with racial classifications, as they were taken before the classification of the races of maize in Peru had been completed. At Áspero, Feldman (1980) found maize cobs in the preceramic context, as was already noted, and these were identified by Grobman as Proto-Confite Morocho.

As for other findings of preceramic corn, we have what Uceda found at Cerro El Calvario in Casma. Grobman classified a cob found in Level 5 as a hybrid specimen between Confite Chavinense and Proto-Confite Morocho. And both cobs from Cerro Guitarra are of Proto-Confite Morocho race.

After reviewing the Ayacucho corn from Pikimachay (Ac 100), Rosamachay (Ac 117), and Big Tambillo Cave (Ac 244), Galinat concluded (1972: 107–108) that most of them were hybrids derived from Confite Morocho and with an introgression of Confite Puneño: "Thus it seems that the oldest cobs represent a common ancestor to these primitive races rather than being hybrid derivative." He posited the name of Ayacucho for this "ancestral race." We saw that Grobman – along with Galinat – examined these cobs in 1973 and reclassified them in the Proto-Confite Morocho and Confite Chavinense races, with their respective hybrids. We must not, however, forget, as was seen in Chapter 5, that of the sites mentioned, only Rosamachay (Ac 117) can be accepted as a secure site.

The Los Cerrillos site in Ica was excavated by Dwight Wallace (1962) in 1961. It dates to the Early Horizon (c. 900–200 BC). Grobman and colleagues (1961: 75–79) prepared the first report of the maize found. At this site, where little difference has been found between the earliest and the latest phase, the Proto-Confite Morocho, Proto-Kculli, and Confite Iqueño races and their interracial hybrids were once again found. Confite Iqueño – as is how it was then called – is now taken to be a local variant of the Confite Chavinense race. Only the size of the ears in the oldest strata at Los Cerrillos coincides with those of the Proto-Confite Morocho from Los Gavilanes. The size of the ears in all of the other strata is larger. This could be taken as evidence of a small increase in size due to the selection of the ears, but it is not significant. The clear increase in the size of the ears and the kernels took place much later, by hybridization with exotic corns that undoubtedly arrived in the Christian era (see Grobman et al., 1961: 61, 63). The explosive increase in the size of maize probably happened in AD 200–500 onward.

Another point to which we must once again return, is the variability of the maize found in early archaeological sites. In all of the sites mentioned here – which date to preceramic times, with the sole exception of Los Cerrillos – we are consistently before the presence of the same complex of primitive races. These exhibit quite definite characteristics in perfectly identifiable typical specimens. In the same context we also – abundantly – find the outcome of hybridization, which segregate among the three races. Only a few specimens from Los Gavilanes and Guitarrero Cave have bigger kernels. This is explained by the apparition of slightly bigger kernels, which points toward roasting; this appeared due to the use of popcorn, which is the direction the development of the use of maize followed in Peru. Yet the size of these kernels falls within the approximate range of variation also found in Bat Cave, New Mexico, for an age of more than 4,000 years (see Mangelsdorf, 1974: figure 14.3, 151). The differentiation in the size of the kernels is conditioned by other additional factors, such as the position of the kernel in the ear, the nutritional state of the plant, possible droughts, and so on.

R. McK. Bird (1990: 832) made other errors of interpretation. He mentions as evidence that the more than 200 cobs from Los Gavilanes have 8–10 rows. Things are actually different. These are almost the same number of cobs that have differential characteristics in each of the races (8–10 rows of kernels in the Proto-Confite Morocho race, and an even larger number of rows – 12–14 – plus fasciation in the Confite Chavinense race). Bird's data in his table 2 (Bird, 1990: 835) only include information from the first excavations made at Los Gavilanes (Grobman et al., 1977; Kelley and Bonavia, 1963) but not from the final report (Bonavia, 1982; Grobman, 1982: tables 11 and 12, 160–161). For Áspero, Bird includes incomplete cobs from which most of his data for length were taken – a useless measure that was not used in the studies Grobman and I made.

Nothing has been found in any early archaeological site, nor in the most ancient racial complexes in Mexico, that even remotely resembles the Confite Chavinense race as regards its morphology and the fasciation of its ears, characteristics that were inherited by many Andean races. Nor has anything been found that resembles the Proto-Kculli. The oldest Peruvian primitive corns are not tripsacoid, whereas the most ancient ones from Mexico are so. The extremely high frequency of anthocyanin color in the vegetable residues of maize has been shown for Los Gavilanes but has not been found in Mexico (Grobman, 1982: 161). This proves it originated in the high Andes before moving down to the coast. Based on these indications, Grobman and I posited that there was a long period of time in the central Andes in which these races of maize formed and developed in a completely independent way from Mesoamerica (Bonavia and Grobman, 1989b: 459–464; Grobman et al., 1961: 337–343).

The aforementioned racial diversity and the environmental adaptations of the races of corn in the Peruvian area could not have been attained in the short span that followed the Preceramic period. On the other hand, the process of formation of the 72 races of native maize in the Peru-Bolivia region, a number that surpasses the 57 races of native Mexican corn (Taba, 1995), must have required a greater time period and a bigger formative genetic base than that available in succeeding periods; it also must have required a period that had the inflow of Mesoamerican corn, just as the process of racial formation in Mexico had introductions from the Andean region. This process was not simultaneous, and in the case of Peru it must have taken place in AD 200–500 (Bonavia and Grobman, 1989b: 463; Grobman et al., 1961: 60–64;), and in AD 600–900 in Mexico (Mangelsdorf, 1974: figure 16.5, 192). We thus see that all of the objections raised by Bird are groundless.

The presence of preceramic maize in the central Andean area cannot be denied, and although it is true that some scholars may not trust the finds made at Las Aldas, Culebras, Guitarrero Cave, and Tambillo Boulder because they have not been properly documented, no objections can be raised against Cerro Guitarra, Cerro Julia, Cerro El Calvario, Tuquillo, Los Gavilanes, Áspero,

Huaricoto, and Rosamachay. Burger and Van der Merwe (1990: 91) noted the following in this regard:

> Some scholars (Vescelius 1981a, 1981b; Bird 1987) have questioned the reliability of the associations and dating of maize fragments from Rosamachay and Guitarrero Cave ... and even the discoveries of maize at Preceramic coastal sites such as Los Gavilanes, Aspero, Culebras, and Haldas [Las Aldas].... The isotopic analysis of the Chaukayán-phase from Huaricoto confirms that maize was already being cultivated in highland Peru during the late Preceramic.

As for Los Gavilanes, we have the position of Gary Vescelius – who always was a harsh critic and a strict follower of the evidence – who said of the site that it had "... ostensibly early corn ... " (Vescelius, 1981b: 10).

It has been said that there is no paleobotanical evidence of the intensive use of maize in northern South America until very late times (Pearsall, 1994b: 122; 1995b). Pearsall does, however, acknowledge that this is a tentative conclusion based on the study of maize carbon isotopes in the Orinoco River basin – where corn only gained significance around AD 400 (Van der Merwe et al., 1981) – and on the highlands of Peru, where it postdates Chavín (Pearsall, 1995b: 129, citing Burger and Van der Merwe, 1990). Tykot and Staller (2002: 671–672) made the same claim but only used data for Early Horizon Peru and fully ignored preceramic sources, which makes their work useless.

Now, it is clear that we still lack the information required to solve this issue. I have long held that the available data on preceramic maize, and in general on domestication and the first uses given to plants, have no statistical value whatsoever because the sites studied are few, and not many of these have been exhaustively studied, particularly as regards their botanical aspects. A large part of the South American continent is still unmapped from an archaeological standpoint. Even so, the evidence that is available on the Peruvian Preceramic period is most significant. For example, we have the storage pits used to store maize at Los Gavilanes, which could hold about 461,128 kg of corn, with an estimated maximum of 712,364 kg (Bonavia, 1982: table 1, 67; Bonavia and Grobman, 1979: table 1, 43).

It has been shown that this storage was not limited to Los Gavilanes, for two more sites have been found in the Huarmey area with the same characteristics – PV35–107 and Gallinazo (PV35–128). It has likewise been established that similar storage facilities existed in Áspero, in the Supe Valley (Bonavia, 1982: 236–242; Bonavia and Grobman, 1979: 37–40). It is hard to believe that these people would have built silos if corn were not widespread. Finally, the analysis of the human coprolites from Los Gavilanes showed that *Zea mays* pollen was present in 27% of the samples from Epoch 2, and 45% in Epoch 3. The pericarp of the kernels has also been found, so there can be no question that maize was used as food (Weir and Bonavia, 1985: 100–101, 106). The llama coprolites found close to the site were also analyzed. Interestingly enough, close to 50% of the samples from Epoch 2 had grains of corn pollen, whereas in Epoch 3 27% and 6% had them. This means the llamas also ate this plant (J. G. Jones and Bonavia, 1992: table 1, 839, 840).

One question many scholars have asked is why if maize was known on the North-Central Coast at the end of the Late Preceramic, was it not known elsewhere on the coastlands. Long ago I made the following statement, which is still valid:

> it is simplistically believed that once a new phenomenon like the arrival of an unknown cultigen takes place, its diffusion has to be rapid and constant. This is not so. Feeding habits usually are the habits that most resist change. The potato is the best example [in this regard]. When it was first introduced in Europe it was initially rejected and a long time passed by not only before it was just accepted, but before it became one of the major staples in many areas of Europe. There is an interesting parallelism with human groups that lived in Central California – which were studied by Heizer and are mentioned by Rowe[10] – who became acquainted with cultivated maize through their barter with other areas but long refused to accept it (see Rowe, 1964: 28). This may well have happened in many coastal parts of the Peruvian Preceramic Epoch. (Bonavia, 1982: 372)

Gremillion has broached this problematic for the North American area, but some of his arguments are not only interesting but also applicable to our case. One of his arguments is that the discussion of a predictive model based on archaeological evidence holds several implications for an understanding of why maize had a limited role hundreds of years after its introduction, and why the dramatic change in its use took place when it did. One of the implications is that the initial rejection or adoption of a plant and its economic role will have different consequences; this is why different types of explanation will probably be required. Gremillion then points out that the exchange networks and storage technology chain precede the intensification of maize cultivation.[11] Once maize was established as a major crop, the role of other plant resources also changed. He gives the example of the differences in the diffusion of the use of maize in the eastern United States and reaches the conclusion that these differences may perhaps reflect separate evolutive pathways, separated by the emergence of maize-based economies in response to regional differences in the productivity of maize, as well as in the efficiency of the farming technology (Gremillion, 1996: 197).

The problematic that pertains to feeding customs must on the other hand be taken into account. Kahn (1987: 48–49) presented some very good examples of how hard it is to change the feeding habits of the people and to use new plants. Crosby (1975: 169) likewise pointed out that "... human beings, in matter of

[10] This was a personal communication.

[11] This also holds in the Peruvian case.

diet, especially of the staples of diet, are very conservative, and will not change unless forced." Interestingly enough, Pearsall (1994a: 246) has accepted that maize may even have been rejected by a group that was acquainted with it. When discussing the diffusion of maize using phytoliths in Panama, Piperno (1995: 141) noted that not all human groups in a given area used the same plants, or identical combinations of them. Heiser (1965: 945) likewise pointed out that when a plant is taken to a new area, it has to compete with the preexisting plant complex. It has been said (Schaaffhausen, 1952) that humans are conservative in their diet; when a native group has a satisfactory diet, it may reject a newly introduced plant, even if it grows well in the new area.

The peoples of the Casma and Huarmey Valleys were in contact with people in the Callejón de Huaylas, one of the areas where maize cultivation probably began. The people of Supe were quite close to the Fortaleza Valley, which is another natural entryway into the Callejón de Huaylas. This explains why this plant was rapidly introduced into the North-Central Coast. The picture remains blank for the coastlands north and south, due to the lack of research. We are before two possibilities: either this plant was rejected at first, or it was used indeed but we have as yet not found any evidence of it.

Another point on which we have to insist, as it is frequently forgotten, is that early maize was a popcorn. This means that it must have been exclusively used by directly eating its kernels. Anderson and Cutler studied the ways in which maize can be popped without using either pottery or metal. They concluded that this can be done in two ways. The simplest method is to cast the kernels into the fire, or to place them in a clean area around the fire. This technique is known in both the Old and the New World. The kernels are picked up as they pop; in Asia they are picked up using bamboo tongs. The second technique, which Anderson and Cutler say is "slightly more sophisticated," consists in popping the kernels by placing them over sand that has been previously heated by placing it below a fire. This procedure is also known in both the Old and the New World (Anderson and Cutler, 1950: 304).

We know that the Indians of Baja California used the second technique (Sauer, 1969a: 11), and that it is still used in India to cook some cereals. We must bear in mind that the Indian tradition is quite ancient, even though archaeological evidence has not yet been found. The technique has not been studied by ethnologists. Rice, maize, peas, peanuts, and many other species are still cooked in warm sand. This practice is common even in Calcutta (Asok K. Ghosh, personal communication, 1977). Mangelsdorf, however, tried a different technique that also does not require sophisticated equipment, with magnificent results. All that is needed is a layer of warm charcoal and a pointed green stick that is stuck at the base of the ear, and that allows it to be held and slowly turned over the charcoal. The kernels thus pop in a short time and can be easily removed to be eaten. The glumes are slightly charred in the process but are not fully carbonized (Mangelsdorf, 1974: 154).

Any of these procedures could have been used in Peru in preceramic times, and it will be very difficult for any of them to be archaeologically identified. The technique presented by Mangelsdorf has the advantage that one does not have to pick up the kernels from the fire or the sand, and they are not dirtied by it, nor do they get buried. It is also more practical when it comes to eating the kernels. Yet the warm sand method remains an open possibility, because it is also quite an easy method to use. I tried to establish what temperatures can be reached by warming the sand using only salt-impregnated wood, of the kind found on the shore cast out by the sea, and that has been dried by the sun. This type of wood was used because it is not easy to get any other type of material for this on the Peruvian coastline. A thermometer was placed in the sand 10 cm below the surface on which the embers lay, some twenty to thirty minutes after lighting the fire. The temperature registered in half a minute was more than 200°C.[12] It therefore follows that Feldman's assumption (1980: 110) that pottery is the most appropriate method for popping corn is groundless.

A most serious problem has recently appeared. It is a quite technical issue and so is not discussed here in depth, but attention is drawn to it for the benefit of those who are not archaeologists and see it from the outside, so that they are at least aware of it. Besides, this issue – the methods used to date the samples – has already been raised in this book.

Archaeology uses two chronologies, which are known as relative and absolute chronologies. A relative chronology is established on-site through a stratigraphic excavation, which allows objects to be arranged in a sequence wherein one knows what comes before and what afterward. An absolute chronology that gives us an age in years that can be correlated with our calendar is later established by specialists using sophisticated methods that are applied to the materials excavated. Many methods are available for this depending on the available materials, as well as the range of the antiquity in question.

Carbon 14 (C14) is the most commonly used method. It establishes the radiocarbon age of the death of an organism, so that the time that has passed between its death and the present day can be determined. This method was discovered thanks to the research on the atom bomb undertaken during the Second World War, and it began to be used in the 1940s. It certainly is the method most used in American archaeology. The method known as AMS dating was discovered in the 1970s. It consists in improving the accuracy of determining the age of the death of an organism through a direct isotopic reading of carbon 14 in a mass accelerator. The advantage of this method is that it requires a very small sample for dating, whereas the traditional C14 method requires a bigger test sample. The materials most commonly used to date a given context using the traditional C14 method were specimens associated with that specific context. In

[12] The exact temperature could not be established, as the thermometer used in this test only reached up to 200°C, which in this case proved more than enough.

AMS dating a part of the very object one wants to date is used, as this method is less destructive than C14 due to the small amount of material needed. For instance, in the case of plants, only a very small part of them is needed.

Several precautions must be taken when collecting the samples and during storage, to avoid contamination. In the laboratory itself, they are subjected to a series of cleaning procedures prior to their use in dating. Besides, there are several other factors that must be taken into account once we have a date and before it can be used. But despite all of the precautions taken, both by archaeologists in the field and by the specialists who process the sample in the laboratory, there always is the danger that distortions and errors will appear. This is precisely when the archaeologist must intervene, being fully aware of how he or she must then proceed.

Archaeology is a science, and as such it has principles and methods that it cannot abandon. Association is one of these principles – and one of the major ones. We can have both primary and secondary association. If a group of archaeological remains is found in the same stratum in an excavation, and if we are sure that there has not been subsequent disturbance, then the remains will all have the same age, and we have a primary association. Whenever there is a possibility that there have been later intrusions – which may be due to various causes – we are faced with a secondary association in which different remains may have different ages, even though they are together in the same stratum. However, any well-prepared and earnest archaeologist realizes this and is able to discriminate the evidence.

Another factor that we must understand is that a date always indicates just one moment in a temporal sequence, so all it does is give us a temporal signpost within a continuum. The only one who can realize whether or not a date is correct is the archaeologist, who has to place it within the context he or she is studying. Dates are, in other words – and this is very important – at the service of the archaeologist and not the other way around. They are not by themselves an absolute truth that has to be accepted as such. And it is also quite common that some dates within a group of dates will turn out to be aberrant and will therefore have to be discarded.[13]

At first the traditional C14 method had problems that specialists gradually solved. Now we have the same situation with AMS dating, whose results in many cases not only disagree with those obtained with traditional C14 but also do not fit the facts. These disagreements were mentioned in Chapter 5 when discussing the archaeological evidence. Here only three of them need be mentioned: the case of Tehuacán, Mexico, which is the most resounding instance (Flannery, 1997; Fritz, 1994a; Long et al., 1989; MacNeish, 1997; MacNeish and Eubanks, 2000); Nanchoc in Peru (Rossen et al.: 1996); and Tiliviche in Chile (Rivera, 2006).

[13] I have already drawn attention to this point in a previous publication (Bonavia, 1996c).

What is most serious here is that some archaeologists, as well as scholars from other fields, believe that only AMS dates are valid, and not those obtained with the traditional C14 method, and they have even gone as far as to reject and practically declare that the principle of association is no longer valid. Let us see some specific cases. Fritz (1994a: 305) is one archaeologist who makes a staunch defence of AMS dating, to the point that she claims it "... eliminated the need to rely on associated material for age determination by its use of milligram-sized samples." The same holds for Blake (2006), who in his table 4–2 (Blake, op. cit., 61) brought together 17 dates "... from Mexico to Peru, where archaeologists have hypothesized that early maize was present and *presumably of some economic importance. Although some of these samples may turn out to be as old as suggested by indirect dating*, I believe we *should hold off* on incorporating them into our distributional models *until their ages can be confirmed with direct dating*" (Blake, 2006: 60; emphasis added).[14] Smith is another scholar who blindly defends this methodology, claiming that "*many of the early dates assigned to South American domesticates on the basis of age associations are, in all likelihood, incorrect*" (B. D. Smith, 1994–1995b: 180; emphasis added).

The arguments used to support this are interesting. It is claimed that AMS dating "... end[ed the] reliance on age estimates obtained by conventional large-sampling dating of charcoal or other organic material found in close proximity to botanical samples and, therefore considered to be contemporaneous with them." These were just "... age estimates ... based on conventional radiocarbon 14 dating of organic material assumed to be contemporaneous," whereas with the new standard in evidence, that is, AMS dating, they began to be "... more widely accepted and applied, [hence] the strength of these secondary classes of evidence has declined, particularly when that evidence is at odds with direct dates ..." (Smith, 2006: 176). In another publication, Smith claimed that because AMS dating allows small early plant specimens to be directly tested, thus avoiding all potential pitfalls inherent to the traditional radiocarbon methods using "... assumedly associated ..." organic materials, it therefore has clear and obvious applications when specifying when it was that a given plant species was first domesticated (Smith, 1997a: 349).

Significantly enough, Smith himself acknowledges that the AMS method is not problem-free, for a careful reading of the very same publication in which he defends this method reveals that Smith also claims that at present we have both a "long," C14-based traditional chronology and a "short" AMS chronology (this is not entirely true; B. D. Smith,1994–1995b: figure 6, 182). Smith is probably unaware that the traditional C14 methodology originally met this same problem when Rowe (1965b) drew attention to the existence of "long" and "short" chronologies. Pearsall and colleagues (2004: 424) pointed out a similar problem for Ecuador.

[14] The reader should be warned that Blake arbitrarily chose these 17 dates.

Blake does not even consider the traditional C14 methods and analyzes only the AMS dates because "... indirect dates have often proven to be unreliable...." He then points out that although individual indirect dates can in some cases be correct, they are not so in many others. Interestingly enough, when he criticizes the Tehuacán dates (Blake, 2006: 56–57) he only mentions Long and colleagues (1989) and ignores the defense made by MacNeish (1997) and Flannery (1997). As for dating by association, Blake claims that "... this is becoming a significant problem because the associations between the materials used for dating, such as wood charcoal, and the maize macrobotanical remains are not always secure" (Blake, 2006: 59, see also 68). Significantly enough, when he excavates, no errors are made, and so the other option is likewise valid, as shown by his claim that "... there are many cases where maize remains are in their primary context and *can be reliably dated by associations* ...," for instance in Chiapas, "... *where my colleagues and I*..." worked (Blake, 2006: 60; emphases added).

We see that none of those who defend the AMS method as the only valid one have presented solid arguments. I insist that archaeology is a science, and all its practitioners have to treat it as such. The rejection of the validity of associations is tantamount to disavowing one of the major tenets of archaeology and distorts the truth. No earnest archaeologist can accept this. As for the problems raised by the difference between traditional C14 and AMS dates, it is not archaeologists but the specialists who must pass judgment. The AMS method is new and has to be adjusted. In the meantime it is not the figures thus obtained that will decide, but the archaeologists themselves, who have to analyze each and every case, pass judgment, and establish which dates can be accepted and which ones are aberrant. What has instead been definitively established is that the correct traditional C14 dates are still valid.

Some archaeologists have actually realized the existence of this issue. Such is the case of Schoenwetter (1974: 301), one of the few palynologists who is aware of the overriding significance of association, over and above botanical identification. To conclude this section we turn now to Pickersgill and Heiser (1976: 60; emphasis added), who wrote thus: "In maize, as in most other crops, understanding the changes which have taken place under domestication *requires a study of the archaeological specimens, not in isolation, but in their archaeological context*, with all the supplementary evidence about diet, processing, storage pests etc. that can be obtained from coprolites, artifact inventories and the like."

Now, if we list all of the sites where corn has been found, we clearly see that it is only in Cueva Cebollita and Bat Cave in the United States, the Ocampo and Tehuacán Caves *sensu lato* in Mexico, and Los Gavilanes in Peru, that significant amounts of remains have been recovered that are not just statistically valid but also allowed a detailed botanical analysis to be performed. This is one thing that is not usually taken into account.

One fact that is worth noting is that thus far no early corn-storage facilities have been found in Mesoamerica. Pearsall (2003b) and Pearsall and colleagues (2004) claim these facilities did exist in the Real Alto site in Ecuador, but we still await a detailed report that lists their characteristics. Actual evidence of these facilities does exist for the north-central Peruvian coast in the Preceramic period, which has already been mentioned (see previously in this chapter and Chapter 5), and which could store a significant amount of maize (Bonavia, 1982; Bonavia and Grobman, 1979). This is something that must be taken into account when appraising the significance of this plant, and its role in the initial development of human groups.

Wilson concludes that when we compare fishing- and agriculture-based subsistence, we clearly see that the "carrying capacity" of agriculture is six times greater than that of fish in the worst possible scenario. Wilson estimates that the carrying capacity of early maize agriculture is 50 individuals per hectare, or 50 individuals per km^2. Using Moseley's estimate (1975) of 2,000 individuals in the Ancón-Chillón zone, without any contribution from the marine subsistence system, we find that they would have necessitated 40 km^2 of cultivated land to support themselves. This figure is equal to 33% of the land cultivated in the Chillón Valley in the 1970s (Wilson, 1981: 107). We should likewise bear in mind that the seeds usually live 3–5 years, and that the maximum period of viability under normal storage conditions rarely exceeds 10 years. According to the Huarmey farmers, corn is viable for 2 years when it is stored in the sand, just like in preceramic times (Bonavia, 1982: 71). The viability of the seeds extends up to 25 years or more when they are placed in cold storage, which was not the case for preceramic times (Mangelsdorf, 1974: 8).

A solid racial identification is of the utmost importance when using the results derived from the remains of early maize – whenever macro-remains are available of course – to apply a comparative approach. The identification is clearly deficient in the case of the Ecuadorean remains, and all the more so in the case of Chile and Argentina. I noted along with Grobman (Bonavia and Grobman, 1989b: 463) that although these specimens are potentially significant (we meant the southern specimens, but this can be extended to the northern ones too), more information, particularly detailed botanical data, is required before they can be included in the scientific literature for comparative purposes. This is important because it so happens that the major problem we have at present is reconciling the chronology with the racial types that have supposedly been found in the aforementioned zones, which presumably are older than those of the central Andean area.

There is one point that is very hard to broach, and that I would rather not touch. It cannot, however, be avoided, as it not only goes against each and every scientific tenet but is also holding back the study of maize. By this I mean the way in which a group of U.S. scholars behave in regard to the publications made by Latin American scholars, and Peruvian ones in particular, vis-à-vis the preceramic remains of maize found in Peru. An objective analysis of the publications made in regard to this issue shows that there have been at least five different ways in which this issue has been approached.

First we have those scholars who accept the validity of the data presented. Such is the case of Harlan (1992: 222; 1995: 185), who believes that Grobman and I have presented "rather compelling evidence" of an independent domestication in the central Andes. The same goes for Flannery (1973: 303 and table 3), who does not accept an independent domestication but believes that the findings made at Los Gavilanes are valid.

Then we have the second group: those who simply ignore everything that Latin American scholars do. The best example here is D. D. Smith (1994–1995b): in his overview of plant domestication in the Americas, the bibliography has 42 entries, only 9 of which (i.e., 21%) are for South America, and all of them were authored by North American scholars. Not a single reference in Spanish is included, and no Latin American scholar is listed, not even when they have published in English.

The third group comprises those scholars who may have eventually used original data, but who in most cases used secondhand information without even bothering to check whether or not it was reliable. Stark (1986), and her study of the origins of food production in the New World, is a good case in point. An examination of her vast bibliography (Stark, op. cit.: 306–321) shows that only 25 studies on Peru are listed, none of them authored by a Latin American scholar. Had Stark bothered to check the literature, she would have realized that just up to 1982 there were about 103 publications on this issue, 11 of them by Peruvian scholars (see Bonavia, 1982: 449–490). Another good example is Pearsall (2003b), which compares plant food resources in Ecuador and the central Andes, and whose bibliography only includes 25 entries for the latter area, only 2 of which were written by Peruvian authors.

The fourth group comprises those who, instead of using the data in an objective fashion, select it arbitrarily so that it will further their interests or support what they want to prove, and who thus go against each and every scientific principle. Here only three instances are mentioned, which I believe are the most striking ones. To avoid any misunderstanding we must point out that in science, each and every critique is welcome as long as it is well intentioned and is supported by very specific evidence. This unfortunately has not been the case with the discovery of preceramic maize in Peru. It was Robert McKelvy Bird (1970: 124, 148) who began this "dirty war." Although he initially accepted that "at about 2000 B.C. maize without pottery, was in the Supe area" – based on Willey and Corbett (1954) – and that there was a "... very early maize from Huarmey ..." – using the initial work done by Bonavia (Kelley and Bonavia, 1963), he later began to systematically reject each and every discovery of preceramic maize without providing at the same time a valid argument (Bird, 1984,

1987; R. McK. Bird and B. Bird, 1980). A detailed examination of his writings has already been presented, and it is not worth going over old ground (see Bonavia and Grobman, 1989a).[15]

The major arguments given by Bird are summarized here to let the reader decide. He claims that "the samples of maize supposedly preceramic (pre-1750 B.C.) from Áspero and Huarmey seem to be considerably later, unless a very variable maize, with some cobs being as large as maize of two millennia later, is erased from the scene between 1750 and 1050 B.C. to be succeeded by a thoroughly different set of types" (R. McK. Bird, 1978: 92). "Interestingly, even by the end of Gallinazo [c. 200 years BC–AD 200], cobs from the Chicama-Virú area had not reached the size of the larger specimens from Áspero and Los Gavilanes.... which may mean these last two sites are not preceramic" (R. McK. Bird and J. B. Bird, 1980: 330). "Maize found in superficial and/or disturbed layers of the large preceramic site of Áspero ... morphologically is an array more typical of the AD 200–1200 period ..." (R. McK. Bird, 1984: 43); "coastal maize purported to predate 1500 B.C. is much more recent in appearance or gives late radiocarbon dates or comes from disturbed contexts ..." (R. McK. Bird, 1984: 49).[16]

It is striking that Bird never criticizes the associations, contexts, stratigraphy, or overall archaeological work undertaken at the sites he mentions. In his 1970 paper Bird furthermore pointed out that "one important problem" was "to integrate data for ears, cobs and cob fragments"; the fact that this had yet to be done raised a problem. He then not only insisted on the importance that a systematic study of "... the morphology of the chromosomes" would have but also pointed out that "one important pattern is quite evident ... which McClintock calls 'Andean' ..." (R. McK. Bird, 1970: 125, 128–129). Yet he avoided mentioning that the first data regarding the differential pattern of chromosomal knobs, which distinguished the primitive – and even the evolved – Andean maize races from those from Mexico and Central America, had been established by Grobman and colleagues (1961). All of these analyses were undertaken in later studies with the preceramic maize from Los Gavilanes (see Grobman, 1982), but Bird never took them into account.

Here it is worth pointing out that besides the present writer, the only archaeologist who analyzed Bird's writings was Lathrap, who then showed "the circular quality of Robert Bird's thought ..." and its inconsistency (Lathrap, 1987: 351–352). Yet almost all of the other U.S. archaeologists who touched on this subject either followed Bird without first making a critical assessment of the original sources or instead acted in a nonscientific fashion for reasons I cannot fathom. The damage Bird has inflicted on Andean archaeology in this regard is

15 R. McK. Bird (1990) replied to the critique Bonavia and Grobman made of his work (Bonavia and Grobman, 1989a), and the latter in turn subsequently made their rebuttal (Bonavia and Grobman, 1999).

16 The botanical aspects have already been discussed, so there is no point in going over them once again.

massive, not only because he derailed research for many years but even more so because he sowed discord among colleagues, and this has still not been ironed out.

Pearsall and Hastorf probably are the two individuals, besides Bird, who best fall into this category. It is clear that not only has Pearsall not analyzed the work I made with Grobman and other colleagues, or has done so only cursorily, but she has even used secondhand sources or based her work on Bird. Years before the final report on Los Gavilanes appeared, Pearsall complained that "many of the Peruvian archaeological maize collections lack racial identification, including some collections of the earliest pre-ceramic maize." When she suggested the movement of maize between Mesoamerica and South America, Pearsall demanded that "… remains … be tested more fully by archaeological excavations in the crucial regions, and by genetic studies of the races of maize involved" (Pearsall, 1978a: 47, 53). In a subsequent study, on wondering "… what constitutes evidence for the presence of maize," she pointed out that

> archaeological maize remains, whether cob fragments, pollen grains, phytoliths, or cooking residues, must be studied by specialists and documented in detail in print. The archaeological context of the occurrence should be described. Identification criteria must be clearly explained and the precision or confidence level of the identification stated. Direct dating of remains should be a priority. No model can be considered adequately tested or supported if systematic recovery methods (flotation, fine sieving, soil testing) have not been carried out. (Pearsall, 1994a: 247)

At the same time she insisted that "the importance of the crop must be documented, not assumed …" (Pearsall, 1996: 2) and suggested (1995c: 21) that "the best way to address issues of diet and subsistence … is to focus on multiple lines of evidence: charred macroremains, phytoliths, pollen, faunal data, settlement pattern data, and chemical and physical analyses of human bone."

It so happens that of all of the conditions set by Pearsall, the only three were not fulfilled by the work done at Los Gavilanes were the study of phytoliths (which was unnecessary, given the abundance and excellent preservation of maize macro-remains), the direct dating of the samples (also unnecessary, given the clear associations of the contexts), and chemical bone analysis (none were associated with corn). All of the information recovered was published (Bonavia, 1982), so the question is, why has Pearsall not considered it? To show she has not done so, we need only turn to two of her studies. In one of them she demands "… specific associations of plant material and imperishable artifacts, features, and stratigraphy." Pearsall also claims (1992b: 190) that "Guitarrero is the notable exception," but fails to mention the severe critique Vescelius (1981a, 1981b) leveled at this site. And why does she omit Los Gavilanes?

Pearsall, however, made a mistake in this same study that clearly shows she has not read the sources. In her table 9.2, Pearsall (1992b: 178) mentions the

North-Central Coast and acknowledges the presence of preceramic maize, giving Kelley and Bonavia (1963) as her source, that is, the first report on Los Gavilanes, when the site was still without a name and was known as Huarmey Norte 1. But when Pearsall mentions the site with its proper name in table 9.6 (Pearsall, 1992b: 184), she claims that I do not accept the direct date of corn and only cites "Bonavia (1982: 73)," when the correct thing to do would have been to cite Mangelsdorf and Cámara-Hernández (1967: 47), Grobman and colleagues (1977: 224), and Bonavia (1982: 73, 275–277), where the reasons why these are not valid are clearly laid out. We can substantiate the fact that Pearsall (1992b: 190) has not actually read the studies she cites with the fact that she claims Feldman (1980) believes the maize from Áspero is intrusive. Now, if she had actually read pages 182–184 of Feldman's dissertation, Pearsall would have realized that this is not so. Besides, why does she fail to mention the research done by Willey and Corbett (1954), or the reports presented by Towle (1954) and Moseley and Willey (1973)?

Pearsall made the same claim regarding Los Gavilanes in a later study (1994a: table 15.2, 258). Had she read Piperno (1994a: 638), she would have realized "... how misleading conclusions drawn solely from macrobotanical data can be." In another of the studies done by Pearsall, we find several mistakes in the Peruvian data she cites, thus showing once more that she is not familiar with the sources and lacks a critical capacity. Here we need only mention some of these mistakes. Pearsall claims that "... detailed dietary and health data are largely lacking for this period [she means the cotton Preceramic]" (Pearsall, 2003b: 245). The question here is why she ignores Feldman (1981), Patrucco and colleagues (1982, 1983), Weir and colleagues (1988), and Weir and Bonavia (1985). And when Pearsall discusses Los Gavilanes and mentions the archaeobotanical remains found on the Peruvian coast in regard to this latter site (Pearsall, 2003b: 236), she only lists Popper (1982). Why not Grobman (1982), Stephens (1982), Morán Val (1982), or Kaplan (1982)?

A look at table 3 (Pearsall, 2003b: 238) shows that the presence of 15 plants is accepted for Los Gavilanes, yet the presence of maize is denied without presenting any argument in this regard, just like in the other studies done by Pearsall. On page 242 of this same study we find that in the "Initial period" the "... potato appeared ...," but on table 3 (Pearsall, 2003b: 238) we find that in Huaynuná there was "potato" in the "Cotton Preceramic" (but Bonavia, 1993, and Ugent et al., 1982, have not been used). In the case of Casma, great significance is attached to the work done by Pozorski and Pozorski (1987), who undertook only limited excavations at Huaynuná, but why is it that no mention is made of Uceda Castillo (1987, 1992), who worked this same valley, under the same conditions, and did find corn at two sites in preceramic contexts? Table 3 (Pearsall, 2003b: 239) also uncritically accepts the finds made at La Galgada, but Pearsall does not point out that Grieder and Bueno Mendeza (1981: 45) claim that these remains also included "mangos" and "bananas" (*sic*). Are these findings secure? Interestingly enough, no study is mentioned throughout this discussion of the "Emergence of Agriculture on the Coast of Peru" that is not authored by U.S. nationals; as for the report on Los Gavilanes (Bonavia, 1982), it does appear in the bibliography but was not taken into account (Pearsall, 2003b: 243–249).

Finally, in a recent study Pearsall makes the following claim when discussing the origins of agriculture: "Storage technology ... *is the key* to making food a usable currency of social networking. Type and extent of storage is something that we can document archeologically and evaluate in relationship to independent indicators of status and crop production (within the limits of preservation)" (Pearsall, 2009: 610; emphasis added). The presence of storage facilities for corn that held the perfectly preserved remains of this plant has been documented at Los Gavilanes, as well as in several other contemporary sites (e.g., La Laguna, Gallinazo; see Bonavia, 1982). The question once again is why Pearsall has never taken them into account.

Hastorf is another case in point. Her work is lacking in objectivity, to say the least, for it includes several statements that fly in the face of all available evidence and are completely unsupported. As for the research undertaken at Los Gavilanes, we find there was a radical shift in what Hastorf believes. This would be acceptable had she made a critique supported by the evidence, but this was not so. Let us look at the record. In her review of the Los Gavilanes report (Bonavia, 1982), Hastorf made the following assessment:

> This book provides a detailed account of each pit and stratum excavated.... For the first time in Andean prehistoric studies an excellent group of specialists have their analyses reported in one volume, making *Los Gavilanes* one of the best documented and promptly published data sets in Andean archaeology ... [the] *presence of preceramic maize is supported clearly by the Los Gavilanes data* ... *Los Gavilanes* is a very important book for Andean scholars for two reasons. It provides a very complete site report of an early coastal site as well as a thorough and up to date presentation of the early Andean prehistoric record. (Hastorf, 1985: 928–929; emphasis added)

And yet, when discussing "the Peruvian case" in a recent study on "crop introduction in Andean prehistory," Hastorf included a note (Hastorf, 1999: 55) that reads thus: "I have not included data from two sites that are still controversial in terms of dates relating to their botanical remains and mixture of levels. These two sites are Los Gavilanes (Bonavia 1982) and the early levels of the Ayacucho Caves (MacNeish 1977)."[17] It is not for me to judge why Hastorf has changed her mind, and in any case she should do the explaining. I personally asked her this (in a letter, 11 August 2001) but never received an answer. And as far as Ayacucho is concerned, we have seen in Chapter 5 that

[17] MacNeish 1977 does not appear in Hastorf's bibliography.

here there are more than 10 caves, so there is no way of knowing which one Hastorf meant.

This way of distorting the data is not restricted to Los Gavilanes and appears everywhere. We examine just the two most striking instances of this distortion. To start, the dates set for the Preceramic VI (i.e., the Late Preceramic) and the Initial period (Hastorf, 1999: 35, note 1, 55) are modified, and the only argument given for this is that "the phases are an updating of Lanning (1967: 25) and Rowe and Menzel (1967: ii)." This being a major issue, the least Hastorf could have done is to briefly explain the reasons for these changes. On reading her work it seems that the only reason why the Initial period is pushed back (it is traditionally taken to extend from 1800/1500 to 900 BC, whereas Hastorf places it in 2100–1400 years BC) is because she wants to show that in the Peruvian zone "... crop production and use ..." appeared in the Initial period (Hastorf, 1999: 53).[18] Hastorf in fact claims that "... substantial agriculture, with a regular array of fifteen to twenty crops growing up and down the coast, occurs only by the end of the Initial phase [*sic*], 2100–1400 BC ..." (Hastorf, op. cit.: 41). This list is incomplete because she has not taken into account that 20 cultivated food plants and 2 industrial ones, likewise cultivated, were known in the coastlands by the end of the Preceramic VI (Bonavia, 1991: 130).

When discussing the diffusion of maize, Hastorf (1999: 43–45) claims that it arrived "quite late into the Andean region" and notes that "its route into the western Andes was either down the western coast and/or over the mountains to the coast from the eastern slopes." First of all, here Hastorf supports her claim that this took place "quite late" by using Benz (1994b), Bush and colleagues (1989), and Pearsall (1994a) as support, but none of these authors has worked on this subject in the central Andean area. Second, no reference is made to the early maize found in Casma (Uceda Castillo, 1987, 1992). Finally, Hastorf does not present any evidence with which to support her proposed path of maize diffusion. It would be worthwhile to have actual data for this.

Of the finding of potato at Cueva Tres Ventanas (Chilca), Hastorf (1999: 44–45) states that she is "... not convinced of the security of the date and stratigraphy of the tubers ...," leaving only Engel (1973) and Martins-Farias (1976) as sources. It should be pointed out here that the dissertation presented by Martins-Farias only mentions an analysis of the samples. As for Engel, it is hard to believe that Hastorf was able to approach this issue based on the 1973 work alone, as this publication is not the most relevant in this regard – she should at least have gone over Engel (1970a, b, c). Yet the only scholar who made a critical review of this issue is the present writer (Bonavia, 1984). Why was this study ignored?

[18] Hastorf is likewise heavily influenced by Robert McKelvy Bird, who we have seen believes that the use of plants is a late occurrence.

When Hastorf (1999: 48) lists the sites with maize in the Preceramic VI, she says it was present in "… the valleys of Viru, Supe, Chancay, and Chilca, as well as in the Ayacucho Caves …," but not a single reference is given. In the case of Supe, she evidently means Áspero, which is correct. But in the other valleys no remains of maize have ever been found in preceramic strata, and the bibliography holds no data in this regard. As for the "Ayacucho Caves," Hastorf has not realized that there are many of them, that the presence of preceramic maize has been claimed only for some of them, and that a critical review only leaves standing the maize found at Rosamachay (Bonavia and Grobman, 1999). It is likewise strange that La Galgada is not included in the list of sites with maize given by Hastorf (1999), yet it does appear in figure 2.4 (Hastorf, 1999: 49). No one has ever made this claim. It is also striking that although the text does not mention Culebras, it is included in this same figure as a site that held corn. Culebras is then mentioned in the text and appears in the map as belonging to the Initial period (Hastorf, 1999: 50, figure 2.5, 51). The question here is where Hastorf found this information, and why it is that she does not cite it. Here we must bear in mind that reference to this site is made only in Lanning (1959, 1960), and that the only scholar who can bear witness to this site is the present writer, as I was personally present when Lanning was excavating there (Bonavia, 1982: 359–362).

Another interesting detail is that Hastorf claims that "manioc is found only at Guitarrero Cave …" (1999: 48). Why does she not point out that remains of this plant were also found in Epochs 2 and 3 at Los Gavilanes (see Bonavia, 1982: table 10, 149; Pearsall, 1992b: table 9.6, 185)? A review of the bibliography listed by Hastorf (1999: 55–58) is quite revealing. All of the pieces cited are by U.S. scholars save for Bonavia and Bueno, and the information has been used in a quite selective fashion. For instance, the most important pieces authored by Engel are not cited, and the same holds true for Lanning, for whom only his handbook is mentioned, not so his other specific papers. Finally, although the 1982 study of Los Gavilanes is included in the bibliography, and as has been shown here is well known by Hastorf, it is not mentioned even once in her paper and appears only tangentially cited in note 2 (Hastorf, 1999: 55). Hastorf would do well to remember what Piperno (1994a: 639) said in regard to the use of "a single sentence in a footnote": "This is not good academic practice, and it will hardly advance the dialogue and consensus building by which most debates arrive at some resolution."[19]

[19] Piperno has correctly noted that "scholarly disagreement plays an important role in the scientific process, and *good debates* often lead to the development or refinement of new techniques with which scientists can evaluate questions that fall outside the purview of more faddish types of analyses. *A requirement of authentic debate however, is that the content of the research under critique is discussed and evaluated in the way that it was originally presented to the scientific community*" (Piperno, 2003b: 832; emphasis added). Unfortunately it is not just the aforementioned colleagues but Piperno herself who has not followed these guidelines vis-à-vis the Peruvian findings.

Finally, the fifth group includes those scholars who out of ignorance have made claims that are entirely unsupported by the evidence, or that are even of their own invention. The best example here is Chevalier (1999). These publications are not worth noting or even considering.

I feel that I must apologize to the readers not just because of this long and tedious analysis but also because this has made me leave behind the specific subject of this book. The intention here was to show the overall inconsistency of these writings, which is not limited to just the corn problematic. Should the trend exhibited by this group of U.S. scholars of systematically ignoring the work undertaken in Latin America – and even distorting the sources on the rare occasions they are used – continue, then far too many issues, particularly that of maize, will take far too long to be solved. Only an open dialogue and an earnest interdisciplinary approach can solve this serious issue, which has persisted since the 1970s. Mangelsdorf (1974: 184) correctly noted that those who reject certain positions "... have attempted to discredit [their colleagues] not by direct criticism but obliquely by implication.... If the purpose of all this was to obfuscate it has succeeded."

Curatola brought up one aspect of maize that has to be discussed, albeit superficially, for on the one hand it should be studied by pathologists, and on the other it still has not received adequate attention. By this I mean the disease known as pellagra. Curatola (1985: 9; this source also appeared in 1990 with only minor changes) points out that maize is completely lacking in niacin (or nicotinic acid) or vitamin PP (i.e., pellagra preventing), as well as tryptophan, an amino acid that human and animal organisms can turn into niacin. Curatola explains that pellagra is a disease that appeared in the Old World with the consumption of maize. The name first appeared in the Padan Plain – it comes from *pell'agra* (rough skin) – and the disease developed when, due to their massive poverty, the people were forced to subsist on corn alone. It is known that in 1784 20% of the population in Lombardy had this disease (Curatola, 1985: 12–13). The hypothesis presented by Curatola is that pellagra was a major disease "... that recurred every year at the end of the dry season, and struck the major regions of Tawantinsuyu." Curatola claims there are several traces of this in historical sources (e.g., Acosta, 1954: 109;[20] see Curatola, 1985: 16). The only thing that Acosta (1954: 109) actually says is that those who eat "... too much [maize] often suffer swellings and scabies." I discussed this point with Uriel García Cáceres[21] (personal communication, 30 April 2007), who pointed out that although the description is inaccurate, it could well be interpreted as indicating pellagra. Curatola (1985: 13–24) claims that this disease is what the Indians knew as *Taki Onqoi*, and that it was the first one to be exorcised during

20 The reference given by Curatola – "Bk. IV, Chap. XVI" – is mistaken and should be book IV, chapter XV.

21 García Cáceres is a renowned Peruvian pathologist.

the major *citua* festival. According to the chroniclers *Taki Onqoy* was a synonym of *Sara Onqoy*, that is, the "sickness of maize." "We [i.e., Curatola] submit the hypothesis that the expression *Sara Onqoy* designated the phase in pellagra that is characterised by skin alterations, gastrointestinal disorders and several other collateral symptoms, whereas *Taki Onqoy* designated the phase wherein mental confusion, hallucinations and psychomotor seizures and madness take place" (Curatola, 1985: 22).

Lastres (1951: 160) pointed out that "pellagra may well have existed" but added that "it would be worthwhile studying the distribution of corn in Tahuantinsuyo; it is possible that pellagra likewise existed in this [same] area" (Lastres, 1951: 268). This latter suggestion is naïve, for it is well known that corn was eaten throughout all of the Andes. Buikstra (1992: 87–89) went over this issue and hinted that although "... it is tempting to link the maize sickness or 'sara oncuy' to a nutritional disease such as pellagra," there is no actual proof of it.

Although no detailed study of this subject has ever been undertaken, and although no more data are available, it is hard to believe that pellagra was widespread in the Andes, for two major reasons. First, in ancient Peru we find a complex that associated three plants – *Zea mays* (maize), *Phaseolus* sp. (beans), and *Cucurbita* sp. (squash). Mangelsdorf (1974: 1–2) noted that

> ... the combination [of these three plants] ... is now recognized as furnishing an adequate, indeed an excellent diet. Corn supplies carbohydrates, small amount of protein, and fat; the beans represent the principal source of protein, but more important still, they contain adequate amounts of some of the 'dietary essential,' amino acids – the building blocks of protein – in which corn is deficient, especially tryptophane and lysine. Beans can also remedy corn's notorious deficiency in two vitamins, riboflavin and nicotinic acid. Squashes are valuable in supplying additional calories as well as vitamin A, and their seeds furnish an increment of wholesome fat in which a diet of corn and beans alone is barely adequate.

Mangelsdorf also believes that this is not the only reason this discovery was just highly beneficial for the American Indians, as it also allowed for an efficient use of the land, particularly bearing in mind the technology they had at their disposal. Beans climb and entwine themselves on the maize stalk, exposing their leaves to the sunlight without dramatically shading the leaves of maize, whereas squash vines spread out over the ground, between the hills of corn, thus choking out the weeds. To this we can add that when its leaves rot, they provide an excellent fertilizer for the corn plants.[22]

Flannery believes (1973: 291), based on the Mexican case, that the *Zea-Phaseolus-Cucurbita* complex "... is not an invention of the Indians; nature

[22] For more information, see Kaplan (1965: 359–360; 1968: 509); Sauer (1969a: 64, 131–132).

provided the model...." Thus in Guerrero, when a maize field is abandoned it is invaded by teosinte. *Phaseolus* and *Cucurbita* naturally occur there, and the beans twine themselves around teosinte.

To this we have to add the extensive use of the potato in the pre-Hispanic period, alongside root crops that have a significant starch content but are oil and protein deficient. We must not underestimate the fact that each potato plant provides more calories and proteins per unit of time and space than any other plant (Bonavia, 2006). This means that the pre-Hispanic population had a balanced diet, and it is therefore hard to believe that pellagra was widespread. Perhaps it did occur in some restricted zones. I consulted Uriel García Cáceres in this regard (personal communication, 21 February 2007), and he concurs, adding that pellagra is very hard to detect histologically. Brenton and Paine (1998: 113) concur, and they furthermore conclude that this disease did not exist in America.

As regards chicha, we need only insist on a specific point that was already mentioned in Chapter 9 – its preparation through salivation, which is known in Quechua as *muko*. This must once have been the most common way in which this beverage was prepared, yet it has gradually been forgotten, so nowadays the predominant type is the *chicha de jora*, that is, a malted chicha prepared with germinated corn. Significantly enough, in both Peru and Argentina this type of chicha is known as "falsa" or "postiza" ("false chicha"; Cámara-Hernández and Arancibia de Cabezas, 1976: 223; Sevilla Panizo, 1994: 223). The places where chicha is still prepared in Peru through salivation[23] cannot at present be pinpointed, for apparently no one has studied this. It probably does not survive in many places. I carried out a small survey among highlanders who have migrated to Lima but still retain their Andean roots: not only was the salivation technique unknown, its mere mention proved revolting. When asked if they would drink this type of chicha, the answer was a resounding no. Mercedes Quispe Palomino, a native of the community of San José de Pucaraqay,[24] is a native Quechua speaker who had no idea what *muko* is. The survey she carried out in her community showed that no one there was aware of the use of insalivation to prepare chicha, and only an old man could recall having heard someone in his family mention this procedure when he was young.

This is also the case with the *claro* or *clarito* chicha of Guadalupe,[25] which is not known anywhere else. It is a variant of the *chicha del año*, which I often had in the 1950s and 1960s. The only explanation then given was that this chicha was stored underground in large clay vessels, to which was added the leg of an ox. This beverage was of a clear yellow color, was quite clean, and had a high

[23] This term is used by Cutler and Cárdenas (1947: 41) and is more accurate than "mastication."

[24] This is close to Vilcashuamán, in the province of Cangallo, in Ayacucho.

[25] This is on the North Coast province of Pacasmayo, in the department of La Libertad.

alcohol content. Its taste was similar to the Italian *prosecco veneto*.[26] These are important Indian customs that are being irrevocably lost, and that may still be saved by studying them.

There is one major aspect of the role the Indian peoples played in the evolution of maize that is often forgotten, but that was described in depth by Grobman and his team, so here we need just summarize their position. First of all we must emphasize both the enormous success Andeans had as farmers and several sociological factors that brought about the intensification of their farming technology. Both on the coast and in the highlands, maize must have become one of the driving forces of Andeans quite early on, as we cannot otherwise understand the intimate relation established with this plant, which clearly surfaced in the art, religion, and other social activities of the various Andean cultures. In Peru this relation was far more than the simple connection established between plant and farmer. This was a deep, empirical knowledge of corn physiology, its morphological variations, and its forms of cultivation.

Both Weatherwax (1942) and Kempton (1937) wrote and showed that Indians were quite experienced farmers who were able to point the evolution of the plants they cultivated toward the desired phenotypic characteristics. Wellhausen and colleagues (1957) discussed this issue in regard to the Central American Indian population. They did not downplay the role the latter had in the evolution of maize, but they did question the idea that the native American populations were able to act as modern agriculturalists, that is, that they were able to conceive of the desired phenotypes beforehand. For Wellhausen and colleagues (1957), the Indians were able to direct the evolution of maize along different lines through selection rather than through hybridization.

Grobman and his team (1961) believe that in the Peruvian case both positions can be reconciled. Andean Indians at first had limited knowledge and less stringent social obligations, so a conscious improvement through selection took place, albeit in a not-too-rigorous manner. This probably was the simple selection of the bigger ears, as well as those with fancy shapes and colors.

The need for better harvests grew as society became more complex, thus probably giving rise to a whole system of state supervision to produce and maintain the output. The most-desired or finest races were devoted to religion and the large state storehouses. This clearly attained its highest development with the Inca Empire. When the Spanish arrived, they found an agricultural technology dedicated to maize, with irrigation, agricultural terraces, row planting, and fertilization of the land. Farmers were thus able to improve cultivation and to better direct the selection of the plants that could increase the yield and the stabilization of racial types, as well as their preferential uses. The Inca state was

[26] It was seen in Chapter 9 that Tschudi also collected data in this regard in the highlands (Tschudi, 1918: 42) and pointed out that a piece of boneless, fatless, and muscleless flesh was placed in each vessel, thus giving out a taste that resembled that of Spanish wine.

responsible in this regard not just for improving the races of the Cuzco region but also for the diffusion of the most productive of these races into the areas conquered during the Inca expansion, both in the highlands and on the coast. The Inca Empire does not, however, seem to have been extensive enough or of sufficient duration to consolidate the variability of maize into a few types, because this clearly is not the current situation.

It seems that the Indians never arrived at the knowledge required to attain hybridization as a cultivation system. It is possible that they experimented with favorable effects by planting together different seeds of one or several races in successive generations. Nor did they reach the level of individual selection of plants, but they did identify the best ones and spread them in Inca times.

The biggest success of the indigenous agriculture in late pre-Hispanic times was having attained a pattern of human selection of maize in what is now Peru that is different in the way it operates and in its results from those found in other primary corn areas (Grobman et al., 1961: 37–39).

Finally, some suggestions made by both this writer and other colleagues are in order, as regards the study of the origins and domestication of maize. First, we must work with significant samples, in regard not just to their size but also to the total sample assembled for comparative purposes. We have seen how Mesoamerican and South American results are often compared using mostly northern specimens and an insignificant number of southern specimens. This clearly distorts the results.

In regard to significant samples, we must bear in mind that many archaeologists have acquired the bad habit of undertaking small-scale excavations and then claiming to have attained results, although it is clear that these will be distorted. Area excavations must be the rule, and no site should be left behind before it has been fully studied. There are far too many preliminary reports that are just a few pages long, and far too few monographs that present the full results of the complete, in-depth study of one site.

Furthermore, it is of the utmost importance that the analyses be undertaken with openness, and they must not be unduly influenced by some hypothesis the researcher likes or has developed. The data must be examined critically, bearing in mind what they are, that is, mere testimonies that may fit in one or another position. And we must also be ready to accept the results, even if they do not match what we believe. Inability to follow this principle has brought about a long-term distortion of reality. Grobman (2004: 448) therefore correctly noted that "the data are not accepted if [the evidence] does not fit with an a priori hypothesis. To a large extent this is the way in which the information that goes against the hypothesis of an early domestication in Mexico and its late movement to South America has been treated."

It is also worth noting that at present very few individuals have the training required to distinguish the various races of maize found in South America. We have seen here that the racial classifications thus far made of Chilean and

Argentinean corn leave much to be desired and have raised serious problems. In this regard it is essential that all botanists or ethnobotanists who dedicate themselves to this task must be trained not just by reading the literature but also by handling and familiarizing themselves with the modern races. Sevilla (1994: 244) correctly notes that to draw conclusions from archaeological corn remains, one must have a vast knowledge of modern races of maize. A comparison of measurements does not suffice, as this can be misleading, since many races may be morphologically similar but phylogenetically different. We need to know the area of diffusion of the races, their adaptive scope, the context in which the race evolved, and their ecogeographical and phylogenetic relationships. Besides, because many traits, such as flexible rachises, shape, wide butts, and the number of rows, are common to various races that could never have been in contact with one another, it is possible that samples of maize may seem similar even though there is no genetic relation between them (Sevilla, 1994: 228).

At present we clearly need more specialists in this problematic. It is true that the Mexican area has been widely studied, yet even there there are several zones that have never been studied. It was seen in Chapter 5 that little work has been done from Mexico to South America, and the same holds true for the Pacific coast up to Ecuador. Some advances have been made in Ecuador and Peru, but this is still far too little, considering the great potential these two areas hold, particularly Peru, due to the exceptional preservation conditions found on the coastlands and to the significance of its midaltitude highland valleys. Goloubinoff and colleagues (1993: 2001) clearly noted this: "A more extensive survey of ancient and modern maize, teosinte, and *Tripsacum* is needed to determine the contribution of each process to the domestication and early evolution of maize."

Another major point is that we must realize that the analysis of modern samples does not suffice for an understanding of past ones, when we do not already have a good comparative base of archaeological samples: "In order to be meaningful, the findings of experimental research must be consistent with the archaeological record documenting maize evolution" (Eubanks, 2001b: 498).

In recent years genetic analyses underwent an impressive development, and DNA studies have an ever-increasing significance. The studies made by Goloubinoff and colleagues (1994: 117) showed that under certain conditions, long nuclear DNA fragments (some thousands of base pairs) can be preserved in the archaeological remains of corn. Yet scholars insist on the high potential for contamination of the samples. For them to remain valid, one must take several precautions, not just in their handling but even in the vessels used, the rapidity with which the samples must be stored once they have been found, and so on. They also point out that the best containers for this are those that have been kept in the dark as much as possible, are as sterile as can be, and are dry (Goloubinoff et al., 1994: 113, 124–125). The archaeologists who have dedicated themselves to ethnobotany should go over these recommendations carefully. But it is a fact that the study of DNA could help solve the issue of the domestication of maize.

This is very important for those plants whose wild ancestors have been lost, as is the case with *Vicia faba* and maize.[27]

As for the specific case of Peru, it is worth insisting here on a point I have often made. Except for some rare and isolated exceptions, ethnobotany began in Peru thanks to the work Junius Bird carried out at Huaca Prieta in the 1940s (see J. B. Bird et al., 1985). Yet to date not one Peruvian university has a real ethnobotany program, and in Peru there are almost no specialists in this major archaeological field. The fact that archaeologists are trained in faculties of the humanities and the social sciences, and the sharp break between these faculties and the natural sciences, means that when on the ground, Peruvian archaeologists do not have the training required to handle not just botanical and zoological remains but even geological phenomena – and these are just some of the major shortcomings of this state of affairs.

This problem is even worse when the period studied is the Preceramic period, for which this knowledge is essential. This is the reason why so much information has been lost or is simply absent in many fieldwork reports. Besides, we need more ethnobotanists rather than pure botanists. The latter do not have a detailed knowledge of archaeology and usually will not understand the real value of the samples, and their interpretation may include some serious mistakes. Archaeologists should, furthermore, get used to interdisciplinary work and collaborate with the highest number of specialists possible, but just assembling them will not suffice. To ask the proper questions and correctly interpret the answers, archaeologists must have a basic grounding in each of the disciplines with which they are working and must have mastered the technical jargon.

This is clearly visible when one analyzes Peruvian studies of the early cultural development in Peru from an ethnobotanical standpoint, as works are the ones that should help us dispel the many questions that still remain concerning the origin and domestication of plants. In the end these are isolated contributions made by a few specialists, and in only a handful of cases do they have really significant and extensive excavations. On the coast, only the Zaña, Casma, Culebras, Huarmey, and Supe Valleys have been studied with these goals in mind. But of these five valleys, Huarmey alone has been the subject of a systematic study, and even so it had two shortcomings that could not be addressed – the study of the *lomas* and the upper valley, where it joins the Callejón de Huaylas, which may yet yield some surprises. All the rest of the coast is as yet unknown.

In the vast expanses of the highlands, only one site in the Callejón de Huaylas and another one in Ayacucho have been studied, with the latter leaving much to be desired. This means that we know little of this huge area, which includes the midaltitude valleys where the earliest domestication of some plants took place. This is the great challenge that future generations will have to face.[28]

27 For more information, see M. Jones and Brown, 2000: 773.

28 Pickersgill made the same point (2007: 828).

Finally, there is another vast expanse of Peru of which we know next to nothing – the upper and the low tropical forest (the *ceja de selva* and the *selva baja*). It is true that there are some signs that some plants, like peanuts and manioc, actually come from this zone, yet we know nothing about this, because no studies have been made in this region. There obviously are some factors that go against any archaeological study undertaken in these lands, that is, the harsh preservation conditions and the many problems any study will face. But we must likewise acknowledge that now there are several resources that were not previously available, which can help to facilitate the task. This is a virgin area that is waiting to be explored and studied.

I have often discussed with Grobman the need to study maize outside Peru,[29] particularly in northern Chile, northwestern Argentina, and Ecuador and Colombia. But to carry out these tasks, and furthermore to analyze the results reached in the research already undertaken in this vast expanse of the central Andes, one must have more than a working knowledge of its geography *sensu lato*. Andean geography is a very difficult one, as has been pointed out in previous chapters, and it is only with an adequate command of its intricacies that one can understand not just what it meant to domesticate plants in it but also the results achieved. We have already seen how many mistakes can be made when comparisons are attempted without having an adequate grasp of the issues. Nor must we forget that each ecology has its own characteristics and that these cannot be generalized. Iltis (1987: 196) drew attention to this and pointed out how the cereals and legumes of the Near East not only can be planted en masse but also can be harvested in the same way. In contrast, maize, beans, and squashes – all large mesophytes – have to be individually planted, tended, harvested, and selected. The anthropological and biological implications of these contrasting agricultural strategies have yet to be fully understood.

As regards the origin and domestication of maize, it is true that very specific information is available, but it is also undeniable that far too many hypotheses have been made that are based only on ideas and that lack any kind of actual support. Randolph (1976: 321) therefore correctly noted that there has been "an unusual amount of speculations." Following Mangelsdorf (1974: 147), we can say that in the absence of facts, the debate may well continue indefinitely and the exchanges may even be heated at times, but at the same time it is very hard to argue against the solid evidence provided by well-preserved archaeological specimens.

On reaching the final section of this book, it would be nice if one could draw some specific conclusions regarding this subject. Yet here it is perhaps best to cite Walton Galinat, one of the scholars who devoted his life's work to the study of this plant: "... More questions than answers remain on the origin and diffusion of maize, [so] the available time and state of knowledge are inadequate to

[29] Grobman also noted this in his most recent publication (2004: 470).

attempt a more comprehensive treatment here" (Galinat, 1985b: 274). More than a quarter of a century has gone by since this statement was made, yet things have not changed significantly.

One finds it very hard to understand why some colleagues have stubbornly rejected the possibility of an independent domestication of maize, despite the specific and irrefutable evidence supporting this claim, all the more so when this has been accepted and has been shown for other plants in the South American continent. *Phaseolus* and *Capsicum* come to mind as just two examples that no one questions now. And this in an epoch – 2000–10000 (13000?) years BP – when we know there were at least 10 independent domestication centers in the world, of which the 4 major ones were in the Near East, and Far East, Mexico, and South America (Balter, 2007: 1831, 1833).

Of all the plants used by humanity, maize probably is the hardest one to study. This is because this plant tends to vary, and even now aboriginal groups lacking the required technology find it hard to restrain this change. This also is the reason why maize has changed so much throughout time (Johannessen, 1982: 97). Johannessen points out that he "know[s] of no other such wind-pollinated crop, anywhere, that has been modified continually over such a long time and now has such extreme variability as does maize" (Johannessen, 1982: 98). This explains the various possibilities that arise when interpreting the changes seen in archaeological remains, which some believe were rapid and others long in developing (we should recall in this regard the work carried out by Goloubinoff et al., 1993).

Goodman (1988: 203) correctly noted that little was known of maize in Latin America until the 1920s and 1930s. In the 1940s the studies undertaken by Vavilov and his colleagues and by Anderson and his team and the pieces published by the Committee for the Preservation of Indigenous Strains of Maize of the National Academy of Sciences–National Research Council all showed the genetic diversity of maize, which quite probably has been better described than that of any other crop. A review of the existing literature easily shows the point Goodman made.

Although it is true that maize is a demanding crop, under the right conditions it also has a productive capacity that is the highest among the major cereals (Tschauner, 1998: 326). This clearly is one of the reasons why humanity used this plant from the moment it established contact with it. On the other hand we must not forget that although it is true that maize does not include all of the component parts required for a complete diet (see previously), according to the study made by S. A. Watson and Ramstad (1987), it still includes on average 73.4% starch, 9.1% proteins, and 4.4% lipids. For Van der Merwe and Tschauner (1999), maize is the most nutritious substance grown in the world in terms of biomass.

When we combine maize with wheat and rice, we get a trio that yields more than 50% of the calories consumed by humanity (Doebley, 2006: 1318). Although I do not have any figure for the number of people who use the *Zea*,

Phaseolus, and *Lagenaria* association that has been described at length, it clearly is of the utmost importance in America.

It is a fact that all cultivated plants are intimately related with man, but there clearly are differences among them. In the case of corn I agree with Tschauner (1998: 326), for whom maize is "… conceptually associated with humankind and culture rather than with nature."

And yet interestingly enough, although with domestication maize underwent a spectacular increase in its size and productivity, and although it became a staple food of pre-Hispanic American populations, it has even so not undergone any substantial transformation in its botanical characteristics in the 7,000 years during which man has been acquainted with it (Mangelsdorf et al., 1967a: 200).

Flannery (1986a: 7) correctly noted that "maize (*Zea mays*) has by far the most enigmatic and controversial origin [and domestication, I would add] of any major cultivated plant."

In 1999, when I reviewed the corn problematic alongside Grobman (Bonavia and Grobman, 1999: 256–257), we concluded our overview by citing a passage taken from an article authored by Moseley and Willey. Being well acquainted with Willey's style, with his ideas and his great modesty, I would like to believe that he wrote the paragraph in question. The passage is on Áspero, but its content and moral are perfectly applicable to the study of maize, which is why it is once again cited here. It reads thus:

> Well formulated hypothesis and a problem orientation are basic to archaeological research. In fact, these are always with us, but they must be overtly spelled out lest they confine rather than orient research. In 1941 the investigators of Aspero operated with certain problems in mind, and with certain covert concepts about the course of Peruvian prehistory. These concepts were never clearly expressed, nor thoroughly assessed. Although the Aspero data pointed to new hypotheses and called for different concepts the pre-extant intellectual framework did not genuflect. A major preceramic settlement – the first systematically excavated in Peru – was fit into a ceramic stage framework. The fit was not good, and to various degrees certain data were overlooked or explained away. Surprisingly, for 30 years the status of Aspero was never seriously challenged, only accepted as somewhat anomalous. Perhaps abler and wiser investigators would have risen above the constraints of their preconceptions; yet hindsight is always better than foresight, and it is difficult to be sure. The only moral to be drawn from Aspero is that *while hypotheses are necessary stimuli for investigation, the archaeologist should not allow himself to be fettered by them.* (Moseley and Willey, 1973: 466–467; emphasis added)

APPENDIX

ORIGIN, DOMESTICATION, AND EVOLUTION OF MAIZE: NEW PERSPECTIVES FROM CYTOGENETIC, GENETIC, AND BIOMOLECULAR RESEARCH COMPLEMENTING ARCHAEOLOGICAL FINDINGS

ALEXANDER GROBMAN

Introduction

The present appendix to the book written by my colleague Duccio Bonavia tries to bridge and interpret the information available on the evolution and domestication of maize, from two main sources: (a) archaeological research and (b) genetic, cytogenetic, and molecular biology research. The first area has been amply covered by Bonavia in the main text of this book. I attempt to cover selectively recent advances in the second fields and to interpret the new findings in retrospect of previous knowledge.

From the vantage point of a reviewer and specialized analyst of the plethora of information now available on the evolution of maize and other species, it behooves me to make sense of it all in painting the most accurate vision I could make, after more than 60 years of exploring maize in the field as a plant breeder and seed producer, geneticist, ethnobotanist, student of its diversity and evolution, and advocate and starter of the collection and analysis of its genetic and molecular variability. In trying to organize the information and presenting it in this document, I attempt to advance an educated and, as far as possible, an impartial set of conclusions. They are reached with the majority of the evidence at hand, on how maize evolved, from where it started, and how it interrelated to its relatives, teosinte and *Tripsacum*. I am aware that this vision is necessarily a changing one and that it is, at this time, a photograph of the present evidence from this vantage point.

I must give credit to a large group of researchers who have advanced our knowledge of maize evolution to this point. Theories that were useful at one time as an explanation may prove – as some have – to be untenable under present facts. The evidences used may be valid at one time but may be subject to a variety of interpretations as mounting evidence accumulates to either upgrade working hypotheses to the level of theories or to dethrone them and thus allow other theories to rise up.

At this stage the theory that maize evolved from teosinte in a direct path through mutation and selection has gained many supporters but is not, in any way, shared by all researchers. One school of maize researchers advanced it from the rank of theory and has pronounced this hypothesis practically as fact. Although we grant that evidence accumulated so far weighs heavily in favor of acceptance of teosinte as the putative parent of maize, through selection of adaptive changes to make it an agricultural crop, we are reluctant to give the theory the category of fact. The reason for this is that the evidence is not factual; rather it is in many cases contrary to the teosinte-to-maize descent theory, and there are many unexplained situations and vacuum areas that this theory does not explain. We deal with them in the course of this analytical review. Although many studies show high synteny in genes of man and chimpanzee, with about 1.11% divergence of coding genes (Chen and Li, 2001[1]), no one postulates, as did Darwin's critics at the time of the presentation of his book, *Origin of Species,* that man descended from chimpanzee only because they are the closest related present living species and share close to 99% of genes. We now know that there are a number of preceding species of apes that point in the direction that both species are part of a gradation of species and subspecies that branched and were subjected to differential selection ending in our present human species and in chimpanzees, through parallel evolution. Man and chimpanzee originated from a common ancestor but did not originate one from the other as a direct branch. A similar situation is valid for maize, in spite of the fact that maize and teosinte cross easily today. We interpret the evidence accumulated thus far – archaeologic, genetic, cytogenetic, and DNA molecular analysis of specific genes – to indicate that, in all likelihood, maize and teosinte, the latter in its multiple versions, originated from a common ancestor and diverged before domestication to find themselves again and interact genetically with each other. We present in the following our evaluation of the multiple evidence now available of how this could have happened and why we think so.

Maize Origin, Domestication, and Evolution

The subject of maize origin, domestication, and evolution has attracted the imagination and scientific skills of many scientists over a period of more than a century. Maize being such an important crop, this knowledge could lead to a better understanding of how it evolved its present genetic structure, and its variability, which could lead to improvement by breeding. From a purely scientific point of view, maize has become one of the best-known plants through the genomic and genetic information that has been accumulated. It can be used as a model organism for understanding both plant and animal evolution. Also, from an ethnological viewpoint, the close association of maize with human beings

[1] Chen, F. C, and W. H. Li. 2001. Genomic divergences between humans and other hominoids and the effective population size of the common ancestor of humans and chimpanzees. *American Journal of Human Genetics*, 68 (2): 444–456.

over several centuries has allowed it to be used as a tracer of the movement of human cultural groups in the Americas.

Maize physical remains, preserved in the archaeological record of the Americas for more than 7,000 years, have yielded through their study a wealth of information. They have been approached from the archaeological and ethnobotanical viewpoints and more recently through archaeobiology, using improved techniques. This has helped recently to follow through the analysis of new data to gain a better insight on maize domestication and evolution, resulting in the reorientation of former hypotheses to propose new schemes or the alternative courses that maize domestication might have followed.

Recently great strides have been made in recognizing the factors that might have led to the emergence of maize as a domesticated plant from a former wild ancestor, and to acquire its extraordinary variation. Maize is the most- and best-studied species in regard to plant evolution. Much has been learned so far, but at each step of the advance of science, new frontiers and questions open up. It is a fallacy to state that the whole picture of maize evolution is now clear. As much as we accept that there have been considerable advances in our knowledge, such as the genomic constitution of maize and its relatives, still there are many voids that need to be filled in regard to its antecedents and evolution.

Through advances in archaeological, genetic, cytogenetic, and, more recently, molecular biology research, new vistas have been added to our knowledge on how maize might have evolved from a wild plant that preceded it, and how it could have proceeded in its domestication and evolution with the formation of more than 300 races in the Americas and some others in other parts of the world in post-Columbian times. How did maize, starting from a single or from multiple centers, develop into a crop plant, at what speed did it pass through several stages, and how was the process of domestication and evolution accomplished to this day, resulting in modern maize, one of the most widely planted crops in the world? From its early beginnings from one or multiple sites, maize spread out and was further reselected for specific adaptations. The whole process is a thrilling scientific story.

I am thankful and honored to have the opportunity to have these pages added as an appendix to the English edition of the present excellent book written by my good friend and colleague Duccio Bonavia. A long working and personal friendship has linked us for more than 40 years. I consider him an outstanding scientist, with whom I have had the privilege to share many days as a collaborator in the fields of maize genetics and ethnobotany, adding extra value to his fine archaeological studies in Peru. These studies have resulted in a series of publications relative to the origin and evolution of maize in Peru; he has put these studies in focus in the context of the larger picture of the archaeology and ethnobotany of maize in the Americas, which he treats with skill and detail in this book.

The field of research on maize origin and evolution had taken in recent years a considerable impulse from the advances made in genetics, genomics, and molecular biology; from new techniques to trace the ancient presence of maize; and from the reevaluation of the age of the sites in which maize was present. Accordingly we felt that these advances had to be reviewed and meshed into the fabric of this book to amplify its treatment of the subject and to afford the opportunity for fresh visions from the synthesis of the multiplicity of information now available.

Bonavia and I have supported all along the presence in Peru of preceramic maize against opposers to that belief. Coastal agriculture in Peru, including peanuts, squash, and cotton being grown with agricultural skills and industry between 5,000 to 10,000 years ago, has been well documented (Dillehay 2011;[2] Dillehay et al., 2007[3]). Recent dated findings push back maize in Peru to approximately 6000 years BP in the Casma Valley (Cerro El Calvario and Cerro Julia). (See the afterword at the end of the appendix.) New early site archaeological digs are needed in Peru to scan the evidence for the beginnings of agriculture and the plants used, especially in the intermediate altitudes in the highlands. This will provide additional hard evidence to confront or support the genetic and molecular evidence on maize domestication.

I have tried to supplement in this appendix Bonavia's ample treatment of the subject of maize origin and evolution in Peru with the presentation of selected advances in various fields by critically reviewing the literature and by integrating these findings, hoping that the excercise will shed new light on the subjects of maize origin, domestication, and evolution. The most recent advances in the areas of genetics, cytogenetics, and genomics and in molecular, population, and evolutionary biology, from selected and relevant papers tracing the origin, domestication, evolution, and differentiation of maize, are treated in a critical review form in this appendix.

Some concepts in this review may not conform to some, even majority, views on the subject. Nevertheless, I have felt obliged to voice scientific opinions supported by published and some new available but yet unpublished data on the origin and domestication of maize. Much of the data that has accumulated from many studies on this subject, when reviewed objectively, renders itself open to more than one way of interpretation. I accept full responsibility for these opinions, and I accept any possible errors of form and substance as my own.

The treatment of information directly linked to archaeology has been retained in the main part of this book. Such areas as the analysis of macro-remains, pollen, phytoliths, and starch grains are included there.

[2] Dillehay, T. D., J. Rossen, T. C. Andres, and D. E. Williams. 2007. Preceramic adoption of peanut, squash, and cotton in northern Peru. *Science*, 316 (5833): 1890–1893.

[3] Dillehay, Tom D. (editor). 2011. *From Foraging to Farming in the Andes: New Perspectives on Food Production and Social Organization*. Cambridge University Press. New York.

Our hope is to be able to help bridge the gaps of knowledge between archaeology and the biological sciences, presenting fresh interpretations with questions and proposals without qualms.

Theories on the Descent of Maize and Its Relatives: I

Various hypotheses have been advanced on the origin of maize. There is no unanimity of acceptance of any of them, as has been established in the main text of this book. It is now well accepted that *Zea mays* L. is an ancient amphidiploid species resulting from the hybridization of two closely related preexisting species with $n = 5$ as the basic chromosome number and duplicating it to $n = 10$. It is also accepted that *Z. mays* and *Tripsacum* have diverged from a common ancestor between 0.5 and 1.2 million years ago and that the genus *Tripsacum* evolved independently. *Tripsacum* and *Zea* species have interacted in the past, as it appears in the case of *T. andersonii* (formerly *Tripsacum laxum* Nash), a sterile species that has evolved with $2n = 54$–72 chromosomes. *T. andersonii* is suggested, on the basis of banding studies, to have originated as a hybrid between diploid *T. latifolium* with *T. laxum* to give triploid *T. latifolium*, which then hybridized with *Zea luxurians* (Barre et al., 1994[4]). Barre and colleagues proposed the hypothesis that the genetic constitution of *T. andersonii* was derived from two hybridization events:

1. *T. latifolium* ($2x$) × *T. laxum* ($2x$) => *T. latifolium* ($3x = 54$)
2. *T. latifolium* ($3x$) × *Zea luxurians* ($2n = 20$) => *T. andersonii* (54 + 10 chromosomes)

Two events may have occurred, given that two *T. latifolium* populations have been found to have different isozyme profiles. That the second event has probably been unique is strongly suggested by at least two lines of evidence: more than 20 different accessions of *T. andersonii* from several South American countries show exactly the same morphology and the same isozyme pattern. *T. andersonii* may, therefore, be an example of how an apparently very improbable set of events can give rise to a new species.

Both maize and *Tripsacum* chromosomes show homoeologous sections, and crosses can be made between them, producing viable progeny. There is also evidence of common DNA polymorphisms in a number of alleles present only in maize and *Tripsacum* and absent from teosinte (Eubanks, 2001[5]), which may be ascribed to genetic interaction with wild maize prior to domestication. They cannot be explained by descent from a common ancestor, if maize was a

4 Barre, M., J. Berthaud, D. González de León, Y. Savidan. 1994. Evidence for the tri-hybrid origin of *Tripsacum andersonii* Gray. *Maize Genetics Cooperation Newsletter*, 68: 58–59.

5 Eubanks, Mary W. 2001. The mysterious origin of maize. *Economic Botany*, 55 (4): 492–514.

domesticate of teosinte. The meaning of these findings has been disputed by adherents to the teosinte-to-maize hypothesis.

The work of Galinat, Rao, Chandravadana, and Tentravahi, summarized by Galinat (1971[6]) has demonstrated that although there are homoeologous regions between maize and *Tripsacum* and crossing over has been observed between maize and *Tripsacum* chromosomes, the assemblage of genes is different. For example, the critical region of chromosome 4, which is required to express the peculiar fruitcase of teosinte, is not found in *Tripsacum* as a single segment of genes; these do exist in *Tripsacum* but are dispersed over the genome. These findings do not lend credibility to the theory reviewed previously that teosinte was formed through the introgression of *Tripsacum* into maize.

HYPOTHESES ON THE DOMESTICATION OF MAIZE

Several hypotheses have been advanced on the origin of domesticated maize, as we know it today. They have been amply presented, and their relative merits have been discussed in a number of reviews: Goodman (1965[7]), Mangelsdorf (1983[8]), and Galinat (1977[9]), among others. They have been thoroughly explained in Chapter 3 in the main text of this book.

MAIZE WAS ALWAYS MAIZE

According to this hypothesis, proposed by Kempton (1936[10]), Mangelsdorf (1974[11]) states that maize was domesticated from a wild ancestor that was essentially a pod corn. It was a plant that had separate tassels and ears, the latter bearing basal pistillate spikelets with almost complete tunicate soft glumes, and the ear ended in a staminate tip, reminiscent of the *Tripsacum* female inflorescences. Such ears – semi-tunicate, with staminate tips – have been found in a relic form in some archaeological sites and in present Andean maize races and are well described photographically from San Marcos Cave cob relics in Mangelsdorf (1974). This primitive maize had a polystichous ear – it was 4-ranked with 8 rows of seeds and had an average number of 12 seeds per row, a slender rachis, and long, navicular, and shallow cupules that did not even partially cover the seeds. The seeds were inserted on long rachillae (and pedicels) in a position perpendicular to the rachis, as found in cobs from the Proto-Confite Morocho race from the Cerro Guitarra,

[6] Galinat, W. C. 1971. The origin of maize. *Annual Review of Genetics*, 5: 447–478.

[7] Goodman, M. M. 1965. *The History and Origin of Maize.* North Carolina Agricultural Experiment Station. Technical Bulletin, No. 170.

[8] Mangelsdorf, P. C. 1983. The mystery of corn: New perspectives. *Proceedings of the American Philosophical Society*, 127 (4): 215–247.

[9] Galinat, Walton C. 1977. The origin of corn. In G. F. Sprague, editor. *Corn and Corn Improvement.* Agronomy Series No. 18. American Society of Agronomy. Madison. pp. 1–47.

[10] Kempton, J. H. 1936. Maize as a measure of Indian skill. *University of New Mexico Bulletin*, 296: 19–28.

[11] Mangelsdorf, Paul C. 1974. *Corn: Its Origin, Evolution and Improvement.* Belknap Press, Harvard University. Cambridge.

Cerro El Calvario, Los Gavilanes, Áspero, and Cueva del Guitarrero preceramic archaeological sites in Peru. It had a similar aspect as in archaeological maize from the Abejas phase of Tehuacán, Mexico (Mangelsdorf, 1974), with an average of 13.6 seeds per row. The seeds of this modern maize ancestor must have been very small, corneous, round, or imbricated and surrounded by soft glumes protecting a good part of the seeds, which were attached to long pedicels, easily breakable at harvest time, with an abscission layer, as occurs in the wild grain progenitors of other species of grasses. There is a possibility that the rachis, itself flexible and breakable in primitive wild maize, could have been an alternative mechanism for seed dispersal. Some ears could have had basal branching, as found in the archaeological maize at the Los Gavilanes site in Peru (Grobman, 1982[12]), and the husk system was loose and semiopen, as perceived from archaeological maize specimens from Mexico and Peru. The husk system from archaeological samples of maize collected in the valley of Huarmey on the coast of Peru by David H. Kelley and Duccio Bonavia was reconstructed and described by Grobman and colleagues (1977: figure 4[13]) as having a long peduncle of the ear, which may have projected the ear and exposed it beyond the husks, which did not fully cover the ear at maturity, thus allowing for dispersal of the seeds.

The domestication process could have selected for adherence of the seeds to the rachis in a manner similar to that of other cereal species, for more seeds per cob, and for the enhancement of the closure of the ear with longer and tighter bracts, especially in more tropical environments under insect pressure. Mummified *Helicoverpa* (*Heliothis*) *zea* Boddie larvae and perforations in stalks and husks have been found in archaeological maize ears on the coast of Peru, attesting to very early insect pressure and coadaptation of host and predator over a long time period.

It appears that there was sufficient pollination – caused either by time or coincidence of maturation of male/female inflorescences or by large enough populations with variable days to flowering – to produce ears with a full complement of seeds formed; this was not the case for teosinte, for which it is commonly found that there are many ears with an incomplete formation of kernels (Wilkes, 1989[14]).

After studying the earliest maize from San Marcos Cave in Mexico, Benz and Iltis (1990[15]) declared that these archaeological specimens exhibit none of the

[12] Grobman, Alexander. 1982. Maíz (*Zea mays*). In Duccio Bonavia, *Precerámico Peruano Los Gavilanes. Mar, desierto y oasis en la historia del hombre.* COFIDE/Instituto Arqueológico Alemán. Lima. pp. 157–179.

[13] Grobman, Alexander, Duccio Bonavia, and David H. Kelley, with Paul C. Mangelsdorf and Julián Cámara-Hernández. 1977. Study of pre-ceramic maize from Huarmey, North Central Coast of Peru. *Botanical Museum Leaflets.* Harvard University, 25 (8): 222–242.

[14] Wilkes, G. H. 1989. Maize: Domestication, racial evolution and spread. In D. R. Harris and G. C. Hillman, editors. *Foraging and Farming: The Evolution of Plant Exploitation.* Unwin Hyman. London. pp. 441–455.

[15] Benz, Bruce F., and Hugh H. Iltis. 1990. Studies in archaeological maize I: The wild maize from San Marcos Cave reexamined. *American Antiquity*, 55 (3): 500–511.

attributes one might expect of an intermediate between a teosinte female spike and a typical maize ear.

As discussed in his 1974 publication, based on experimental facts, by introducing the t^d component of the dominant allele *Tu* on chromosome 4 of maize and the inhibitor gene *Ti,* Mangelsdorf was able to reconstruct a type of ear similar to those found in archaeological specimens from the Tehuacán caves.

Objections to this theory are based on the inability to conceive of how the maize ear could have possibly allowed natural seed scattering. The objectors are visualizing the present-day monstrous maize ear, which has undergone centuries of selection, and fail to consider the points mentioned previously explaining how wild maize could have scattered its seeds without human intervention. We have added a further suggestion – that is, the possibility that birds could have eaten the tiny maize seeds and that many of the seeds were able to pass though their digestive system and spread along a larger territory, as we have explained using the present-day time example of sorghum in Peru (Bonavia and Grobman, 1989[16]).

Again, the nonexistence of a wild maize relic population has been invoked as an argument to dismiss this hypothesis (Galinat, 1971). The reply is easy: such a wild population, in all likelihood, was initially small, somewhat distinct from annual teosinte, and geographically isolated. It would have been swamped out and discarded by human and natural selection after the new domesticated population(s) moved onward. Another possibility is that it became extinct by the grazing of imported animals after the Spanish conquest (Mangelsdorf et al., 1964[17]).

MULTIPLE DOMESTICATION

Mangelsdorf (1974) has promoted the concept that there were at least six primitive wild popcorn races of maize that originated through independent domestication and later interaction with teosinte, leading to the abundance – about 300 – of races of maize that exists today. In this context it is interesting to review the conclusions of Goloubinoff and colleagues (1993[18]). They studied an *Adh2* (alcohol dehydrogenase) segment sequence between positions 85 and 403 from the transcription start site of genes, from archaeological maize from Peru and Chile of different ages, as well as genes from the present primitive races Kculli and Proto-Confite Morocho of Peru, present maize races from

[16] Bonavia, Duccio, and Alexander Grobman. 1989. Andean maize: Its origin and domestication. In D. R. Harris and G. C. Hillman, editors. *Foraging and Farming: The Evolution of Plant Exploitation*. Unwin Hyman. London. pp. 456–470.

[17] Mangelsdorf, P. C., R. S. MacNeish, and W. C. Galinat. 1964. Domestication of corn. *Science*, 143: 538–545.

[18] Goloubinoff, Pierre, Svante Pääbo, and Allan C. Wilson. 1993. Evolution of maize inferred from sequence diversity of an *Adh2* gene segment from archaeological specimens. *Proceedings of the National Academy of Sciences USA*, 90: 1997–2001.

Mexico (Tabloncillo) and the United States, *parviglumis, mexicana, diploperennis*, and *luxurians* teosintes and *Tripsacum pilosum*. They arrived at several conclusions: (1) The sequence diversity of archaeological maize (Proto-Confite Morocho, 4700 years BP) equals that of contemporary maize. (2) Ancient *Adh2* alleles are identical to contemporary alleles, indicating that there has not been an accelerated change and that the sequence diversity has become constant, as is seen from the fact that ancient alleles are more similar to modern alleles than to one another. (3) The extent of sequence differences remained constant from a period about halfway between domestication and the present day. (4) If maize had originated from one domestication event and subsequently evolved at an accelerated pace, it would be predicted that ancient alleles would be less diverse than modern alleles, and they are not. (5) The gene pool of maize must be several million years old and must have preceded the domestication era. (6) Because there is no presence of teosinte in South America and no evidence there ever was, a suggested explanation of the findings, listed previously, is that a constant flow of genes to maize from teosinte occurred after the introduction of the Andean maize races to South America, that is, in very early maize history. (7) Many maize alleles appear as more closely related to teosinte than to other maize alleles and vice versa, and this applies not only to races of teosinte related to maize but to *Z. diploperennis* and *Z. luxurians*, which do not commonly cross-pollinate with maize. (8) There is no evidence to support the notion that modern races of maize emerged from a single common ancestor, such as a specific line of *Z. mays* ssp. *parviglumis*, which is in clear contradiction to the hypothesis of Matsuoka and colleagues (2002[19]) or to a hypothetical line of wild maize. (9) Domestic maize, in spite of its morphological variability, remains indistinguishable from teosinte in the *Adh2* allele variation. (10) Ancient maize alleles have not evolved faster than modern maize alleles relative to *Tripsacum*, therefore there is no significant acceleration of evolution in maize.

These conclusions have implications for the development of three scenarios for domestication that can fit the information obtained, namely, that maize was domesticated: (1) from a single wild ancestor that was subsequently introgressed by teosinte prior to the early movement of maize to South America, (2) from a population of wild ancestors that initially contained and later perpetuated a high degree of allelic polymorphism, or (3) independently from several distinct wild ancestors that have subsequently interbred among themselves and with wild teosinte. The first scenario (1) is not in consonance with the archaeological and cytogenetic data, which point out that, whatever origin maize may have had, early maize arrivals in the Andean region did not have evident morphological signs of teosinte introgression. Nor do their descendant races have

[19] Matsuoka, Y., S. E. Mitchell, S. Kresovich, M. Goodman, and J. Doebley. 2002. Microsatellites in *Zea* – variability, patterns of mutations, and use for evolutionary studies. Theoretical applied. *Genetics*, 104: 436–450.

them now – that is, they exhibit low or no rachis and glume induration, low condensation index of tassels, long rachillae and no tillers, as well as low frequencies or absence of chromosome knobs; the last characteristic is a decidedly critical marker. This evidence does not support the existence of teosinte introgression in the first migrating populations. Grobman and colleagues (1961[20]) have entertained the possibility that later contacts may have produced migration of Mexican races to the Andean region and of Andean races to Mexico, which is also accepted by Wellhausen and colleagues (1952[21]).

The independent origin of maize, teosinte, and *Tripsacum* from a common ancestral species was advanced by Montgomery (1906[22]) and Weatherwax (1918,[23] 1935[24]). Both teosinte and *Tripsacum* have a two-ranking condition in their ears, a trait that is controlled, according to Rogers (1950[25]), by two genes, one in chromosome 1 and one in chromosome 2 – the latter was also confirmed by Galinat (1988[26]) – and that corresponds to the *Tripsacum* two-ranking gene in partial homoeologue to chromosome 9 of *Tripsacum* (Galinat, 1988).

THE TEOSINTE-TO-MAIZE HYPOTHESIS

The description of the various teosinte taxa has been provided in the main text of this book. Since the publication of the Spanish edition, *Zea nicaraguensis* has been separated as a new species from the teosinte section Luxuriantes, and three new teosintes have been found (Sánchez et al., 2011[27]). These newly discovered populations are distinct from one another and from other *Zea* species. They represent three new entities based on their unique combinations of morphological, ecological, ploidy, and DNA markers. A perennial diploid population from Nayarit is distinguished by its early maturation plants and by having male inflorescences with few tassel branches and long spikelets, unlike Balsas teosinte. A perennial tetraploid population from Michoacán is characterized by tall and late-maturing plants, and by having male inflorescences with many branches. An annual diploid population from Oaxaca is characterized by

[20] Grobman, A., W. Salhuana, and R. Sevilla, in collaboration with Paul C. Mangelsdorf. 1961. *Races of Maize in Peru: Their Origins, Evolution and Classification.* National Academy of Sciences/National Research Council. Publication 915. Washington, D.C.

[21] Wellhausen, E. J., L. M. Roberts, and X. E. Hernández, in collaboration with P. C. Mangelsdorf. 1952. *Races of Maize in Mexico.* Bussey Institute, Harvard University. Cambridge.

[22] Montgomery, E. G. 1906. What is an ear of corn? *Popular Science Monthly*, 68: 55–62.

[23] Weatherwax, P. 1918. The evolution of maize. *Torrey Botanical Club*, 45: 309–342.

[24] Weatherwax, P. 1935. The phylogeny of *Zea mays*. *American Midland Naturalist*, 16: 1–71.

[25] Rogers, J. S. 1950. The inheritance of inflorescence characters in maize-teosinte hybrids. *Genetics*, 35: 342–358.

[26] Galinat, W. C. 1988. The origin of corn. In G. F. Sprague and J. W. Dudley, editors. *Corn and Corn Improvement.* 3rd ed. American Society of Agronomy. Madison. pp. 1–31.

[27] Sánchez-G., J. J., L. De La Cruz L., V. A. Vidal M., J. Ron P., S. Taba, F. Santacruz-Ruvalcaba, S. Sood, J. B. Holland, J. A. Ruíz C., S. Carvajal, F. Aragón C., V. H. Chávez T., M. M. Morales R., and R. Barba-González. 2011. Three new teosintes (*Zea* spp., Poaceae) from Mexico. *American Journal of Botany*, 98 (9): 1537–1548.

having male inflorescences with fewer branches and longer spikelets than those found in the sister taxa *Z. luxurians* and *Z. nicaraguensis*, plants with high thermal requirements, and very long seed dormancy. Sánchez and colleagues have placed the three new populations of teosinte as distinct entities within the section Luxuriantes of the genus *Zea*.

The basis of this theory, which has prompted a large number of followers, treats the domestication of maize as a direct descent process from teosinte through the application of strong artificial selection pressure to overcome the centrifugal and stabilizing natural selection on the original wild population(s). Backcrosses to teosinte producing new variation, with further selection for the domesticated traits, took place in later periods. Initially it was postulated (Galinat, 1971) that *Z. mays* ssp. *mexicana* was the putative parent of maize, because of the great affinity of its morphological traits with the upland races of maize of Mexico. Recent advances in molecular genetics have pointed in the direction of the teosinte race Balsas or *Z. mays* ssp. *parviglumis* as the most likely ancestor, according to Doebley and colleagues (1987[28]), Doebley (1990a[29]) and Matsuoka and colleagues (2002).

The origin of maize as a direct domesticate from teosinte was initially advanced by Ascherson (1975,[30] 1880[31]), Beadle (1939,[32] 1972[33]), Langham (1940[34]), Iltis (1983[35]), and Galinat (1977[36], 1985a[37]), and in recent years, molecular information on the polymorphisms present in some genes responsible for expression of different traits in maize and teosinte have been exhibited as proof of the hypothesis of maize having been domesticated from a teosinte ancestor (Clark et al., 2005;[38] Doebley 1990a; Matsuoka et al. 2002).

[28] Doebley, J. F., W. T. Renfroe, and A. Blanton. 1987. Restriction site variation in the *Zea* chloroplast genome. *Genetics*, 117: 139–147.

[29] Doebley, J. F. 1990a. Molecular evidence and the evolution of maize. *Economic Botany*, 44 (suppl. 3): 6–27.

[30] Ascherson, P. 1975. Über *Euchlaena mexicana* Schräd. *Verhandlungen des Botanischen Vereins der Provinz Brandenburg*,17: 76–80.

[31] Ascherson, P. 1880. Bemerkungen über ästigen Maiskolben. *Verhandlungen des Botanischen Vereins der Provinz Brandenburg*, 21: 133–138.

[32] Beadle, G. W. 1939. Teosinte and the origin of maize. *Journal of Heredity*, 30: 245–247.

[33] Beadle, G. W. 1972. The mystery of maize. *Field Museum of Natural History Bulletin*, 43: 2–11.

[34] Langham, D. G. 1940. The inheritance of intergeneric differences in *Zea-Euchlaena* hybrids. *Genetics*, 25: 88–107.

[35] Iltis, H. H. 1983. From teosinte to maize: The catastrophic sexual transmutation. *Science*, 22z: 886–894.

[36] Galinat, W. C. 1977. The origin of corn. In G. W. Sprague, editor. *Corn and Corn Improvement.* 2nd ed. American Society of Agronomy. Madison. pp. 1–47.

[37] Galinat, W. C. 1985a. Teosinte. The ancestor of maize: Perspectives for its use for maize breeding for the tropics. In A. Brandolini and F. Salamini, editors. *Breeding Strategies for Maize Production Improvement in the Tropics.* Monograph: 1–11. FAO. Firenze.

[38] Clark, R. M., S. Tavaré, and J. Doebley. 2005. Estimating a nucleotide substitution rate for maize from polymorphism at a major domestication locus. *Molecular Biology Evolution*, 22: 2304–2312.

The latter have presented information based on single nucleotide polymorphisms (SNPs) of some specific quantitative trait loci (QTLs), genes capable of producing large morphological effects, such as *tb1*, that relate different taxa through greater affinities in polymorphisms. Ending a long series of other studies, they sustain the hypothesis (which has inappropriately been elevated to the rank of a consensus) that teosinte ssp. *parviglumis* was the undisputed ancestor of maize and that domestication occurred at a single location in the lowland Balsas River basin of southwestern Mexico. Because of the geography of the Balsas River basin, it is more likely that domestication of maize would have taken place at intermediate altitudes where teosinte populations are still found. It has been suggested by Wang and colleagues (1999[39]) and Doebley (2004[40]) that strong selection was exercised to change the regulatory segments of chromosome 1 upstream of the transcriptional unit of gene *tb1* without any major changes in the gene itself. Sequencing studies by Wang and colleagues (2005[41]) on another QTL, this time *tga1*, disclosed that selection has altered nucleotide diversity, reducing it considerably at exon 1 but strangely not at exons 2 and 3 of this gene. Vigouroux, McMullen, and colleagues (2002[42]) worked with microsatellites (SSRs [single sequence repeats]) rather than nucleotides on 501 genes of a series of maize inbred lines. Their evidence shows that 15 SSRs show signs of selection in maize, and that 10 SSRs are candidates for being under selection during maize domestication and improvement. It is discussed that, at the *tb1* locus, domestication produced a reduction of diversity in the promoter region but much less so in the coding region of the gene.

A caveat to this hypothesis, in our opinion, is that the inferences resulting from the SNP analysis that have prompted the justification of the direct descent of maize from Balsas teosinte could just as well be explained as follows: a wild maize exhibiting maize-like plant and ear characteristics, distinct from but closer to teosinte than present-day maize, was recognized as a potential food source and underwent selection pressure by human plant gatherers not only during the initial domestication process but also, very importantly, during almost 9,000 years of crop improvement by farmers and diversification into races. The initial wild maize populations – which were allopatric to teosinte populations and derived, as has been dated, some 60,000 years ago from a common ancestor – were subjected to domestication, and at a later date they interacted by reciprocal introgression and

[39] Wang, R. L., A. Stec, J. Hey, L. Lukens, and J. Doebley. 1999. The limits of selection during maize domestication. *Nature*, 398: 236–239.

[40] Doebley, J. F. 2004. The genetics of maize evolution. *Annual Review of Genetics*, 38: 37–59.

[41] Wang, H., T. Nussbaum-Wagler, B. Li, Q. Zhao, Y. Vigouroux, M. Faller, K. Bomblies, L. Lukens, and J. Doebley. 2005. The origin of the naked grains of maize. *Nature*, 436: 714–719.

[42] Vigouroux, Y., M. McMullen, C. T. Hittinger, K. Houchins, L. Schulz, S. Kresovich, Y. Matsuoka, and J. Doebley. 2002. Identifying genes of agronomic importance in maize by screening microsatellites for evidence of selection during domestication. *Proceedings of the National Academy of Sciences USA*, 99: 9650–9655.

exchanged genes with teosinte. A reduction of variation of SNPs could result, in the case of domestication according to either theory, from population bottlenecks and genetic drift in small initial populations of selected and domesticated maize, prior to its radiation from an original location or locations.

Unless new archaeological findings of very early maize appear and molecular and fingerprinting analyses are conducted on them, the question of maize domestication will remain open. Archaeological precision dating and new finds of both macro- and micro-remains of maize and teosinte must be brought in line with genetic, cytogenetic, and DNA micromolecular evidence. Opposing the possible existence of a wild maize population just because it is not found today does not take into consideration that some teosinte populations are on the verge of extinction with only few individuals left (Wilkes, 1967[43]), and the indication is that some bridging teosinte populations between various species, which may have existed in the past, are now extinct. A similar occurrence with wild maize could not be discounted.

The differences of morphology of teosinte and maize are of such enormous magnitude (Galinat, 1977), not seen in other cases of crop domestication, that the differences in morphology would require a long period of selection or a traumatic change, such as was advocated by Iltis (1983), that is, a single event for the origin of maize from a teosinte original parent. Because the early maize cobs from the Tehuacán caves do not exhibit a distichous condition or induration or other ear characteristics typical of teosinte introgression, the proponents of the teosinte hypothesis have tried to present a number of possible genetic explanations to account for an accessory hypothetical reversal from teosinte-introgressed cob types to the early nonintrogressed cob types found in the Tehuacán caves of Mexico and in the earliest dated archaeological maize specimens that we have examined in the early dated archaeological sites of Peru (Cerro Guitarra, Cerro El Calvario, Los Gavilanes, Áspero, and Cueva del Guitarrero). One such proposal by Beadle (1974[44]) at a seminar, which this author attended, attempted to explain the lack of induration in early archaeological cobs as due to the mutation to a tunicate gene in a teosinte population, from which domestication may have begun and which, conveniently for this hypothesis, disappeared later. It is far simpler, on the other hand, to account for the preexisting tunicate gene and its expression in a moderate form in ancestral maize. It is still present in archaeological maize relics, notwithstanding the fact that a mutant tunicate form was found by Randolph (1976[45]) in teosinte in a chimeric section of a teosinte plant. Randolph, however, discounted teosinte as the originator of maize.

[43] Wilkes, E. G. 1967. *Teosinte: The Closest Relative of Maize.* The Bussey Institution of Harvard University. Cambridge.

[44] Beadle, G. W. 1974. Proceedings of a symposium on the origin of *Zea mays* and its relatives. Harvard University (unpublished). Copy in possession of the author.

[45] Randolph, L. F. 1976. Contributions of wild relatives of maize to the evolutionary history of domesticated maize: A synthesis of divergent hypotheses I. *Economic Botany*, 30: 321–345.

A segmented cob was found at Tehuacán that shows no teosinte introgression but a teosinte-like abscission layer that might have provided a seed dispersal mechanism. More important still, the long spikelet pedicels and rachillae of the primitive Proto-Confite Morocho popcorn race of Peru and the Bat Cave maize (graphic reconstruction provided in Mangelsdorf, 1974: figure 14.2), with a fragile abscission layer, could well have been the original wild maize seed dispersal mechanism.

THE ROLE OF THE PEDICEL IN THE SHATTERING OF SEEDS OF WILD MAIZE

The role of the pedicel is crucial as a system of attachment of the developing caryopsis of maize to the mother plant and the adaptations it has for transport of photosynthates to the endosperm of the seed and later for separation of the seed from the plant in a proposed wild maize ancestor. Doebley (1990a[46]) proposed that domestication of maize from teosinte had made an abscission layer, present in the teosinte rachis, disappear in the maize caryopsis. Dermastia and colleagues (2009[47]) have advanced our knowledge on the similarities and differences in cellular traits of developing caryopses of maize and of teosinte (*Zea mays* sp. *parviglumis*). These features, each with a possible role in development, include (1) an early programmed cell death in the maternal placenta-chalazal (P-C) layer; (2) accumulation of phenolics and flavonoids in the P-C layer that may be related to antimicrobial activity; (3) formation of wall ingrowths in the basal endosperm transfer layer (BETL); (4) localization of cell wall invertase in the BETL, which is attributed to the increased transport capacity of photosynthates to the sink; and (5) endo-reduplication in endosperm nuclei, which is suggested to contribute to increased gene expression and greater sink capacity of the developing seed.

Programmed cell death (PCD) in maize was characterized by the autolytic rupture of the vacuole and then an almost complete loss of cell content and was more apoptotic-like in the integumental part of the P-C layer. However, in teosinte, apoptotic PCD took place in both subdomains of the P-C layer.

As in maize, the development of the teosinte P-C layer was accompanied by differential deposition of different phenolic compounds in the remaining cell walls of the nucellar and integumental P-C layer. An outcome of maize domestication resulting in a nonshattering ear at maturity is the loss of the abscission layer (Doebley et al., 1990[48]) at the base of the pedicel, whose equivalent is

[46] Doebley, J. F. 1990. Molecular evidence and the evolution of maize. *Economic Botany*, 44: 6–27.

[47] Dermastia, Marina, Ales Kladnik, Jasna Dolenc Koce, and Prem S. Chourey. 2009. A cellular study of teosinte *Zea mays* subsp. *parviglumis* (Poaceae) caryopsis development showing several processes conserved in maize. *American Journal of Botany*, 96 (10): 1798–1807.

[48] Doebley, J. F., A. Stec, J. Wendel, and M. Edwards. 1990. Genetic and morphological analysis of a maize-teosinte F2 population: Implications for the origin of maize. *Proceedings of the National Academy of Sciences USA*, 87: 9888–9989.

found in teosinte as cells between adjacent cupulate fruitcases. Quite to the contrary, Kiesselbach (1949: figures 57 and 58[49]) describes a black layer in the integumental P-C layer of the maize caryopsis pedicel that he defines as a brown abscission layer and a well-developed pedicel. A prominent pedicel, and a long rachilla have been described and diagrammed based on dissected parts from an archaeological Bat Cave cob fragment (Mangelsdorf, 1974: figure 14.2). This condition of relatively long rachillae and pedicels for grain attachment is found in primitive as well as in Andean highland floury kernel maize races; the pedicel is much reduced in the Corn Belt Dent race, which has been the preferred maize material for studies on this subject. Kernels of the Cuzco race of maize are attached by rather long, fragile rachillae; they are easily shelled nowadays out of their cobs by humans trampling the corn ears with their feet. Similarly, developed transport layers appear in maize and teosinte and most likely in sorghum kernels, indicating conservation in the transport system to their respective caryopses in their primitive races. Phenolic compounds are deposited to strengthen the cell walls and to protect them against decay and pathogens in both maize and teosinte. The appearance of a separate black layer in maize according to Dermastia and colleagues (2009) is thus merely the result of differences in the phenolic compounds in the integumental layer from those in the nucellar P-C layer. They concluded that the essential developmental cellular processes in the caryopsis evolved before maize domestication and do not contribute to the striking changes in the structure of the maize caryopsis phenotype compared with that in Balsas teosinte. Notably, only the specific distribution of large cells with very large endopolyploid nuclei in the upper central endosperm of maize, which are not observed in teosinte, might contribute to more effective storage of starch in maize.

Dermastia and colleagues (2009) state that these cellular traits as present in the maize caryopsis have been previously attributed to domestication and selection for larger seed size and vigor. On the basis of these results, which show conservation of the entire cellular program present previously in maize, it is suggested that these features evolved independently of human selection pressure and domestication in the developing maize and teosinte caryopses. In other words, they have been present in maize-like *Zea mays* populations prior to domestication, thus giving credibility to the hypothesis of the existence of a wild maize plant that shattered its seeds via the presence of an abscission layer in the pedicels of the caryopsis, and that was independent of and predated human selection.

The evolution of morphological traits has been explained by mutation. However, the evolution of discrete new characters may require multiple gene changes, to bring the organism forward to a new threshold level, prior to achieving the fixation and permanence of a new character. Testing this hypothesis,

[49] Kiesselbach, T. A. 1949. *The Structure and Reproduction of Corn*. Research Bulletin N. 161. Agricultural Experiment Station, University of Nebraska. Lincoln.

Lauter and Doebley (2002[50]) identified cryptic genetic variation in teosinte for traits that are invariant to teosinte. These authors agreed that the preexistence of cryptic (not expressed) genes is necessary to achieve not only qualitative but also quantitative new traits. Such cryptic genes are overtly expressed in maize, a fact that indicates that maize could have carried them all along, having obtained them from a common ancestor, without requiring the hypothesis of a change from teosinte to maize. The opposite is equivalent to assuming that given the genetic similarity of chimpanzee and humans, direct descent from the first to the latter must have existed in these two species because they share the same cryptic genes.

Smith and Lester observed close association in serological, double diffusion, immuno-electrophoresis and albumin protein electrophoresis data (Stephen et al., 1980[51]) between Mexican and north Guatemalan teosinte with maize; they thus concluded that maize might have been domesticated from teosinte. They also underlined the lack of similarity of both maize and Mexican teosinte to southern Guatemalan teosinte, which did not exhibit any greater similarity to *Tripsacum* either in their data, which otherwise do not correspond to the great plant morphological similarity of southern Guatemalan teosinte (known also as Florida teosinte) with *Tripsacum*. They did not account for the long previous and pervasive introgression between maize and teosinte, as explanatory of their data association.

THE DESCENT OF MAIZE AND TEOSINTE FROM A COMMON ANCESTOR

Montgomery (1906[52]), Weatherwax (1935[53]), and Randolph (1976[54]) had expressed the opinion that maize and teosinte were closely related and that both derived from a common ancestor. In our view, present-day maize descends directly, in all likelihood, from an extinct wild maize ancestor phylogenetically linked to teosinte though a common ancestor and separated from it thousands of years before the presence of humans in the American continent. Very recent molecular evidence from gene nucleotide polymorphisms presented elsewhere in this appendix support this position. Its morphological characteristics can be projected backward in time from the archaeological record and from present genetic knowledge. The wild maize populations were, most likely, initially present in some isolated location(s) in Mesoamerica or Mexico, although it is not ruled out that they could have been present also in South America. At

50 Lauter, Nick, and John Doebley. 2002. Genetic variation for genotypically invariant traits detected in teosinte: Implications for the evaluation of novel forms. *Genetics*, 160: 333–342.

51 Stephen, J., C. Smith, and Richard N. Lester. 1980. Biochemical systematics and evolution of *Zea*, *Tripsacum* and related genera. *Economic Botany*, 34 (3): 201–218.

52 Montgomery, E. G. 1906. What is an ear of corn? *Popular Science Monthly*, 68: 55–62.

53 Weatherwax, P. 1935. The phylogeny of *Zea mays*. *American Midland Naturalist*, 16: 1–71.

54 Randolph, L. F. 1976. Contributions of wild relatives of maize to the evolutionary history of domesticated maize: A synthesis of divergent hypothesis I. *Economic Botany*, 30: 321–345.

some point in time, probably around 9,000 to 10,000 years ago, this wild maize started being utilized by humans, who gathered its seeds and consumed them as popped grain, starting a long domestication process from a maize-like parent population, which was obviously different from present-day maize but also different from teosinte. Some domesticated populations were transported either to or within Mexico and ended up growing in isolated pockets linked to human settled groups. Bretting and colleagues (1990[55]) have suggested that divergent combinations of isozymatic, karyotypic, and morphological features have evolved in local maize races from Mexico, Guatemala, and Bolivia, perhaps as the result of the different selective regimens that indigenous cultivators have imposed on different regional phylogenetic lineages. We do not support the claim to karyotypic diversity per se, in spite of clear evidence of the association of chromosome knob number with the racial distribution in altitude above sea level (see, for example, Grobman et al., 1961; McClintock 1978;[56] Wellhausen et al., 1957[57]). Rather, we support the opportunities of introgression or isolation from teosinte and further migration of the initial races.

It is difficult to conceive that a process of slow adaptation to a domesticate plant morphology could have occurred if the new selected plants had emerged from teosinte and were grown in the vicinity of the parent population; the latter would have swamped them out. It is striking that the morphology of the ear of maize has not changed much during the early agricultural period in Mexico, where supposedly artificial selection pressure would have been at maximum vigor if separation from teosinte ear architecture was the objective target. Such a slow evolution also took place during the Late Preceramic and Formative period in Mexico between 2500 BC and AD 150, which is the period when agriculture became the principal mode of subsistence in Mesoamerica (Benz and Long, 2000[58]; Flannery et al., 1981[59]; MacNeish, 1967[60]; McClung et al., 2001[61]).

[55] Bretting, P. K., M. M. Goodman, and C. W. Stuber. 1990. Isozymatic variation in Guatemalan races of maize. *American Journal of Botany*, 77 (2): 211–225.

[56] McClintock, Barbara. 1978. Significance of chromosome constitutions in tracing the origin and migration of raices of maize in the Americas. In D. B. Walden, editor. *Maize Breeding and Genetics*. John Wiley and Sons. New York. pp. 159–189.

[57] Wellhausen, E. J., A. Fuentes O., and A. Hernandez C., in collaboration with Paul C. Mangelsdorf. 1957. *Races of Maize in Central America*. National Academy of Sciences–National Research Council. Publication 511. Washington, D.C.

[58] Benz, B. F., and A. Long. 2000. Prehistoric maize evolution in the Tehuacán valley. *Current Anthropology*, 41 (3): 459–465.

[59] Flannery, K. V., J. Marcus, and Stephen A. Kowaleski. 1981. The Preceramic and Formative in the valley of Oaxaca. In J. A. Sabbloff, editor. *Supplement to the Handbook of Middle American Indians. Vol. I. Archaeology*. University of Texas Press. Austin. pp. 48–93.

[60] MacNeish, R. S. 1967. A summary of the subsistence. In D. S. Buyers, editor. *The Prehistory of the Tehuacán Valley*. Vol. 1. University of Texas Press. Austin. pp. 290–309.

[61] McClung de Tapia, Emily, Diana Martínez Yrizar, Guillermo Acosta, Francisca Zalaquet, and Eleonor A. Robitaille. 2001. Nuevos fechamientos para las plantas domesticadas en el México prehispánico. *Anales de Antropología*, 35, IIA, UNAM: 125–156.

The very early archaeological maize does not exhibit, in the case of Mexico, external signs of teosinte introgression. Such primitive ancestral maize in an early stage of domestication, at a remote time period that may be at least 7000 years BP, based on the findings of pollen and microfossils of maize in Panama (Piperno and Pearsall, 1998[62]), may have migrated to South America before becoming coetaneous with the teosinte colonies, having been previously isolated from them. Similarities among maize cobs of the Cerro El Calvario (6070 years BP – Cal. 2Σ 6667–7154 BP) and Los Gavilanes sites in Peru and Bat Cave maize cobs in New Mexico, both examined by the present author, are striking. It is clear, in our view, that the lack of teosinte introgression signals in the early maize of both Mexico and Peru is a strong signal in the direction of an independent evolution of both lineages of domesticated maize from wild maize in the absence of prior teosinte descent or introgression. Such introgression of teosinte, which is evident later in Mexican archaeological maize, and much later in Peru (Grobman et al., 1961), could only have happened when maize and teosinte were brought together in the same sympatric areas and when there was maize movement by commerce or occupation resulting from war. Therefore, maize evolved under new ecologic and agronomic conditions imposed by the first farmers, acting on preexisting maize, with teosinte hybridization and introgression following later. This would explain the lack of visible teosinte morphological signs in early archaeological maize in both Mexico and the Andean region, and it would suggest that molecular data could be reinterpreted, according to this view. Perceived similarities are the result of relatively later introgression both ways: maize into teosinte and teosinte into maize in Mexico. In the same way, reduction of variability of SNPs in maize could be the result of drift of early domesticated populations and a gradually increasing human selection pressure in the process of maize improvement.

TESTS OF THE VARIOUS HYPOTHESES

The major differences in morphology between maize and teosinte, with their respective interpretations, are as follows.

Tassel. The teosinte terminal tassel of the main stalk has no central spike in Guatemalan teosinte and few tertiary branches, whereas maize has a central spike with a higher condensation of spikelets. Some Mexican teosinte races (Chalco and Central Plateau) have acquired the maize tassel characteristic through gene flow from maize. Balsas teosinte or ssp. *parviglumis* has tertiary branches but a very small central spike.

Leaves. Maize leaves are wide, whereas Balsas teosinte has narrow leaves.

Stalks. Maize stalks are generally strong and tiller less, except in Mexican highland races. In hybrids between the Mexican race Chalqueño and high-altitude races of maize from Peru, the tiller-less condition of the Peruvian races is

[62] Piperno, D. R., and D. M. Pearsall. 1998. *The Origin of Agriculture in the Lowland Neotropics.* Academic Press. San Diego.

dominant over the tillering condition of the Mexican races introduced into them from teosinte. Balsas teosinte, in addition to having many tillers, has very slender stalks (Wilkes, 1967).

Ears. Teosinte properly has no ears. The lateral branch of teosinte is a deceptive structure (Weatherwax, 1955[63]). Externally it resembles an ear of corn, but when dissected it reveals a branched structure with ramifications that end in spikes, which are the homologues of ears, each enclosed in a spathe. The homologues of these multiple spikes are found in early maize at Los Gavilanes in Peru, where small, branched – but fully maize-type – ears have been found, and also in the race Quicheño of Guatemala; it appeared and had been selected in ancient times as fasciated ears in the ancient cultures of Peru, which attributed to these ears, as charms, the properties of fecundity. The differentiating traits between maize and teosinte ears are as follows: teosinte has single spikelets, two-ranked ears, and sessile spikelets, and maize has double spikelets, four-ranked phyllotaxy, and pedicellate pistillate spikelets, all of which are recessive in teosinte. Although Galinat (1971[64]) has dismissed a genetic complexity in determining these differences, assigning them to just one region in the short arm of chromosome 4, Mangelsdorf (1947[65]), Mangelsdorf and Reeves (1939[66]), Rogers (1950), and Galinat himself (1971) credited the whole short arm of chromosome 4 of maize as acting in a partial dominant form in controlling some but not all of the basic differences enumerated previously. Chromosome 4 is strongly prevented from exchanging genes by crossing over in some crosses of teosinte and maize by several factors that may be compounded: (1) linkage to gametophyte genes, promoting some sterility in the female parent; (2) incomplete chromosome 4 pairing, as observed by Wilkes (1967); and (3) an inversion found in the chromosome 4 of some teosinte, such as Nobogame. Chromosomes 3 and 7 and possibly others are involved in the expression of the single-spikelet trait, which has been found as a rare mutant in maize. Regarding sessile spikelets in maize, Galinat (1971) assures that he had extracted this trait from the primitive maize race Confite Morocho of Peru.

The earliest archaeological maize cobs found at Tehuacán are uniform and small – 2.0 to 2.5 cm in length – and have eight rows of kernels, and some have four rows, with slender cobs and seeds covered by long, soft glumes (Mangelsdorf et al., 1964, 1967[67]), characteristics that are not at all what would

[63] Weatherwax, Paul. 1955. Structure and development of reproductive organs. In G. F. Sprague, editor. *Corn and Corn Improvement.* Academic Press. New York. pp. 89–121.

[64] Galinat, W. C. 1971. The origin of maize. *Annual Review of Genetics*, 5: 447–478.

[65] Mangelsdorf, P. C. 1947. The origin and evolution of maize. *Advances in Genetics*, 1: 161–207.

[66] Mangelsdorf, P. C., and R. G. Reeves. 1939. *The Origin of Indian Corn and Its Relatives.* Texas Agricultural Experiment College Station. Bulletin 574.

[67] Mangelsdorf, P. C., R. S. MacNeish, and W. C. Galinat. 1967. Prehistoric wild and cultivated maize. In D. S. Byers, editor. *The Prehistory of the Tehuacán Valley.* Environment and Subsistence. University of Texas Press. Austin. pp. 178–200.

be expected had they been derived from teosinte or had they experienced teosinte introgression. Eubanks (2001: figures 13 and 14) has obtained by crossing *Zea diploperennis* teosinte × *Tripsacum dactyloides*, in an F_2 advanced generation, two-ranked, four-rowed ears that are very close replicas of the ears of the same types found in Oaxaca and Tehuacán, Mexico.

We have carefully examined the biomolecular and enzymatic information supplied by a number of researchers, which was interpreted as pointing to a direct descent of maize from teosinte. We see no problem of adjustment if the information we reinterpreted according to the hypothesis that is being put forward here: that maize interacted with teosinte only after it had been domesticated from a maize-like plant. Graphic reconstructions of an ear of maize prior to or during early domestication were presented by Grobman (1982), Mangelsdorf (1974), and Eubanks (2001).

The hypothesis of direct descent of maize from annual teosinte – which has gained many adherents basically through the demonstration of isozyme, genetic, and nucleotide polymorphism affinities and divergences of maize with teosinte, especially with the *parviglumis* subspecies (Matsuoka et al., 2002) – could also be reinterpreted as resulting from the hybridization and pervasive reciprocal introgression of the two species over a period of several thousand years, interspersed with a strong human selection concentrating and stabilizing genes of human use and agronomic interest in maize, while teosinte was left free to continue maintaining its wild sort of variability, increasing at the nucleotide level of genes, and also at the gene and chromosome levels (as can be seen in the stable inversion of teosinte chromosome 8). The maize that first interacted with teosinte when their ranges met had already been acted on by human domestication, starting from a type of plant that was quite different from present-day maize, in its capability to disperse seeds without human assistance.

There is evidence at the genetic level that the differentiation of teosinte and maize is not simply based on five genes. Each one of them is supported by a complex of modifier genes in adjacent chromosomal regions, and some may be duplicated in other chromosomal regions, according to Mangelsdorf (1974) and to Doebley and colleagues (1995[68]). Mangelsdorf (1947) has demonstrated, by using a multiple gene linkage tester, that a number of genes controlling fundamental morphological traits differentiating maize and teosinte are located in blocks found in chromosomes 1, 3, 4, and 9. He had no linkage tester at the time for chromosome 5. Rogers (1950), in later studies, found added differences in segments of chromosomes 1, 3, 4, 5, 6, 8, 9, and 10. There are genes controlling the two-ranked condition of either the ear or the central spike in chromosomes 1, 2, 6, 8, and 9 of Nobogame teosinte and in chromosomes 2, 3, 4, 8, and 9 of Durango teosinte, whereas genes controlling paired versus single

[68] Doebley, J., A. Stec, and C. Gustus, 1995. *Teosinte branched 1* and the origin of maize: Evidence for epistasis and the evolution of dominance. *Genetics*, 141: 335–346.

spikelets are located in chromosome 4 of Durango and also in Nobogame teosinte (Mangelsdorf, 1974). Lower glumes, important for determining teosinte introgression through induration, are controlled by numerous genes located in chromosomes 4, 6, 7, 8, 9, and 10. These are not simple Mendelian genes but are rather gene blocks, the most prominent of which is in chromosome 4. Collins and Kempton (1920[69]) and Mangelsdorf (1974) have pointed out that there are strong correlations of teosinte characters with one another, suggesting that several blocks of genes differentiate simultaneously and in coordination, maize from teosinte.

Doebley and Stec (1993[70]) evaluated the results of using molecular marker loci (MMLs) to map QTLs in an F_2 population derived from a cross of maize (*Zea mays* ssp. *mays*) and its suggested progenitor, teosinte (*Z. mays* ssp. *parviglumis*). A total of 50 significant associations (putative QTLs) between the MMLs and nine key traits that distinguish maize and teosinte were identified. When compared with a previous study of another subspecies of teosinte (*Z. mays* ssp. *mexicana*) for traits that measure the plant architectural differences between maize and teosinte, the two F_2 populations possessed similar suites of QTLs. For traits that measure components of yield, substantially different suites of QTLs were identified in the two populations. In a previously published analysis of the maize × ssp. *mexicana* teosinte population, these authors identified *five regions* of the genome that control most of the key morphological differences between maize and teosinte. These same *five regions* also control most of the differences in the maize × ssp. *parviglumis* teosinte population. The authors established in their conclusions that results from both populations support the hypothesis that a relatively small number of loci with large effects were involved in the early evolution of the key traits that distinguish maize and teosinte. They suggested that loci with large effects on morphology may not be a specific feature of crop evolution, but rather a common phenomenon in plant evolution whenever a species invades a new niche with reduced competition. This would be the case if wild maize had been brought to coinhabit with teosinte a certain ecological niche.

A more radical process of modification of the fruitcase of teosinte, as produced by a single gene *tga1*, located in the short arm of chromosome 4 near the centromere, has been advocated. Such an active region on chromosome 4 had already been determined by Mangelsdorf (1974) and Rogers (1950). According to Dorweiler and colleagues (1993[71]), this gene has dramatic effects on the glumes and would be responsible, singly, for the differences between maize and teosinte in this respect. At that time, these authors supported the model, first

[69] Collins, G. N., and J. H. Kempton. 1920. A teosinte-maize hybrid. *Journal of Agricultural Research*, 19: 1–38.

[70] Doebley, J., and A. Stec. 1993. Inheritance of the morphological differences between maize and teosinte: Comparison of results for two F2 populations. *Genetics*, 134: 559–570.

[71] Dorweiler, J., A. Stec, J. Kermicle, and J. Doebley. 1993. Teosinte glume architecture 1: A genetic locus controlling a key step in maize evolution. *Science*, 262: 233–235.

advocated by Beadle (1939), that changes through mutations in a few genes and selection of their modified phenotypes had been sufficient to transform teosinte into maize. It is doubtful that this gene *tga1,* obtained from race Reventador of Nayarit and used by Dorweiler and colleagues (1993), represents the original maize controlling region, because the Reventador maize race collections from Nayarit examined by Kato-Yamakake and McClintock (1981[72]) are teosintoid, as reflected by the many medium and large knobs in chromosomes 1, 2, 4, and 8 that indicate teosinte introgression into this race and possible modification of the ancestral maize *tg1* controlling region. Dorweiler and colleagues (1993), furthermore, had selected out, through RFLP elimination, the adjoining nucleotides to the *tga1* gene. It is good to remember that Mangelsdorf (1974), in his studies on different maize backgrounds, found a more complex inheritance of the lower glume of maize.

A reasonable conclusion, based on recent findings, according to Sang (2009[73]), is that in most cases a single gene played a pivotal role in a key domestication transition. QTLs of smaller effect or modifier genes played relatively minor but necessary roles in the course of the optimization of a domestication trait. This observation seems remarkable given that there are multiple regulators in a developmental pathway that could be potentially targeted by domestication selection. Equally intriguing is the question of the distance that has been conserved among the evolutionary lineages of the targets of domestication selection. The recent findings from fine mapping and cloning of domestication QTLs of cereals seem to indicate that such conservation has a more stringent phylogenetic constraint than previously thought. Additionally, recent evidence suggests that genes involved in the major domestication transitions are regulatory genes whose mutations can generate substantial phenotypic modifications that serve as suitable targets for strong artificial selection in crop evolution (Doebley and Lukens, 1998;[74] Doebley et al., 2006[75]). Taking this theory one step farther, one can imagine that knocking out or drastically altering the function of a regulatory gene without severely negative pleiotropic effect may not be an easy genetic modification to engineer. It is thus not surprising to find that repeated selection experiments performed independently by early farmers in independent domestication trails ended up at the same target genes.

72 Kato-Yamakake, T. A., in collaboration with B. McClintock. 1981. The chromosome constitution of races of maize in North and Middle America. Part 2. In B. McClintock, T. A. Kato-Yamakake, and A. Blumenschein, editors. *Chromosome Constitution of Races of Maize: Its Significance in the Interpretation and Relationship between Races and Varieties of the Americas.* Colegio de Postgraduados. Chapingo.

73 Sang, Tao. 2009. Genes and mutations underlying domestication transitions in grasses. *Plant Physiology*, 149: 63–70.

74 Doebley, J. F., and L. Lukens.1998. Transcriptional regulators and the evolution of plant form. *Plant Cell*, 10: 1075–1082.

75 Doebley, J. F., B. S. Gaut, and B. D. Smith. 2006. The molecular genetics of crop domestication. *Cell*, 127: 1309–1321.

Examples include the origins of white rice and six-rowed barley, in which where the same phenotypic modification in each case was accomplished through independent selection for the loss-of-function mutations of the same gene. Sang (2009) states that this demonstrates that there were a limited number of suitable targets in the developmental pathway for the artificial selection that aimed at developing the most desirable phenotype for cultivation in the cereal species. Furthermore, strong artificial selection coupled with introgression could drive the fixation of the most beneficial gene for a key domestication transition of a crop. Even for cultivars with different origins and partial reproductive isolation, gene flow could spread domestication genes across the entire gene pool of a crop and could provide opportunities for replacing less favorable genes with the most beneficial ones, especially when there was negative epistasis between them. This eventually led to the fixation of a gene of large phenotypic effect for a domestication trait, such as *sh4* for non-shattering rice and *nud* for naked barley. This mechanism, however, does not work between crops that are reproductively isolated.

This view opens up a number of interesting challenges in the analysis of the initial suitable targets in maize domestication that created the new phenotype of maize. However, the new maize phenotype does not exclude a large amount of new variability that transcends that which existed in teosinte itself in adaptation to a wide array of new habitats. Examples of extreme forms of adaptation fall within two categories: (1) the type of plants and their adaptation from the primitive Chococeño race of Colombia – which resembles a maize × teosinte (or *Tripsacum*) hybrid in plant type with many tillers and branches and which grows without cultivation in a high rainfall area – to the primitive race Enano – which grows an ear in the lower nodes of a tiny plant and which is partially covered with soil by the practice of hilling by farmers, allowing it to grow well and produce seeds in cold climate at an altitude of 3,800 masl on the shores of Lake Titicaca of Peru and Bolivia, which have scanty precipitation – and (2) in the large variation of ear and kernel sizes, shapes, textures, and number of rows of kernels.

The simplistic view that the differences between maize and teosinte are due to a few genes and that their mutation and selection has been responsible for a rather rapid domestication process by the empirical action of teosinte spike and caryopses gatherers is no longer tenable. It is not clear how an alleged domestication process of maize starting from teosinte could have derived into developing and maintaining so many morphologically critical differences if only as few as five genes – now increased to five regions, but involving many more QTL and modifier genes and recently found controlling regions and active preexisting transposable elements with controlling characteristics, existing before domestication – were orchestrated and intervened in the domestication process to produce a quantum change in a short period of time. The archaeological evidence does not lend witness to such a short process, as phytolith evidence (Piperno et al., 2009[76]) shows maize presence 8700 years BP in the presumed general area of maize domestication. The teosinte-to-maize hypothesis would require us to accept pushing back the date of initiation of the domestication of maize to a time long prior to 9000 years BP and closer to 10,000 years ago. We should take into consideration that the earliest macroarchaeological record in Mexico indicates the domestication of squash in the 9,000- to 10,000-year period but shows no record of domesticated maize or teosinte in that time period. For teosinte to have originated maize, domestication would have had to begin around 10,000 years ago, which is rather doubtful, considering our present knowledge. Had wild maize been the origin of domesticated maize, no such time constraint would apply, as no major changes in the genetic structure of the species would have been required.

The orthodox teosinte hypothesis had the ear of maize derived from an ear of teosinte through a series of mutations (Beadle, 1980;[77] Galinat, 1983[78]). This hypothesis has been criticized on the basis of the many changes that would have needed to occur by progressive mutations for the early farmers to have achieved a straight forward objective. What was such an objective, if it ever existed? The puzzle is still not solved; a vacuum of knowledge exists on how and where and when these changes occurred and what the motivations of the farmers were to develop a new model of plant if that model did not exist. The motivations of the farmers in the domestication process of maize are a big question mark (Wilkes, 1989). It is far easier to conceive that the plant model essentially preexisted before artificial domestication started and was improved on.

Considering this riddle, Iltis (1983[79]) conceived the catastrophic sexual transmutation theory in which, by a few mutations occurring over a short period of time and resulting in the feminization and condensation of its central spike, the tassel of teosinte would have turned into an ear of maize. Iltis (1988[80]), realizing that there were no incentives for farmers to utilize the hard fruitcase-covered seeds of teosinte as food, has suggested that early farmers would have used teosinte stalks with a high sugar content for chewing, basing his work on the finding of some "chews" of maize in archaeological deposits at Tehuacán. If this is so, why would early farmers have bothered to change the size and quality of teosinte seeds for direct consumption, when the plant already had a good production of seeds (more than 100 spikes per plant) for already-easy propagation?

76 Piperno, D. R., A. J. Ranere, I. Holst, R. Dickau, and J. Iriarte. 2009. Starch grain and phytolith evidence for early ninth millennium B.P. maize from the central Balsas River valley, Mexico. *Proceedings of the National Academy of Sciences USA*, 106: 5019–5024.

77 Beadle, G. W. 1980. The ancestry of corn. *Scientific American*, 242 (1): 96–103.

78 Galinat, W. C. 1983. The origin of maize as shown by key morphological traits of its ancestor, teosinte. *Maydica*, 28: 121–138.

79 Iltis, H. H. 1983. From teosinte to maize: The catastrophic sexual transmutation. *Science*, 222: 886–894.

80 Iltis, H. H. 1988. Maize evolution and agricultural origins. *International Symposium on Grass Systematics and Evolution*. Smithsonian Institution Press. Washington, D.C. pp. 195–220.

Seed weight differences between maize and ssp. *mexicana* teosinte have been studied by Doebley and colleagues (1994[81]); they found up to 10 QTLs in two different population crosses, accounting for up to 34% of the variation of kernel weight. There might be additional genes, thus making selection for seed size a complex and lengthy rather than an easy process.

A further complication of the teosinte-to-maize hypothesis lies in the recognized problem of the requirement of spatial isolation in the formative period of a maize subspecies. It has been recognized (Wilkes, 1989) that without such spatial or geographic isolation or genetic barriers, early populations in the process of domestication would have been swamped out if they had grown next to the original teosinte parental population. Stebbins (1950[82]), following Darlington (1940[83]), Stebbins (1942[84]), and Muller (1942[85]), has explicated the various types of reproductive isolation mechanism required for speciation as external and internal barriers, or, following Dobzhansky (1941[86]): (1) spatial isolation and (2) physiological isolation, which we may also apply to subspeciation. Internal barriers such as a temporal isolation at the same location, hybrid inviability, or hybrid sterility can be counted out, as maize and teosinte are cross-fertile, and any first- or advanced-generation hybrids between clines or subpopulations, without strong selection in the initial process of domestication, would indefectibly revert to the more adapted wild parent. It does not appear that the assumption that the many changes required for differentiation of maize and teosinte would have worked out in geographic isolation immediately after the first attempts at domestication, following a series of successive genetic and morphological changes. On the contrary, if a wild maize population or populations with essentially the same characteristics as later domesticated maize had existed, branching out recently from a common ancestor, in isolation from teosinte, no major problems would arise in explaining a gradual process of domestication within its own species (or subspecies). The introduction of domesticated maize at a later day in teosinte growing areas and subsequent gene exchange between both subspecies would go a long way toward explaining the inconsistencies in the archaeological records of Mexico and Peru with the teosinte to maize domestication hypothesis. It would also help to explain

81 Doebley, J., A. Bacigalupo, and A. Stec. 1994. Inheritance of kernel weight in two maize-teosinte hybrid populations: Implications for crop evolution. *Journal of Heredity*, 85: 191–195.

82 Stebbins, G. L., Jr. 1950. *Variation and Evolution in Plants.* Columbia University Press. New York.

83 Darlington, C. D. 1940. Taxonomic species and genetic systems. In J. Huxley, editor. *The New Systematics.* Clarendon Press. Oxford. pp. 137–160.

84 Stebbins, G. L., Jr. 1942. The role of isolation in the differentiation of plant species. *Biological Symposia*, 6: 217–233.

85 Muller, H. J. 1942. Isolating mechanisms, evolution and temperature. *Biological Symposia*, 6: 71–125.

86 Dobzhansky, Th. 1941. *Genetics and the Origin of Species.* Columbia University Press. New York.

the contrasting karyotypes: low chromosome knob number in early highland race Palomero Toluqueño in Mexico, which had a low chromosome knob number (2.1 knobs), in comparison to its contemporary Andean early maize races, such as Confite Morocho, Kculli, and derivatives of the archaeological Confite Chavinense maize races in Peru, which were early maize domesticate relics (0 to 1 or 2 knobs) prior to teosinte introgression into maize. Such introgression has increased the knob number and positions in Mexican maize races and their derivatives to karyotypes of anywhere between 8 to 10 knobs. Chromosome knobs are permanent features of chromosomes and are useful identifiers of maize evolution, as has been demonstrated by Brown (1949[87]) when he traced the Corn Belt Dent race to its hybrid origin from northern flints and southern dents from the United States, which diverged in chromosome knob numbers.

The lack of surviving populations of original maize, either wild or semidomesticated populations, is of no special concern to this hypothesis.

THE TRIPARTITE HYPOTHESIS

Annual teosinte was postulated in the tripartite hypothesis as originating in the hybridization of wild maize and *Tripsacum* (Mangelsdorf and Reeves, 1939). This hypothesis was later abandoned and substituted by a new hypothesis of the origin of annual teosinte from the hybridization of maize with a diploid perennial teosinte, *Zea diploperennis*, first expressed by Garrison Wilkes in a letter to Paul C. Mangelsdorf and then formally advocated by Wilkes (1979[88]). This latter hypothesis was confronted recently with new cytogenetic and biomolecular evidence that has rendered it as highly implausible.

The hypothesis of direct descent of maize as a domesticate from the ssp. *parviglumis* teosinte that occurred in the Balsas River valley, less than 10,000 years ago, has been upheld by new evidence obtained in the field of molecular genetics and interpreted in the frame of a preordained hypothesis. However, there are vacuums, contradictions, and a series of unexplained facts that lead to skepticism about this hypothesis. Although the inference from the molecular biology data presumes to demonstrate a direct descent of maize from teosinte, nevertheless, cytogenetic, genetic, and archaeological information are not in synchrony. If proven to be flawed or subject to serious criticisms, then the aforementioned number-one hypothesis remains as a serious alternative.

There are a number of considerations that require further analysis and contradict the direct-descent hypothesis.

First, the archaeological evidence in Mexico contains an incongruity in the fact that the Guilá Naquitz maize cobs that evidence teosinte introgression were AMS dated at an average of 5415 years BP (6250 calibrated calendar years BP)

[87] Brown, W. L. 1949. Numbers and distribution of chromosome knobs in U.S. maize. *Genetics*, 34: 524–536.

[88] Wilkes, R. G. 1979. Mexico and Central America as a centre for the origin of agriculture and the evolution of maize. *Crop Improvement (India)*, 6: 1–18.

(Benz, 2001[89]), whereas Tehuacán early maize dated 750 years later exhibits no signs of teosinte introgression, which appears in later periods. Second, these early cobs of maize from Mexico and southwestern United States exhibit polystichous spikes and paired pistillate spikelets, as well as longer pedicels and soft rachises and glumes. Introgression of teosinte is evident only in specimens dated much later. This is apparent in remains from San Marcos Cave in Puebla and Romero, La Perra Caves in Tamaulipas, and caves in the Tehuacán Valley; Tau, Slab, and Olla Caves in northeastern Mexico; Lister and Tonto Caves in Arizona; and Bat Cave and Cebollitas Cave in New Mexico (Mangelsdorf, 1974).

DISCUSSION OF INCOMPLETE OR MISSING EVIDENCE ON THE INTERPRETATION OF THE HYPOTHESES OF THE ORIGIN OF MAIZE AS A DOMESTICATE FROM TEOSINTE

There is no evidence of teosinte introgression in the abundant early archaeological cob specimens that we have examined in the Cerro Guitarra, Cerro El Calvario, Los Gavilanes, and Áspero preceramic coastal sites and also in the Guitarrero Cave in the highlands of Peru. The site of Cerro El Calvario is radiocarbon dated to approximately 6000 BP, which coincides with the maize in morphological characteristics in Mexico that do not exhibit teosinte introgression. Guilá Naquitz maize cobs exhibit teosinte introgression in a time period when there is none in Peru, with information that, based on phytolith evidence, maize had already moved to Panama 7400 years BP, and there was already maize in Ecuador around that same time period. Furthermore, direct evidence of the archaeological presence of teosinte appears at only two sites, and both are late (Romero Cave, Tamaulipas, and the cave of Guilá Naquitz, Oaxaca). Piperno and Flannery (2001[90]) have established AMS dates for domestication of *Cucurbita pepo* squash at Guilá Naquitz in a period of 6980 to 8990 C14 years BP. If teosinte were in existence at that time, it is strange that there are no maize macrofossils earlier than 5415 AMS years BP, considering that thousands of years before there was already a capacity for plant selection and domestication of the local dwellers. However, Piperno and colleagues (2009) have reported starch grain and phytolith data from the Xihuatoxtla shelter, located in the central Balsas Valley of Mexico, that indicate that maize was present by 8,700 calendar years ago (cal. BP). The firm evidence of maize phytoliths and starch grains would back the existence of actual maize rather than an early modified form of teosinte, which presumably would not have changed its cell characteristics at that time.

89 Benz, B. F. 2001. Archaeological evidence of teosinte domestication from Guilá Naquitz, Oaxaca. *Proceedings of the National Academy of Sciences USA*, 98 (4): 2104–2106.

90 Piperno, D. R., and K. V. Flannery. 2001. The earliest archaeological maize (*Zea mays* L.) from highland Mexico: New accelerator mass spectrometry dates and their implications. *Proceedings of the National Academy of Sciences USA*, 98: 2101–2103.

The differential presence of chromosome knob positions in Andean and in Mexican maize and teosinte is a key distinguishing feature. Andean maize has three patterns present in the early maize races: it is knobless, or has one small knob in either chromosomes 6L or 7L, or has knobs in both. Mexican and Guatemalan maize exhibit knobs at many of the 23 different knob positions known in teosinte. The Tepecintle race, which is presumed to have teosinte as one of its putative parents, has a range of 11 to 16 knob positions with an average of 13.3 in Guatemala (Wellhausen et al., 1957[91]).

Incomplete analysis suggests bottleneck effects to explain the absence of variation in genes *tb1* and *pbf* from early archaeological maize (Ocampo Cave, 2300–4300 years BP, and Tularosa Cave, 650–1900 BP) to present-day maize, whereas much more variation is found in teosinte. The *su1* gene is a different story, as there is a burst of variation appearing in the New Mexico maize archaeological remains at 650–1870 years BP and then a decrease in modern maize. The New Mexico maize carries a *su1* allele, present in teosinte but not in maize of the early period (Jaenicke-Deprés et al., 2003[92]). This would indicate, most likely, a late introgression of teosinte, rather than the explanation of the influence of New England maize. Jaenicke-Deprés and colleagues (2003) did not consider in their discussion that *su1* emerged in corn in the Chullpi race, descended from Confite Chavinense, a primitive race from the south-central Andean region of Peru that was in existence there at least 6,000 years ago, and from where it radiated to other geographical locations north and south (Mangelsdorf, 1974).

Alloplasmic lines obtained from *Zea mays* × *Z. mays* ssp. *mexicana* crosses, the latter as female, have a regular meiotic behavior of 10 bivalent chromosomes but with two different groups of 5 bivalents. This interaction of maize nucleus with teosinte cytoplasm conditions a special distribution of the chromosomes in the nucleus, promoting the separation of the two asynchronous ancestral chromosome groups of five each. If teosinte were the ancestor of maize, such behavior would not need to take place. However, if teosinte and maize represent different branches of a common ancestor, and if their respective nuclei and cytoplasms had been separated for several thousand years, they would have lost synchrony, and more recent introgression of teosinte into maize, or reciprocally, would explain the differential interaction.

Studer and colleagues (2011[93]) have informed us that one of the critical genes, *tb1*, which has been found to be associated with a DNA region that is in

91 Wellhausen, E. J., A. O. Fuentes, and A. C. Hernández X., in collaboration with P. C. Mangelsdorf. 1957. *Races of Maize in Central America.* National Academy of Sciences–National Research Council Publication 511. Washington, D.C.

92 Jaenicke-Deprés, V. J., E. S. Buckler, B. D. Smith, M. T. P. Gilbert, A. Cooper, J. Doebley, and S. Pääbo. 2003. Early allelic selection in maize as revealed by ancient DNA. *Science*, 302: 1206–1208.

93 Studer, Anthony, Qiong Zhao, Jeffrey Ross-Ibarra, and John Doebley. 2011. Identification of a functional transposon insertion in the maize domestication gene *tb1*. *Nature Genetics*, 43: 1160–1163.

a transposable element located near it, enhances the expression of this allele to the trait typical of maize and is not found in teosinte in the same level of expression. They consider that this association predates by many thousands of years the beginning of agriculture, thus early agriculturalists had to work on domesticating a plant that already exhibited the most typical maize traits. If this plant is referred to as a teosinte with maize characteristics, we might agree, but it would be simpler and easier to call it undomesticated or wild maize.

Gene flow from *Zea mays* ssp. *mexicana* has been proven to occur with a high frequency to maize. Because ssp. *mexicana* and ssp. *parviglumis* are much more closely related to each other than to maize, it is expected that the gene flow has had the potential to affect the similarity between maize and ssp. *parviglumis*, making present-day Mexican maize races phenotypically resemble ssp. *Mexicana*, even though, according to data produced and interpreted by Matsuoka and colleagues (2002), maize originated from the domestication of ssp. *parviglumis* of teosinte in the Balsas Valley lowlands some 9,000 years ago. Lowland Mexican maize races tend to be more similar to the putative parent *parviglumis* teosinte, whose early derived maize would have grown at low altitudes. The profound adaptive differences found in Mexico between lowland and highland maize today may not have existed at the time of maize domestication. Evidence presented by van Heerwaarden and colleagues (2010[94]) indicates that *Z. mays* ssp. *parviglumis* teosinte introgression into maize races in the Mexican lowlands is on the order of 1%, and *Z. mays* ssp. *mexicana* teosinte has an introgression pressure on the order of 20% in the highland Mexican races of maize. These researchers tried to bridge the gap of knowledge concerning the fact that the origin of the highland races of Mexican maize is unknown. These races are more closely related to ssp. *mexicana* than to ssp. *parviglumis*, being that the former is postulated as the putative ancestor of maize. To solve this point, they resorted to a reevaluation of SNP information by way of using estimates of differentiation from ancestral gene frequencies, inferred from extant maize populations as a measure of genetic distance from domesticated populations. By separating the lowland west Mexican group of races, they showed allele frequencies closer to ssp. *parviglumis* for this group if current frequencies of modern races can be taken as valid for past frequencies. These assumptions in the Mexican case may not be really valid because of the interference of the reciprocal introgression effects of teosinte and maize in Mexico, which has been a pervasive process over thousands of years. In any event, the study is a case of adjusting and partitioning the data to fit it into the prevalent theory. Because the evidence indicates that the radiation of maize into Mesoamerica and North America originated in the highland races of Mexico, it was important to preserve the initial theory of the origin of maize in the lowlands of the Balsas River basin.

[94] Van Heerwaarden, J., J. Doebley, B. Briggs, J. Glaubitz, M. M. Goodman, J. J. Sánchez-G., and J. Ross-Ibarra. 2010. A second look at the cradle of maize cultivation. *Proceedings of the National Academy of Sciences USA*, 108: 1088–1092.

We find a similar situation of two groups of early maize races in the highland and lowland (coastal) archaeological deposits in Peru, with a pervading anthocyanin pattern present in the archaeological material found on the coast of Peru, which was very well preserved because of the dry desert conditions. This indicator signals that the early coastal maize of Peru, around 6000 years BP, originated in the Peruvian highlands, rather than being transported from lowland maize in Mesoamerica. This presumed highland origin is in contrast to the evidence that lowland maize had existed in Tabasco, Mexico (7300 years BP), and in Panama (7400 years BP) and might have radiated to South America at that time. This apparently incongruous situation might be resolved if we assume that three early and primitive maize races were grown in the intermediate highlands of Peru between 1500 and 2500 masl. These were popcorn types with short, early plants adapted to a reduced rainfall pattern and had a wider range of adaptability than later, more evolved races. They would have been related to an early maize domesticate that arrived from the Mexican highlands and had yet not been introgressed with teosinte. In fact, in the three-dimensional figures presented by van Heerwaarden and colleagues (2010), when using allele frequencies from present races, the Andean group of races shows the widest distance apart of any group of races, in spatial coordinates indicating differentiation from teosinte.

Maize Domestication and the *tb1* Gene

Studer and colleagues (2011) and Studer (2011[95]) have advanced additional recent information on the regulatory region of the *teosinte branched* (*tb1*) gene. There is a genetic complex that includes transposable elements that control and affect the expression of the host gene. In a maize background, it reinforces the maize-type expression of apical dominance as compared to teosinte. Studer conducted a thorough study for his Ph.D. dissertation in Doebley's laboratory on the upstream region to gene *tb1*, which, in turn, had been identified as a QTL located in the long arms of chromosome 1 by Doebley and colleagues (1995[96]). Studer (2011) proposes that the difference between maize and teosinte plant architecture is controlled by a pair of transposable element (TE) insertions upstream of *tb1* (*Hopscotch*, a retrotransposon, and *Tourist*, a MITE [miniature inverted transposable element]), which are found in the maize haplotype but not in the teosinte haplotype. This conclusion was based on studying a large sample of maize races from Mexico, U.S. lines, and South and Central American races (including Andean races) and a large sample of teosinte accessions that present uniformly the same situation indicated previously. They established that a

[95] Studer, A. J. 2011. The genetic, molecular and evolutionary dissection of the teosinte branched gene. A dissertation submitted in partial fulfillment of the requirements for the degree of doctor of philosophy (genetics) at the University of Wisconsin. Madison.

[96] Doebley, J., A. Stec, and C. Gustus. 1995. *Teosinte branched 1* and the origin of maize: Evidence for epistasis and the evolution of dominance. *Genetics*, 141: 333–346.

complex regulation takes place, including cis-regulatory regions and four multiple-linked QTLs, two of them with epistatic interactions. A 12 kb control region was established previously by Clark and colleagues (2006[97]) to be located 58.7 to 69.5 kb upstream of the open reading frame (ORF) of *tb1*. These previous studies indicated that the change of expression between the maize and teosinte trait caused by *tb1* was based on the change of regulation rather than a change in the coding region. The expression of *tb1* to produce a maize phenotype is enhanced by the *Hopscotch* TE, acting on repressing axillary buds (Hubbard et al., 2002[98]). Furthermore, *tb1* has been found to encode a transcription factor that is a member of the TCP family of transcription regulators. This is a family of TCP plant transcription factors. TCP proteins were named after the first characterized members (TB1, CYC, and PCFs), and they are involved in multiple developmental control pathways. The TE insertions near *tb1* elicit a twofold increase in expression in maize relative to teosinte.

INTERPRETATION OF THESE FINDINGS

Through molecular dating, the TE *Hopscotch* was found to predate maize domestication by at least 10,000 years; in fact it was dated as 23,300 years, and the *Tourist* TE insertion is even older. These TEs could not have been transferred to maize from teosinte, because teosinte does not carry them now except in 5% of tested chromosomes, and they are probably acquired by gene flow from maize, so it is most likely teosinte did not carry them prior to or at the time of maize domestication. We rationalize that if teosinte had been the ancestor of maize, according to the orthodox teosinte theory, the transposable element insertions must have been present in teosinte before the domestication process began, to enhance the expression of the traits characteristic to maize in the domestication process. Because the finding of Studer (2011) points in the opposite direction, that maize had the regulatory transposon insertions long before domestication occurred, then it follows that the hypothesis that maize was domesticated from a wild maize ancestor rather than from teosinte becomes plausible. How would domestication from teosinte have occurred otherwise, if it happened on standing genetic variation, and if teosinte did not have this transposon insertion that regulates and enhances a maize phenotype? Where would such a cryptic control system come from if not from wild maize?

A selective sweep during domestication is postulated to explain the difference in variation between maize and teosinte in the *tb1* region. Such an explanation

[97] Clark, R. M., T. Nusbaum Wagler, P. Quijada, and J. Doebley. 2006. A distant upsteam enhancer at the maize domestication gene *tb1* has pleiotropic effects on plant and inflorescence architecture. *Nature Genetics*, 38: 594–597.

[98] Hubbard, L., P. J. McSteen, J. Doebley, and S. Hake. 2002. Expression patterns and mutant phenotype of *teosinte branched*, correlate with growth suppression in maize and teosinte. *Genetics*, 162: 1927–1935.

would not need to be resorted to if wild maize had been all along the source of origin of domesticated maize. The study reports, furthermore, that tests for past selection disclose that components of the region do not depart from neutral expectations except for one segment in the middle of the control region, which still requires explanation.

By producing *tb1* insertions in an isogenic maize background from nine different teosinte sources (Studer, 2011) it was found that morphological differences were encountered in the expression of *tb1*. This finding suggests that the *tb1* gene has suffered modifications over a long span of time in evolution and that it may also have been implied in the differentiation of various teosinte races.

Theories on the Descent of Maize and Its Relatives: II

Mangelsdorf and Reeves (1939) had resolved that the differences between maize and teosinte could be assigned to four blocks of genes, closely linked and located in different chromosomes. These differences were too large to have come about if domestication had taken place in a short period of time. Beadle (1939[99]) responded to Mangelsdorf and Reeves's tripartite hypothesis of 1939 with his affirmation that maize was directly derived through domestication from teosinte. Following a classical field study carried out in Mexico, Beadle (1972[100]) reported that, after classifying 50,000 plants in the segregating population of a cross between an eight-rowed maize and teosinte, he could recover one typical plant out of about 500 as corresponding to one of the parents. He thus claimed that the difference between both parents could be resolved through five major genes and a few additional modifier genes and that the difference was based on simple Mendelian inheritance. Doebley and Stec (1993[101]), through an analysis and mapping of QTLs of two F_2 segregating populations of maize × teosinte crosses, concluded that genes separating the two taxa were scattered throughout the genome. They also claimed that the differences were attributed mostly to the expression of six blocks of multiple linked genes, agreeing in this respect with Mangelsdorf and Reeves (1939). Plant architecture was associated with chromosome 1L; ear rank, with chromosome 2S; cupulate fruitcase, with chromosome 4S; and other large effects, with chromosomes 1S, 3L, and 5S. The inheritance of ear disarticulation was complex, and no less than nine genes were involved. Ten QTLs operate in differentiating singles against paired spikelets, with major effects in this respect localized in chromosomes 1L, 1S, and 3L. None of these genes segregated in a true Mendelian fashion (Doebley, 2004[102]).

[99] Beadle, G. W. 1939. Teosinte and the origin of maize. *Journal of Heredity*, 30: 245–247.
[100] Beadle, G. W. 1972. The mystery of maize. *Field Museum of Natural History Bulletin*, 43: 2–11.
[101] Doebley, J. F., and A. Stec. 1993. Inheritance of the morphological differences between maize and teosinte: Comparison of resuts for two F2 populations. *Genetics*, 134: 559–570.
[102] Doebley, J. 2004. The genetics of maize evolution. *Annual Review of Genetics*, 38: 37–59.

These findings dismiss the simple theory of Beadle that five gene mutations had been sufficient to explain the transformation of teosinte into maize. Even so, Beadle had admitted that perhaps thousands of additional genes would have been needed to shape modern maze in its present form.

Doebley and his group of researchers (Doebley, 2004) transferred a QTL located near the centromere in chromosome 4 reciprocally from maize to teosinte and from teosinte to maize and found that this QTL opened up the closed fruitcase of teosinte and allowed the kernel to be extracted freely; they saw this as a possible step toward domestication.

These experiments were conducted using maize alleles from Mexican maize, which may have been modified in their controlling regions by teosinte introgression during thousands of years of reciprocal gene exchange. Therefore, they may produce biased results and may not simulate the genome of their precursor. Similar experiments ought to be conducted with some primitive Andean maize races that are unmodified by teosinte and checked through chromosome knob absence or other means, to select a genome that would be much closer to the primitive ancestry of maize.

A group of researchers employing the techniques of analysis of gene polymorphisms of microsatellites have affirmed that maize originated by domestication some 9,000 to 10,000 years ago from the teosinte race Balsas or *Z. mays* ssp. *parviglumis* at a single location in the Balsas River basin (Doebley et al., 1987; Matsuoka et al., 2002). Matsuoka and colleagues (2002) have also interpreted their data to indicate that the oldest surviving maize types are those of the Mexican highlands, with maize spreading from this region over the Americas along two major paths. This conclusion is subject to serious doubts, because there has undoubtedly been a gene flow from teosinte to highland maize in Mexico, as evidenced by its high chromosome knob number, a clear indication of teosinte introgression, which the authors of this publication are at a loss to explain. They also postulate that their phylogenetic work is consistent with a model based on the archaeological record, suggesting that maize diversified in the highlands of Mexico before spreading to the lowlands. To arrive at consistent conclusions, we need to integrate not only the Mexican archaeological evidence but also that of Panama and South America with its new more remote dating in a single, large picture to gain a better understanding of the meaning of these correlations.

Maize arrived in South America at a very early date, as judged by micro- and macrofossil evidence. Additional cytogenetic and molecular studies of present primitive landraces that appear to be direct descendants of archaeologically determined races confirm a very interesting and diverse pattern from Mesoamerican maize. The pattern of the spread of maize was clarified by McClintock and colleagues (1981[103]), who, on the basis of an ample survey of chromosome knobs

[103] McClintock, Barbara, T. A. Kato-Yamakake, and A. Blumenschein. 1981. *Chromosome Constitution of the Races of Maize: Its Significance in the Interpretation of Relationship between Races and Varieties in the Americas.* Colegio de Postgraduados. Chapingo.

as tracers, concluded that maize was initially introduced from Mesoamerica to the central Andean region and that from there it spread through the highlands and to the lowlands. It was not until much later that other maize was reintroduced into the area on the West Coast (Grobman et al., 1961: see figure 18) some 1,000 years ago, and on the east coast of South America, in fairly recent times. Contrasting this view, Matsuoka and colleagues (2002) basing their analysis on microsatellite variation, have advocated the view that maize was first introduced into the lowlands of South America, at an early date, reaching the Andes highlands in a later stage.

Recent – soon-to-be-published – evidence supports a highland-to-lowland movement of early maize within Peru. Additionally, Lia, Confalonieri, Ratto, and colleagues (2007[104]) have studied cobs and kernels from high elevations at Catamarca, Argentina, dating from 400 to 1320 AMS years BP; three microsatellite loci were examined: *phi127*, *phi029*, and *phi059* (the nomenclature is from Matsuoka and colleagues, 2002, and they are on linkage groups 2, 3, and 10, respectively). As expected, extracted DNA was of low molecular weight, but some 9 amplicons out of 52 archaeological samples were obtained. At the *phi127* locus, all the nine archaeological specimens that gave results were homozygous for allelic variant 112. This allele was also present in modern populations, along with four others ranging in size from 114 to 126 bp. Analysis of 37 clones from archaeological sequences revealed sequence variations at a total of 23 nucleotide positions. The two ancient specimens typed at locus *phi029* were homozygous for allelic variant 154, which was also present in all the modern populations examined. A total of eight alleles were detected in contemporary specimens, with allele 154 being found at high frequency in several modern check populations from Argentina's landraces, which were obtained from Catamarca, Jujuy, Salta, and Misiones at medium to high elevations.

Determination was made of the genetic affiliations between modern and archaeological specimens. Despite being cultivated in the same region (northwestern Argentina), the eight landraces can be placed in three groups according to morphological and cytogenetic evidence (Poggio et al., 1998[105]): (1) the Andean complex (Altiplano populations 6473 and 6167; Amarillo Chico populations 6476 and 6484; Amarillo Grande population 6480, and Blanco population 6485), (2) South American popcorns (Pisingallo population 6313), and (3) incipient races derived from the introduction of commercial germplasm into local varieties approximately 40 years ago (Orgullo Cuarentón population 6482). Generally, the results of the assignment test indicate that the

[104] Lia, Veronica V., Viviana A. Confalonieri, Norma Ratto, Julián A. Cámara-Hernández, Ana M. Miante Alzogaray, Lidia Poggio, and Terence A. Brown. 2007. Microsatellite typing of ancient maize: Insights into the history of agriculture in southern South America. *Proceedings of the Royal Society. B. Biological Sciences*, 274 (1609): 545–554.

[105] Poggio, L., M. Rosato, A. M. Chiavarino, and C. A. Naranjo. 1998. Genome size and environmental correlations in maize (*Zea mays* ssp. *mays*, Poaceae). *Annals of Botany*, 82: 107–115.

archaeological specimens are more closely affiliated to the races of the Andean complex than to South American popcorns or incipient races, even though these are all currently cultivated in the same area, or to populations from the lowland regions of South America. All the three microsatellite loci examined in the archaeological specimens exhibited a single allelic variant, identical in size to the allele found most frequently in contemporary populations belonging to the races Amarillo Chico (6476 and 6484), Amarillo Grande (6480), Blanco (6485), and Altiplano (6167). This genetic homogeneity is remarkable when considering the diversity of the archaeological sites included in this study. These not only encompass a time period of nearly 1,000 years, but also cover different sociohistorical periods, each characterized by a distinctive pattern of agricultural production and interregional exchange. Furthermore, the specimens from Punta Colorada and Lorohuasi were found in association with funerary artifacts, whereas the Tebenquiche specimens were retrieved from households.

The conclusion of the assignment tests is that each of the nine archaeological specimens for which aDNA sequences were obtained belongs to the Andean complex and that this gene pool has therefore predominated in the western regions of southern South America for at least the last 1,300 years. Lia, Confalonieri, Ratto, and colleagues' (2007) interpretations of their genetic data from aDNA analysis are pertinent to the competing hypotheses regarding the spread of maize cultivation into and through South America. Taking into account the location of northwestern Argentina at the extreme southern range of maize distribution, it appears likely that the genetic ancestors of the Andean complex became established in northwestern Argentina, soon after the first arrival of maize cultivation to South America. The prevailing view that the Andean complex is a highland rather than lowland population therefore supports a highland origin for maize cultivation in this region.

The inferences of Lia, Confalonieri, Ratto, and colleagues (2007) indicate that the antiquity of the Andean complex is not compatible with the interpretation of Matsuoka and colleagues (2002) that maize cultivation reached the Andes at a presumably late stage, only after its initial introduction to the lowlands of South America. If the races from the lowlands of South America were ancestral to those of the Andean complex, then at least some indication of their presence might be expected at archaeological sites from northwestern Argentina, but no evidence for the presence of germplasm from sources other than the Andean complex was found within the samples that were analyzed.

One of the riddles of the theory of the Balsas River basin as the place of origin of domesticated maize is that the maize cultivars that are most closely related to Balsas teosinte are found mainly in the Mexican highlands, where ssp. *parviglumis* does not grow. Balsas teosinte has short plants and is morphologically the least maize-like teosinte. Genetic data appears to point to the primary diffusion of domesticated maize from the highlands rather than from the region of the purported place of initial domestication. Recent evidence, based on starch grains and phytoliths but not on macrofossils of maize itself, points to the early lowland presence of maize close to 9000 years BP in Mexico (Piperno et al., 2009[106]). The archaeological evidence of early lowland maize is based not on actual maize macro parts but on phytoliths and starch granules on stone artifacts. The latter is a rare occurrence, because, at that time, maize had small kernels that were most likely used by being popped on hot stones or sand in the Preceramic period, rather than, as was done much later, grinding seeds for use in tortillas or pozole (which requires large grains such as those from the race Cacahuacintle in Mexico). That evidence, after all, left the issue of the origin of highland maize in Mexico unresolved.

Van Heerwaarden and colleagues (2011[107]) claim that their new study presents a resolution of the former paradox. They have compared SNPs from sampling a single plant representing accessions of *parviglumis* and of *mexicana* teosinte, and of a single (presumably) representative plant from each of the 351 races described in the American continent. Their results would suggest that the west Mexican lowland maize is more similar to the inferred maize ancestor than is highland maize and that it is also more closely related to other extant populations based on gene frequency analysis. Introgression of ssp. *mexicana* teosinte to highland races is suggested as having caused the similarity, in contradiction to the earlier findings of Matsuoka and colleagues (2002), which dismissed introgression of teosinte as unimportant. They suggest that their data show that previous genetic evidence for an apparent highland origin of modern maize is best explained by gene flow from *mexicana* teosinte and demonstrate that admixture with a related nonancestral wild relative (*mexicana* teosinte) can interfere with analyses based on straightforward comparisons with the claimed known ancestor (*parviglumis* teosinte).

A sideline to the report of van Heerwaarden and colleagues (2011) is the wide separation observed in terms of posterior densities of the drift parameter F for 10 genetic groups studied that appears between the 2 groups of Andean maize and lowland Bolivian maize, which has essentially derived from highland Andean races; among all other eight racial groups of maize; and to an even greater extent between *mexicana* and *parviglumis* teosinte, indicating a prolonged divergence of the two South American groups of races from Mesoamerican and Mexican maize races and an even greater divergence from teosinte. (It is unfortunate that no descriptions are given of the races involved, especially of those labeled South American lowlands because recent imports from North America, especially from the Caribbean, have spread widely in fairly recent historical times in the lowland

[106] Piperno, D. R., A. J. Ranere, I. Holst, J. Iriarte, and R. Dickau. 2009. Starch grain and phytolith evidence for early ninth millennium B.P. maize from the central Balsas River valley, Mexico. *Proceedings of the National Academy of Sciences USA,* 106: 5019–5024.

[107] Van Heerwaarden, Joost, John Doebley, William H. Briggs, Jeffrey C. Laubitz, Major M. Goodman, José de Jesús Sánchez-González, and Jeffrey Ross-Ibarra. 2011. Genetic signals of origin, spread, and introgression in a large sample of maize landraces. *Proceedings of the National Academy of Sciences USA*, 108 (3): 1088–1092.

tropics.) They state that gene flow between maize and its wild relatives meaningfully impacts their inference of geographic origins.

One of the problems with the estimations of divergence of maize and teosinte through the use of microsatellite polymorphisms by Matsuoka and colleagues (2002), which they date at 9188 years BP, is that the phytolith archaeological evidence would present maize as being already in existence at 8500 years BP. Such a short span of time from purported domestication is questionable, because of the many changes that teosinte, if it was the parent of maize, would have had to undergo.

Evidence of past selection of the *tga1* allele, responsible for glume architecture in teosinte was presented by Wang and colleagues (2005[108]). They report that this allele, which allows for exposure of the grain of maize, previously concealed and covered under a fruitcase in teosinte, experienced strong selection, resulting in a reduction of variability. They accounted for six single-base pair polymorphisms that lie close to the coding sequence and that might affect the *tga1* gene, but a selective sweep appears not to have extended over the whole gene. The 6 kb region of gene *tga1* appears to be homologous to SPB (squamosa-promoter-binding protein), which is a transcriptional regulator, this was determined by identifying matching ESTs (expressed-sequence-tags) from other cereals. A *tga1* variant allele obtained by mutagenesis appears to differ from the SPB gene (supposedly the dominant *tga1* allele in maize) through a substitution in the former of a phenylalanine for a leucine amino acid in position 5. The differences in expression of the *tga1* alleles are suggested to be due to the different proteins resulting from this gene, which accounts for the morphological differences in the ear and to a lesser extent in the husk. The assumption is that a radical change of morphology in glume architecture – not only a change of the size of glumes and orientation but also lignification and accumulation of silica in the epidermal cells of the glumes and internodes of teosinte, against the small and soft glumes of maize – is conditioned by a single gene *Tga1,* which in its dominant form, present in maize, has not been found in teosinte. This situation is different from the *tb1* allele, responsible for plant architecture differences, which has been found in both maize and teosinte.

ALTERNATIVE TRIPARTITE HYPOTHESIS

Eubanks (1995,[109] 1997,[110] and 2001[111]) presented a new tripartite hypothesis in which *Tripsacum dactyloides* × *Zea diploperennis* could have originated maize

[108] Wang, H., T. Nussbaum-Wagler, B. Li, Q. Zhao, Y. Vigouroux, M. Faller, K. Bomblies, L. Lukens, and J. Doebley. 2005. The origin of the naked grains of maize. *Nature*, 436: 714–719.

[109] Eubanks, Mary. 1995. A cross between two maize relatives *Tripsacum dactyloides* and *Zea diploperennis. Economic Botany*, 49: 172–182.

[110] Eubanks, Mary. 1997. Molecular analysis of crosses between *Tripsacum dactyloides* and *Zea diploperennis* (Poaceae). *Theoretical and Applied Genetics*, 94: 707–712.

[111] Eubanks, Mary. 2001. The mysterious origin of maize. *Economic Botany*, 55: 492–514.

from a segregating generation of their hybridization. Eubanks was able to produce fertile segregates; one of them she named Sundance when *Z. diploperennis* was the female parent and Tripsacorn when *Tripsacum* was the female parent, both with $2n = 20$ chromosomes. She was able to transfer resistance to root insects from *Tripsacum* to maize through the genetic bridge provided by *Z. diploperennis.* Buckler and Stevens (2006[112]) have criticized her work on the basis of the improbability that a $2n = 20$ hybrid could have resulted from the cross, suggesting that the two grasses were not successfully hybridized.

MULTIPLE DOMESTICATION

Multiple domestication based on at least two races of maize or of teosinte was first suggested by Randolph (1959[113]) on the basis of cytological, morphological, and physiological characteristics. McClintock (1960[114]) was also of the same opinion, indicating that cultivated maize may have had several independent origins based on the evidence of knob-forming regions in the chromosomes of maize races and teosinte. Mangelsdorf and Sanoja (1965[115]) and Mangelsdorf and Galinat (1964[116]) concluded that there had been at least two races of wild maize in Mexico. Mangelsdorf (1974) expanded the number to six, when comparing the races of primitive popcorn that could have been independently domesticated from their respective wild maize progenitors.

Interracial hybridization has contributed in more recent periods to expand the variability of corn. Grobman and colleagues (1961) have argued that the complete genealogy of the Corn Belt Dent race, for example, when traced back, may have involved up to 12 races, not simply the northern flints and southern dents proposed by Anderson and Brown (1950[117]).

Origin and Preservation of Maize Genes

Imperfectly concatenated introns and exons arising from genes present throughout the genome have formed "pseudogenes" through the action of transposable elements such as *helitrons*, which create and move them around (Brunner,

[112] Buckler, E. S., and N. M. Stevens. 2006. Maize origins, domestication and selection. In T. Motley, N. Zerega, and H. Cross, editors. *Darwin's Harvest: New Approaches to the Origins, Evolution, and Conservation of Crops.* Columbia University Press. New York.

[113] Randolph, L. F. 1959. The origin of Indian maize. *Journal of Genetics and Plant Breeding*, 19: 1–12.

[114] McClintock, Barbara. 1960. Chromosome constitution of Mexican and Guatemalan races of maize. Annual Report of the Department of Genetics. 59: 461–472.

[115] Mangelsdorf, P. C., and M. Sanoja O. 1965. Early archaeological maize from Venezuela. *Botanical Museum Leaflets.* Harvard University, 21: 105–112.

[116] Mangelsdorf, P. C., and W. C. Galinat. 1964. The tunicate locus in maize dissected and reconstituted. *Proceedings of the National Academy of Sciences USA*, 51: 147–150.

[117] Anderson, E., and W. L. Brown. 1950. The history of common maize varieties in the United States Corn Belt. *Journal of the New York Botanical Garden*, 51: 242–267.

Pea, and Rafalski, 2005[118]). Over time, through accidental concatenation and expression of exons from different genes, effects on plant phenotype may have arisen in the course of evolution, and then been selected (Brunner, Fengier, et al., 2005[119]).

Chromosome 2 exhibits a reduced genetic diversity over one-third of its length, as demonstrated (Anderson et al., 2004;[120] Fengler et al., 2007[121]) using the pi factor of Tajima (1983[122]), which has been used to measure average differences and their variability of nucleotide differences within a population or between populations. High diversity regions on chromosome 2 appear to flank the region of low diversity (near the centromere), as has been reported by Rafalsky and Ananiev (2009[123]). Stabilization of diversity in some chromosomal regions has to be considered against a generalization that domestication carries a reduction of variation, in spite of the fact that maize appears to have some 75% of the variation included in teosinte. Nevertheless, we must take into consideration that probably only 5–6% of the maize genome is made up of protein-coding genes.

Insertions and deletions or transpositions of complete repeat clusters may be a mechanism whereby disruption of entire gene linear order can occur, creating large-scale polymorphisms differentiating individual plants or inbreds in a given population and adding to variability (Rafalsky and Ananiev, 2009).

Allelic diversity is important in maize breeding if the objective is to build populations that will interact heterotically in the production of hybrids. Heterosis in maize appears to be the result of accumulation of loci with alleles showing partial to complete dominance with additive effects (Gardner and Lonnquist, 1959;[124] Gardner et al., 1953;[125] Robinson et al., 1958[126]). Hallauer and

[118] Brunner, S., G. Pea, and A. Rafalski. 2005. Origin, genetic organization and transcription of a family of non-autonomous helitron elements in maize. *The Plant Journal*, 43: 799–810.

[119] Brunner, S., K. Fengler, M. Morgante, S. Tingey, and A. Rafalski. 2005. Evolution of DNA sequence nonhomologies among maize inbreds. *Plant Cell*, 17: 343–360.

[120] Anderson, L. K., N. Salameh, H. W. Bass, L. C. Harper, W. Z. Conde, G. Weber, and S. M. Stack. 2004. Integrating genetic linkage maps with pachytene chromosome structure in maize. *Genetics*, 166: 1923–1933.

[121] Fengler, K., S. M. Allen, B. Li, and A. Rafalski. 2007. Distribution of genes, recombination, and repetitive elements in the maize genome. *The Plant Genome: A Supplement to Crop Science*, 46: S83–S95.

[122] Tajima, F. 1983. Evolutionary relationship of DNA sequences in finite populations. *Genetics*, 106 (2): 437–460.

[123] Rafalsky, A., and E. Ananiev. 2009. Genetic diversity, linkage disequilibrium and association mapping. In S. L. Bennetzen and S. Hake, editors. *Handbook of Maize: Genetics and Genomics*. Springer Science and Business Media. New York. pp. 201–220.

[124] Gardner, C. O., and J. H. Lonnquist. 1959. Linkage and the degree of dominance of genes controlling quantitative characters in maize. *Agronomy Journal*, 51: 524–528.

[125] Gardner, C. O., P. H. Harvey, R. E. Comstock, and H. F. Robinson. 1953. Dominance of genes controlling quantitative characters in maize. *Agronomy Journal*, 45: 186–191.

[126] Robinson, H. F., C. C. Cockerham, and R. H. Moll. 1958. Studies on the estimation of dominance variance and effects of linkage bias. In O. Kempthorne, editor. *Biometical Genetics*. Pergamon Press. New York. pp. 171–177.

colleagues (1988[127]) have suggested surveying the available maize populations for adequate genetic variability.

Past evolution of maize under grower's selection for several millennia did not greatly advance the productivity in terms of number and weight of grains per plant. It was only after populations of different origins with different gene frequencies that were capable of additive genetic effects met and came into contact that explosive yield increases through hybridization and further selection took place (Grobman et al., 1961). Recurrent selection under hybridization is three or more times more effective than pedigree selection, such as selecting for single ears and following through their progeny (Sprague, 1952[128]); this explains why size of ears and their yield were stagnant for such a long time in the archaeological record.

The inclusion of exotic germplasm from the Andean region into the germplasm from the Mexican region and North American regions, and reciprocally within type of use phenotypes, may be very useful in improving heterosis and yields, after adaptation through backcrossing and selection is effected. Such efforts have been marked by relative success in the past. The availability of new molecular techniques and the understanding of evolutionary phylogeny should be helpful in bringing about a much more refined and closer monitoring of relevant gene-rich areas to be exploited and followed through by associated effects.

Allelic Diversity in Maize Gene Sequences

Sequencing of genic regions (including coding regions, introns, untranslated regions, and single-copy DNA surrounding genes) from multiple strains or varieties of maize has documented the existence of considerable allelic variation in the species. This variation had been already known at the macro gene level, but now additional variation is becoming amply known at the molecular level inside the genes.

Springer and Stupar (2007[129]) have documented how this information is becoming useful to select complementing maize inbred lines for maximizing heterosis. An example is given with inbred lines Mo 17 and B73, whose hybrid exhibits 64.7% heterosis for yield of grain, 46% heterosis for seed number, and 101% heterosis for plant height. An investigation by Vroh Bi and colleagues (2005[130]) of randomly selected sequences in the maize inbred B73 relative to

[127] Hallauer, A. B., W. B. Russell, and K. R. Lamkey. 1988. Corn breeding. In G. F. Sprague and J. W. Dudley, editors. *Corn and Corn Improvement*. 3rd ed. Agronomy Series No. 18. American Society of Agronomy. Madison. pp. 463–564.

[128] Sprague, G. F. 1952. Additional studies of the relative effectiveness of two systems of selection for oil content in the corn kernel. *Agronomy Journal*, 44: 329–331.

[129] Springer, N. M., and R. M. Stupar. 2007. Allelic variation and heterosis in maize: How do two halves make more than a whole? *Genome Research*, 17 (3): 264–275.

[130] Vroh Bi, I., M. D. McMullen, H. Sanchez-Villeda, S. Schroeder, J. Gardiner, M. Polacco, C. Soderlund, R. Wing, Z. Fang, and E. H. Coe Jr. 2005. Single nucleotide polymorphisms and

inbred Mo 17 found that, on average, indel polymorphisms are present every 309 bp and that SNPs occurred every 79 bp. The analysis of 300–500-bp amplicons found that 44% of the sequences contained at least one polymorphism in B73 relative to Mo 17. In general, it is estimated that there is one polymorphism every 100 bp in any two randomly chosen maize inbred lines (Tenaillon et al., 2001[131]). Ching and colleagues (2002[132]) examined the frequency and distribution of DNA polymorphisms at 18 maize genes in 36 maize elite U.S. inbreds, representing the genetic diversity in the breeding pool. The frequency of nucleotide changes is high: on average 1 polymorphism per 31 bp in non-coding regions and 1 polymorphism per 124 bp in coding regions. Insertions and deletions are frequent in noncoding regions (1 per 85 bp) but are rare in coding regions. A small number (2–8) of distinct and highly diverse haplotypes can be distinguished at all loci examined. Within genes, SNP loci comprising the haplotypes are in linkage disequilibrium with each other.

No decline of linkage disequilibrium within a few hundred base pairs was found in the elite maize germplasm. This finding, as well as the small number of haplotypes, relative to neutral expectation, is consistent with the effects of breeding-induced bottlenecks and selection on the elite germplasm pool. The genetic distance between haplotypes is large and indicative of an ancient gene pool and of possible interspecific hybridization events in maize ancestry.

Collectively, these studies indicate that maize has a relatively high level of sequence polymorphism compared to many other species. For example, the level of sequence diversity in genic sequences within maize is estimated to be higher than the level of diversity between humans and chimpanzees (Buckler, Gaut, and McMullen et al., 2006[133]). Recent research efforts have made tremendous strides toward characterizing this diversity: structural diversity appears to be largely mediated by *helitron* transposable elements. Patterns of diversity are yielding insights into the number and type of genes involved in maize domestication and improvement, and functional diversity experiments are leading to allele mining for future crop improvement.

There is growing evidence that structural variation in the form of copy number variation (CNV) and presence–absence variation (PAV) can lead to variation in the genome content of individuals within a species. Array comparative genomic hybridization (CGH) was used to compare gene content and

insertion-deletions for genetic markers and anchoring the maize fingerprint contig physical map. *Crop Science*, 46: 12–21.

131 Tenaillon, M. I., M. C. Sawkins, A. D. Long, R. L. Gaut, J. F. Doebley, and B. S. Gaut. 2001. Patterns of DNA sequence polymorphism along chromosome 1 of maize (*Zea mays* ssp. *mays* L.). *Proceedings of the National Academy of Sciences USA*, 98: 9161–9166.

132 Ching, A., K. S. Caldwell, M. Jung, M. Dolan, O. S. Smith, S.Tingey, M. Morgante, and A. Rafalski. 2002. SNP frequency, haplotype structure and linkage disequilibrium in elite maize inbred lines. *BMC Genetics*, 3: 1.

133 Buckler, E. S., B. S. Gaut, and M. D. McMullen. 2006. Molecular and functional diversity of maize: Current opinion plant. *Biology*, 9: 172–176.

copy number variation among 19 diverse maize inbreds and 14 genotypes of teosinte. Identification was made of 479 genes exhibiting higher copy number in some genotypes (UpCNV) and 3,410 genes that have either fewer copies or are missing in the genome of at least one genotype relative to maize inbred line B73 (DownCNV/PAV). Many of these DownCNV/PAV are examples of genes present in maize inbred line B73, but missing from other genotypes. More than 70% of the CNV/PAV examples are identified in multiple genotypes, and the majority of events are observed in both maize and teosinte, suggesting that these variants predate domestication and that there is not strong selection acting against them. Many of the genes affected by CNV/PAV are either maize specific (which the authors anchored in the teosinte-to-maize hypothesis of descent characterize as possible annotation artifacts) or members of large gene families, suggesting that the gene loss can be tolerated through buffering by redundant functions encoded elsewhere in the genome. Although this structural variation may not result in major qualitative variation due to genetic buffering, it has been considered that it may significantly contribute to quantitative variation (Swanson-Wagner et al., 2010[134]).

The significance of CNV and PAV on the results of studies made with small plant samples, even with one plant from an accession representing a maize landrace, lends low credibility to the inferences from such studies, which should have taken into consideration intrapopulation variation and its reduction by statistically sound designs before coming to final conclusions. Unfortunately, some experiments on molecular genetics have been conducted without such precautions.

The Early Phases of Maize Domestication

Considering this new information, it is interesting to note that during the first two to three millennia after domestication the variation in maize yield capacity did not evolve rapidly. Ear size was conservatively small for hundreds or thousands of generations and increased slowly, as is reflected in the archaeological record. As an example, the number of seeds in a Proto-Confite Morocho ear in the third millennia in Peru was about 96 (8 rows by 12 seeds per row), and it stayed so in that race throughout hundreds of years. Ear size increased gradually both in length and through the development of ear fasciation (widening and flattening of the cob), which allowed the fitting of more rows of kernels on the cob by increasing the number of vascular bundles reaching the ear and through other physiological mechanisms that allowed a greater sink for the deposition of more starch, oil, and protein in the aggregate of kernels of the ear. The early

[134] Swanson-Wagner, Ruth A., Steven R. Eichten, Sunita Kumari, Peter Tiffin, Joshua C. Stein, Doreen Ware, and Nathan M. Springer. 2010. Pervasive gene content variation and copy number variation in maize and its undomesticated progenitor. *Genome Research*, 20: 1689–1699.

development of this "fasciation syndrome" is an exclusive characteristic of maize evolution in the early period of the central Andean region. Ear fasciation appears very early with a precursor race that we have named Confite Chavinenese, which is morphologically distinct in ear shape and size from the cylindrical, eight-rowed Proto-Confite Morocho archaeological race, which spread widely in Peru and neighboring areas. There is little difference in ear size and shape between ears of the maize races Proto-Confite Morocho, Chapalote/Nal-Tel complex, and Pollo. The race Pira is also a primitive one and a later derivative from this group. At the same time, Confite Chavinenese evolved to form a number of races, with a globular, hand grenade shape, unlike the conical ears of the Mexican region.

During the early evolution of maize in the Andean region, the rate of growth of ear size was low, but at some time around 1000 BP a noticeable increase in ear length and row number appeared followed by an even steeper rate of expansion of ear size beginning around AD 400 to 800 (Grobman et al., 1961: figure 18).

The reasons for the improvement of yield of maize under artificial selection, we believe, are not just based on the accumulation of mutations leading to allelic variation and indel polymorphisms. The coevolution of a large number of small-effect genes or QTLs must have taken place, leading to the formation of a certain type of plant architecture, growth habit, and resilience in its changing habitats, which was driven more by a stabilizing selection required for survival in the initial domestication phase than by a disruptive forward selection driven by humans. These new alleles and polymorphisms would accumulate over time, driven also by the activity of transposable elements, and by gene duplication and specialization, as expressed in other parts of this appendix, and finally leading to the establishment of populations with certain gene pools. Insertion of new genes into this pool by hybridization with other maize populations and hybridization with teosinte and *Tripsacum* in some regions may have undoubtedly started differentiating populations in regard to frequencies of certain alleles. There are two forces that rule the physiology of populations, not of individuals, and that could lead to the production of differences between populations in gene frequencies, as presented by Stebbins (1950[135]): the chance random fixation of variation and the directive action of selection, either natural or artificial. Sewall Wright (in an oral communication quoted by Stebbins, 1950), has characterized evolution as the "statistical transformation of populations" (Stekkines, 1950: 104).

It is likely that growing conditions, aggression of insect plagues, diseases, weeds, soil moisture limitations, and unskilled farmers conspired against the full utilization of the early potential variability in maize. Selection of plants that are depauperate cannot be made at a fast rate, as a certain productivity threshold must be crossed. This is what may have happened in the early phases of maize

[135] Stebbins, G. Ledyard, Jr. 1950. *Variation and Evolution in Plants.* Columbia University Press. New York.

domestication. Concomitantly, the inherent variability was limited, because in the early evolution of maize under domestication there had not been an opportunity for an expansion of variability. It is not until later, with migrations from initial and from secondary diversification nuclear sites to others, that a range of climatic, soil, and farming conditions exerted differential selection pressures and allowed for the fixation and coevolution of a series of mutations and gene polymorphisms.

The transhumance of populations from highlands to lowlands and vice versa in Mexico, Mesoamerica, and the central Andes – and later migration movements, followed much later by military campaigns, as in the case of the Inca Empire – brought populations of early maize in contact, creating shifts in gene frequencies by hybridization leading to heterosis, when the effects (not, at that time, the causes) of these shifts were discovered, they led to the accelerated creation of new races of maize by empirical observation and selection.

Reduction of the Variability of Maize after Domestication

A number of researchers have observed a reduced variability in maize as compared to its putative wild ancestors, when working on polymorphisms at the subgenic level. A genetic bottleneck, similar to the one found in other cases of presumed direct evolution of a domesticated species from a wild ancestor, has been also postulated to occur in maize. However, the effects of an assumed short-period, domestication bottleneck in which a small group of individual maize plants participated in the beginning stages of domestication cannot be separated from a long selection process conducted over thousands of years by farmers and in the past 100 years by professional breeders, which may have resulted in present maize appearing less variable than its wild relatives at the genic and subgenic levels. For example, Buckner and colleagues (1990,[136] 1998[137]) and Palaisa and colleagues (2003[138]) have estimated that the strong selection for the *Y1* allele, which encodes a phytoene synthase and conditions yellow endosperm, which is a preferred trait in many maize grain markets because of the added color value of yellowish chicken meat and higher beta-carotene content of the grain, results also in a more than 10-time reduction in diversity. This reduction sweep was found extending several hundred kilobases from the *Y1* locus (Palaisa et al., 2004[139]). In a survey of 1,000

[136] Buckner, B., T. L. Kelson, and D. S. Robertson. 1990. Cloning of the y1 locus of maize, a gene involved in the biosynthesis of carotenoids. *The Plant Cell*, 2: 867–876.

[137] Buckner, B., P. San Miguel, D. Janick-Buckner, and J. L. Bennetzen. 1998. The y1 gene of maize codes for pytoene synthase. *Genetics*, 143: 479–488.

[138] Palaisa, K., M. Morgante, M. Williams, and A. Rafalski. 2003. Contrasting effects of selection on sequence diversity and linkage disequilibrium at two phytoene synthase loci. *The Plant Cell*, 15: 1795–1806.

[139] Palaisa, K., M. Morganta, S. Tingey, and A. Rafalski. 2004. Long-rage patterns of diversity and linkage disequilibrium surrounding the maize Y1 gene are indicative of an asymmetric selective sweep. *Proceedings of the National Academy of Sciences USA*, 101: 9885–9890.

genes, Yamasaki and colleagues (2005[140]) found a great reduction of diversity in eight genes that had been subjected to selection.

Wright and colleagues (2005[141]) have estimated that maize has retained 57% of the diversity of its presumed progenitor, teosinte, on the basis of a study of genic SNPs. An analysis of 21 genes of chromosome 1 made by Tenaillon and colleagues (2001) reduced it to 77% when referring to races of maize. Liu and colleagues (2007[142]) obtained a similar estimate on the basis of microsatellites. At the *tb1* locus, which was initially discovered as a mutant in maize and considered by some to be a basic link to maize ear architecture and domestication from an ancestor such as teosinte *parviglumis*, Wang and colleagues (1999[143]) found reduced diversity as compared to teosinte. Estimates by Vigouroux, and colleagues (2002,[144] 2005[145]) based on SSRs (single sequence repeats) have disclosed a much lower difference in variability with this type of estimation. These studies are in their beginnings and should be pursued over many more gene loci in the future.

Anthocyanin Synthesis and Its Relation to Maize Evolution

Understanding which genes contribute to evolutionary change and the nature of the alterations in them are fundamental challenges in evolution studies in maize.

Hanson and colleagues (1996[146]) analyzed regulatory and enzymatic genes in the maize anthocyanin pathway as related to the evolution of anthocyanin-pigmented kernels in maize. Genetic tests indicate that teosinte, which has colorless kernels, possesses functional color alleles at all enzymatic loci. At two

[140] Yamasaki, M., M. I. Tenaillon, I. V. Bi, S. G. Schroeder, H. Sanchez-Villeda, J. F. Doebley, B. S. Gaut, and M. D. McMullen. 2005. A large scale screen for artificial selection in maize identifies candidate agronomic loci for domestication and crop improvement. *Plant Cell*, 17: 2859–2872.

[141] Wright, S. I., I. V. Bi, S. G. Schroeder, M. Yamasaki, J. F. Doebley, M. D. McMullen, and B. S. Gaut. 2005. The effects of artificial selection on the maize genome. *Science*, 308: 1310–1314.

[142] Liu, R., C. Vitte, J. Ma, A. A. Mahama, T. Dhliwayo, M. Lee, and J. L. Bennetzen. 2007. A gene trek analysis of the maize genome. *Proceedings of the National Academy of Sciences USA*, 104: 11844–11849.

[143] Wang, R. l., A. Stec, J. Hey, I. Lukens, and J. Doebley. 1999. The limits of selection during maize domestication. *Nature*, 398: 236–239.

[144] Vigouroux, Y., J. S. Jaqueth, Y. Matsuoka, O. S. Smith, W. D. Beavis, J. S. Smith, and J. Doebley. 2002. Rate and pattern of mutation of microsatellite loci in maize. *Molecular Biology and Evolution*, 19: 1251–1260.

[145] Vigouroux, Y., S. Mitchell, Y. Matsuoka, M. Hamblin, S. Kresovich, J. S. Smith, J. Jaqueth, O. S. Smith, and J. Doebley. 2005. An analysis of genetic diversity across the main genome using microsatellites. *Genetics*, 169: 1617–1630.

[146] Hanson, M. A., B. S. Gaut, A. O. Stec, S. I. Fuerstenberg, M. M. Goodman, E. H. Coe, and J. F. Doebley. 1996. Evolution of anthocyanin biosynthesis in maize kernels: The role of regulatory and enzymatic loci. *Genetics*, 143: 1395–1407.

regulatory loci, most teosintes possess alleles that encode the promoter element necessary for the activation of the anthocyanin pathway during kernel development. On the other hand, their genetic tests indicate that teosinte *c1* alleles are not active during kernel development. The authors claim that their analyses suggest that the evolution of the purple kernel color resulted from changes in cis-regulatory elements at regulatory loci and not changes in either regulatory protein function or the enzymatic loci.

If the deductions of the previous authors are correct, we are at a loss to explain how, in the course of less than 1,000 years, a complex system of anthocyanin expression in blue or black aleurone color could have arisen and become uniformly expressed in all the ancient races of Peru and in some races of other countries, such as Güirua in Colombia and Negro de Chimaltenango and its subraces Negro de Tierra Fría and Negro de Tierra Caliente in Guatemala. Pericarp color is a manifestation of plant color in the kernel, and the deep purple, almost black color of the Proto-Kculli race found archaeologically and manifested in the highlands of Peru is accompanied by red pericarp color in the kernels of all primitive and subsequent evolutionary races in archaeological contexts from the very early period in Peru. It is hard to explain how the deep purple color of highland maize in Peru could have had such a fast development of all the control loci required for the expression of anthocyanin color in the pericarp and aleurone. Some 97.83% of all corn cobs found in the Los Gavilanes coastal site in Peru (Grobman, 1982: cuadro 13) exhibit purple color, which would have originated in maize that migrated from the Peruvian highlands, where it was previously cultivated.

Maize is a plant species that through evolution has developed multiple genetic polymorphisms, expressed as allelic series at different loci. At the molecular level these originate as single nucleotide polymorphisms (SNPs); nucleotide deletions or insertions (indels); tandem repeats; gene duplications, of which maize has a large number; nucleotide rearrangements; and other repeats, especially of transposable elements (transposons) present in many areas of the genome. Maize nucleotide polymorphisms occur at an astounding rate. One nucleotide in every 28 bp is polymorphic (a frequency of 3.5%). The overall nucleotide diversity in maize is 1.3%, whereas in teosinte it reaches 2%. It is strange to note that diversity in maize at the nucleotide level would be lower than in teosinte if it evolved from it and moved at least some 7,500 years ago from Mesoamerica, and occupied an extensive territory with many ecological niches, underwent selection for many characteristics in a territory much larger than that occupied by teosinte.

The prevalence of high anthocyanin coloring in the stems and husks and the pericarp purple color in maize races from the high-elevation Andean region could have developed through natural selection and in accommodation to environmental restrictions. The series of four different structural (A, B, Pl, and r) and two regulatory genes (C and I) for anthocyanin synthesis could have interacted with low mean temperatures and high UV radiation to enhance the synthesis of anthocyanins and thus to imprint a natural selection pressure for alleles that would confer a deeper color in the ancient high-altitude Andean races of maize. That such a regulatory effect of natural environment is indeed active in promoting anthocyanin production in the secondary metabolic pathway has been demonstrated by Yingqing and colleagues (2009[147]). Anthocyanin is a flavonoid that may confer protection to plants from UV radiation (Koes et al., 1993[148]).

Another advantage of deep purple color in high-altitude maize would be greater absorption of heat, which stimulates enhanced metabolic processes in cells in low-temperature or high-altitude climates. This observation and implication has been tested by applying thermocouple temperature sensors to plants with different depths of anthocyanin colors in a test population of maize with stalks exhibiting deep purple plant color in the high-elevation (3200 masl) Mantaro valley in Peru, where it was found that that deeper-purple-colored stalks had a higher temperature than green or lighter-colored ones (Greenblat, 1968[149]). The high intensity of purple color of maize plants in the high-elevation Central Plateau of Mexico corresponds to the same situation in the deep purple color of maize in the high-elevation valleys and slopes of the Andes range of mountains.

Whitt and colleagues (2002[150]) found that although maize exhibits a great amount of variability, several loci that have been subjected to strong artificial selection, such as *c1* and *tb1*, have low levels of genetic variation. Six nonselected genes of maize and *Zea mays* ssp. *parviglumis*, a teosinte relative of maize, were compared for silent diversity by sampling some 500–2,700 silent bases for each locus. The genes were *Adh1, Adh2, glb1, hm1, hm2,* and *tm1; tb1* is the only one that was defined as a domesticated gene. Maize exhibited less variation than teosinte. These findings, if confirmed with more detailed studies, would be contrary to the theory of direct descent of maize from teosinte. If maize were derived from teosinte, it would be strange that neutral genes exhibit less variation in the derived subspecies than in its progenitor, given that maize and teosinte diverged according to most hypotheses of domestication about 9,000 years ago, and then the area covered by maize expanded to all the American continent, where it was subjected to a great variety of environments. It would be expected that – in the course of the explosion of variability that ensued, with more than 350 races that

[147] Yingqing, Lu, Jin Du, Jingyu Tang, Fang Wang, Jie Zhang, Jinxia Huang, Weifeng Liang, and Liangsheng Wang. 2009. Environmental regulation of floral anthocyanin synthesis in Ipomoea purpurea. *Molecular Ecology*, 18 (18): 3857–3871.

[148] Koes, Ronald E., Francesca Quattrocchio, and Joseph N. M. Mol. 1993. The flavonoid biosynthetic pathway in plants: Function and evolution. *Bio Essays*, 16: 123–132.

[149] Greenblat, Irwin M. 1968. A possible selective advantage of plant color at high altitudes. *Maize Genetics Cooperation Newsletter*, 42: 144–145.

[150] Whitt, Sherry R., Larissa M. Wilson, Maud Tenaillon, Brandon S. Gaut, and Edward S. Buckler IV. 2002. Genetic diversity and selection in the maize starch pathway. *Proceedings of the National Academy of Sciences USA*, 99 (20): 12959–12962.

grew in locations that ranged from 3,800 meters of altitude to deserts, jungles, and intermountain valleys and slopes – would have acquired a greater variability than teosinte, which is primarily confined to a few locations in Mexico and Guatemala. The results of the study may be due to the fact that the maize that was compared is a single, mostly uniform corn race, Corn Belt Dent.

The Evolution of Inflorescence Development in Maize and Teosinte

Male inflorescences in both maize and teosinte are distichous (two ranked). Through condensation, some maize male inflorescences appear as polystichous (multi-ranked) in terms of rows of spikelets. Female inflorescences in the teosintes are always distichous, whereas in maize they are always polystichous, and this is a main characteristic that separates the ear phenotypes of teosinte and maize.

Studies made by Orr and Sundberg (1994[151]) revealed that femininity and masculinity in teosinte and maize were derived from a common developmental background. Again, these authors (Orr and Sundberg, 2004[152]), have been able to prove by means of scanning electron micrographs that tassel and ear primordia in maize are similar and that they both may be subject to the same mechanism of development in their early stages, and they proposed that this mechanism is general to the Andropogoneae. Sundberg and Orr (1996: 1264–1265[153]) proposed that "... the primary mechanism for a shift from distichy to polystichy during the evolution of maize may have involved the transition from distichous to spiral phyllotaxy, a process that occurs ontogenetically during the development of the very young inflorescence meristems." Sundberg and colleagues (2008[154]) identified successive waves of cambial development as associated with primordial development resulting from possible canalized auxin flow. They propose that transition from distichous to spiral phyllotaxy in the ear branch of maize occurs in the vegetative region of ear husk leaf production well before the transition to vegetative growth. Necessary and complex changes would have had to occur in vascular development and in patterns of procambial development in the ear shoots, which are genetically controlled, if a transition from teosinte to maize had taken place.

In a study of a recently discovered teosinte species, *Zea nicaraguensis* Iltis and Benz, which is an almost extinct population from the Nicaraguan lowlands of Chinandega, near the Gulf of Fonseca, Iltis and Benz found the same situation as in other *Zea* taxa. Spikelet primordia produce both sessile and

[151] Orr, A. R., and M. D. Sundberg. 1994. Inflorescence development in a perennial teosinte: *Zea perennis* (Poaceae). *American Journal of Botany*, 81: 598–608.

[152] Orr, A. R., and M. D. Sundberg. 2004. Inflorescence development in a new teosinte: *Zea nicaraguensis* (Poaceae). *American Journal of Botany*, 91: 165–173.

[153] Sundberg, M. D., and A. B. Orr. 1996. Early inflorescence and floral development in *Zea mays* race Chapalote (Poaceae). *American Journal of Botany*, 83: 1255–1265.

[154] Sundberg, M. D., A. R. Orr, and T. D. Pizzolato. 2008. Phyllotactic pattern is altered in the transition to flowering in the early ears of *Zea mays* landrace Chapalote. *American Journal of Botany*, 95 (8): 903–913.

pedicellate spikelets; although they remain so in the male inflorescence, in the female inflorescence the pedicellate spikelet growth is arrested, and only the sessile spikelet develops. In maize, both spikelets develop, and this makes for a basic difference in the formation of duplicate grains in maize, whereas only one grain remains in teosinte at each formative point. These authors report that they found a polystichous trait (multiple rows of spikelets) in some individuals of this species of teosinte. In the 80–90 *Z. nicaraguensis* male and female inflorescences that they examined, all of them exhibited a distichous (two-rank) condition, except for two male tassels that developed four ranks (polystichous) of spikelet pair primordia along the axis of the central spike. This situation was also found although rarely in male inflorescences of teosinte from Toluca, Mexico (Orr et al., 2002[155]).

The presence of a rare polystichous novel male inflorescence phenotype in wild populations of teosinte is interesting and should be further explored. Is this an ancient characteristic maintained by hidden genes in the Andropogoneae that could be exhibited by wild plants in the absence of human selection, as it occurred in *Z. nicaraguensis*? There was no maize present for miles in the area where *Z. nicaraguensis* was found. This case opens the possibility that a wild polystichous maize could also have existed. Populations of wild plants disappear – as this one, which has only 6,000 individuals, is about to do, if it is not preserved for its extremely interesting characteristics of growing under flooded conditions, which maize is incapable of supporting for long periods of time. It should be noted that *Z. nicaraguensis*, an annual species, has been found to be basic to the Luxurians group of teosintes, which are perennial.

The Directional Evolution of Microsatellite Size in Maize

Microsatellites, also known as single sequence repeats (SSRs) or short tandem repeats (STRs), are repeating sequences of 2–6 bp of DNA. They are used as molecular markers in genetic research.

Microsatellite loci have been subjected to analysis as a reference to evolutionary processes in plants. Their use in this type of application presupposes adjustment to a determined mutation model, which may not always be followed, introducing a certain bias in the results. One well-documented bias in microsatellite mutation is the tendency for new mutations to cause an increase in the size of the allele, leading to "directional evolution" in plants and animals (Vigouroux, Jaqueth, et al., 2002). Through mutational bias and differential mutation rate, there may be an increase in microsatellite size. Vigouroux et al. (2003[156]) reported both a

[155] Orr, A. R., Mullen D. Klaahsen, and M. D. Sundberg. 2002. Inflorescence development in a high altitude annual teosinte of Mexico. *American Journal of Botany*, 89: 1730–1740.

[156] Vigouroux, Y., Y. Matsuoka, and J. Doebley. 2003. Directional evolution for microsatellite size in maize. *Molecular Biology and Evolution*, 20: 1480–1483.

"directional increase" in microsatellite size in geographically derived maize racial groups and a negative correlation of allele size and altitude that arose independently in North and South America.

Their statement – "We have previously analyzed a data set of 193 pre-Columbian maize land races genotyped at 99 microsatellite loci (Matsuoka et al., 2002)" (Vigouroux et al., 2003: 1480) – contains an error. These are not pre-Columbian landraces but modern successors of primitive but presently evolved races that may have changed considerably from their previous ancestral makeup.

Comparisons were made of average allele size of the three groups NA (North American), ME (Mexican), and SA (South American) – and it was found that both NA and SA (geographically derived groups) were significantly larger than the ancestral group ME ($T = 5.65$, $P < 0.0001$, and $t = 3.74$, $P < 0.0003$, respectively). This would demonstrate directional evolution according to the authors. No significant differences were established between the NA and SA groups.

The authors state: "The SA group was derived from low altitude maize in Guatemala and the NA group was derived from maize of Northern Mexico (Matsuoka et al. 2002)" (Vigouroux et al., 2003: 1481). The first part of the statement is subject to debate, whereas the second is more likely to be correct. There is no evidence that points to a relationship of putative origin of the earlier Andean races of maize from either lower- or higher-altitude Guatemalan races. Quite to the contrary, several high-altitude races of maize in Guatemala are considered to be derived from South American races (Wellhausen et al., 1957).

Based on the tests, they found that (1) there is a difference in average size of alleles between groups and (2) there is a significant correlation between average individual size and altitude.

The mean difference in size of alleles between SA and NA with ME is 4.1 and 3.3 bp per locus, respectively. Therefore, their conclusion is that allele size has increased and has not remained in equilibrium from ancestral to the geographically derived populations. This conclusion is subject to skepticism. Why would the ME group, which is a reflection of present-day races, have stopped increasing in average allele size for the last 10,000 years, whereas the other two groups maintained a directional evolution away from the "ancestral" region? What keeps the Mexican races from increasing the size of their alleles? Is there a stabilizing selection force in operation? Is it due to the high penetration of teosinte DNA in the maize genome?

The authors try to explain this directional evolution in terms of a mutation bias in the new environments that happened twice independently: the transfer of maize to a new environment changed the size of alleles in North and South America. The demographic explanation made by Vigouroux and colleagues (2003) resorts to hypothetical conditions that would require a more detailed demonstration. For their inferences to hold, there would be a need to demonstrate that the sizes of their genomes in the present Andean primitive and essentially derived races are greater than the comparatively smaller genome sizes in the Mexican primitive maize races. Among primitive maize races, only the Palomero Toluqueño genome has been sequenced, and it has been found to be smaller by 25% than the genome of the B73 North American Corn Belt maize inbred line (Vielle-Calzada et al., 2009a,[157] 2009b[158]).

Taking into consideration directional evolution, Matsuoka and colleagues (2002) decided to evaluate time of divergence between teosinte and maize in a single environment. The dynamics of microsatellite evolution are not yet fully understood; therefore, the estimations of these authors should be taken with caution.

We may recall that there was a long-standing controversy between the advocates of directed or "orthogenetic evolution," which was supported by Goldschmidt (1940[159]) and Willis (1940[160]), whereas Dobzhansky (1941[161]), Mayr (1942[162]), Simpson (1944[163]), and Wright (1941[164]) criticized this approach, indicating that there was no selective basis for certain evolutionary changes. Observations of natural and induced mutations have been found to be random, at least in respect to species differences and evolutionary trends (Stebbins, 1950).

Evidence of Teosinte Introgression

A classification of the genus *Zea* appears in the first part of this book.

The dramatic morphological differences between maize and teosinte have made taxonomists separate them in the past as two different species. However,

157 Vielle-Calzada, Jean-Philippe, Octavio Martínez de la Vega, Gustavo Hernández-Guzmán, Enrique Ibarra-Laclette, Cesar Alvarez-Mejía, Julio C. Vega-Arreguín, Beatriz Jiménez-Moraila, Araceli Fernández-Cortés, Guillermo Corona-Armenta, Luis Herrera-Estrella, and Alfredo Herrera-Estrella. 2009a. The Palomero Toluqueño genome suggests metal effects on domestication. *Science*, 326 (5956): 1078.

158 Vielle-Calzada, Jean-Philippe, Octavio Martínez de la Vega, Gustavo Hernández-Guzmán, Enrique Ibarra-Laclette, Cesar Alvarez-Mejía, Julio C. Vega-Arreguín, Beatriz Jiménez-Moraila, Araceli Fernández-Cortés, Guillermo Corona-Armenta, Luis Herrera-Estrella, and Alfredo Herrera-Estrella. 2009b. El genoma de la raza Palomero Toluqueño y sus implicaciones para el entendimiento del proceso de domesticación del maíz. Simposio, Fronteras en la Biotecnología Agricola. XIII Congreso Nacional de Biotecnología y Bioingeniería. Acapulco.

159 Goldschmidt, R. 1940. *The Material Basis of Evolution*. Yale University Press. New Haven.

160 Willis, J. C. 1940. *The Course of Evolution by Differentiation or Divergent Mutation Rather Than by Selection*. Cambridge University Press. Cambridge.

161 Dobzhansky, Th. 1941. *Genetics and the Origin of Species*. Columbia University Press. New York.

162 Mayr, Ernst. 1942. *Systematics and the Origin of Species*. Columbia University Press. New York.

163 Simpson, G. G. 1944. *Tempo and Mode in Evolution*. Columbia University Press. New York.

164 Wright, S. 1941. The material basis of evolution. *Scientific Monthly*, 53: 165–170.

the fact that they are cross-compatible, and that their chromosomes are able to pair in meiosis and produce often-fertile offspring, brought Reeves and Mangelsdorf (1942[165]) to be the first to propose a revision in taxonomy and to establish annual teosinte and maize to be co-specific in the genus *Zea*.

The genus *Tripsacum* is the closest relative to the genus *Zea*. The species of teosinte belonging to the Luxuriantes group are the most primitive and approach in morphological characteristics of various species of the genus *Tripsacum*.

The major differences between maize and teosinte lie on the following characteristics: (1) The central spike of the tassel in maize is more compact and differentiated from the branches, which in teosinte look all alike. (2) The ear of maize is abnormal in all the Andropogoneae in being polystichous, whereas all other species are distichous and spread their seeds by breakage of their rachis, as in teosinte, or have an abscission layer in the rachillae supporting the seeds, vestiges of which appear in the pedicels of modern maize as relics from an ancestor that might have had such a seed-scattering mechanism. (3) The seeds of teosinte are enclosed by hard glume extensions, whereas the seeds of maize are exposed. (4) Teosinte produces lateral branches terminating in tassels. Tillering is found in teosinte under certain ecological conditions and not others for the same taxa, and (5) teosinte has chromosome knob positions unknown in maize.

Maize cobs that have teosinte genes, due to introgression, have lower indurated glumes (a *tripsacoid* or *teosintoid* expression), usually extending at an angle.

Mangelsdorf and Smith (1949[166]) considered that primitive archaeological maize discovered in New Mexico was a pod corn; its kernels were partially or totally enclosed in glumes and were not a teosinte derivative. The earliest maize found at Bat Cave has slender ears that were not entirely enclosed by husks and is either entirely a pod corn or a weekly pod corn.

Archaeological evidence of teosinte introgression into maize is very strong. The indirect evidence comes from a sequence of corn cobs at Tehuacán, Mexico, extending back to 3400 BC (Mangelsdorf et al., 1964). The direct evidence of teosinte and maize × teosinte hybrids, apparently in the F_1 generation, appears from 770 to 500 BC in Mitla, Oaxaca, and in Romero's Cave (900–400 BC) in western Tamaulipas, Mexico (Wilkes, 1972[167]). The latter dates coincide with the period when maize showing teosinte introgression was the most abundant type in the remains at Tehuacán.

[165] Reeves, R. G., and P. C. Mangelsdorf. 1942. A proposed taxonomic change in the tribe Maydeae (family Gramineae). *American Journal of Botany*, 29: 815–817.

[166] Mangelsdorf, P. C., and C. E. Smith. 1949. New archaeological evidence on the evolution of maize. *Botanical Museum Leaflets*, 13: 213–247.

[167] Wilkes, H. G. 1972. Maize and its wild relatives. *Science*, 177: 1071–1077.

On one hand, Kato-Yamakake (1976,[168] 1984[169]) and Kato-Yamakake and Sánchez (2002[170]) consider that no introgression occurs between maize and teosinte. On the other hand, Mangelsdorf (1974) and Wilkes (1967, 1972) have the opposite view. Wilkes (1977[171]) has observed numerous fields in Mexico that provide evidence of teosinte introgression, for example, in the case of the race Olotillo with teosinte in the Balsas River basin. Introgressed Olotillo by teosinte was also mapped by Wellhausen et al. (1952[172]) for Guerrero and Oaxaca. A modified Olotillo has tassel morphology very similar to the teosinte in the region (Wilkes, 1977).

Wellhausen and colleagues (1952: 21) described teosinte and maize relations in Mexico some 60 years ago as follows:

> There is no doubt (and least on the part of those who have studied the problem intensely) that there is in Mexico a constant and reciprocal introgression of maize and teosinte. This is easily seen in Chalco, a village 34 kilometers southeast of Mexico City. In this region teosinte grows in profusion in the maize fields. Its flowering period overlaps the flowering period of maize and natural hybridization between the two species is naturally occurring. There is a constant fraction of plants which are first generation hybrids and which intercross both ways. The teosinte of Chalco has acquired phenotypic characteristics of maize such as strongly pigmented and pubescent leaf sheaths. Even yellow endosperm, confined to maize, can be found in teosinte kernels (Mangelsdorf, 1947), and colored aleurone also occurs. The maize in the same region shows unmistakable evidence of teosinte introgression in a number of characteristics, particularly the induration of the rachis and glumes.

There are barriers to natural crossing between teosinte and maize, but there is no doubt that teosinte introgression into maize is occurring at the present time in Mexico, and there is little doubt that it has occurred in the past for as long a period and as often as the two species have been in contact (Wellhausen et al., 1952). In Chalco teosinte, hybrids between maize and teosinte represent a sizable part of the population (Wilkes, 1977). Reciprocal introgression also occurs from maize to teosinte. Varying amounts of maize germplasm are identified in

[168] Kato-Yamakake, T. A. 1976. *Cytological Studies of Maize* (Zea Mays L.) *and Teosinte* (Zea mexicana *Schrader Kuntze*) *in Relation to Their Origin and Evolution*. Massachusetts Agricultural Experiment Station Bulletin, 635.

[169] Kato-Yamakake, T. A. 1984. Chromosome morphology and the origin of maize and its races. *Evolutionary Biology*, 17: 219–253.

[170] Kato-Yamakake, T. A., and J. J. Sánchez-G. 2002. Introgression of chromosome knobs from *Zea diploperennis* into maize. *Maydica*, 47: 33–50.

[171] Wilkes, H. G. 1977. Hybridization of maize and teosinte in Mexico and Guatemala and the improvement of maize. *Economic Botany*, 31: 254–293.

[172] Wellhausen, E. J., L. M. Roberts, and F. Hernandez-Xolocotzi, in collaboration with P. C. Mangelsdorf. 1952. *Races of Maize in Mexico*. The Bussey Institution of Harvard University. Cambridge.

six races of teosinte (Nobogame Central Plateau, Chalco, and Balsas) in Mexico and in teosinte races Huehuetenango and Guatemala in Guatemala (Wilkes, 1977).

Van Heerwaarden and colleagues (2010) analyzed a large SNP dataset obtained from maize (*Z. mays* ssp. *mays*) and two teosinte subspecies, *Z. mays* ssp. *parviglumis* and Z. *mays* ssp. *mexicana*. Their analysis indicated a large amount of introgression between highland maize and ssp. *mexicana*, which grow associated in farmers' fields in the highlands of central Mexico. Thus it is possible that the apparent close relationship of highland maize to ssp. *parviglumis* may be an artifact of later introgression of teosinte alleles from ssp. *mexicana*.

Hufford, Lubinsky, and colleagues (2011[173]) have studied sympatric hybrid swarms of maize and of ssp. *mexicana* of teosinte, genotyping more than 40,000 SNPs. They arrived at the conclusion that, although they retain their individual morphologies, there has been in the past and continues to exist a gene flow into maize from "preadapted" ssp. *mexicana* teosinte for alleles for high-altitude adaptation. This explanation is needed for the march of early adapted maize from the lowlands to the Mexican highlands, if indeed it was domesticated in the Balsas Valley area. This process of a lowland to highland march appears to be the reverse from what we have found in Peru, where early lowland archaeological races are the same ones as early highland races and basically exhibit the morphological signs, of deep anthocyanin pigmentation, of a previous highland adaptation, and, in the absence of teosinte introgression, of cytological or morphological signals.

Teosinte has undoubtedly participated in the formation of maize races in Mexico. One of them, Reventador, which has characteristics that appear to originate in teosinte, has been postulated to be the product of the hybridization of Chapalote and teosinte (Wellhausen et al., 1952). Tepecintle, another highly teosintoid race, could have evolved from Olotillo, which exhibits characteristics related to strong teosinte introgression, although with a medium chromosome knob number. The modern race Tuxpeño shows teosinte introgression, supposedly coming from both of its putative parents, Olotillo and Tepecintle (Wellhausen et al., 1952). More advanced modern races, such as Corn Belt Dent, trace their ancestry directly to Mexican and southwestern U.S. races, and to teosinte, and more distantly to South American races, as proposed by Grobman and colleagues (1961).

The fact that no teosinte introgression is evident in primitive Andean maize races and their earlier derived races – either archaeologically or in present times – would tend to support the hypothesis that early maize in Mexico or Mesoamerica, where it presumably originated, did not share teosintoid characteristics with

[173] Hufford, Matthew B., Pesach Lubinsky, Tanja Pyhäjärvi, Norman C. Ellstrand, and Jeffrey Ross-Ibarra. 2011. Dueling genomes: Reciprocal gene flow in hybrid swarms of maize and its wild relative *Zea mays* ssp. *mexicana*. Oral paper. University of California, Davis. http://www.evolutionmeeting.org/engine/search/index.php?func=detail&aid=711.

the teosinte subspecies when, during early times, it was transported to South America in a pure maize form. This assumption is supported by chromosome knob number and location patterns of the primitive popcorn races. The most primitive maize races from Peru and Bolivia, forming the so-called Andean complex (McClintock, 1960[174]), range from knobless to more frequently having one small knob in chromosomes 6 and/or 7. Zero or low chromosome knob numbers (one or two small knobs in chromosomes 6L and 7L) are also found in the South American primitive races Confite Morocho, Kculli, Enano, Confite Puntiagudo, and Pisinkalla, which share the "Andean" low-knob pattern, and in the derived races of the archaeological maize Confite Chavinense, such as Granada, Chullpi, Paro, and Huayleño (Grobman et al., 1961). On the other hand, the early Mexican popcorn races Chapalote and Nal-Tel (whose ancestors have been identified in archaeological sites in Mexico) have an average chromosome knob number of 6 and 5.5, respectively (Wellhausen et al., 1952), or 10 and 11.7 knobs, respectively, according to Longley and Kato-Yamakake (1965[175]). This indicates that the latter maize races received their knobs from teosinte at a later date.

It is more likely that new knobs are acquired rather than lost, and teosinte is their purveyor. Loss by mutation or negative selection against knobs is unlikely, as they have tended to be more ubiquitous in many races of maize and teosinte. The primitive maize Palomero Toluqueño has retained a low chromosome knob number probably because of genetic isolating mechanisms present in that race (*Ga* genes) that work against teosinte introgression.

Knobs are formed by a large number of 180-bp and some 350-bp DNA tandem repeats that are condensed as a resource to allow proper chromosome pairing of the rest of the chromosome. They are orders of magnitude larger than genes, and although they are inherited in a Mendelian fashion, they are not liable to be lost by mutations, as has been speculated. Their absence in some races of maize is a strong tracer of a knobless precondition in the remote origins of maize, and unless evidence appears of a knobless teosinte, similar to knobless *Tripsacum australe*, there is no choice but to reopen the theory of the origin of maize and teosinte from a common ancestor.

The evidence at hand would indicate that early maize was essentially knobless. It encountered teosinte and interacted with it much later in the process of maize evolution. Maize, if this concept holds, could not, therefore, be directly descended from teosinte but must come from some wild plant more akin to maize. The introgression activity between maize and teosinte in later periods of

[174] McClintock, B. 1960. Chromosome constitutions of Mexican and Guatemalan races of maize. Annual Report of the Department of Genetics. *Carnegie Institution of Washington Yearbook*, 59: 461–472.

[175] Longley, A. E., and Kato-Yamakake, T. A. 1965. Chromosome morphology of certain races of maize in Latin America. International Maize and Wheat Improvement Center (CIMMYT). *Research Bulletin*, 1: 1–112.

time would explain the commonality of polymorphisms in some selected genes. Explanations that try to make room for teosinte where it does not appear in the archaeological record, such as the suggestion that it was not used as grain but was chewed, are not convincing and are simply speculative.

Biochemical Techniques Used in the Taxonomy of the Maydeae

Biochemical characters have been used extensively in taxonomic and evolutionary studies. Waines (1972[176]) compared the alcohol- and salt-soluble proteins of Chalco, Balsas, and perennial teosintes with primitive maize races from Mexico and Peru and found them identical, with identical bands, whereas few common bands were found with *Tripsacum*.

Serological and electrophoretic techniques have been used in taxonomy of the Maydeae by Stephen and colleagues (1980[177]), who devised tests based on injecting an extract of seeds of maize, teosinte, and *Tripsacum* into rabbits and then studying difference in antibodies reacting to the respective antigens through this "immunization." Immunological and immunoelectrophoretic tests can be run on different seed extracts put on a starch gel by applying an electric current, which makes them migrate in patterns that differ according to their various different components' relative migrating speeds. Their test included 8 primitive races of maize, 6 races of teosinte with samples from various localities, and 44 entries of *Tripsacum* species and other Maydeae (*Coix, Andropogon, Bothriochloa, Elyonurus, Manisuris*, and *Dichanthium*). In addition, they tested other Poaceae, such as *Panicum, Triticum*, and *Hordeum*. Races of maize from Mexico and northern Guatemala had electrophoretograms similar to those of teosinte. Mexican teosintes differed from Guatemalan teosintes, but much more from *Tripsacum*. The four *Tripsacum* species were identical. None of the *Tripsacum* species shared all the bands with maize. Differences with the other species were much greater.

Gene Evolution and Species Evolution

The earliest farmers, who were hunter-gatherers, or their women recognized the useful variation of first wild and then semiwild plants that grew in the garbage refuse patches near their habitation, which would have the potential to express some hidden variation in the new semicultivation environment. They gradually developed improved populations, gradually endowed with a range of new and desirable traits. The domesticated forms would not have been very different

[176] Waines, J. G. 1972. Protein electrophoretic patterns of *maize*, teosinte and *Tripsacum dactyloides*. *Maize Genetics Cooperation Newsletter*, 46: 164–165.

[177] Stephen, J., C. Smith, and R. N. Lester. 1980. Biochemical systematics and evolution of *Zea, Tripsacum* and related genera. *Economic Botany*, 34 (3): 201–218.

from the wild forms, and their selection was enhanced by the expression of domestication genes.

With reference to plant evolution, a few traits make the difference between domesticates and wild plants. These traits are controlled by a few major genes. But the process of adjustment of the domesticate to the new environment under selection must have put into action thousands of minor genes or quantitative genes (QTLs) whose action is regulated by major genes, and interaction becomes orchestrated for the expression of certain important traits between key genes.

The discovery of such domestication genes has been facilitated by the development of molecular markers. Discovering the location of these makers in individual genes or chromosome regions was made possible by studying segregating polymorphisms at the molecular level, and in segregating populations exhibiting visual phenotypic traits (Hancock, 2005;[178] Paterson 2002[179]). Some closely linked genes or genes with pleiotropic effects may be able to express various traits that appear simultaneously. In these cases it would be difficult to assign the effects to single gene pleiotropy or to close linkage of the genes in control, as indicated by Bomblies and Doebley (2006[180]). Genes that may have been involved in early phases of domestication have been cloned and studied in their molecular composition. During domestication, population genetic diversity is reduced as a consequence of selection. Domestication-related genes experience a more severe genetic bottleneck due to selection than neutral genes, as discussed by Doebley and colleagues (2006[181]). An estimate of the severity of the genetic bottleneck of domestication is about 80% in maize (Wright and Gaut, 2005[182]).

The causes of a reduced genetic variation include not only the bottleneck effect due to selection but also the "founder effect," which arises from genetic drift caused by a small population establishing itself in a new habitat with few individuals in isolation of the original population, or after the disappearance of it.

QTL analysis has permitted the detection of previously undetected domestication-related genes across the genome. Selective sweeps enable hidden domestication genes to be detected based on the selection profile of comparative sequences. Genomic comparison of crops and their wild progenitors for hidden

[178] Hancock, J. F. 2005. Contributions of domesticated plant studies to our understanding of plant evolution. *Annals of Botany*, 96: 953–963.

[179] Paterson, A. H. 2002. What has QTL mapping taught us about plant domestication? *New Phytologist*, 154: 591–608.

[180] Bomblies, K., and J. F. Doebley. 2006. Pleiotropic effects of the duplicate maize FLORICAULA/LEAFY genes zfl1 and zfl2 on traits under selection during maize domestication. *Genetics*, 172: 519–531.

[181] Doebley, J., B. S. Gaut, and B. D. Smith. 2006. The molecular genetics of crop domestication. *Cell*, 127: 1309–1321.

[182] Wright, S. I., and B. S. Gaut. 2005. Molecular population genetics and the search for adaptive evolution in plants. *Molecular Biology and Evolution*, 22: 506–519.

domestication-related genomic regions is a new approach to detecting potentially useful diversity in wild progenitors for crop improvement.

Genome sequencing has enabled the evolution of domestication-related genes to be elucidated. By the end of the twentieth century, classical methods of genetics and cytogenetics had revealed the different routes by which genomic variation took place. This variation included macro changes, such as chromosome deletions, deficiencies, inversions, and supernumerary chromosomes that were normally transmitted, such as the B chromosomes in maize and polyploidization. Genetic information on transposable element insertions, paramutation, and controlling genes and epigenetic control added to the potential for genome evolution.

The study of the *Hardness* (*Ha*) locus, which controls grain hardness in hexaploid wheat (*Triticum aestivum*) and its relatives (*Triticum* and *Aegilops* species), represents a classical example of a trait whose variation arose from gene loss after polyploidization. The investigation carried out at the molecular level provided evidence for the basis of the evolutionary events observed at this locus. It revealed numerous genomic rearrangements, such as transposable element insertions, genomic deletions, duplications, and inversions (Chantret et al., 2005[183]). Genomic rearrangements at the *Ha* locus in wheat were believed to be mainly caused by illegitimate recombination, in which DNA sequences not originally attached to one another become joined, and this type of recombination is considered a major evolutionary mechanism in wheat species (Chantret et al., 2005).

Murat and colleagues (2010[184]) in paleobotanical studies of the evolution of species of the Poaceae (= Gramineae), compared genomes of rice, sorghum, maize, and *Brachypodium* regarding the changes in synteny of gene blocks. It appeared to them that centromeric/telomeric illegitimate recombination between nonhomologous chromosomes led to nested chromosome fusions (NCFs) and synteny breakpoints (SBPs). When intervals comprising NCFs were compared in their structure, they concluded that SBPs (1) were meiotic recombination hot spots, (2) corresponded to high-sequence turnover loci through repeat invasion, and (3) might be considered as hot spots of evolutionary novelty that could act as a reservoir for producing adaptive phenotypes.

Many studies have shown that in large plant genomes, long terminal repeat (LTR)–retrotransposon families often contain thousands (or tens of thousands)

[183] Chantret, N., J. Salse, F. Sabot, S. Rahman, A. Bellec, B. Laubin, I. Dubois, C. Dossat C., P. Sourdille, P. Joudrier, M. F. Gautier, L. Cattolico, M. Beckert, S. Aubourg, J. Weissenbach, M. Caboche, M. Bernard, P. Lerog, and B. Chalhoub. 2005. Molecular basis of evolutionary events that shaped the hardness locus in diploid and polyploid wheat species (*Triticum* and *Aegilops*). *The Plant Cell*, 17: 1033–1045.

[184] Murat, F., J. H. Xu, E. Tannier, M. Abrouk, N. Guilhot, C. Pont, J. Messing, and J. Salse. 2010. Ancestral grass karyotype reconstruction unravels new mechanisms of genome shuffling as a source of plant evolution. *Genome Research*, 20 (11): 1545–1557.

of copies with high-sequence identity, which suggests that they originate from a recent massive retrotransposition event. In the case of maize, San Miguel and colleagues (1998[185]) found, by comparison of LTR divergences with the sequence divergence between the *Adh1* locus in maize and sorghum, that all retrotransposons examined have inserted within the last 6 million years, most in the last 3 million years. The structure of the *Adh1* region appears to be standard relative to the other gene-containing regions of the maize genome, thus suggesting that retrotransposon insertions have increased the size of the maize genome from approximately 1,200 Mb to 2,400 Mb in the last 3 million years. Furthermore, the results indicate an increased mutation rate in retrotransposons compared with genes.

Piequ and colleagues (2006[186]) studied the case of the *Oryza australiensis* genome, which has accumulated more than 90,000 retrotransposon copies during the last 3 million years, leading to a rapid twofold increase of its size. Bursts of activity of the *Wallabi*, *Kangourou*, and *RIRE1* LTR-retrotransposon families in *Oryza australiensis* in recent periods have impacted the genome of this species by doubling its size relative to other diploid *Oryza*.

Illegitimate recombination mechanisms targeting LTR retrotransposons have been identified as inducing considerable loss of DNA and contributing to genome size reduction in *Arabidopsis* and rice. Two balanced and competing forces – increase, induced by retrotransposition, and decrease, caused by recombinations and deletions (Petrov, 2002;[187] Vitte and Panaud, 2005[188]) – act in shaping the evolution of genome size.

An important consideration in crop evolution in the last 10,000 years of agriculture is the role of weeds in speciation and their coevolution with crop plants. Common bread wheat is a classical example of a crop species being formed by spontaneous hybridization, in the case of tetraploid wheat (*Triticum turgidum* ssp. *dicoccum*, $2n = 4x = 28$) and diploid goat grass (*Aegilops tauschii*, $2n = 2x = 14$), a weed of early wheat fields. As with crops, the emergence of new weedy rice forms in the last decade is an example of evolution of weeds under selection (Cao et al., 2006[189]). The coevolution of teosinte with maize, which mimetized it, is very similar to the evolution of red rice in rice fields.

[185] San Miguel, P., B. S. Gaut, A. Tikhonov, Y. Nakajima, and J. L. Bennetzen. 1998. The paleontology of intergene retrotransposons of maize. *Nature Genetics*, 20: 43–45.

[186] Piequ, B., R. Guyot, N. Picault, A. Roulin, A. Saniyal, H. Kim, K. Collura, D. S. Brar, S. Jackson, R. A. Wing, and O. Panaud. 2006. Doubling genome size without polyploidization: Dynamics of retrotransposition-driven genomic expansions in *Oryza australiensis*, a wild relative of rice. *Genome Research*, 16 (10): 1262–1269.

[187] Petrov, D. A. 2002. Mutational equilibrium model of genome size evolution. *Theoretical Popular Biology*, 61: 531–544.

[188] Vitte, C., and O. Panaud. 2005. LTR retrotransposons and plant genome size: Emergence of the increase/decrease model. *Cytogenetic and Genome Research*, 110: 91–107.

[189] Cao, Q., B. R. Lu, H. Xia, J. Rong, F. Sala, A. Spada, and F. Grassi. 2006. Genetic diversity and origin of weedy rice (*Oryza sativa* f. *spontanea*) populations found in north-eastern China revealed by simple sequence repeat (SSR) markers. *Annals of Botany*, 98: 1241–1252.

Domestication-related genes have been cloned, and the molecular basis has been clarified for changes on cryptic genes that selection acted on in several species. For example, the two genes that are most important in relation to spikelet shattering in rice (*sh4* and *qSH1*) have been cloned (Konishi et al., 2006;[190] Li et al., 2006[191]). The reduction of seed shattering is a fundamental difference between a wild species or a weedy form of an early cultivated species and a domesticated plant. In the Poaceae, shattering is controlled by changes in the abscission layer, where the pedicel attaches to the seed; *sh4* is the key shattering gene that distinguishes cultivated rice from wild rice, whereas the *qSH1* gene controls the difference in the degree of shattering between some *indica* and *japonica* varieties of rice. Moreover, *sh4* is a transcription regulator, and a single amino acid substitution results in reduced shattering. For *qSH1*, a single nucleotide in the regulatory region of this gene results in the altered level of seed shattering, and *sh4* activates the abscission process, whereas *qSH1* regulates abscission-layer formation. Sequence analysis of *sh4* has revealed a single base-pair mutation that is responsible for non-shattering characteristics, and this change is the same in both *indica* and *japonica* rice varieties (Lin et al., 2007[192]). This result raises doubts about whether Asian rice was domesticated more than once, as has been suggested in several recent papers (for a review, see Sang and Ge, 2007[193]). In contrast, studies sequencing and comparing seven loci in wild and landrace barley have provided strong evidence that barley was domesticated once in the Fertile Crescent and a second time in a location between 1,500 and 3,000 km to the east (Morrell and Clegg, 2007[194]). In the case of maize, if domestication started from teosinte, the process had to be more complex and unlike any other grass species in which a change in the abscission layer of the pedicel insertion was the key factor. In teosinte, the whole rachis breaks down, and the structural changes that maize had to undergo through domestication to reduce shattering involved not only the disappearance of the abscission layer but major structural modification of the insertion of the seed. If maize evolved from a wild maize parent by a process similar to those in other grasses, the changes required would have been of lower complexity. This subject is discussed in more detail elsewhere in this publication.

The increase of row number in maize ears is due to the *zfl2* gene, which has pleiotropic effects on yield, such as reduction of number of ears and lower

[190] Konishi, S., T. Izawa, S. Y. Lin, K. Ebana, Y. Fukuta, T. Sasaki, and M. Yano. 2006. An SNP caused loss of seed shattering during rice domestication. *Science*, 312: 1392–1396.

[191] Li, C., A. Zhou, and T. Sang. 2006. Rice domestication by reduced shattering. *Science*, 311: 1936–1939.

[192] Lin, Z., M. E. Griffith, X. Li, Z. Zhu, L. Tan, Y. Fu, W. Zhang, X. Wang, D. Xie, and C. Sun. 2007. Origin of seed shattering in rice (*Oryza sativa* L.). *Planta*, 226: 11–20.

[193] Sang, T., and S. Ge. 2007. The puzzle of rice domestication. *Journal of Integrative Plant Biology*, 49: 760–768.

[194] Morrell, P. L., and M. T. Clegg. 2007. Genetic evidence for a second domestication of barley (*Hordeum vulgare*) east of the Fertile Crescent. *Proceedings of the National Academy of Sciences USA*, 104: 3289–3294.

position in the stalk, as well as earlier flowering (Bomblies and Doebley, 2006). Bomblies and Doebley's suggestion is that pleiotropic effects such as those associated with the domestication gene *zfl2* could produce secondary undesirable effects limiting selection for favorable "domestication alleles" during early stages of the differentiation of a crop from its wild progenitor. On the other hand, selection for beneficial traits controlled by pleiotropic genes could result in associated neutral or even detrimental traits being concurrently selected. This may explain, at least partially, the presence, in wild populations, of alleles for traits of the domestication syndrome that apparently evolved prior to domestication and survived despite their possibly deleterious effects in the wild.

Furthermore, the duplicate genes *zfl1* and *zfl2* of maize are orthologous to the *FLORICAULA/LEAFY* (*FLO/LFY*) genes of the species of *Antirrhinum* and *Arabidopsis*, among others (Bomblies and Doebley, 2006). These belong to a large assortment of orthologous genes that belong to families of transcriptional regulators in plants involved in domestication, as discussed by Doebley and colleagues (2006). Within a given family of transcriptional regulators, gene structure may be sufficiently conserved for similarities to be identified not just between genera of the same plant family but between taxonomically very distantly related species. Thus, *monoculm1* in maize shares similarities with *LATERAL SUPPRESSOR* from *Arabidopsis thaliana* and tomato (Doust, 2007[195]), and *Q* in wheat is similar to *APETALA2* (*AP2*) of *Arabidopsis* (Simons et al., 2006[196]). Therefore, care should be exercised when using similarities between taxa, invoking the presence of genes or alleles that may be widely dispersed in higher plants.

Plant Molecular Genetics and the Need for Additional Research

Plant domestication is the genetic modification of a wild species to create a new form of a plant altered to meet human needs. A common suite of traits – known as the "domestication syndrome" – distinguishes most seed and fruit crops from their progenitors (Hammer, 1984[197]). During the domestication process changes in plant architecture take place to adapt the plant and its usable parts to human needs. Basic to the process is the avoidance of seed scattering, which would diminish yields. For many crops, domestication has rendered the plant completely dependent on humans, such that it is no longer capable of propagating itself in nature. Maize and cauliflower are good examples of such highly modified forms. However, other crops, such as hemp, carrot, and lettuce,

[195] Doust, A. 2007. Architectural evolution and its implications for domestication in grasses. *Annals of Botany*, 100: 941–950.

[196] Simons, K. J., J. P. Fellers, H. N. Trick, Z. Zhang, Y. S. Tai, B. S. Gill, J. D. Faris. 2006. Molecular characterization of the major wheat domestication gene Q. *Genetics*, 172: 547–555.

[197] Hammer, K. 1984. Das Domestikationssyndrom. *Kulturpflanze*, 32: 11–34.

have been more modestly modified compared to their progenitors, and they can either revert to their wild forms or become self-propagating weeds.

Plant molecular population genetics has come to the aid of the study of domestication and plant evolution. The first papers on plant molecular population genetics were published approximately 10 years ago. Since that time, well more than 50 additional studies of plant nucleotide polymorphism have been published, and many of these studies focused on detecting the signature of balancing or positive selection at a locus. In two well-studied taxa (maize and *Arabidopsis*), more than 20% of studied genes have been interpreted as containing the signature of selection. This is probably an overstatement of the prevalence of natural selection in plant genomes, for two reasons. First, demographic effects are difficult to incorporate and have generally not been well integrated into the plant population genetics literature. Second, the genes studied to date are not a random sample, so selected genes may be overrepresented. The next generation of studies in plant molecular population genetics requires additional sampling of local populations, explicit comparisons among loci, and improved theoretical methods to control for demography (Wright and Gaut, 2005).

More attention should be given to early periods of maize evolution in South America. Most of the research has concentrated in the past in the Mexican and Mesoamerican areas, with a complacency derived from the assumption that all basic problems on how domestication and evolution of maize have proceeded are now well understood. This is not so. Many of the arguments stemming from research data are interpreted in one sense only – that maize was domesticated from teosinte, excluding any other possible interpretation, such as that teosinte has been strongly influenced by maize. The data showing that maize could have retained a fraction of the original variability that was present in teosinte is simplistic. There are options to be considered other than the bottleneck explanation to justify a reduction in variability in domesticated maize. Modern maize has a much greater variability than that found in teosinte today. Therefore, open minds are needed to conduct a quiet and sober revision of the present standing of the question of how, and where, maize domestication and increase of variability took place.

Estimation of Gene Number in Maize

The estimation of gene number from draft whole-genome sequence and finished individual chromosomes in maize has varied from approximately 32,000 to approximately 70,000 (Liu et al., 2007[198]). The estimated gene number in maize using BACs was 37,000. The arrival at this figure is particularly

[198] Liu, Renyi, Clémentine Vitte, Jianxin Ma, A. Assibi Mahama, Thanda Dhliwayo, Michael Lee, and Jeffrey L. Bennetzen. 2007. A gene trek analysis of the maize genome. *Proceedings of the National Academy of Sciences USA*, 104 (28): 11844–11849.

challenging, because the complete genome sequence was not available, and the majority of the genome consists of nested LTR retrotransposons (San Miguel et al., 1996[199]). The estimation of maize gene number (37,000) is similar to the gene number estimated in the nearly completed rice genome sequence (approximately 32,000). Although maize has a fairly recent history as a tetraploid, it is approaching a diploid status, because 50–90% of the duplicated copies of genes have been deleted at least partially in one of the homoeologous regions. Haberer and colleagues (2005[200]) estimated 42,000 genes for maize but identified as genes what would be sequences of actually truncated gene fragments and/or sequences within transposable elements. Morgante and colleagues (2005[201]) have predicted that approximately 20% of annotated maize genes are actually gene fragments within *helitrons*. They identified putative autonomous *helitron* elements and found evidence for their transcription. *Helitrons* in maize seem to continually produce new nonautonomous elements responsible for the duplicative insertion of gene segments into new locations and for the unprecedented genic diversity. The maize genome is in constant flux, as transposable elements continue to change both the genic and nongenic fractions of the genome, profoundly affecting genetic diversity. Genic content polymorphisms involve as many as 10,000 sequences and are mainly generated by DNA insertions. The ends of eight of the nine genic insertions that were analyzed shared the structural hallmarks of *helitrons*, which are rolling-circle transposons. The effect that transposons have had and continue to have in increasing the diversity of maize cannot be minimized.

It has also been found that blocks of gene-free repetitive DNA of more than 100 kb in size appear to be common in the maize genome and are likely to be intermixed with genic blocks.

These results indicate that genes are found in small islands that are unevenly distributed around the genome, and that different families of transposable elements (TEs) preferentially associate with gene-containing or gene-free regions. Mapping of these regions suggests that most of these gene-free regions are not associated with known heterochromatic features of the genome.

TEs are the major components of genomes of most plant species. TEs have various families or types that proliferate at different rates in the genome.

[199] San Miguel, Phillip, Alexander Tikhonov, Young-Kwan Jin, Natasha Motchoulskaia, Dimitrii Zakharov, Admasu Melake-Berhan, Patricia S. Springer, Keith J. Edwards, Michael Lee, Zoya Avramova, and Jeffrey L. Bennetzen. 1996. Nested retrotransposons in the intergenic regions of the maize genome. *Science*, 274 (5288): 765–768.

[200] Haberer, George, Sarah Young, Arvind K. Bhart, Heidrun Gundlach, Christina Raymond, Galina Fuks, Ed Butler, Rod A. Wing, Steve Rounsley, Bruce Birren, Chad Nusbaum, Klaus F. X. Maye, and Joachim Messing. 2005. Structure and architecture of the maize genome. *Plant Physiology*, 139: 1612–1624.

[201] Morgante, Michele, Stephan Brunner, Giorgio Pea, Kevin Fengler, Andrea Zuccolo, and Antoni Rafalski. 2005. Gene duplication and exon shuffling by helitron-like transposons generate intraspecies diversity in maize. *Nature Genetics*, 37: 997–1002.

Proliferation activity is counteracted by TE removal via recombination and population processes driven by natural selection, creating opportunities for genetic variation.

Amazingly, *Zea* shows as many or more coding genes as humans do.

B Chromosomes and the Evolution of Maize

B chromosomes are abnormal or supernumerary, and highly heterochromatic chromosomes found in many species have been more intensely studied in maize. They have no correlation with the normal or A chromosomes and are essentially devoid of coding genes. They are usually small and vary in size and number in maize and are considered to be nonessential in the genome. They may or may not be present in the maize genome, and if present, they vary in number (usually between zero and four), in the maize genome. They fail to pair with the A chromosomes at meiosis. They are equally transmitted by male and female gametes.

B chromosomes are club shaped due to the accumulation of chromatin at their distal end, forming one clearly defined arm. Unlike A chromosomes, which have knobs located distally from the centromere, they have a small knob next to the centromere. Two-armed B chromosomes have been reported in the literature but are rare, if they are indeed real.

B chromosomes are found in a large number of plant and animal species (Jones and Rees, 1982[202]). They are supposed to be genetically inert, although a specific case is known of their direct action on external phenotypes in the case of striped leaves in maize (Staub, 1987[203]). B chromosomes in maize can compose up to 4% of the total DNA volume. There appears to be a negative correlation between B chromosome number and number of large knobs (Rosato et al., 1998[204]). B chromosomes are shorter and unlike any of the A chromosomes. The fact that B chromosomes do not synapse with A chromosomes would be an indication of their remote origin.

Randolph (1941[205]) found that B chromosomes in maize vary in number from one generation to the next: they may increase or decrease. It is very easy to increase the number of Bs by inbreeding and selection up to 20 or more per nucleus. In crosses of a plant with zero Bs by a plant with one B, one would get

[202] Jones, R. N., and H. Rees. 1982. *B Chromosomes*. Academic Press. New York.

[203] Staub, R. W. 1987. Leaf striping correlated with the presence of B chromosomes in maize. *Journal of Heredity*, 78: 71–74.

[204] Rosato, M., A. M. Chiavarino, C. A. Naranjo, J. Cámara-Hernández, and L. Poggio. 1998. Genome size and numerical polymorphism for the B chromosome in races of maize (*Zea mays* ssp. *mays*, Poaceae). *American Journal of Botany*, 85 (2): 168–174.

[205] Randolph, L. A. 1941. Genetic characteristics of the B chromosomes of maize. *Genetics*, 16: 608–831.

one-third of the plants would have one B chromosome, and two-thirds would have zero Bs, which is the approximate ratio in which they are found in Andean maize populations. In small numbers they do not seem to produce any phenotypic or deleterious effects, although in high numbers they may induce larger cells and nuclei, larger pollen grains, and some infertility. Their evolutionary significance is not certain, although it would appear that they have been capable of coevolution with the rest of the genome.

Studies on sequencing the B and A chromosomes of maize have found the existence of a great homology of sequences between them. A more detailed study was made by Cheng and Lin (2003[206]), who dissected the B chromosome from microsporocytes and obtained 19 B sequences, 18 of which share homology with the A chromosomes. Their results confirmed the previous conclusions of similarity between B and A chromosomes. A total of 19 B sequences were isolated, all of which are repetitive and, with one exception, are homologous to the A chromosome(s). Three sequences have strong homology to maize sequences that include 2 knob repeats and 1 *zein* gene (noncoding region), and 10 others are homologous to the noncoding regions of *adh1*, *bz1*, *gag*, *zein*, and B centromere to a lesser degree. Six sequences have no homology to any gene. The B-specific sequence and another partially B-specific one were also mapped, by 7 newly characterized TB-10L translocations, to a similar location on the central portion of the distal heterochromatic region, spreading over a region of about one-third of the B chromosome. Those two specific areas in the B chromosome are required for nondisjunction.

This information, in addition, added to other data, would tend to confirm that the origin of B chromosomes may be found in their evolution from A chromosomes. They could have originated as a simple by-product of the evolution of the A chromosomes from centric fragments, fusions, or amplification of the paracentromeric region of an A chromosome in plants, although in animals they could have originated from sex chromosomes as well (Camacho et al., 2000[207]).

Evidence for an alternative, interspecific origin of B chromosomes has been advanced by Perfectti and Warren (2001[208]). They claimed, citing the case of the grasshopper *Nasonia*, that new chromosomes acquired through interspecific hybridization could be heterochromatized and inactivated, creating in the process instability for a number of generations.

B chromosomes have been found to contain no coding genes. However, in some intriguing way, they control their own fate and interact with the genome. Although they try to avoid being lost, they can still be lost through the formation

[206] Cheng, Ya-Ming, and Bor-Yaw Lin. 2003. Cloning and characterization of maize B chromosome sequences derived from microdissection. *Genetics*, 164: 299–310.

[207] Camacho, J. P. M., T. F. Sharbel, and L. W. Beukeboom. 2000. B-chromosome evolution. *Philosophical Transactions of the Royal Society*, 355: 163–178.

[208] Perfectti, Francisco, and John H. Warren. 2001. The interspecific origin of B chromosomes: Experimental evidence. *Evolution*, 55 (5): 1069–1073.

of univalents at meiosis. The loss avoidance may come through several mechanisms: mitotic nondisjunction, reduction of meiotic loss, preferential fertilization, and possibly, but rarely, conferring a selective advantage to the host.

B chromosomes may have been absent in primitive maize and may have appeared later in evolution as parasitic chromosomes generated by "selfish genes" that found an additional way of self-propagation, which then had extreme success in self-propagation in the *Ab10* chromosome.

B chromosomes have some common DNA sequences with some A chromosomes but also have some sequences that are uniquely of their own (Cheng and Lin, 2003[209]). In the second meiotic division, they exhibit nondisjunction. In addition, gametes with the nondisjoined B chromosomes have a preferential fertilization of the eggs as compared with gametes without B chromosomes (Roman, 1947[210]); therefore, the maize gametes that carry B chromosomes will tend to perpetuate, through such a preference, the number of B chromosomes (Rusche et al., 1997[211]), but there will always be fertilization from gametes without B chromosomes. The tendency toward an increase in the frequency of Bs has been observed in some of the more modern Peruvian races of maize. Some of these races have progressively increased from zero Bs, which may have been the situation in the earliest race, Proto-Confite Morocho.

A meiotic drive impacting positively on preservation has been found in rye and in maize. In maize it is based not only on nondisjunction of the B chromosomes at the second pollen grain mitosis but especially on preferential fertilization with the gametes that carry the B chromosomes, which brings an advantage of about 70% in the rate of fertilization. But the interesting part of the scheme is that the preference for the B gametes is regulated by A genes (González-Sánchez et al., 2003;[212] Lin, 1978[213]). Lin and González-Sánchez and colleagues found a single locus, which was called *mBt* (male B transmission), controlling B preferential fertilization in maize. The egg cells control which one of the sperm nuclei from the maize pollen will fertilize them. Preferential fertilization is carried out on the mBt^{h} egg cells by the sperm nucleus carrying the supernumerary B chromosomes (Bs). A hypothesis was formulated in the sense that the *mBt* gene is involved in the normal fertilization of maize, but the parasitic Bs take advantage

[209] Cheng, Ya Ming, and Bor-Yaw Lin. 2003. Molecular organization of large fragments in the maize B chromosome: Indication of a novel repeat. *Genetics*, 166: 1947–1961.

[210] Roman, H. 1947. Mitotic nondisjunction in the case of interchanges involving the B-type chromosome in maize. *Genetics*, 32: 391–409.

[211] Rusche, M. L., H. L. Mogensen, L. Shi, P. Keim, M. Rougier, A. Chaboud, and C. Dumas. 1997. B chromosome behavior in maize pollen as determined by a molecular probe. *Genetics*, 147: 1915–1921.

[212] Gonzáles-Sánchez, M., E. Gonzáles-Sánchez, F. Molina, A. M. Chiavarino, M. Rosato, and M. J. Puertas. 2003. One gene determines maize B chromosome accumulation by preferential fertilization; another gene(s) determines their meiotic loss. *Heredity*, 90: 122–129.

[213] Lin, B. Y. 1978. Regional control of non-disjunction in the B chromosome of maize. *Genetics*, 90: 613–627.

of the *mBt*^{h} allele to increase their own transmission. To complete the picture, the gene(s) that González-Sánchez and colleagues (2003) call *fBt* (female B transmission), which controls female transmission of B chromosomes, is located on the A chromosomes acting at diploid level, the *fBt*^{l} allele(s) for low transmission being dominant. This allele causes the loss of Bs at meiosis. Therefore, in the case of the B chromosome survivability, there is an intragenome conflict, because the *mBt* and *fBt* loci constitute a polymorphic system of attack and defense between the A and B chromosomes.

One characteristic of B chromosomes that is especially interesting in evolution is that they selfishly avoid being eliminated in the meiotic process, by a particular transmission dynamic, and instead try to accumulate themselves in plants. They do not follow a Mendelian segregation rate, and B chromosomes would appear not to be essential, as maize plants can function without them. However, when they accumulate in relatively large numbers, it is possible that they may have functions that they have acquired through selection. Their selfish selective advantage, nevertheless, has eluded B chromosome researchers in spite of their polymorphisms and widespread presence in plants and animals. That they, somehow, may have important, although yet unknown, effects on the genome expression is well known, but it is also known that their increase in excessive numbers in plant cells may become deleterious to the organism.

B chromosomes differ from the normal complements (A chromosomes) in several aspects: they are mitotically telocentric and highly heterochromatic and have no detectable genetic effects on the plant, except in high numbers. In addition, they enhance recombination on the A chromosomes and undergo nondisjunction at the second pollen mitosis. Yet, B and A chromosomes are not molecularly divergent (Jones et al., 2008[214]).

Longley and Kato-Yamakake (1965[215]), Kato-Yamakake (1976,[216] 1984[217]), and McClintock and colleagues (1981[218]) have reviewed the distribution of knobs, abnormal chromosome 10, and B chromosomes in maize and teosinte in the Americas, with the exception of Peru, where the work was carried out by a Peruvian team who had previously been trained by McClintock in maize cytogenetic techniques during her stay in Peru (Grobman et al., 1961).

[214] Jones, R. Neil, Wanda Viegas, and Andreas Houben. 2008. A century of B chromosomes in plants: So what? *Annals of Botany*, 101: 767–775.

[215] Longley, A. E., and T. A. Kato-Yamakake. 1965. *Chromosome Morphology in Certain Races of Maize in Latin America.* Research Bulletin 1. CIMMYT. Chapingo.

[216] Kato-Yamakake, T. A. 1976. *Cytological Studies of Maize* (Zea Mays *L.*) *and Teosinte* (Zea mexicana *Schrader Kuntze*) *in Relation to Their Origin and Evolution.* Massachussetts Agricultural Experiment Station Research Bulletin 635.

[217] Kato Yamakake, T. A. 1984. Chromosome morphology and the origin of maize and its races. *Evolutionary Biology*, 17: 219–253.

[218] McClintock, Barbara, T. A. Kato-Yamakake, and A. Blumenschein. 1981. *Chromosome Constitution of the Races of Maize: Its significance in the Interpretation of Relationship between Races and Varieties in the Americas.* Colegio de Postgraduados. Chapingo.

B chromosomes have been found in maize from Mexico, Central America, the various tribes of the United States, Colombia, Venezuela, the Caribbean, and the Andean region, and they extend as far as Argentina and Chile. Although they are peculiarly absent from maize in the eastern coast of South America, they are present in teosinte (McClintock et al., 1981). In the Peruvian races of maize, they are found in frequencies of zero to five per nucleus in races with low knob numbers (Grobman et al., 1961). They have been found absent in Corn Belt Dent maize, but present in the northeastern flints (Randolph, 1941).

B chromosomes are found in both maize and teosinte. In some of the Peruvian races of maize, they have been identified in various numbers (Grobman et al., 1961). They are found generally in low frequencies in maize, with zero B chromosomes per plant being the highest frequency in both Mexico and Peru. The same is true for teosinte in Mexico, except the teosinte collections sampled in the states of Michoacán and Guerrero, which exhibit higher frequencies of B chromosomes (see table and map in McClintock et al., 1981).

B chromosomes are absent or present only in frequencies of one or two in some 30% of the plants of primitive and highland- and lowland-derived races of maize in Peru. They have also been found in the highland region of Bolivia (McClintock et al., 1981). This situation is also found in the races of the eastern slope of the Andes and their derived races in the Amazonian plain, where there is a tendency to higher frequencies of plants with low numbers of B chromosomes (Grobman et al., 1961).

In the Peruvian races of maize, they are found in frequencies of zero to five per nucleus in races with low chromosome knob numbers. For example, the race Ancashino features a small knob subterminal on the long arm of chromosome 7 with a frequency of 78% and a small knob subterminal on the long arm of chromosome 6 with a frequency of 22%; knobless plants were found with a frequency of 22%, which may be the original condition of early maize in the central Andean region. At the same time, B chromosomes were found with a frequency of 22% in that same race. In the maize race Piricinco of Peru, which is named Entrelazado (Interlocked) in Brazil and Coroico in Bolivia, and which has the most extended distribution of any race in the South American lowlands, the number of knobs is variable from zero to four and the number of B chromosomes, which are more likely to be found, is from zero to two. The ancestor race of Piricinco – which we believe, on the basis of genetic, morphological, and archaeological evidence, is the race Rabo de Zorro of Peru – features zero to three knobs, with the "Andean type" of knobs in chromosomes 6 and 7 being the most frequent; it has B chromosomes ranging from zero to four, with zero knobs as the most frequent number. The most ancient race of this group is Proto-Confite Morocho. Granada, a race with small, globular ears grown at a mean altitude of 3,200 masl in Cuzco, exhibits the Andean type of knobs and also has B chromosomes in 37.5% of plants examined. Huayleño, a race postulated as directly descended from the race Confite Chavinense, which was amply distributed in the past in the highlands and the coastal belt of Peru in the Preceramic period, also exhibits the Andean type of knobs and a frequency range of zero to two B chromosomes. Kculli is considered another of the ancient precursor races of maize in Peru. It exhibits the Andean knob complex: a higher frequency of a knob 88% subterminal on the long arm of chromosome 7 and a knob subterminal in the long arms of chromosome 6 with a frequency of 38%, and occasionally on chromosome 4L, with plants exhibiting between two and five B chromosomes per cell. Confite Morocho, which is the present-day successor of the ubiquitous race Proto-Confite Morocho, found in all lowland and highland early archaeological sites, exhibits small knobs in the typical Andean knob complex positions on chromosomes 7L and 6L, with frequencies of 100% and 25%, respectively, and with no B chromosomes (Grobman et al., 1961).

Levings and colleagues (1975[219]) have reported the presence of three to four B chromosomes in *Tripsacum.*

A significant negative correlation between A-DNA content and altitude of cultivation and between A-DNA content and mean number of Bs was found in native populations of maize of northern Argentina by Rosato and colleagues (1998[220]) and Chiavarino (1988[221]). This indicates that there is a close interrelationship between the DNA content of A chromosomes and doses of Bs. Lia, Confalonieri, and Poggio (2007[222]) expanded the work of the previous authors in Argentina to study the adaptive significance of the altitudinal cline of B chromosomes by using molecular markers (SSRs) at 18 loci, from 7 maize races of northern Argentina, for which the altitude of the origin of the collections ranged between 910 and 3,000 masl. They studied the association of genetic differentiation at the SSR level and B chromosome differentiation, altitudinal distance, and geographic distance. Of the 18 loci, 17 were polymorphic. As expected, the populations exhibited a significant degree of genetic subdivision. On analyzing the genetic differentiation of the 183 alleles found with the clinal variation (altitude variation series) of the B chromosome

[219] Levings, C. S., D. H. Timothy, and W. W. L. Hu. 1975. Cytological characteristics and nuclear DNA buoyant densities of corn, teosinte, *Tripsacum* and corn-*Tripsacum* hybrids. *Crop Science*, 16 (1): 63–66.

[220] Rosato, Marcela, Amilcar Chiavarino, Carlos A. Naranjo, Julian A. Cámara-Hernández, and Lidia Poggio. 1998. Genome size and numerical polymorphism for the B chromosome in races of maize (*Zea mays* ssp. *mays,* Poaceae). *American Journal of Botany,* 85 (2): 168–174.

[221] Chiavarino, Amilcar M., Marcela Rosato, Pabo Rosi, Lidia Poggio, and Carlos A. Naranjo. 1988. Localization of the genes controlling B chromosome transmission rate in maize (*Zea my* spp. *mays,* Poaceae). *American Journal of Botany,* 85 (1): 1581–1585.

[222] Lia, Veronica V., Viviana A. Canfalonieri, and Lidia Poggio. 2007. B chromosome polymorphism in maize land races: Adaptive vs. demographic hypothesis of clinal variation. *Genetics,* 177: 895–904.

cline, they found no association between allele frequency and mean number of Bs per plant.

B chromosome research at the molecular level, at present, is concentrated on using the three model plants maize, rye, and *Brachycome* as the three lead species. Questions that are being asked are whether the increase of the quantity of DNA in cells due to the B chromosomes has any further evolutionary meaning, whether there are genes in the B chromosomes, and what further regulatory functions they have in the genome.

Viotti and colleagues (1985[223]) characterized clones from a family of highly repeated sequences present in a heterochromatin-rich maize line by sequencing and chromosome location. By means of in situ hybridization experiments, they found that the repeats are mainly located in the knob heterochromatin of the A chromosomes and the centromeric heterochromatin of the B chromosome. However, some copies are also distributed in euchromatic regions of the A chromosomes and in the distal heterochromatic block of the B chromosome.

Ananiev and colleagues (1998[224]) isolated a class of tandemly repeated DNA sequences (TR-1) of 350-bp unit length from the knob DNA of chromosome 9 of *Zea mays* L. Comparative fluorescence in situ hybridization revealed that TR-1 elements are also present in cytologically detectable knobs on other maize chromosomes in different proportions relative to the previously described 180-bp repeats. At least one knob on chromosome 4 is composed predominantly of the TR-1 repeat. In addition, several small clusters of the TR-1 and 180-bp repeats have been found in different chromosomes, some not located in obvious knob heterochromatin. Variation in restriction fragment fingerprints and copy number of the TR-1 elements was found among maize lines and among maize chromosomes. TR-1 tandem arrays up to 70 kb in length can be interspersed with stretches of 180-bp tandem repeat arrays. DNA sequence analysis and restriction mapping of one particular stretch of tandemly arranged TR-1 units indicate that these elements may be organized in the form of fold-back DNA segments. The TR-1 repeat shares two short segments of homology with the 180-bp repeat. The longest of these segments (31 bp; 64% identity) corresponds to the conserved region among 180-bp repeats. The polymorphism and complex structure of knob DNA suggest that, similar to the fold-back DNA-containing giant transposons in *Drosophila*, maize knob DNA may have some properties of transposable elements.

[223] Viotti, A., E. Priviterra, E. Sala, and N. Pogna. 1985. Distribution and clustering of two highly repeated sequences in the A and B chromosomes of maize. *Theoretical and Applied Genetics*, 70: 234–239.

[224] Ananiev, E. V., R. L. Phillips, and H. W. Rines. 1998a. A knob-associated tandem repeat in maize capable of forming fold-back DNA segments: Are chromosome knobs megatransposons? *Proceedings of the National Academy of Sciences USA*, 95 (18): 10785–10790.

miRNA in Maize

Knowledge of and interest in the biogenesis and activity of diverse classes of small noncoding RNAs (sRNAs) has boomed recently. These include microRNAs (miRNAs), small interfering RNAs (siRNAs), transacting siRNAs (ta-siRNAs), and others. A great deal of interest has been placed on miRNAs due to their ability to act post-transcriptionally, regulating gene expression. The critical regulatory behavior of miRNAs is evident at key positions in a variety of pathways, such as in root, shoot, leaf, and flower development and cell fate. Additionally, they also include responses to phytohormones, limited nutrient availability, and other environmental stresses.

Zhang and colleagues (2009[225]) predicted miRNA targets computationally based on the most recent maize protein annotations. Analysis of the predicted functions of target genes, on the basis of gene ontology, supported their roles in regulatory processes. By analyzing the synteny of orthologs of sorghum, they found that maize-homoeologous miRNA genes were retained more frequently than expected. They also explored miRNA nucleotide diversity among many maize inbred lines and partially inbred teosinte lines. The results indicated that mature miRNA genes were highly conserved during their evolution.

Taken together, it is apparent that miRNA regulation is intertwined with key plant development processes. Thus, there is considerable interest in taking advantage of the complete genome sequence of maize B73 reference genome version 1 (Schnable et al., 2009[226]) to systematically identify miRNA genes, and their corresponding targets, and to decipher their regulatory roles.

225 Zhang, Lifang, C. Jer-Ming, K. Sunita, Joshua C. Stein, L. Zhijie, Apurva Narechania, Christopher A. Maher, Katherine Guill, Michael D. McMullen, and Doreen Ware. 2009. A genome-wide characterization of micro RNA genes in maize. *PLoS Genetics*. November.

226 Schnable, P. S., D. Waren, R. S. Fulton, J. C. Stein, F. Wei, S. Pasternack, C. Liang, J. Zhang, L. Fulton, T. A. Graves, P. Minx, A. D. Reily, L. Courtney, S. S. Kruchowski, C. Tomlinson, C. Strong, K. Delehaunty, C. Fronick, B. Courtney, S. M. Rock, E. Belter, F. Du, K. Kim, R. M. Abbott, M. Cotton, A. Levy, P. Marchetto, K. Ochoa, S. M. Jackson, B. Gillam, W. Chen, Le Yan, J. Hihhinbotham, M. Cardenas, J. Walogorski, E. Applebaum, L. Phelps, J. Falcone, K. Kanchi, T. Thane, A. Scimone, N. Thane, J. Henke, T. Wang, J. Ruppert, N. Shah, K. Rotter, J. Hodges, E. Ingenthron, M. Cordes, S. Kohlberg, J. Sgro, B. Delgado, K. Mead, A. Chinwalla, S. Leonard, K. Crouse, K. Collura, D. Kudma, J. Currie, R. He, A. Angelova, S. Rajasekar, T. Mueller, R. Lomely, G. Scara, A. Ko, K. Delaney, M. Wissotski, G. Lopez, D. Campos, M. Braidotti, E. Ashley, W. Golser, H. Kim, S. Lee, J. Lin, Z. Dujmic, W. Kim, J. Talag, A. Zuccolo, C. Fan, A. Sebastian, M. Kramer, L. Spiegel, L. Nascimento, T. Zutavern, B. Miller, C. Ambroise, S. Muller, W. Spooner, A. Narechania, L. Ren, S. Wei, S. Kuman, B. Faga, M. J. Levy, L. McMahan, P. Van Buren, M. W. Vaughn, K. Ying, C. Yeh, S. J. Emrich, Y. Jia, A. Kalyanaraman, A. Hsia, W. B. Barbazuk, R. S. Baucom, T. P. Brutnell, N. C. Carpita, C. Chaparro, J. Chia, J. M. Deragon, J. C. Estill, Y. Fu, J. A. Jeddeloh, Y. Han, H. Lee, P. Li, D. R. Lisch, S. Liu, Z. Liu, D. Holligan Hagel, M. C. McCann, P. SanMiguel, A. M. Myers, D. Nettleton, J. Nguyen, B. W. Penning, L. Ponnala, K. L. Schneider, D. C. Schwartz, A. Sharma, C. Soderlung, N. M. Springer, Q. Sun, H. Wang, M. Waterman, R. Westerman, T. K. Wolfgruben, L. Yang, Y. Yu, L. Zhang, S. Zhou, Q. Zhu, J. L. Bennetzen, R. Kelly Dawe,

To investigate the evolution of miRNA loci, Zhang and colleagues (2009) sequenced 28 loci and flanking regions in panels of inbred and teosinte lines. There was no polymorphism detected within the mature miRNA sequences. Their conservation within maize throughout its evolution is expected given the importance of miRNA genes in suppressing target gene expression during development and stress. The flanking regions displayed diversity levels similar to protein coding genes, indicating that purifying selection is limited to mature miRNAs. None of the 28 loci tested exhibited the extreme reductions in diversity in inbreds relative to teosinte accessions that would be indicative of artificial selection during domestication or crop improvement. MicroRNA loci may control such fundamental processes in development; thus alterations of sequence or expression are not tolerated.

The Structure of the Maize Plant

The concept of plant structure in grasses has evolved from morphological and histological studies and has led to the concept of the phytomer as a morphogenetic basic unit in the formation of the structure of the grass plant. The maize plant bears leaves and buds on opposite sides of successive nodes of the main axis of the culm or plant axis. This repetition, although with modifications, is perceived in the tassel branches and the ear, where phytomers are present as basic units, equivalent to those of the main axis, with their respective morphological modifications (Cutler and Cutler, 1948[227]).

The classical concept of the vegetative phytomer is that it is a basic anatomical unit of grasses composed of an internode with a leaf at its upper end and a bud at its lower end in a position opposite to the leaf (Sharman, 1942;[228] Weatherwax, 1923[229]). Adventitious roots arise from the basal plate of the phytomer (Cutler and Cutler 1948). Another view of the phytomer is based on vascular and physiological associations and states that it is composed of a leaf, lateral bud, and adventitious roots, all connected to the lower end of the associated internode (Bossinger et al., 1992[230]). The internode of the grasses is intercalated between two meristematic plates (Cutler and Cutler, 1948).

For a long time it has been recognized and reaffirmed that there is homology of the ear and tassel. Both the central spike of the tassel and the ear originate

J. Jiang, N. Jiang, G. G. Presting, S. R. Wessier, S. Aluru, R. A. Martienssen, S. W. Clifton, W. R. McCombie, R. A. Wing, and R. K. Wilson. 2009. The B73 maize genome: Complexity, diversity, and dynamics. *Science*, 326 (5956): 1112–1115.

[227] Cutler, H. C., and P. C. Cutler. 1948. Studies on the structure of the maize plant. *Annals of the Missouri Botanical Garden*, 35: 301–316.

[228] Sharman, B. C. 1942. Developmental anatomy of the shoot of *Zea mays* L. *Annals of Botany*, 6: 245–282.

[229] Weatherwax, P. 1923. *The Story of the Maize Plant*. University of Chicago Press. Chicago.

[230] Bossinger, G., M. Maddaloni, M. Motto, and F. Salamini. 1992. Formation and cell lineage patterns of the shoot apex of maize. *The Plant Journal*, 2: 311–320.

from the reduction of branches of a panicle to one pair of spikelets for each member (Mangelsdorf and Reeves, 1939; Reeves, 1953[231]). The lateral spikelet initials were shown to be homologous with branch initials in maize (Bonnett 1940,[232] 1948,[233] 1953[234]).

It would be expected that if there is a change in one part of the plant, similar changes would be reflected in other parts of the plant according to a new genetically based plan. If there is condensation in the tassel, something similar would be experienced in the stalk or the ears. For example, it was observed that increased condensation of the tassel is accompanied by an increase in row number in the ears of Mexican and North American varieties of maize (Anderson, 1944;[235] Nickerson, 1954[236]).

Organs that are not homologous may thus be subject to change due to the genetic change in the overall pattern of growth of the plant. But these changes are easily observable in the tassel and ear, whereas the leaves will not be affected in their distichous disposition as the growth pattern changes.

Galinat (1956[237]) has contributed to this discussion with a study of the significance of the cupulate fruitcase in its evolution from a common ancestor to teosinte and maize and in posterior evolution. Mangelsdorf and Reeves (1939), in their extensive work on maize and its relatives, concluded that the maize ear has evolved from an ancestor with a perfect flower in which the pistillate inflorescences aborted in the upper half and the staminate inflorescences remained in the lower half. Such double-structured ears are found both archaeologically and in present races of maize in Peru with relatively high frequencies. We have depicted an ideotype of such inflorescences based on our archaeological observations at the Los Gavilanes site (Grobman, 1982: figure 60, 167). Although this view has been challenged by Iltis (1983) and others, who postulate a chain of successive modifications or a sudden catastrophic occurrence that changed the spike of teosinte into the ear of maize, we still hold that a simple hormonal influence area in parts of the dioecious inflorescence of the ancient precursor of maize led to the monoecious actual differentiation of the inflorescences of maize, without having to resort to the complex

[231] Reeves, R. G. 1953. Comparative morphology of the American Maydeae. *Texas Agricultural Experiment Station Bulletin*, 761: 3–26.

[232] Bonnett, O. T. 1940. The development of the staminate and pistillate inflorescences of sweet corn. *Journal of Agricultural Research*, 60: 25–37.

[233] Bonnett, O. T. 1948. Ear and tassel development in maize. *Annals of the Missouri Botanical Garden*, 25: 260–287.

[234] Bonnett, O. T. 1953. *Developmental Morphology of the Vegetative and Floral Shoots of Maize.* Illinois Agricultural Experiment Station Bulletin 568. 47 pp.

[235] Anderson, E. 1944. Homologies of the ear and tassel in *Zea mays*. *Annals of the Missouri Botanical Garden*, 31: 325–340.

[236] Nickerson, N. N. 1954. Morphological analysis of the maize ear. *American Journal of Botany*, 41: 87–92.

[237] Galinat, W. C. 1956. Evolution leading to the formation of the cupulate fruit case in the American Maydeae. *Botanical Museum Leaflets*. Harvard Univeristy, 17: 217–239.

lucubrations and hypothesizing required for explaining the evolution of the ear of maize starting from teosinte.

The cobs of archaeological maize found in Peru at Los Gavilanes, at Cerro El Calvario, and at the highland Guitarrero Cave, among other sites, have dates that are similar to the earliest archaeological maize in Mexico, with no indications of teosinte morphological characteristics present at a date as early as about 6700 years BP (see the afterword at the end of the appendix).

Key Genes Involved and Their Variation in the Process of Maize Domestication

A number of key genes have been involved in the process of domestication of maize through a process of directional selection. These are *tb1*, which changes plant habit (Wang et al., 1999); *c1*, which regulates anthocyanin formation (Hanson et al., 1996); the group *bt2*, *ae1*, and *su1*, which is involved in the starch pathway (Whitt et al., 2002); *zagl1*, a transcription factor (Vigouroux, McMullen, et al., 2002); *d8*, which is involved in sex determination (Tenaillon et al., 2001[238]); and *ts2*, which is involved in sex determination (Harberd and Freeling, 1989[239]). Through their investigation on genes *tb1, d8, ts2*, and *zagl1*, Tenaillon and colleagues (2004[240]) found a loss of nucleotide diversity of 38%, but it was skewed downward for the four selected genes. A bottleneck effect was likely, but use of statistical approaches did not rule it as conclusive. Gene s*ts2* and *d8* appear more likely to have been selected during the process of breeding after domestication.

One additional important QTL is *tga1* or *glume architecture 1*. This gene, which is recessive to *tga1* in maize, controls the depth of the cavity in which is inserted the grain of teosinte, the formation of a cupule that grows and extends as a modified glume and encloses the grain, and the induration and silification of the glume and rachis segment of teosinte. Wang and colleagues (2005[241]) propose that it belongs to the SBP-domain family of transcriptional regulators and that it controls the key phenotypic difference between maize and teosinte in regard to their respective grain structures: enclosed or open, indurated or not, different angles of insertion, and fragmentable rachis versus firm rachis. *Tga1* maps to a 1-kb region, within which maize and teosinte show only seven fixed

[238] Tenaillon, M., M. Sawkins, A. Long, R. Gaut, J. Doebley, and B. Gaut. 2001. Patterns of DNA sequence polymorphism along chromosome 1 of maize (*Zea mays* ssp. *mays*). *Proceedings of the National Academy of Sciences USA*, 98: 9161–9166.

[239] Harberd, N. P., and M. Freeling. 1989. Genetics of dominant gibberellins-insensitive dwarfism in maize. *Genetics*, 121: 827–838.

[240] Tenaillon, M. I., J. U'Ren, O. Tenaillon, and B. S. Gaut. 2004. Selection versus demography: A multilocus investigation of the domestication process in maize. *Molecular Biology and Evolution*, 21: 1214–1225.

[241] Wang, Huai, Tina Nussbaum-Wagler, Bailin Li, Qiong Zhao, Yves Vigouroux, Marianna Faller, Kirsten Bomblies, Lewis Lukens, and John F. Doebley. 2005. The origin of the naked grains of maize. *Nature*, 436: 714–719.

differences in their DNA sequences. One of these differences encodes a non-conservative amino acid substitution and may affect protein function, and the other six differences potentially affect gene regulation. This region according to interpretation of molecular analysis by the authors, could have been the target of selection during maize domestication.

Higher levels of genetic diversity in terms of DNA polymorphism were found in teosinte ssp. *parviglumis* and ssp. *mexicana* as compared to maize inbred line B73 in the *tb1* to gene 3 region (Clark et al., 2004[242]). The study of this region is important because the *tb1* gene is associated with the changes in plant habit that might have occurred during domestication of maize from an ancestor that could have been teosinte or a closely associated wild maize, related to but morphologically different from teosinte. It appears that during the domestication process, through a bottleneck effect derived from a limited initial population and direction over which the selection process for agronomically valuable genes was enacted (Whitt et al., 2002; Zhang et al., 2002[243]), there would have followed a loss of variation. However, it is important to note that inbred line B73 – a Corn Belt Dent race – has suffered a very strong selection process, first in the development of that race; second by farmers, who subjected several varieties of Corn Belt Dent to strong selection for ear type, number of ears per plant, and ear size; and finally by the selection of breeders of B73 as an inbred line derived from Stiff Stalk Synthetic, a restricted population developed by Dr. George F. Sprague at Arlington Farms, Virginia, and then improved at Iowa State University by him and his collaborators. We must note that the comparison of variation of the *tb1* region in teosinte is being made with an extreme type of modern maize. A more meaningful comparison to establish primary differences in variability of DNA polymorphisms at the *tb1* region, would be needed. It is suggested that it be made with primitive maize races from the Peruvian Andean region which, through isolation from teosinte for thousands of years, would possibly yield results more indicative of the extent of change that took place in the original setting of domestication. Even better comparisons would be those made with archaeological maize specimens from Peru, which do not exhibit teosinte introgression.

A test was made of the hypothesis that transcription factors are involved in the evolution of morphological characteristics of plants than other genes (Zhao et al., 2011[244]). Zhao and colleagues selected a family of transcription

[242] Clark, R. M., E. Linton, J. Messing, and J. F. Doebley. 2004. Pattern of diversity in the genomic region near the maize domestication gene *tb1*. *Proceedings of the National Academy of Sciences USA*, 101: 700–707.

[243] Zhang, L., A. S. Peek, D. Dunams, and B. S. Gaut. 2002. Population genetics of duplicated disease-defense genes, *hm1* and *hm2*, in maize (*Z. mays* ssp. *mays* L.) and its wild ancestor (*Z. mays* ssp. *parviglumis*). *Genetics*, 162: 851–860.

[244] Zhao, Q., A. Weber, M. McMullen, K. Guil, and J. F. Doebley. 2011. MADS-box genes of maize: Frequent targets of selection during domestication. *Genetics Research*, 93: 65–75.

factors – the MADS-box genes of maize, which are key regulators of vegetative and floral development – and sequenced 32 of them from a diverse set of maize and teosinte. They came to the conclusion that more MADS-genes were under selection during maize domestication and improvement than a randomly chosen set of 32 other genes with which they were compared, and the differences were statistically significant. Both inbred lines and landraces had fewer segregating sites (SNPs) than teosinte, and – as expected – they had expanded lengths, which might be seen as new nucleotide additions since domestication.

Although it is now believed that MADS-box genes were of importance in the evolution of maize, it is still too early to ascertain what their specific role in the process of domestication was.

A study conducted on two perennial teosinte species, *Zea diploperennis* (diploid) and *Zea perennis* (tetraploid), on the basis of their relative DNA sequence diversity of a number of selected genes (*Adh1, glb1, c1*, and *waxy*) showed that the two species did not differ much in terms of genetic diversity. It is likely that *Z. perennis* is of autotetraploid origin, and based on the different genes selected, the time of its formation as a species is very recent (Tiffin and Gaut, 2001[245]). The estimated divergence times between these species could be estimated, on the basis of nucleotide substitution rates for the genes studied, as 381,000–563,500 years (*Adh1*), 184,700–1,168,000 years (*glb1*), 41,800–251,000 years (*c1*), and 24,000–142,000 years (*waxy*). In evolutionary histories this is a relatively recent time period. There is no evidence for the occurrence of a selection bottleneck in the development of the tetraploid *Z. perennis.* Also, because of the wide distribution of differential alleles in the genealogies of *Z. perennis* studied, there is the possibility that there could have been multiple origin events. At the four loci studied, the genetic diversity of both *Z. diploperennis* and *Z. perennis* is lower than *Z. mays* ssp. *parviglumis* and the domesticate *Z. mays* ssp. *mays. Z. perennis* has higher diversity than that found in other studies on *Z. luxurians.* At the *c1* locus, *Z. mays* ssp. *mays* has the lowest diversity when compared to the other species. It has been hypothesized that the gene *c1*, which is involved in anthocyanin synthesis of the aleurone layer of the maize seed, may be of recent domesticate origin.

Isozyme and genetic sequence studies provide evidence that *Z. mays* ssp. *parviglumis* has the greatest amount of genetic diversity, whereas *Z. luxurians* has the lowest. Why this is so, that is, a diploid having more variability than a tetraploid, is puzzling, because theoretically a tetraploid has a larger gene population, and the reverse should be true. The data of the Tiffin and Gaut (2001) study, especially at the *c1* locus, provide evidence of introgression between taxa. In the case of the *c1* gene, there is strong evidence for

[245] Tiffin, P., and B. S. Gaut. 2001. Sequence diversity in the tetraploid *Zea perennis* and the closely related diploid *Z. diploperennis*: Insights from four nuclear loci. *Genetics*, 158: 401–412.

Zea mays having had introgression with *Z. diploperennis*. This is supported by isozyme data (Doebley et al., 1984[246]) and the isolation of maize-like ITS sequences from both *Z. diploperennis* and *Z. perennis* by Buckler and Holtsford (1996[247]). Furthermore, it is known that maize and *Z. diploperennis* cross easily in areas in Mexico where both are present. Cámara-Hernández and Mangelsdorf (1981[248]) and Mangelsdorf and colleagues (1981[249]) obtained fertile crosses between the maize primitive race Palomero Toluqueño of Mexico and *Zea diploperennis*, which in the F_1 and backcross generations resembled annual teosinte. The research material they used was the popcorn race Palomero Toluqueño from Jalisco, Mexico. Given the chromosome knob position counts made by McClintock and colleagues (1981) on this race, which they listed as Reventador in their chromosome knob position tables, and given the high number of basal tillers, it is evident that this maize race has experienced considerable teosinte introgression. Even so F_2 ears of this cross had a strong dominance of the maize parent phenotype. It is a fact that most maize alleles in teosinte × maize crosses show dominance over their teosinte counterparts.

Five chloroplast haplotypes have been described (Buckler, Goodman, et al., 2006[250]). Haplotype 1 appears to be basal and most likely ancestral to all teosinte species and found in *Zea luxurians, Zea perennis*, and *Zea diploperennis*. Haplotype 2 is found only in *Zea mays* ssp. *huhuetenangensis*. Haplotypes 3, 4, and 5 have been derived more recently. There is a significant difference of hapolotype 1 and 2 frequency between *Zea mays* ssp. *parviglumis* and *mexicana*. Derived haplotypes 4 and 5 are only found in ssp. *mexicana*. Haplotype 3 is ubiquitous. Because average linkage cluster analysis had established that the two subspecies are well differentiated and the isozyme analysis establishes that in all likelihood an eastern population of ssp. *parviglumis* is basal to ssp. *mexicana*, using the chloroplast haplotype data, we can get a better indicator of differentiation, because chloroplast genes tend to be conserved. The chloroplast data indicates clearly that ssp. *parviglumis* and *mexicana* have a different origin and diverged a long time ago. A crossability barrier between ssp. *parviglumis* and ssp. *mexicana*, but no barrier between ssp. *parviglumis* and

[246] Doebley, J. F., M. M. Goodman, and C. W. Stuber. 1984. Isoenzymatic variation in *Zea* (Gramineae). *Systematic Botany*, 9: 203–218.

[247] Buckler, E. S., and T. P. Holtsford. 1996. *Zea* systematic: Ribosomal ITS evidence. *Molecular Biology and Evolution*, 13 (4): 612–622.

[248] Cámara-Hernández, J., and P. C., Mangelsdorf. 1981. *Perennial Corn and Annual Teosinte Phenotypes in Crosses of Zea Diploperennis and Maize*. Publication No. 10. The Bussey Institution of Harvard University. Cambridge. pp. 3–37.

[249] Mangelsdorf, P. C., L. M. Roberts, and J. S. Rogers. 1981. *The Probable Origin of Annual Teosintes*. Publication No. 10. The Bussey Institution of Harvard University, Cambridge. pp. 39–69.

[250] Buckler, E. S., M. M. Goodman, T. P. Holtsford, J. F. Doebley, and J. Sanchez. 2006. Phylogeography of the wild subspecies of *Zea mays*. *Maydica*, 51: 123–134.

maize in Mexico, has been reported by Kermicle (1997[251]). Unless chloroplast haplotype data are further analyzed in different maize backgrounds, this riddle will not be solved.

Present-day races or subspecies of teosinte *parviglumis* and *mexicana* may be considered in some sense "anthropogenic artifacts" in the same way that Doebley and Iltis (1980[252]) considered maize to be such an anthropogenic artifact, because both maize and teosinte have introgressed each other, as postulated by Mangelsdorf and colleagues (1981: 51) as follows: "... in the sense that the activities of man in changing the relationship of their ancestors from allopatric to sympatric made possible the hybridization that gave them birth." Hufford, Xu, and colleagues (2011[253]) have proposed that their data establish much stronger selection during domestication than during the improvement period, with a more pronounced bottleneck on diversity observed between the putative wild ancestor (*Zea mays* ssp. *parviglumis*) and landraces than between landraces and improved lines. Until their full data is published, we must consider the alternative that *Zea mays* ssp. *parviglumis* may show an increased variability over maize through the result of hybridization of a wild maize ancestor with a wild teosinte precursor, rather than through a bottleneck effect, which would lead to a different interpretation of the results.

In the process of studying, through a fine mapping approach, the effect of QTLs and specifically the region of the teosinte branched *tb1* gene, which conditions regulatory changes of large effect in morphological differences that distinguish maize (*Zea mays* ssp. *mays*) from teosintes (*Z. mays* ssp. *parviglumis* and *mexicana*), Clark and colleagues (2006) showed that sequences more than 41 kb upstream to *tb1* act in cis to alter *tb1* transcription. They also established that their findings show that large stretches of the noncoding DNA that comprise the majority of many plant genomes can be a source of variation affecting gene expression and quantitative phenotypes, and that maternal and environmental influences affected trait values of individuals when the effect of the *tb1* region of teosinte was isolated into a maize background. The hypothesis that the whole region was carried in a selection sweep is required to explain this situation.

It is well known that some of these noncoding stretches of DNA are composed of transposons that have been present in maize for thousands of years

[251] Kermicle, J. 1997. Cross compatibility within the genus *Zea*. In J. A. Serratos, M. C. Willcox, and F. Castillo Gonzáles, editors. *Gene Flow among Maize Land Races, Improved Maize Varieties and Teosinte*. CIMMYT Forum, Mexico City, D.F. pp. 40–43.

[252] Doebley, J. F., and H. H. Iltis. 1980. Taxonomy of *Zea* (Gramineae). 1. A subgeneric classification with a key to taxa. *American Journal of Botany*, 67: 982–993.

[253] Hufford, M. B., X. Xu, J. van Heerwarden, T. Pyhäjärvi, J. M. Chia, R. A. Cartwright, R. I. Elshiere, J. C. Glandbitz, R. E. Grill, S. Kaeppler, J. Lai, L. M. Shamon, C. Song, N. M. Springer, R. A. Swanson-Wagner, P. Tiffin, J. Wang, G. Zhang, J. Doebley, M. D. McMullen, E. S. Buckler, D. Ware, S. Yung, and J. Ross-Ibarra. 2011. Genome-wide effects of domestication and improvement in landraces and modern maize. *Maize Genetics Conference Abstracts*, 53: T06.

previous to its domestication and that the maize plant morphology could have been patterned long before it encountered teosinte in a sympatric condition. The presence of QTLs and cis-regulatory sequences at relatively distant positions from the *tb1* genes in an essentially noncoding region is quite interesting. This is evidence that the hypothesis of the construction of the maize genome based on the alteration of some major genes of teosinte to transform it into maize is too simplistic, is not completely understood, and needs to be continuously reassessed.

Pickersgill (2007[254]) established that in spite of difficulties in defining domestication (see Casas et al., 1999;[255] Gepts, 2004[256]), most workers agree that there were several independent regions of plant domestication in the Americas and that, quite frequently, different species of the same genus were domesticated independently, in different regions and by different peoples. Therefore, the transfer of a wild or semidomesticate maize from Mesoamerica to South America and its independent full domestication cannot be ruled out, when all the evidence is assembled and examined. Among other examples covering the various species that she exposed, in support of this position, the fact that sweet corn could have arisen independently in North America, Mexico (Whitt et al., 2002), and South America, specifically the Peruvian central Andes region (Mangelsdorf, 1974), is a distinct possibility that reinforces the possibility of early independent evolution after domestications in the Andean region, selected separately from Mesoamerica and Mexico.

GAMETOPHYTE GENES AS AN ISOLATING MECHANISM IN *ZEA MAYS*

Pollination in wind-pollinated plants requires physiological interaction between pollen and pistil to regulate hybridization. Incompatibility factors act in some cases to prevent hybridization as a stabilizing selection mechanism. In many outcrossing species, genetic mechanisms exist to prevent self-fertilization (self-incompatibility [SI]) and crossing among individuals of the same species or subspecies (cross-incompatibility [CI]), minimizing inbreeding and furthermore isolating taxa.

Studies based on physiological and molecular biology analysis have revealed the mechanisms of physiological self-incompatibility and genetic self-incompatibility that are present in many species, including the *Zea mays* species, both within the ssp. *mays* and within its teosinte relatives. Incompatibility may be total in the presence of viable pollen and receptive stigmas in some cases, whereas it

254 Pickersgill, B. 2007. Domestication of plants in the Americas: Insides from Mendelian and molecular genetics. *Annals of Botany*, 100 (5): 925–940.

255 Casas, A., J. Caballero, A. Valiente-Banuet, J. A. Soriano, and P. Dávila. 1999. Morphological variation and the process of domestication of *Stenocereus stellatus* (Cactaceae) in central Mexico. *American Journal of Botany*, 86: 522–533.

256 Gepts, P. 2004. Crop domestication as a long-term selection experiment. *Plant Breeding Review*, 24: 1–44.

might be partial in other cases. These mechanisms have emerged as a system of ensuring against inbreeding, but evidence is mounting that they may also have emerged as a mechanism of natural stabilizing selection for protection against crossing between closely related taxa and against different emerging populations in the course of plant evolution.

Although genetic systems for self-incompatibility have been found in many plant species, it is strange that they should be present in maize. Maize is a monoecious plant with separate functional staminate flowers borne in the tassel and with pistillate flowers borne on ears, branching out in numbers of one or two or as many as four, growing from axils of leaves. A few popcorn races, such as Argentine popcorn, may have many more ears, up to eight. As a comparison, a teosinte plant may have as many as 100 spikes or female "ears" on a single plant. A single maize plant was estimated by Sturtevant (1881[257]) to produce 18,000,000 pollen grains, whereas Kiesselbach (1949[258]) reckons that a maize plant of the Corn Belt Dent race, growing in Kansas, will produce about 25,000,000 pollen grains. Such a large quantity of pollen, if compared to, at most, 1,000 silks (stigmas) per ear, would have a ratio of 25,000 pollen grains available to pollinate one stigma, and thus pollination is ensured except under the most drastic drought conditions.

Maize is essentially an outcrossing species, and it is estimated that no more than 10% of the seeds of a plant, and perhaps even less in plants that exhibit protandry – which is a more frequent trait in maize, in which pollen is shed over a period of time ahead of the emergence of the silks – are formed with pollen from the same plant. Outcrossing ensures that populations of maize plants avoid inbreeding, which is the normal case in other grain cereals. Being easy outcrossers, teosinte and maize have been able to pollinate each other freely and have to some extent exchanged genes in a restricted but permanent reciprocal gene flow. The tremendous outcrossing capacity of maize has undoubtedly been one of the reasons for increasing and maintaining its explosive variability, resulting in some 400 cultivated races registered at present around the Americas and other continents.

Given the physiological evolution of outcrossing in maize, there arises the intriguing question of why maize has developed genetic self-incompatibility systems to avoid inbreeding and other systems, to be detailed subsequently, which would appear redundant. The alternative explanation is that the self-incompatibility mechanisms built a genetic barrier that preceded the domestication of maize and was present in maize, teosinte, and *Tripsacum*.

There are a number of nuclear genes that condition sterility in maize when in a heterozygous condition. In addition, there are some 850 translocations identified that result in some 1,700 terminal chromosome segments generating

[257] Sturtevant, E. L. 1881. The superabundance of pollen in Indian corn. *American Naturalist*, 15: 1000.

[258] Kiesselbach, T. A. 1949. *The Structure and Reproduction of Corn*. University of Nebraska, College of Agriculture, Agricultural Experiment Station, Bulletin No. 161.

deficiency-bearing gametophytes, resulting in pollen sterility. This indicates that a very large number of genes are vital for the development of the gametophyte in maize (Kindiger, 1986[259]).

GAMETOPHYTE ISOLATION BARRIERS DUE TO NUCLEAR GENES

The existence of nuclear genes, called gametophyte genes (*Ga* series genes), which induce partial sterility and segregation ratios in crosses between plants differentiated into races or subspecies of a number of species, including rice, barley, maize, and so on, has been known for a long period of time. Nakagahra and colleagues (1972[260]) found a series of three *Ga* genes in rice subspecies, making it difficult to cross rice varieties from different subspecies or geographical origins.

There is evidence of the existence of a set of genes, the gametophyte (*Ga*) gene series, which acts as a barrier to reproduction and which is recognized also by distortion of Mendelian ratios in crosses between plants that possess these factors (Emerson, 1934;[261] Jiménez and Nelson, 1965;[262] Jones, 1920;[263] Mangelsdorf and Jones, 1926;[264] Schwartz, 1950[265]). The incompatibility barrier is exercised by the male gametophyte. In maize it is frequently present in modern popcorn races, which are all directly derived from primitive precursors. They tend to foster genetic isolation of these races from other races, and this may be one of the reasons for their persistence in spite of their lower yield. An interesting observation is that *Ga* genes are present in greater frequency in maize from the Andean region (Mangelsdorf, 1974). One additional point of interest is that the *Ga1* gene locus is in chromosome 4S.32, where the most important block of genes differentiating maize from teosinte is also located, and it is linked with *su1* gene at 4S.66. This monolocus gene identified by Schwartz (1950) has an allele from popcorn that is partially dominant (*Ga1-s*). This allele is partially dominant such that *Ga1-s/Ga1-s* silks are commonly nonreceptive to *ga1* pollen, but *Ga1-s/ga1-s* silks are partially receptive or fail to be fertilized by *ga1* pollen even in the absence of competing *Ga1-s* pollen, according to Nelson (1952[266]). On *Ga1-s/Ga1-s* pistils, *ga1* pollen, fails to effect fertilization, even

259 Kindiger, B. 1986. Development abnormalities in hypoploid pollen grains of *Zea mays* L. Ph.D. dissertation. University of Missouri. Columbia.

260 Nakagahra, Masahiro, Takeshi Omura, and Naburo Iwata. 1972. Gametophyte genes and their loci on the eleventh linkage group of cultivated rice. *Japanese Journal of Breeding*, 22 (6): 305–312.

261 Emerson, R. A. 1934. Relation of the differential fertilization gene *Ga ga* on certain other genes of the *Su-Tu* linkage group in maize. *Genetics*, 19: 137–156.

262 Jiménez, J. R., and O. E. Nelson. 1965. A new fourth chromosome gametophyte locus in maize. *Journal of Heredity*, 23: 259–263.

263 Jones, D. F. 1920. Selective fertilization in pollen mixtures. *Biology Bulletin*, 38: 251–289.

264 Mangelsdorf, P. C., and D. F. Jones. 1926. The expression of Mendelian factors in the gametophyte of maize. *Genetics*, 11: 423–455.

265 Schwartz, D. 1950. The analysis of a case of cross-sterility in maize. *Proceedings of the National Academy of Sciences USA*, 36: 719–724.

266 Nelson, O. E. 1952. Non-receptive cross-sterility in maize. *Genetics*, 37: 121–124.

in the absence of competing *Ga1-s* pollen, resulting in barrenness. The same is true for the *Tcb1-s* allele of the *teosinte crossing barrier 1* locus discovered in a population of annual teosinte (Kermicle and Allen, 1990[267]).

It has been found, furthermore, that pollen-pistil incompatibility distorts transmission of alleles at six or more loci (Nelson, 1993[268]). Gametophyte factors *ga2*, *ga7*, *ga8*, *ga9*, and *ga10* have been identified on chromosomes 3, 5, 9, and 1. They act by enhancing the competition of pollen of, for example, *Ga1*, over *ga1* on *ga1* maize female stigmas (silks). This is a potent system that was evolved in maize plants as a barrier to prevent outcrossing among some races and may help explain why primitive races of maize have persisted in the presence of other races, for example, popcorn races and the race Chullpi, the sweet-grained maize race from Peru whose *su* gene is linked in chromosome 4 (see previously). Its higher frequency in the Andean region where there is no presence of teosinte might indicate that the genetic barrier was maintained by natural selection in some primitive popcorn races from the time of their origin, as a barrier to teosinte introgression, which was retained when other races evolved from hybridization with teosinte and possibly with introgression from *Tripsacum*.

Reproductive isolation between maize and teosinte may be conditioned by the presence of the *Ga2-s* allele, excluding therefore the presence of hybrids that might be less fit. It has been known that some weedy populations of teosinte, especially in the Mexican highlands, resemble local maize in plant color, vegetative period, and general plant morphology to the point of mimicking maize very well. However, hybridization and its products have been found to occur less frequently than in the case where wild populations of teosinte grow in dense masses of plants in close proximity to maize fields, as happens in the Balsas River region with *parviglumis* teosinte (Wilkes, 1977). Additionally, Wilkes (1967) suggested that the lower than expected frequency of hybrids between maize and teosinte in fields where teosinte grew next to maize as a weed might be based on pollen-pistil incompatibility of the sort associated with the *Ga1-s* gene.

The allele *Ga1-s* was recently reported in annual teosinte populations (Kermicle et al., 2006[269]). However, all of the associated landrace maize populations carried the cross-neutral allele *Ga1-m*, which fertilizes *Ga1-s* but accepts *ga1* pollen. Thus it is not obvious how *Ga1-s* could serve as a primary barrier to crossing in this circumstance.

An analogous gene, *Tcb1-s*, was found in some teosinte populations but not in sympatric or parapatric maize. Pistils carrying *Tcb1-s* are unreceptive to pollen carrying the *tcb1* allele of this locus and not containing the same allele

[267] Kermicle, J. L., and J. O. Allen. 1990. Cross incompatibility between maize and teosinte. *Maydica*, 35: 399–408.

[268] Nelson, O. E. 1993. The gametophyte factors of maize. In M. Freeling and V. Walbot, editors. *The Maize Handbook*. Springer-Verlag. New York/Berlin/Heidelberg. pp. 496–503.

[269] Kermicle, J. L., S. Taba, and M. M. S. Evans. 2006. The *gametophyte-1* locus and reproductive isolation among *Zea mays* subspecies. *Maydica*, 51: 219–225.

Tcb1-s (Evans and Kermicle, 2001[270]). Kermicle (2006[271]) studied 14 teosinte populations and their *tcb1* alleles among 13 of these populations. His findings implicate *Tcb1-s* as providing reproductive isolation while promoting its own propagation. The crossing barrier gene *Tcb1-s* was found as predominant in the five weedy teosinte populations tested, was present in one population and one other plant of four weedy populations, and was found in one of four *parviglumis* populations, all classified as wild, indicating an association of *Tcb1-s* with an isolating ecology of the teosinte populations. Pollen carrying the introgressed *Tcb1-s* segment fertilizes *tcb1/tcb1* maize less efficiently than *tcb1* pollen. This reduction could be significant in preventing *Tcb1-s* from becoming established in open-pollinated maize landraces that grow sympatrically with teosinte.

Kermicle (2006) suggested that assortative fertilization spurs postzygotic ecological isolation but without involving the classical reinforcement sequence. In typical cases, prezygotic isolation, such as that found between sympatric populations of *Drosophila* subspecies (Ehrman, 1965;[272] Noor, 1995[273]), is considered to come into play subsequent to nascent postzygotic isolation. He considers that, alternatively, gratuitous sexual-incompatibility genes may preexist in local populations. This seems to be the case presently within ssp. *parviglumis* where its Teloloapan population contains *Tcb1-s*, although this population grows wild, occurring only incidentally as a weed in maize fields. Ellstrand and colleagues (2007[274]), by means of field crossing experiments in California, found that maize and *Z. mays* ssp. *mexicana* naturally hybridize at a low rate (less than 1%), whereas *Z. mays* ssp. *parviglumis* hybridizes with the crop at a high rate (more than 50%) (Ellstrand et al., 2007[275]). It would thus appear that the gene flow is primarily one way, from *parviglumis* teosinte to maize. This does not exclude the existence of reciprocal crosses at lower frequencies.

The presence of the *teosinte crossing barrier 1* alleles *Tcb1-s* and *Tcb1-m* only in teosinte and not in maize is a mystery. The most logical explanation passes by accepting the possibility that maize and teosinte have evolved separately from a common ancestor, at which time of separation the gametophyte gene barriers evolved. This is further evidence of modern maize being derived from preexisting wild maize, with later gene flow from teosinte.

270 Evans, M. M. S., and J. L. Kermicle. 2001. *Teosinte crossing barrier 1*, a locus governing hybridization of teosinte with maize. *Theoretical and Applied Genetics*, 103: 259–265.

271 Kermicle, J. L. 2006. A selfish gene governing pollen-pistil compatibility confers reproductive isolation between maize relatives. *Genetics*, 172 (1): 499–506.

272 Ehrman, L. 1965. Direct observation of sexual isolation between allopatric and between sympatric strains of different *Drosophila paulistorum* races. *Evolution* 19: 459–464.

273 Noor, M. A. 1995. Speciation driven by natural selection in *Drosophila*. *Nature*, 375: 674–675.

274 Ellstrand, N., L. C. Garner, S. G. Hegde, R. Guadagnuolo, and L. Blancas. 2007. Spontaneous hybridization between maize and teosinte. *Journal of Heredity*, 98: 183–187.

275 Ellstrand, Norman C., Lauren Garner C., Subray Hegde, Roberto Guadagnuolo, and Lesley Blancas. 2007. Spontaneous hybridization between maize and teosinte. *Journal of Heredity*, 98 (2): 183–187.

Kermicle and colleagues (2006) and Kermicle and Alleman (1990[276]) have noted that *Ga2-s* prevents teosinte from being fertilized by maize. The *ga1* locus acts similarly where the maize landraces are sympatric to the four *Ga1-s* teosinte populations identified as *Ga1-m*. The former gene distribution contrasts with genes *teosinte crossing barrier 1* in which the alleles *Tcb1-s* and *Tcb1-m* were reported only in teosinte. *Ga1-s* and *Ga2-s* could have been effective in isolating teosinte from *ga* maize in the past.

Kermicle and Evans (2010[277]) have reported that pollen simultaneously carrying both *ga2* and *Ga2* was functional on *Ga2* silks, which have the pistil barrier, indicating that *Ga2* conditions acceptance of the pollen grain, rather than *ga2* conditioning rejection of the pollen grain by *Ga2* silks. The strong allele (*Ga2-s*), a weaker one such as reported among maize genetic stocks (*Ga2-w*), and an allele having only pollen competence (*Ga2-m*), or some combination of these, was found in all 13 of the teosinte populations sampled. Sympatric and parapatric maize landraces carried *Ga2-m* or the presumed null allele *ga2*, but *Ga2-s* or *Ga2-w* was not found. The combination of exclusively *Ga2-s* teosinte with *ga2* maize, which could provide strong reproductive isolation, was not characteristic of the five paired populations tested. Again the following question is posed: if maize were derived from teosinte by domestication and human selection, why is there no presence of the *Ga2-s* allele in maize at all? It is unlikely that the self-incompatibility mechanism is of recent origin and was developed after or in the course of domestication. Further research is needed on this key question.

MALE STERILITY AS AN ISOLATION MECHANISM

Male sterility in maize is of two types: genetic and cytoplasmic. Genetic male sterility is determined by a series of *Ms* genes that, when in a double recessive *ms* allele condition, produce pollen sterility. Genetic sterility in maize causing abortion of anthers or nonfunctional pollen was reported by various early maize geneticists (Burnham, Emerson, Eyster, Singleton and Jones, Sprague, and Wiggans), and these sources of sterility were studied and reported by Beadle (1932[278]) as a series of 16 genes named *sterile-1* (m_{s1}), then *sterile-2* (v_{a2}), *wa*, and *sterile-4* to *sterile-16* (m_{s4} to m_{s16}); at least three of these sources came from crosses with South American maize or directly from South American maize. This series is now increased to 22 male sterile genes, of which 2 are dominant and the rest recessive, with rearrangement of the early notation (Coe et al., 1988[279]). A new male sterility gene dominant Ms_{30} was reported to be

[276] Kermicle, J. L., and M. Alleman. 1990. Gametic imprinting in maize in relation to the angiosperm life. *Development Supplement*: 9–14.

[277] Kermicle, J. L., and M. M. S. Evans. 2010. The *Zea mays* sexual compatibility gene *ga2*: Naturally occurring alleles, their distribution, and role in reproductive isolation. *Journal of Heredity*, 101 (6): 737–749.

[278] Beadle, G. W. 1932. Genes in maize for pollen sterility. *Genetics*, 17: 413–431.

[279] Coe, E. H., M. G. Neuffer, and D. H. Hoisington. 1988. The genetics of corn. In G. G. Sprague and J. W. Dudley, editors. *Corn and Corn Improvement*. 3rd ed. American Society of Agronomy. Agronomy 18. Madison. pp. 83–256.

located in chromosome 4 (Liang et al., 1999[280]). It is important to note that chromosome 4 is where blocks of genes differentiating teosinte and maize are also located.

Cytoplasmic male sterility is conditioned by the presence of extra-chromosomal factors. It is obtained in the presence of *rf* alleles (*rf1* and *rf2*) in a homozygous state in the appropriate cytoplasm. The dominant *Rf1* and *Rf2* alleles restore pollen fertility. This knowledge was used to produce hybrid seed in maize without the need for detasseling.

Cytoplasmic male sterility, another form of self-sterility, which in effect enhances outcrossing, was first found by Rhoades (1931[281]) to be transmitted by the cytoplasm, in a maize collection from Arequipa, Peru, collected by R. A. Emerson and F. D. Richey. Since then, a number of cytoplasmic male sterility cases have been found. The most extensively used for commercial maize hybrid seed production was *cms-T*, discovered by Rogers and Edwardson (1952[282]) in the Texas variety Golden June in 1952, followed later by *cms-S* and *cms-C*. The respective nuclear gene restorer series are *Rf1Rf2* for T, *Rf3* for S, and *Rf4* is C. The mitochondria of the respective types of cytoplasms produce some polypeptides that differ in molecular size from 13,000 M in the case of *cms-T* to 42,000–88,000 M in *cms-S* (Forde and Leaver, 1980[283]). The control of the production of these polypeptides is achieved by the nuclear gene *Rf1*, which suppresses the polypeptide production.

The present author has found several lines with *cms-T* cytoplasm, identified as such in the Peruvian maize race Perla, which were restored with the U.S. tester line Keys, which carries the allele restorer composition *Rf1Rf1Rf2Rf2*. Plants with *cms-T* exhibit extreme sensitivity to *Helminthosporium maydis* Nisik and Miy (now *Bipolaris maydis* Nisik) of their mitochondria, where the genes conditioning cytoplasmic sterility are located. The discovery of *cms-T* in a modern race such as Perla from the coastal belt of Peru; of mitochondrial genes similar to those found in the Cuzco race collected in Arequipa and in a Texas variety, the latter of which was probably descended from the variety Mexican June; and of the set of dominant restorer genes indicates that cytoplasmic male sterility is widespread and probably evolved a long time ago. Genes in the homozygous condition *rf1rf2* must be present to allow for the expression of such cytoplasmic sterility. Their presence in some genotypes and their fertility restoration by other genotypes with *Rf* dominant genes ensures greater variability through forced crossing and at the same time enhances the effects of

[280] Liang, Y., H. Zhou, and W. Jiang. 1999. Molecular mapping of a male sterile gene (ms30) in maize. *Maize Genetics Cooperation Newsletter*, 73: 5–6.

[281] Rhoades, M. M. 1931. Cytoplasmic inheritance of male sterility in maize. *Science*, 31: 340–341.

[282] Rogers, J. S., and J. R. Edwardson. 1952. The utilization of cytoplasmic male sterile inbreeds in the production of maize hybrids. *Agronomy Journal*, 44: 8–13.

[283] Forde, B. G., and C. J. Leaver. 1980. Nuclear and cytoplasmic genes controlling synthesis of variant mitochondrial polypeptides in male-sterile maize. *Proceedings of the National Academy of Sciences USA*, 77 (1): 418–422.

heterosis. It is unknown to what extent cytoplasmic male sterility is an ancient trait in maize and teosinte and whether it could have acted as a reciprocal barrier to crossing.

A gene identified as *urf13* has been located upstream of a conserved mitochondrial gene called *orf221*, which encodes a membrane-bound protein, and which causes pollen disruption in maize. The fertility restorer gene *Rf2* for *cms-T* has been cloned and found to exhibit activity similar to aldehyde dehydrogenases. It is considered to be a biochemical restorer for a residual effect left from incomplete action of *RF1* on restoration of cytoplasmic male fertility in maize. Plants that are *Rf2Rf2rf1rf1* are sterile. Apparently *Rf2* has an effect on elimination of the toxicity of aldehydes remaining from the *Rf1* down-regulation of *urf13* expression. *Rf2* is capable of reducing aliphatic and aromatic aldehydes. Thus nuclear fertility restorer genes in maize may act by compensating a toxic protein generated by the mtDNA associated cytoplasmic male sterility (CMS) pollen sterility (Hanson and Bentolita, 2004[284]). The *cms-T* URF13 protein has been subjected to considerable research because it is associated with extreme susceptibility to the T-toxin produced by the fungus *Cochliobolus heterostrophus* and is associated also with susceptibility to the insecticide methomyl.

The *cms-S* system discovered by the U.S. Department of Agriculture, requiring the *rf3* allele for expression and *Rf3* for restoration, has an *atp9* and adenosime triphosphate (ATP) synthase gene. Unlike other systems, the *cms-S* pollen fertility-restoration system is gametophytic, but its cytoplasmic revertants occur rather frequently, which is a reason why this system has not been reliable for hybrid seed production.

The evolutionary implications of the development of CMS have not been developed yet. There are many species of plants that have CMS systems with conserved mitochondrial genes in species as varied as *Brassica*, *Petunia*, maize, rice, or sorghum. However, there are sequences that are not identified as similar from one species to another. An advantage that a CMS cytoplasm may report is that of enhancing crossability, because fertile progeny, offsetting the maternal CMS in the progeny, will necessarily be all hybrid. The origin of these regulatory genes in the mitochondrial genome probably arose from invasion of nuclear or chloroplast or other unidentified DNA into the mitochondrial genome and its insertion. To offset possible detrimental effects of these new arrivals, one of a duplicate pair of genes in the nuclear genome became specialized to control the expression of the new mitochondrial gene (Hanson and Bentolita, 2004).

A study of genes regulating the growth of the female gametophyte before and after pollination in maize found the series of genes *zmES 1–4*, which produce a type of defensins (small proteins) that are expressed only during the formation of the embryo sac of the female gametophyte and that control and also

[284] Hanson, M., and S. Bentolita. 2004. Interactions of mitochondrial and nuclear genes that affect male gametophyte development. *The Plant Cell*, 16: S154–S169.

direct pollen tube growth with its two gametes to the egg cell and the secondary cell of the embryo sac. Two maize inbred lines, A158 and B73, had similar electrophoretic patterns after digestion by four enzymes for the defensins, and both were very different from *Tripsacum pilosum* (Cordts et al., 2001[285]). The difference in production of similar regulatory molecules may be one of the biochemical mechanisms that would prevent the successful fertilization of maize and *Tripsacum*.

Supergenes

Supergenes are defined as a group of closely linked genes occupying a large chromosomal segment, and they frequently function as a genetic unit and are related in an evolutionary sense, although they are rarely coregulated genetically. Supergenes have their own regulatory region and may produce simultaneous and pleiotropic effects. Supergenes have cis-effects due to multiple loci, which are located either within a gene or within a single gene's regulatory region, and which exhibit tight linkage. Other duplications may give rise to tightly linked gene complexes, in which duplicate genes diverge through evolution into different specializations and have independent action.

The presence of minute extra loci in a teosinte chromosome, as compared to a maize chromosome, or in different maize populations or races could produce, through a mechanism of unequal crossing over, deficiencies and duplications. The reciprocal crossover, if it carries a deletion, could give rise to deficient seeds and other unusual types of mutations. On the other hand, these duplications, which were named "paraduplications" or "supergenes" by Brink and associates, when created by hybridization of parents that lack collinearity in their allele arrangements (Fu and Dooner, 2002,[286] found cases in which genetic microlinearity in maize was violated), would, through their enhanced action and close linkage, be able to exercise strong action in the manifestation of some traits. These genes, as mentioned previously, are different from the duplicates arising through a slow evolutionary process (Mangelsdorf, 1974).

The presence of supergenes in chromosomal segments of maize was demonstrated by Mangelsdorf (1974: 127–131) in his experiments on the effect of the introduction of chromosome segments – of various teosinte sources, of maize derivatives from teosinte from Mexico and Central America, and of four South American accessions representing different races with tripsacoid characteristics – into two control inbred lines of the Corn Belt Dent race. The tripsacoid effects

[285] Cordts, Simone, Jörg Bantin, Peter E. Wittich, Erhard Kranz, Horst Lörz, and T. Dresselhaus. 2001. *ZmES* genes encode peptides with structural homology to defensins and are specifically expressed in the female gametophyte of maize. *The Plant Journal*, 25 (1): 103–114.

[286] Fu, Huihua, and H. K. Dooner. 2002. Intraspecific violation of genetic collinearity and its implications in maize. *Proceedings of the National Academy of Sciences USA*, 99 (14): 9573–9578.

that were observed are characterized by glume and rachis induration due to tissue sclerenchymatization and were found to be similar regardless of whether the chromosome segments originated from teosinte or from Mexican or Central American maize races derived from teosinte, or from tripsacoid South American maize races. The effects were clearly established as coming from closely linked genes or possible supergenes in chromosome segments that Mangelsdorf attributes as of teosinte origin, and in the case of the selected tripsacoid South American maize accessions, as originating most likely from *Tripsacum*, because teosinte is not found in South America. It was shown that the tripsacoid effects could be produced by chromosome gene blocks from South American maize races and were perceived as equal in effect to those of teosinte and of teosinte-derived races of maize. It is noteworthy that the South American races selected by Mangelsdorf for his experiments are all tripsacoid, lowland tropical flint maize types, with the exception of the Bolivia 1157 accession, which is a floury maize of the Piricinco-Coroico race – the most widely distributed maize race in South America, covering the whole Amazonian basin. One of them, Maíz Amargo from Argentina, which was originally studied by Rosbaco (1951[287]), exhibits extreme tripsacoid characteristics, which are possible evidence of chromosome segments of *Tripsacum* inserted in this type of maize and acting as supergenes (see Mangelsdorf, 1974: figure 11.5, 128). Teosinte introgression is conclusive in the case of races of maize in Mexico and Central America, but less so in ancient maize races in South America, of which Piricinco-Coroico, a virtually knobless type descended from Andean maize, is one example. Its tripsacoid chromosome segments may have come from South American *Tripsacum australe*, which has been found to be knobless (Graner and Addison, 1944[288]), pointing to the origin of such segments as of probable *Tripsacum* introgression. Initially natural and later human selection have been active in those cases – for example in Maíz Amargo, a highly tripsacoid maize from Argentina – in conditioning resistance to biotic pressure, in this case to grasshopper damage.

Domestication Genes

Some important distinguishing characteristics separating domesticated maize from teosinte are the differences in branching and in spikelet suppression. Both traits are controlled by the *tb1* allele and attributed to up-regulation of *tb1* in maize. Doebley and colleagues (1995) and Hubbard and colleagues (2002), comparing maize with teosinte, established that maize has one dominant axis of growth, whereas teosinte is highly branched; they concluded in their studies

[287] Rosbaco, U. F. 1951. Consideraciones sobre "maíces amargos" con especial referencia a su cultivo en la provincia de Entre Rios. *Idia*, 46: 1–12.

[288] Graner, G. A., and G. Addison. 1944. Meiose em *Tripsacum australe* Cutler and Anderson. *Anais da Escola Superior de Agricultura "Luiz de Quiroz,"* Universidade de São Paulo, 9: 213–224.

that the axillary branches in maize are short and feminized, whereas the axillary branches of teosinte are long and end in a male inflorescence under normal growing conditions. Previous QTL and molecular analysis suggested that the *teosinte branched 1* (*tb1*) gene of maize contributed to the architectural difference between maize and teosinte, and *tb1* mutants of maize resemble teosinte in their overall architecture. They analyzed the *tb1* mutant phenotype in more detail and showed that the highly branched phenotype was due to the presence of secondary and tertiary axillary branching, as well as to an increase in the length of each node, rather than to an increase in the number of nodes. Double-mutant analysis with *anther ear1* and *tasselseed2* revealed that the sex of the axillary inflorescence was not correlated with its length. RNA in situ hybridization showed that *tb1* was expressed in maize axillary meristems and in stamens of ear primordia, consistent with a function of suppressing growth of these tissues. Expression in teosinte inflorescence development suggests a role in pedicellate spikelet suppression. These results provide support for a role for *tb1* in growth suppression and reveal the specific tissues where suppression may occur.

Most of the known domestication genes that have been cloned are diverse transcription factors that are usually functional, thus placing the role of human selection on wild populations during crop domestication at the gene level, as being modification rather than elimination of gene function, as postulated by Consonni and colleagues (2005[289]) and Doebley and colleagues (2006). These views would relegate gene mutations to be inconsequential per se in the course of domestication of plants and would presuppose that the major role would be adaptation of genes to produce new structures. The aforementioned study of Hubbard would indicate that the mutant *tb1* of maize is not identical with the gene "*tb1*" that was present in teosinte, on which the proponents of the orthodox hypothesis of maize domestication directly from teosinte base their theory, because their effects on branching are not the same. Their supporting argument is that there is a relative rarity of mutations, leading to new structural or functional genes and the short time span of crop domestication. The reality is that maize exhibits a tremendous variation in its genome due to mutations at the visual level and polymorphisms that may have helped shape more differences than those that have been studied up to now with a few so-called domestication genes.

The origin of the maize ear, according to Galinat (1983, 1985b[290]), derives from the pistillate inflorescence of teosinte through a series of mutations that were selected for. The genes involved are (1) *Tr* for two-ranked arrangement of seeds in teosinte versus four-ranked arrangement in maize; (2) *Pd* for single

[289] Consonni, G., G. Gavazzi, and S. Dolfini. 2005. Genetic analysis as a tool to investigate the molecular mechanisms underlying seed development in maize. *Annals of Botany*, 96: 353–362.

[290] Galinat, W. C. 1985b. The missing links between teosinte and maize: A review. *Maydica*, 30: 137–160.

arrangement, in teosinte versus paired spikelets in maize; (3) *Ab* for presence against absence of abscission layer, which allowed easy shattering of the seeds (including fruitcases) of teosinte against the nonshattering condition of cultivated maize; and (4) *Tu*, which controls the expression of soft outer glumes and soft rachis in maize versus the indurated glumes and hardened rachis of teosinte. A different hypothesis for the evolution of the maize ear was proposed by Iltis (1983, 1987[291]). His proposal is that the terminal tassel of the lateral branch of teosinte had its central spike feminized and the staminate florets transformed into pistillate ones by a "transmutation" process. This process must have proceeded in a very fast manner according to this hypothesis; human selection followed in the stabilization of the expression of the new maize ear. Under the Iltis hypothesis, the action of the *Pd* gene is not required, because the teosinte tassel already has paired spikelets, whereas the ear of teosinte has single spiklets.

Genes that exhibit pleiotropic effects in maize such as those associated with *zfl2*, which increases row number, in all likelihood, would also select for earlier flowering and fewer ears placed lower on the plant (Bomblies and Doebley, 2006). This led Bomblies and Doebley to suggest that, in general, undesirable secondary effects associated with pleiotropic genes could limit selection for favorable "domestication alleles" during early stages of the differentiation of a crop from its wild progenitor. On the other hand, selection for beneficial traits controlled by pleiotropic genes could result in associated neutral or even detrimental traits being concurrently selected. This may explain, at least partially, the presence in wild populations of alleles for traits of the domestication syndrome that apparently evolved prior to domestication and survived despite their possibly deleterious effects in the wild. Alternatively, the explanation could be that those genes came through the introduction of maize genes into the wild plants by early hybridization and by successive introgression of maize.

The former discussion applies to some key genes, such as *tb1*, responsible for some differences between maize and teosinte. The gene *teosinte branched 1* (*tb1*) mutant of maize has pleiotropic effects on apical dominance, length of lateral branches, growth of blades of leaves on lateral branches, and development of the pedicillate spikelet in the female (Hubbard et al., 2002). In teosinte ssp. *parviglumis*, a *tb1* region haplotype with sequences identical to that of the major maize *tb1* haplotype was found. This result suggested that haplotypes that confer maize-like phenotypes could predate domestication (Clark et al., 2004).

Recently, however, Clark and colleagues (2006[292]) located a factor or factors controlling the levels of the message produced by the transcriptional

[291] Iltis, H. H. 1987. Maize evolution and agricultural origins. In T. Soderstrom, K. Hilu, C. Campbell, and M. Barkworth, editors. *Grass Systematics and Evolution*. Smithsonian Institution Press. Washington, D.C. pp. 195–213.

[292] Clark, R. M., T. N. Wagler, P. Quijada, and J. Doebley. 2006. A distant upstream enhancer at the maize domestication gene *tb1* has pleiotropic effects on plant and inflorescent architecture. *Nature Genetics*, 38: 594–597.

regulator *teosinte branched 1* (*tb1*) in maize, and hence the phenotypic differences between maize and teosinte associated with *tb1* are assigned to an intergenic region upstream from *tb1*. This region consists of a mixture of repetitive and unique sequences not previously considered to contribute to phenotypic variation. Doebley and Lukens (1998) had proposed earlier that modifications in cis-regulatory regions of transcriptional regulators would prove a predominant means for the evolution of novel forms. The findings of Clark and colleagues (2006) appear to provide a supporting example, due to the effects of up- and down-regulating all transcription factors in a given genome. Or if one is open to accept other explanations, it might again indicate the possibility that the up- and down-regulation effects could be the result of the introduction of regulatory segments in teosinte via maize introgression. This explanation in no way interferes with the acceptance of strong selection pressures on preexisting variation.

Protracted Age of Plant Domestication

The field of archaeobotany is fast producing evidence that undermines the quick development model of plant domestication. Robin Allaby of the University of Warwick's plant research arm – Warwick HRI – and his associates have found genetic evidence that supports a revision of the age of initiation of plant domestication. They claim that the emergence of agriculture in prehistory took much longer than originally thought (Allaby et al., 2008[293]).

There were three stages in domestication: (1) a wild gathering stage that was very long, longer in age than the sum of the following two stages; (2) a predomestication cultivation stage (not envisioned in the case of maize, if maize were domesticated out of teosinte); and (3) an equally long – as the preceding one – domestication syndrome fixation stage, as we can interpret from figure 1 in Allaby and colleagues (2008). Until recently, domestication in the Near East had been viewed as a rapid process in the aforementioned three principal steps, which closely followed the climatic transition between the Pleistocene and Holocene, with little predomestication cultivation, a rapid rise of domesticated crops, and an explosive expansion of agriculturists and agriculture from the centers of origin.

The invention, place of origin, and spread of agriculture is a subject of a long, ongoing debate. The most plausible interpretation of the available data is that some agriculturally useful plant species were domesticated no more than a few times, and perhaps only once, in the Near East, Mexico, Mesoamerica, and the Andean region. Multiple domestications of some other species may

[293] Allaby, Robin G., Dorian Q. Fuller, and Terence A. Brown. 2008. The genetic expectations of a protracted model for the origins of domesticated crops. *Proceedings of the National Academy of Sciences USA*, 105 (37): 13982–13986.

have taken place in the central Andean region independently from Mexico and Mesoamerica, but it is impossible to say whether any such cases were truly independent, because of the incomplete archaeological as well as botanical records. In the case of maize, the dearth of such records extends not only to Mexico and the Andean region but to other agricultural centers such as Colombia, the tropical South American lowlands, and Mesomerica.

Our accumulated evidence does not negate the conclusion that multiple independent domestications of the same species or genus did occur in different regions. In fact the evidence strongly points in that direction, with several examples of plant domestication (squash, beans, lima beans, *Amaranthus*, *Chenopodium*, *Capsicum*, and cotton) having taken place independently in Mexico/Mesoamerica and the central Andean region (see Pickersgill, 2007, and extensive references). It is important to note also that there is evidence of very early plant domestication in the tropical lowlands of South America (peanuts and cassava).

Genetic evidence by itself, isolated from the context of the archaeological record, is insufficient to mark the sequence of short-term evolutionary events (in the hundreds or thousands of years). It is by now evident that in the case of maize and its relatives, the genetic background – in terms of existing major alleles capable of discrete changes and series of QTLs capable of accumulated additive action and capable of initially subtle morphological changes, which were selected by early gatherers/farmers – was already present in the genomes of the preagricultural plants. The question is, what were these plants like?

The human mind has tried to order nature into boxes and to classify the components of its proximal environment into them, starting with Aristotle, passing through Linnaeus, and ending in modern taxonomists. Although this is a useful mental organizational procedure, it may overlook the reality that nature works in a continuous evolutionary flow, not in leaps and jolts. Gaps in the evolutionary process are most likely due to accidental missing links but not to their inexistence. Plant populations are more likely to have had enough variation already present in them to allow domestication possibilities in many directions, such as may have happened with *Zea mays* in early periods of its contact with humans.

A major underlying assumption has been that artificial selection pressures were substantially stronger than natural selection pressures in the early stages, resulting in genetic patterns of diversity that reflect a major initial genetic impact, independent of the human selection pressures applied at a later stage in various geographic localities and over a much longer period of time. Recent archaeobotanical evidence has overturned the notion of a rapid domestication transition, resulting in a protracted time model that undermines these assumptions. Conclusions of genome-wide multilocus studies that support a rapid-transition model, by indicating that domesticated crops appear to be associated by monophyly with only a single geographic locality, may have questionable and perhaps biased interpretations and remain problematic.

New information, including simulations presented by Allaby and colleagues (2008), tries to resolve this conflict, indicating that the results observed in such plant domestication studies are inevitable over time at a rate that is largely influenced by the long-term population size. Counterintuitively, they state, multiple-origin crops are shown to be more likely to produce monophyletic clades than crops of a single origin. Under the protracted transition, the importance of the rise of the domestication syndrome becomes paramount in producing the patterns of genetic diversity from which crop origins may be deduced. They propose to identify four different interacting levels of organization that now need to be considered to track crop origins from modern genetic diversity, making crop origins a problem that could be addressed through system-based approaches.

Mangelsdorf (1974) has proposed the possibility of multiple locations for the domestication of maize, starting from preexisting wild maize, which would have had a different plant architecture and grain dispersal system than present cultivated maize, in several locations. The archaeological data points in that direction. These views are disputed by a vast majority of students of maize evolution, based on the interpretations of molecular genetic data of some key studies (see, for example, Matsuoka et al., 2002), who present evidence for a single location and time period for the domestication of maize. There is modern evidence, however, which we have dealt with elsewhere in this appendix, that requires that we revise and take a second look at the data and that behooves that we reinterpret such data when the evidence so demands. Kato and colleagues (2009[294]) also dispute the single center of origin in Mexico and advance a position of four locations in that territory.

The appearance of agriculture and plant cultivation in the Near East was thought to have begun around 10000 years BP. New genetic evidence disputes that model. Robin Allaby's team developed a new mathematical model that shows how plant agriculture actually began much earlier than first thought, well before the Younger Dryas (the last "big freeze" with glacial conditions in the higher latitudes of the Northern Hemisphere). It also shows that useful gene types could have actually taken thousands of years to become stable.

A similar situation is starting to emerge with the domestication of plants in the Americas. New discoveries are adding to the age of plant domestication. There is a scarcity of information from the central Andean area, especially at intermediate altitudes in the Andean region, where early maize would have found favorable conditions for development of an early agriculture. Some evidence comes from the studies at the Rosamachay Cave site in Ayacucho. There are some 26 caves in Ayacucho and others in Huanuco in Peru that should be explored for evidence of early human settlement and domestication, now that local pacification of terrorist group activities has taken place in that area. The present author pointed out this opportunity to MacNeish, and that information

[294] Kato, T. A., C. Mapes, L. M. Mera, J. A. Serratos, and R. A. Bye. 2009. Origen y diversificación del maíz: una revisión analítica. Universidad Nacional Autónoma de México. Comisión Nacional para el Conocimiento y Uso de la Biodiversidad. Mexico, D.F.

led to the initial archaeological studies of three Ayacucho caves, reported in the main body of this book by Bonavia. It is unfortunate that much of the work done was lost due to the poor handling of the material by the workers and that a botanical report was never published. Extensive new digs are required in undisturbed sites in that general area, where the presence of early man in a long pre-agricultural period has been sufficiently well established (Cardich, 1964[295]).

It is important to notice that evidence of early domestication may have proceeded from plants such as gourds and beans, which are different from Mesoamerican and Mexican beans and apparently occurred earlier in the Andean region. The sparse evidence from Peru, where more archaeological research needs to be done at the early preceramic levels, already points out to the presence of maize with more evolved racial variation than at similar early periods in Mexico, and with clear evidence of its having been brought from the highlands to the sites where it was found in later periods in the coastal locations of Peru. The high anthocyanin pigmentation frequency in cob cupules in early maize from coastal locations such as Cerro El Calvario, Los Gavilanes, or Áspero and morphological affinity to early maize from archaeological sites such as Rosamachay and Guitarrero Caves, in the highlands in Peru, point to a long period of previous adaptation to cultivation in the middle-altitude, inter-Andean valleys of Peru, before moving down to the coastal locations.

On the other hand, archaeological sequences should not be interpreted without reference to genetic contexts, and based only on in situ data. That would allow researches to jump to wrong conclusions on the linkages of the site and its timing and botanical relations to other sites. Such was the incongruity of the presence of early Tehuacán maize as a domesticated derivative of teosinte in areas where teosinte is not prevalent today. It also makes it difficult to establish a rational interpretation of primitive maize from Mexico as being a unique early development, independent of similar events that were taking place at the same approximate time (as judged from existing AMS dating that was published while this book was in publication; see the afterword at the end of the appendix) and in different directions in the central Andean region.

Our proposal is that a wider and multidisciplinary approach emphasizing archaeobiology should be adopted, aiming at expanding our knowledge on the rise of the domestication syndrome. A new systematic and concerted approach to the discovery of early domestication and crop origins is essential. Those who take up this challenge should heed the warning that tracking crop origins from modern genetic diversity, although useful, is risky because of the blurring effect of teosinte introgression into maize.

It has been our objective, in reexamining in this appendix some of the most critical scientific information relevant to the origin of maize as a crop, to interpret the information, without prejudice, following the logic emanating from the

[295] Cardich, A. 1964. Lauricocha. Fundamentos para una prehistoria de los Andes Centrales. *Studia Pehistorica*, III. Centro Argentino de Estudios Prehistóricos. Buenos Aires.

available data. Suffice it to say, at this point, that the theme is still subject to dispute and more information on the evolution of plants and of maize is required to shed light on how the process of domestication began, where it started from, how long it continued, and what were the forces that propelled it.

The Origin of Genome Diversity in Maize

The diversity of maize can be easily observed in the range of plant, ear, and grain phenotypes of its 359 races. The origin of such diversity may have resulted initially from very early isolation and human selection on at least six primitive maize popcorn races, as proposed by Mangelsdorf (1974). Considerable information has been gathered that allows us to explain the maize diversity by a conjunction of the following processes:

1. Ancient polyploidization, gene duplication, and gene specialization of members of gene pairs
2. Unequal crossing over, producing gene duplication and gene deletions, followed by gene functional diversification of one member of the pair
3. Point mutations due to nucleotide changes, deletions, or insertions, including SSRs
4. Mutations due to transposon insertions into gene-coding regions and transposon losses after insertion
5. Transport of segments of genes by certain types of transposons, such as helitrons
6. Insertion of genes from related species (teosinte) and genera (*Tripsacum*)
7. Migration of nuclear genes to the mitochondrial genome

Gene Duplication

Gene duplication is an important and common evolutionary phenomenon in plants. The ultimate fate of a duplicated gene is that either it ends up being silenced through inactivating mutations or both copies are maintained by selection. The survival of both copies can occur via "neofunctionalization," wherein one copy acquires a new function, or by "subfunctionalization," wherein the original function of the gene is partitioned across both copies. The relative probabilities of these three different fates involve often very subtle interactions between population size, mutation rate, and selection. All three of these fates are critical to the expansion and diversification of gene families (Monson, 2003;[296] Walsh, 2003[297]).

[296] Monson, R. K. 2003. Gene duplication, neo-functionalization, and the evolution of C4 photosynthesis. *International Journal of Plant Science*, 164: S43–S54.

[297] Walsh, B. 2003. Population-genetic models of the fates of duplicate genes. *Genetica*, 118 (2–3): 279–294.

The complexity of living organisms is attributed to evolving new gene functions that arise from gene duplications. Gene duplication is believed to be the primary source of new genes (Ohno, 1970[298]), and evolution by gene duplication has emerged as a general principle of biological evolution, because evidence has accumulated from the prevalence of duplicate genes in all sequenced genomes of *Bacteria*, *Archaea*, and *Eukarya*: for example, the percentage of duplicate genes range from 17% in the bacterium *Helicobacter pylori*, to 66% in the plant *Arabidopsis thaliana*, and to 38% in *Homo sapiens* (reviewed in Zhang, 2003[299]).

It is generally agreed in the classical model for the evolution of duplicate genes that an ancestral function of a progenitor gene will be retained in at least one of the daughter genes after duplication, and that shared functions between duplicates are ancestral functions. There are two possible fates: one copy evolves a new beneficial function, and one member of the duplicated pair usually degenerates within a few million years by accumulating deleterious mutations, or, in the alternative case, the other duplicate under this model retains the original function (Fisher, 1935;[300] Haldane, 1933[301]), and others. This model further predicts that on rare occasions, one duplicate gene may acquire a new adaptive function, resulting in the preservation of both members of the pair, one with the new function and the other retaining the old. However, empirical data suggest (Force et al., 1999[302]) that a much greater proportion of gene duplicates is preserved than predicted by the classical model.

A duplication-degeneration-complementation (DDC) model has been proposed that predicts that (1) degenerative mutations in regulatory elements can increase rather than reduce the probability of duplicate gene preservation and (2) the usual mechanism of duplicate gene preservation advances partitioning of ancestral functions rather than the evolution of new functions.

According to Force and colleagues (1999), a newly duplicated paralog (twin gene) will survive if it is capable of acquiring by chance an advantageous regulatory mutation. It may fix an advantageous allele, giving it a slightly different, and selectable, function from its original copy. This initial fixation provides substantial protection against future fixation of null mutations, allowing additional mutations to accumulate that refine functional differentiation. Alternatively, a duplicate locus can instead first fix a null allele, becoming a pseudogene (Walsh 1995: 426[303]). Duplicated genes persist only if mutations

[298] Ohno, S. 1970. *Evolution by Gene Duplication*. Springer-Verlag. New York.

[299] Zhang, J. 2003. Evolution by gene duplication – an update. *Trends in Ecology and Evolution*, 18: 292–298.

[300] Fisher, R. A. 1935. The sheltering of lethals. *American Naturalist*, 69: 446–455.

[301] Haldane, H. B. S. 1933. The part played by recurrent mutation in evolution. *American Naturalist*, 67: 5–9.

[302] Force A., M. Lynch, F. B. Pickett, A. Amores, Y. L. Yan, and J. Postlethwait. 1999. Preservation of duplicate genes by complementary degenerative mutations. *Genetics*, 151 (4): 1531–1545.

[303] Walsh, J. B. 1995. How often do duplicated genes evolve new functions? *Genetics*, 110: 345–364.

create new and essential protein functions, an event that is predicted to occur rarely (Nadeau and Sankoff, 1997[304]). Thus, overall, with higher plants and complex metazoans, the major mechanism for retention of ancient gene duplicates would appear to have been the acquisition of novel expression sites for developmental genes, with its accompanying opportunity for new gene roles underlying the progressive extension of development itself. If a new allele comprising duplicate genes is selectively neutral, compared with preexisting alleles, it only has a small probability, 1/2N, of being fixed in a diploid population, where N is the effective population size. This situation suggests that many duplicated genes will be lost. For those that do become fixed, fixation is time consuming, because it takes, on average, 4N generations for a neutral allele to become fixed (Kimura, 1983[305]).

The neofunctionalization (NF) hypothesis asserts that after duplication one daughter gene retains the ancestral function while the other acquires new functions. In contrast, the subfunctionalization (SF) hypothesis argues that duplicate genes experience degenerate mutations that reduce their joint levels and patterns of activity to that of the single ancestral gene. Force and colleagues (1999) hypothesize that duplicate genes that are preserved by neofunctionalization will tend to be unlinked, whereas those preserved by subfunctionalization (or silencing of the ancestral gene) will tend to be more closely linked (at least during the period of preservation). Neofunctionalizing mutations are more likely to be established in large populations. If the neofunctional allele, in the process of fixation, has a very small selective advantage or is a degenerative mutation, it is assured that it will be silenced by the time the mutation is fixed, contributing thus to an increase in genome size without affecting the original organism's functions or phenotype. Many of the duplicate genes of maize fall into this category. Silent nucleotide sites of duplicate loci may provide a means of ascertaining the time of duplication. Rates of substitution at silenced against replacement sites may provide information on whether different gene regions are evolving in a neutral manner, are being maintained by selection, or are in the process of being transformed to new functions that have selective advantage.

Rapid SF, accompanied by prolonged and substantial NF in a large proportion of duplicate genes, has suggested a new model termed sub-neofunctionalization (SNF). He and Zhang (2005[306]), who proposed the new model, claimed that their results demonstrate that enormous numbers of new functions have originated via gene duplication.

[304] Nadeau, J. H., and D. Sankoff. 1997. Landmarks in the Rosetta Stone of comparative mammalian comparative maps. *Nature Genetics*, 15: 6–7.

[305] Kimura, M. 1983. *The Neutral Theory of Molecular Evolution*. Cambridge University Press. Cambridge.

[306] He, Xionglei, and Jianzhi Zhang. 2005. Rapid subfunctionalization accompanied by prolonged and substantial neofunctionalization in duplicate gene evolution. *Genetics*, 169: 1157–1164.

A phylogenetical method for testing whether pairs of genes evolve in a similar manner over their protein-coding domains has been proposed by Dermitzakis and Clark (2001[307]).

Gene duplication is often accompanied by genetic map changes, and it is a common and ongoing feature of all genomes. This raises the possibility that the differential expansion or contraction of various genomic sequences may be just as important a mechanism of phenotypic evolution as changes at the nucleotide level. Considering that the population-genetic mechanisms responsible for the success or failure of newly arisen gene duplicates are poorly understood, Lynch and colleagues (2001[308]) have examined the influence of various aspects of gene structure, mutation rates, degree of linkage, and population size (N) on the joint fate of a newly arisen duplicate gene and its ancestral locus.

In maize it has been estimated that about a third of genes are tandem duplicates due to unequal recombination or transposition events that have involved gene fragments (Emrich et al., 2007[309]). Rondeau and colleagues (2005[310]) have shown duplication and subsequent functional specialization of the NADH-MDH genes in some, but not all, grasses with C_4 photosynthesis. NADP-MDH genes have been characterized in maize (Metzler et al., 1989[311]) and in sorghum (Luchetta et al., 1991[312]). Only one NADP-MDH nuclear gene was identified in maize and in the complete genome sequence of rice, whereas two NADP-MDH tandemly repeated encoding genes have been found in sorghum.

Enzymes involved in C_4 photosynthesis have been selected for their high expression level in the mesophyll cells (as is probably the case in *Zea* in which only one NADP-MDH gene has been identified). New mutations during the evolution from C_3 to C_4 NADP-MDH are likely to have occurred, and selective pressures have favored their establishment and fixation. Gene duplication offers an opportunity to partition the original functions of NADP-MDHs across different copies.

307 Dermitzakis, F. T., and A. G. Clark. 2001. Differential selection after duplication in mammalian developmental genes. *Molecular Biology and Evolution*, 18: 557–562.

308 Lynch, M., M. O'Hely, B. Walsh, and A. Force. 2001. The probability of preservation of a newly arisen gene duplicate. *Genetics*, 159 (4): 1789–1804.

309 Emrich, S. J., L. Li, T. J. Wen, M. D. Yandeau-Nelson, Y. Fu, L. Guo, H. H. Chou, S. Aluru, D. A. Ashlock, and P. S. Schnable. 2007. Nearly identical paralogs: Implications for maize (*Zea mays* L.) genome evolution. *Genetics*, 175: 429–439.

310 Rondeau P., C. Rouch, and G. Besnard. 2005. NADP-malate dehyrogenase gene evolution in Andropogonae (Poaceae): Gene duplication followed by sub-functionalization. *Annals of Botany*, 96: 1307–1314.

311 Metzler, M. C., B. A. Rothermet, and T. Nelson. 1989. Maize NADP malate dehydrogenase: CDNA cloning, sequence and mRNA characterization. *Plant Molecular Biology*, 12: 713–722.

312 Luchetta, P., C. Crétin, and P. Gadal. 1991. Organization and expression of the two homologous genes encoding the NADP-malate dehydrogenase in *Sorghum vulgare* leaves. *Molecular Genetics and Genomics*, 228: 473–481.

In *Sorghum bicolor*, NMDH-I was reported to have a high transcription level in green leaves and is likely to be involved in C_4 photosynthesis, whereas NMDH-II, which displays a lower transcription level, could be involved in reducing equivalent export (Luchetta et al., 1991). Genes for NMDH-I and NMDH-II have different expression levels, sustaining the hypothesis that these two genes were maintained in some Andropogoneae due to subfunctionalization (Rondeau et al., 2005).

The theoretical model presented previously predicts that maintenance of duplicate genes could be associated with selective pressures via neofunctionalization, in which one copy acquires a new function, or by subfunctionalization, in which the original function is partitioned across both copies (Walsh, 2003). Maize has maintained the single copy and does not seem to have required in the process of its evolution a second copy of this gene. On the other hand, sorghum is well known for its high photosynthetic efficiency in terms of rate of building of biomass per unit time, and this gene duplication and subsequent gene specialization may have contributed to it. Maize has a high C_4 efficiency, but it does not match the C_4 efficiency of sorghum, which has required the coadaptation of other genes.

Among duplicated genes are a class called nearly identical paralogs (NIPs) that appear to be of recent origin. This class of duplicate gene shares greater than or equal to 98% identity. Many NIPs in maize are differentially expressed. This has led to the suggestion that the variation in this class of duplicate gene provides new variation that may have had a selective advantage during domestication and improvement of maize. It is of evolutionary significance that members of many NIP families also exhibit differential expression. The finding that some families of maize NIPs are closely linked genetically whereas others are genetically unlinked is coherent with multiple modes of origin. NIPs provide a mechanism for the maize genome to circumvent the inherent limitation that diploid genomes can carry at most two "alleles" per "locus." They may have played important roles during the evolution and domestication of maize. Emrich and colleagues (2007), from their sequence analysis of the genome of the inbred maize line B73's "gene space," conclude that, as an ancient segmental tetraploid, maize contains large numbers of paralogs that are expected to have diverged by a minimum of 10% over time. NIPs are defined as paralogous genes that exhibit greater than or equal to 98% identity. Sequence analyses have revealed that, conservatively, at least approximately 1% of maize genes are NIPs, which is a rate substantially higher than that present in *Arabidopsis*. In most instances, both members of maize NIP pairs are expressed and are therefore at least potentially functional.

An entirely redundant duplicate copy cannot be maintained in the genome for a long time, according to population genetic theories. These predict that as deleterious mutations accumulate, they may render the gene nonfunctional. The only exception may be the concerted evolution among certain duplicate genes for which a larger amount of gene product is beneficial (Zhang, 2003). In other words, functional divergence between duplicates is usually required for their long-term retention in the genome.

The *a1*, *a2*, and *a3* series; the *B-b*, *B–Bolivia*, *B–Peru*, *B′*, *B–v* series; and the *Rg-rg*, *R–r*, *R–mb*, *R–nj*, *r–r:Pu*, *R–scm*, *R–st* series of alleles in maize, all conditioning anthocyanin synthesis with variations in phenotypic outcome in different plant organs, might be considered examples of such a case.

The *Pr* gene for aleurone color, which has no visible effect in the presence of *A*, *C*, and *R*, has a high frequency (more than 80%) in the Andean region (Bolivia, Peru, Ecuador, and Colombia) and a lower incidence in Mexico and Central America. The *I* or *C1-I* allele at *C1* locus in chromosome 9 inhibits aleurone color regardless of the presence of the other alleles of the anthocyanin series of genes. The development of a strong inhibitor and dominant allele in a heterozygous state, when the simple presence of *c* in a homozygous state would be sufficient to produce colorless aleurone, has to be viewed in the context of possible human selection, because *C1-I* has a very low frequency in the central and southern Andean region (Bolivia and Peru, and also in Brazil) but a high frequency in Mexico and Central America (Mangelsdorf, 1974).

One important advance reported recently (Schnable et al., 2009) is a new, improved draft nucleotide sequence of the 2.3-Gb genome of maize with a prediction of more than 32,000 genes. Some 99.8% of the genes were placed on reference chromosomes. It was confirmed that nearly 85% of the genome is composed of hundreds of families of TEs. These are dispersed in a nonuniform manner across the genome and are considered responsible for the capture and amplification of numerous gene fragments. They are presumed to affect the composition, sizes, and positions of centromeres. The authors reported the correlation of methylation-poor regions with *Mu* transposon insertions and recombination, and copy number variants with insertions and/or deletions, as well as how uneven gene losses between duplicated regions were involved in returning an ancient allotetraploid to a genetically diploid state.

Recent advances in genotyping maize are not only disclosing the genetic composition of maize at the molecular level but also helping to discern the existence of a great variation in conservatism of some chromosomal regions. Several million sequence polymorphisms were identified and genotyped among 27 diverse maize inbred lines (Gore et al., 2009[313]). In their research, Gore and colleagues discovered that the maize genome, as reflected in the collection of inbred lines of maize being used, is characterized by highly divergent haplotypes, which show a 10- to 30-fold variation in recombination rates. Most chromosomes were found to have pericentromeric regions in which there is

[313] Gore, Michael, A. Jer-Ming Chia, Robert J. Elshire, Qi Sun, Elhan S. Ersoz, Bonnie L. Hurwitz, Jason A. Peiffer, Michael D. McMullen, George S. Grills, Jeffrey Ross-Ibarra, Doreen H. Ware, and Edward S. Buckler. 2009. A first-generation haplotype map of maize. *Science*, 326 (5956): 1115–1117.

highly suppressed recombination. By having low crossing over and recombination, these regions of the maize chromosomes, close to the centromeres, may have influenced the effectiveness of selection by stabilizing the association of blocks of genes in the vicinity of the centromeres against disruptive selection. They found hundreds of selective sweeps and highly differentiated regions that contain loci that probably are keys to geographic adaptation.

Ancient tetraploids are found throughout the eukaryotes. After duplication, one copy of each duplicate gene pair is free to explore modification, be lost (fractionate), or be modified in function, whereas the mirror partner of the pair stands in maintaining its original function. For all studied tetraploidies, the loss of duplicated genes, known as homoeologues, ohnologs, or syntenic paralogs, is uneven between duplicate regions. Maize, as a species, experienced a karyotype modification from diploidy to tetraploidy 5–12 million years ago.

Schnable and colleagues (2011[314]) have postulated that, in addition to uneven ancient gene loss, the two complete genomes contained within maize are differentiated by ongoing fractionation, as expressed among diverse inbreds, as well as by a pattern of overexpression of genes from the original genome of the pair that has experienced less gene loss. In ancient tetraploids, and perhaps in all tetraploids, there appears to be selection against loss of the gene (of the original pair in the two previous diploid genomes), which ends up being responsible in the tetraploid for the majority of total expression for a duplicate gene pair. Although the tetraploidy of maize is ancient, biased gene loss and expression continue today and explain, at least in part, the remarkable genetic diversity found among modern maize cultivars.

Lynch and colleagues (2001) have argued that unless there is active selection against duplicate genes, the probability of permanent establishment of such genes is usually no less than 1/(4N) (half of the neutral expectation), and it can be orders of magnitude greater if neofunctionalizing mutations are common. The probability of a map change (reassignment of a key function of an ancestral locus to a new chromosomal location) induced by a newly arisen duplicate is also generally greater than 1/(4N) for unlinked duplicates, suggesting that recurrent gene duplication and alternative silencing may be common mechanisms for generating the microchromosomal rearrangements responsible for postreproductive isolating barriers among species.

As an example, Mangelsdorf (1974) found defective seeds due to mutations that appeared in the cross of maize × Florida teosinte and of maize × Nobogame teosinte, when teosinte chromosome 4 was introduced, which he designated as de^{t1} and de^{t2}. It is noteworthy that chromosome 4 is considered – on the basis of genetic tests and molecular analysis – to be an active macro-differentiating part of the genome between the teosinte and maize

[314] Schnable, James C., Nathan M. Springer, and Michael Freeling. 2011. Differentiation of the maize subgenomes by genome dominance and both ancient and ongoing gene loss. *Proceedings of the National Academy of Sciences USA*, 108 (10): 4069–4074.

subspecies. Bianchi (1957[315]) found that the two loci are located in chromosome 4 with a 14% crossover value between them. Further studies by Mangelsdorf (1974: 134) and one of his students, Surinder Sehgal, established that *De* was 30 crossover units from *Ga*, a gametophyte gene also located on the short arm of chromosome 4. This gametophyte gene is different from the *Ga* series of genes found in popcorns, which enable them to preferentially transmit their genome in crosses. The fact that *Ga* and *De* are located on chromosome 4, which is a differentiating chromosome, had already been noted by Mangelsdorf and Jones (1926). The crosses also led to an enhanced mutability in the filial population.

Because maize and teosinte separated through mutation and adaptation of mutated genes to new functions, although the timing of the event is in dispute, redundancy has been built into the critical chromosome 4 for interspecies (or subspecies according to how the taxa are considered) cross-sterility plus seed lethality, leading to the isolation between them. Although chromosomes are of the same length and there is apparently complete pairing, there may be minute differences, which are due to gene duplications and other major rearrangements in specific parts of the genome, where genes with new functions that are acquired do not lead to the maintenance of the previous metabolic blueprint. The accumulation of barriers to cross-fertility between maize and teosinte races or species that do or do not grow in the vicinity of maize appears evident. On the other hand, where homogenization of the respective genomes has taken place over thousands of years of crosses and backcrosses, cross-fertility reduction genes, through mutation and acquisition of new functions, may have enabled maize to be a bridge for gene flow with *parviglumis* and *mexicana* teosintes. The flow of wild genes to cultivated maize and the reverse flow of alleles originating from domesticated maize, which are adapted to cultivation, to teosinte that grows in a wild habitat – in spite of the accumulation of a high frequency of common alleles – may tend to diminish, in the absence of strong selection, the respective fitness of each one in their respective habitats. Therefore, it is easy to realize why a gametophyte series of genes has arisen as an isolating mechanism between teosinte and maize. Their frequency is higher in Mexico and Central America, as expected, and is very low or nonexistent in the Andean region. This is indirect evidence of the early separation of maize from Mexico and the Andean region, according to Mangelsdorf (1974).

The high proportion of mutants arising in plants whose morphological characteristics are defined as tripsacoid (resembling segregates from maize × teosinte or maize × *Tripsacum* crosses), as described by Mangelsdorf (1974: 139), points to the existence of a large number of micro-rearrangements differentiating the genomes of teosinte and maize, which in a state of linkage disequilibrium,

[315] Bianchi, A. 1957. Defective caryopsis factors from maize teosinte derivatives. I. Origin, description and segregation. *Genetica Agraria*, 7: 1–38.

through crossing over, may originate new mutations. The mutagenic effect of such hybridization may have induced the formation of additional duplicate genes and their selection in the wild and under cultivation, respectively, as a survival mechanism in both maize and teosinte.

A relevant example of the species divergence and specialization of duplicate genes comes from the research results of Bomblies and Doebley (2006), who worked on identifying pleiotropic quantitative effects of regulatory genes, which may have been directly involved in evolution. They assessed how trait correlations may arise. Significant associations were found between several quantitative traits and copy numbers of both *zfl* genes in several maize genetic backgrounds. Despite overlap in traits associated with these duplicate genes, *zfl1* showed stronger associations with flowering time, whereas *zfl2* associated more strongly with branching and inflorescence structure traits, suggesting some divergence of function. Considering that *zfl2* associates with quantitative variation for ear rank and also that its position maps near a QTL on chromosome 2, controlling ear rank differences between maize and teosinte, tests were made as to whether *zfl2* might have been involved in the evolution of this trait by using a QTL complementation test. The results suggest that *zfl2* activity is important for the QTL effect, supporting *zfl2* as a candidate gene for a role in the morphological evolution of maize.

It is now accepted that recurrent gene duplication and alternative gene silencing may be a common mechanism for generating microchromosomal rearrangements, which may become builders of postreproductive isolating barriers among species. Relative to subfunctionalization, neofunctionalization is expected to become a progressively more important mechanism of duplicate gene preservation in populations with increasing size. However, even in large populations, the probability of neofunctionalization scales only with the square of the selective advantage. Tight linkage also influences the probability of duplicate gene preservation, increasing the probability of subfunctionalization but decreasing the probability of neofunctionalization.

The Role of Gene Flow in Plant Speciation

The genetic mechanisms of speciation and how species diverge and exchange genes after their divergence from a common ancestor are subjects of primary interest in evolutionary biology. Recent proposals are that chromosomal rearrangements are the first step of population differentiation. They allow the formation and persistence of alleles that promote isolation. Reduced recombination permits the accumulation of alleles contributing to isolation and adaptive differentiation and protects existing differences from the homogenizing effects of introgression between incipient species.

In the genus *Zea*, the existence of introgression has been amply documented between several subspecies or races of teosinte and maize (Wilkes, 1977), and gene exchanges between maize and *Tripsacum* have also been proposed to have occurred (Eubanks, 2001;[316] Mangelsdorf, 1974).

Genes that are considered to be major contributors to the morphological differences between maize and teosinte have been studied in regard to their single nucleotide polymorphisms as well as, to some extent, to their DNA segmental genomic sequencing analysis, and deductions have been inferred from these studies on differentiation and rates of evolution. Most studies have looked at very few loci (10 at most), and within the reference accessions studied, few plants have been sampled, in some cases just one plant (Matsuoka et al., 2002); therefore there are doubts that the effect of sampling variance may be as large as or larger than the effect of the biological processes. Additionally, introgression can obscure historical phylogenetic relationships. For example, Jaenicke-Deprés and colleagues (2003), report that, regarding the *tb1* gene, maize carries a single allele, whereas teosinte is estimated to carry 11 alleles, and for the *tbf* segment that was sequenced, maize is expected to carry 6 alleles and teosinte 16 alleles, the latter when increasing the number of analyzed samples. Nucleotide diversity in maize was found to be 11-fold lower than in teosinte. How this finding squares with the known large extent of gene flow between teosinte and maize, which should equalize differences in the variability of nucleotide polymorphisms, remains a mystery.

It is important to consider that methylation of DNA and epigenetic control may be a very important source of variation that needs to be explored in maize. Working with *Arabidopsis*, Roux and colleagues (2011[317]) have recently found that artificially induced DNA methylation not only caused heritable phenotypic diversity but also produced heritability patterns closely resembling those of the natural accessions. Their findings indicate that epigenetically induced variation and naturally occurring variation in complex traits share part of their polygenic architecture and may offer complementary adaptation routes in ecological settings.

Ross-Ibarra and colleagues (2009[318]), in a study of divergence and gene flow in the genus *Zea*, state that "substantial uncertainty remains about the evolutionary history of the genus *Zea* due in part to the complicating effects of hybridization and introgression" as expressed also by Wilkes (1977), Doebley (1990b[319]), and Fukunaga and colleagues (2005[320]). Consideration of its effects

[316] Eubanks, Mary. 2001. The mysterious origin of maize. *Economic Botany*, 55: 492–514.

[317] Roux, Fabrice, Maria Colomé-Tatché, Cécile Edelist, René Wardenaar, Philippe Guerche, Frédéric Hospital, Vincent Colot, Ritsert C. Jansen, and Frank Johannes. 2011. Genome-wide epigenetic perturbation jump-starts patterns of heritable variation found in nature. *Genetics*, 188: 1015–1017.

[318] Ross-Ibarra, J., M. Tenaillon, and B. S. Gaut. 2009. Historical divergence and gene flow in the genus *Zea*. *Genetics*, 181 (4): 1399–1413.

[319] Doebley, J. F. 1990b. Molecular evidence of gene flow among *Zea* species. *Bioscience*, 40: 43–448.

[320] Fukunaga, K., J. Hill, Y. Vigouroux, Y. Matsuoka, J. Sánchez, G. K. Liu, E. S. Buckler, and J. Doebley. 2005. Genetic diversity and population structure of teosinte. *Genetics*, 169: 2241–2254.

brought attention on maize domestication from Eyre-Walker and colleagues (1998[321]), Matsuoka and colleagues (2002), Tenaillon and colleagues (2004), and Wright and colleagues (2005); despite their discussion, the relation of gene divergence in the process of divergence among other lineages of *Zea* remains obscure. It is necessary to note with Wright and Gaut (2005) that multiple factors may contribute to nucleotide variation across plant species, including assembling representative samples for the studies, mutation rates, demography, and selection. Hilton and Gaut (1998[322]) indicated that nucleotide diversity in rice differentiation between species was only 23–46% of that found in *Zea* species. When comparing results obtained on nucleotide polymorphisms from a number of Angiosperm species based on multi loci samples, Ling-Bin and Ge (2007[323]) note in their table 6 that *Zea mays* ssp. *parviglumis* has about twice the variation exhibited by *Zea mays* ssp. *mays*, by *Zea diploperennis*, and *Zea perennis*, and has even much higher variation than other cross-pollinated species in the genera *Oryza*, *Helianthus*, and *Quercus*. Therefore the four reasons for such a difference between maize and teosinte ssp. *parviglumis* in nucleotide polymorphism need to be researched more thoroughly and should not be judged only in the context of a possible descent of maize from teosinte ssp. *parviglumis*.

To ascertain the effect of gene flow on divergence, Ross-Ibarra and colleagues (2009) assembled a set of polymorphisms combined with new resequencing of 26 genes of *Zea luxurians*, and three *Zea mays* subspecies: ssp. *mays*, ssp. *parviglumis*, and ssp. *mexicana*. They studied divergence at three levels: recent domestication, subspecies differentiation, and speciation. They applied in their calculations a mutation rate of 3×10^{-8} per bp, based on Clark and colleagues' (2005) analysis of postdomestication mutations calibrated from archaeological data, although this rate might be higher than other researchers' estimates. They found some measure of isolation between species but not so among subspecies. Very significantly the only subspecies that showed a fixed difference between them were ssp. *parviglumis* and maize, of one SNP at locus *asg65*, when theory of direct descent would preclude that being so. Approximately half of the loci analyzed show introgression effects of sequence segments, with about 70% being of recent introgression between subspecies of *Zea*, and six loci provide evidence of gene flow also with *Z. luxurians*.

The following models were reported the most likely explanations: (1) For ssp. *parviglumis–luxurians*, there could have been a sympatric history of ancestral gene flow during divergence. (2) Recent gene flow has higher probability if

[321] Eyre-Walker, A., R. L. Gaut, H. Hilton, D. Feldman, and B. S. Gaut. 1998. Investigations on the bottleneck leading to the domestication of maize. *Proceedings of the National Academy of Sciences USA*, 95: 4441–4446.

[322] Hilton, H., and B. S. Gaut. 1998. Speciation and domestication in maize and its wild relatives: Evidence from the *Globulin-1* gene. *Genetics*, 150: 863–872.

[323] Ling-Bin, Zhang, and Song Ge. 2007. Multilocus analysis of nucleotide variation and speciation in *Oryza officinalis* and its close relatives. *Molecular Biology and Evolution*, 24 (3): 769–783.

the isolation model has been rejected. (3) The *parviglumis*–maize comparison models of recent gene flow (island and allopatry) have the lowest probability if only the island effect is rejected. They estimate a *time of divergence* of maize from teosinte ssp. *parviglumis* at 55,000 years and maize from teosinte ssp. *mexicana* at 60,000 years ago. These figures are several times higher than those used by other research workers (Pohl et al., 2007[324]). The estimated *time of isolation* of *parviglumis* teosinte from maize is 27,000 years, which the authors, trapped in their dogma of maize being domesticated from teosinte, cannot fathom and thus record as archaeologically implausible. It would be plausible if maize and teosinte had diverged from a common ancestor and, thus, divergent wild maize was the ancestor of modern maize. Divergence of *parviglumis* from *luxurians* took place about 149,000 years ago, and cessation of gene flow between them happened about 60,000 years ago. Between *parviglumis* and *mexicana* a recent divergence and gene flow with introgression occurring between them are also suggested. Hanson and colleagues (1996[325]) have proposed the divergence of these two taxa 61,000 years ago. Also, recent gene flow is suggested between maize and teosinte ssp. *mexicana*, which would account for statistics that indicate a closer relationship of these two taxa, with recent gene flow as detected at several loci. This confirms other observations on this point (Blancas et al., 2002[326]). These authors, using allozyme data, propose that introgression between maize and Mexican teosinte may be common.

Zea luxurians and *Zea diploperennis* may be phylogenetically farther apart from each other than maize is from either one of them, based on unpublished data by Hanson and colleagues (1966), Blancas and colleagues (2002), and Ross-Ibarra and colleagues (2009). *Tripsacum* and ssp. *parviglumis* teosinte may have diverged long before, about 1 to 1.2 million years ago. White and Doebley (1999[327]) have estimated this time of divergence at a shorter 0.5 million years. An intriguing possibility was offered by Ross-Ibarra and colleagues (2009) that maize may have served as a bridge for gene flow among all four taxa.

The number of individuals involved in the separation between *parviglumis* teosinte and *Zea luxurians* and between *parviglumis* and *mexicana* teosintes was estimated for both cases at 120,000–160,000 individuals; there were

[324] Pohl, M. E., D. R. Piperno, K. O. Pope, and J. G. Jones. 2007. Microfossil evidence for pre-Columbian maize dispersals in the neo-tropics from San Andrés, Tabasco, Mexico. *Proceedings of the National Academy of Sciences USA*, 104: 6970–6875.

[325] Hanson, M. A., B. S. Gaut, A. O. Stec, S. I. Fuerstenberg, M. M. Goodman, E. H. Coe, and J. F. Doebley. 1996. Evolution of anthocyanin biosynthesis in maize kernels: The role of regulatory and enzymatic loci. *Genetics*, 143: 1395–1407.

[326] Blancas, L., D. L. Arias, and N. C. Ellstrand. 2002. Patterns of genetic diversity in sympatric and allopatric populations of maize and its wild relative teosinte in Mexico: Evidence for hybridization. In A. Snow, editor. *Scientific Methods Workshop: Ecological and Agronomic Consequences of Gene Flow from Transgenic Crops to Wild Relatives.* Ohio State University. Columbus. pp. 31–38.

[327] White, S. E., and J. F. Doebley. 1999. The molecular evolution of *terminal ear 1*, a regulatory gene in the genus *Zea*. *Genetics*, 153: 1455–1462.

50,000 individuals for *luxurians* and 45,000 individuals for maize. It is suggested from these studies that the various species of *Zea* may have arisen at an almost contemporaneous period in the range of 100,000 to 300,000 years ago. *Zea mays* ssp. *parviglumis* and *Zea diploperennis* would seem to be the more closely related taxa, not including maize.

Ross-Ibarra and colleagues (2009) state that the estimates of divergence may be complicated by historical introgression and that domesticated maize may have acted as a genetic go-between with wild populations. They consider that nuclear linked data from multiple unlinked loci are not appropriate for phylogenetic reconstruction because introgression can obscure historical relationships. This is why phylogenetic analysis based on Mexican maize races that have been subjected to a long period of mutual introgression with teosinte and have been also subjected to different selection pressures may not be a good subject for molecular analysis to reconstruct ancient relationships. It is also important to consider that the geographic distribution of all the *Zea* taxa and of *Tripsacum* may have been different in the past from what it is at present.

The Effect of Cytoplasm on Evolution of *Zea*

The effect of cytoplasm on heredity is based on organelle genomes: plastids and mitochondria. All organelle genomes found in mitochondria of plant or animal cells are considered to have originated from an endosymbiotic form of α-Proteobacteria, and to have given rise to the emerging eukaryotic cell more than 10^9 years ago. Animal mitogenomes are compact – about 20 kb. Plant mitochondrial genomes vary in size from 187 kb (Oda et al. 1992[328]) to more than 2,400 kb (Ward et al., 1981[329]) and are less compact than their animal counterparts due to the occurrence of noncoding sequences and duplicated fragments.

The mitochondrial genome in higher plants has a complex organization. It can undergo homologous recombination that results in variation within species. The total genetic information of the plant mitochondrial genome can be arranged into a single circular molecule that is referred to as the master chromosome. This circular DNA molecule contains repeated sequences that can generate, via intramolecular recombination, either isomeric forms of the master chromosome or smaller subgenomic circular DNA molecules.

The maize mitochondrial genome is the most complex and largest mitochondrial genome for which a physical map is presently available. Its organization

[328] Oda, K., K. Yamato, E. Ohta, Y. Nakamura, M. Takemura, N. Nozato, T. Kohchi, Y. Ogura, T. Kanegae, K. Akashi, and K. Ohyama. 1992. Gene organization deduced from the complete sequence of liverwort *Marchantia polymorpha* mitochondrial DNA: A primitive form of plant mitochondrial genome. *Journal of Molecular Biology*, 223: 1–7.

[329] Ward, B. L., R. S. Anderson, and A. J. Bendich. 1981. The mitochondrial genome is large and variable in a family of plants (*Cucurbitaceae*). *Cell*, 25: 793–803.

varies considerably among the different maize cytotypes. Fauron, Casper, Gao, and Moore (1995[330]) have proposed a general model of genome evolution that can explain a multitude of genomic rearrangements, for the maize mitochondrial DNA but also applicable to other higher plant mitochondrial genomes.

Recombination has occurred at the intraspecific level of the organelle genome through small repeats accounting for large gene-order shuffling and the emergence of new ORFs, some of which have been involved in CMS in maize. It has been known that the linear order of their genes can be highly variable among cytotypes even within a species (reviewed in Fauron, Casper, Gao, and Moore 1995; Fauron, Moore, and Casper 1995[331]). However, it is not clear how plant mitochondrial genomes rearrange so readily or how their genome sizes can increase or decrease dramatically over relatively short evolutionary times. Circular master genome maps have been generated for most of the plant mitochondrial genomes sequenced to date (e.g., Fauron et al., 2004[332]).

Plant mitochondria have a number of distinctive features, including considerable variation in genome size and organization, which can occur even within a single species (Fauron, Casper, Gao, and Moore, 1995).

The movement of DNA between cellular compartments is responsible for some of the variation in the known gene sets of different plants and appears to be an ongoing evolutionary process in plants (Palmer et al., 2000[333]). An additional curiosity is that, although plant mitochondrial genes are translated according to the universal code, transcripts of many genes require editing in order for that to occur (Brennicke et al., 1999[334]).

The NB mitochondrial genome found in most fertile varieties of commercial maize (*Zea mays* ssp. *mays*) was sequenced (Clifton et al., 2004[335]). The final assembly of the maize NB mitochondrial sequences generated a single circular map of 569,630 bp, larger than any of the previously sequenced plant mitochondrial genomes. It should be noted that a circular map does not mean that

[330] Fauron, C., M. Casper, Y. Gao, and B. Moore. 1995. The maize mitochondrial genome: Dynamic, yet functional. *Trends in Genetics*, 11 (6): 228–235.

[331] Fauron, C., B. Moore, and M. Casper. 1995. Maize as a model of higher plant plasticity. *Plant Science*, 112: 11–32.

[332] Fauron C., J. O. Allen, S. Clifton, and K. J. Newton. 2004. Plant mitochondrial genomes. In H. Daniell and C. Chase, editors. *Molecular Biology and Biotechnology of Plant Organelles.* Kluwer Academic. Dordrecht. pp. 151–177.

[333] Palmer, J. D., K. L. Adams, Y. Cho, C. L. Parkinson, Y. L. Qiu, and K. Song. 2000. Dynamic evolution of plant mitochondrial genomes: Mobile genes and introns and highly variable mutation rates. *Proceedings of the National Academy of Sciences USA*, 97: 6960–6966.

[334] Brennicke, A., E. Zabaleta, S. Dombrowski, M. Hoffmann, and S. Binder. 1999. Transcription signals of mitochondrial and nuclear genes for mitochondrial proteins in dicot plants. *Journal of Heredity*, 90: 345–350.

[335] Clifton, Sandra W., Patrick Minx, Christiane M. R. Fauron, Michael Gibson, James O. Allen, Hui Sun, Melissa Thompson, W. Brad Barbazuk, Suman Kanuganti, Catherine Tayloe, Louis Meyer, Richard K. Wilson, and Kathleen J. Newton. 2004. Sequence and comparative analysis of the maize NB mitochondrial genome. *Plant Physiology*, 136 (3): 3486–3503.

the genome is actually composed of a circular molecule in vivo. Indeed, a number of studies suggest that there are alternative physical structures.

The maize NB main mitochondrial genome contains 58 identified genes, including 34 genes coding for 33 known proteins.

To ascertain the contribution of organelle genes to plant phenotype, a study was made by Allen (2005[336]). Because variation in the plastid and mitochondrial genomes is reduced because mutations affect only a few of them, whereas the rest remain unchanged, the study was on how cytoplasms of the various *Zea* species affect variation, and it was conducted by transferring cytoplasms from various teosinte species and subspecies to a single inbred line A619 (Corn Belt Dent race line developed in Minnesota). Maize lines were created in which the maize organelle genomes were replaced through serial backcrossing by those representing the entire genus, yielding alloplasmic sublines, or cytolines. They found that the effects of cytoplasmic substitution can be substantial on the 58 characters that were observed or calculated in this study. More than 90% were different at a significance level of at least $\rho < 0.01$ in at least one cytotype and more than half at $\rho < 0.0001$.

Given that the organelle genomes of the cytolines are evolutionarily diverged from those that they replaced and are assumed to be a less optimal match to the maize nuclear genome, it was anticipated that characters such as growth rate would be negatively affected. Mazoti (1954[337]) in Argentina observed that maize into which teosinte cytoplasm had been substituted had slower growth and development. Consistent with this observation, growth in the Allen experiment was retarded in several *Z. luxurians* and *Z. diploperennis* cytotypes. Almost all of the characters that were selected to be observed in the study varied significantly, and, of critical importance, most of them did so independently of the other characters. These results suggest that there are many phenotypically important genes among the approximately 60 protein-coding genes in the organellar genomes whose interactions with the nuclear genome, or with the genome of the other type of organelle, are varied and extensive.

The cytolines of each of the three section Luxuriantes species contain a broad range of phenotypic diversity. This is somewhat surprising in light of the apparent consistency within the group by a variety of other measures. For instance, the authors found in this study that *Z. luxurians* cytolines were the most heterogeneous phenotypically, yet *Z. luxurians* teosinte itself is the most homogeneous of the *Zea* taxa both phenotypically and isoenzymatically (Doebley et al., 1984). Looking more broadly, phenotypic variation within and among section Luxuriantes cytolines was widespread and substantial, but variation within section *Zea* cytolines was minimal.

[336] Allen, J. O. 2005. Effect of tesosinte cytoplasmic genomes on maize phenotype. *Genetics*, 169: 863–880.

[337] Mazoti, L. B. 1954. Caracteres citoplasmáticos heredables derivados del híbrido de *Euchlaena* por *Zea*. *Revista Investigaciones Agrícolas*, 8: 175–183.

A survey of the plastid genome was conducted with 22 restriction enzymes and yielded only three differences separating the annual from the perennial teosintes; only one restriction site polymorphism distinguished the two perennial species *Z. diploperennis* and *Z. perennis*, and all three taxa were monotypic (Doebley et al., 1987). On the other hand, within *Z. mays* at least five types were observed.

Given the apparently highly conserved nature of plant organelle genomes, it is surprising that cytoline phenotypes should be so diverse within species. There have been no reports of this level of cytoplasmically influenced phenotypic diversity within any other species in which cytoplasmic effects have been investigated using non-CMS cytoplasms.

Zea mitochondrial and plastid genome sequencing should shed a more precise light on sequence divergence rates among the taxa involved in this reference study, as well as providing candidate genes and alleles to account for specific phenotypic differences.

Z. luxurians cytotype 8 was so maize-like that it is actually more similar to *Z. mays* cytotype 6 than were the other *Z. mays* cytotypes. This result is not consistent with studies of plastid DNA (Doebley et al., 1987) or with the mitochondrial RFLP groupings used in the study by Allen (2005).

Panayotov (1983[338]) surmised that the variation that he observed was due to an alien cytoplasm in the source and not to inherent differences within the species. The existence of such "phenotypic series" in the *Zea* cytolines could be explained by roughly simultaneous divergence of all three species, *Z. mays*, *Z. luxurians*, and *Z. diploperennis*, from a common ancestor, with only some *Z. luxurians* and *Z. diploperennis* cytoplasms subsequently accumulating phenotypically important mutations (at least in the context of the Allen study). Alternatively, because all of the *Zea* species are interfertile, there may be interspecific mitochondrial gene flow. Native Mexicans occasionally cross maize with *Z. diploperennis* for crop improvement purposes (Benz et al. 1990[339]), and the reciprocal cross can also occur. Mitochondria, but not plastids, are known to take up exogenous DNA, and such a mechanism may also be responsible for the surprising diversity in these genomes and perhaps for the seemingly chimeric mitochondrial genome of *Z. perennis* cytotype.

Darracq and colleagues (2010[340]) analyzed the whole sequences of eight mitochondrial genomes from maize and teosintes to comprehend the events that led to their structural features, that is, the order of genes, tRNAs, rRNAs,

[338] Panayotov, I. 1983. The cytoplasm in *Triticinae*. In S. Sakamoto, editor. *The Sixth International Wheat Genetics Symposium*. Plant Germplasm Institute, Faculty of Agriculture, Kyoto University. Kyoto. pp. 481–497.

[339] Benz, B. F., L. R. Sanchez-Velásquez, and F. J. Santana Michel. 1990. Ecology and ethnobotany of *Zea diploperennis*: Preliminary investigations. *Maydica*, 35: 85–98.

[340] Darracq, Aude, Jean-Stéphane Varré, and Pascal Touzet. 2010. A scenario of mitochondrial genome evolution in maize based on rearrangement events. *BMC Genomics*, 11:233.

ORFs, pseudogenes, and noncoding sequences shared by all mitogenomes and duplicate occurrences. They suggested a tandem duplication model similar to the one described in animals, except that some duplicates can remain. Their analysis of maize and teosinte mitogenomes revealed the occurrence of duplications. Duplication length varied from 0.54 kbp to 120 kbp. Duplicated fragments were an important part of the total genome length for the longest genomes – 23.4% for NA, 31.5% for CMS-C, and 21.2% for *Zea mays* ssp. *parviglumis* – and more generally were the main cause of size differences among maize mitogenomes (Allen et al., 2007[341]). Six duplicated fragments were shared between maize and *Zea mays* ssp. *parviglumis* mitogenomes: {NA, NB, CMS-C, CMS-S, and *Zea mays* ssp. *parviglumis*} shared two duplications (11 and 17 kbp), {NA, NB, CMS-S, and *Zea mays* ssp. *parviglumis*} shared a 0.7 kbp duplication; {NA, CMS-S, CMS-T, and *Zea mays* ssp. *parviglumis*} shared a 5.3 kbp duplication; {NA, NB and *Zea mays* ssp. *parviglumis*} shared another 5.3 kbp duplication; and {NA and *Zea mays* ssp. *parviglumis*} shared a 0.6 kbp duplication.

CMS-C and NA cytoplasm (the latter the most widespread one in North American maize) had a maximal duplication length of 105 and 120 kbp, respectively, as compared to 55.0, 10.1, and 13.6 kpb of *parviglumis*, *luxurians*, and *perennis* teosintes, respectively. In median duplication length CMS-C was 31.00 kpb, four times larger than *parviglumis* and three times larger than *perennis* teosintes. This might be interpreted as a large duplication occurrence in this maize cytoplasm, whereas the other maize cytoplasms do not reflect such duplication.

Maize Chromosome Divergence

Chromosomes are the condensed filaments of DNA, which in a discontinuous form constitute the genome of a plant or animal and are located in every nucleus of a cell.

Maize chromosomes have a basic number $n = 10$, which proceeds from an ancestral basic number of 5 through polyploidization by addition of two different complements of 5 pairs some 5 million years ago.

Maize chromosomes are best identified by observing their relative total and arm lengths while in the Leptotene and Pachytene stages of meiosis or in the Prophase of mitosis, which precedes cell division. They have unequal lengths of two arms, which are at both sides of the centromere; when the chromosome is stained with special dyes, the centromere does not show coloration. The centromere is a region where the chromosome attaches to fibers that form the spindle and that pull each chromosome to a cell pole in the processes of either mitosis – or cell division with maintenance of the same number of chromosomes

[341] Allen, J. O., C. M. Fauron, P. Minx, S. Oddiraju, L. Westgate, G. N. Liu, M. Gibbon, J. Cifrese, L. Meyers, H. Sun, K. Kim, C. Wang, F. Du, D. Xu, S. Welifton, and K. J. Newton. 2007. Comparisons among two fertile and three male-sterile mitochondrial genomes of maize. *Genetics*, 177: 1173–1192.

in the daughter cells, with the double complement of DNA originating in both parents of the plant – or meiosis, in which the chromosome number is reduced in half to form cells that carry one set of genes or gametes.

The centromere is in a relatively constant position in the chromosome. With further increase of thickness of the chromosomes by condensation of the spirals of DNA, the centromere appears as a constriction. Centromeres have repeat sequences of DNA usually of 150–180 bp in grasses, with great variation among grass species. For example, in barley and maize, the ZmBs repeat is 9 Mb in length (Jin et al., 2005[342]). The major tandem repeat sequence of maize centromeres is 156 bp in size, and is called CentC (Ananiev et al., 1998b[343]). The total length of CentC arrays vary from as little as 100 kb to as much as several thousand kilobases (Jin et al., 2004[344]). Studies should be made of centromere CentC arrays in maize races and their ancestors – because they appear as conserved structures with minor changes. They would shed light on the evolution of maize and its relatives under domestication and prior to domestication. Centromeric retroelements are a class of retroelements present in the centromeres of grasses with a large number of repeats that are full portions of the Ty3-*Gypsy* family of retroelements. Zhong and colleagues (2002[345]) have shown that these centromeric retroelements are conserved up to 85% in cereal species that diverged 60 million years ago.

The centromere insertion point in the DNA chain has been set precisely, so far, only on chromosome 8 of maize (Luce et al., 2006[346]). A conserved histone C3-like protein, CENH3, is associated in the centromere with the kinetochore formation for movement of the chromosomes in the spindle in cell division and is conserved in all grasses. In the centromeres it is associated with the chromatin coloring protein and some retrotransposons. The repetitive DNA in centromeres is megabases in length and is often left out in genomic analysis (Jin et al., 2005).

Genes located in centromeres tend to maintain linkage equilibrium. Nondisjunction of genes located in loci in the vicinity of the centromeres could

342 Jin, W., J. C. Lamb, J. M. Vega, R. K. Dawe, J. A. Birchler, and J. Jiang. 2005. Molecular and functional dissection of the maize B chromosome centromere. *Plant Cell*, 17: 1412–1423.

343 Ananiev, E. V., R. L. Phillips, and H. W. Rines. 1998b. Complex structure of knob DNA on maize chromosome 9: Retrotransposon invasion into heterochromatin. *Genetics*, 149: 2025–2037.

344 Jin, W., J. R. Melo, K. Nagaki, P. B. Talbert, S. Henikoff, R. K. Dawe, and J. Jiang. 2004. Maize centromeres: Organization and functional adaptation in the genetic background of oat. *Plant Cell*, 16: 571–581.

345 Zhong, C. X., J. B. Marshall, C. Topp, R. Mroczek, A. Kato, K. Nagaki, J. A. Birchler, J. Jiang, and R. K. Dawe. 2002. Centromeric retroelements and satellites interact with maize kinetochore protein CENH3. *Plant Cell*, 14: 2825–2836.

346 Luce, A. C., A. Sharma, O. S. Mollere, T. K. Wolfgruber, K. Nagaki, J. Jiang, G. G. Presting, and R. K. Dawe. 2006. Precise centromere mapping using a combination of repeat junction markers and chromatin immunoprecipitation-polymerase chain reaction. *Genetics*, 174: 1057–1061.

be a mechanism for conservation of some genes. Except for B chromosomes, no knobs are located near the centromeres in maize or teosinte.

Knobs, contrary to centromeres, are deep-staining locations in the chromosomes of maize. McClintock identified 23 positions, whereas Kato-Yamakake (1976) has identified 34 positions in maize and teosinte, of which 13 from teosinte are not shared by maize. Recently, Sánchez-González (2011[347]) has reported up to 47 knob positions in maize and its relatives. The knobs can be large or small and exhibit a marked constancy of position at the tips or interstitial positions in specific chromosomes in the various races of maize and their wild relatives (Grobman et al., 1961; Kato-Yamakake, 1976). Their sizes also vary from very small to large (Adawy et al., 2004;[348] Kato-Yamakake, 1976). Maize knobs contain a basic 180-bp repeat in many thousands of bases per knob, and two families of tandemly repeated DNA sequences have been identified within the maize knobs (Adawy et al., 2004). Peacock and colleagues (1981[349]), working in Australia, showed that a 180-bp tandem repeat is the main component in maize knobs and also in abnormal chromosome 10. Another repeat called TR-1 has been identified by Ananiev and colleagues (1998[350]) in many but not all knobs. Transposable elements are found in small numbers in the knobs but much more infrequently than in other parts of the genome.

Fluorescence in situ hybridization (FISH) analyses were conducted by Adawy and colleagues (2004) to examine the presence or absence of the 180- and 350-bp knob-associated tandem repeats in maize strains previously defined as one-knobbed or knobless. Multiple loci were found to hybridize to these two repeats in all maize lines analyzed. They concluded that the number of 180- and 350-bp repeat loci fails to correlate with the number of knobs in maize and that these tandem repeats are not independently sufficient to confer knob heterochromatin, even when present at megabase sizes.

Knobs are extremely important as markers in tracing the evolution of maize. They appear in terminal chromosome positions in annual teosinte, which are not identified in maize, where they are mostly interstitial. This fact provokes thoughts as to why such positions, if they were transmitted by a presumed teosinte ancestor, are not found in maize.

[347] Sánchez-González, J. J. 2011. Diversidad del Maíz y el Teocintle. Informe preparado para el proyecto: "Recopilación, generación, actualización y análisis de información acerca de la diversidad genética de maices y sus parientes silvestres en México." Comisión Nacional para el Conocimiento y Uso de la Biodiversidad. Manuscrito.

[348] Adawy, S. S., R. M. Stupar, and J. Jiang. 2004. Fluorescence in situ hybridization analysis reveals multiple loci of knob-associated DNA elements in one-knob and knobless maize lines. *Journal of Histochemistry and Cytochemistry*, 52: 1113–1116.

[349] Peacock W. J., E. S. Dennis, M. M. Roades, and A. Pryor. 1981. Highly repeated DNA sequence, limited to knob heterochromatin in maize. *Proceedings of the National Academy of Sciences USA*, 78: 4490–4494.

[350] Ananiev, E. V., R. L. Phillips, and H. W. Rines. 1998c. Chromosome-specific molecular organization of maize (*Zea mays* L.) centromeric regions. *Proceedings of the National Academy of Sciences USA*, 95, 13073–13078.

One of the critical differences between primitive and immediately derived maize races from the Andean region is that, contrary to most Mexican and Mesoamerican races, they are either knobless or contain very few knobs, basically only a large one and a small one in chromosomes 6 and 7 (Grobman et al., 1961). Because chromosomes are essentially conserved, this would be used as evidence either of independent domestication or separate evolution after early divergence of Andean and Mexican/Mesoamerican races of maize, prior to teosinte introgression.

The abnormal chromosome 10 (*Ab10*), first observed by Longley (1938[351]), is present in races of maize from Mexico and southwestern United States and in teosinte (Longley, 1937,[352] 1938), and in northern and eastern South America (Kato-Yamakake and McClintock, 1981[353]). In Peru it has been found only in recently introduced races or races resulting from hybridization with them, but not in primitive races of the Andean type or their direct derivatives (Grobman et al., 1961). *Ab10* is identified by extreme condensation and large heterochromatic regions as compared to normal chromosome 10. It appears that *Ab10* is capable of moving the whole chromosome faster to the poles in Anaphase I and II of meiosis, resulting in preferential segregation in a ratio of 3:1 (Rhoades, 1942[354]). This happens when *Ab10* is in a heterozygous condition. The process is also called meiotic drive and is a segregation distorter that may have arisen in evolution to give some genes called selfish genes – possibly parasitic genes – a transmission advantage over the whole genome. This characteristic of selective transmission is shared by B chromosomes (see subsequently). Meiotic drive has recently been shown to be controlled by four genes (Hiatt and Dawe, 2003[355]).

Meiotic drive may drive evolution through the *Ab 10* chromosome, which promotes its own transmission, carrying along with it some knobs regarded to host fitness (Ardlie, 1998[356]). At least three other knobs show the same levels of preferential segregation when *Ab10* is present (Longley 1945;[357] Rhoades

[351] Longley, A. E. 1938. Chromosomes of maize from North American Indians. *Journal of Agricultural Research*, 56: 177–196.

[352] Longley, A. E. 1937. Morphological characters of teosinte chromosomes. *Journal of Agricultural Research*, 54: 835–862.

[353] Kato-Yamakake, T. A., in collaboration with B. McClintock. 1981. The chromosome constitution of races of maize in North and Middle America. Part 2. In B. McClintock, T. A. Kato-Yamakake, and A. Blumenschein. *Chromosome Constitution of Races of Maize: Its Significance in the Interpretation and Relationship between Races and Varieties of the Americas.* Colegio de Postgraduados. Chapingo.

[354] Rhoades, M. M. 1942. Preferential segregation in maize. *Genetics*, 27: 395–407.

[355] Hiatt, E. N., and R. K. Dawe. 2003. Four loci on abnormal chromosome 10 contribute to meiotic drive in maize. *Genetics*, 164: 699–709.

[356] Ardlie, K. G. 1998. Putting the brake on drive: Meiotic drive of *t* haplotypes in natural populations of mice. *Trends in Genetics*, 14: 189–193.

[357] Longley, A. E. 1945. Abnormal segregation during megasporogenesis in maize. *Genetics*, 30: 100–113.

and Dempsey, 1985[358]). These data suggest that all 23 knobs are preferentially segregated, but only when *Ab10* is present.

Preferential segregation produces a "genomic conflict" (Burt and Strivers 2006[359]), in which the selfish interests of the DNA, in this case, are at odds with the interests of the organism. Any allele linked to a knob is constrained in evolutionary terms, because it is slated to increase in the population whether or not it is a fit allele. As the majority of the maize genome is linked to knobs, meiotic drive is presumed to have had a major impact on the makeup of maize (Buckler et al., 1999;[360] Dawe, 2009[361]). In retrospect, thus, a knobless condition of maize would be a primitive one.

Furthermore, *Ab10* is capable of generating an abnormal number of mutations. The results of some work done earlier with the inbred line W23 – such as the work of the mutation genetics group at Kiev, Ukraine, whom I visited in 1972, and who were working on mutation research with radiation and chemical mutagenesis – may have been affected by self-mutation capabilities of that line, which carried *Ab10* (Brink, personal communication), something that I pointed out to them at that time. It may be speculated that in maize, *Ab10* may have been a late contributing factor for its permanence and generation of variability, which could then have been selected for adaptation to the many different environments found in altitude, mean temperature, length of growing season, soil type, and precipitation regime variations, which are present within a short altitudinal distance in the slopes of mountain ranges in the area within the tropical lines. There also appears to be a preferential segregation for all knobs when *Ab 10* is present.

Ab10 has not been found in the basic Andean maize races but is found in recent introductions or their derived races, such as Jora and Perla in Peru.

The Evolution of the Maize Nuclear Genome

Maize is the product of a polyploid event that occurred some 11 million years ago. The first evidence came from McClintock's observations (1933) that the 10 maize chromosomes paired in two sets of 5 nonhomologous chromosomes in the meiosis of haploid cells. The polyploid event occurred after the divergence between sorghum and maize. The DNA content of maize expanded

[358] Rhoades, M. M., and E. Dempsey. 1985. Structural heterogeneity of chromosome 10 in races of maize and teosinte. In M. Freeling, editor. *Plant Seretics.* Alan R. Liss Inc. New York. pp. 1–18.

[359] Burt, A., and R. Strivers. 2006. *Genes in Conflict: The Biology of Selfish Genetic Elements.* Harvard University Press. Cambridge.

[360] Buckler, E. S. I., T. L. Phelps-Durr, C. S. K. Buckler, R. K. Dawe, J. F. Doebley, and T. P. Holsford. 1999. Meiotic drive of chromosomal knobs reshaped the maize genome. *Genetics,* 153: 415–426.

[361] Dawe, Kelly R. 2009. Maize centromeres and knobs (neocentromeres). In J. L. Bennetzen and S. C. Hake, editors. *Handbook of Maize.* Vol. 2. Springer. New York. Chap. 12. pp. 239–250.

considerably, and hence the polyploid event explains some of the difference in DNA content between these two species. Genomic rearrangement and diploidization followed the polyploid event. Most of the repetitive DNA in the maize genome is made up of retrotransposable elements, and they comprise more than 75% of the genome. Retrotransposon multiplication has been relatively recent – within the last 5–6 million years – suggesting that the proliferation of retrotransposons has also contributed to differences in DNA content between sorghum and maize.

The recent studies of diversity in the wild relatives of maize indicate the course that different genes have taken and also show that domestication and intensive breeding on the one hand and natural selection on the other have produced heterogeneous effects on genetic diversity across genes.

To infer the mechanisms of evolution that have shaped maize is a Herculean task. With the aid of a multisystematic approach from various scientific vantage points, researchers have produced an enormous quantity of data that are not easy to handle and organize into coherent solutions. These data, which shed light on the organization and structure of the genomes of maize and its relatives, range from extensive marker-based genetic maps, to "chromosome paintings" based on fluorescent in situ hybridization, and to complete genomic DNA sequences.

Maize is a member of the grass family (Poaceae). The grasses represent a range of genome size and structural complexity, with rice on one extreme. A diploid with 12 chromosomes ($2n = 24$), rice has one of the smallest plant genomes, with only 0.9 pg of DNA per 2C nucleus. Other grass species exhibit far larger genomes. Wheat, for example, is a hexaploid with 21 chromosomes ($2n = 42$) and a haploid DNA content of 33.1 pg. Genera like *Saccharum* (sugarcane) and *Festuca* are even more complicated, displaying wide variation in ploidy level and more than 100 chromosomes in some species. As a diploid with 10 chromosomes ($2n = 20$) and a 2C genome content roughly six-fold larger than rice, maize lies somewhere in the middle of grass genome size and structural complexity (Figure A.1).

The segmental allotetraploid event would have predicted a two-fold variation in DNA content between sorghum and maize, but it does not account for the actual 3.5-fold variation in DNA content (Figure A.1). Based on this information, differences in DNA content probably reflect the allopolyploid event *and* additional evolutionary changes, such as the accumulation of repetitive DNA.

Rhoades (1955[362]) noted that some regions of linkage maps did not contain mutants, and he proposed that the lack of mutants reflected genetic redundancy caused by chromosomal duplication. Rhoades's proposal has since been supported by molecular and isozyme data that have documented the presence of duplicated, linked loci in maize.

[362] Rhoades, M. M. 1955. The cytogenetics of maize. In G. F. Sprague, editor. *Corn and Corn Improvement*. Academic Press. New York. pp. 123–219.

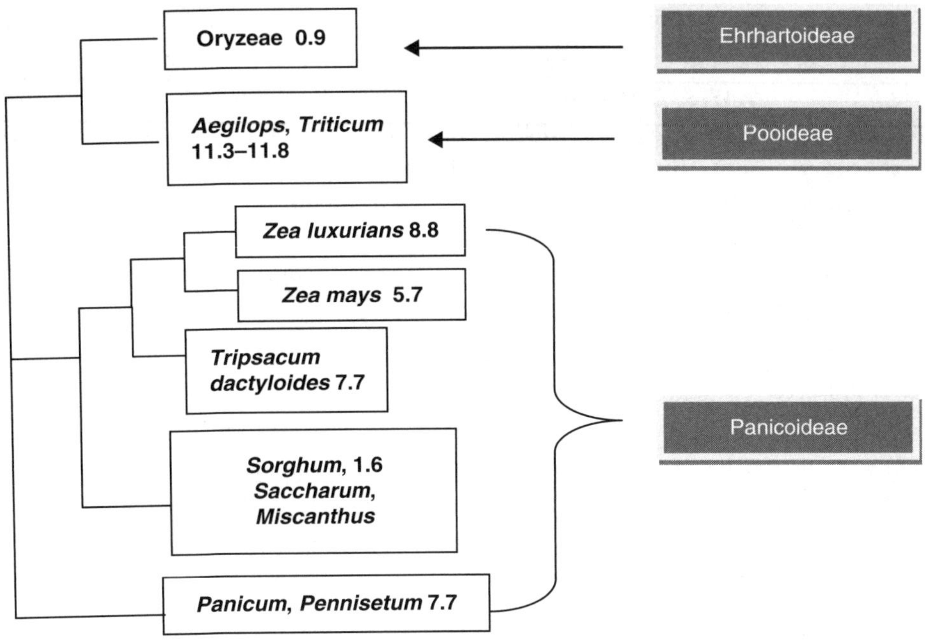

A.1. A phylogeny of selected grass tribes and species. The diagram is based on information drawn from data from the Grass Phylogeny Working Group and several authors and the 2c genome content of the species in pictograms shown after respective species from J. L. Bennetzen and E.A. Kellogg's 1997 article "Do Plants Have a One-Way Ticket to Genomic Obesity?" (*Plant Cell*, 9 [9]: 1509–1514). The formation of present *Zea* and *Tripsacum* species was preceded by a polyploidization event and by retrotransposon invasion.

Rearrangements of genes in the chromosomes is known to occur after allopolyploid events but does not lead exactly to duplication of genes. Mapping studies have documented regions of chromosome duplication in maize. There is a consensus about some chromosomal pairs having rearrangements; portions of chromosome 1 are duplicated on chromosomes 5 and 9, meaning that the process of diploidization rearranged one copy of chromosome 1. Alternatively, chromosome 1 could be an amalgamation of regions from different parental chromosomes. Chromosome 2 had a similar fate in that portions of chromosome 2 are also found on chromosomes 7, 10, and perhaps 4, and so on. In this sense, maize is not an exception. A great number of species contain chromosomal duplications as inferred from gene maps. The recently sequenced maize genome should disclose the extent of the gene duplication.

These duplications may account for the maintenance of reproducibility of phenotypes in maize in spite of the lack of collinearity found in gene positions in the maize genome.

Song and Messing (2003[363]) found significant the diversity of the maize germplasm within a small chromosomal region and examined the impact of such

[363] Song, Rentao, and Joachim Messing. 2003. Gene expression of a gene family in maize based on noncollinear haplotypes. *Proceedings of the National Academy of Sciences USA*, 100 (15): 9055–9060.

diversity on gene expression in hybrids. Because maize was domesticated only 10,000 years ago, it is quite unusual for the sequence differences between the two major haplotypes of the interval common to two inbred lines – BSSS53 and B73 – to be so extensive, given that allelic 22-kDa zein sequences are conserved between 97% and 100%. Although haplotypes of loci in other species consist primarily of single nucleotide polymorphisms or small insertions and deletions (indels), the same genome interval of the two maize inbreds differs substantially in size and content. Genes are missing or added as whole sequence segments that contain more than one gene. A previous phylogenetic analysis of the z1C-1 zein cluster in BSSS53 showed that the cluster arose not by unequal crossing over between genes but by the amplification of segments containing at least two genes, a finding confirmed by comparing the inbred lines as well. The two extra genes downstream of the BSSS53 *z1C* gene cluster also represent a segmental indel. In a recent study comparing an interval containing the *bz* gene and two different maize lines, researchers found that four extra genes in one inbred line are also clustered within a single segment. Conceivably, therefore, segmental indels could dramatically change gene content and size of the same interval in maize inbred lines.

In the same way, transposable elements, particularly DNA transposons and retrotransposons of relatively large size, affect sequence content and size of the same interval of different maize inbred lines. DNA transposable elements contribute to sequence divergence, and retrotransposons also contribute to the size variation of the same interval. Because active retrotransposons reach a length of between 8 and 12 kb and tend to insert on top of one another (the "hotspot" effect), reiterative retrotranspositions result in the large retrotransposon blocks observed in the maize genome. Retrotransposons also see movement around the genome and "hitchhike" with segmental duplications.

San Miguel and Bennetzen (1998[364]) established that the z1C-1 interval in maize and sorghum appears more different than that between maize inbreds alone, whereas sorghum and rice appear to be much more conserved downstream of the *z1C* gene cluster. When compared to rice or sorghum, maize seemed to be much more active in segmental rearrangements and transpositions of its strains. By accumulating just a few "large unit" mutations, maize lines rapidly yield divergent intervals from a common ancestral chromosomal region.

The highly methylated retrotransposon clusters are probably heterochromatic as are similar blocks in the knobs of maize (Ananiev et al., 1998b), and most likely affect recombination. Genes next to retrotransposon clusters may be less recombinogenic, because the more condensed chromatin state of the retrotransposon cluster may interfere with the access of the recombination machinery to

[364] San Miguel, P., and J. I. Bennetzen. 1998. Evidence that a recent increase in maize genome size was caused by a massive amplification of intergene retrotransposons. *Annals of Botany*, 82 (Suppl. A): 37–44.

the adjacent euchromatic regions. Because they had available three different *bz1* locus haplotypes, McC, B73, and W22, in the same genetic background, Dooner and He (2008[365]) were able to examine the effect of retrotransposon heterozygosity on recombination in the adjacent *bz1* and *stc1* genes. They analyzed recombination between the *bz1* and *stc1* markers in heterozygotes that differ by the presence and absence of a 26-kb intergenic retrotransposon cluster. The genetic distance between the markers was twofold smaller in the presence of the retrotransposon cluster, in spite of its being inert. These findings imply that haplotype structure will profoundly affect the correlation between genetic and physical distance for the same interval in maize.

Genomic Imprinting

Researchers have discovered a new modality of control of plant and animal phenotypic expression that operates as epigenetic control, and it has been named genomic imprinting.

Genomic imprinting, the allele-specific expression of a gene dependent on its parent of origin, means that both parents do not have the same hereditary effect on the progeny. Genomic imprinting has independently evolved in flowering plants and mammals. Parental genomic imprinting is characterized by the expression of a selected panel of genes from one of the two parental alleles (Feil and Berger, 2007[366]). Recent evidence shows that DNA methylation and histone modifications are responsible for this parent-of-origin-dependent expression of imprinted genes. Because similar epigenetic marks have been recruited independently in plants and mammals, the only organisms in which imprinted gene loci have been identified so far, this phenomenon represents a case for convergent evolution.

Genomic imprinting is an epigenetic mechanism that results in mono-allelic gene expression that is parent-of-origin dependent (Kinoshita, 2007[367]). In plants it is observed as a control of flow of nutrients to the grain endosperm tissue under control of the mother plant. In *Arabidopsis*, recent studies of several imprinted gene loci have identified the epigenetic mechanisms that determine genomic imprinting. A crucial feature of genomic imprinting is that the maternally and paternally derived imprinted genes must carry some form of differential mark, usually DNA methylation and/or histone modification. Although the epigenetic marks should be complementary on maternally and paternally imprinted genes within a single species, it is possible that neither the patterns of

[365] Dooner, Hugo K., and Limei He. 2008. Maize genome structure variation: Interplay between retrotransposon polymorphisms and genic recombination. *The Plant Cell*, 20: 249–258.

[366] Feil, R., and F. Berger. 2007. Convergent evolution of genomic imprinting in plants and mammals. *Trends in Genetics*, 23 (4): 192–199.

[367] Kinoshita, T. 2007. Reproductive barrier and genomic imprinting in the endosperm of flowering plants. *Genes & Genetic Systems*, 82 (3): 177–186.

epigenetic marks nor the expression of imprinted genes are the same in different species. The regulation of expression of imprinted genes in seeds and endosperm tissue of hybrid seeds can be affected by upstream regulatory mechanisms in the male and female gametophytes. Species-specific variations in epigenetic marks, the copy number of imprinted genes, and the epigenetic regulation of imprinted genes in hybrids might all play a role in the reproductive barriers observed in the endosperm of interspecific and interploidy crosses. These predicted molecular mechanisms might be related to earlier models such as the endosperm balance number (EBN) and polar nuclei activation (PNA, which also refers to peptide nucleic acid) hypotheses. The determination of the type, form, and composition of the grain of hybrids in interspecific crosses between maize and its related species, teosinte and *Tripsacum*, could be subject to such subtle epigenetic regulation in addition to the effect of major genes and QTL genes.

Köhler and Weinhofer-Molisch (2010[368]) have advanced the concept that imprinting might have evolved as a by-product of a defense mechanism destined to control transposon activity in gametes (the defense hypothesis). Recent studies provide substantial evidence for the defense hypothesis, by showing that imprinted genes in plants are located in the vicinity of transposons or repeat sequences, suggesting that the insertion of transposons or repeat sequences was a prerequisite for imprinting evolution. DNA methylation causes silencing of neighboring genes in vegetative tissues, and transposons might be thus silenced. However, because of genome-wide DNA demethylation in the central cell, genes located in the vicinity of transposon or repeat sequences will be active in the central cell and the maternal alleles will remain unmethylated and active in the descendent endosperm, assuming an imprinted expression. Consequently, many imprinted genes are likely to have an endosperm-restricted function, or, alternatively, they have no functional role in the endosperm and are on the trajectory to convert to pseudogenes. Thus, the defense hypothesis and the kinship theory together can explain the origin of genomic imprinting; whereas the first hypothesis explains how imprinting originates, the latter explains how imprinting is manifested and maintained.

One theory worth examining is the Parental conflict theory; maternally expressed imprinted genes act by enhancing the flow of nutrients to the seed endosperm, while paternally expressed imprinted genes repress endosperm development. It has been proposed that the methylation process in *Arabidopsis* is controlled in the central cell of the gametophyte. A balance between paternal and maternal contributions is required for prezygotic development in interspecific crosses to avoid endosperm breakdown of the seed and loss of germinating capacity. The desynchronization of endosperm development cells in interploidy crosses has been characterized as affecting the development of the seed

[368] Köhler, C., and I. Weinhofer-Molisch. 2010. Mechanisms and evolution of genomic imprinting in plants. *Heredity*, 105: 57–63.

endosperm. In maize, this destabilizing effect is called the ploidy barrier and applies whenever crosses are made with unequal chromosome number parents (Kinoshita et al., 2008[369]).

Sixteen imprinted genes have since been identified in maize and *Arabidopsis*, and these are expressed primarily in the endosperm, which nurtures embryo development. Imprinting results from the regulation of transcriptional silencing by DNA methylation or by Polycomb Group (PoG) complex-mediated histone methylation (Berger and Chaudhury, 2009[370]).

AtFH5, an imprinting gene in *Arabidopsis*, has been recently identified and could shed light on how seed regulation may be altered in crosses between relatives in maize to stabilize some outcomes in seed morphology and viability of seed in interspecific crosses. AtFH5 is not specifically activated in the female gamete during female gametogenesis as a requirement for imprinting. Imprinting status is defined by the silencing of the paternal allele followed by zygotic activation of the maternal copy. The silencing takes place by PcG activity in vegetative tissues and the endosperm. PcG complexes act by depositing silencing histone modifications on homoeotic genes that regulate the patterning of other transcription factors. The PcG complex establishes the pattern of cell fate. The PcG complexes regulate genes that control transcription factors and structural molecules that establish the patterns of the development of the plant or animal (Fitz Gerald et al., 2009[371]).

Lin (1984[372]), working in Brazil, studied various ploidy levels as they affected maize endosperm development. Maize kernels inheriting the indeterminate gametophyte mutant (*ig*) on the female side had endosperms that ranged in ploidy level from diploid (2x) to nonaploid (9x). In crosses with diploid males, only kernels of the triploid endosperm class developed normally. Most endosperms started to degenerate soon after pollination and remained in an arrested state. Hexaploid endosperm was exceptional; it developed normally during the sequence of stages studied and accounted for plump kernels on mature ears. Because such kernels have diploid maternal tissues (pericarps) but triploid embryos, the present finding favors the view that endosperm failure or success in such circumstances is governed by conditions within the endosperm itself. Whereas tetraploid endosperm, consisting of three maternal genomes and one paternal genome, is slightly reduced in size but supports viable seed

[369] Kinoshita, T., Y. Hikeda, and R. Ishikawa. 2008. Genomic imprinting: A balance between antagonistic roles of parental chromosomes. *Seminary in Cell and Developmental Biology*, 19: 574–579.

[370] Berger, Fred, and Abed Chaudhury. 2009. Parental memories shape seeds. *Trends in Plant Science*, 14 (10): 550–556.

[371] Fitz Gerald, J. N., P. S. Hul, and F. Berger. 2009. Polycomb group dependent imprinting of the actin regulator *AtFH5* regulates morphogenesis in *Arabidopsis thaliana*. *Development*, 136: 3399–3404.

[372] Lin, Bor-Yaw. 1984. Ploidy barrier to endosperm development in maize. *Genetics*, 107: 103–115.

development, the endosperm that has two maternal and two paternal chromosome sets was highly defective and conditioned abortion. Thus, evidently, development of maize endosperm is affected by the parental source of its sets of chromosomes.

This observation would be a mechanism for defense of maize from foreign pollen of the *Tripsacum* genus or from tetraploid teosinte, depending on in which direction the hybridization is made and the ploidy level of the zygote and endosperm in the F_1 and succeeding generations.

Successful mating between species results in the presence of different genomes within a cell (hybridization), which can lead to incompatibility in cellular events due to adverse genetic interactions. In addition to such genetic interactions, recent studies have shown that the epigenetic control of the genome, silencing of transposons, control of nonadditive gene expression, and genomic imprinting might also contribute to reproductive barriers in plant and animal species.

The hybridization process is strictly limited by a number of reproductive barriers that can act either before fertilization or after fertilization. Postzygotic barriers can be stimulated in zygotes formed through the fusion of genetically divergent genomes; embryonic lethality, seed abortion, adult invariability, or sterility is often exhibited due to incompatible interactions between parental alleles.

Accumulating evidence suggests that epigenetic control is an important mechanism in reproductive barriers between species. Epigenetic control intervenes by altering gene expression without changes in DNA sequence. This regulatory change is mediated by DNA methylation, histone modifications, and small RNAs, which control the structural folding of the nucleosomal array and render the chromatin state as active or silent, thereby controlling gene expression (Henderson and Jacobsen, 2007;[373] Henikoff, 2008;[374] Kouzarides, 2007[375]). A case of epigenetic control in *Arabidopsis* by means of *FWA* is not a postzygotic barrier; similar mechanisms may also induce mating time variation in other plant species. Therefore, this is a potential prefertilization barrier through control of flowering time. Similarly, epigenetic mechanisms may be involved in the post-fertilization barrier.

Whereas hybrid incompatibility only emerges after fusion of two different genomes, the epigenetic program or memory, in order to be functional, must have already been set up before fertilization, especially in cases of species hybridization.

Repetitive elements or transposons may have beneficial effects on organisms, but more often they produce deleterious effects. Epigenetic control of

[373] Henderson, I. R., and S. E. Jacobsen. 2007. Epigenetic inheritance in plants. *Nature*, 447: 418–424.

[374] Henikoff, S. 2008. Nucleosome destabilization in the epigenetic regulation of gene expression. *National Review of Genetics*, 9: 15–26.

[375] Kouzarides, T. 2007. Chromatin modifications and their function. *Cell*, 128: 693–705.

transposons has surged during evolution to moderate their effect. Reciprocally, in addition to mobilization, transposons can also affect expression of neighboring genes through alteration of their epigenetic status. Host organisms have developed genetic and epigenetic mechanisms to silence the activities of transposons, such as transcriptional silencing by modification of DNA or histones. Evidence that has been accumulated also suggests the importance of small RNAs recruited in defense of the organism against transposons.

The striking finding that transposon proliferation may contribute to speciation came from a study of the sunflower and its species hybrids (Rieseberg et al., 1995[376]). Comparative linkage mapping demonstrated that extensive genomic reorganization had occurred in the hybrid species relative to its parents. Recently, it was shown that genome expansion in three hybrid sunflower species compared to the parental species was the result of retrotransposon proliferation (Ungerer et al., 2006[377]). A number of studies presented by Ishikawa and Kenoshita (2009[378]) have provided a framework for understanding how transposons are constitutively suppressed in the germ line to prevent their mobilization, which could have strikingly adverse effects in the next generation.

Recent Research on the Races of Maize

Progress in maize racial analysis has taken place in Peru through the development of additional information to amplify the previous information on Peruvian races of maize. Salhuana (2004[379]) has updated the information on the races of maize in Peru, which number 52 out of a total number of 259 in the American continent, as defined at that time.

Other recent studies have focussed on studying different indicators and applying statistical techniques to evaluate their use in differentiating the races of maize in more detail. Ortiz and Sevilla (1997[380]) identified in their studies the most useful descriptors for classifying races of maize in Peru. Their bases were heritability (H), repeatability (R), and coefficients of variation (CV). When the interaction of the genotype with the environment was high and the H value

376 Rieseberg, L. H., C. V. Fossen, and A. M. Desrochers. 1995. Hybrid speciation accompanied by genomic reorganization in wild sunflowers. *Nature*, 375: 313–316.

377 Ungerer, M. C., S. C. Strakosh, and Y. Zhen. 2006. Genome expansion in three hybrid sunflower species is associated with retrotransposon proliferation. *Current Biology*, 16 (20): R872–R873.

378 Ishikawa, R., and T. Kenoshita. 2009. Epigenetic programming: The challenge to species hybridization. *Molecular Plant*, 2 (4): 589–599.

379 Salhuana, Wilfredo. 2004. Diversidad y descripción de las razas de maíz en el Perú. In W. Salhuana, A. Valdez, F. Sheuch, and J. Davelouis, editors. *Cincuenta Años del Programa Cooperativo de Investigaciones en Maíz (PCIM)*. Universidad Nacional Agraria La Molina. Lima. pp. 204–251.

380 Ortiz, R., and R. Sevilla. 1997. Quantitative descriptors for classification and characterization of highland Peruvian maize. *Plant Genetic Resources Newsletter*, 110: 49–52.

was low, the descriptors were not recommended for classification or characterization. Moreover, descriptors were significantly affected by the environment, and examples with low R should also be discarded for characterization. The best descriptors were reproductive traits such as ear length, number of rows of kernels, cob diameter, and kernel width. These descriptors show high H and R, and low CV.

Abu Alrob and colleagues (2004[381]) concluded that kernel traits are the best descriptors for Peruvian highland maize germplasm, followed by ear traits. Tassel traits are not reliable descriptors for classifying this germplasm. Likewise, PCA (principle componet analysis) biplots are better than dendrograms from average linkage cluster analysis for grouping accessions within races of this Peruvian highland maize germplasm. Overlap of accessions from distinct races might reflect interaction and gene flow among races of Peruvian highland maize.

Similar conclusions were derived from a study of 47 Latin American accessions of maize and 30 F_2 populations from certain pairs of those parents by Martínez et al. (1983[382]) in the United States. They attempted to measure racial differentiation or divergence by six different statistical procedures. Classical taxonomic methods and multivariate analysis were in general harmony in measuring racial divergence with the statistical analysis methods used (Euclidean distance, Mahalanobis distance, generalized distance, modified generalized distance, approximate Dempster's distance, and Dempster's distance). Morphological analysis of F_2 populations can be useful in understanding the variability and relationships of races.

The maize of Latin America, with its enormous diversity, has played an important role in the development of modern maize cultivars of the American continent. Peruvian highland maize shows a high degree of variation stemming from its long history of cultivation by Andean farmers. Multivariate statistical methods for classifying accessions have become powerful tools for classifying genetic resources conservation and the formation of core subsets.

A study was undertaken by Ortiz, Crossa, and colleagues (2008[383]) in Peru with two objectives: (1) to use a numerical classification strategy for classifying eight Peruvian highland races of maize based on six vegetative traits evaluated in two years, and (2) to compare this classification with the existing racial classification. The numerical classification maintained the main structure of the eight races but reclassified parts of the races into new groups (Gi). The new groups are more separated and well defined, with a decreasing accession within group

[381] Abu Alrob, I., J. L. Christiansen, S. Madsen, R. Sevilla, and R. Ortiz. 2004. Assessing variation in Peruvian highland maize: Tassel, kernel and ear descriptors. *Plant Genetic Resources Newsletter*, 137, 34–41.

[382] Martínez, W. O. J., M. M. Goodman, and D. H. Timothy. 1983. Measuring racial differentiation in maize using multivariate distance measures standardized by variation in F2 populations. *Crop Science*, 23: 775–781.

[383] Ortiz, R., J. Crossa, J. Franco, R. Sevilla, and J. Burgueño. 2008. Classification of Peruvian highland maize races with plant traits. *Genetic Resources and Crop Evolution*, 55: 151–162.

"environment interaction." Most of the accessions from G1 are from Cuzco Gigante, all (except one) of the accessions from G3 are from Confite Morocho, and all of the accessions from G7 are from Chullpi. Group G2 has four accessions from Huayleño and four accessions from Paro, whereas G4 has four accessions from Huayleño and five accessions from San Geronimo. Group G5 has accessions from four races, and G6 and G8 formed small groups with two and one accessions each, respectively. These groups can be used for forming core subsets for the purpose of germplasm enhancement and assembling gene pools for further breeding.

Maize landraces are an important source of germplasm for the genetic improvement of the crop. Classification of genetic resources requires both appropriate descriptors and sound numerical and statistical methods. Another research project was undertaken by Ortiz, Sevilia, Alvarado, and Crossa (2008[384]) to assess the use of six internal ear traits for classifying a set of four related Peruvian highland maize races comprising a total of 24 accessions. Several accessions of the four races were included in field trials planted in Peru's inter-Andean valleys. The trials were sown on two planting dates (normal and late) in two consecutive years. Variance components among races and among accessions with races were used to estimate broad-sense heritability and repeatability for each internal ear trait. The Ward-modified location model (Ward MLM) and canonical analysis were undertaken for clustering the 24 accessions.

For most traits, the variance components among races were more important than the accession within races, and the variance components for race by environment or accession within race by environment were, for the most part, negligible. Results suggest that internal ear traits such as cob and pith diameter, as well as cupule sizes and glume texture, are among the most appropriate for clustering these materials in their respective races. The numerical classification maintained the structure of the more differentiated races but identified two distinct accessions in one race and separated them into a homogeneous group. The Ward-MLM numerical method produced groups with distinct characteristics in terms of internal ear variables. Racial classification carried out in Peru by the use of descriptors as presented by Grobman and colleagues (1961) coincided with the new findings.

Ortiz, Sevilla, and colleagues (2008[385]) investigated the use of variance components to calculate total phenotypic variation for 12 vegetative and reproductive maize traits. A set of 59 accessions, belonging to nine Peruvian highland maize races, were grown at two consecutive planting seasons in two years at one

[384] Ortiz, R., R. Sevilla, G. Alvarado, and J. Crossa. 2008. Numerical classification of related Peruvian highland maize races using internal ear traits. *Genetic Resources and Crop Evolution*, 55: 1055–1064.

[385] Ortiz, R., R. Sevilla, and J. Crossa. 2008. Minimum resources for phenotyping morphological traits of maize (*Zea mays* L.) genetic resources. *Plant Genetic Resources: Characterization and Utilization*, 6: 195–200.

inter-Andean site in northern Peru. The trial data provided means for calculating the variance components using the restricted maximum-likelihood method. The variance components were assumed to be stable while the number of environments and replications varied to simulate phenotypic variation for each trait. The lowest number of environments and replications that do not affect the precision of phenotyping was selected for assessing each trait. Tabulated data provide the number of environments and replications that can be used as a reference for Peruvian highland trials to assess quantitative variation in plant and reproductive traits. The results suggest that fewer environments and replications are needed for reproductive than for vegetative plant traits because the former show higher heritability than vegetative traits.

A visual grouping exercise of the Andean races of maize from Colombia, Ecuador, Peru, Bolivia, and Chile was attempted by Edgar Anderson and the present author on the basis of morphological characters of typical collections of the various described races in 1959. Notes were taken at the agricultural experiment station Tulio Ospina in Medellín, Colombia, where ears of typical collections of the maize races were dispayed at the same time in a large patio. We took notice of the resemblance of races across countries and wider racial groupings by visual ear characters. Unfortunately, the majority of these notes were in the possession of Dr. Anderson, and they were not published before his death.

Goodman and Bird (1977[386]) grouped 219 races of Latin American maize using both principal component and cluster analysis and using seven characters previously used by Goodman and Paterniani (1969[387]), which had a ratio of variance due to races divided by variance due to environment greater or equal to 3. They were able to delimit 14 groups of races. They found minimum overlap of the groups when compared to classical morphological analysis.

Various studies have been undertaken in Mexico on amplifying maize and teosinte collections, evaluating them, and grouping them into races, employing classical measures and using biometric and statistical tools. Teosinte collections were studied by Sánchez-Gonzáles and Ordáz-Sánchez (1987[388]), Sánchez-Gonzáles and Ruiz Corral (2002[389]), Ruiz Corral and colleagues (2001[390]), and Sánchez-Gonzáles (2011). New and former maize collections

[386] Goodman, M. M., and R. McKelvy Bird. 1977. The races of maize IV: Tentative grouping of 219 Latin American races. *Economic Botany*, 31 (2): 204–211.

[387] Goodman, M. M., and E. Paterniani. 1969. The races of maize III: Choices of appropriate characters for racial classification. *Economic Botany*, 33: 265–273.

[388] Sánchez-Gonzáles, J. J., and J. Ordáz-Sánchez. 1987. *El teocintle en México. Distribución y situación actual de las poblaciones.* Systematic and Ecogeographic Studies on Crop Gene Pools: 2. International Board for Plant Genetic Resources. Rome.

[389] Sánchez-Gonzáles, J. J., and J. A. Ruiz Corral. 2002. Distribución del teocintle en México. In Antonio Senatos, Martha C. Wilcox, and Fernando Castillo, editors. *Memoria del Foro Flujo genético entre maíz criollo, maíz mejorado y teocintle. Implicaciones para el maíz transgénico.* INIFAP, CIMMYT. Mexico City. pp. 20–38.

[390] Ruiz Corral, J. A., J. J. Sánchez-Gonzáles, and M. Aguilar. 2001. Potential geographical distribution of teosinte in Mexico: A GIS approach. *Maydica*, 46: 105–110.

were studied by Ortega (1985[391]) and by Sánchez-Gonzáles (2011), who increased the number of Mexican races of maize from the original 32 races, established by Wellhausen and colleagues (1952), to some 56 races, which are now considered to exist within the territory of Mexico. The new data on the classification of the Mexican races of maize essentially confirms the original classification of Wellhausen and colleagues (1952) by adding new derived races and subraces.

Reif and colleagues (2003[392]) conducted a study on Mexican races of maize (*Zea mays* L.) by applying SSR markers to characterize 25 accessions. The objectives were to (1) study the molecular genetic diversity within and among these accessions and (2) examine their relationships as had been previously established on the basis of morphological data. A total of 497 individuals were fingerprinted with 25 SSR markers. A high total number of alleles (7.84 alleles per locus) and total gene diversity (0.61) were observed. The broad genetic base of the maize races from Mexico was thus confirmed. The maize accessions were grouped into distinct racial complexes on the basis of a model-based clustering approach. The principal coordinate analyses of the four Modern Incipient hybrids corroborated the proposed parental races of Chalqueño, Cónico Norteño, Celaya, and Bolita on the basis of the morphological data. Consequently, for some of the accessions, at least, hybridizations provide a clue that can be used to further explain the associations among the Mexican races of maize.

An analysis of the diversity of maize races in Colombia, which were described by Roberts and colleagues (1957), was undertaken by Cardona (2010[393]), employing the Ward-MLM (conglomerate) method. The 3 racial groups established by Roberts and colleagues (1957) were revalidated. They proposed an increase in the number of primitive races from 2 to 5 (Pollo, Pira, Pira Naranja, Clavo, and Imbricado) and a reduction of the number of hybrid Colombian races from 12 to 9. The number of introduced races in the previous study was maintained in the new one. In general, save the changes indicated, the groupings made by using morphological markers in the previous study were essentially confirmed with the new analytical strategy, proving that the initial methodology for racial descriptors was a sound one.

Evaluations were conducted on the Latin American races of maize by CIMMYT based on 12,406 accessions. The Latin America Maize Project (LAMP) organized a series of studies and increased the amount of seed of those accessions

[391] Ortega, P. R. 1985. Variedades y razas mexicanas de maíz y su evaluación en cruzamiento con líneas de clima templado o material de partida para fitomejoramiento. Abbreviated Spanish translation of a Ph.D. dissertation at the N.I. Vavilov National Institute of Plants, Leningrad, USSR.

[392] Reif, J. C., M. L. Warburton, X. C. Xia, D. A. Hoisington, J. Crossa, S. Taba, J. Muminovic, M. Bohn, M. Fritsch, and A. E. Melchinger. 2003. Grouping of accessions of Mexican races of maize revisited with SSR markers. *Theoretical and Applied Genetics*, 113 (2): 177–185.

[393] Cardona, J. O. 2010. Análisis de diversidad genética de las razas colombianas de maíz a partir de datos Roberts et al., (1957) usando la estrategia Ward-MLM. *CienciAgro*, 2 (1): 199–207.

that was available at that time in seed banks. For reports on the South American area, see LAMP Report v. June (1995[394]), Salhuana and Sevilla (1995[395]), and Salhuana and colleagues (1995[396]). Salhuana and Pollak (2006[397]) related the already-collected and studied germplasm to its potential use in breeding.

Transposons or Transposable Elements

Transposable elements (TEs) are the major components of genomes of most plant species. TEs have various families or types that proliferate at different rates in the genome. Proliferation activity is counteracted by TE removal via recombination and population processes driven by natural selection, creating opportunities for genetic variation.

Transposons or TEs are a kind of mobile genetic element (MGE); MGEs include (1) DNA transposons, which are transposed without RNA intermediates; (2) retrotransposons; (3) insertion sequences; (4) *helitrons*; and (5) episomal replicons, including plasmids of archaea and bacteria, circular single-stranded DNA (ssDNA) bacteriophages, and geminiviruses (circular ssDNA viruses replicating in plant cells).

Transposable elements were first discovered in plants and specifically in maize by McClintock (1948[398]) in the 1940s and 1950s. She had built, by that time, a reputation as a serious research worker on maize cytogenetics, including the analysis of crossing over during meiosis in maize chromosomes – a mechanism by which chromosomes exchange information. She produced the first genetic map of maize, linking regions of the chromosome with physical traits, and demonstrated the role of the telomere and centromere regions of the chromosome that are important in the conservation of genetic information.

When McClintock published her information on the Ac/Ds transposable system in maize, she encountered widespread skepticism about her research and its implications. At that time the thinking was that genes were continuous and unmovable in their relation to one another in the chromosome, like a string of beads, and that order could only be altered by breakage and fusion of blocks of genes by chromosome translocations or inversions. The present

[394] LAMP (Latin American Maize Project). 1995. Proyecto Latinoamericano de Maíz. Versión Junio de 1995. CIMMYT. Mexico City, D.F.

[395] Salhuana, W., and R. Sevilla. (Editors). 1995. Latin American Maize Project (LAMP). Stage 4 results from homologous areas 1 and 5. National Seed Storage Laboratory. Fort Collins.

[396] Salhuana, W., R. Sevilla, and S. A. Eberhart. (Editors). 1995. Latin American Maize Project (LAMP). Stage 4. Results from Homologous areas 2, 3, and 4. National Seed Storage Laboratory. Fort Collins.

[397] Salhuana, W., and L. Pollak. 2006. Latin American Maize Project (LAMP) and Germplasm Enhancement of Maize (GEM) project: Generating useful breeding germplasm. *Maydica*, 51: 339–335.

[398] McClintock, B. 1948. Mutable loci in maize. *Carnegie Institution of Washington Year Book*, 47: 155–169.

author witnessed that universal skepticism when McClintock informed me of her research and conclusions on transposons. At that time she had a stage at the Cooperative Maize Research Program in Peru in the late 1950s to train the present senior author and his associates in cytogenctic techniques, which we later used to develop the study and understanding of the chromosome knobs constitution of the maize races of Peru (Grobman et al., 1961). She stopped publishing her data in 1953 because of the "puzzlement and hostility" that her findings created in the scientific community at that time.

The development of the concept of gene regulation must also be credited to her. She identified the *Dissociator* and *Activator*, and she later discovered the *Spm* element as a class of controlling elements to explain how cells with different tissues and organs but identical genomes have different functions.

Later, she made an extensive study of the chromosomes of maize races from South America, excluding Peru, which had already been studied by the present writer and associates, especially Ulises Moreno. McClintock's research became well and widely understood only later, in the 1960s and 1970s. She was awarded the Nobel Prize in Physiology or Medicine in 1983 for her discovery of genetic transposition.

DNA transposons are sequences of DNA that can move around and are capable of intragenomic multiplication by transferring a DNA segment from one genomic site to another; that is, they *transpose* themselves to new positions within the genome of a single cell. The mechanism of transposition can be described as either copy and paste or cut and paste. The first type of transposons are called Class I transposons, whereas the second type are Class II transposons.

Transposons can have tremendous effects on genome structure and gene function. Although only a few or no elements may be active within a genome at any time in any individual, the genomic alterations they cause can have major outcomes for a species. They cause unstable mutations, with reversions to previous states caused by excision of the transposon. Their action is similar to that of parasites in that they selfishly benefit from the cell mechanisms to multiply themselves; however, their persistence indicates that they may have come to terms with their hosts over thousands of years and may benefit them through some sort of symbiosis and the acquisition of some regulatory functions.

All major element types appear to be present in all plant species and animals, but their quantitative and qualitative contributions are enormously variable, even between closely related lineages. In some large-genome plants, mobile DNAs make up the majority of the nuclear genome. They can rearrange genomes and alter individual gene structure and may exercise gene regulation through any of the activities they promote: transposition, insertion, excision, chromosome breakage, and ectopic recombination. Many genes may have been assembled or amplified through the action of transposable elements, and it is likely that most plant genes contain legacies of multiple transposable element insertions into promoters. Transposons may be more likely to activate themselves under conditions of stress of the plant, thus becoming a powerful set of elements for genomic change, setting up mechanisms of adaptation and plant evolution.

Because chromosomal rearrangements such as inversions can lead to infertility in heterozygous progeny, transposable elements may be responsible for the rate at which such incompatibility is generated in separated populations, which may eventually lead to development of a different species. For these reasons, understanding plant gene and genome evolution is only possible if we comprehend the contributions of transposable elements (Bennetzen, 2000[399]).

TRANSPOSONS IN PERUVIAN RACES OF MAIZE

External characteristics of the kernels of the Peruvian races of maize Pisccorunto, Huancavelicano, Paro, and Cuzco, from the south-central high-altitude region of the Andes mountains of Peru, exhibit clear evidence of the action of transposons on aleurone and pericarp, causing the development of mosaic kernel colors. Aleurone color requires deep anthocyanin pigmentation (cyanidine and pelargonin glucosides). The color patterns of maize seed that exhibit aleurone color mosaicism are under the unstable inheritance of two dominant and interacting genetic loci that McClintock (1950,[400] 1953[401]) named *Dissociator* (*Ds*) and *Activator* (*Ac*). *Dissociator* inhibits the synthesis of anthocyanin, and the seeds are colorless when there is no *Ac* element present. When *Ac* is present in one, two, or three doses, different patterns of color appear.

McClintock found that the *Dissociator* did not just dissociate or cause the chromosome to break; it also had a variety of effects on neighboring genes when the *Activator* was also present, and in early 1948, she made the surprising discovery that both *Dissociator* and *Activator* could transpose, or change position, on the chromosome. The presence of the Ac/Ds system is suspect in the aforementioned races, one of which is specifically selected – the race Pisccorunto – for the dotted mosaic pattern, whereas in the other races the mosaic pattern appears in some ears through gene flow. The possibility that other genes also may be acting in developing a similar pattern, such as etched (*et*) and blotching (*Bh*), should be investigated.

We have examined (unpublished) kernels of archaeological maize from the Huaca Prieta site in northern Peru (collected by Duccio Bonavia and Tom Dillihay in 2007), under stereomicroscope at high resolution; during the process we saw that there was a possibility that some blotches of clear pericarp color on otherwise brown/red color could be due to transposon effects. The suspicion is established, but we have no definite means to prove it.

[399] Bennetzen, J. L. 2000. Transposable element contributions to plant gene and genome evolution. *Plant Moleculat Biology*, 42 (1): 251–269.

[400] McClintock, B. 1950. The origin and behavior of mutable loci in maize. *Proceedings of the National Academy of Sciences USA*, 36 (6): 344–355.

[401] McClintock, B. 1953. Induction of instability at selected loci in maize. *Genetics*, 38 (6): 579–599.

RETROTRANSPOSONS

Retrotransposons are also called transposons via RNA intermediates and are a subclass of transposons that are genetic elements that can amplify themselves in a genome and are ubiquitous components of the DNA of many eukaryotic organisms. They are a principal component of the nuclear DNA in many plants. In maize, 49–78% of the genome is made up of retrotransposons (San Miguel and Bennetzen, 1998[402]).

The chromosomes of higher plants are littered with retrotransposons that, in many cases, constitute as much as 80% of the plant genomes. Long terminal repeat retrotransposons have been especially successful colonizers of the chromosomes of higher plants. Examinations of their function, evolution, and dispersal are essential to understanding the evolution of eukaryotic genomes. Retrotransposons copy themselves to RNA and then back to DNA that may integrate back to the genome. The second step of forming DNA may be carried out by a reverse transcriptase encoded by the retrotransposon. Retrotransposon as well as host genome encoded factors regulate the retrotransposon movement and insertion, thus avoiding deleterious effects. There is evidence that retrotransposons have existed for many millions of years in plants and animals, where they encompass a significant proportion of the genome.

San Miguel and Bennetzen (1998) estimated the size and copy number of the retrotransposons in a 240-kb region flanking the *Adh1* gene of maize. Their data suggest that 33–62% of the maize genome is composed of the high-copy-number retrotransposons (LTR retrotransposons) found in this region. An additional 16% of the maize genome is estimated to be composed of middle- and low-copy-number retrotransposons (non-LTR retrotransposons). The sorghum genome, which is more than one-third the size of the maize genome, does not have any detected copies of the maize retrotransposons in a region orthologous to that of maize locus *Adh1* in the long arm of chromosome 1. Thus, it appears that retrotransposons have increased the size of the maize genome two- to five-fold since the divergence of maize and sorghum from a common ancestor about 16 million years ago.

Retrotransposons and retroviruses are related genetic elements that replicate through a cycle of successive transcription, reverse transcription, and integration into the genome. Retroviruses differ from retrotransposons in their being infective. The infectivity of mammalian retroviruses depends critically on their encoded envelope (ENV) glycoproteins, which recognize receptor proteins on the surface of host cells, allowing adsorption into them, and help to mediate subsequent penetration of the plasma membrane. Retroviruses are more similar to one particular class of plant, fungal, and invertebrate retrotransposons, those

[402] San Miguel, P., and J. L. Bennetzen. 1998. Evidence that a recent increase in maize genome size was caused by the massive amplification of intergene retrotransposons. *Annals of Botany*, 82 (Suppl. A): 37–44.

resembling the type-element *gypsy* of *Drosophila melanogaster*. The strong internal sequence similarities, respectively, in the *copia*-like and *gypsy*-like groups of retrotransposons suggest that they are lineages that have been separated since early in eukaryote evolution. The results of research by Vicient and colleagues (2001[403]) indicate that *env*-containing elements were identified in maize by PCR (polymerase chain reaction) amplification (see Vicient et al. 2001: figure 1).

A distinct class of *gypsy*-like, *env*-class retrotransposons related to *Athila* is widespread, transcribed in flowering plants, and probably ubiquitous and active in the grasses and other species of Angiosperms. *Athila4* is a so-called envelope gene. The primary distinguishing feature between the retrotransposons and retroviruses is that the latter have a third gene called *envelope* (*env*). The *env* gene encodes a transmembrane protein that associates with the cell membrane. *Athila4* is an infectious element classified as a retrovirus, found in *Arabidopsis thaliana* and later in a related element in soybean (*Calypso*), in *BAGY*-2 from barley, and in a degenerate element from rice, identified from the rice genome sequence data, as well as in other plant species (Wright and Voytas, 2002[404]). *Athila4* is a degenerate, centromere-associated retroelement and a Ty3-*gypsy* group retrotransposon with an *env*-like ORF (Wright and Voytas, 1998[405]). Their ubiquitous nature and infectiousness indicate that these endogenous retroviruses may be important vehicles for plant genome evolution. Analyses of *env*-like genes from the various retroelement groups suggests that *env* was independently acquired from viruses multiple times during evolution. The loss of an envelope-like coding domain suggests that noninfectious retrotransposons could swiftly evolve from infectious retroviruses, possibly by anomalous splicing of genomic RNA (Yano et al., 2005[406]).

These types of *env* elements were not recovered from a gymnosperm (pine) and from maize, teosinte, and *Tripsacum*. It may be that the endogenous retroviruses are not present in the genomes of these plants or that they are divergent and cannot be amplified by the primers used. The fact that they have been found in other cereal species antedating the Maydeae in evolution, such as oat, rye, and barley, may lead to their future discovery.

TEs in maize have increased considerably the physical length of the genome but not the size of the genetic map, relative to the previously studied genome

[403] Vicient, Carlos M., Ruslan Kalendar, and Alan H. Schulman. 2001. Envelope-class, retrovirus-like elements are widespread, transcribed and spliced and insertionally polymorphic in plants. *Genome Research*, 11: 2041–2049.

[404] Wright, David A., and Daniel F. Voytas. 2002. *Athila4* of *Arabidopsis* and *Calypso* of soybean define a lineage of endogenous plant retroviruses. *Genome Research*, 12 (1): 122–131.

[405] Wright, David A., and Daniel F. Voytas. 1998. Potential retroviruses in plants: Tat1 is related to a group of *Arabidopsis thaliana* Ty3/gypsy retrotransposons that encode envelope-like proteins. *Genetics*, 149: 703–715.

[406] Yano, S. T., B. Panbehi, A. Das, and H. M. Laten. 2005. Diaspora, a large family of Ty3-gypsy retrotransposons in Glycine max, is an envelope-less member of an endogenous plant retrovirus lineage. *BMC Evolutionary Biology*, 5 (1): 30.

of rice, by adding intergene space or repetitive noncoding sequences of nucleotides.

Different families of TEs exist that have repeats in different parts of the genome but have their preferred sites of insertion. For example, the Ty1-*copia* elements were first identified as insertions near maize genes, whereas the highly repetitive Ty3-*gypsy* elements prefer to insert into or near other repetitive elements (Bennetzen, 1996;[407] Kumar and Bennetzen, 2000[408]). In maize and other plant species, the Class II TEs such as *Ac/Ds*, *En/Spm*, *Mu*, and miniature inverted-repeat TEs insert preferentially into genes and low-copy-number DNA, which are relatively hypomethylated.

HELITRON TRANSPOSONS

Helitron transposons, which are also rolling-circle eukaryotic transposons, were first discovered in plants (*Arabidopsis thaliana* and *Oryza sativa*) and in the nematode *Caenorhabditis elegans*. To date, *helitrons* have been identified in a diverse range of species, from protists to mammals. They represent a major class of eukaryotic transposons and are fundamentally different from classical transposons in terms of their structure and mechanism of transposition. *Helitrons* seem to have a major role in the evolution of host genomes (Kapitonov and Jurka, 2001,[409] 2007[410]). They frequently capture diverse host genes, some of which can evolve into novel host genes or become essential for *helitron* transposition. *Helitrons* have been transposed recently in the rice genome, where they are represented by just a few copies. *Helitrons* reside in heterochromatin regions, which are underrepresented in the available sequence data. Moreover, multiple highly divergent families of nonautonomous *helitrons* are represented by one or two copies per genome, and only some of them can be detected on the basis of distant similarity to known TEs. In their host genome, *helitron* transposons act as a powerful tool of evolution. They have recruited host genes, modified them to an extent that is unreachable by the Mendelian process, and multiplied them in the host genomes.

EPIGENETIC GENE REGULATION BALANCING TRANSPOSONS

Epigenetic regulation involves the stable propagation of gene activity states through mitotic, and sometimes even meiotic, cell divisions without changes in DNA sequence. Transposons replicate, increase in copy number, and persist by

[407] Bennetzen, Jeffrey L. 1996. The contributions of retroelements to plant genome organization, function and evolution. *Trends in Microbiology*, 4 (9): 347–353.

[408] Kumar, Amar, and Jeffrey L. Bennetzen. 2000. Retrotransposons: Central players in the structure, evolution and function of plant genomes. *Trends in Plant Science*, 5 (12): 509–510.

[409] Kapitonov, Vladimir V., and Jerzy Jurka. 2001. Rolling circle transposons in eukaryots. *Proceedings of the National Academy of Sciences USA*, 98 (15): 8714–8719.

[410] Kapitonov, Vladimir V., and Jerzy Jurka. 2007. Helitrons on a roll: Eukaryotic rolling circle tranposons. *Trends in Genetics*, 23: 521–529.

moving around in the genome, and insertion into genes occurs and is potentially and generally mutagenic. To avoid damage and loss of fitness, there must be an offsetting and delicate mechanism. A strong selection for transposons must be present that can achieve a balance between their own replication and minimal damage to their host. A widespread way to achieve this balance is by means of epigenetic gene regulation, which quiets transposition but also allows for reversions (Weil and Martienssen, 2008[411]).

A TE database has been recently published, containing exemplar sequences of 1,526 TE families and subfamilies (Schnable et al., 2009). Overall, TEs constitute more than 85% of the maize reference (B73) genome (Schnable et al., 2009), of which the 20 most common TE families comprise approximately 70% (Baucom et al. 2009[412]).

These 20 "common" families are all members of the Class I LTR retrotransposons, such as the *Gypsy* and *Copia* superfamilies. In the genus *Zea*, amplification of LTR retrotransposons has been particularly dramatic during the last 3 million years, leading to a doubling of genome size. Baucom and colleagues (2009) found that retroelements occupy the majority (more than 75%) of the nuclear genome in maize inbred B73. Unprecedented genetic diversity was discovered in the LTR-retrotransposon class of retroelements, with more than 400 families (more than 350 newly discovered) contributing more than 31,000 intact elements. The two other classes of retroelements, SINEs (4 families) and LINEs (at least 30 families), were observed to contribute 1,991 and approximately 35,000 copies, respectively, or a combined approximately 1% of the B73 nuclear genome. The maize genome provides a great number of different niches for the survival and procreation of a great variety of retroelements that have evolved to differentially occupy and exploit this genomic diversity.

The genome of maize (*Zea mays* ssp. *mays*) consists mostly of transposable elements (TEs) and varies in size among lines. This variation extends to other species in the genus *Zea*. Although maize and *Zea luxurians* diverged only approximately 140,000 years ago, their genomes differ in size by about 50%. Tenaillon and colleagues (2011[413]) found that Class II DNA transposable elements were present significantly more often in genic regions than Class I RNA transposable elements, but Class 1 elements were found more often near other TEs. Overall, both Class I and II TE families account for approximately 70%

[411] Weil, Cliff, and Rob Martienssen. 2008. Epigenetic interactions between transposons and genes: Lessons from plants. *Current Opinion in Genetcs & Development*, 18 (2): 188–192.

[412] Baucom, Regina S., James C. Estill, Cristian Chaparro, Naadira Upshaw, Ansuya Jogi, Jean-Marc Deragon, Richard P. Westerman, Phillip J. San Miguel, and Jeffrey L. Bennetzen. 2009. Exceptional diversity, non-random distribution, and rapid evolution of retroelements in the B73 maize genome. *PLoS Genetics*, 5 N°(11), Article ID e1000732.

[413] Tenaillion, Maud, I. Matthew, B. Hufford, Brandon S. Gaut, and Jeffrey Ross-Ibarra. 2011. Genome size and transposable element content as determined by high-throughput sequencing in maize and *Zea luxurians*. *Genome Biology and Evolution*, 3: 219–229.

of the genome size difference between the maize inbred line B73 and *luxurians* teosinte. Interestingly, the relative abundance of TE families was conserved between species (r = 0.97), suggesting genome-wide control of TE content rather than family-specific effects. This confirms the previous findings that TEs may have a significant effect on the evolution of genome size and that, because of their ability to transfer DNA segments, they may intervene in substantial gene modification. A genome size difference of 22% has been reported in the primitive popcorn maize race Palomero from Mexico when compared to the larger Corn Belt inbred line B73 genome. It would appear that the increase in genome size has proceeded unabated after domestication and selection. Class II TEs would be – because of their concentration in genic regions – potential carriers of gene fragments, thus modifying them as they move along the genome.

Pooling all the archaeological, genetic, and cytogenetic information of plant anatomy and physiology, and of identification of variability at the subgene level through molecular analysis of its components, of maize and its relatives, should help us obtain an integrated view of how, when, where, and from what maize domestication proceeded. The assembly of new evidence is required, and a critical reexamination of all the evidence is essential. An analysis of the present situation renders the quest unended. More research in all these fields is required. Periodic reexamination of the advances and the integration of such information will bring us closer to the truth, and understanding this information will help us better understand the foundations of maize domestication and evolution. It will also help in positioning crop improvement efforts on a sounder basis.

Paramutation

Paramutation is an epigenetic phenomenon involving changes in gene expression that are stably transmitted from one allele of a gene to another through mitosis as well as meiosis to establish a heritable state of gene expression across generations. These heritable changes are mediated by trans-interactions between homologous DNA sequences on different chromosomes, such that epigenetic information is carried as a new expression state to subsequent generations in spite of the fact that the allele or the DNA sequences that issued the instructions are not transmitted. Furthermore the locus that has been altered by paramutation continues to issue instructions similar to the ones originally received to homologous sequences, and there are no associated DNA changes in the affected DNA or in the sequences of the active gene resulting from the instructions.

Although paramutation was initially discovered in plants, it has recently been observed in mammals as well, suggesting that the mechanisms underlying paramutation might be evolutionarily conserved. Recent findings point to a crucial role for small RNAs in the paramutation process. In mice, small RNAs appear sufficient to induce paramutation, whereas in maize, it seems not to be the only player in the process. Stam (2009[414]) discusses potential mechanisms in relation to the various paramutation phenomena.

This form of inheritance affecting the *R* locus of maize was discovered by Brink (1956[415]). The *R* locus experiences an inherited reduction on the functional capacity of a sensitive or "paramutable" allele after it has been affected in a heterozygous condition by a "paramutagenic" allele, which remains unchanged. *R* alleles continue to maintain the changed expression state through several generations in the homozygous state but may regain the original state depending on the homologous type of allele found in the heterozygous state. A similar situation was found by Coe (1959[416]) in the *B* locus of maize.

How does this phenomenon change the genetic paradigm of phenotypic changes with no changes in the coding DNA?

The case of the *B-1* locus in maize is a good example. Deep purple color due to a high level of anthocyanin synthesis results from the *B-1/B-1* homozygous state, whereas the expression of another allele *B'* in the homozygous state conditions a light purple color. When the sequences of both alleles are compared, they are found to be identical. In plants that are heterozygous *B-1/ B'*, the *B-1* allele is converted (paramutated) to *B'*. This paramutated allele *B'* can in subsequent generations convert newly encountered *B-1* alleles to *B'* when in a heterozygous state and can continue that capability in subsequent generations. The changes are not due to a permanent mutation, because there are no sequence changes, and under certain conditions reversions to prior states can take place (Chandler, 2007[417]). The paramutation to *B'* is extremely stable and has 100% penetrance. The key sequences required for paramutation are tandem repeats of noncoding DNA that are located about 100 kb upstream of the *B-1* transcription start site. The *B-1* DNA is in a different pattern of methylation than the *B'* DNA, although their sequences are identical.

Paramutation is associated with a 10- to 20-fold reduction in transcription of *B'* relative to *B-1*. Recombination experiments have shown that sequences required for paramutation are tightly linked to *B'* and map upstream of the transcribed region (Patterson et al., 1995[418]).

The *B-1* chromatin is in a more open state than the *B'* chromatin (Stam et al., 2002[419]).

414 Stam, M. 2009. Paramutation: A heritable change in gene expression by allelic interactions in trans. *Molecular Plant*, 2 (4): 578–588.

415 Brink, R. A. 1956. A genetic change associated with the R locus in maize which is directed and potentially reversible. *Genetics*, 41: 872–889.

416 Coe, E. H., Jr. 1959. A regular and continuing conversion–type phenomenon at the B locus in maize. *Proceedings of the National Academy of Sciences USA*, 45: 828–832.

417 Chandler, V. L. 2007. Paramutation: From maize to mice. *Cell*, 128: 641–645.

418 Patterson, G. I., K. M. Kubo, T. Shroyer, and V. L. Chandler. 1995. Sequences required for paramutation of the maize b gene map to a region containing the promoter and upstream sequences. *Genetics*, 140: 1389–1406.

419 Stam, M., C. Belele, J. E. Dorweiler, and V. L. Chandler. 2002. Differential chromatin structure within a tandem array 100 kb upstream of the maize *b1* locus is associated with paramutation. *Genes and Development*, 15: 1906–1918.

RNA intervenes by the activity of transcription of the two strands of the tandem repeats upstream of the *b-1* locus, which may result in the formation of double-stranded RNA (dsRNA). There is also an RNA-dependent RNA polymerase called mediator of paramutation1 (*mop1*), which is essential for silencing of *B-1* by *B'* and for paramutation to take place in other maize genes. Small interfering RNAs (siRNAs) from the repeats have been detected later in all the three different genotypes but not in the mutant lines lacking *mop1*. The conditions required for paramutation have been established previously.

Penetrance of paramutantion in other genes may be different. The case of an easily observable alteration such as the one produced at the *B-1/B'* locus may not be the same for many other possible paramutation loci that have other types of expression. Chandler (2007) believes that paramutation, observed also in mice, may be a fundamental mechanism of gene regulation and heredity. She has suggested that silencing of genes by paramutation has to do with methylation. The case of the *B-1* gene of maize may be the extreme case of an allele that is highly sensitized to become silenced because it is recognized as a foreign allele by a cellular defense system.

Paramutation has also been described for four genes in maize (*b1*, *pr1*, *pl1*, and *p1*), all of which contribute to the synthesis of flavonoid pigments (Brink, 1973[420]).

The great variability of anthocyanin states of expression in Andean-region maize suggests that paramutation has been active in Andean maize races for a very long time, contributing to acceleration of the evolution of the species in the region.

Heterochromatin

Heterochromatin is a deeply stainable form of chromatin, as opposed to euchromatin. It is made up of densely packed and repetitive DNA at the molecular level and is genetically inert. In maize chromosomes it appears in neighboring centromere regions, chromomeres, knobs, nucleolar organizer regions (NORs), and chromosome satellites, as well as in B chromosomes and abnormal chromosome 10. It is clonally inherited in the division of sister chromatids during cell division. Facultative heterochromatin is the result of genes that are silenced through a mechanism such as histone methylation or siRNA through RNAi. Genes found in heterochromatin and coding for proteins provide us with recent information on the structure and function of heterochromatin.

Heterochromatin has the capacity to silence nearby genes, a phenomenon called position effect variegation (PEV), which is reported in *Drosophila melanogaster* and suspected in plants, and results from translocations in which an euchromatic gene is placed in a heterochromatic environment, or from ectopic

[420] Brink, R. A. 1973. Paramutation. *Annual Review of Genetics*, 7: 129–152.

expression of transgenically introduced genes (Avramova, 2002;[421] Eisenberg and Elgin, 2000[422]). Failure of transgenic expression in plants may be attributed to a PEV effect (Matzke and Matzke, 1998[423]). Gene silencing with methylated DNA or with deacetylated histones is not synonymous with heterochromatin. Despite the fact that the heterochromatin of many species contains densely methylated DNA, it is not known whether methylated DNA can provoke the assembly of heterochromatin (Avramova, 2002). Plants appear to have heterochromatin silencing complexes. Solitary coding genes have been found in maize surrounded by blocks of highly methylated transposons, making the methylation paradigma of silencing of genes not totally applicable (San Miguel et al., 1996;[424] Tikhonov et al., 1999[425]).

DNA and proteins required for certain protection functions may have coevolved for repair purposes in cells, for telomere repair, or for defense against foreign proteins. Forming DNA/protein complexes, of which heterochromatin is a basic component, may be one such example.

Rhoades (1978[426]) has proposed that certain genetic effects are attributable to the heterochromatin of B chromosomes, such as the loss of chromosomal segments from knobbed A chromosomes at the second microspore mitosis when two or more than two B chromosomes are present. This and the B chromosome nondisjunction in the second microspore mitosis would be caused by delayed replication of centric B chromatin.

Chromosome Knobs

Chromosome knobs are important markers for the study of the evolution of maize. Maize and teosinte chromosomes at the pachytene stage of meiosis exhibit prominent chromomeres and larger heterochromatic knobs; the latter can be seen in 23 known positions. The number, size, and position of knobs on the chromosomes vary among landraces of maize, and among various species of teosinte and *Tripsacum*. These knobs are powerful phylogenetic tracers of

[421] Avramova, Zoya. 2002. Heterochromatin in animals and plants: Similarities and differences. *Plant Physiology*, 129: 40–49.

[422] Eisenberg, J. C., and S. C. R. Elgin. 2000. The HP1 protein family getting a grip on chromatin. *Current Opinion in Genetics & Development*, 10: 204–210.

[423] Matzke, A., and M. Matzke. 1998. Position effect and epigenetic silencing of maize transgenes. *Current Opinion in Plant Biology*, 1: 142–148.

[424] San Miguel, P., A. Tikhonov, Y. K. Jin, N. Motchoulskaya, D. Zakharov, A. Melake-Berhan, P. S. Springer, K. J. Edwards, M. Lee, Z. Avramova, and J. L. Bennetzen. 1996. Nested retrotransposons in the intergenic regions of the maize genome. *Science*, 274: 765–768.

[425] Tikhonov, A. P., P. J. San Miguel, Y. Nakajima, N. Gorenstein, J. L. Bennetzen, and Z. V. Avramova. 1999. Colinearity and its exceptions in orthologous *adh* regions of maize and sorghum. *Proceedings of the National Academy of Sciences USA*, 96: 7409–7414.

[426] Rhoades, M. M. 1978. Genetic effects of heterochromatin in maize. In D. B. Walden, editor. *Maize Breeding and Genetics*. John Wiley & Sons. New York. pp. 641–671.

the evolution of maize. Knobs may affect the frequency and position of genetic recombination (Rhoades, 1978).

Knobs tend to replicate later than other heterochromatic regions and may interfere with crossing over between genes in their immediate vicinity, contributing thus to conservation of some gene blocks.

Peacock and colleagues (1981[427]) found that a repeating unit of 180 bp is the major component of knob heterochromatin. Some polymorphisms have been detected in these 180-bp sequences by Dennis and Peacock (1985[428]). Viotti and colleagues (1985[429]) characterized clones from a family of highly repeated sequences present in a heterochromatin-rich maize line by sequencing and by chromosome location. By means of in situ hybridization experiments, they found that the tandem DNA repeats are mainly located in the knob heterochromatin of the A chromosomes and the centromeric heterochromatin of the B chromosome. However, some copies are also distributed in euchromatic regions of the A chromosomes and in the distal heterochromatic block of the B chromosome.

Ananiev and colleagues (1998a[430]) isolated a class of tandemly repeated DNA sequences (TR-1) of 350-bp unit length from the knob DNA of chromosome 9 of *Zea mays* L. Comparative fluorescence in situ hybridization revealed that TR-1 elements are also present in cytologically detectable knobs on other maize chromosomes in different proportions relative to the previously described 180-bp repeats. At least one knob on chromosome 4 is composed predominantly of the TR-1 repeat. In addition, several small clusters of the TR-1 and 180-bp repeats have been found in different chromosomes, some not located in obvious knob heterochromatin. Variation in restriction fragment fingerprints and copy number of the TR-1 elements was found among maize lines and among maize chromosomes. TR-1 tandem arrays up to 70 kb in length can be interspersed with stretches of 180-bp tandem repeat arrays. DNA sequence analysis and restriction mapping of one particular stretch of tandemly arranged TR-1 units indicate that these elements may be organized in the form of fold-back DNA segments. The TR-1 repeat shares two short segments of homology with the 180-bp repeat. The longest of these segments (31 bp; 64% identity) corresponds to the conserved region among 180-bp repeats. The polymorphism and complex

[427] Peacock, W. J., E. S. Dennis, M. M. Rhoades, and A. Pryor. 1981. Highly repeated DNA sequence limited to knob heterochromatin in maize. *Proceedings of the National Academy of Sciences USA*, 78: 4490–4494.

[428] Dennis, E. S., and W. J. Peacock. 1985. Maize heterochromatin homology in maize and its relatives. *Journal of Molecular Evolution*, 20: 341–350.

[429] Viotti, A., E. Privitera, E. Sala, and N. Pogna. 1985. Distribution and clustering of two highly repeated sequences in the A and B chromosomes of maize. *Theoretical and Applied Genetics*, 70: 234–239.

[430] Ananiev, E. V., R. L. Phillips, and H. W. Rines. 1998a. A knob-associated tandem repeat in maize capable of forming fold-back DNA segments: Are chromosome knobs megatransposons? *Proceedings of the National Academy of Sciences USA*, 95 (18): 10785–10790.

structure of knob DNA suggest that, similar to the fold-back DNA-containing giant transposons in *Drosophila*, maize knob DNA may have some properties of transposable elements.

Members of the TR-1 family have been found by Hsu and colleagues (2003[431]) to be composed of three basic sequences: A (67 bp); B (184 bp); its variants B' (184 bp), 2/3B (115 bp), and 2/3B' (115 bp); and C (108 bp). The B components appear to arise through mutation during evolution from the 180-bp basic number. B' may arise through lateral amplification plus base changes. Fluorescence in situ hybridization localized the B repeat to the B centromere and the 180-bp and TR-1 repeats to the proximal heterochromatin knob on the B chromosome. Hsu and colleagues (2003) report that a comparison of the nucleotide sequences of the 180-bp and TR-1 repeats revealed two regions of homology between the 180-bp repeat and the B component of the TR-1. In the 180-bp repeat, the two regions are separated by 26 bp, whereas in the B component of TR-1, one region is included in the other. It is suggested that the loss of internal repetition due to mutation during evolution may be the cause for the origin of the B component of TR-1. This hypothesis is supported by the fact that *Tripsacum* has the 180-bp repeats in its chromosome knobs but lacks the TR-1 elements that are present in maize and teosinte (Buckler and Holtsford, 1996;[432] Kellogg and Birchler, 1993[433]). The knobs became differentiated in their composition during divergence between *Zea* and *Tripsacum*.

Knob composition is a valid argument against the recent participation of *Tripsacum* in the formation of teosinte. The fact that both teosinte and maize share the TR-1 family of 350-bp nucleotides found by Ananiev in the Seneca 60 variety of maize indicates that both separated from the phylogenetic branch in which *Tripsacum* diverged. However, it does not signal that one may have originated the other. A consideration of major importance is that knob presence or absence is a critical marker in the tracing of the evolutionary path and is discussed in other sections of this appendix.

THE CHROMOSOME KNOB EVIDENCE IN MAIZE EVOLUTION

The Andean type or complex of knobs in the chromosome knob-forming regions – typified by the early maize race Confite Morocho, which is grown in Peru – is the basic knobless karyotype. It may have, alternatively, either one small knob subterminal on the chromosome 7 long arm (7L) and/or one small

[431] Hsu, F. C., C. J. Wang, C. M. Chen, and C. C. Chen. 2003. Molecular characterization of a family of tandemly repeated DNA sequences, TR-1, in heterochromatic knobs of maize and its relatives. *Genetics*, 164: 1087–1097.

[432] Buckler, E. S., IV, and T. P. Holtsford. 1996. *Zea* systematics: Ribosomal ITS evidence. *Molecular Biology and Evolution*, 13: 612–622.

[433] Kellogg, E. A., and J. E. Birchler. 1993. Linking phylogeny and genetics: *Zea mays* as a tool for phylogenetic studies. *Systematic Biology*, 42: 415–439.

knob subterminal on 6L (Grobman et al., 1961; McClintock, 1978). This finding coincides with the studies of Reeves (1944[434]), who stated that almost all of the Peruvian varieties that he studied had knobless chromosomes, whereas eastern South American varieties – which are in the Caingang and Cateto racial groups from Brazil, Uruguay, and Argentina and originate from Caribbean flint races – had many knobs (Mangelsdorf, 1983[435]). Mangelsdorf and Reeves (1939) noted the low number of chromosome knobs in Andean maize. They considered Andean maize "uncontaminated" by teosinte, and maize with a high number of chromosome knobs was, according to them, contaminated by *Tripsacum*, as was posited in their tripartite hypothesis. If we change *Tripsacum* by teosinte, we have the same effect: the higher the knob number, the more teosinte genetic components are signaled as having been introgressed into maize.

There is every reason to assume that wild maize and/or early domesticated maize followed the basic Andean knobless and the 6L/7L small-knob pattern. Such a pattern has been conserved in Andean maize due to the absence of teosinte influence on early maize in South America. In passing, we must again emphasize that knobs are complex permanent chromosome structures formed by a large number of folded DNA segment repeats. Their number, position, and size are, therefore, important tracers of maize dispersal and evolution.

We must also state clearly that all available archaeological evidence does not point, either in Mexico or in Peru, to an early influence of teosinte on maize, but rather to its incursion into maize by introgression at a later stage. This fact has been recognized by many maize evolution students, who, while still adhering to the teosinte hypothesis of origin of maize for lack of any other, cannot make the archaeological facts fit into their theory. The archaeological facts as of early 2012, either in Mexico or Peru, do not support – in our opinion – direct evolution of maize from teosinte, but rather a later influence of teosinte on domesticated maize.

A number of studies – some of which have been reviewed previously – show that teosinte and maize share a high number of alleles and allelic polymorphisms. They also share some chromosome structural rearrangements, such as an inversion in chromosome 8 of Chalco teosinte, as reported by Ting (1964[436]) and a similar inversion reported by McClintock (1933[437]) that is terminal in maize chromosome 8S.

434 Reeves, R. G. 1944. Chromosome knobs in relation to the origin of maize. *Genetics*, 29: 141–147.

435 Mangelsdorf, P. C. 1983. The mystery of corn: New perspectives. *Proceedings of the American Philosophical Society*, 127 (4): 215–247.

436 Ting, Y. C. 1964. *Chromosomes of Maize-Teosinte Hybrids*. Bussey Institute of Harvard University. Cambridge.

437 McClintock, Barbara. 1933. The association of non-homologous pairs of chromosomes in a mid-prophase of meiosis in *Zea mays*. *Zeitschrift für Zellforschung und microskopishe Anatomie*, 19: 191–237.

Chromosome knob size, position, and number are powerful discriminators of maize races as related to their putative ancestry and also to their close relatives teosinte and *Tripsacum*. They have been used to establish possible origins and geographical migration routes (McClintock, 1978).

Longley (1938[438]), in studies of knob frequency and positions of maize from 33 Indian tribes, found a low number of chromosome knobs in maize from northern Indian varieties in the United States; more knobs were observed in that of the southeastern tribes, and many knobs were found in the maize of the New Mexico and Arizona tribes. Longley suggested that knob number and position could give a clue as to the geographical origins of maize. Brown (1949[439]) studied 171 strains of maize (varieties and inbred lines) from the United States in regard to chromosome knob frequency. He again found the northern flint maize varieties, with cylindrical and eight-rowed ears, to be low in chromosome knobs (0 to 4), whereas southern dent varieties with a higher row number and pyramidal (conical) ears with a high row number had a higher chromosome knob number (usually 7 to 9 and up to 12). Corn Belt Dent varieties had an intermediate number of knobs, as expected based on their hybrid origin between southern dents and northern flints. It was easy to ascribe the high knob number of southern U.S. dents to the influence of Mexican races of maize. However, Longley was at a loss to explain the origin of the low knob number association of northern flints. They could be of Guatemalan or other origin, but not from Mexico. Their tripsacoid nature moved Brown to suggest that this trait could have been introduced from introgression of South American *Tripsacum* species that are knobless. This explanation would, therefore, necessitate that the primary origin of North American northern flints be traced back to eight-rowed maize with cylindrical ears and a low chromosome knob number, which are found in South America, uncontaminated by either teosinte or Mexican and Central American *Tripsacum*, both of which have a high chromosome knob number.

McClintock and colleagues (1981) confirmed the low chromosome knob number of maize from the Andean region, which is quite different from the high chromosome knob numbers found in the karyotypes of almost all Mexican races of maize, which closely follow the chromosome knob patterns of their neighboring teosinte populations.

It is very interesting to note that the Confite Morocho chromosome knob structure – 0, 1, or 2 small knobs in the chromosomes – is the simplest, and this pattern might be a relic of the most primitive early maize or wild maize itself. Confite Morocho is also the most primitive among maize races in terms

[438] Longley, A. E. 1938. Chromosomes of maize from North American Indians. *Journal of Agricultural Research*, 56: 177–195.

[439] Brown, W. L. 1949. Numbers and distribution of chromosome knobs in United States maize. *Genetics*, 34: 524–536.

of structure and basic cob morphology, as well as one of the oldest according to dating of archaeological finds in Peru (approximately 7000 BP). It is a primitive race of maize more basic in ear morphology as well as chromosome knob composition than either of the races Nal-Tel or Chapalote, which are the present races of maize in Mexico most closely linked to the earliest archaeological maize found in that area. In its present form, very likely due to introgression from teosinte, Nal-Tel has the presence of knobs of different sizes in positions at least in chromosomes 1S, 1L, 2S, 2L, 3S, 3L, 4L, 5L, 6L, 7L, 8L, and 9L, whereas Chapalote, a sister primitive race of Mexico has knobs in positions 1S, 1L, 2S, 2L, 3L, 4S, 4L, 5L, 6L, 7S, 7L, 8L, 9S, and 9L (McClintock et al., 1981: table 6).

The situation regarding knob numbers of the ancient derived races Nal-Tel and Chapalote is quite complex. Both races, considered the most ancient races of maize in Mexico, and their derived races (Zapalote Grande, Zapalote Chico, Nal-Tel, Tepecintle, and Comiteco) in southwest Mexico, exhibit a high number of chromosome knobs located in 21 positions on the chomosomes. The Pacific coastal maize races that also exhibit a high number of large knobs are Chapalote, Reventador, Harinoso de Ocho, Tabloncillo, Jala, and Celaya (Kato-Yamakake, 1981). It is important to stress that the distribution of knob sizes and positions is not random but relatively constant in each geographical region. Therefore chromosome knobs are stable and transmissible.

From these earlier observations, reanalyzed in view of the later complementary work carried out by McClintock and colleagues (1981) on cytological analysis of maize from other Latin American countries, we may infer that the original chromosome knob number per maize nucleus may have been a total of zero, with variations of zero to one knobs in chromosome 6 and zero to one knobs in chromosome 7, with no additional knobs present in other positions in the remaining chromosomes. I do not share the view advanced by McClintock that the maize that arrived very early in the Andean region from Mexico was loaded with a full complement of knobs in most chromosomes, as exhibited today by Mexican maize races and teosinte, and then lost them through selection and adaptation. Considering the large repetitive structure of DNA basic units that compose the knobs and the fact that they persist through many generations in the same positions as fixed chromosomal structures, it is difficult to imagine how they would have been massively eliminated by selection against them. If anything, it is much more logical and easier to accept the hypothesis of an essentially primitive, knobless maize whose range may have extended at a very early period across Mexico, Central America, and South America's Andean region. At a later period, this primitive maize was subjected in Mexico to introgression from annual teosinte races, when they entered in a sympatric association. This introgression provoked a gradual accumulation of chromosome knobs, which are now present in the modern Mexican maize races. There are still some isolated locations where low chromosome knob numbers are found in Mexican maize, such as the race Arrocillo Amarillo (Wellhausen et al., 1952). This case may represent a relic status of a primitive knobless maize race that existed in Mexico and was more extended geographically than it is today, before introgression from teosinte.

Recently introduced races of maize such as Perla of the coast of Peru, which may be related to the Amagaceño or Común maize race of Colombia and has a high chromosome knob number (Roberts et al., 1957[440]), have had at least 100 years, if not more, of cultivation on the coast of Peru as open pollinated cultivars. The Perla maize race retains the highest number of knobs, ranging from 6 to 13 (Grobman et al., 1961), even though it was subjected to the same selection pressures in the same areas as other races such as Pagaladroga, which has almost zero knobs. The persistence of high chromosome knob numbers in most Colombian and Venezuelan maize races, which are teosintoid, also fits the picture. These facts make it more than dubious that chromosome knobs were lost in maize evolution through migration to the Andean region in a very early time period. This is a powerful argument for the existence of wild maize, which may have been domesticated and diffused before it came into contact with teosinte.

Kato-Yamakake (1976), in his studies of maize and teosinte from the area encompassing the south of Guanajuato to the south-central state of Guerrero, found that teosinte possessed each of the chromosome knob positions found in the maize of the area and some more. It is the persistent introgression of teosinte into maize in Mexico that has produced the high number of knob positions in Mexican maize. The absence of teosinte in South America was the reason for the preservation of the originally knobless or 6L–7L cytological condition of maize in the Andean region. Chromosome knob position numbers and their frequencies are consequently a most powerful tracer of the routes of evolution of maize. If we accept their persistence at specific positions in the chromosomes, as observation and experimentation indicate, knob numbers are also indicators that Andean maize from Ecuador, Peru, and Bolivia, as well as northern Chile and northwestern Argentina, have evolved independently from the Mexican and Central American maize in prolonged isolation. They are also indicators of what a wild maize precursor karyotype was like.

There is every reason to assume that wild maize or early domesticated maize followed the Andean complex chromosome knob pattern, which has not changed and is preserved in Andean maize due to the absence of teosinte in South America. It is also notable that the Andean chromosome pattern has spread widely along the rivers flowing from the highlands of Peru, Ecuador,

[440] Roberts, L., U. J. Grant, R. Ramírez, W. H. Hatheway, and D. L. Smith, in collaboration with P. C. Mangelsdorf. 1957. *Races of Maize in Colombia*. National Academy of Sciences, National Research Council, Publication 510. Washington, D.C.

Bolivia, and Colombia into the Amazon basin. The Andean chromosome complex can be traced as a precursor of the most widely geographically dispersed ancient race in the Americas, the race Piricinco (Grobman et al., 1961), also named Coroico in Bolivia and Entrelazado (Interlocked) in Brazil, with the influence of the Andean complex ranging as far as Argentina and Paraguay, according to McClintock (1978).

The fact that Andean maize exhibits a knobless complex, and that there is total absence of evidence of teosinte introgression in early archaeological maize (approximately 6000 years BP) in Peru, could be interpreted as an indicator that wild maize or early semidomesticated, essentially knobless maize entered the South American continent at a very early prehistorical period, as already evidenced by archaeological findings in Panama, Ecuador, and Peru. A possible form of its mobilization without human help has been suggested by Bonavia and Grobman (1989), through the carriage of seeds by migrating birds. Seeds of sorghum of similar size and hardness to early popcorn seeds are found scattered, and sorghum plants are sprouting and growing in the northern Peruvian deserts after some El Niño phenomenon rains, at great distances from grain sorghum farmers' fields (Bonavia and Grobman, 1989[441]). Seed dispersal by endemic and migratory birds (e.g., *Spiza americana*) has thus been tested and considered possible.

It is interesting to note that *Tripsacum* species found in Peru also have knobless chromosomes.

Direct descent from a wild form of maize precursor to domesticated maize constitutes a simple theory. Further introgression of early domesticated maize with teosinte ssp. *parviglumis* conforms to facts present in the archaeological record in Mexico. It would conform also to the finding of ancient maize pollen in Mexico and Panama and the fact that the earliest archaeological evidence in Mexico has been identified as that of maize starch and not of teosinte starch granules. Introgression of teosinte into maize after more than 7,000 years of sympatric establishment and exchange of genes could very easily explain both the similarities in the molecular data and the differences between maize and teosinte presented by Doebley's laboratory scientists. Present patterns of coincidence and divergence as presented in molecular studies by Matsuoka and colleagues (2002) and others and by Goodman (1969[442]) on the similarities and divergences of maize races and teosinte could be easily explained by this new interpretation of the available information.

There is little doubt that migration of maize races from Mexico and Central America that were already loaded with teosinte genes took place at a later date to the Andean region, the Caribbean, and the eastern coast of South America.

[441] Bonavia, Duccio, and Alexander Grobman. 1989. Andean maize: Its origin and domestication. In David R. Harris and Gordon C. Hillman, editors. *Foraging and Farming: The Evolution of Plant Exploitation*. Unwin Hyman. London. pp. 456–470.

[442] Goodman, M. M. 1969. Measuring evolutionary divergence. *Japanese Journal of Genetics*, 44 (Suppl. 1): 310–316.

Their influence is found to be more prevalent in lowland races. They have been of marginal importance in shaping the present highland races in the Andean region, except in Venezuela and northern Colombia. The reciprocal effect of influence of South American races on the highland races of Guatemala in more recent times is very great and has been explained by Mangelsdorf and Cameron (1942[443]). The Guatemalan races in turn have had great influence on eastern North American races and in the Southwest of the United States starting with the Pueblo period. Some exotic Mexican maize races were classified by Wellhausen and colleagues (1952) as originating in South America. Finally, in the last 500 years, after Spanish and Portuguese presence, a great deal of maize interchange must have taken place, followed in more recent years by introductions of improved varieties that have started to hybridize with local maize races, creating new variability.

We believe that the field is still open for more exploration into the evolution of maize. We are just entering into a better understanding of the ramifications and complexities of interpretation of the data. More archaeological explorations in the Andean and Mexican region are needed. So are other studies at the chromosome, genetic, and molecular levels of maize and its relatives. Only with such studies will it be possible to entertain a clearer picture and better understanding of the mysteries of maize origin and evolution.

The Time of Arrival of Maize in South America

If maize, as a species, originated in the general Mexican area, as most researchers presume from evidence available so far, when did it arrive in South America? According to Piperno and Pearsall (1998) it arrived between 7,700 and 6,000 years ago. C14 dating evidence in Peru available at this time clearly indicates the presence of maize, which is dated approximately 6000 BP years in two coastal archaeological sites in Casma (Cerro Julia and Cerro El Calvario) on the coast of Peru. These maize cob samples were examined by the present author and exhibit no evidence of teosinte introgression. Comparatively, it is interesting to observe that the oldest actual cob samples from Guilá Naquitz, Oaxaca, Mexico, are dated 5410 and 5420 C14 years BP (about 6,250 calendar years ago) (Piperno and Flannery, 2001[444]).

We are convinced that the earliest coastal maize in Peru was preceded by maize cultivated at intermediate highland altitudes in the Andes of Peru. We are supported by observations on the high frequency of the presence of anthocyanin

[443] Mangelsdorf, P. C., and J. W. Cameron. 1942. Western Guatemala: A secondary center of origin of cultivated maize varieties. *Botanical Museum Leaflets*, 10: 217–252.

[444] Piperno, D. R., and K. V. Flannery. 2001. The earliest archaeological maize (*Zea mays*) from highland Mexico: New accelerator mass spectrometry dates and their implications. *Proceedings of the National Academy of Sciences USA*, 98 (4): 2101–2103.

pigmentation in the cupules of large numbers of well-preserved maize cobs obtained from archaeologically precise diggings. Maize macrofossils obtained on the Peruvian coast are notoriouslky well preserved by the desert environment, except under conditions of close proximity to the ocean or in cases of underground water or nearness to swamps, which occur least frequently.

Some archaeologists have been reluctant to accept the antiquity of the dates attached to macrofossils obtained in Peru, because they did not fit into their preconceived framework of perception of the antiquity of maize in Peru. Staller and Thompson (2002[445]) proposed a more recent introduction (2200–1200 BC) based on archaeological and paleoethnobotanical grounds. Robert McKelvy Bird and others (see Chapters 5 and 10 of the present text) have rejected the presence of preceramic maize in Peru simply on the argument of size of ears, without realizing that ear size is a racial characteristic that is subject to change by hybridization or preservation by human selection. Lia, Confalonieri, Ratto, and colleagues (2007[446]), on the other hand, have provided evidence, from microsatellite studies of Argentina's present and archaeological races of maize, that exhibits a great uniformity of alleles of the microsatellites that were studied. They concluded that their studies point to an association of the Argentine maize with races of maize from the western part of South America. They support the hypothesis that maize was first introduced to South America via a highland route.

The distance of the movement of maize from the highlands to the coast of Peru is very short in the transversal valleys that link these areas. The association of the early archaeological maize on the coast to highland races also found in archaeological sites (Guitarrero Cave in Ancash and the Ayacucho caves) has been firmly established. Early maize races were evidently widely adapted at an early period of cultivation in both the highlands and on the coast of Peru, as is evidenced by the fact that the same three primitive races, Proto-Confite Morocho, Confite Chavinenese, and Proto-Kculli, have been found in the highlands and on the coast of Peru in early periods.

From another source of evidence, that is, the chromosome knob constitution of maize races in the Andean complex, it has been inferred by McClintock (1978) and McClintock and colleagues (1981) that there was an early spread of maize cultivation into South America. Her extensive cytological studies of maize races of the Americas, together with those of her associates Kato-Yamakake and Blumenschein, brought her to the conclusion that maize was initially introduced

[445] Staller, J. E., and R. G. Thompson. 2002. A multidisciplinary approach to understanding the initial introduction of maize into coastal Ecuador. *Journal of Archaeological Science*, 29: 33–50.

[446] Lia, Veronica V., Viviana A. Confalonieri, Norma Ratto, Julián A. Cámara-Hernández, Ana M. Miante Alzogaray, Lidia Poggio, and Terence Brown. 2007. Microsatellite typing of ancient maize: Insights into the history of agriculture in southern South America. *Proceedings of the Royal Society*, 274 (1609): 545–554.

into the central Andes, from where it moved to other highland and lowland regions of the continent. Later on, new races spread from the north along the eastern coast of South America in relatively recent times. The ample archaeological and cytological evidence stands in contrast to the molecular information interpretation of Matsuoka and colleagues (2002), which states, on the basis of microsatellite variation, that maize reached the lowland of South America at a later stage.

Bringing all this information together, and looking at the case of the similarities between chimpanzees and humans, which have some 98% of genes in common, similarities in gene composition and chromosome structure are not decisive proof of direct descent. Species or subspecies of *Zea* could have evolved in parallel in the wild, from a common ancestor at a recent time, not long enough ago to have experienced early major chromosomal changes, although some of them have appeared later. Teosinte and maize did not drift genetically apart in the wild to the point of establishing strong barriers to intercrossing between maize and the *parviglumis* and *mexicana* subspecies. This is precisely so because initial reciprocal intercrossing has tended to obliterate allelic differences in the last 8,000 years of sympatric coexistence. This situation is entirely different between maize and other species of teosinte, as has been found by Mangelsdorf and Reeves (1939), Langham (1940), and Rogers (1950).

Final Thoughts

It is now accepted that *Zea mays* L. ssp. *mays* evolved by polyploidization resulting from the reassembly of two very closely related ancestors of the new species that belonged in the tribe Andropogoneae, each one with a chromosome number $n = 5$ but with many common gene elements, such that, in fact, present maize appears as having many duplicated genes (Goodman et al. 1980[447]). Swigonova and colleagues (2004[448]) have calculated that the tetraploidy event took place almost simultaneously with the divergence of the two maize progenitors from sorghum about 11.9 million years ago.

We may imagine that maize as we call it today had a wild plant ancestor, thousands of years ago, that was different from how both maize and teosinte appear to us today. Continuing its evolution as a wild plant, it adapted to several different environments, and from it several new subspecies arose. We do not know if this wild maize ancestor was, a few million years ago, an annual or a perennial plant. We can envision a branching out of the initial phylogenetic

[447] Goodman, M. M., C. W. Stuber, K. Newton, and H. H. Weisinger. 1980. Linkage relationships of 19 enzyme loci in maize. *Genetics*, 96: 697–710.

[448] Swigonova, Zuzana, Jinsheng Lai, Jianxin Ma, Wusirika Ramakrishna, Victor Llaca, Jeffrey L. Bennetzen, and Joachim Messing. 2004. On the tetraploid origin of the maize genome. *Genomics* 5 (3): 281–284.

line into a tree of populations from which annual teosinte and a modern maize precursor arose. Some of these populations, growing near river banks, were subjected to seasonal flooding and evolved branching of their inflorescences from buds located along the stem in leaf axils, projecting branches that would sustain small ears and tassels above the water level, whenever water levels rose, and adapting in a way that allowed the ears to form and mature when the water receded. Some teosinte and *Tripsacum* populations continue to grow presently along river banks. This evolutionary pattern duplicates, in some respects, the branching habit that developed in flooded rice. For other populations of the common wild maize and teosinte ancestor, which grew in areas that were not subjected to seasonal flooding but were dryer, upland areas with occasional seasonal rains, natural selection would have preferred a more efficient plant architecture, distributing the final allocation of photosynthates to a better balance between vegetative and reproductive biomass, increasing the harvest index, and favoring a single stalk over a number of tillers and an annual rather than a perennial growth habit. The annual growth habit would confer advantages of genetic plasticity under selection as compared to the perennial plants, which tend to become conserved as polyploids. These various populations diverged in gene frequencies and stabilized themselves through the establishment of allopatric groupings, which gradually lead to speciation in a taxonomic sense, but without fully establishing intercrossing barriers.

Primitive men may have recognized that maize was a potential food product according to the following scenario. They may have used dry maize stalks as fuel for cooking meat and have accidentally discovered that the kernels popped and could be eaten. Collection of stalks of wild and early maize with their ears attached appears to have been the harvest system in ancient times in Peru (and continues to be done today in both coastal and highland locations). Popping was the first use for the grain of maize, as the seeds were very small and hard and had very little starch, with their cellular starch granules surrounded by a hard, proteinaceous layer. The popping of maize grains could have evolved into a communal social ceremony. The large underground maize silos at the Los Gavilanes site in Huarmey, Peru, in addition to being used as grain storage facilities, appear to have been used as communal grain poppers in preceramic periods, that is, the inhabitants popped shelled or unshelled grain over sand heated by hot stones, as evidenced by the charred stones that cover their walls (Bonavia, 1982[449]). In the very early periods of maize domestication and cultivation, maize kernels were very tiny and of the popping type. It is very likely that they were surrounded by soft glumes, as they appear in a semi-tunicate phenotype today and in the archaeological specimens from Bat Cave and other locations.

[449] Bonavia, Duccio. 1982. *Precerámico Peruano. Los Gavilanes. Mar, desierto y oasis en la historia del hombre.* Corporación Financiera de Desarrollo S.A. COFIDE, Instituto Arqueológico Alemán, Comisión de Arqueología General y Comparada. Lima.

Maize and teosinte grow in proximity in the same general geographical region in Mexico and northern Central America. Annual teosinte races, or subspecies of *Zea mays* (as one might decide to call them), and maize itself have morphological resemblances in plant type, which are accentuated in the teosinte ssp. *mexicana* of the high-altitude Mexican plateau, which mimics maize. Teosinte crosses with maize in agricultural fields and produces fertile offspring; presently the flow of genes is much stronger from maize to teosinte in the former region, whereas in the Balsas River valley, where teosinte grows in large, pure, contiguous tracts, opportunities for gene exchange are more limited. The karyotypes (chromosome number and characteristics) of maize and annual teosinte are similar and their homologous chromosomes pair at meiosis, although differential translocations have been found in teosinte.

These coincidences have led to a rational conclusion, first advocated by Mangelsdorf and Reeves (1945[450]), that from the point of view of their placement in a taxonomical classification, maize and annual teosinte could well belong in the same species: *Zea mays* L. Later on, the different teosintes were further classified into subspecies or races, and into separate species, thus ordaining them from the taxonomical perspective.

As a caveat, we must emphasize that, taxonomically, a species is a human psychological construct that the human mind requires for classifying and ordering living beings into discrete groups. Nature does not work strictly as human minds depict it, and variation discreteness in nature is a human mental construct. Rather, the real interpretation of the concept of species involves the existence of variable populations that have bell-type variability curves that overlap or touch at their ends or else have separated among themselves over time. Crossability is no longer an absolute discriminator for species separation, as wide crosses are found to be possible between different species, for example, rye × wheat and *Zea diploperennis* × *Tripsacum laxum*.

As a consequence of their taxonomic proximity, Beadle (1939) resurrected the hypothesis, which had been shelved for many years, of the origin of maize proceeding directly from the domestication of teosinte. He has been supported in this theory by Doebley (1990a), Galinat (1992[451]) Iltis (2000[452]), Bennetzen and colleagues (2001[453]), Benz (2001), Matsuoka and colleagues (2002), and others.

As Mangelsdorf had abandoned in 1974 the part of the tripartite hypothesis that included *Tripsacum* as one of the direct ancestors of teosinte, which he had

[450] Mangelsdorf, P. C., and R. G. Reeves. 1945. The origin of maize: Present status of the problem. *American Anthropologist*, 47: 235–243.

[451] Galinat, W. C. 1992. Evolution of corn. *Advances in Agronomy*, 47: 203–231.

[452] Iltis, H. H. 2000. Homeotic sexual translocation and the origin of maize (*Zea mays*, Poaceae): A new look at an old problem. *Economic Botany*, 54: 7–42.

[453] Bennetzen, J., E. Buckler, V. Chandler, J. Doebley, J. Dorweiler, B. Gaut, M. Freeling, S. Hake, E. Kellog, R. Scott Poethig, V. Walbot, and S. Wessler. 2001. Genetic evidence and the origin of maize. *Latin American Antiquity*, 12: 84–86.

elaborated with Reeves in 1939 on the basis of a succession of brilliant experiments, the field was wide open for new hypothesis to fill in the evolutionary picture in the vacuum that had been left. It was then that Iltis (1983) advanced his hypothesis of the emergence of maize as a species, from a single catastrophic event that comprised many quick changes in teosinte that took place in a short period of time. The incompatibility of intermediate forms between maize and teosinte as survivors in nature before domestication, as is required by this theory, presents serious objections. His theory did not carry enough evidence to support it and is no longer in favor.

The field was then open for teosinte to regain center stage as the hypothetical direct parent of maize. This theory had many elements in its favor. Mangelsdorf in his research had already detected blocks of genes in chromosome 4 of maize that he advocated were major components of its differentiation from teosinte. Others analyzed the effect of other genes and mutant alleles that could have been instrumental in the transformation of teosinte into maize – specifically, the change of the multi-stemmed, multi-branched teosinte plant morphology changing to a single stalked plant in maize (in most present-day races and varieties); the momentous transformation of the lateral-branched inflorescence of teosinte into a maize ear; and the change of the closed fruitcase of teosinte into a cupule and exposed seeds with a different angle of disposition, along with the transformation of the distichous condition of the arrangement of seeds of teosinte into a polystichous arrangement in maize.

These changes have been hypothesized as having occurred by concentration, through human selection, of a set of preexistent alleles into teosinte in only about five major genes or gene blocks, which lead to the transformation of teosinte into maize. This theory is simple and elegant, and many but not all students of evolution have accepted it as a fact that would close our gap of knowledge.

To support this theory further, isozyme analysis and molecular analysis of polymorphisms of sequences of nucleotides at the subgene level for certain genes (Doebley, 1990b, 2004, and Matsuoka et al., 2002) have been interpreted as providing additional confirmation that maize originated in a single domestication event and at a single site. Furthermore, the greater similarity of SNPs for some genes between present-day Mexican maize races and present-day *parviglumis* ssp. of teosinte has been presented as evidence that they are the closest existing relatives and, ergo, that one must have been the progenitor of the other.

This theory, which appears plausible on the basis of accumulated genetic evidence, did not take into consideration major difficulties that still exist and have to be explained to account for this supposedly simple transformation of teosinte to maize. Undoubtedly there are thousands of genes with minor effects, which work in cascades to accomplish steps in development that would have required coordination and stabilization, if maize were derived from teosinte, to produce the morphological and physiological changes required to move one species into another. Doebley (2004) has recently admitted that the available facts favor the view that although QTLs of major effects are likely to account for most of the morphological changes brought about in the process of maize domestication, many other QTLs of additive action are required to explain the changes that have occurred, in spite of the effects that some major QTLs, such as *tga1* and *tb1*, may have exercised on the morphological characters that they control. What does not follow was Beadle's (1939) support of the teosinte-to-maize hypothesis by some five simple mutations – in spite of his own observations that few traits separating maize and teosinte follow a simple Mendelian inheritance, which does not back the hypothesis that only a few genes are responsible for the transformation of teosinte into maize.

Studies on the development of the leaf of maize (Li et al., 2010;[454] Majeran et al., 2010;[455] Nelson et al., 1984[456]) have disclosed the extraordinary coordination needed between hundreds of genes of the nuclear and chloroplast genomes that operate sequentially in large sets under close coordination to cause a shift in control of the development of a leaf of maize. These studies support the notion of the close relationship and coordination required between large series of genes for expression of complex traits such as those that differentiate teosinte from maize.

Another theory that was advocated and tested in the early 1980s, after the discovery of *Zea diploperennis*, a perennial diploid teosinte unknown until then in Mexico, prompted the resurrection of the tripartite hypothesis, substituting *Tripsacum* with *Z. diploperennis* and again proposing that the end result was not maize but annual teosinte. The hypothesis was first advocated by Wilkes (1979), an experienced student of teosinte, and tested by Mangelsdorf and his former students (Cámara-Hernández and Mangelsdorf, 1981; Mangelsdorf et al., 1981). Although annual teosinte-like plants have been recovered in segregating populations of *Z. diploperennis* crossed with the Mexican popcorn maize race Palomero Toluqueño, subrace Jaliscience (from the state of Jalisco), this hypothesis carries some weaknesses. One of them is the lack of explanation for the unlikely chromosome knob outcome in the present annual teosinte races of Mexico, which have a large number of knobs in different positions, as compared

[454] Li, Pinghua, Lalit Ponnala, Neeru Gandotra, Lin Wang, Yaqing Si, S. Lori Tausta, Tesfamichael H. Kebrom, Nicholas Provart, Rohan Patel, Christopher R. Myers, Edwin J. Reidel, Robert Turgeon, Peng Liu, Qi Sun, Timothy Nelson, and Thomas P. Brutnell. 2010. The developmental dynamics of the maize leaf transcriptome. *Nature Genetics*, 42: 1060–1067.

[455] Majeran, Wojciech, Giulia Frisor, Lalit Ponnalar, Brian Connolly, Mingshu Huang, Edwin Reidel, Cankui Zhang, Yukari Asakura, Nazmul H. Bhuiyan, Qi Sun, Robert Turgeon, and Klaas J. van Wijk. 2010. Structural and metabolic transitions of C_4 leaf development and differentiation defined by microscopy and quantitative proteomics in maize. *The Plant Cell*, 22 (11): 3509–3542.

[456] Nelson, Timothy, Mark H. Hapter, Stephen P. Mayfield, and William Taylor. 1984. Light-regulated gene expression during maize leaf development. *The Journal of Cell Biology*, 98: 558–564.

to their putative parentage under this new tripartite hypothesis. Furthermore, the maize parent at the time of crossing would not have had chromosome knobs like those of Palomero Toluqueño, which is multi-knobbed and exhibits now. At the presumed ancient time of such formative hybridization, the maize parent would have been essentially knobless (or with one or two small knobs); furthermore, the chromosome knob positions of annual teosinte and the perennial *Z. diploperennis* do not coincide.

Mangelsdorf (1974) had expressed strong reservations about the theory of maize originating from teosinte in a time period of only 10,000 years ago, which is about the maximum period of time in which domestication could have taken place on the basis of agricultural origins. He advocated that maize was originally a wild plant before domestication, similar to other Andropogonaceous plants, and to the reconstructed pod-popcorn he had obtained, except that it had pistillate and staminate spikelets that were separated but in the same inflorescence. The change to the monoecious flowering condition has occurred in many plants and could have also occurred in maize. Remains of the change are still seen in present maize. Maize and teosinte must have diverged relatively recently, initially as independent but related species, and after the formation of a number of domesticated maize races, they came together and produced through mutual introgression some of the variability in maize we find today.

A novel theory has been formulated by Eubanks in which teosinte and *Tripsacum* hybridization may have transformed the simple spike of their progenitors into the maize ear (Eubanks, 2001). According to her, comparative genomic analysis of maize, teosinte, and *Tripsacum* confirms that maize has inherited unique polymorphisms from a *Tripsacum* ancestor and other unique polymorphisms from a teosinte progenitor. This would support the hypothesis that *Tripsacum* introgression provided the mutagenic action for the transformation of the teosinte spike into the maize ear. On the basis of pollen morphology examined by scanning electron microscopy of archaeological maize in Peru, the possibility of *Tripsacum* introgression into maize had been expressed by Grobman (1982).

A remaining possible hypothesis is that wild maize, annual teosinte, and other teosintes diverged at different periods of time, rather recently, from a common ancestor. This wild maize would have been unlike modern teosinte in that it would not have had the long, lateral branches that are peculiar to teosinte, with a pronounced apical meristem exhibiting dominance over the lateral axillary meristems; a polystichous ear terminated in a diminute staminate tassel; small, naked, hard kernels surrounded by soft glumes, with brittle rachillae and an abscission layer that allowed for easy dispersal; and a loose husk system. These have all been found in archaeological and in a vestigial form in some present-day maize races.

The opposition against this hypothesis surges from the fact that such postulated predomestication maize, although reconstructed genetically, has never been found growing presently in the wild, although Mangelsdorf (1974) was in favor of considering the Tehuacán Cave cobs as being examples of wild maize. A common additional argument is that the maize ear, as known today, has seeds that do not shatter and, furthermore, that its seeds are enclosed by husk leaves (teosinte ears are also enclosed by a husk), which make present-day maize unable to disperse seeds without the intervention of man.

These objections are not insurmountable. Wild *Zea mays*, being endemic to a certain location, could have been swamped out by interbreeding, under human selection, with plants with phenotypes that were acquiring higher grain yield and a domesticate phenotype. This would have happened by the thickening of the cob from a thin rachis, which is first evident in the early ancient maize race Proto-Confite Morocho, and later through ear fasciation as a permanent racial characteristic in the primitive race Confite Chavinense; both races mentioned are from Peru.

Through continued evolution, most domesticated forms of formerly wild species have had their gene pools modified gradually such that present-day forms of the species are very different in size but retain most of the former phenotypic characteristics of the wild plant, as is the case of tomato or barley. The original seed dispersal mechanism in wild maize could have been similar to that of other cereals. Wild maize could have had longer and more fragile rachillae with an abscission layer, a protruding ear, and relatively open husks that were lower in number and opened up when the ear dried out, as do some races of maize today.

The fact that no wild *Zea mays* ssp. *mays* populations are found today does not preclude that they may have existed in the past with characteristics that resemble those of the ears of pure maize found in the Tehuacán and Bat Caves in Mexico, and with cobs that were similar to the small cobs (2 cm long) of very early maize at the Los Gavilanes site in Peru (Grobman, 1982: photo 47, 159), which are probably very early domesticates. These particular ear cobs exhibit no teosinte introgression, and furthermore they exhibit a morphology that is totally different from that of teosinte ears. In addition to being polystichous, some of them have a staminate ear tip terminating the basal pistillate inflorescence of the ear where seeds are formed, resembling the bisexual inflorescences of *Tripsacum*, which is totally different from the type of inflorescences of teosinte. These same types of inflorescences have been described by Mangelsdorf (1974: figure 15.24, 180; figure 8.7, 98; figure 8.8, 96; figure 8.10, 98; and figure 11.4, 128) and by Grobman (1982: figure 60, 167) from the Los Gavilanes archaeological maize excavated by Bonavia in Peru. The maize plants from Los Gavilanes, furthermore, show the regular presence of small, lateral ears covered by their own husks, originating from accessory buds at the base of insertion of the main ear on its shank (Grobman, 1982: photo 52, 167). These accessory branched ears, we hypothesize, may have been under control of the *ramose 1* and *ramose 2* genes, which Vollbrecht and colleagues (2005[457]) assume may have played an important role in maize domestication.

These coincidences are further carried on to the present maize races in Peru. Many of them, indigenous to the Andean region, exhibit the staminate inflorescence characteristic in their ears as a relic. Some cryptic genes may exist that show up as a staminate spike terminating the ear in crosses of maize from South America with maize from North America (Mangelsdorf, 1974: figure 11.4, 128).

When examining the archaeological race that we identify as Proto-Confite Morocho, from which the present primitive Confite Morocho race has descended, we find the most basic structure of a maize ear of any race. It is basically a modified central, slender spike with very long, superficial, navicular shaped cupules with eight rows of kernels (Grobman et al., 1961: figure 49, 143). This basic cob structure is very similar to that of an elongated ear of Guarani pod corn, which is expressed due to either the *Tu* (tunicate) or tu^h (half tunicate) genes, obtained by Mangelsdorf (1974: figure 7.4, 80). Wild maize ears could have been similar to either one of these present relic ears, of which the race Proto-Confite Morocho of Peru is the closest exponent.

Morphologically, maize and teosinte differ in some striking and distinguishing characteristics. The most visible ones are the extreme difference in structure of the female inflorescence or pistillate spike of the two species. The main differences between the pistillate spikes of maize and teosinte include paired versus single spikelets, a many-ranked versus a two-ranked arrangement of the spikelets, and inconspicuous soft glumes versus prominent horny glumes, respectively. Granting that there is variation among different types of pistillate spikelets within each of the two species, none of the types in one subspecies approach any of the types found in the other subspecies in pure maize or teosinte, whether early archaeological or modern. These characteristics are discontinuous and belong to two different morphological classes in maize and teosinte.

The question that requires an answer is whether, from the condition of single-ranked, single spikelets of the teosinte female inflorescence, the two-ranked, double-spikelet traits could have arisen by mutations in the course of the domestication process, or whether those latter traits preexisted independently in some populations of a maize precursor, which could then accurately be identified as wild maize. We must note that double spikelets in maize was considered to be a simple dominant trait over single spikelets in genetic experiments conducted by Collins and Kempton (1920) on the hybridization of Tom Thumb popcorn (a primitive maize race) with Florida teosinte. Mutations from a wild allele are usually recessive, therefore the double-spikelet condition trait of

[457] Vollbrecht, E., P. S. Springer, L. Goh, and E. S. Buckler IV. 2005. Architecture of floral branch systems in maize and related grasses. *Nature*, 436: 1119–1126.

the female inflorescence of maize is more likely the wild state of maize, and not a mutation from teosinte. They measured 33 plant characters in an F_2 population of the indicated cross but found no instance in which a strictly Mendelian inheritance occurred.

Rogers (1950) made a number of crosses of maize by several teosinte races, and the maize-teosinte F_1 hybrids were consistent in their behavior for one character; all plants produced paired pistillate spikelets, attesting to maize dominance.

Concluding Statement

The discussions that we have conducted in this appendix to the main portion of this book have been associated with the problem of resolving the origin, domestication, and evolution of maize. They have evolved into the following hypothesis, which results from the integration of available scientific information up to the end of 2011 and its reinterpretation in the light of the author's 60 years of experience and research in maize, comprising a variety of studies (in fields such as ethnobotany, evolution, breeding, cytogenetics, and genetics).

Wild maize, in all likelihood, originated together with teosinte, or it branched out from a common ancestor before the presence of humans in the American continent. Wild maize populations were not initially sympatric to teosinte populations and remained so until human intervention and many years after it. Wild maize fundamentally exhibited the plant and ear characteristics of maize. Its ears were small, with semiopen husk systems, and were polystichous and capable of shattering their small, horny seeds, through an abscission layer in the pedicel. Each ear terminated in a staminate tip, and possibly the ears were branched. The plants may have been single-stalked, although tiller production is not discarded, and were essentially a maize plant all along. These plants had basically knobless chromosomes.

Domestication of maize took place from a now-extinct wild population that resembled maize more than teosinte, in an unknown location or locations. These locations possibly were in Mexico or northern Mesoamerica, but other locations cannot be ruled out. Maize domestication, under this hypothesis, is not time constrained, as the changes in the genetics, physiology, and morphology of the plants from wild to domesticate could have been gradual, not explosive, as demanded by the teosinte-to-maize hypothesis, and were minor, following a course similar to what occurred in rice or barley domestication. Wild maize populations were, very likely, swamped out by their semidomesticates in a gradual process and became formally extinct but not really so, as their cultivated descendants continued to be improved under human selection. At a more advanced stage of domestication, seeds were carried out of the primary center(s) to locations in Central and South America, where the gradual process of selection for yield improvement and for a variety of uses was continued.

Present archaeological evidence (recently published by Bonavia and me; see the afterword) dates that initial movement of a primitive maize, not introgressed by teosinte, to the coast of Peru at no later than 7000 years BP. That maize was in cultivation in the higher or middle elevations of the Andes at an earlier period of time.

As domesticated maize expanded in area, its incipient cultivated populations came into contact with teosinte ssp. *parviglumis* and ssp. *mexicana*, and gene flow, primarily from teosinte to maize, was initiated and has continued unabated for several thousand years. That initial re-encounter of the two related subspecies (or more meaningfully, "sister species," as modern taxonomy should permit it), while allowing for mutual introgression, also created some genetic mechanisms of defense in both maize and teosinte against disruptive or centripetal selection acting independently on them. They then had different selective forces acting on each one: human selection in a pampered agricultural environment on maize, and natural selection acting on the wild teosinte populations. But one new factor was present. Both had been introgressed by DNA from the other, which had evolved separately, and they were no longer the same as when they branched out many thousands of years ago (one provocative study mentioned in this appendix refers to a separation between maize and teosinte happening no earlier than 23,000 years ago).

The variability of maize increased considerably since domestication. Some 400 races of maize worldwide are witness to this fact. Allelic analysis and sub-allelic variation analysis have disclosed that maize appears to be less variable in this respect than teosinte. New evidence indicates that maize after domestication has increased its variation (Hufford et al., 2011[458]). We must reflect that present maize has gone not only through an apparent domestication bottleneck, but also through the filter of a continuous process of selection under genetic drift. Over thousands of years, few seeds from each generation were saved, and many small populations of maize were lost by accidents, migration, war, climatic changes, and so on, while others survived. Finally, maize underwent stabilizing selection to maintain its present races. Intensive breeding followed in the last century, and this may have been responsible for further reduction in variation at the nucleotide assembly level.

Maize movements have continued to take place among centers of variation at varying times and intensities. These centers are located in Mexico, the United States, Mesoamerica, the Colombian region, the extended Andean region (from

[458] Hufford, Matthew B., Xun Xu, Joost van Heerwaarden, Tanja Pyhäjärvi, Jer-Ming Chia, Reed A. Cartwright, Robert J. Elshire, Jeffrey C. Glaubitz, Kate E. Guill, Shawn M. Kaeppler, Jinsheng Lai, Peter L. Morrell, Laura M. Shannon, Chi Song, Nathan M. Springer, Ruth A. Swanson-Wagner, Peter Tiffin, Jun Wang, Gengyun Zhang, John Doebley, Michael D. McMullen, Doreen Ware, Edward S. Buckler, Shuang Yang, and Jeffrey Ross-Ibarra. 2012. Comparative population genomics of maize domestication and improvement. *Nature Genetics*, 44: 808–811.

Pasto in Colombia to northern Argentina, Chile, and the Bolivian highlands), the Amazonian lowland region, and the Venezuelan/Caribbean/eastern South American region.

The present hypothesis of maize origin and domestication, like some others, has strong and weak points. In advancing it, we believe it has fewer weaknesses than others at this time. Venturing to posit it now, when the scientific press admits some papers mentioning the word "consensus" as regards the teosinte-to-maize hypothesis, has required a long and careful evaluation of the evidence at hand.

We are aware that interpreting present evidence and reformulating once again a bold hypothesis on the origin and the course of evolution of modern maize from a precursor that was akin to maize, rather than to teosinte, is a risky proposition, but we are taking that risk, if this book, avoiding complacency, opens up the subject for further analysis. We do not believe that the analysis on the domestication of maize has been terminated and that the conclusions are carved in stone. New archaeological digs, interpretation, and research should be resumed not only in Mexico, Mesoamerica, and the Andean region but all over the Americas. Very little has been really done up to now, in relation to what needs to be done in archaeological exploration of the early agricultural period of the Americas. Molecular biology is a powerful tool and has provided new important insights into the evolution of maize, which we mostly follow, accept, and respect. This tool, when applied to modern archaeology, should be able to exact new, valuable information from the early actors: primitive farmers with their archaeological maize and its relatives. At any rate, it should provide better information than that obtained with the modern maize races, which have been modified by teosinte introgression, and reciprocally, with teosinte, which has been modified by maize.

In all isozyme and molecular studies regarding maize evolution, regardless of the method used, it can be seen that early derived Andean maize races stand apart as a group from maize races modified by teosinte introgression in the Mexican and Mesoamerican regions. We start our journey from a deep insight into these differences.

The hypothesis presented here could and should be tested by further research in archaeology, genetics, cytogenetics, and molecular biology. We have presented a number of arguments in all these fields in the course of the discussions in the various sections of this book and this appendix. We hope that new vistas will be opened from our modest contribution and that they will stimulate further research and open discussions. Time will tell whether or not they prove correct.

Afterword

While the present book was in publication, newly discovered macrobotanical and microbotanical remains of maize were reported that shed significant light on the chronology, landrace variation, and cultural contexts associated with the crop's evolution in South America (Grobman et al. 2012*). The evidence comes from the coastal Peruvian sites of Paredones and Huaca Prieta. Dates from a series of middle- and late-preceramic periods and early ceramic periods (between c. 6700 and 3000 calibrated years before the present) were based on accelerator mass spectrometry radiocarbon determinations carried out directly on different structures of preserved maize plants – cobs, husks, stalks, and tassels; these findings represent some of the earliest known specimens in the American continent. Indirect dating of the maize remains at the Paredones site points to dates around 7000 years BP. These dates, based on macrofossils, coincide with or are earlier than the earliest macrofossil finds in Mexico.

The macrobotanical record indicates that a diversity of racial complexes characteristic of the Andean region emerged during the preceramic era in an early development that was largely independent from that of the Mexican and Mesoamerican region. Because of high frequency of anthocyanin pigmentation in the cupules of the earliest maize cobs, it is presumed that the origin of these coastal landraces could be traced to the highlands of Peru, where derived races similar to those found at these sites are still found growing at the present time.

The earliest maize cobs from Paredones and Huaca Prieta in Peru have no phenotypic evidence of teosinte introgression and differ in this respect from the earliest maize cobs found in Mexico at Guilá Naquitz, Oaxaca, dated at 6300 BP.

This evidence constitutes added support to the hypothesis on the evolution of maize presented in this book and especially in this appendix.

Alexander Grobman
November 2012

* Grobman, Alexander, Duccio Bonavia, Tom D. Dillehay, Dolores R. Piperno, José Iriarte, and Irene Holst. 2012. Preceramic maize from Paredones and Huaca Prieta, Peru. *Proceedings of the National Academy of Sciences USA*, 109 (5): 1755–1759.

Alexander Grobman was educated at the Anglo-Peruvian School in Lima; Escuela Nacional de Agricultura La Molina, Peru; Ohio State University; and Harvard University, where his Ph.D. dissertation was on the origin, evolution, and classification of the races of maize in Peru. He has worked on corn breeding at the international level and has been a research administrator and an agricultural consultant in 20 countries. He has teamed up with Duccio Bonavia throughout more than 40 years of ethnobotanical research on maize. He is an emeritus professor at Universidad Nacional Agraria La Molina and president of PERUBIOTEC, the National Peruvian Association for the Development of Biotechnology. Grobman is the author of two books, four book chapters, and more than 100 scientific articles on agricultural research, agricultural development, genetics, and ethnobotany.

Bibliography

ACOSTA, Padre José de

(1590) 1954. *Historia Natural y Moral de las Indias. Escritos Menores. De Procuranda Indorum Salute o Predicación del Evangelio en las Indias.* Obras del P. José de Acosta de la Compañía de Jesús. Estudio preliminar y edición del P. Francisco Mateos de la misma compañía. Biblioteca de Autores Españoles desde la Formación del Lenguaje hasta nuestros días. Tomo LXXIII. Ediciones Atlas. Madrid. pp. 1–247.

ADAIR, Mary J.

1994. Corn and culture history in central plains. In: *Corn and Culture in the Prehistoric New World*, edited by Sissel Johannessen and Christine A. Hastorf. Westview Press. Boulder, San Francisco, Oxford. pp. 315–334.

ADAMS, K. R.

1994. A regional synthesis of *Zea mays* in the prehistoric American Southwest. In: *Corn and Culture in the Prehistoric New World*, edited by Sissel Johannessen and Christine A. Hastorf. Westview Press. Boulder, San Francisco, Oxford. pp. 273–302.

AGUERRE, Ana M., Alicia A. FERNÁNDEZ DISTEL, and Carlos A. ASCHERO

1973. Hallazgo de un sitio acerámico en la quebrada de Inca Cueva (Prov. de Jujuy). *Relaciones de la Sociedad Argentina de Antropología*, 7, N.S.: 197–231.

1975. Comentarios sobre nuevas fechas en la cronología arqueológica precerámica en la provincia de Jujuy. *Relaciones de la Sociedad Argentina de Antropología*, IX, Nueva Serie: 211–214.

AIKENS, C. Melvin

1981. Review of T.F. Lynch (ed.), *Guitarrero Cave: Early Man in the Andes. American Anthropologist*, 83 (1): 224–226.

ALCORN, Janis B., Barbara EDMONSON, and Cándido HERNÁNDEZ VIDALES

2006. Thipaak and the origins of maize in northern Mesoamerica. In: *Histories of Maize: Multidisciplinary Approaches to the Prehistory, Linguistics, Biogeography, Domestication and Evolution of Maize*, edited by John Staller, Robert Tykot, and Bruce Benz. Academic Press, Elsevier. San Diego, London. pp. 599–609.

AMPUERO BRITO, Gonzalo, and Mario RIVERA DÍAZ

1971. Secuencia arqueológica del alero rocoso de San Pedro Viejo-Pichasca. *Boletín*, 14: 45–69 (Publicaciones del Museo Arqueológico de La Serena, Chile).

ANDAGOYA, Pascual de

([1829]1880) 1954. Relación de los sucesos de Pedrarias Dávila en las provincias de Tierra firme ó Castilla del oro, y de lo ocurrido en el descubrimiento de la mar del Sur y costa del Perú y Nicaragua, escrita por el Adelantado Pascual de Andagoya. In: *Pascual de Andagoya. Ein Mensch elrebt die Conquista*, edited by Hermann Trimborn. Universität Hamburg. Abhandlungen aus dem Gebiet der Auslandskunde. Band 59. Reihe B. Völkerkunde, Kulturgeschichte und Sprachen. Band 33. Cram, de Gruyter & Co. Hamburg. pp. 224–261.

ANDERSON, Edgar, and Hugh C. CUTLER

1942. Races of *Zea mays*: I. Their recognition and classification. *Annals of the Missouri Botanical Garden*, 29 (2): 69–86, 88.

1950. Methods of corn popping and their historical significance. *Southwestern Journal of Anthropology*, 6 : 303–308.

ANDERSON, Edgar, and R. O. ERICKSON

1941. Antithetical dominance in North American maize. *Proceedings of the National Academy of Sciences of the United States of America*, 27: 436–440.

ANDRES, J. M.

1950. Granos semivestidos, restos de un carácter ancestral del maíz. *Revista Argentina de Agronomía*, 17: 252–256.

ANGLERÍA, Pedro Mártir de (ANGHIERA, Pietro Martire di)

(1530) 1944. *Décadas del Nuevo Mundo*. Editorial Bajel. Buenos Aires.

(Note: A first volume of *De Orbe novo decades* was published in 1911. The complete collection was published in 1530.)

Anonymous

1982. Maíz evolution: A manuscript by L.F. Randolph. *Economic Botany*, 36 (2): 193–194.

(Note: See Randolph, 1976.)

Anonymous

(1534) 1992. *Nouvelles certaines des Isles du Peru* (Facsimilé de l'édition originale). Texte rapproché du Français moderne par Hélène Cazes. Suivi de *Nouvelle du Pérou* de Miguel de Estete. Texte présenté par Isabel De Soto, traduit de l'espagnol par Angeles Muñoz et Rosine Gars. Amiot. Lenganey. Thaon. pp. XII–XLII.

ARFORD, M. R., and S. P. HORN

2004. Pollen evidence of the earliest maize agriculture in Costa Rica. *Journal of Latin American Geography*, 3: 108–115.

ARONA, Juan de (pseudonym of Pedro Paz Soldán y Unanue)

1938. *Diccionario de peruanismos*. Biblioteca de Cultura Peruana, primera serie, N°10. Paris.

ASCHERSON, P.

1875. Über *Euchlaena mexicana* Schrad. *Botanischen Vereins* (Prov. Brandeburg), 17: 76–80.

ATHENS, J. S.
1990. *Prehistoric Agricultural Expansion and Population Growth in Northern Highland Ecuador: Interim Report for 1989 Fieldwork.* International Archaeological Research Institute, Inc. Honolulu.
1991. Early agriculture in northern highland Ecuador. Paper presented at the 56th Annual Meeting of the Society for American Archaeology. New Orleans.
BABOT, M.P.
2004. Tecnología y utilización de artefactos en el noroeste prehispánico. Ph.D. dissertation. Universidad Nacional de Tucumán. Tucumán.
2005. Granos de almidón en contextos arqueológicos: Posibilidades y perspectivas a partir de casos del Noroeste Argentino. In: *Paleoetnobotánica del Cono Sur: Estudios de casos y propuestas metodológicas*, edited by B. Marconetto, P. Babot, and N. Oliszewski. Museo de Antropología, FF y H-UNC. Córdoba. pp. 95–125.
BAKER, Herbert G.
1968. *Las plantas y la civilización.* Herrero Hermanos Sucesores, S.A. Mexico City.
BALTER, Michael
2001. Did plaster hold neolithic society together? *Science*, 294 (5550): 2278–2279, 2281.
2007. Seeking agriculture's ancient roots. *Science*, 316 (5833): 1830–1835.
BANERJEE, Umesh Chandra
1973. Morphology and fine structure of the pollen grains of maize and its relatives. A thesis presented by ... to the Department of Biology in partial fulfillment of the requirements for the degree of Doctor of Philosophy in the Subject of Biology. Harvard University. Cambridge.
BANERJEE, Umesh Chandra, and Elso S. BARGHOORN
1972. Fine structure of pollen grain ektexine of maize, teosinte and *Tripsacum.* Reprinted from *30th Annual Proceedings of Electron Microscopy Society of America.* C. J. Arcenaux, editor. Los Angeles. pp. 226–227.
1973a. The oldest convincing evidence for natural introgression between *Tripsacum* and *Zea. Maize Genetics Cooperation Newsletter*, 47: 47–49.
1973b. Palynological evidence for natural introgression between *Tripsacum* and *Zea. American Journal of Botany*, 60 (4) (Suppl.): 34.
1977. Scanning electron microscopy of fossil pollen from Mexico. Abstract presented at the Annual Meeting of the Botanical Society of America at Michigan State University at East Lansing, 21–26 August.
BARBERENA, Santiago I.
1894. *Quicheísmos. Contribución al estudio del folklore americano por el Dr. Santiago I. Barberena.* Tipografía La Luz. San Salvador.
(Note: Valdizán, 1990, cites this, giving an incomplete and mistaken reference.)
BÁRCENA, Roberto J.
2001. Prehistoria del Centro Oeste Argentino. In: *Historia Argentina Prehispánica*, edited by E. Berberian and A. Nielsen. Ediciones Brujas. Córdoba. pp. 561–634.
BARGHOORN, Elso S., M. K. WOLFE, and K. H. CLISBY
1954. Fossil maize from the valley of Mexico. *Botanical Museum Leaflets.* Harvard University, 16 (9): 229–240.

BARTLETT, Alexandra, and Elso S. BARGHOORN

1973. Phytogeographic history of the isthmus of Panama during the past 12,000 years. (A history of vegetation, climate, and sea-level change). In: *Vegetation and Vegetational History of Northern Latin America*, edited by Alan Graham. Elsevier. Amsterdam. pp. 203–299.

BARTLETT, Alexandra, Elso S. BARGHOORN, and Rainer BERGER

1969. Fossil maize from Panama. *Science*, 165 (3891): 389–390.

BEADLE, George W.

1932. Studies of *Euchlaena* and its hybrids with *Zea*: Chromosome behavior in *Euchlaena mexicana* and its hybrids with *Zea mays. Zeitschrift fuer Absendunggen der Vererbungswissenschaft*, 62: 291–304.

1939. Teosinte and the origin of maize. *Journal of Heredity*, 30: 245–247.

1972. The mystery of maize. *Field Museum of Natural History Bulletin*, 43 (10): 2–11.

1977. The origin of *Zea mays*. In: *Origin of Agriculture*, edited by C. E. Reed. Mouton. The Hague. pp. 615–655.

1978. Teosinte and the origin of maize. In: *Maize Breeding and Genetics*, edited by D. B. Walden. John Wiley & Sons. New York. pp. 113–128.

1980. The ancestry of corn. *Scientific American*, 242 (1): 112–119.

1981. Origin of corn: Pollen evidence. *Science*, 213 (4510): 890–892.

BENNETT, Wendell C.

1948. The peruvian co-tradition. *A Reappraisal of Peruvian Archaeology*, assembled by Wendell C. Bennett. Memoirs of the Society for American Archaeology. *American Antiquity*, XIII (4, Part 2) (suppl): 1–7.

BENNETT, Wendell C., and Junius B. BIRD

(1949) 1960. *Andean Culture History*. American Museum of Natural History, Handbook Series No. 15. New York.

BENNETZEN, Jeff, Edward BUCKLER, Vicki CHANDLER, John F. DOEBLEY, Jane DORWEILER, Brandon GAUT, Michael FREELING, Sarah HAKE, Elizabeth KELLOG, R. Scott POETHIG, and Virginia WALBOT

2001. Genetic evidence and the origin of maize. *Latin American Antiquity*, 12 (1): 84–86.

BENZ, Bruce F.

1986. Taxonomy and evolution of Mexican maize. Ph.D. dissertation. University of Wisconsin. Madison.

1989. On the origin, evolution, and dispersal of maize. Paper presented at the Circum-Pacific Prehistory Conference. Seattle.

1994a. Can prehistoric racial diversification be deciphered from burned corn cobs? In: *Corn and Culture in the Prehistoric New World*, edited by Sissel Johannessen and Christine A. Hastorf. Westview Press. Boulder, San Francisco, Oxford. pp. 23–33.

1994b. Reconstructing the racial phylogeny of Mexican maize: Where do we stand? In: *Corn and Culture in the Prehistoric New World*, edited by Sissel Johannessen and Christine A. Hastorf. Westview Press. Boulder, San Francisco, Oxford. pp. 157–179.

1999. On the origin, evolution, and dispersal of maize. In: *Pacific Latin America in Prehistory: The Evolution of Archaic and Formative Culture*, edited by M. Blacke. Washington State University Press. Pullman. pp. 25–38.

2000. Review of Eubanks 1999. *Latin American Antiquity*, 11 (1): 106–107.

2001. Archaeological evidence of teosinte domestication from Guilá Naquitz, Oaxaca. *Proceedings of the National Academy of Sciences of United States of America*, 98(4): 2104–2106.

2006. Maize in the Americas. In: *Histories of Maize: Multidisciplinary Approaches to the Prehistory, Linguistics, Biogeography, Domestication and Evolution of Maize*, edited by John Staller, Robert Tykot, and Bruce Benz. Academic Press, Elsevier. San Diego, London. pp. 9–20.

BENZ, Bruce F., Li CHENG, Steven W. LEAVITT, and Chris EASTOE

2006. El Riego and early maize agricultural evolution. In: *Histories of Maize: Multidisciplinary Approaches to the Prehistory, Linguistics, Biogeography, Domestication and Evolution of Maize*, edited by John Staller, Robert Tykot, and Bruce Benz. Academic Press, Elsevier. San Diego, London. pp. 73–82.

BENZ, Bruce F., and Hugh H. ILTIS

1990. Studies on archaeological maize I: The "wild" maize from San Marcos Cave reexamined. *American Antiquity*, 55 (3): 500–511.

BENZ, Bruce F., and Austin LONG

2000. Prehistoric maize evolution in the Tehuacán Valley. *Current Anthropology*, 41 (3): 459–465.

BENZ, Bruce F., and John E. STALLER

2006. The antiquity, biogeography, and culture history of maize in the Americas. In: *Histories of Maize: Multidisciplinary Approaches to the Prehistory, Linguistics, Biogeography, Domestication and Evolution of Maize*, edited by John Staller, Robert Tykot, and Bruce Benz. Academic Press, Elsevier. San Diego, London. pp. 665–673.

BERRAONDO, Ramón de

1927. El maíz. *Revista Internacional de los Estudios Vascos. Revue Internationale des Études Basques*, 18: 305–306.

BERRY, Michael S.

1980. Time, space and transition in Anasazi prehistory. Ph.D. dissertation. Department of Anthropology, University of Utah.

1985. The age of maize in the greater Southwest: A critical review. In: *Prehistoric Food Production in North America*, edited by Richard I. Ford. Anthropological Papers, Museum of Anthropology, University of Michigan N° 75. Ann Arbor. pp. 279–307.

BERTONIO, S. J. Ludovico

(1612) 1956. *Vocabvlario de la lengva Aymara*. Litografía Don Bosco. La Paz.

BETANZOS, Juan de

(1551) 1968. *Suma y narración de los Incas*. Biblioteca Peruana. Primera Serie, Tomo III. Editores Técnicos Asociados, S.A. Lima. pp. 197–293.

(1551) 1987. *Suma y narración de los Incas*. Ediciones Atlas. Madrid.

(1551) 1996. *Narrative of the Incas*. Translated and edited by Roland Hamilton and Dana Buchanan from the Palma de Mallorca manuscript, Madrid, Spain. University of Texas Press. Austin.

BIRD, Junius B.

1943. Excavations in northern Chile. *Anthropological Papers*, 38 (4): 171–318. The American Museum of Natural History. New York.

1946. The cultural sequence of the north Chilean coast. In: *Handbook of South American Indians. Vol. 2. The Andean Civilizations*, edited by Julian H. Steward. Smithsonian Institution. Bureau of American Ethnology. Bulletin 143. Washington, D.C. pp. 587–594.

1960. Preface to second edition. In: *Andean Culture History*. American Museum of Natural History, Handbook Series No. 15. New York. pp. 1–6.

1970. Muestras de Radiocarbono de un Basural Precerámico de Quiani, Arica. *Boletín de la Sociedad Arqueológica de Santiago*, 4: 13–15.

BIRD, Junius B., John HYSLOP, and Milica DIMITRIJEVIC SKINNER

1985. *The Preceramic Excavations at the Huaca Prieta: Chicama Valley, Peru. Vol. 62. Part 1*. Anthropological Papers of the American Museum of Natural History. New York.

BIRD, Junius B., and Mario A. RIVERA

1988. *Excavaciones en el Norte de Chile*. Ediciones de la Universidad de Tarapacá. Arica.

BIRD, Robert McKelvy

1970. Maize and its cultural and natural environment in the Sierra of Huánuco, Peru. Unpublished Ph.D. dissertation. Department of Botany, University of California. Berkeley.

1978. Archaeological maize from Peru. *Maize Genetics Cooperation Newsletter*, 52: 90–92.

1979a. The evolution of maize: A new model for the early stages. *Maize Genetics Cooperation Newsletter*, 53: 53–54.

Ms. 1979b. Report of plant remains from the Tiliviche site, Tarapacá Province, Chile. Smithsonian Institution. Washington, D.C.

1980. Maize evolution from 500 B.C. to the present. *Biotropica*, 12: 30–41.

1984. South American maize in Central America? In: *Pre-Columbian Plant Migration*, edited by Doris Stone. Papers of the Peabody Museum of Archaeology and Ethnology. Vol. 76. Harvard University Press. Cambridge. pp. 39–65.

1987. A postulated tsunami and its effects on cultural development in the Peruvian Early Horizon. *American Antiquity*, 52 (2): 285–303.

1990. What are the chances of finding maize in Peru dating before 1000 B.C.?: Reply to Bonavia and Grobman. *American Antiquity*, 55 (4): 828–840.

BIRD, Robert McKelvy, and Junius B. BIRD

1980. Gallinazo maize from the Chicama Valley, Peru. *American Antiquity*, 45 (2): 325–332.

BIRKET-SMITH, K.

1943. The origin of maize cultivation. *Kongelige Danske Videnskabernes Selskab Historisk Filologiske Meddelelser*, XXIX (3): 1–49.

BLAKE, Michael

2006. Dating the initial spread of *Zea mays*. In: *Histories of Maize: Multidisciplinary Approaches to the Prehistory, Linguistics, Biogeography, Domestication and Evolution of Maize*, edited by John Staller, Robert Tykot, and Bruce Benz. Academic Press, Elsevier. San Diego, London. pp. 55–72.

BONAVIA, Duccio

1972. Factores ecológicos que han intervenido en la transformación urbana a través de los últimos siglos de la época precolombina. In: *Urbanización y proceso social en América*, R. P. Schaedel, W. Borah, H. L. Browning, D. Bonavia, R. Cortés Conde, N. López de Nisvovich, F. Mauro, A. Rofman, J. E. Hardoy, A. Moreno Toscano, M. Kaplan, B. J. Prices, J. R. Scobie, M. Mamalakis, E. E. Calneck, R. A. Gakenheimer, G. Gasparini, E. W. Palm, and R. M. Morse. Instituto de Estudios Peruanos. Lima. pp. 79–97.

1981. Los Gavilanes, ein später präkeramischer Fundplatz im Tal von Huarmey, Dpto. Ancash, Peru. In: *Beiträge zur Allgemeinen und Vergleichenden Archäologie*. Band 3. Sonderdruck. C.H. Beck'sche Verlagsbuchlandlung. Munich. pp. 391–413.

1982. *Precerámico peruano. Los Gavilanes. Mar, desierto y oasis en la historia del hombre.* With the collaboration of Ramiro Castro de la Mata, Félix Caycho Quispe, Alexander Grobman, Lawrence Kaplan, César A. Morán Val, Raúl Patrucco, Mario Peña, Virginia Popper, Elizabeth J. Reitz, Stanley George Stephens, Raúl Tello, and Elizabeth S. Wing. Corporación Financiera de Desarrollo S.A. COFIDE, Instituto Arqueológico Alemán, Comisión de Arqueología General y Comparada. Lima.

1984. La importancia de restos de papas y camote de época precerámica halladas en el valle de Casma. *Journal de la Société des Américanistes*, LXX: 7–20.

1990a. La storia del mais andino. *L'Umana Aventura*, 5 (14): 73–79.

1990b. La domestication des plantes dans le monde andin. In: *Inca-Perú. 3000 Ans d'Histoire*. Musée Royaux d'Art et d'Histoire. Imschoot, Uitgevers. Gent. pp. 78–89.

1991. *Perú: Hombre e Historia. De los Orígenes al siglo XV.* Edubanco. Lima.

1991–1992. Historia del maíz. *Acta Herediana*, 12: 6–16.

1993. La papa: apuntes sobre sus orígenes y su domesticación. *Journal de la Société des Américanistes*, LXXIX: 173–187.

1996a. *Los Camélidos Sudamericanos. Una introducción a su estudio.* IFEA, UPCH, Conservation International. Lima.

1996b. De la caza-recolección a la agricultura: una perspectiva local. *Bulletin de l'Institut Français d'Études Andines*, 25 (2): 169–186.

1996c. Letter to the editor. *SAA* [Society for American Archaeology] *Bulletin*, 14 (4): 3, 30.

(1993–1995) 1997. La domesticación de las plantas y los orígenes de la agricultura en los Andes Centrales. *Revista Histórica*, XXXVIII: 77–107.

1998. ¿Bases marítimas o desarrollo agrícola? In: *50 Años de Estudios Americanistas en la Universidad de Bonn. 50 Years of Americanist Studies at the University of Bonn. I. Contribuciones arqueológicas. Contribuciones etnohistóricas. Archaeological Contributions. Ethnohistorical Contributions*, edited by Sabine Dedebanch-Salazar Sáenz, Carmen Arellano Hoffmann, Eva König, and Heiko Prümers. Bonner Amerikanistische Studien, 30. Verlag Anton Saurwein. Markt Schwaben. pp. 45–62.

2000. Almacenamiento en arena: una vieja técnica que se ha perdido. *Arkinka*, 5 (59): 84–92.

2002a. Orígenes de la agricultura en el Área Central Andina. In: *Enciclopedia Temática del Perú. Ecología Prehistórica Andina y Ciencia en el Perú.* Vol. V. Dirección, coordinación, revisión, ilustraciones, epígrafes, diagramación y edición Carlos Milla Batres. Editorial Milla Batres. (Universidad Ricardo Palma). Lima. pp. 139–173.

2002b. Del Precerámico a hoy: un raro caso de continuidad cultural. In: *El Hombre y los Andes. Homenaje a Franklin Pease G.Y.* Vol. I, edited by J. Flores Espinoza and R. Varón Gabai. Pontificia Universidad Católica del Perú. Fondo Editorial. IFEA, BCP, Fundación Telefónica. Lima. Volume I, pp. 421–435.

2006. Origen y domesticación de la papa. *Boletín Jaka Tinkuy*, Edición Especial, 4 (7): [12]. (The pages are not numbered.)

BONAVIA, Duccio, and Claudia GRIMALDO, with the collaboration of Jimi ESPINOZA

2001. *Bibliografía del Período Precerámico Peruano.* Academia Nacional de la Historia. I. Pontificia Universidad Católica del Perú. Fondo Editorial. Academia Nacional de la Historia. Lima.

BONAVIA, Duccio, and Alexander GROBMAN

1978. El origen del maíz andino. In: *Amerikanistische Studien, Estudios Americanistas*, edited by Roswith Hartman and Udo Oberem. Festschrift für Hermann Trimborn, Libro Jubilar en homenaje a Hermann Trimborn con motivo de su septuagésimo aniversario. I. Collectanea Instituti Anthropos, Vol. 20. Haus Völker uns Kulturen, Anthropos-Institut, D-5205 St. Agustin l. pp. 82–91.

1979. Sistema de depósitos y almacenamiento durante el período precerámico en la costa del Perú. *Journal de la Société des Américanistes*, LXVI: 21–43.

1989a. Preceramic maize in the central Andes: A necessary clarification. *American Antiquity*, 54 (4): 836–840.

1989b. Andean maize: Its origin and domestication. In: *Foraging and Farming: The Evolution of Plant Exploitation*, edited by D. R. Harris and G. C. Hillman. Unwin Hyman. London. pp. 456–470.

1998. Review of evidence on preceramic maize in the central Andean region. In: *Proceedings of XIII International Congress of Prehistoric and Protohistoric Sciences.* Vol. 5, edited by G. Bermond Montanari, R. Francovich, F. Mori, P. Pensabene, S. Salvatori, M. Tosi, and C. Peretto. Forlì. A.B.A.C.O. Edizioni, M.A.C. Srl. Forlì. pp. 403–406.

1999. Revisión de las pruebas de la existencia de maíz precerámico en los Andes Centrales. *Boletín de Arqueología P.U.C.P. El Período Arcaico en el Perú: Hacia una definición de los orígenes.* Departamento de Humanidades. Especialidad de Arqueología. Pontificia Universidad Católica del Perú. Lima. 3: 239–261.

Ms. 2012. El maíz de Huaca Prieta, Paredones y de la Unidad 16. (In preparation.)

BONAVIA, Duccio, Alexander GROBMAN, Laura W. JOHSON-KELLY, John G. JONES, Inés R. ORTEGA, Raúl PATRUCCO, Alberto PUMAYALLA D., Elizabeth J. REITZ, Raúl TELLO, Glendon H. WEIR, Elizabeth S. WING, and Angel ZARATE ZAVALETA

2009. Historia de un campamento del Horizonte Medio de Huarmey, Perú (PV35-4). *Bulletin de l'Institut Français d'Études Andines*, 39 (2): 237–287.

BONAVIA, Duccio, and Lawrence KAPLAN

1990. Bibliography of American archaeological plant remains (II). *Economic Botany*, 44 (1): 114–128.

BONAVIA, Duccio, Carlos M. OCHOA, Oscar TOVAR S., and Rodolfo CERRÓN PALOMINO

2004. Archaeological evidence of Cherimoya (*Annona cherimolia* Mill.) and Guanabana (*Annona muricata*) in ancient Peru. *Economic Botany*, 58 (4): 509–522.

BONZANI, René M., and Augusto OYUELA-CAYCEDO

2006. The gift of the variation and dispersion of maize: Social and technological context in Ameridian societies. In: *Histories of Maize: Multidisciplinary Approaches to the Prehistory, Linguistics, Biogeography, Domestication and Evolution of Maize*, edited by John Staller, Robert Tykot, and Bruce Benz. Academic Press, Elsevier. San Diego, London. pp. 343–356.

BOPP, Mónica

1961. El análisis de pollen con referencia especial a dos perfiles polínicos de la cuenca de México. In: *Homenaje a Pablo Martínez del Rio en el XXV Aniversario de la Edición de Los orígenes Americanos*. Instituto Nacional de Antropología e Historia. Mexico City. pp. 49–56.

BORREGÁN, Alonso de

(1565) 1968. *Crónica de la Conquista del Perú*. Biblioteca Peruana. Primera Serie, Tomo II. Editores Técnicos Asociados S.A. Lima. pp. 415–473.

BRADLEY, J. E., and T. VIEJA

1994. An Archaic and Early Formative site in the Arenal region. In: *Archaeology, Volcanism, and Remote Sensing in the Arenal Region, Costa Rica*, edited by P. D. Sheets and B. R. McKee. University of Texas Press. Austin. pp. 73–86.

BRANDOLINI, A.

1970. Maize. In: *Genetic Resources in Plants: Their Exploration and Conservations*, edited by O. H. Frankel and E. Bennett. F. A. Davis. Philadelphia. pp. 273–309.

BRAY, Warwick, Leonor HERRERA, Marianne CARDALE SCHRIMPFF, Pedro BOTERO, and José G. MONSALVE

1987. The ancient agricultural landscape of Calima, Colombia. In: *Prehispanic Agricultural Fields in the Andean Region*. Vol. 2, edited by William M. Denevan, Kent Mathewson, and George Knapp. BAR International Series, 359. Oxford. pp. 443–481.

BRENTON, Barnet P., and Robert R. PAINE

1998. The skeletal biology of Pellagra with intensive maize horticulture in the New World. *American Journal of Physical Anthropology*, 26 (Suppl.): 113–114.

BRETTING, P., and Major GOODMAN

1989. Karyotypic variation in Mesoamerican traces of maize and its systematic significance. *Economic Botany*, 43: 107–124.

BRIEGER, F. G.

1961. Collections and evaluation of indigenous races of maize. Food and Agriculture Organization of the United Nations. Technical Meeting on Plant Exploration and Introduction. Rome.

1968. The main ethnobotanical regions of Central and South America. In: *Actas y Memorias*, Vol. II. XXXVII Congreso Internacional de Americanistas. 1966. Buenos Aires. pp. 547–558.

BRIEGER, F. G., J. T. A. GURGEL, E. PATERNIANI, A. BLUMENSCHEIN, and M. R. ALLEONI

1958. *Races of Maize in Brazil and Other Eastern South American Countries.* National Academy of Sciences. National Research Council. Publication 593. Washington, D.C.

BROCHADO, José J. J. P.

1984. An ecological model of the spread of pottery and agriculture into eastern South America. Unpublished Ph.D. dissertation. Department of Anthropology, University of Illinois. Urbana.

BROWN, Cecil H.

2006. Glottochronology and the chronology of maize in the Americas. In: *Histories of Maize: Multidisciplinary Approaches to the Prehistory, Linguistics, Biogeography, Domestication and Evolution of Maize*, edited by John Staller, Robert Tykot, and Bruce Benz. Academic Press, Elsevier. San Diego, London. pp. 647–663.

BROWN, Kathryn

2001. New trips through the back alleys of agriculture. *Science*, 292 (5517): 631–633.

BROWN, William L.

1949. Number and distribution of chromosome knobs in United States maize. *Genetics*, 34: 524–536.

1953. Maize of the West Indies. *Tropical Agriculture*, 30: 141–170.

1974. Review of *Corn: Its Origin, Evolution and Improvement*, by Paul C. Mangelsdorf. *Science*, 185 (4152): 687–688.

BROWN, William L., and E. ANDERSON

1947. The northern flint corns. *Annals of the Missouri Botanical Garden*, 34 (1): 1–28.

BROWN, William L., Ricardo BRESSANI, David V. GLOVER, Arnel R. HALLAUER, Virgil A. JOHNSON, and Calvin O. QUALSET

1988. *Quality-Protein Maize.* National Academy Press. Washington, D.C.

BROWN, William L., and Paul C. MANGELSDORF

1951. The effects on yield and morphology of the addition of teosinte germ plasm to maize. *Genetics*, 36: 544.

BRUHNS, Karen Olsen

1994. *Ancient South America.* Cambridge University Press. Cambridge.

2003. Social and cultural development in the Ecuadorian highlands and eastern lowlands during the Formative. In: *Archaeology of Formative Ecuador*, edited by J. Scott Raymond and Richard L. Burger. Jeffrey Quilter, general editor. Dumbarton Oaks Research Library and Collection. Washington, D.C. pp. 125–174.

BUCKLER, Edward S., IV, and Timothy P. HOLTSFORD

1996. *Zea* systematics: Ribosomal ITS evidence. *Molecular Biology and Evolution*, 13: 612–622.

BUCKLER, Edward S., IV, Deborah M. PEARSALL, and Timothy P. HOLTSFORD
1998. Climate, plant ecology and central Mexican archaic subsistence. *Current Anthropology*, 39 (1): 152–164.
BUGÉ, David E.
1974. Maize in South America. *Journal of the Steward Anthropological Society*, 5 (2): 29–65.
BUIKSTRA, Jane E.
1992. Diet and disease in Late Prehistory. In: *Disease and Demography in the Americas*, edited by John W. Verano and Douglas H. Ubelaker. Smithsonian Institution Press. Washington, D.C. pp. 87–101.
BURGER, Richard L.
1989. Andean South America. *American Antiquity*, Current Research, 54 (1): 187–194.
BURGER, Richard L., and Lucy SALAZAR-BURGER
1980. Ritual and religion at Huaricoto. *Archaeology*, 33: 26–32.
BURGER, Richard L., and Nikolaas J. van der MERWE
1990. Maize and the origin of highland Chavín civilization: An isotopic perspective. *American Anthropologist*, 92 (1): 85–95.
BURLEIGH, Richard, and Don BROTHWELL
1978. Studies on American dogs, 1: Carbon isotopes in relation to maize in the diet of domestic dogs from early Peru and Ecuador. *Journal of Archaeological Science*, 5 (4): 355–362.
BUSH, Mark B., Dolores R. PIPERNO, and Paul A. COLINVAUX
1989. A 6000 year history of Amazonian maize cultivation. *Nature*, 340 (6231): 303–305.
BUSH, Mark B., Miles R. SILMAN, Mauro B. de TOLEDO, Claudia LISTOPAD, William D. GOSLING, Christopher WILLIAMS, Paulo E. de OLIVEIRA, and Carolyn KRISEL
2007. Holocene fire and occupation in Amazonia: Records from two lake districts. *Philosophical Transactions of the Royal Society*, B, 362: 209–218.
CABELLO VALBOA, Miguel
(1840) 1951. *Miscelánea antártica. Una historia del Perú Antiguo.* Universidad Nacional Mayor de San Marcos. Facultad de Letras. Instituto de Etnología. Lima.
CADDEO, Rinaldo
1930. Studio Introduttivo. In: *Le Historie della Vita e dei Fatti di Cristoforo Colombo*, per D. Fernando Colombo suo figlio. Vol. I. Viaggi e Scoperte di Navigatori ed Esploratori Italiani. 11. Edizioni "Alpis." Milan. pp. I–LXXXVII.
CALLEN, Eric O.
1965. Food habits of some pre-Columbian Mexican Indians. *Economic Botany*, 19 (4): 335–343.
1967a. The first New World cereal. *American Antiquity*, 32 (4): 535–538.
1967b. Analysis of the Tehuacán coprolites. In: *The Prehistory of the Tehuacan Valley. Vol. 1. Environment and Subsistence*, edited by Byers Douglas. University of Texas Press. Austin. pp. 261–289.

CÁMARA-HERNÁNDEZ, Julián.

Ms. n.d. *Informe preliminar sobre el material arqueológico de maíz coleccionado por la Lic.* A. Fernández Distel en Huachichocana, Qda. De Purmamarca, Prov. de Jujuy.

1989. Restos arqueológicos de maíz (*Zea mays* L.) de León Huasi, Provincia de Jujuy, Argentina. In: *Comunicaciones Científicas*. Dirección Provincial de Antropología e Historia, Año 1, N° 1. San Salvador de Jujuy. pp. 18–26.

CÁMARA-HERNÁNDEZ, Julián, and D. ARANCIBIA DE CABEZAS

1976. El maíz y sus usos en la Quebrada de Humahuaca. *Jujuy Cultural*, 5, 1er Cuatrimestre. Jujuy.

CÁMARA-HERNÁNDEZ, Julián, and Paul C. MANGELSDORF

1981. Perennial corn and annual teosinte phenotypes in crosses of *Zea diploperennis* and maize. *Publication N° 10.* The Bussey Institution of Harvard University. Cambridge. pp. 3–37.

CAMINO, Lupe

1987. *Chicha de maíz. Bebida y vida del pueblo de Catacaos.* Centro de Investigación y Promoción del Campesinado. Piura.

CÁRDENAS, M.

1969. *Manual de Plantas Económicas de Bolivia.* Imprenta Iethus. Cochabamba.

CARDICH, Augusto

1964. Lauricocha. Fundamentos para una pre-historia de los Andes Centrales. *Studia Praehistorica*, III. Centro Argentino de Estudios Prehistóricos. Buenos Aires.

CARTER, George F.

1946. Origins of American Indian agriculture. *American Anthropologist*, 48 (1): 1–21.

CASTAÑEDA CASTAÑEDA, Benjamin, Lucy Adelina IBAÑEZ, and Renán MANRIQUE MEJÍA

2005. Estudio Fitoquímico Farmacológico del *Zea mays* L. Amilaceae st. (maíz morado). *Cultura*, XXIII (19): 105–130.

CASTRO DE LA MATA, Ramiro, and Duccio BONAVIA

1980. Lumbosacral malformation and spina bifida in a Peruvian preceramic child. *Current Anthropology*, 21 (4): 515–516.

CAVERO CARRASCO, Arnulfo

1986. *Maíz, chicha y religiosidad andina.* Universidad Nacional de San Cristóbal de Huamanga. Ayacucho.

CENDRERO, Orestes

1943. *Curso elemental de Historia Natural. Botánica.* Imprenta López. Buenos Aires.

CHÁVEZ, Sergio J.

2003. Comments. *Current Anthropology*, 44 (5): 689–690.

2006. Native Aymara and Quechua botanical terminologies of *Zea mays* in the Lake Titicaca and Cuzco regions. In: *Histories of Maize: Multidisciplinary Approaches to the Prehistory, Linguistics, Biogeography, Domestication and Evolution of Maize*, edited by John Staller, Robert Tykot, and Bruce Benz. Academic Press, Elsevier. San Diego, London. pp. 623–629.

CHEVALIER, Alexandre
1999. Archaeology + ethnobotany = palaeoethnobotany? *Bulletin* (Société Suisse des Américanistes), 63: 149–162.
CIEZA DE LEÓN, Pedro de
(1873) 1967. *El Señorío de los Incas.* Instituto de Estudios Peruanos. Lima.
(1553) 1984. *Crónica del Perú.* Primera Parte. Pontificia Universidad Católica del Perú, Academia Nacional de la Historia. Lima.
1985. *Crónica del Perú.* Segunda Parte. Pontificia Universidad Católica del Perú, Academia Nacional de la Historia. Lima.
1987. *Crónica del Perú.* Tercera Parte. Pontificia Universidad Católica del Perú, Academia Nacional de la Historia. Lima.
CLARK, Richard M., Eric LINTON, Joachim MESSING, and John F. DOEBLEY
2004. Pattern of diversity in the genomic region near the maize domestication gene *tb1*. *Proceedings of the National Academy of Sciences of the United States of America*, 101 (3): 704–707.
COBO, P. Bernabé
(1890–1893) 1964a. *Historia del Nuevo Mundo.* I. Biblioteca de Autores Españoles desde la formación del lenguaje hasta nuestros días. Tomo XCI. Obras del ... Ediciones Atlas. Madrid.
1964b. *Historia del Nuevo Mundo.* II. Biblioteca de Autores Españoles desde la formación del lenguaje hasta nuestros días. Tomo XCII. Obras del ... Ediciones Atlas. Madrid. pp. 1–275.
COE, Michael D., and Kent V. FLANNERY
1967. *Early Cultures and Human Ecology in South Coastal Guatemala.* Smithsonian Contributions to Anthropology, Vol. 3. Smithsonian Institute. Washington, D.C.
COHEN, Mark Nathan
1978. *The Food Crisis in Prehistory: Overpopulation and the Origins of Agriculture.* Yale University Press. New Haven and London.
COLLIER, Donald
1961. Agriculture and civilization on the coast of Peru. In: *Anthropologica Supplement Publication, N° 2. The Evolution of Horticultural Systems in Native South America: Causes and Consequences. A Symposium*, edited by J. Wilbert. Sociedad de Ciencias Naturales. La Salle, Caracas. September. Editorial Sucre. pp. 101–109.
COLLINS, G. N.
1912. The origin of maize. *Journal of the Washington Academy of Sciences*, 2: 520–530.
1918. Maize, its origin and relationships. *Journal of the Washington Academy of Sciences*, 8: 42–43.
COLLINS, G. N., and J. H. KEMPTON
1920. A teosinte-maize hybrid. *Journal of Agricultural Research*, XIX (1): 1–38.
COLOMBO, Fernando (COLUMBUS, Ferdinand)
(1571) 1930. *Le Historie di Cristoforo Colombo.* Le Historie della vita e dei fatti di Cristoforo Colombo per D. Fernando Colombo suo figlio. Due volumi a cura di Rinaldo Caddeo con studio introduttivo, note, appendici e numerose incisioni.

Vol. I. Viaggi e Scoperte di Navigatori ed Esploratori Italiani. 11. Edizioni "Alpes." Milan.

COLÓN, Cristóbal (COLOMBUS, Christopher)
(1492) 1984. See VARELA, Consuelo.

COLTRAIN, Joan Brenner, Joel C. JANETSKI, and Shawn W. CARLYLE
2006. The stable and radio-isotope chemistry of eastern Basketmaker and Pueblo groups in the four corners region of the American Southwest: Implications for Anasazi diets, origins, and abandonment in Southwestern Colorado. In: *Histories of Maize: Multidisciplinary Approaches to the Prehistory, Linguistics, Biogeography, Domestication and Evolution of Maize*, edited by John Staller, Robert Tykot, and Bruce Benz. Academic Press, Elsevier. San Diego, London. pp. 275–287.

CONARD, Nicholas, David L. ASCH, Nancy B. ASCH, David ELMORE, Harry GOVE, Meyer RUBIN, James A. BROWN, Michael D. WIANT, Kenneth B. FARNSWORTH, and Thomas G. COOK
1984. Accelerator radiocarbon dating of evidence for prehistoric horticulture in Illinois. *Nature*, 308 (5958): 443–446.

COOK, O. F.
1925. Peru as a center of domestication. *Journal of Heredity*, 16: 33–46, 95–110.

COOPER, John M.
1949. Stimulants and narcotics. In: *Handbook of South American Indians*. Vol. 5, edited by Julian H. Steward. Smithsonian Institution, Bureau of American Ethnology, Bulletin 143. Washington, D.C. pp. 525–558.

COROMINAS, J., and J. A. PASCUAL
1989. *Diccionario Crítico Etimológico Castellano e Hispánico*. Vol. I–V. Editorial Gredos. Madrid.

CORREAL URREGO, Gonzalo, and María PINTO NOLLA
1983. *Investigación Arqueológica en el Municipio de Zipacón, Cundinamarca*. Fundación de Investigaciones Arqueológicas Nacionales. Banco de la República. Bogotá.

COSTANTIN, J., and D. BOIS
1910. Sur les graines et tubercules des tombeaux péruviennes de la Période Incasique. *Revue Général de Botanique*, 22: 242–265.

CRAWFORD, Gary W., Della SAUNDERS, and David G. SMITH
2006. Pre-contact maize from Ontario, Canada: Context, chronology, variation, and plant association. In: *Histories of Maize: Multidisciplinary Approaches to the Prehistory, Linguistics, Biogeography, Domestication and Evolution of Maize*, edited by John Staller, Robert Tykot, and Bruce Benz. Academic Press, Elsevier. San Diego, London. pp. 549–559.

CREEL, Darrell, and Austin LONG
1986. Radiocarbon dating of corn. *American Antiquity*, 51 (4): 826–837.

CROCE, Benedetto
1960. *La historia como hazaña de la libertad*. Fondo de Cultura Económica. Mexico City.

CROSBY, Alfred W., Jr.,
1975. *The Columbian Exchange: Biological and Cultural Consequences of 1492*. Greenwood Press. Westport.

CURATOLA, Marco

1985. "Mal de Canto" y "Mal de maíz," opresión, enfermedad, muerte y búsqueda de salvación en los Andes (ss. XV–XVI). Trabajo presentado al 45° Congreso Internacional de Americanistas. Bogotá. (Mimeographed.)

1990. "Mal de Canto" y "Mal del Maíz." Etiología de un movimiento milenarista. Offprint of *Antropológica*, 8: [120]–144. (The first page is not numbered.)

(Note: This is the same paper as that of 1985, with just a few changes.)

CUTLER, Hugh C.

1946. Races of maize in South America. *Botanical Museum Leaflets*, 12 (8): 257–291.

(Note: There is a 1949 Spanish translation in the *Revista de Agricultura*, VI (5): 3–19, Cochabamba, Bolivia.)

1952. A preliminary survey of plant remains of Tularosa Cave. *Fieldiana Anthropology*, 40: 461–479.

CUTLER, Hugh C., and E. ANDERSON

1941. A preliminary survey of the genus *Tripsacum*. *Annals of the Missouri Botanical Garden*, 28: 249–269.

CUTLER, Hugh C., and Leonard W. BLAKE

1971. Travels of corn and squash. In: *Man across the Sea: Problems of Pre-Columbian Contacts*, edited by Carrol L. Riley, J. Charles Kelley, Campbell W. Pernigton, and Robert Rands. University of Texas Press. Austin. pp. 366–375.

1973. *Plants from Archaeological Sites East of the Rocky Mountains*. Missouri Botanical Garden. St. Louis. (Mimeographed.)

CUTLER, Hugh W., and Martín CÁRDENAS

1947. Chicha, a native South American beer. *Botanical Museum Leaflets*, 13 (3): 33–60.

DAMON, P., J. C. LERMAN, and A. LONG

1978. Temporal flucuations of atmospheric 14C: Casual factors and implications. *Annual Review of Earth and Planetary Sciences*, 6: 457–494.

DANNENFELDT, Karl H.

1968. *Leonhard Rauwolf.* Harvard University Press. Cambridge.

DARLINGTON, C. D.

1963. *Chromosome Botany and the Origin of Cultivated Plants.* 2nd ed. George Allen and Unwin. London.

DARWIN, Charles

1839. *Voyages of Adventure and Beagle. Vol. III. Journal and Remarks 1832–1836. Adventure and Beagle between the Years 1826 and 1836 Describing of the Southern Shores of South America and the Beagle's Circumnavigations of the Globe.* Henry Colburn, Great Malborough Street. London.

(1839) 1921. *Diario del viaje de un naturalista alrededor del mundo en el navío de S.M. "Beagle."* Tomo 1 (Pp. X + 361) and Tomo 2 (Pp. VIII + 359). Espasa Calpe, S.A. Madrid.

De AZARA, F.

1850. *Viajes por la América del Sur.* 2nd ed. Montevideo.

De BOER, Warren R.

2003. Comments. *Current Anthropology*, 44 (5): 690–691.

De CANDOLLE, Alphonse

(1886) 1959. *Origin of Cultivated Plants.* Hafner. London.

(Note: The 1959 edition is the reprint of the second edition. Published by Hafner Publishing Company, London.)

DEWALD, C. L., B. L. BURSON, J. M. J. de WET, and Jack R. HARLAN

1987. Morphology, inheritance and evolutionary significance of sex reversal in *Tripsacum dactyloides* (Poaceae). *American Journal of Botany*, 74: 1055–1059.

De WET, J. M. J., L. M. ENGLE, C. A. GRANT, and S. T. TANAKA

1972. Cytology of maize-*Tripsacum* introgression. *American Journal of Botany*, 59: 1026–1029.

De WET, J. M. J., and Jack R. HARLAN

1972. Origin of maize: The tripartite hypothesis. *Euphytica*, 21: 271–279.

1976. Cytogenetic evidence for the origin of teosinte (*Zea mays* ssp. *mexicana*). *Euphytica*, 25: 447–455.

De WET, J. M. J., Jack R. HARLAN, and C. A. GRANT

1971. Origin and evolution of teosinte (*Zea mexicana* [Schrad] Kuntze). *Euphytica*, 20: 255–265.

De WET, J. M. J., Jack R. HARLAN, and A. V. RADRIANASOLO

1978. Morphology of teosintoid and tripsacoid maize (*Zea mays* L.). *American Journal of Botany*, 65 (7): 741–747.

De WET, J. M. J., Jack R. HARLAN, H. T. STALKER, and A. V. RADRIANASOLO

1978. The origin of tripsacoid maize (*Zea mays* L.). *Evolution*, 32 (2): 233–244.

De WET, J. M. J., J. LAMBERT, Jack R. HARLAN, and S. M. NAIK

1970. Stable triploid hybrid among *Zea-Tripsacum-Zea* backcross populations. *Caryologia*, 23: 183–187.

DICKAU, Ruth, Anthony J. RANERE, and Richard G. COOKE

2007. Starch grain evidence for the preceramic dispersal of maize and root crops into tropical dry and humid forests of Panama. *Proceedings of the National Academy of Sciences of the United States of America*, 104 (9): 3651–3656.

DIETLER, M.

1996. Feast and commensal politics in the political economy: Food, power, and status in prehistoric Europe. In: *Food and the Status Quest: An Interdisciplinary Perspective*, edited by W. Weissner and W. Schiefenhovel. Berghahn Books. Oxford. pp. 87–125.

DILLEHAY, Tom Dalton

2003. El colonialismo Inka, el consumo de chicha y los festines desde una perspectiva de banquetes políticos. In: *Identidad y transformación en el Tawantinsuyu y en los Andes Coloniales. Perspectivas Arqueológicas y Etnohistóricas. Segunda parte*, edited by P. Kaulicke, G. Urton, and I. Farrington. *Boletín de Arqueología PUCP*, 7: 355–363.

(editor) 2011. *From Foraging to Farming in the Andes: New Perspectives on Food Production and Social Organization.* Cambridge University Press. Cambridge.

DILLEHAY, Tom Dalton, Herbert H. ELING, and Jack ROSSEN

2005. Preceramic irrigation canals in the Peruvian Andes. *Proceedings of the National Academy of Sciences of the United States of America*, 102 (47): 17241–17244.

DILLEHAY, Tom Dalton, Patricia J. NETHERLY, and Jack ROSSEN

1989. Middle preceramic public and residential sites on the florestal slope of the western Andes, northern Peru. *American Antiquity*, 54 (4): 733–759.

2011. Preceramic mounds and hillside villages. In: *From Foraging to Farming in the Andes: New Perspectives on Food Production and Social Organization*, edited by Tom D. Dillehay. Cambridge University Press. Cambridge. pp. 135–161.

DILLEHAY, Tom Dalton, Jack ROSSEN, and Patricia J. NETHERLY

1997. The Nanchoc tradition: The beginnings of Andean civilization. *American Scientist*, 85 (1): 46–55.

DILLEHAY, Tom Dalton, Jack ROSSEN, and David E. WILLIAMS

2007. Preceramic adoption of peanut, squash, and cotton in northern Peru. *Science*, 316 (5833): 1890–1893.

DOBRIZHOFFER, M.

1822. *An Account of the Abipones, an Equestrian People of Paraguay*. Vol. 1. John Murray. London.

DODOENS (DODONAEUS), Rembert

1578. *A niewe herball; or Histories of Plantes: Wherein Is Contayned the Whole Discourse and Perfect Description of All Sortes of Herbes and Plantes: Their Divers & Sundry Kindes ... and That Not Onely of Those Whiche Are Here Growing in This Our Countrie of Englande, but of All Others Also of Forrayne Realmes Commonly Used in Physicke. First Set Foorth in the Doutche or Almaigne Tongue. And Nowe First Translated Out French into English by Henry Lyte*. Gerard Dewes. London.

DOEBLEY, John F.

1983a. The maize × teosinte inflorescence: A numerical taxonomical study. *Annals of the Missouri Botanical Garden*, 70: 32–70.

1983b. The taxonomy and evolution of *Tripsacum* and teosinte, the closest relatives of maize. In: *Proceedings International Maize Virus Disease Colloquium and Workshop*, edited by D. T. Gordon, J. K. Knope, L. R. Nault, and R. M. Ritter. Ohio Agricultural Research Development Center. Wooster. pp. 15–28.

1984. Maize introgression into teosinte: A reappraisal. *Annals of the Missouri Botanical Garden*, 71: 1100–1113.

1990. Molecular evidence for the evolution of maize. *Economic Botany*, 44 (3) (Suppl.): 6–27. New Perspectives on the Origin and Evolution of New World Domesticated Plants, edited by Peter K. Brettin. New York.

1994. Morphology, molecules, and maize. In: *Corn and Culture in the Prehistoric New World*, edited by Sissel Johannesen and Christine A. Hastorf. Westview Press. Boulder. pp. 101–112.

2004. The genetics of maize evolution. *Annual Review of Genetics*, 38: 37–59.

2006. Unfallen grains: How ancient farmers turned weeds into crops. *Science*, 312 (5778): 1318–1319.

DOEBLEY, John F., Major M. GOODMAN, and C.W. STUBER

1984. Isoenzymatic variation in *Zea* (Gramineae). *Systematic Botany*, 9: 203–218.

1987. Patterns of isozyme variation between maize and Mexican annual teosinte. *Economic Botany*, 41: 234–246.

DOEBLEY, John F., and Hugh H. ILTIS
1980. Taxonomy of *Zea* (Gramineae). I. Subgeneric classification with key to taxa. *American Journal of Botany*, 67: 986–993.

DOEBLEY, John F., W. RENFROE, and A. BLANTON
1987. Restriction site variation in the *Zea* chloroplast genome. *Genetics*, 117: 139–147.

DOEBLEY, John F., Adrian STEC, and L. HUBBARD
1997. The evolution of apical dominance in maize. *Nature*, 386 (6624): 485–488.

DOEBLEY, John F., Adrian STEC, Jonathan WENDEL, and Marlin EDWARDS
1990. Genetic and morphological analysis of a maize-teosinte F2 population. Implications for the origin of maize. *Proceedings of the National Academy of Sciences of the United States of America*, 87 (24): 9888–9892.

DOOLITTLE, William E., and C. D. FREDERICK
1991. Phytoliths as indicators of prehistoric maize (*Zea mays* subsp. *mays*, Poaceae) cultivation. *Plant Systematics and Evolution*, 177 (3–4): 175–184.

DOOLITTLE, William E., and Jonathan B. MABRY
2006. Environmental mosaics, agricultural diversity, and the evolutionary adoption of maize in the American Southwest. In: *Histories of Maize: Multidisciplinary Approaches to the Prehistory, Linguistics, Biogeography, Domestication and Evolution of Maize*, edited by John Staller, Robert Tykot, and Bruce Benz. Academic Press, Elsevier. San Diego, London. pp. 109–121.

DORWEILER, Jane E., and John F. DOEBLEY
1997. Developmental analysis of teosinte glume architecture. 1: A key locus in the evolution of maize (Poaceae). *American Journal of Botany*, 84 (10): 1313–1322.

DORWEILER, Jane E., Adrian STEC, Jerry KERMICLE, and John F. DOEBLEY
1993. Teosinte glume architecture. 1: A genetic locus controlling a key step in maize evolution. *Science*, 262 (5131): 233–235.

DOWSWELL, C. D., R. L. PALIWAL, and R. P. CANTRELL
1996. *Maize in the Third World*. Westview Press. Boulder.

DULL, Robert A.
2006. The maize revolution. A view from El Salvador. In: *Histories of Maize: Multidisciplinary Approaches to the Prehistory, Linguistics, Biogeography, Domestication and Evolution of Maize*, edited by John Staller, Robert Tykot, and Bruce Benz. Academic Press, Elsevier. San Diego, London. pp. 357–365.

DUNN, M. E.
1983. Phytolith analysis in archaeology. *Midcontinental Journal of Archaeology*, 8 (2): 287–297.

DUNN, Oliver, and James E. KELLEY Jr. (editors)
1988. *The Diario of Christopher Columbus's First Voyage to America 1492–1493. Abstracted by Bartolomé de las Casas.* Described and translated into English, with notes and concordance of the Spanish by Dunn and Kelley. University of Oklahoma Press. Norman.

ELIOT, Charles
1965. *Turkey in Europe*. Barnes and Noble. New York.

ENGEL, Frédéric A.

1958. Algunos datos con referencia a los sitios precerámicos de la costa peruana. *Arqueológicas*, 3. Publicaciones del Instituto de Investigaciones Antropológicas. Museo Nacional de Antropología y Arqueología. Pueblo Libre. Lima.

1970a. La Grotte du Mégathérium à Chilca et les Écologies du Haut-Holocène Péruvien. In: *Échanges et communications*, Mélanges offerts à Claude Lévi-Strauss à l'occasion de son 60ème anniversaire. Mouton. La Haye. pp. 413–436.

1970b. Explorations of the Chilca Canyon, Peru. *Current Anthropology*, 11 (1): 55–58.

1970c. *Las lomas de Iguanil y el Complejo de Haldas.* Universidad Nacional Agraria. Lima.

1973. New facts about pre-Columbian life in the Andean lomas. *Current Anthropology*, 14 (3): 271–280.

ERICSON, Jonathan E., Michael WEST, Charles H. SULLIVAN, and Harold W. KRUEGER

1989. The development of maize agriculture in the Viru Valley, Peru. *The Chemistry of Prehistoric Human Bone*, edited by T. D. Price. Cambridge University Press. Cambridge. pp. 68–104.

ESTETE, Miguel de (Anonymous?)

(1535?) 1968. *Noticia del Perú.* Biblioteca Peruana, Primera Serie, Tomo I. Editores Técnicos Asociados, S.A. Lima. pp. 345–402.

ESTETE, Miguel de

(1534) 1968. La relación del viaje que hizo el señor capitán Hernando Pizarro por mandado del señor Gobernador su hermano, desde el pueblo de Caxamalca a Parcama, y de allí a Jauja. In: *Verdadera Relación de la Conquista del Perú y provincia del Cuzco, llamada la Nueva Castilla.* Francisco de Jerez. Biblioteca Peruana, Primera Serie. Tomo I. Editores Técnicos Asociados, S.A. Lima. pp. 242–257.

EUBANKS, Mary Wilkes

1995. A cross between two maize relatives: *Tripsacum dactyloides* and *Zea diploperennis* (Poaceae). *Economic Botany*, 49 (2): 172–182.

1997a. Molecular analysis of crosses between *Tripsacum dactyloides* and *Zea diploperennis* (Poaceae). *Theoretical and Applied Genetics*, 94: 707–712.

1997b. Reevaluation of the identification of ancient maize pollen from Alabama. *American Antiquity*, 62 (1): 139–145.

1999a. Comparative analysis of the genomes of *Zea* and *Tripsacum. Maize Genetics Cooperation Newsletter*, 73: 30–32.

1999b. *Corn in Clay: Maize Paleoethnobotany in Pre-Columbian Art.* University Press of Florida. Gainesville.

2001a. The origin of maize: Evidence for *Tripsacum* ancestry. *Plant Breeding Review*, 20: 15–61.

2001b. The mysterious origin of maize. *Economic Botany*, 55 (4): 492–514.

2001c. An interdisciplinary perspective on the origin of maize. *Latin American Antiquity*, 12 (1): 91–98.

2003. Comments. *Current Anthropology*, 44 (5): 691–692.

EUBANKS DUNN, Mary

1979. Ceramic depictions of maize: A basis for classification of prehistoric races. *American Antiquity*, 44 (4): 757–774.

EYRE-WALKER, Adam, Rebecca L. GANT, Holly HILTON, Dawn L. FELDMAN, and Bernedon S. GAUT

1998. Investigations of the bottleneck leading to the domestication of maize. *Proceedings of the National Academy of Sciences of the United States of America*, 95 (8): 4441–4446.

FARFÁN, J. M. B.

1941. La clave del lenguaje quechua del Cuzco. *Revista del Museo Nacional*, 10 (2): 215–239.

FARNSWORTH, Paul, James E. BRADY, Michael J. DeNIRO, and Richard S. MacNEISH

1985. A re-evaluation of the isotopic and archaeological reconstruction of diet in the Tehuacan Valley. *American Antiquity*, 50 (1): 102–116.

FEARN, Miriam L., and Kam-biu LIU

1995. Maize pollen of 3500 B.P. from Southern Alabama. *American Antiquity*, 60 (1): 109–117.

FEDERMAN, Nicolás de

(1579) 1958. *Historia Indiana*. Translated from the German by Juan Friede. Aro Artes Gráficas. Madrid.

FEDOROFF, Nina V.

2003. Prehistoric GM corn. *Science*, 302 (5648): 1158–1159.

FELDMAN, Robert Alan

1980. Aspero, Peru: Architecture subsistence economy and other artifacts of a Preceramic maritime chiefdom. A thesis presented to Department of Anthropology in partial fulfillment of the requirements for the degree of doctor of philosophy in the subject of anthropology. Harvard University. Cambridge.

1981. Two additional cases of lumbar malformation from the Peruvian coastal preceramic. *Current Anthropology*, 22 (3): 286–287.

1992. Preceramic architectural and subsistence traditions. *Andean Past*, 3: 67–86.

FERNÁNDEZ DE OVIEDO Y VALDÉZ, Gonzalo

(1535) 1959. *Historia General y Natural de las Indias*. I. Biblioteca de Autores Españoles desde la formación del lenguaje hasta nuestros días. Tomo CXVII. Ediciones Atlas. Madrid.

FERNÁNDEZ DISTEL, Alicia A.

1974. Excavaciones arqueológicas en las cuevas de Huachichocana, dep. de Tumbaya, prov. de Jujuy, Argentina. *Relaciones*, VIII (Nueva Serie): 101–127.

1975. Restos vegetales de etapas arcaicas en yacimientos del N.O. de la República Argentina (Pcia. De Jujuy). *Etnia*, 22 (artículo 86): 11–24.

1980. Las fechas radiocarbónicas en la arqueología de la provincia de Jujuy. Fechas radiocarbónicas de la Cueva CH III de Huachichocana, Tiniyan e Inca-Cueva. In: *Argentina. Radiocarbono en Arqueología*. Museo de Historica Sam Rafael. Tomo I, Nos. 4–5. Mendoza. pp. 89–100.

1986. Las cuevas de Huachichocana, su posición dentro del precerámico con agricultura incipiente del Noroeste Argentino. In: *Beiträge zur Allgemeinen und Vergleichenden Archäologie*, Vol. 8. Verlag Philipp von Zabern. Mainz. pp. 353–430.

1989. Una nueva cueva con maíz acerámico en el N.O. argentino: León Huasi I, Excavación. *Comunicaciones Científicas. Dirección Provincial de Antropología e Historia*, 1 (1): 4–17.

FICCARELLI, G., M. COLTORTI, M. MORENO-ESPINOSA, P. L. PIERUCCINI, L. ROOK, and D. TORRE

2003. A model for the Holocene extinction of the mamal megafauna in Ecuador. *Journal of South American Earth Sciences*, 15: 835–845.

FINAN, John J.

1950. *Maize in the Great Herbals*. Chronica Botanica Company. Waltham.

FINUCANE, Brian Clifton

2007. Mummies, maize, and manure: Multi-tissue stable isotope analysis of late prehistoric human remains from the Ayacucho Valley, Peru. *Journal of Archaeological Science*, 34: 2115–2124.

2009. Maize and sociopolitical complexity in the Ayacucho Valley, Peru. *Current Anthropology*, 50 (4): 535–545.

FINUCANE, Brian Clifton, Patricia MAITA AGURTO, and William H. ISBELL

2006. Human and animal diet at Conchopata, Peru: Stable isotope evidence for maize agriculture and animal management practices during the Middle Horizon. *Journal of Archaeological Science*, 33: 1766–1776.

FLANNERY, Kent V.

n.d. [1970]. Preliminary archaeological investigations in the valley of Oaxaca, Mexico 1966–1969. Report to the National Science Foundation and I.N.A.H.

(Note: This report was not published. Joyce Marcus, personal communication, 11 October 2006.)

1973. The origins of agriculture. *Annual Review of Anthropology*. Vol. II, edited by B. J. Siegel, A. R. Beals, and S. B. Tyler. Annual Reviews. Palo Alto. pp. 271–310.

1985. Los orígenes de la agricultura en México: las teorías y la evidencia. In: *Historia de la agricultura. Época prehispánica – siglo XVI*, edited by Teresa Rojas Rabiela and William T. Sanders. Colección Biblioteca del Instituto Nacional de Antropología e Historia. Mexico City. pp. 237–266.

(editor) 1986a. *Guilá Naquitz: Archaic Foraging and Early Agriculture in Oaxaca, México*. Academic Press. Orlando.

1986b. The research problem. In: *Guilá Naquitz. Archaic Foraging and Early Agriculture in Oaxaca, México*, edited by Kent V. Flannery. Academic Press. Orlando. pp. 3–18.

1997. In defense of the Tehuacán Project. *Current Anthropology*, 38 (4): 660–662.

FORD, Richard I.

1968. An ecological analysis involving the population of San Juan Pueblo, New Mexico. Ph.D.dissertation. University of Michigan. Ann Arbor.

1994. Corn is our mother. In: *Corn and Culture in the Prehistoric New World*, edited by Sissel Johannesen and Christine A. Hastorf. Westview Press. Boulder. pp. 513–525.

FORNASIN, Alessio

1999. Diffusione del mais e alimentazione nelle campagne del Seicento. In: *Vivere in Friuli. Saggi di demografía storica*. A cura di M. Breschi. Forum. Udine. pp. 21–42.

FREITAS, Fabio Oliveira, Gerhard BENDEL, Robin G. ALLABY, and Terence A. BROWM

2003. DNA from primitive maize landraces and archaeological remains: Implications for the domestication of maize and its expansion into South America. *Journal of Archaeological Science*, 30 (7): 901–908.

FRITZ, Gayle J.

1990. Multiple pathways to farming in precontact eastern North America. *Journal of World Prehistory*, 4: 387–435.

1993. Early and Middle Woodland period paleoethnobotany. In: *Foraging and Farming in the Eastern Woodland*, edited by C. Margarets Searry. University Press of Florida. Gainesville. pp. 39–56.

1994a. Are the first American farmers getting younger? *Current Anthropology*, 35 (3): 305–308.

1994b. Reply. *Current Anthropology*, 35 (5): 639–643.

FUKUNAGA, K., J. HILL, Yves VIGOROUX, Y. MATSUOKA, G. J. SÁNCHEZ, K. LIU, E. S. BUKLER, and John F. DOEBLEY

2005. Genetic diversity and population structure of teosinte. *Genetics*, 169: 2241–2254.

FUNG PINEDA, Rosa

1969. Las Aldas: su ubicación dentro del proceso histórico del Perú antiguo. *Dédalo*, 9–10: 5–207.

2004. *Quehaceres de la Arqueología Peruana. Compilación de escritos.* Museo de Arqueología y Antropología. Centro Cultural de San Marcos. Universidad Nacional Mayor de San Marcos. Lima.

GALINAT, Walton C.

1954. Corn grass. I. Corn grass as a possible prototype of a false progeny of maize. *American Naturalist*, 88 (839): 101–104.

1957. The effects of certain genes on the outer pistillate glume of maize. *Botanical Museum Leaflets*, 18 (2): 57–76.

1963. Form and function of plant structures in the American Maydae and their significance for breeding. *Economic Botany*, 17: 51–59.

1964. *Tripsacum* a possible amphidiploid of *Manisuris* and wild maize. *Maize Genetic Cooperation Newsletter*, 38: 50.

1965. The evolution of corn and culture in North America. *Economic Botany*, 19: 350–357.

(Note: This was published with the same title in 1971 in *Prehistoric Agriculture*, edited by Stuart Struever. American Sourcebook in Anthropology. New York. pp. 534–543.)

1970. The cupule and its role in the origin and evolution of maize. *University of Massachusetts Agricultural Experiment Station Bulletin*, 585: 1–20.

1971a. The origin of maize, edited by H. Roman. *Annual Review of Genetics*, 5: 447–478.

1971b. Identificación de muestras arqueológicas de maíz de San Pedro Viejo (Apéndice 1). In: *Nuevos enfoques de la teoría arqueológica aplicada al Norte Chico*. Mario Rivera. Separata de Actas del VI Congreso de Arqueología Chilena. Universidad de Chile. Departamento de Ciencias Antropológicas y Arqueología. Sociedad Chilena de Arqueología. Santiago. pp. 306–307.

1972. Common ancestry of the primitive races of maize indigenous to the Ayacucho area in Peru. *Maize Genetic Cooperation Newsletter*, 46: 107–108.

1972–1973. Identificación de muestras arqueológicas de maíz de San Pedro Viejo. In: *VI Congreso de Arqueología Chilena*, Sociedad Chilena de Arqueología. Universidad de Chile. pp. 306–307.

1974a. Intergenomic mapping of maize, teosinte and *Tripsacum*. *Evolution*, 27: 644–655.

1974b. The domestication and genetic erosion of maize. *Economic Botany*, 28: 31–37.

1975a. The evolutionary emergence of maize. *Bulletin of the Torrey Botanical Club*, 102 (6): 313–324.

Ms. 1975b. Identificación del maíz de Tiliviche 1.B.

1977. The origin of corn. In: *Corn and Corn Improvement*, edited by G. F. Sprague. Agronomy 18. American Society of Agronomy. Madison. pp. 1–47.

1978. The inheritance of some traits essential to maize and teosinte. In: *Maize Breeding and Genetics*, edited by D. B. Walden. John Wiley & Sons. New York. pp. 93–111.

1981. Corn's origin by means of domestication. Paper presented at the XIII International Botanical Congress. Sydney.

1983. The origin of maize as shown by key morphological traits of its ancestor, teosinte. *Maydica*, 28: 121–138.

1985a. The missing links between teosinte and maize: A review. *Maydica*, 30: 137–160.

1985b. Domestication and diffusion of maize. In: *Prehistoric Food Production in North America*, edited by Richard I. Ford. Anthropological Papers, N° 75, Museum of Anthropology, University of Michigan. Ann Arbor. pp. 245–278.

1988a. The origin of corn. In: *Corn and Corn Improvement*, edited by G. F. Sprague and J. W. Dudley. 3rd ed. Agronomy Monographs 18, American Society of Agronomy, Inc., Crop Science Society of America, Inc. Soil Science Society of America, Inc. Madison. pp. 1–31.

1988b. The origin of maiz de ocho. *American Anthropologist*, 90 (3): 682–683.

1988c. Palomero Toluqueño and certain Andean maize carry the short rachillae and reduced cupule traits probably descended from an independent domestication of teosinte. *Maize Genetic Cooperation Newsletter*, 62: 111.

1992. Corn, Columbus and culture. *Perspectives in Biology and Medicine*, 36: 1–12.

2001a. A scenario for one of the teosinte origins of maize. *Maize Genetics Cooperation Newletter*, 75: 77–78.

2001b. A reconstruction of a possible role of crucial transformation of wild teosinte into first maize. *Economic Botany*, 55 (4): 570–574.

GALINAT, Walton C., and J. H. GUNNERSON
1963. Spread of eight-rowed maize from the prehistoric southwest. *Botanical Museum Leaflets*, 20 (5): 117–160.
GALINAT, Walton C., Paul C. MANGELSDORF, and L. PIERSON
1956. Estimates of teosinte introgression in archaeological maize. *Botanical Museum Leaflets*, 17 (4): 101–124.
GARCÍA, A.
1992. Hacia un ordenamiento preliminar de las ocupaciones prehistóricas agrícolas precerámicas y agroalfareras en el NO de Mendoza. *Revista de Estudios Regionales*, 10: 7–34.
GARCÍA COOK, Ángel
1974. El origen del sedentarismo en el área de Ayacucho, Perú. *Boletín del Instituto Nacional de Antropología e Historia, Época II*, 11: 15–30.
GARCÍA COOK, Ángel, and Richard S. MacNEISH
1981. The stratigraphy of Puente, Ac 158. In: *Prehistory of the Ayacucho Basin, Vol. II. Excavations and Chronology*, R. S. MacNeish, A. García Cook, L. G. Lumbreras, R. K. Vierra, and A. Nelken-Terner. University of Michigan Press. Ann Arbor. pp. 80–112.
GARCILASO DE LA VEGA, Inca
(1609) 1959a. *Comentarios Reales de los Incas.* Tomo I. Universidad Nacional Mayor de San Marcos. Patronato del Libro Universitario. Lima.
1959b. *Comentarios Reales de los Incas.* Tomo II. Universidad Nacional Mayor de San Marcos. Patronato del Libro Universitario. Lima.
1959c. *Comentarios Reales de los Incas.* Tomo III. *Universidad Nacional Mayor de San Marcos.* Patronato del Libro Universitario. Lima.
1966. *Royal Commentaries of the Incas and General History of Peru [1604].* Part One. Translated by H. V. Livermore. University of Texas Press. Austin.
GAUT, B. S., and M. T. CLEGG
1993. Molecular evolution of the *Adh1* locus in genus *Zea*. *Proceedings of the National Academy of Sciences of the United States of America*, 90 (11): 5095–5099.
GAUT, B. S., and John F. DOEBLEY
1997. DNA sequence evidence for the segmental allotetraploid origin of maize. *Proceedings of the National Academy of Sciences of the United States of America*, 94 (13): 6809–6814.
GAY, Jean Pierre
1987. Le maïs. *La Recherche*, 187: 458–466.
GIL, Adolfo F.
2003. *Zea mays* on the South American periphery: Chronology and dietary importance. *Current Anthropology*, 44 (2): 295–300.
GIL, Adolfo F., Gustavo A. NEME, Robert H. TYKOT, Paula NOVELLINO, V. CORTEGOSO, and Víctor DURÁN
2009. Stable isotope and maize consumption in central western Argentina. *International Journal of Osteoarchaeology*, 19: 215–236.
GIL, Adolfo F., Robert H. TYKOT, Gustavo A. NEME, and Nicole R. SCHELNUT
2006. Maize on the frontier: Isotopic and macrobotanical data from central-western Argentina. In: *Histories of Maize: Multidisciplinary Approaches to the Prehistory,*

Linguistics, Biogeography, Domestication and Evolution of Maize, edited by John Staller, Robert Tykot, and Bruce Benz. Academic Press, Elsevier. San Diego, London. pp. 199–214.

GIL, Juan

1984. Introducción. In: *Cristóbal Colón. Textos y documentos.* Prólogo y notas de Consuelo Varela. Alianza Universidad (Editorial). Madrid. pp. IX–LXVIII.

GILLIN, John

1947. *Moche: A Peruvian Coastal Community.* Smithsonian Institution. Institute of Social Anthropology. Publication N° 3. Washington, D.C.

GOETHE, Johann Wolfgang von

1962. *Italian Journey, 1786–1788.* Translated by W. H. Anden and Elizabeth Mayer. Pantheon Press. New York.

GOLOUBINOFF, Pierre, Svante PÄÄBO, and Allan C. WILSON

1993. Evolution of maize inferred from sequence diversity of an *Adh2* gene segment from archaeological specimens. *Proceedings of the National Academy of Sciences of the United States of America*, 90 (5): 1997–2001.

1994. Molecular characterization of ancient maize: Potentials and pitfalls. In: *Corn and Culture in the Prehistoric New World*, edited by Sissel Johannessen and Christine Hastorf. Westview Press. Boulder. pp. 113–125.

GÓMEZ HUAMÁN, Nilo

1966. Importancia social de la chicha como bebida popular en Huamanga. *Wamani*, 1 (1): 33–57.

GONÇALEZ (GONZÁLEZ) HOLGUÍN, Diego

(1608) 1989. *Vocabulario de la Lengua general de todo el Perú llamada Lengua Quechua o del Inca.* Universidad Nacional Mayor de San Marcos. Editorial de la Universidad. Lima.

GOODMAN, Major M.

1976. Maize: *Zea mays* (Gramineae, Maydeae). In: *Evolution of Crop Plants*, edited by N. W. Simmond. Longman. London. pp. 128–136.

1978. Historia e origen do milho. In: *Melhoramento e Produção do Milho no Brasil*, edited by E. Paterniani. Fundação Cargil. São Paulo. pp. 30–65.

1988. The history and evolution of maize. *CRC Critical Review in Plant Sciences*, 7 (3): 197–220.

GOODMAN, Major M., and Robert McKelvy BIRD

1977. The races of maize. IV: Tentative grouping of 219 Latin American races. *Economic Botany*, 31 (2): 204–221.

GOODMAN, Major M., and W. L. BROWN

1988. Races of corn. In: *Corn and Corn Improvement*, edited by G. F. Sprague and J. W. Dubley. 3rd ed. Agronomy Monograph 18, American Society of Agronomy, Inc., Crop Science Society of America, Inc., Soil Science Society of America, Inc. Madison. pp. 33–79.

GOODMAN, Major M., and C. W. STUBER

1980. Genetic identification of lines and crosses using isoenzyme electrophoresis. *35th Annual Corn and Sorghum International Research Conference. Proceedings*, 35: 10–31.

GOTTLIEB, L. D.

1984. Genetics and morphological evolution in plants. *The American Naturalist*, 123: 681–709.

GOULD, Stephen Jay

1984. A short way to corn. *Natural History*, 93 (3): 12–20.

GOYHENECHE, E.

1966. L'Introduction du maïs en Eskual Herria et en Europe (1523). In: *Homenaje a don José Miguel de Barandiarán*. Tomo II. Diputación de Vizcaya. Bilbao. pp. 109–120.

GRANER, E. A., and G. ADDISON

1944. Meiose em *Tripsacum australe* Cutler e Anderson. *Anais da Escola Superior de Agricultura "Luiz de Queiroz"* (Universidade de São Paulo), 9: 213–224.

GRANT, U. J., W. H. HATHEWAY, D. H. TIMOTHY, C. CASSALETT, and L. M. ROBERTS

1963. *Races of Maize in Venezuela*. Publication 1136, National Academy of Sciences, National Research Council. Washington, D.C.

GREENBLAT, Irwin M.

1968. A possible selective advantage of plant color at high altitudes. *Maize Genetics Cooperation Newsletter*, 42: 144–145.

GREMILLION, Kristen J.

1996. Diffusion and adoption of crops in evolutionary perspective. *Journal of Anthropological Archaeology*, 15: 183–204.

2003. Comments. *Current Anthropology*, 44 (5): 692.

GRIEDER, Terence

1988a. Radiocarbon measurements. In: *La Galgada, Peru: A Preceramic Culture in Transition*, Terence Grieder, Alberto Bueno Mendoza, C. Earle Smith Jr., and Robert M. Malina. University of Texas Press. Austin. pp. 68–72.

1988b. Burial patterns and offerings. In: *La Galgada, Peru: A Preceramic Culture in Transition*, Terence Grieder, Alberto Bueno Mendoza, C. Earle Smith Jr., and Robert M. Malina. University of Texas Press. Austin. pp. 73–102.

GRIEDER, Terence, and Alberto BUENO MENDOZA

1981. La Galgada: Peru before pottery. *Archaeology*, 34 (2): 44–51.

GROBMAN T., Alexander

1967. *Tripsacum* in Peru. *Botanical Museum Leaflets*, 21 (9): 285–287.

1974. Conceptos actuales sobre evolución del maíz. *Informativo del Maíz*, 3: [3–4]. (The pages are not numbered.)

Ms. 1980. Informe sobre el maíz precerámico de Áspero.

(Note: The report was sent to Robert Feldman. Lima, August 1980. A copy is in the possession of Duccio Bonavia.)

1982. Maíz (*Zea mays*). In: *Precerámico peruano. Los Gavilanes. Mar, desierto y oasis en la historia del hombre*, D. Bonavia, with the collaboration of R. Castro de la Mata, F. Caycho Quispe, A. Grobman, L. Kaplan, C. A. Morán Val, R. Patrucco, M. Peña, V. Popper, E. J. Reitz, S. G. Stephens, R. Tello, and E. S. Wing. Corporación Financiera de Desarrollo S.A. COFIDE, Instituto Arqueológico Alemán, Comisión de Arqueología General y Comparada. Lima. pp. 157–180.

Ms. 1992. Nota referente a muestras de maíz de Puémape.

(Note: A copy of the report is in the possession of Duccio Bonavia.)

2004. El origen del maíz. In: *Cincuenta Años del Programa Cooperativo de Investigaciones en Maíz (PCIM). Logros y perspectivas*, edited by Wilfredo Salhuana M., Américo Valdéz M., Federico Scheuch H., and José Davelouis M. Universidad Nacional Agraria La Molina. Lima. pp. 426–470 (corrigenda in VII–XIII).

GROBMAN T., Alexander, and Duccio BONAVIA

1978. Preceramic maize on the north-central coast of Peru. *Nature*, 276 (5686): 386–387.

1979–1980. Maíz precerámico en la costa nor-central peruana: análisis preliminar. *Informativo del maíz.* Número extraordinario de investigación. Vol. III. Universidad Nacional Agraria, Programa Cooperativo de Investigación en maíz. Lima. pp. 134–135.

GROBMAN A., D. BONAVIA, T. D. DILLEHAY, D. R. PIPERNO, J. IRIARTE, and I. HOLST

2012. Preceramic maize from Paredones and Huaca Prieta, Peru. *Proceedings of the National Academy of Sciences of the United States of America*, 109 (5): 1755–1759.

GROBMAN T., Alexander, Duccio BONAVIA, David H. KELLEY, Paul C. MANGELSDORF, and Julián CÁMARA-HERNÁNDEZ

1977. Study of preceramic maize from Huarmey, North Central Coast of Peru. *Botanical Museum Leaflets*, 25 (8): 221–242.

GROBMAN T., Alexander, and Wilfredo SALHUANA, in collaboration with Paul C. MANGELSDORF

1956. *Races of Maize in Peru.* Escuela Nacional de Agricultura. Genetic Cooperation News Letter 30. Lima.

GROBMAN T., Alexander, Wilfredo SALHUANA, and Ricardo SEVILLA, in collaboration with Paul C. MANGELSDORF

1961. *Races of Maize in Peru.* National Academy of Sciences. National Research Council. Publication 915. Washington, D.C.

GROHNE, Udelgard

1957. The importance of phase-contrast microscopy in pollen analysis, demonstrated on corn-type Gramineae pollen. *Photographische Forschung*, 7 (8): 237–249.

GUAMAN POMA DE AYALA, Felipe

1936. *Nueva Corónica y Buen Gobierno (Codex péruvien illustré).* Université de Paris. Travaux et Mémoires de l'Institut d'Ethnologie. XXIII. Institut d'Ethnologie. Paris.

GUMERMAN, George, IV

1994. Corn for the dead: The significance of *Zea mays* in Moche burial offerings. In: *Corn and Culture in the Prehistoric New World*, edited by Sissel Johannessen and Christine A. Hastorf. Westview Press. Boulder. pp. 399–426.

GUMILLA, J.

(1791) 1963. *El Orinoco ilustrado y defendido.* Biblioteca de la Academia Nacional de la Historia. Serie: Fuentes para la historia Colonial de Venezuela, N°68. Caracas.

GUTIÉRREZ DE SANTA CLARA, Pedro

(1904–1929) 1963. *Quinquenarios o Historia de las Guerras Civiles en el Perú.* Biblioteca de Autores Españoles desde la formación del lenguaje hasta nuestros días. Crónicas del Perú. III. Ediciones Atlas. Madrid.

HALPERN, Joel Martín

1958. *A Serbian Village*. Columbia University Press. New York.

HARLAN, Jack R.

1951. Anatomy of gene centers. *American Naturalist*, 85: 97–103.

1956. Distribution and utilization of natural variability in cultivated plants. In: *Genetics in Plant Breeding*. Brookhaven Symposia in Biology, N° 9. Brookhaven National Laboratory. New York. pp. 191–206.

1992. *Crops and Man*. American Society of Agronomy, Inc. Crop Science Society of America, Inc. Madison.

1995. *The Living Fields: Our Agricultural Heritage*. Cambridge University Press. Cambridge.

HARLAN, Jack R., and J. M. J. de WET

1971. Toward a rational classification of cultivated plants. *Taxon*, 19: 509–517.

1972. Origin of maize: The tripartite hypothesis. *Euphytica*, 21: 271–279.

1977. Pathways of genetic transfer from *Tripsacum* to *Zea mays*. *Proceedings of the National Academy of Sciences of the United States of America*, 74 (8): 3494–3497.

HARMS, Herman von

1922. Ubersicht der bisher in altperuanishen Graben gefundenen Pflanzenreste. *Festschrift Eduard Seler*. Strecker und Schröeder. Stuttgart. pp. 157–186.

HAROWITZ, S., and A. H. MARCHIONI

1940. Herencia de la resistencia a la langosta en el maíz "Amargo." *Anales del Instituto de Fitotecnia de Santa Catalina*, 2: 27–52.

HARSHBERGER, J. W.

1893. Maize, a botanical and economic study. *Contributions of the Botanical Laboratory at the University of Pennsylvania*, 1: 75–202.

1896. Fertile crosses of teosinthe and maize. *Garden and Forest*, 9 (462): 522–523.

1899. Cruzamiento Fecundo del Teosinte y del Maize. *Boletín de la Sociedad Agraria Mexicana*, 23: 263–267.

HART, John P., Hetty Jo BRUMBACH, and Robert LUSTECK

2007. Extending the phytoliths evidence for early maize (*Zea mays* ssp. *mays*) and squash (*Cucurbita* sp.) in central New York. *American Antiquity*, 72 (3): 563–583.

HART, John P., Robert G. THOMPSON, and Hetty Jo BRUMBACH

2003. Phytolith evidence for early maize (*Zea mays*) in the northern Finger Lake region of New York. *American Antiquity*, 68 (4): 619–640.

HASTORF, Christine A.

1985. (Review) "Precerámico peruano: Los Gavilanes. Mar, desierto y oasis en la historia del hombre." Duccio Bonavia. Corporación Financiera de Desarrollo S.A. COFIDE e Instituto Arqueológico Alemán. Lima, Perú, 1982. xxiii + 512 pp. Illus., biblio., index. Cloth. *American Antiquity*, 50 (4): 927–929.

1999. Cultural implications of crop introduction in Andean prehistory. In: *The Prehistory of Food: Appetites for Change*, edited by Chris Gosden and Jon Hather. Routledge. London and New York. pp. 35–58.

2009. Rio Balsas most likely region for maize domestication. *Proceedings of the National Academy of Sciences of the United States of America*, 106 (13): 4957–4958.

HASTORF, Christine A., and Sissel JOHANNESSEN

1993. Pre-Hispanic political change and the role of maize in the central Andes of Peru. *American Anthropologist*, 95: 115–138.

1994. Becoming corn eaters in prehistoric America. In: *Corn and Culture in the Prehistoric New World*, edited by Sissel Johannessen and Christine A. Hastorf. Westview Press. Boulder. pp. 427–443.

HAUDRICOURT, André G., and Louis HÉDIN

1987. *L'Homme et les plantes cultivées*. Éditions A.M. Métailié. Paris.

HAURY, Emil W.

1962. The greater American Southwest. In: *Courses Toward Urban Life*, edited by Robert J. Braydwood and Gordon R. Willey. Aldine. Chicago. pp. 106–131.

HEISER, Charles B., Jr.

1965. Cultivated plants and cultural diffusion in nuclear America. *American Anthropologist*, 67 (4): 930–949.

HELBAEK, Hans

1953. *Archeology and Agricultural Botany*. Annual Report. Institute of Archaeology. London.

HERNÁNDEZ, Xolocotzi Efraím, and F. G. ALANÍS

1970. Estudio morfológico de cinco nuevas razas de maíz de la Sierra Madre Occidental de México: Implicaciones filogenéticas y fitogeográficas. *Agrociencia*, 5: 3–30.

HERRERA, L. F., I. CAVELIER, C. RODRIGUEZ, and S. MORA

1992. The technical transformation of an agricultural system in the Colombian Amazon. *World Archaeology*, 24: 98–113.

HILL, Jane H.

2006. The historical linguistics of maize cultivation in Mesoamerica and North America. In: *Histories of Maize: Multidisciplinary Approaches to the Prehistory, Linguistics, Biogeography, Domestication and Evolution of Maize*, edited by John Staller, Robert Tykot, and Bruce Benz. Academic Press, Elsevier. San Diego, London. pp. 631–645.

HILTON, Halley, and Brandon S. GAUT

1998. Speciation and domestication in maize and its wild relatives: Evidence from the *Globuline-1* gene. *Genetic*, 150: 863–872.

HO, P. T.

1956. The introduction of American food plants into China. *American Anthropologist*, 57: 191–201.

HOCQUENGHEM, Anne-Marie, and Susana MONZÓN

1995. *La cocina piurana. Ensayo de antropología de la alimentación*. Tomo 80 de Travaux de l'Institut Français d'Études Andines y N°8 de la serie Miscelánea del Instituto de Estudios Peruanos. CNRS Pics 125, IFEA, IEP. Lima.

HOLDRIDGE, Leslie R.

1967. *Life Zone Ecology*. Rev. ed. Tropical Science Center. San José.

HORKHEIMER, Hans

1958. *La alimentación en el Perú Prehispánico y su interdependencia con la agricultura*. UNESCO, Programa de la Zona Arida Peruana. Lima. (Mimeographed.)

(Note: There are two later editions of this study – 1960 and 1973 – both of which include serious errors. See Bonavia, 1996a: 167, note 2.)

HORN, Sally P.

2006. Pre-Columbian maize agriculture in Costa Rica: Pollen and other evidence from lake swamp sediments. In: *Histories of Maize: Multidisciplinary Approaches to the Prehistory, Linguistics, Biogeography, Domestication and Evolution of Maize*, edited by John Staller, Robert Tykot, and Bruce Benz. Academic Press, Elsevier. San Diego, London. pp. 367–380.

HOROWITZ, S., and A. H. MARCHIONI

1940. Herencia de la resistencia a la langosta en el maíz Amargo. *Anales del Instituto de Fitología*, 2: 27–52.

HUCKELL, Lisa W.

2006. Ancient maize in the American Southwest: What does it look like and what can it tell us? In: *Histories of Maize: Multidisciplinary Approaches to the Prehistory, Linguistics, Biogeography, Domestication and Evolution of Maize*, edited by John Staller, Robert Tykot, and Bruce Benz. Academic Press, Elsevier. San Diego, London. pp. 97–107.

ILTIS, Hugh H.

1969 [1970]. The maize mystique – a reappraisal of the origin of corn. Lecture presented at the University of Illinois, Urbana (1969), at Iowa State University, Ames (1970). (Mimeographed.)

(Note: The lecture was republished in 1985 under the same title in *Contribution to the University of Wisconsin Herbarium*.)

1972. The Taxonomy of *Zea mays* L. (Gramineae). *Phytologia*, 23: 248–249.

1981. The catastrophic transmutation theory (CSTT): From the teosinte tassel spike to the ear of corn. Paper presented at the XIII International Botanical Congress. Sidney.

1983a. The catastrophic sexual transmutation theory (CSTT): From teosinte tassel spike to the ear of corn. *Maize Genetics Cooperation Newsletter*, 57: 81–91.

1983b. From teosinte to maize: The catastrophic sexual transmutation. *Science*, 222 (4626): 886–894.

1984. Letter to the editor, in reply to: The Origin of Maize. *Science*, 225 (4667): 1094, 1096.

1987. Maize evolution and agricultural origins. In: *Grass Systematics and Evolution*, edited by Thomas R. Soderstrom, Khidir W. Hilu, Christopher S. Campbell, and Mary E. Barkworth. An international symposium held at the Smithsonian Institution. Smithsonian Institution Press. Washington, D.C. pp. 195–213.

2000. Homeotic sexual translocation and the origin of maize (*Zea mays*, Poaceae): A new look at an old problem. *Economic Botany*, 54 (1): 7–42.

2006. Origin of polystichy in maize. In: *Histories of Maize: Multidisciplinary Approaches to the Prehistory, Linguistics, Biogeography, Domestication and Evolution of Maize*, edited by John Staller, Robert Tykot, and Bruce Benz. Academic Press, Elsevier. San Diego, London. pp. 21–53.

ILTIS, Hugh H., and John F. DOEBLEY

1980. Taxonomy of *Zea*. II. Subspecific categories in the *Zea mays* complex and a generic synopsis. *American Journal of Botany*, 67: 994–1004.

1984. *Zea* – a biosystematical odyssey. In: *Plant Biosystematics*, edited by W. Grant. Academic Press. Montreal. pp. 587–616.

ILTIS, Hugh H., John F. DOEBLEY, Rafael M. GUZMÁN, and Batia PAZY

1979. *Zea diploperennis* (Gramineae): A new teosinte from Mexico. *Science*, 203 (4376): 186–188.

IRIARTE, José

2003. Assessing the feasibility of identifying maize through the analysis of cross-shaped size and three-dimensional morphology of phytoliths in the grasslands of southeastern South America. *Journal of Archaeological Science*, 30 (9): 1085–1094.

2006. Vegetation and climate change since 14,810 14C yr B.P. in southeastern Uruguay and implications for the rise of early Formative societies. *Quaternary Research*, 65 (1): 20–32.

IRIARTE, José, Irene HOLST, Óscar MAROZZI, Claudia LISTOPAD, Eduardo ALONSO, Andrés RINDERKNECHT, and Juan MONTAÑA

2004. Evidence for cultivar adoption and emerging complexity during the mid-Holocene in the La Plata basin. *Nature*, 432 (7017): 614–617.

IRWIN, H., and Elso S. BARGHOORN

1965. Identification of the pollen of maize, teosinte, and *Tripsacum* by phase contrast microscopy. *Botanical Museum Leaflets*, 21 (2): 37–56.

JAENICKE-DEPRÉS, Viviane R., Edward. S. BUCKLER IV, Bruce D. SMITH, M. Thomas GILBERT, Alan COOPER, John F. DOEBLEY, and Svante PÄÄBO

2003. Early allelic selection in maize as revealed by ancient DNA. *Science*, 302 (5648): 1206–1208.

JAENICKE-DEPRÉS, Viviane R., and Bruce D. SMITH

2006. Ancient DNA and the integration of archaeological and genetic approaches to the study of maize domestication. In: *Histories of Maize: Multidisciplinary Approaches to the Prehistory, Linguistics, Biogeography, Domestication and Evolution of Maize*, edited by John Staller, Robert Tykot, and Bruce Benz. Academic Press, Elsevier. San Diego, London. pp. 83–95.

JEFFREYS, M. D. W.

1971. Pre-Columbian maize in Asia. In: *Man across the Sea*, edited by Caroll L. Riley, J. Charles Kelley, Campbell W. Pennington, and Robert L. Rands. University of Texas Press. Austin. pp. 376–400.

JEREZ (XEREZ), Francisco de

(1534) 1968. *Verdadera relación de la Conquista del Perú y provincia del Cuzco llamada la Nueva Castilla*. Biblioteca Peruana. Primera Serie. Tomo I. Editores Técnicos Asociados S.A. Lima. pp. 191–272.

JOHANNESSEN, Carl L.

1982. Domestication process of maize continues in Guatemala. *Economic Botany*, 36 (1): 89–99.

JOHANNESSEN, Carl L., Michael R. WILSON, and William A. DAVENPORT

1970. The domestication of maize: Process or event? *The Geographical Review*, LX (3): 393–413.

JOHANNESSEN, Sissel, and Christine A. HASTORF (editors)

1994. *Corn and Culture in the Prehistoric New World*. Westview Press. Boulder.

JOHNSON, Elmer C.

1977. Arquitectura de la planta de maíz. *Informativo del maíz*, 17: [5–8]. (The pages are not numbered.)

JOHNSON, F., and Richard S. MacNEISH

1972. Chronometric dating. In: *The Prehistory of the Tehuacán Valley: Chronology and Irrigation*. Vol. 4, edited by Frederick Johnson. University of Texas Press. Austin. pp. 3–55.

JONES, John G.

1988. Middle to Late Preceramic (6000–3000 BP) subsistence patterns on the central coast of Peru: The coprolite evidence. Master's thesis. Department of Anthropology, Texas A & M University. College Station.

1991. Palinology. In: *The Agroecological Evolution of Cobweb Swamp, Belize*, edited by J. S. Jacob. Preliminary Report to the National Geographic Society, Palinology Laboratory, Texas A&M University. College Station. pp. 109–131.

JONES, John G., and Duccio BONAVIA

1992. Análisis de coprolitos de llama (*Lama glama*) del Precerámico tardío de la costa Nor-Central del Perú. *Bulletin de l'Institut Français d'Études Andines*, 21 (3): 835–852.

JONES, Martin, and Terry BROWN

2000. Agricultural origins: The evidence of modern and ancient DNA. *The Holocene*, 10 (6): 769–776.

KAHN, E. J., Jr.

1987. *Granos y Raíces. Fuentes de Vida de la Humanidad*. Ediciones Bellaterra, S.A. Barcelona.

KAPLAN, Lawrence

1965. Archeology and domestication in American *Phaseolus* (beans). *Economic Botany*, 19 (4): 358–368.

1968. Archeology and domestication in American *Phaseolus*. In: *Actas y Memorias*. XXXVII Congreso Internacional de Americanistas (República Argentina, 1966), Tomo II. Buenos Aires. p. 509.

1980. Variation in the cultivated beans. In: *Guitarrero Cave: Early Man in the Andes*, edited by Thomas F. Lynch. Studies in Archaeology. Academic Press. New York. pp. 145–148.

1982. Pallar (*Phaseolus lunatus*). In: *Precerámico peruano. Los Gavilanes. Mar, desierto y oasis en la historia del hombre*, Duccio Bonavia, wth the collaboration of Ramiro Castro de la Mata, Félix Caycho Quispe, Alexander Grobman, Lawrence Kaplan, César A. Morán Val, Raúl Patrucco, Mario Peña, Virginia Popper, Elizabeth J. Reitz, Stanley George Stephens, Raúl Tello, and Elizabeth S. Wing. Corporación Financiera de Desarrollo S.A. COFIDE, Instituto Arqueológico Alemán, Comisión de Arqueología General y Comparada. Lima. pp. 181–182.

KATO-YAMAKAKE, Takeo A.

1975. Cytological studies of maize and teosinte in relation to their origin and evolution. Ph.D. dissertation. University of Massachusetts. Amherst.

1976. Cytological studies of maize (*Zea mays* L.) and teosinte (*Zea mexicana* [Schrader] Kuntze) in relation to their origin and evolution. *University of Massachusetts Agricultural Experiment Station Bulletin*, 635: 1–185.

1984. Chromosome morphology and the origin of maize and its races, edited by M. K. Hecht, B. Wallace, and G. T. Prance. *Evolutionary Biology*, 17: 219–253.

1988. Cytological classfication of maize race populations and its potential use. In: *Recent Advances in the Conservation and Utilization of Genetic Resources.* Proceedings of the Global Maize Germplasm Workshop. CIMMYT. Mexico City, D.F. pp. 106–117.

KATZ, S. H., M. L. HEDIGER, and L. A. VALLEROY

1974. Traditional maize processing techniques in the New World. *Science*, 184 (4138): 765–773.

KATZENBERG, Anne M.

2003. Comments. *Current Anthropology*, 44 (5): 692–693.

KAUTZ, Robert R.

1980. Pollen analysis and paleoethnobotany. In: *Guitarrero Cave: Early Man in the Andes*, edited by Thomas F. Lynch. Academic Press. New York. pp. 45–59.

KELLEY, David H., and Duccio BONAVIA

1963. New evidence for pre-ceramic maize in the coast of Peru. *Ñawpa Pacha*, 1: 39–41.

KEMPTON, J. H.

1937. *Maize: Our Heritage from the Indians.* Report of the Smithsonian Institution. Washington, D.C. pp. 385–408.

KEMPTON, J. H., and W. POPENOE

1937. *Teosinte in Guatemala.* Carnegie Institution. Publication 483: 199–218. Washington, D.C.

KIESSELBACH, T. A.

1949. The structure and reproduction of corn. *Nebraska Agriculture Experiment Station Research Bulletin*, 161: 1–96.

KIRKBY, Anne V. T.

1973. *The Use of Land and Water Resources in the Past and Present Valley of Oaxaca, Mexico.* University of Michigan Museum of Anthropology, Memoirs N° 5. Ann Arbor.

KOSOK, Paul

1965. *Life, Land and Water in Ancient Peru.* Long Island University Press. New York.

KUHRY, P.

1988. Palaeobotanical-palaeoecological studies of tropical high Andean peat-bog sections (Cordillera oriental, Colombia). Thesis, University of Amsterdam (Dissertationes Botanicae, 116). Cramer. Berlin.

KULESHOV, N. N.

1929. The geographical distribution of the varietal diversity of maize in the world. *Bulletin of Applied Botany, Genetics and Plant Breeding (Trudy Po Prikladnoi Botanike, Genetike I Selektsii)*, 20: 506–510.

1933. World's diversity of phenotypes of maize. *Journal of the American Society of Agronomy*, 25: 688–700.

(1930) 1981. Maíces de México, Guatemala, Cuba, Panamá y Colombia (según las colecciones de N.S. Bukasov). In: *Las Plantas cultivadas de México, Guatemala y Colombia*. Translated by Jorge León. CATIE. Turrialba. pp. 40–53.

KURTZ, Edwin B., Jr., James L. LIVERMAN, and Henry TUCKER

1960. Some problems concerning fossil and modern corn pollen. *Bulletin of the Torrey Botanical Club*, 87 (2): 85–94.

KURTZ, Edwin B., Jr., Henry TUCKER, and James L. LIVERMAN

1960. Reliability of identification of fossil pollen as corn. *American Antiquity*, 25 (4): 605–606.

LAGIGLIA, Humberto A.

1980. El proceso de agriculturación del sur de Cuyo. La cultura del Atuel II. In: *Actas del V Congreso Nacional de Arqueología Argentina*. Vol. 1. pp. 231–252.

2001. Los orígenes de la agricultura en la Argentina. In: *Historia Argentina Prehispánica*, edited by Eduardo E. Berberian and Axel E. Nielsen. Tomo I. Editorial Brujas. Córdoba. pp. 41–81.

LANDSTROM, B.

1967. *Columbus*. Macmillan. New York.

LANE, Chad S., Claudia I. MORA, Sally P. HORN, and Kenneth H. ORVIS

2008. Sensitivity of bulk sedimentary stable carbon isotopes to prehistoric forest clearance and maize agriculture. *Journal of Archaeological Science*, 35: 2119–2132.

LANGHAM, D. G.

1940. The inheritance of intergenomic difference in *Zea-Euchlaena* hybrids. *Genetics*, 25: 88–107.

LANNING, Edward Putnam

1959. Early ceramic chronology of the Peruvian coast. Berkeley. (Mimeographed.)

1960. Chronological and cultural relationship of early pottery styles in ancient Peru. Dissertation. Submitted in partial satisfaction of the requirements for the degree of doctor of philosophy in anthropology in the graduate division of the University of California. Berkeley.

1963. A pre-agricultural occupation on the central coast of Peru. *American Antiquity*, 28 (3): 360–371.

1967. *Peru before the Incas*. Prentice-Hall. Englewood Cliffs.

LARRAMENDI, Manuel R. P. de

1882. Corografía o descripción general de la muy noble y muy leal provincia de Guipúzcoa. In: *Verdadera Ciencia Española*, Vol. XIX. Imprenta de la Viuda e Hijos de I. Subirana. Barcelona.

(Note: This work is reprinted in Vol. VI of the Biblioteca de Autores Vascongados, San Sebastián, 1897. There is a 1950 facsimile edition, and another one was published by the Sociedad Guipuzcoana de Ediciones y Publicaciones, San Sebastián, 1969.)

LASTRES, Juan B.

1951. *Historia de la medicina peruana. Vol. 1. La medicina Incaica*. Historia de la Universidad, Tomo V. Universidad Nacional Mayor de San Marcos. Publicación del Cuarto Centenario. Lima.

LATCHAM, Ricardo

1936. *La agricultura precolombina en Chile y los países vecinos.* Ediciones Universidad de Chile 16. Vol. I. Santiago.

LATHRAP, Donald W.

1960. (Review) Cultura Valdivia. Clifford Evans, Betty Meggers and Emilio Estrada. N° 6. Guayaquil. 1959. 128 pp., 81 figs, 5 tables. *American Antiquity*, 26 (1): 125–127.

1968a. The "hunting" economies of the tropical forest zone of South America: An attempt at historical perspective. In: *Man the Hunter*, edited by Richard B. Lee and Irven De Vore. Aldine. Chicago. pp. 23–29.

1968b. Aboriginal occupation and changes in river channel on the central Ucayali, Peru. *American Antiquity*, 33 (1): 62–79.

1970. *The Upper Amazon.* Thames and Hudson. London.

1975. *Ancient Ecuador: Culture, Clay and Creativity, 3000–300 BC. El Ecuador Antiguo. Cultura, Cerámica y Creatividad. 3000–300 AC.* Field Museum of Natural History. Chicago. pp. 13–71.

1987. The introduction of maize in prehistoric eastern North America: The view from Amazonia and the Santa Elena Peninsula. In: *Emergent Horticultural Economies of the Eastern Woodlands*, edited by William I. Keegan. Center for Archaeological Investigations. Occasional Paper, N° 7. Board of Trustees Southern Illinois University. Carbondale. pp. 345–371.

LATHRAP, Donald W., and Jorge G. MARCOS

1975. Informe preliminar sobre las excavaciones del sitio Real Alto por la Misión Antropológica de la Universidad de Illinois. *Revista de la Universidad Católica*, Centro de publicaciones de la Pontificia Universidad Católica del Ecuador, III (10): 41–64.

LAUTER, Nick, and John F. DOEBLEY

2002. Genetic variation for phenotypically invariant traits detected in teosinte: Implications for the evolution of novel forms. *Genetics*, 160: 333–342.

LEFEBVRE, Th.

1933. *Les modes de vie dans les Pyrénées atlantiques orientales.* Armand Colin. Paris.

LI, Wen-Hsiung, and Dan GRAUR

1991. *Fundamentals of Molecular Evolution.* Sinaner Associates. Sunderland.

LIPPI, Ronald, Robert McKelvy BIRD, and David M. STEMPER

1984. Maize recovered at La Ponga, an early Ecuadorian site. *American Antiquity*, 49 (1): 118–124.

LLERENA LANDA, J. Enrique

1957. *Calculos, reducciones y equivalencias.* Esso. International Petroleum Company, Ltd. División de Ventas. Lima.

LOCKE, John

1953. *Locke's Travels in France, 1675–1679.* Cambridge University Press. Cambridge.

LONG, Austin, Bruce F. BENZ, D. J. DONAHUE, A. J. T. JULL, and L. J. TOOLIN

1989. First direct AMS dates on early maize from Tehuacán, México. *Radiocarbon*, 31 (3): 1035–1040.

LONG, Austin, and Gayle J. FRITZ

2001. Validity of AMS dates on maize from the Tehuacán Valley: A comment on MacNeish and Eubanks. *Latin American Antiquity*, 12 (1): 87–90.

LONGLEY, A. E.

1924. Chromosomes in maize and maize relatives. *Journal of Agriculture Research*, 28: 673–682.

1937. Morphological characters of teosinte chromosomes. *Journal of Agriculture Research*, 54: 835–862.

1938. Chromosomes of maize from North American Indians. *Journal of Agriculture Research*, 56: 177–195.

1941. Chromosome morphology in maize and its relatives. *Botanical Review*, 7: 263–289.

LONGLEY, A. E., and Takeo A. KATO-YAMAKAKE

1965. Chromosome morphology of certain races of maize in Latin America. International Maize and Wheat Improvement Center (CIMMYT). *Research Bulletin*, 1: 1–112.

LORENZO, José Luis, and Lauro GONZÁLES QUINTERO

1970. El más antiguo teosinte. *Boletín del Instituto Nacional de Antropología e Historia*, 42: 41–43.

LYNCH, Thomas F.

1978. Andean South America: Current research, edited by Thomas P. Myers. *American Antiquity*, 43 (3): 524–526.

(editor) 1980a. *Guitarrero Cave: Early Man in the Andes.* Academic Press. New York.

1980b. Stratigraphy and chronology. In: *Guitarrero Cave: Early Man in the Andes*, edited by Thomas F. Lynch. Academic Press. New York. pp. 29–43.

1980c. Guitarrero Cave in its Andean context. In: *Guitarrero Cave: Early Man in the Andes*, edited by Thomas F. Lynch. Academic Press. New York. pp. 293–320.

1983. The paleo-Indians. In: *Ancient South Americans*, edited by Jesse D. Jenning. W.F. Freeman. San Francisco. pp. 87–137.

LYNCH, Thomas F., R. GILLESPIE, John A. J. GOWLETT, and R. E. M. HEDGES

1985. Chronology of Guitarrero Cave, Peru. *Science*, 229 (4716): 864–867.

MacNEISH, Richard S.

1969. *First Annual Report of the Ayacucho Archaeological Botanical Project.* Robert S. Peabody Foundation for Archaeology, Phillips Academy. Published by the Foundation. N° 1. Andover.

1977. The beginning of agriculture in central Peru. In: *Origins of Agriculture*, edited by Charles A. Reed. World Anthropology. Sol Tax, general editor. Mouton. The Hague, Paris. pp. 753–802.

1981a. The stratigraphy of Pikimachay, Ac 100. In: *Prehistory of the Ayacucho Basin, Peru. Vol. II. Excavations and Chronology*, Richard S. MacNeish, Ángel García Cook, Luis G. Lumbreras, Robert K. Vierra, and Antoinette Nelken-Terner. University of Michigan Press. Ann Arbor. pp. 19–56.

1981b. Synthesis and conclusions. In: *Prehistory of the Ayacucho Basin, Peru. Vol. II. Excavations and Chronology*, Richard S. MacNeish, Ángel García Cook, Luis G. Lumbreras, Robert K. Vierra, and Antoinette Nelken-Terner. University of Michigan Press. Ann Arbor. pp. 199–257.

1981c. Seasonality of the components. In: *Prehistory of the Ayacucho Basin, Peru. Vol. II. Excavations and Chronology*, Richard S. MacNeish, Ángel García Cook, Luis G. Lumbreras, Robert K. Vierra, and Antoinette Nelken-Terner. University of Michigan Press. Ann Arbor. pp. 149–166.

1992. *The Origins of Agriculture and Settled Life*. University of Oklahoma Press. Norman.

1997. In defense of the Tehuacán project. *Current Anthropology*, 38 (4): 663–672.

2001. A response to Long's radiocarbon determinations that attempt to put acceptable chronology on the Fritz. *Latin American Antiquity*, 12 (1): 99–104.

MacNEISH, Richard S., and Mary Wilkes EUBANKS

2000. Comparative analysis of the Río Balsas and Tehuacán models for the origin of maize. *Latin American Antiquity*, 11 (1): 3–20.

MacNEISH, Richard S., and Ángel GARCÍA COOK

1972. Excavations in the locality of the Riego Oasis. In: *The Prehistory of the Tehuacán Valley*. Vol. 5, edited by R. MacNeish, M. Fowler, A. García Cook, F. Peterson, and A. Nelken-Terner. University of Texas Press. Austin. pp. 14–65.

1981. Rosamachay, Ac 117. In: *Prehistory of the Ayacucho Basin, Peru. Vol. II. Excavations and Chronology*, Richard S. MacNeish, Ángel García Cook, Luis G. Lumbreras, Robert K. Vierra, and Antoinette Nelken-Terner. University of Michigan Press. Ann Arbor. pp. 121–124.

MacNEISH, Richard S., Ángel GARCÍA COOK, Luis G. LUMBRERAS, Robert K. VIERRA, and Antoinette NELKEN-TERNER

1981. *Prehistory of the Ayacucho Basin, Peru. Vol. II. Excavations and Chronology*. University of Michigan Press. Ann Arbor.

MacNEISH, Richard S., and Antoinette NELKEN-TERNER

1983. Introduction to preceramic contextual studies. In: *Prehistory of the Ayacucho Basin, Peru. Vol. IV. The Preceramic Way of Life*, Richard S. MacNeish, Robert K.Vierra, Antoinette Nelken-Terner, Rochelle Lurie, and Ángel García Cook. University of Michigan Press. Ann Arbor. pp. 1–15.

MacNEISH, Richard S., Antoinette NELKEN-TERNER, and Ángel GARCÍA COOK

1970. *Second Annual Report of the Ayacucho Archaeological Botanical Project*. Robert S. Peabody Foundation for Archaeology, Phillips Academy. Published by the Foundation. Andover.

MacNEISH, Richard S., Antoinette NELKEN-TERNER, and Robert K. VIERRA

1980. Introduction. In: *Prehistory of the Ayacucho Basin, Peru. Vol. III. Nonceramic Artifacts*, Richard S. MacNeish, Robert K. Vierra, Antoinette Nelken-Terner, and Carl J. Phagan. University of Michigan Press. Ann Arbor. pp. 1–34.

MacNEISH, Richard S., Thomas C. PATTERSON, and David L. BROWMAN

1975. *The Central Peruvian Prehistoric Interaction Sphere*. Papers of the Robert S. Peabody Foundation for Archaeology, Vol. 7. Phillips Academy. Andover.

MacNEISH, Richard S., and Robert K. VIERRA

1983. The preceramic way of life in the thorn forest scrub ecozone. In: *Prehistory of the Ayacucho Basin, Peru. Vol. IV. The Preceramic Way of Life*, Richard S. MacNeish, Robert K. Vierra, Antoinette Nelken-Terner, Rochelle Lurie, and Ángel García Cook. University of Michigan Press. Ann Arbor. pp. 130–187.

MacNEISH, Richard S., Robert K. VIERRA, Antoinette NELKEN-TERNER, Rochelle LURIE, and Ángel GARCÍA COOK

1983. *Prehistory of the Ayacucho Basin, Peru. Vol. IV. The Preceramic Way of Life.* University of Michigan Press. Ann Arbor.

MacNEISH, Richard S., Robert K. VIERRA, Antoinette NELKEN-TERNER, and Carl L PHAGAN

1980. *Prehistory of the Ayacucho Basin, Peru. Vol. III. Nonceramic Artifacts.* The University of Michigan Press. Ann Arbor.

MacNEISH, Richard S., and Wayne WIERSUM

1981. Tambillo Boulder Cave, Ac-240. In: *Prehistory of the Ayacucho Basin, Peru. Vol. II. Excavations and Chronology*, Richard S. MacNeish, Ángel García Cook, Luis G. Lumbreras, Robert K. Vierra, and Antoinette Nelken-Terner. University of Michigan Press. Ann Arbor. pp. 128–129.

MAGUIRE, M. P.

1962. Common loci in corn and *Tripsacum*. *Journal of Heredity*, 53: 87–88.

MANGELSDORF, Paul Christoph

1947. The origin and evolution of maize. In: *Advances in Genetics.* Vol. I, edited by M. Demerec. Academic Press. New York. pp. 161–207.

1958a. The mutagenic effect of hybridizing maize and teosinte. *Cold Spring Harbor Symposium in Quantitative Biology*, 13: 409–421.

1958b. Reconstructing the ancestor of corn. *Proceedings of the American Philosophical Society*, 102: 454–463.

1961. Introgression in maize. *Euphytica*, 10: 157–168.

1967. Report of mineralized corncobs and other prehistoric specimens from Salinas La Blanca. In: *Early Culture and Human Ecology in South Coastal Guatemala*, edited by M. D. Coe and Kent V. Flannery. Smithsonian Contribution to Anthropology. Vol. 3. Smithsonian Institution Press. Washington, D.C. pp. 127–128.

1968. Cryptic genes for "tripsacoid" characteristics in maíz Amargo of Argentina and other Latin American varieties. *Boletín de la Sociedad Argentina de Botánica*, 12: 180–187.

Ms. 1973. Letter to the editor Field Museum of Natural History [Chicago]. Comments on "The Mystery of Maize."

(Note: This letter was sent after the paper by Beadle, 1972, appeared, and it was not published. A copy is in the possession of Duccio Bonavia.)

1974. *Corn: Its Origin, Evolution and Improvement.* Belknap Press of Harvard University Press. Cambridge.

Ms. 1977. More on the San Pablo corn kernel.

(Note: This letter was sent to *Science* after the publication of the paper by Zevallos Menéndez et al., 1977, and it was not published. A copy is in the possession of Duccio Bonavia.)

1983a. The search for wild corn. *Maydica*, 28: 89–96.

1983b. The mystery of corn: New perspectives. *Proceedings of the American Phylosophical Society*, 127 (4): 215–247.

1986. The origin of corn. *Scientific American*, 255 (2): 80–86.

MANGELSDORF, Paul Christoph, Elso J. BARGHOORN, and Umesh C. BANERJEE

1978. Fossil pollen and the origin of corn. *Botanical Museum Leaflets*, 26 (7): 237–255.

MANGELSDORF, Paul Christoph, and Julián CÁMARA-HERNÁNDEZ

1967. Prehistoric maize from a site near Huarmey, Peru. *Maize Genetics Cooperation Newsletter*, 41: 47–48.

MANGELSDORF, Paul Christoph, and J. W. CAMERON

1942. Western Guatemala: A secondary center of origin of cultivated maize varieties. *Botanical Museum Leaflets*, 10 (8): 217–252.

MANGELSDORF, Paul Christoph, H. W. DICK, and Julián CÁMARA-HERNÁNDEZ

1967. Bat Cave revisited. *Botanical Museum Leaflets*, 22 (1): 1–31.

MANGELSDORF, Paul Christoph, and Walton C. GALINAT

1964. The tunicate locus in maize dissected and reconstituted. *Proceedings of the National Academy of Sciences of the United States of America*, 51 (2): 147–150.

MANGELSDORF, Paul Christoph, and Robert H. LISTER

1956. Archeological evidence of the evolution of maize in northwestern Mexico. *Botanical Museum Leaflets*, 17 (6): 151–178.

MANGELSDORF, Paul Christoph, Richard S. MacNEISH, and Walton C. GALINAT

1956. Archaeological evidence on the diffusion and evolution of maize in northeastern Mexico. *Botanical Museum Leaflets*, 17 (5): 125–150.

1964. Domestication of corn. *Science*, 143 (3606): 538–545.

1967a. Prehistoric wild and cultivated maize. In: *The Prehistory of the Tehuacán Valley. I. Environment and Subsistence*, edited by D. S. Byers. University of Texas Press. Austin. pp. 178–200.

1967b. Prehistoric maize, teosinte and *Tripsacum* from Tamaulipas, Mexico. *Botanical Museum Leaflets*, 22 (2): 33–63.

MANGELSDORF, Paul Christoph, Richard S. MacNEISH, and Gordon R. WILLEY

1964. Origin of agriculture in Middle America. In: *Handbook of Middle American Indians. Vol. 1. Natural Environment and Early Cultures*, edited by Robert Wauchope and Robert C. West. University of Texas Press. Austin. pp. 427–445.

MANGELSDORF, Paul Christoph, and R. G. REEVES

1939. The origin of Indian corn and its relatives. *Texas Agricultural Experiment Station Bulletin*, 574: 1–315.

1945. The origin of maize: Present status of the problem. *American Anthropologist*, 47 (2): 235–243.

1959a. The origin of corn. I. Pod corn, the ancestral form. *Botanical Museum Leaflets*, 18 (7): 329–356.

1959b. The origin of corn. III. Modern races, the product of teosinte introgression. *Botanical Museum Leaflets*, 18 (9): 389–411.

1959c. The origin of corn. IV. Place and time of origin. *Botanical Museum Leaflets*, 18 (10): 413–427.

MANGELSDORF, Paul Christoph, Lewis M. ROBERTS, and John S. ROGERS

1981. *The probable origin of annual teosintes.* Publication N° 10, The Bussey Institution of Harvard University. Cambridge. pp. 39–69.

MANGELSDORF, Paul Christoph, and C. SMITH

1949. New archaeological evidences on the evolution of maize. *Botanical Museum Leaflets*, 13 (8): 213–247.

MARCUS, Joyce

1982. The plant world of the sixteenth- and seventeenth-century lowland Maya. In: *Maya Subsistence*, edited by Kent V. Flannery. Academic Press. New York. pp. 239–273.

2006. The roles of ritual and technology in Mesoamerican water management. In: *Agricultural Strategies*, edited by Joyce Marcus and Charles Stanish. Monograph 50 in the Cotsen Institute of Archaeology Series, UCLA. Los Angeles. pp. 221–254.

MARTIENSSEN, Rob

1997. The origin of maize branches out. *Nature*, 386 (6624): 443–445.

MARTINS-FARIAS, Rena

1976. New archaeological techniques for the study of ancient root crops in Peru. Ph.D. dissertation. University of Birmingham.

MATOS MENDIETA, Ramiro

1966. El período cerámico inicial en la costa central del Perú. In: *Actas y Memorias*, XXXVI Congreso Internacional de Americanistas. España, 1964. Vol. 1. Sevilla. pp. 509–518.

MATSUOKA, Yoshihiro, Yves VIGOROUX, Major M. GOODMAN, Jesús SÁNCHEZ-G., Edward BUCKLER, and John F. DOEBLEY

2002. A single domestication of maize shown by multilocus microsatellite genotyping. *Proceedings of the National Academy of Sciences of the United States of America*, 99 (9): 6080–6084.

MAYR, E.

1942. *Systematics and the Origin of Species.* Columbia University Press. New York.

McCLINTOCK, Barbara

1959. Chromosome constitution of some South American races of maize. *Annual Report Department of Genetics, Carnegie Institution Washington Yearbook*, 58: 454–456.

1960. Chromosome constitution of Mexican and Guatemalan races of maize. *Annual Report Department of Genetics, Carnegie Institution Washington Yearbook*, 59: 461–472.

1978. Significance of chromosome constitutions in tracing the origin and migration of races of maize in the Americas. In: *Maize Breeding and Genetics*, edited by D. B. Walden. John Willey and Sons. New York. pp. 159–184.

McCLINTOCK, Barbara, Takeo A. KATO-YAMAKAKE, and A. BLUMENSCHEIN

1981. *Chromosome Constitution of the Races of Maize: Its Significance in the Interpretation of Relationship between Races and Varieties in the Americas.* Colegio de Postgraduados. Chapingo.

McCLUNG de TAPIA, E.
1992. The origins of agriculture in Mesoamerica and Central America. In: *The Origins of Agriculture: An International Perspective*, edited by C. W. Cowan and P. J. Watson. Smithsonian Institution Press. Washington, D.C. pp. 143–171.
MEJÍA XESSPE, M. Toribio
1931. Alimentación de los indios. *Wira Kocha*, 1 (1): 9–24.
MENA, Cristóbal de
(1534) 1968. *La Conquista del Perú.* Biblioteca Peruana, Primera Serie, Tomo I. Editores Técnicos Asociados S.A. Lima. pp. 133–169.
MESA BERNAL, D.
1955. ¿De donde es originario el maíz? III. Colombia considerado como centro de origen. *Agricultura Tropical*, 11 (9): 753–758.
1957. Historia natural del maíz. *Revista de la Academia Colombiana de Ciencias Exactas, Físico-Químicas y Naturales*, X (39): 13–106.
MESSEDAGLIA, Luigi
1927. *Il mais e la vita rurale italiana.* Federazione Italiana dei Consorzi Agrari. Piacenza.
METCALFE, C.R.
1971. *Anatomy of Monocotyledons: V Cyperaceae.* Clarendon Press. Oxford.
MIDDENDORF, Ernst W.
(1894) 1973. *Perú.* Tomo II. Universidad Nacional Mayor de San Marcos. Lima.
MIRACLE, M. P.
1966. *Maize in Tropical Africa.* University of Wisconsin Press. Madison.
MIRANDA, Colín Salvador
1966. Discusión sobre el origen y la evolución del maíz. In: *Memorias del Segundo Congreso Nacional de Fitogenética.* Sociedad Mexicana de Fitogenética. Monterrey. Escuela Nacional de Agricultura. Colegio de Post-graduados. Chapingo. Monterrey. pp. 233–251.
MOLINA, Cristóbal de (El Chileno)
(1552) 1968. *Conquista y población del Perú o destrucción del Perú.* Biblioteca Peruana, Primera Serie. Tomo III. Editores Técnicos Asociados, S.A. Lima. pp. 297–364.
MOLINA, Cristóbal de (El Párroco Cuzqueño)
(1873) 1916. *Relación de las Fabulas y Ritos de los Incas.* Colección de Libros y Documentos referentes a la Historia del Perú. Tomo I. Imprenta y Librería Sanmarti y Cía. Lima. pp. 1–103.
MONSALVE, J. G.
1985. A pollen core from the Hacienda Lusitania. *Pro-Calima*, 4: 40–44.
MONTGOMERY, E. C.
1906. What is an ear of corn? *Popular Science Monthly*, 68: 55–62.
MOORE, Jerry D.
1989. Pre-hispanic beer in coastal Peru: Technology and social context of prehistoric production. *American Anthropologist*, 91 (3): 682–695.
MOORE, Peter D.
1998. Getting to the roots of tubers. *Nature*, 395 (6700): 330–331.

MORA, S.

2003. *Early Inhabitants of the Amazonian Tropical Rain Forest: A Study of Humans and Environmental Dynamics.* University of Pittsburgh Latin American Archaeology Reports, N° 3. University of Pittsburgh. Pittsburgh.

MORÁN VAL, César A.

1982. Descripción de las muestras 2. In: *Precerámico peruano. Los Gavilanes. Mar, desierto y oasis en la historia del hombre*, D. Bonavia, with the collaboration of R. Castro de la Mata, F. Caycho Quispe, A. Grobman, L. Kaplan, C. A. Morán Val, R. Patrucco, M. Peña, V. Popper, E. J. Reitz, S. G. Stephens, R. Tello, and E. S. Wing. Corporación Financiera de Desarrollo S.A. COFIDE, Instituto Arqueológico Alemán, Comisión de Arqueología General y Comparada. Lima. pp. 181.

MORENO, Ulises, Alexander GROBMAN, and Barbara McCLINTOCK

1959. Study of chromosome morphology of races of maize in Peru. *Maize Genetic Cooperation Newsletter*, 33: 27–28.

MORRIS, Craig

1979. Maize beer in the economics, politics and religion of the Inca Empire. In: *Fermented Food Beverages in Nutrition*, edited by Clifford F. Gastineau, William J. Darby, and Thomas B. Turner. Academic Press. New York. pp. 21–34.

1982. The infrastructure of Inka control in the Peruvian central highlands. In: *The Inca and Aztec States, 1400–1800: Anthropology and History*, edited by George A. Collier, Renato I. Rosaldo, and John D. Wirth. Academic Press. New York. pp. 153–171.

1993. Value, investment and mobilization in the Inca economy. In: *The Politics of Production and Consumption*, edited by H. Henderson and Patricia Netherly. Cornell University Press. Ithaca. pp. 34–55.

MORRIS, Craig, and Donald E. THOMPSON

1985. *Huánuco Pampa: An Inca City and Its Hinterland.* Thames and Hudson. London.

MOSELEY, Michael Edward

1975. *The Maritime Foundations of Andean Civilization.* Cumming. Menlo Park.

1978. *Pre-Agricultural Coastal Civilization in Peru.* Carolina Biology Readers. J. J. Head, editor. 90. Scientific Publications Division. Burlington.

1992. Maritime foundations and multilinear evolution: Retrospect and prospect. *Andean Past*, 3: 5–42.

MOSELEY, Michael Edward, and Gordon R. WILLEY

1973. Aspero, Peru: A reexamination of the site and its implications. *American Antiquity*, 38 (4): 452–468.

MUELLE, Jorge C.

1945. La chicha en el distrito de San Sebastián. *Revista del Museo Nacional*, XIV: 144–152.

MURRA, John V.

1960. Rite and crop in the Inca state. In: *Culture in History*, edited by S. Diamond. Columbia University Press. New York. pp. 390–407.

(Note: This paper was later republished without any changes; see Murra, 1973.)

1968. La papa, el maíz y los ritos agrícolas del Tawantinsuyu. *Amaru*, 8: 58–70.

1973. Rite and crop in the Inca state. In: *Peoples and Cultures of Native South America*, edited by D. R. Gross. Natural History Press. Garden City. pp. 377–389.
(Note: This is the same as the 1960 version. There are two Spanish translations – 1968 and 1975.)
1975. Maíz, tubérculos y ritos agrícolas. In: *Formaciones económicas y políticas del mundo andino*, John V. Murra. Instituto de Estudios Peruanos. Lima. pp. 45–57.
1980. *The Economic Organization of the Inca State*. Research in Economic Anthropology. Supplement 1. JAI Press. Greenwich.

MUTIS (BOSIO), José Celestino Bruno
(1760–1790) 1957. *Diario y observaciones de José Celestino Mutis*. Vol. 1. Instituto Colombiano de Cultura Hispánica. Editorial Minerva, Ltda. Bogotá.

NEFF, H., B. ARROYO, J. G. JONES, Deborah M. PEARSALL, and D. E. FREIDEL
2002. Nueva evidencia pertinente a la ocupación temprana del sur de Mesoamérica. Paper presented at the XII Encuentro Internacional: Los Investigadores de la Cultura Maya, Campeche. 10–14 November 2002. Universidad de Campeche.

NETHERLY, Patricia
1977. Local level lords on the North Coast of Peru. Ph.D. dissertation. Department of Anthropology, Cornell University.

NEWSOM, Lee A.
2006. Caribbean maize: First farmers to Columbus. In: *Histories of Maize: Multidisciplinary Approaches to the Prehistory, Linguistics, Biogeography, Domestication and Evolution of Maize*, edited by John Staller, Robert Tykot, and Bruce Benz. Academic Press, Elsevier. San Diego, London. pp. 325–335.

NEWSOM, Lee A., and Kathleen A. DEAGAN
1994. *Zea mays* in the West Indies: The archaeological and early historic record. In: *Corn and Culture in the Prehistoric New World*, edited by Sissel Johannessen and Christine A. Hastorf. Westview Press. Boulder. pp. 203–217.

NEWSOM, Lee A., and Deborah M. PEARSALL
2003. Trends in Caribbean Island archaeobotany. In: *People and Plants in Ancient Eastern North America*, edited by P. E. Minnis. Smithsonian Institution. Washington, D.C. pp. 347–412.

NICHOLSON, G. Edward
1960. Chicha maize types and chicha manufacture in Peru. *Economic Botany*, 14 (4): 290–299.

NICKERSON, Northon M.
1953. Variation in cob morphology among certain archaeological and ethnological races of maize. *Annals of Missouri Botanical Garden*, 40: 79–111.

NIEDERBERGER, Christine
1979. Early sedentary economy in the basin of Mexico. *Science*, 203 (4376): 131–142.

NORR, Lynette
1991. Nutritional consequences of prehistoric subsistence strategies in lower Central America. Ph.D. dissertation. University of Illinois. Urbana.

1995. Interpreting dietary maize from stable isotopes in the American tropics: The state of the art. In: *Archaeology in the Lowland American Tropics: Current Analytical Methods and Recent Applications*, edited by P. W. Stall. Cambridge University Press. Cambridge. pp. 198–223.

NORTHROP, Lisa A., and Sally P. HORN

1996. Pre-Columbian agriculture and forest disturbance in Costa Rica: Palaeoecological evidence from two lowland rainforest lakes. *The Holocene*, 6 (3): 289–299.

NÚÑEZ A., Lautaro

1965. Desarrollo cultural prehispánico del norte de Chile. *Estudios Arqueológicos*, 1: 37–114.

1976. Registro regional de fechas radiocarbónicas del Norte de Chile. *Estudios Atacameños*, 4: 74–123.

1986. Evidencias arcaicas de maíces y cuyes en Tiliviche: Hacia el semisedentarismo en el litoral fértil y quebradas del norte de Chile. *Revista Chungará*, 16–17: 25–47.

NÚÑEZ A., Lautaro, and W. Cora MORAGAS

1976. Revaluación de los primeros poblamientos en las tierras bajas: nuevas evidencias de maíz temprano en el N. de Chile. Paper presented at the IV Congreso de Arqueología Argentina. San Rafael.

1978. Ocupación arcaica temprana en Tiliviche, Norte de Chile. (1 Región). *Boletín* (Museo Arqueológico de La Serena), 16: 53–76.

NÚÑEZ ENRÍQUEZ, Patricio, and Vjera ZLATAR MONTAN

1978a. Tiliviche 1-b y Aragón-1 (Estrato V): Dos comunidades precerámicas coexistentes en Pampa de Tamarugal, Pisagua, N. De Chile. *III Congreso Peruano. El Hombre y la Cultura Andina.* Tomo II, edited by Ramiro Matos M. Lima. pp. 734–756.

1978b. Coexistencia de comunidades cazadoras-recolectoras. Paper presented at the V Congreso de Arqueología Argentina. San Juan.

OLSEN, K. M., and B. A. SCHAAL

1999. Evidence on the origin of cassava: Phylogeography of *Manihot esculenta. Proceedings of the National Academy of Sciences of the United States of America*, 96 (10): 5586–5591.

2001. Microsatellite variation in cassava (*Manihot esculenta*, Euphorbiaceae) and its wild relatives: Further evidence for a southern Amazonian origin of domestication. *American Journal of Botany*, 88: 131–142.

ONERN

1976. *Mapa ecológico del Perú. Guía explicativa.* Oficina Nacional de Evaluación de Recursos Naturales. Lima.

ORR, Alan R., and Marshall SUNDBERG

2001. Inflorescence development in a new teosinte: *Zea nicaraguensis* (Poaceae). *American Journal of Botany*, 91 (2): 165–173.

ORTEGA, E., and J. GUERRERO

1981. *Cuatro nuevos sitios paleoarcaicos en las islas de Santo Domingo.* Ediciones del Museo del Hombre Dominicano. Santo Domingo.

ORTEGA, Ynes R., and Duccio BONAVIA
2003. *Cryptosporidium*, *Giardia*, and *Cyclospora* in ancient Peruvians. *Journal of Parasitology*, 89 (3): 635–636.
ORTIZ, Alfonso
1994. Some cultural meanings of corn in aboriginal North America. In: *Corn and Culture in the Prehistoric New World*, edited by Sissel Johannesen and Christine A. Hastorf. Westview Press. Boulder. pp. 527–544.
OSBORN, Alan J.
1977. Strandloopers, mermaids and other fairy tales: Ecological determinants of marine resources utilization. The Peruvian case. In: *For Theory Building in Archaeology: Essays on Faunal Remains, Aquatic Resources, Spatial Analysis, and Systematic Modeling*, edited by L. R. Binford. Academic Press. New York. pp. 157–205.
OVIEDOS, J.
1824. *Historia de la conquista y población de la provincia de Venezuela*. Reimpreso en Caracas. Imprenta Navas Spinola. Caracas.
OYUELA-CAYCEDO, Augusto
1996. The study of collector variability in the transition to sedentary food producers in North Colombia. *Journal of World Prehistory*, 10 (1): 49–93.
2003. Comments. *Current Anthropology*, 44 (5): 693–694.
PÄÄBO, Swante
1999. Neolithic genetic engeenering. *Nature*, 398 (6724): 194–195.
PAGÁN-JIMÉNEZ, J. R., M. A. RODRIGUEZ LÓPEZ, L. A. CHANLATTE BAIK, and Y. NARGANES STORDE
2005. La temprana introducción y uso de algunas plantas domésticas, silvestres y cultivos en las Antillas precolombinas. *Diálogo Antropológico*, 3: 7–33.
PALAZZI, Fernando
1940. *Novíssimo Dizionario della Lingua Italiana*. Casa Editrice Ceschina. Milan.
PALIWAL, R. L.
2006. Origen, evolución y difusión del maíz. Accessed at www.fao.org/docrep/003/X7650S/x7650s03.htm.
PARMENTIER, A. A.
1984. *Le maïs ou blé de Turquie*. Reprinted by the Association générale des producteurs de maïs. Paris.
PARODI, R. L.
1935. *Relaciones de la agricultura prehispánica con la agricultura argentina actual*. Anales de la Academia Nacional de Agricultura y Veterinaria, 1. Buenos Aires.
1966. *La Agricultura Aborigen Argentina*. Editorial Universitaria. Buenos Aires.
PATIÑO, V. M.
1964. *Plantas Cultivadas y Animales Domésticos en América*. Equinoccial Vol. 2. Plantas Alimenticias. Imprenta Departamental. Cali.
PATRUCCO, Raul, Raúl TELLO, and Duccio BONAVIA
1982. *Homo sapiens sapiens*. In: *Precerámico peruano. Los Gavilanes. Mar, desierto y oasis en la historia del hombre*, D. Bonavia with the collaboration of R. Castro de la Mata, F. Caycho Quispe, A. Grobman, L. Kaplan, C. A. Morán

Val, R. Patrucco, M. Peña, V. Popper, E. J. Reitz, S. G. Stephens, R. Tello, and E. S. Wing. Corporación Financiera de Desarrollo S.A. COFIDE, Instituto Arqueológico Alemán, Comisión de Arqueología General y Comparada. Lima. pp. 226–232.

1983. Parasitological studies of coprolites of pre-Hispanic Peruvian populations. *Current Anthropology*, 24 (3): 393–394.

PATTERSON, Thomas C.

1971. Central Peru: Its population and economy. (Villages developed in Central Peru long before the introduction of agriculture.) *Archaeology*, 24 (4): 316–321.

PEARSALL, Deborah M.

1978a. Early movement of maize between Mesoamerica and South America. *Journal of the Steward Anthropological Society*, 9 (1–2): 41–75.

1978b. Phytolith analysis of archaeological soils: Evidence for maize cultivation in formative ecuador. *Science*, 199 (4325): 177–178.

1978c. Paleoethnobotany in western South America: Progress and problems. In: *The Nature and Status of Ethnobotany*, edited by Richard I. Ford, H. Hodge, and W. L. Merril. Anthropological Papers, N° 67. Museum of Anthropology, University of Michigan. Ann Arbor. pp. 389–416.

1979. The application of ethnobotanical techniques to the problem of subsistence in the Ecuadorian Formative. Unpublished Ph.D. dissertation. University of Illinois at Urbana-Champaign.

1987. Evidence for prehistoric maize cultivation on raised fields at Peñón del Río, Guayas, Ecuador. In: *Pre-Hispanic Agricultural Fields in the Andean Region*, edited by William M. Denevan, Kent Mathewson, and Gregory Knapp. Part 2: BAR International Series 359. pp. 279–296.

1988a. An overview of Formative period subsistence in Ecuador: Paleoethnobotanical data and perspectives. In: *Diet and Subsistence: Current Archaeological Perspectives*, edited by B. V. Kennedy and G. M. Le Moine. Proceedings of the 19th Annual Chacmool Conference. Archaeological Association of the University of Calgary. Calgary. pp. 149–164.

1988b. *La producción de alimentos en Real Alto: la aplicación de las técnicas al problema de la subsistencia en el Período Formativo ecuatoriano.* Centro de Estudios Arqueológicos y Antropológicos, ESPOL. Guayaquil.

1989. *Paleoethnobotany: A Handbook of Procedures.* Academic Press. San Diego.

1992a. Prehistoric subsistence and agricultural evolution in the Jama River valley Manabí Province. *Journal of the Steward Anthropological Society*, 20 (1–2): 181–207.

1992b. The origin of plant cultivation in South America. In: *The Origins of Agriculture: An International Perspective*, edited by C. Wesley Cowan and P. J. Watson, with the assistance of N. L. Benco. Smithsonian Institution Press. Washington, D.C. pp. 173–205.

1993. Contributions of the phytolith analysis for reconstructing subsistence: Example from research in coastal Ecuador. In: *Current Research in Phytolith Analysis: Applications in Archaeology and Paleoecology*, edited by Deborah M. Pearsall and Dolores R. Piperno. MASCA, Research Papers in Science and Archaeology. Vol.

10. Museum Applied Science Center for Archaeology. University of Pennsylvania Museum. Philadelphia. pp. 109–122.

1994a. Issues in the analysis and interpretation of archaeological maize in South America. In: *Corn and Culture in the Prehistoric New World*, edited by Sissel Johannessen and Christine A. Hastorf. Westview Press. Boulder. pp. 245–272.

1994b. Investigating New World tropical agriculture: Contribution from phytolith analysis. In: *Tropical Archaeobotany: Application and New Development*, edited by Kohn G. Hather. Routledge. London. pp. 115–138.

1995a. Domestication and agriculture in the New Worlds tropics. In: *Last Hunters–First Farmers: New Perspectives on the Prehistoric Transition to Agriculture*, edited by T. D. Price and A. B. Gebauer. School of American Research Press. Santa Fe. pp. 157–192.

1995b. "Doing" paleoethnobotany in the tropical lowlands: Adaptation and innovation in methodology. In: *Archaeology in the Lowland Tropics: Current Analytical Methods and Recent Application*, edited by Peter W. Stahl. Cambridge University Press. Cambridge. pp. 113–129.

1995c. Subsistence in the Ecuadorian Formative: Overview and comparison to the central Andes. Paper presented at the Dumbarton Oaks Conference on the Ecuadorian Formative.

1996. The impact of maize on subsistence systems in South America: An example from Jama River valley, coastal Ecuador. Paper presented at WAC-3 (1994). New Delhi.

(Note: The paper was delivered in 1994, and the draft distributed was dated January 1996. This same draft was published in 1999 in *The Prehistory of Food: Appetite for Change*, edited by John Hather. Routledge. London and New York. pp. 408–426.)

2000. *Paleoethnobotany: A Handbook of Procedures.* Academic Press. San Diego.

2002. Maize is *still* ancient in prehistoric Ecuador: The view from Real Alto, with comments on Staller and Thompson. *Journal of Archaeological Science*, 29 (1): 51–55.

2003a. Comments. *Current Anthropology*, 44 (5): 694.

2003b. Plant food resources of the Ecuadorian Formative: An overview and comparison to the central Andes. In: *Archaeology of Formative Ecuador*, edited by J. Scott Raymond and Richard L. Burger. Jeffrey Quilter, general editor. Dumbarton Oaks Research Library and Collection. Washington, D.C. pp. 213–257.

(Note: A copy of this paper was distributed when it was delivered in the 1995 meeting, under the title of "Subsistence in the Ecuadorian Formative: Overview and Comparison to the Central Andes.")

2008. Plant domestication and the shift to agriculture in the Andes. In: *The Handbook of South American Archaeology*, edited by Helaine Silverman and William H. Isbell. Springer. New York. pp. 105–120.

2009. Investigating the transition to Agriculture. *Current Anthropology*, 50 (5): 609–613.

PEARSALL, Deborah M., Karol CHANDLER-EZELL, and Alex CHANDLER-EZELL

2003. Identifying maize in neotropical sediments and soils using cob phytoliths. *Journal of Archaeological Science*, 30 (5): 611–627.

PEARSALL, Deborah M., Karol CHANDLER-EZELL, and J. A. ZEIDLER

2004. Maize in ancient Ecuador: Results of residue analysis of stone tools from the Real Alto. *Journal of Archaeological Science*, 31 (4): 423–442.

PEARSALL, Deborah M., and Dolores R. PIPERNO

1990. Antiquity of maize cultivation in Ecuador: Summary and reevaluation of the evidence. *American Antiquity*, 55 (2): 324–337.

(editors) 1993a. *Current Research in Phytolith Analysis: Applications in Archaeology and Paleoecology*. MASCA Research Papers in Science and Archaeology, Vol. 10, Museum Applied Science Center for Archaeology. University of Pennsylvania Museum. Philadelphia.

1993b. The nature and status of phytolith analysis. In: *Current Research in Phytolith Analysis: Applications in Archaeology and Paleoecology*, edited by Deborah M. Pearsall and Dolores R. Piperno. MASCA Research Papers in Science and Archeaology, Vol. 10, Museum Applied Science Center for Archaeology. University of Pennsylvania Museum. Philadelphia. pp. 9–18.

PERRY, Linda

2004. Starch analysis reveals the relationship between tool type and function: An example from the Orinoco Valley of Venezuela. *Journal of Archaeological Science*, 31: 1069–1081.

PERRY, Linda, Daniel SANDWEISS, Dolores PIPERNO, Kurt RADEMAKER, Michael A. MALPASS, Adán UMIRE, and Pablo de la VERA

2006. Early maize agriculture and interzonal interaction in southern Peru. *Nature*, 440 (7080): 76–79.

PICKERSGILL, Barbara

1969. The archaeological record of chili peppers (*Capsicum* spp.) and the sequence of plant domestication in Peru. *American Antiquity*, 34 (1): 54–61.

1972. Cultivated plants as evidence for cultural contacts. *American Antiquity*, 37 (1): 97–104.

1977. Taxonomy and the origin and evolution of cultivated plants in the New World. *Nature*, 268 (5621): 591–595.

1983. Dispersal and distribution in crop plants. *Sonderband des naturwissenschaft-tlichen Vereis in Hamburg*, 7: 285–301.

1989. Cytological and genetical evidence on the domestication and diffusion of crops within the Americas. In: *Foraging and Farming: The Evolution of Plant Exploitation*, edited by D. R. Harris and G. C. Hillman. Unwin Hyman. London. pp. 426–439.

2007. Domestication of plants in the Americas: Insights from Mendelian and molecular genetics. *Annals of Botany*, 100: 925–940.

2009. Domestication of plants revisited – Darwin to the present day. *Botanical Journal of the Linnean Society*, 161: 203–212.

PICKERSGILL, Barbara, and Charles B. HEISER Jr.

1976. Cytogenetics and evolutionary change under domestication. *Philosophical Transactions of the Royal Society*, 275: 55–69.

1978. Origins and distribution of plants domesticated in the New World tropics. In: *Advances in Andean Archaeology*, edited by David L. Browman. Mouton. The Hague, Paris. pp. 133–165.

PIPERNO, Dolores R.

1984. A comparison and differentiation of phytoliths from maize and wild grasses: Use of morphological criteria. *American Antiquity*, 49 (2): 361–383.

1985a. Phytolith analysis and tropical paleo-ecology: Production and taxonomic significance of siliceous forms in New World plant domesticates and wild species. *Review of Paleobotany and Palynology*, 45: 185–228.

1985b. Phytolithic analysis of geological sediments from Panama. *Antiquity*, LIX: 13–19.

1988a. *Phytolith Analysis: An Archaeological and Geological Perspective*. Academic Press. San Diego.

1988b. Primer informe sobre los fitolitos de las plantas de OGSE-80 y la evidencia del cultivo de maíz en el Ecuador. In: *La Prehistoria Temprana de la Península de Santa Elena, Ecuador: Cultura Las Vegas*, edited by Karen E. Stothert. Banco Central del Ecuador. Miscelánea Antropológica Ecuatoriana, Serie Monográfica 10. Guayaquil. pp. 203–214.

1989. The occurrence of phytoliths in the reproductive structure of selected tropical angiosperm and their significance in tropical paleoecology, Paleo-ethnobotany and systematics. *Review of Paleobotany and Palynology*, 61: 147–173.

1990. Aboriginal agriculture and land usage in the Amazon basin, Ecuador. *Journal of Archaeological Science*, 17: 665–677.

1991. The status of phytolith analysis in the American tropics. *Journal of World Prehistory*, 5 (2): 155–191.

1994a. On the emergence of agriculture in the New World. *Current Anthropology*, 35 (5): 637–639.

1994b. Phytolith and charcoal evidence for prehistoric slash-and-burn agriculture in the Darien rain forest of Panama. *The Holocene*, 4: 321–325.

1995. Plant microfossils and their application in the New World tropics. In: *Archaeology in the Lowlands American Tropics: Current Analytical Methods and Recent Applications*, edited by P. W. Stahl. Cambridge University Press. Cambridge. pp. 130–153.

1998. Paleoethnobotany in the neotropics from microfossils: New insights into ancient plant use and agricultural origins in the tropical forest. *Journal of World Prehistory*, 12: 393–449.

2003a. Comments. *Current Anthropology*, 44 (5): 694–695.

2003b. A few kernels short of a cob: On the Staller and Thompson late entry scenario for the introduction of maize into northern South America. *Journal of Archaeological Science*, 30: 831–836.

2006. *Phytoliths: A Comprehensive Guide for Archaeologists and Paleoecologists*. AltaMira Press. Lanham.

2009. Identifying crop plants with phytoliths (and starch grains) in Central and South America: A review and update of the evidence. *Quaternary International*, 193: 146–159.

2011. Northern Peruvian Early and Middle Preceramic agriculture in Central and South American contexts. In: *From Foraging to Farming in the Andes: New*

Perspectives on Food Production and Social Organization, edited by Tom D. Dillehay. Cambridge University Press. Cambridge. pp. 275–284.

PIPERNO, Dolores R., Mark B. BUSH, and Paul A. COLINVAUX

1990. Paleoenvironments and human occupation in Late Glacial Panama. *Quaternary Research*, 33: 108–116.

1991. Paleoecological perspectives on human adaptation in central Panama. II. The Holocene. *Geoarchaeology*, 6: 227–250.

PIPERNO, Dolores R., and Karen H. CLARY

1984. Early plant use and cultivation in the Santa María basin, Panama: Data from phytoliths and pollen. In: *Recent Advances in the Isthmian Archaeology*, edited by F. Lange. Proceedings of the 44th International Congress of Americanists. British Archaeological Reports. International Series 212. Oxford. pp. 85–121.

PIPERNO, Dolores R., Karen H. CLARY, Richard G. COOK, Anthony J. RANERE, and Doris WEILAND

1985. Preceramic maize in central Panama: Phytolith and pollen evidence. *American Anthropologist*, 87 (4): 871–878.

PIPERNO, Dolores R., and Kent V. FLANNERY

2001. The earliest archaeological maize (*Zea mays* L.) from highland Mexico: New accelerator mass spectrometry dates and their implications. *Proceedings of the National Academy of Sciences of the United States of America*, 98 (4): 2101–2103.

PIPERNO, Dolores R., and I. J. HOLST

1998. The presence of starch grains on prehistoric stone tools from the humid neotropics: Indications of early use and agriculture in Panama. *Journal of Archaeological Science*, 25: 765–776.

PIPERNO, Dolores R., I. J. HOLST, A. J. RANERE, P. HANSELL, and Karen E. STOTHERT

2001. The occurrence of genetically controlled phytoliths from maize cobs and starch grains from maize kernels on archaeological stone tools and human teeth, and in archaeological sediments from southern Central America and northern South America. *The Phytolitharien*, 13: 1–7.

PIPERNO, Dolores R., and John G. JONES

2003. Paleoecological and archaeological implications of a Late Pleistocene/Early Holocene record of vegetation and climate from the Pacific coastal plain of Panama. *Quaternary Research*, 59: 79–87.

PIPERNO, Dolores R., J. E. MORENO, José IRIARTE, I. J. HOLST, M. LACHNIET, John G. JONES, Anthony J. RANERE, and R. CASTANZO

2007. Late Pleistocene and Holocene environmental history of the Iguala Valley, Central Balsas watershed of Mexico. *Proceedings of the National Academy of Sciences of the United States of America*, 104 (29): 11874–11881.

PIPERNO, Dolores R., and Deborah M. PEARSALL

1993. Phytoliths in the reproductive structures of maize and teosinte: Implication for the study of maize evolution. *Journal of Archaeological Science*, 20: 337–362.

1998. *The Origin of Agriculture in the Lowland Neotropics.* Academic Press. San Diego.

PIPERNO, Dolores R., Anthony J. RANERE, Irene HOLST, and Patricia HANSELL

2000. Starch grains reveal early root crop horticulture in the Panamian tropical forest. *Nature*, 407 (6806): 894–897.

PIPERNO, Dolores R., Anthony J. RANERE, Irene HOLST, José IRIARTE, and Ruth DICKAU

2009. Starch grain and phytoliths evidence for early ninth millenium B.P. maize from the Central Balsas River valley, Mexico. *Proceedings of the National Academy of Sciences of the United States of America*, 106 (13): 5919–5024.

PIPERNO, Dolores R., and Karen E. STOTHERT

2003. Phytolith evidence for Early Holocene *Cucurbita* domestication in southern Ecuador. *Science*, 299 (5609): 1054–1057.

PIZARRO, Hernando

(1533) 1968. *Carta de Hernando Pizarro.* Biblioteca Peruana, Primera Serie. Tomo I. Editores Técnicos Asociado, S.A. Lima. pp. 117–130.

PIZARRO, Pedro

(1571) 1968. *Relación del descubrimiento y conquista de los reinos del Perú.* Biblioteca Peruana, Primera Serie. Tomo I. Editores Técnicos Asociados, S.A. Lima. pp. 439–586.

POGGIO, L., V. CONFALONIERI, C. COMAS, A. CUADRADO, N. JOUVE, and C. A. NARANJO

1999. Genomic in situ hybridization (GISH) of *Tripsacum dactyloides* and *Zea mays* spp. *mays*. *Genome*, 42 (4): 687–691.

POHL, Mary E. D., Dolores R. PIPERNO, Kevin O. POPE, and John G. JONES

2007. Microfossil evidence for the pre-Columbian maize dispersals in the neotropics from San Andrés, Tabasco, México. *Proceedings of the National Academy of Sciences of the United States of America*, 104 (16): 6870–6875.

POHL, Mary, Kevin O. POPE, John G. JONES, John S. JACOB, Dolores R. PIPERNO, Susan D. de FRANCE, David L. LENTZ, John A. GIFFORD, Marie E. DANFORTH, and Kathryn J. JOSSERAND

1996. Early agriculture in the Maya lowlands. *Latin American Antiquity*, 7 (4): 355–372.

POLO DE ONDEGARDO, Juan

(1561) 1940. Informe del Licenciado Juan Polo de Ondegardo al Licenciado Briviesca de Muñatones sobre la perpetuidad de las encomiendas en el Perú. *Revista Histórica*, XIII: 128–196.

POPE, Kevin O., Mary E. D. POHL, John G. JONES, David L. LENZ, Christopher von NAGY, Francisco J. VEGA, and Irvy R. QUITMYER

2001. Origin and environmental setting of ancient agriculture in the lowlands of Mesoamerica. *Science*, 292 (5520): 1370–1373.

POPPER, Virginia

1982. Análisis general de las muestras. In: *Precerámico peruano. Los Gavilanes. Mar, desierto y oasis en la historia del hombre*, D. Bonavia, with the collaboration of R. Castro de la Mata, F. Caycho Quispe, A. Grobman, L. Kaplan, C. A. Morán Val, R. Patrucco, M. Peña, V. Popper, E. J. Reitz, S. G. Stephens, R. Tello, and E. S. Wing. Corporación Financiera de Desarrollo S.A. COFIDE, Instituto

Arqueológico Alemán, Comisión de Arqueología General y Comparada. Lima. pp. 148–156.

PORRAS BARRENECHEA, Raúl

1951. Prólogo. In: *Lexicon y vocabulario de la lengua general del Perú, por el maestro fray Santo Tomás, O.P. Domingo de.* Facsimile edition, with a prologue by ... Instituto de Historia (Facultad de Letras en el IV Centenario de la Universidad Nacional de San Marcos). Lima. pp. V–XXXII.

1986. *Los cronistas del Perú (1528–1650) y otros ensayos.* Biblioteca Clásicos del Perú 2. Banco de Crédito del Perú. Lima.

POZORSKI, Shelia G.

1979. Prehistoric diet and subsistence of the Moche Valley, Peru. *World Archaeology*, 11 (2): 163–184.

POZORSKI, Shelia G., and Thomas POZORSKI

1987. *Early Settlement and Subsistence in the Casma Valley, Peru.* University of Iowa Press. Iowa City.

1997. Cherimoya and guanabana in the archaeological record of Peru. *Journal of Ethnobiology*, 17 (2): 235–248.

PRESCOTT, Guillermo H.

1995. *Historia de la Conquista del Perú.* Ediciones Imán. Buenos Aires.

PURSEGLOVE, J. W.

1972. *Tropical Crops: Monocotyledones 1.* Halsted Press. New York.

PRYWER, L. C.

1960. Estudios citológicos sobre algunas especies del género *Tripsacum*. *Boletín de la Sociedad Botánica Mexicana*, 25: 1–21.

QUILTER, Jeffrey

1992. To fish in the afternoon: Beyond subsistence economies in the study of early Andean civilizations. *Andean Past*, 3: 111–125.

QUILTER, Jeffrey, Bernardino OJEDA, Deborah M. PEARSALL, Daniel SANDWEISS, John G. JONES, and Elizabeth S. WING

1991. The subsistence economy of El Paraiso, an early Peruvian site. *Science*, 251 (4991): 277–283.

RAIMONDI, Antonio

(1859–1861) 1942. *Notas de Viajes para su obra "El Perú."* 1er Volumen. Imprenta Torres Aguirre. Lima.

(Note: "Viaje al Norte y a la región del Amazonas. Años 1859–1861" corresponds to the Libretas Nos. 10 and 11.)

RAMÍREZ, E. Ricardo, David H. TIMOTHY, B. Efraín DÍAZ, and U. J. GRANT, in collaboration with G. Edward NICHOLSON, Edgar ANDERSON, and William L. BROWN

1960. *Race of Maize in Bolivia.* National Academy of Sciences, National Research Council. Publication 747. Washington, D.C.

RANDOLPH, L. F.

1952. New evidence on the origin of maize. *American Naturalist*, 86: 193–202.

1955. History and origin of corn. II. Cytogenetic aspects of the origin and evolutionary history of corn. In: *Corn and Corn Improvement*, edited by G. F. Sprague. Academic Press. New York. pp. 16–61.

1959. The origin of maize. *Indian Journal of Genetics and Plant Breeding*, 19: 1–12.

1972. Retention of *Euchlaena* as a genus separate from Zea. *Maize Genetics Cooperation Newsletter*, 46: 8–18.

1976. Contributions of wild relatives of maize to the evolutionary history of domesticated maize: A synthesis of divergent hypothesis I. *Economic Botany*, 30: 321–345.

(Note: Randolph did not publish the second part. See Anonymous, 1982. Goodman, 1988, gives the title of this second part, "Contributions of Wild Relatives of Maize to the Evolutionary History of Different Hypothesis," as an "unpublished manuscript.")

RANERE, A. J.

1980. Preceramic shelters in the Talamancan range. In: *Adaptive Radiations in Prehistoric Panama*, edited by D. Linares and A. J. Ranere. Monograph 5. Peabody Museum of Archaeology and Ethnology. Harvard University. Cambridge. pp. 16–43.

RANERE, A. J., and P. HANSELL

1978. Early subsistence patterns along the Pacific coast of central Panama. In: *Prehistoric Coastal Adaptations: The Economy and Ecology of Maritime Middle America*, edited by B. L. Stark and B. Voorhies. Academic Press. New York. pp. 43–59.

RANERE, Anthony J., Dolores R. PIPERNO, Irene HOLST, Ruth DICKAU, and José IRIARTE

2009. The cultural and chronological context of early Holocene maize and squash domestication in the Central Balsas River valley, Mexico. *Proceedings of the National Academy of Sciences of the United States of America*, 106 (13): 5014–5018.

RAYMOND, J. Scott, and Warren R. DE BOER

2006. Maize on the move. In: *Histories of Maize: Multidisciplinary Approaches to the Prehistory, Linguistics, Biogeography, Domestication and Evolution of Maize*, edited by John Staller, Robert Tykot, and Bruce Benz. Academic Press, Elsevier. San Diego, London. pp. 337–342.

REAL ACADEMIA ESPAÑOLA

2001. *Diccionario de la Lengua Española*. Impreso en Mateu Cromo. Artes Gráficas S.A. Madrid.

REBER, Eleanora A., and Richard P. EVERSHED

2004. Identification of maize in absorbed organic residues: A cautionary tale. *Journal of Archaeological Science*, 31: 399–410.

REEVES, R. G.

1944. Chromosome knobs in relation to the origin of maize. *Genetics*, 29: 141–147.

REEVES, R. G., and Paul C. MANGELSDORF

1959. The origin of corn. II. Teosinte a hybrid of corn and *Tripsacum*. *Botanical Museum Leaflets*, 18 (8): 357–387.

RILEY, Thomas J., Gregory R. WALZ, Charles J. BAREIS, Andrew C. FORTIER, and Kathryn E. PARKER

1994. Accelerator mass spectrometry (AMS) dates confirm early *Zea mays* in the Mississippi River valley. *American Antiquity*, 59 (3): 490–498.

RINDOS, D.

1984. *The Origin of Agriculture: An Evolutionary Perspective.* Academic Press. San Diego.

RIVAL, Laura, and Doyle McKEY

2008. Domestication and diversity in manioc (*Manihot esculenta* Crantz ssp. *esculenta*, Euphorbiaceae). *Current Anthropology*, 49 (6): 1119–1128.

RIVERA, Mario A.

1971. Nuevos enfoques de la teoría arqueológica aplicada al Norte Chico. Offprint of *Actas del VI Congreso de Arqueología Chilena.* Universidad de Chile. Departamento de Ciencias Antropológicas y Arqueología. Sociedad de Arqueología. Santiago. pp. 295–310.

1978a. La agriculturización del maíz en el Norte de Chile. Actualización de Problemas y Metodología de Investigación. *V Congreso Nacional de Arqueología Argentina.* Museo de Historia Natural. San Juan. pp. 157–180.

(Note: This paper was republished in 1980 without any changes at all. See Rivera, 1980b.)

1978b. Cronología absoluta y periodificación en la arqueología chilena. *Boletín del Museo Arqueológico de La Serena*, 16: 13–21.

1980a. *Temas antropológicos del Norte de Chile.* Ediciones de la Universidad de Chile. Antofagasta.

1980b. La agriculturización del maíz en el norte de Chile: actualización de problemas y metodología de investigación. *Estudios Arqueológicos.* Número especial. Temas antropológicos del Norte de Chile. Antofagasta. pp. 105–129.

(Note: This paper was published with the same title and text in 1978. See Rivera, 1978a. There is a manuscript dating to this same year that was distributed among specialists, which has footnotes that were not included in the published text.)

1980c. La agriculturización del Maíz. *Creces*, 2 (11): 36–42.

1991. The prehistory of northern chile: A synthesis. *Journal of World Prehistory*, 5 (1): 1–47.

2006. Prehistoric maize from northern Chile. In: *Histories of Maize: Multidisciplinary Approaches to the Prehistory, Linguistics, Biogeography, Domestication and Evolution of Maize*, edited by John Staller, Robert Tykot, and Bruce Benz. Academic Press, Elsevier. San Diego, London. pp. 403–413.

ROBERTS, L. M., U. J. GRANT, E. R. RAMÍREZ, W. H. HATHEWAY, and D.L. SMITH, in collaboration with Paul C. MANGELSDORF

1957. *Races of Maize in Colombia.* National Academy of Sciences. National Research Council. Publication N° 510. Washington, D.C.

ROCHEBRUNE, Alphonse Tremau de

1879. Recherches d'ethnographie botanique sur la flores des sepultures Péruviennes d'Ancon. *Acte de la Société Linneaus*, Bordeaux, 3: 343–358.

RODRÍGUEZ, A., M. ROMERO, J. QUIROGA, A. ÁVILA, and A. BRANDOLINI

1968. *Maíces Bolivianos.* Food and Agriculture Organization. Rome.

RODRÍGUEZ, María Fernanda, and Carlos Alberto ASCHERO

2007. Archaeological evidence of *Zea mays* L. (Poaceae) in the southern Argentinian puna (Antofagasta de la Sierra, Catamarca). *Journal of Ethnobiology*, 27 (2): 256–271.

ROE, Daphne A.
1976. *A Plague of Corn: The Social History of Pellagra*. Cornell University Press. Ithaca and London.
ROGERS, J. S.
1950. The inheritance of inflorescence characters in maize-teosinte hybrids. *Genetics*, 35: 541–558.
ROIG, Fidel A., Virgilio G. ROIG, and Roberto J. BÁRCENA
1985. Aportes Arqueofito-Zoológicos para la prehistoria del N.O. de la provincia de Mendoza: La excavación de Agua de la Tinaja 1. *Trabajos de Prehistoria*, 42 (1): 311–363.
ROOSEVELT, Anna Curtenius
1980. *Parmana: Prehistoric Maize and Manioc Subsistence along the Amazon and Orinoco*. Academic Press. New York.
1984. Problems interpreting the diffusion of cultivated plants. In: *Pre-Columbian Plant Migration*, edited by Doris Stone. Papers of the Peabody Museum of Archaeology and Ethnology. Vol. 76. Harvard University Press. Cambridge. pp. 1–18.
ROSSEN, Jack T.
2011. Preceramic plant gathering, gardening and farming. In: *From Foraging to Farming in the Andes: New Perspectives on Food Production and Social Organization*, edited by Tom D. Dillehay. Cambridge University Press. Cambridge. pp. 177–192.
ROSSEN, Jack T., Tom Dalton DILLEHAY, and Donald UGENT
1996. Ancient cultigens or modern intrusion? Evaluating plant remains in Andean study. *Journal of Archaeological Science*, 23: 391–407.
ROSTWOROWSKI DE DIEZ CANSECO, Maria
1977. *Etnia y sociedad: Costa Peruana prehispánica*. Instituto de Estudios Peruanos. Lima.
1983. *Estructuras andinas de poder: ideología religiosa y política*. Instituto de Estudios Peruanos. Lima.
ROUSELL, Théophile
1845. *De la pellagre. De son origine, de ses progrès, de son existence en France, de ses causes et de son traitment curatif et préservatif.* Hennuyer et Turpin. Paris.
ROUSH, Wade
1996. Corn: A lot of change from a little DNA. *Science*, 272 (5270): 1873.
ROVNER, Irwin
1995. Mien, mean, and meaning: The limits of typology in phytolith analysis. Paper presented at the 60th Annual Meeting of the Society for American Archaeology. Minneapolis.
1996. (Review) Current Research in Phytolith Analysis: Applications in Archaeology and Paleoecology. Deborah M. Pearsall and Dolores R. Piperno, editors. MASCA Research Papers in Science and Archaeology. Vol. 10. Museum Applied Science Center for Archaeology, University Museum, University of Pennsylvania, 1993. *American Antiquity*, 61 (2): 430–431.
1999. Phytolith analysis. *Science*, 283 (5401): 488–489.

ROVNER, Irwin, and John C. RUSS

1992. Darwin and design in phytolith systematics: Morphometric methods for mitigating redundancy. In: *Phytolith Systematics: Emerging Issue*, edited by G. Rapp Jr. and S. C. Mulholland. Advances in Archaeological and Museum Science. Vol. 1. Plenum Press. New York. pp. 253–276.

ROWE, John Howland

1946. Inca culture at the time of the Spanish conquest. In: *Handbook of South American Indians. Vol. 2. The Andean Civilizations*, edited by Julian H. Steward. Smithsonian Institution. Bulletin 43. Washington, D.C. pp. 183–330.

(assembled by) 1960. Highland South America. In: Notes and News, Nathalie F. S. Woodbury, editor. *American Antiquity*, 26 (1): 140–142.

1962. Stage and periods in archaeological interpretation. *Southwestern Journal of Anthropology*, 18 (1): 40–54.

1963. Urban settlement in ancient Perú. *Ñawpa Pacha*, 1: 1–27.

1964. A review of radiocarbon measurements from Peru and Bolivia. (Ditto copy.)

1965a. The origin of plant cultivation in ancient Peru. Read at the 64th Annual Meeting of the American Anthropological Association. Denver.

(Note: This paper was not published.)

1965b. An interpretation of radiocarbon measurements on archaeological samples from Peru. In: *Proceedings of the Sixth International Conference, Radiocarbon and Tritium Dating*, held at Washington State University, Pullman. Washington, D.C. pp. 187–198.

ROWE, John Howland, and Dorothy MENZEL

1967. Introduction. In: *Peruvian Archaeology: Selected Readings*, edited by John Howland Rowe and Dorothy Menzel. Peek. Palo Alto. pp. v–x.

ROWLEY, J. R.

1960. The exine structure of "Cereal" and "wild" type grass pollen. *Grana Palynologica*, 2 (2): 9–15.

RUE, D. J.

1989. Archaic Middle American agriculture and settlement: Recent pollen data from Honduras. *Journal of Field Archaeology*, 16: 177–184.

RUIZ DE ARCE (or ALBUQUERQUE), Juan

(1545) 1968. *Advertencias.* Biblioteca Peruana. Primera Serie. Tomo I. Editores Técnicos Asociados, S.A. Lima. pp. 405–437.

RUSS, John C., and Irwin ROVNER

1989. Stereological identification of opal phytolith populations from wild and cultivated *Zea*. *American Antiquity*, 54 (4): 784–792.

SAFFORD, William Edwin

1917. Food-plants and textiles of ancient America. *Proceedings of the 19th International Congress of Americanists.* Washington, D.C. pp. 12–30.

SAHAGÚN, Fray Bernardino

1963. *General History of the Things of New Spain: Florentine Codex.* Book 11. Translated by C. E. Dibble and A. J. O. Anderson. Monographs of the School of American Research and the Museum of New Mexico, N° 14, Pts. 12. School of American Research, Santa Fe, and the University of Utah. Salt Lake City.

SAINT-HILAIRE, A. de

1829. Lettre sur une varieté remarquable de maïs de Brésil. *Annales des Sciences Naturelles*, 16: 143–145.

SALHUANA M., Wilfredo

2004. Diversidad y descripción de las razas de maíz en el Perú. In: *Cincuenta Años del Programa Cooperativo de Investigaciones en Maíz (PCIM). Logros y perspectivas*, edited by Wilfredo Salhuana M., Américo Valdéz M., Federico Scheuch H., and José Davelouis M. Universidad Nacional Agraria La Molina. Lima. pp. 204–251.

SALHUANA M., Wilfredo, and Americo VALDÉZ M.

2004. Usos del maíz en el Perú. In: *Cincuenta Años del Programa Cooperativo de Investigaciones en Maíz (PCIM). Logros y perspectivas*, edited by Wilfredo Salhuana M., Américo Valdéz M., Federico Scheuch H., and José Davelouis M. Universidad Nacional Agraria La Molina. Lima. pp. 472–476.

SÁNCHEZ GONZÁLES, José Luis

1994. Modern variability and patterns of maize movement in Mesoamerica. In: *Corn and Culture in the Prehistoric New World*, edited by Sissel Johannessen and Christine A. Hastorf. Westview Press. Boulder. pp. 135–156.

SANCHO DE LA HOZ, Pedro

(1534) 1968. *Relación para Su Majestad*. Biblioteca Peruana, Primera Serie. Tomo I. Editores Técnicos Asociados, S.A. Lima. pp. 275–343.

SANOJA, Mario

1981. *Los hombres de la yuca y el maíz*. Monte Avila Editores C.A. Caracas.

1989. From foraging to food production in northeastern Venezuela and the Caribbean. In: *Foraging and Farming: The Evolution of Plant Exploitation*, edited by David R. Harris and Gordon C. Hillman. Unwin Hyman. London. pp. 523–537.

SANTILLÁN, Hernando de

(1563) 1968. *Relación del origen, descendencia política y gobierno de los Incas*. Biblioteca Peruana, Primera Serie. Tomo III. Editores Técnicos Asociados, S.A. Lima. pp. 375–463.

SANTO TOMÁS, O.P., Domingo de

(1560) 1951. *Lexicón o vocabulario de la lengua general del Perú*, por el maestro fray … Facsimile edition, with a prologue by Raúl Porras Barrenechea. Instituto de Historia de la Facultad de Letras en el IV Centenario de la Universidad Nacional Mayor de San Marcos, II. Lima.

SASS, J. E.

1955. Vegetative morphology. In: *Corn and Corn Improvement*, edited by G. F. Sprague. Academic Press. New York. pp. 63–97.

SAUER, Carl O.

1950. Cultivated plants of South and Central America. In: *Hanbook of South American Indians. Vol. 6. Physical Anthropology, Linguistics and Cultural Geography of South American Indians*, edited by Julian H. Steward. Smithsonian Institution, Bureau of American Ethnology. Bulletin 143. Washington, D.C. pp. 487–543.

1952. *Agricultural Origins and Dispersals.* American Geographical Society. New York.

1969a. *Seeds, Spades, Hearths, and Herds: The Domestication of Animals and Foodstuff.* MIT Press. Cambridge.

1969b [1960]. Maize into Europe. In: *Seeds, Spades, Hearths, and Herds: The Domestication of Animals and Foodstuff.* MIT Press. Cambridge. pp. 147–167.

SCHAAFFHAUSEN, R. V.

1952. Adlay or Job's tears. *Economic Botany*, 6: 216–227.

SCHIAPACASSE, V.

1988. Maíces y llamas. *Boletín de la Sociedad Chilena de Arqueología*, 8: 7–8.

SCHIAPACASSE, V., and H. NIEMEYER

1984. *Descripción y análisis interpretativo de un sitio arcaico temprano en la quebrada de Camarones.* Publicación Ocasional 41. Ediciones del Museo Nacional de Historia Natural de Santiago. Santiago.

SCHOENINGER, Margaret J.

2009. Stable isotope evidence for the adoption of maize agriculture. *Current Anthropology*, 50 (5): 633–640.

SCHOENWETTER, James

1974. Pollen records of Guilá Naquitz Cave. *American Antiquity*, 39 (2): 292–303.

SCHOENWETTER, James, and Landon Douglas SMITH

1986. Pollen analysis of the Oaxacan Archaic. In: *Guilá Naquitz: Archaic Foraging and Early Agriculture in Oaxaca, México*, edited by Kent V. Flannery. Academic Press. Orlando. pp. 179–237.

SCHWARCZ, Henry P.

2006. Stable carbon isotope analysis and human diet: A synthesis. In: *Histories of Maize: Multidisciplinary Approaches to the Prehistory, Linguistics, Biogeography, Domestication and Evolution of Maize*, edited by John Staller, Robert Tykot, and Bruce Benz. Academic Press, Elsevier. San Diego, London. pp. 315–321.

SEARS, Paul B.

1952. Palynology in southern North America. 1. Archaeological horizons in the basin of México. *Geological Society of America Bulletin*, 63 (3): 241–254.

1982. Fosil maize pollen in México. *Science*, 216 (4549): 932–934.

SEARS, Paul B., and K. H. CLISBY

1952. Two long climatic records. *Science*, 116 (3007): 176–178.

SEDEROFF, R. R., C. S. LEVINGS III, D. H. TIMOTHY, and W. W. L. HU

1981. Evolution of DNA sequence organization in mithochondrial genomes of *Zea*. *Proceedings of the National Academy of Sciences of the United States of America*, 78 (10): 5953–5957.

SEGOVIA, Víctor, Aribek MADIM, Mario PÉREZ, and Francia FUENMAYOR

1999. *Origen, evolución e historia del maíz venezolano.* Centro Nacional de Investigaciones Agropecuarias. Caracas.

SEVILLA PANIZO, Ricardo

1994. Variation in modern Andean maize and its implications for prehistoric patterns. In: *Corn and Culture in the Prehistoric New World*, edited by Sissel Johannessen and Christine A. Hastorf. Westview Press. Boulder. pp. 219–244.

SEVILLA PANIZO, Ricardo, and Miguel HOLLE OSTENDORF

2004. *Recursos genéticos vegetales.* Luis León Asociados S.R.L. Editores. Lima.

SHADY SOLIS, Ruth

2001. Caral-Supe y la Costa Norcentral del Perú: La cuna de la civilización y la formación del estado prístino. In: *Historia de la Cultura peruana I.* Fondo Editorial del Congreso del Perú. Lima. pp. 45–87.

(2000) 2003. Sustento socioeconómico del Estado prístino de Supe-Perú: las evidencias de Caral-Supe. In: *La Ciudad sagrada de Caral-Supe. Los orígenes de la civilización andina y la formación del Estado prístino en el antiguo Perú*, edited by Ruth Shady Solis and Carlos Leyva. Instituto Nacional de Cultura. Proyecto Especial Arqueológico Caral-Supe. Lima. pp. 107–122.

(Note: This paper was first published without any changes in 2000, in *Arqueología y Sociedad*, 13: 49–66.)

2004. *Caral. La Ciudad del Fuego Sagrado. The City of the Sacred Fire.* Interbank. Lima.

2006. Caral-Supe and the north-central area of Peru: The history of maize in the land where civilization came into being. In: *Histories of Maize: Multidisciplinary Approaches to the Prehistory, Linguistics, Biogeography, Domestication and Evolution of Maize*, edited by John Staller, Robert Tykot, and Bruce Benz. Academic Press, Elsevier. San Diego, London. pp. 381–402.

SHADY SOLIS, Ruth, Camilo DOLORIER, Fanny MONTESINOS, and Lyda CASAS

(2000) 2003. Los orígenes de la civilización en el Perú: el área norcentral y el valle de Supe durante el Arcaico Tardío. In: *La Ciudad sagrada de Caral-Supe. Los orígenes de la civilización andina y la formación del Estado prístino en el antiguo Perú*, edited by Ruth Shady Solis and Carlos Leyva. Instituto Nacional de Cultura. Proyecto Especial Arqueológico Caral-Supe. Lima. pp. 51–91.

(Note: This paper was first published without any changes in 2000 in *Arqueología y Sociedad*, 13: 13–48.)

SHADY SOLIS, Ruth, Jonathan HAAS, and Winifred CREAMER

2001. Dating Caral, a preceramic site in the Supe Valley on the coast of Peru. *Science*, 292 (5517): 723–726.

SHADY SOLIS, Ruth, and Carlos LEYVA (editors)

2003. *La Ciudad sagrada de Caral-Supe. Los orígenes de la civilización andina y la formación del Estado prístino en el antiguo Perú.* Instituto Nacional de Cultura. Proyecto Especial Arqueológico Caral-Supe. Lima.

SHADY SOLIS, Ruth, and Marco MACHACUAY

(2000) 2003. El Altar del Fuego Sagrado del Templo Mayor de la Ciudad Sagrada de Caral-Supe. In: *La Ciudad sagrada de Caral-Supe. Los orígenes de la civilización andina y la formación del Estado prístino en el antiguo Perú*, edited by Ruth Shady Solis and Carlos Leyva. Instituto Nacional de Cultura. Proyecto Especial Arqueológico Caral-Supe. Lima. pp. 169–185.

(Note: This paper was first published without any changes in 2000 in *Boletín del Museo de Arqueología y Antropología*, UNMSM, 3 [12]: 2–18.)

SHEETS, P. D.

1994. The proyecto prehistórico Arenal: An introduction. In: *Archaeology, Volcanism and Remote Sensing in the Arenal Region, Costa Rica*, edited by P. D. Sheets and B. R. McKen. University of Texas Press. Austin. pp. 1–23.

SILVA Y GUZMÁN, Diego de

(1538) 1968. *La crónica rimada*. (Versión de F. Rand Morton, ediciones de Andrea. México, D.F. 1963), Biblioteca Peruana, Primera Serie. Tomo I. Editores Técnicos Asociados, S.A. Lima. pp. 17–115.

SIMMONS, Alan H.

1986. New evidence for the early use of cultigens in the American Southwest. *American Antiquity*, 51 (1): 73–89.

SINGLETON, W. R.

1951. Inheritance of corn grass, a macromutation in maize and its posible significance as ancestral type. *American Naturalist*, 85: 81–86.

SLUYTER, Andrew, and Gabriela DOMÍNGUEZ

2006. Early maize (*Zea mays* L.) cultivation in México: Dating sedimentary pollen records and its implications. *Proceedings of the National Academy of Sciences of United States of America*, 103 (4): 1147–1151.

SMALLEY, John, and Michael BLAKE

2003. Sweet beginnings: Stalk sugar and the domestication of maize. *Current Anthropology*, 44 (5): 675–689, 696–703.

SMITH, Bruce D.

1995a. *The Emergence of Agriculture*. Scientific American Library. W.H. Freeman & Co. New York.

1994–1995b. The origins of agriculture in the Americas. *Evolutionary Anthropology*, 3 (5): 174–184.

1997a. Reconsidering the Ocampo Caves and the era of incipient cultivation in Mesoamerica. *Latin American Antiquity*, 8 (4): 342–383.

1997b. The initial domestication of *Cucurbita pepo* in the Americas 10,000 years ago. *Science*, 276 (5314): 932–934.

1998. *The Emergence of Agriculture*. Scientific Library Publication. W.H. Freeman & Co. New York.

2005. Reassessing Coxcatlan Cave and the early history of domesticated plants in Mesoamerica. *Proceedings of the National Academy of Sciences of the United States of America*, 102 (27): 9438–9445.

2006. Eastern North America as an independent center of plant domestication. *Proceedings of the National Academy of Sciences of the United States of America*, 103 (33): 12223–12228.

SMITH, C. Earle, Jr.

1967. Plant remains. In: *The Prehistory of Tehuacan Valley. Vol. 1. Environment and Subsistence*, edited by Douglas Byers. Peabody Foundation. Philips Academy, Andover. University of Texas Press. Austin and London. pp. 220–255.

1968. The New World centers of origin of cultivated plants and the archaeological evidence. *Economic Botany*, 22 (3): 253–266.

1980a. Plant remains from Guitarrero Cave. In: *Guitarrero Cave: Early Man in the Andes*, edited by Thomas F. Lynch. Academic Press. New York. pp. 87–119.

1980b. Ancient Peruvian highland maize. In: *Guitarrero Cave: Early Man in the Andes*, edited by Thomas F. Lynch. Academic Press. New York. pp. 121–143.

1988. Floral remains. In: *La Galgada, Peru: A Preceramic Culture in Transition*, Terence Grieder, Alberto Bueno Mendoza, C. Earle Smith Jr., and Robert M. Malina. University of Texas Press. Austin. pp. 125–151.

SMITH, J. S. C., Major M. GOODMAN, and Takeo A. KATO-YAMAKAKE

1982. Variation within teosinte. II. Numerical analysis of chromosome knob data. *Economic Botany*, 36: 100–112.

SMITH, J. S. C., Major M. GOODMAN, and C. W. STUBER

1984. Variation within teosinte. III. Numerical analysis of allozyme data. *Economic Botany*, 38: 97–113.

1985. Relationship between maize and teosinte of Mexico and Guatemala: Numerical analysis of isozyme data. *Economic Botany*, 39: 12–24.

SMITH, J. S. C., and R. N. LESTER

1980. Biochemical systematics and evolution of Zea, *Tripsacum* and related genera. *Economic Botany*, 34: 201–218.

SOARES DE SOUSA, G.

n.d. (Written in 1578). *Noticia do Brasil.* Vol. 1. Livraría Martins Editora. São Paulo.

SOUKUP, SDB, Jaroslav

n.d. [1987]. *Vocabulario de los nombres vulgares de la Flora peruana y Catálogo de los géneros.* Editorial Salesiana. Lima.

STAAL, C.

1974. Excavación arqueológica en el sitio pre-Agrícola Tiliviche 1-b. *Serie Documentos de Trabajo. Universidad de Chile*, 5: 14–26.

STALKER, H. T., Jack R. HARLAN, and J. M. J. de WET

1977. Cytology and morphology of maize-*Tripsacum* introgression. *Crop Science*, 17: 745–748.

STALLER, John E.

2003. An examination of the palaeobotanical and chronological evidence for an early introduction of maize (*Zea mays* L.) into South America: A response to Pearsall. *Journal of Archaeological Science*, 30 (3): 373–380.

2006. The social, symbolic, and economic significance of *Zea mays* L. in the Late Horizon period. In: *Histories of Maize: Multidisciplinary Approaches to the Prehistory, Linguistics, Biogeography, Domestication and Evolution of Maize*, edited by John Staller, Robert Tykot, and Bruce Benz. Academic Press, Elsevier. San Diego, London. pp. 449–467.

2010. *Maize Cobs and Cultures: History of Zea mays* L. Springer. Heidelberg, Dordrecht, London, New York.

STALLER, John E., and Robert G. THOMPSON

2001. Reconsiderando la introducción del maíz en el occidente de América del Sur. *Bulletin de l'Institut Français d'Études Andines*, 30 (1): 123–156.

2002. A multidisciplinary approach to understanding the initial introduction of maize into coastal Ecuador. *Journal of Archaeological Science*, 29: 33–50.

STALLER, John E., Robert H. TYKOT, and Bruce F. BENZ (editors)

2006. *Histories of Maize: Multidisciplinary Approaches to the Prehistory, Linguistics, Biogeography, Domestication and Evolution of Maize.* Academic Press, Elsevier. San Diego, London.

STARK, Barbara L.

1986. Origins of food production in the New World. In: *American Archaeology Past and Future*, edited by David J. Meltzer, Don D. Fowler, and Jeremy A. Sabloff. Published for the Society for American Archaeology by the Smithsonian Institution Press. Washington, D.C. pp. 277–321.

STEBBINS, G. L.

1950. *Variation and Evolution in Plants.* Columbia University Press. New York.

STEPHENS, Stanley George

1982. Algodón (*Gossypium barbadense*). In: *Precerámico peruano. Los Gavilanes. Mar, desierto y oasis en la historia del hombre*, D. Bonavia, with the collaboration of R. Castro de la Mata, F. Caycho Quispe, A. Grobman, L. Kaplan, C. A. Morán Val, R. Patrucco, M. Peña, V. Popper, E. J. Reitz, S. G. Stephens, R. Tello, and E. S. Wing. Corporación Financiera de Desarrollo S.A. COFIDE, Instituto Arqueológico Alemán, Comisión de Arqueología General y Comparada. Lima. pp. 179–180.

STOTHERT, Karen E.

1985. The preceramic Las Vegas culture of coastal Ecuador. *American Antiquity*, 50 (3): 613–637.

(editor) 1988. *La Prehistoria Temprana de la Península de Santa Elena, Ecuador: Cultura Las Vegas.* Miscelánea Antropológica Ecuatoriana. Serie Monográfica 10. Museo Antropológico, Banco Central del Ecuador. Guayaquil.

STOTHERT, Karen E., Dolores R. PIPERNO, and Thomas C. ANDRES

2003. Terminal Pleistocene/Early Holocene human adaptations in coastal Ecuador: The Las Vegas evidence. *Quaternary International*, 109–110: 23–43.

STROSS, Brian

2006. Maize in word and image in southeastern Mesoamerica. In: *Histories of Maize: Multidisciplinary Approaches to the Prehistory, Linguistics, Biogeography, Domestication and Evolution of Maize*, edited by John Staller, Robert Tykot, and Bruce Benz. Academic Press, Elsevier. San Diego, London. pp. 577–598.

STULER, E.

1928. *Leonard Fuchs.* Munich.

STURTEVANT, E. L.

1899. Varieties of corn. United States Department of Agriculture Office of Experiment Station. Government Printing Office. *Bulletin*, 57: 1–108.

SUTO, T., and Y. YOSHIDA

1956. Characteristics of the oriental maize. In: *Land and Crops of Nepal, Himalaya.* Vol. 2, edited by H. Kihara. Fauna and Flora Research Society. Kyoto University. Kyoto. pp. 375–530.

TABA, S.

1995. Maize germplasm: Its spread, use and strategies for conservation. In: *Maize Genetic Resources*, edited by S. Taba. Maize Program Special Reports. CIMMYT. Mexico City. pp. 7–58.

TABÍO, Ernesto E.
1977. *Prehistoria de la Costa del Perú*. Academia de Ciencias de Cuba. La Habana.

TALBERT, L. E., John F. DOEBLEY, S. LARSON, and V. L. CHANDLER
1990. *Tripsacum andersonii* is a natural hybrid involving *Zea* and *Tripsacum*: Molecular evidence. *American Journal of Botany*, 77: 722–726.

TARRAGÓ, Myriam Noemí
1980. El proceso de agriculturización en el Noroeste Argentino, zona Valliserrana. In: *Actas del V Congreso Nacional de Arqueología Argentina*. Tomo I. Instituto de Investigaciones Arqueológicas y Museo. Universidad Nacional de San Juan. San Juan. pp. 181–217.

TAYLOR, Royal Erwin
1987. *Radiocarbon Dating: An Archaeological Perspective*. Academic Press. New York.

TELLO, Julio C.
1943. Discovery of the Chavin culture in Peru. *American Antiquity*, 9 (1): 135–160.

THACHER, John Boyd
1903. *Christopher Columbus: His Life, His Work, His Remains*. G.P. Putnam's Sons. New York and London.

THOMPSON, R. G., and John E. STALLER
2000. An analysis of opal phytoliths from food residues of selected sherds and dental calculus from excavations at the site of La Emerenciana, El Oro Province, Ecuador. *The Phytolitharien*, 13 (3): 8–16.

TIMOTHY, David H., W. H. HATHEWAY, U. J. GRANT, M. TORREGROZA, D. SARRIA, and D. VARELA
1963. *Races of Maize in Ecuador*. National Academy of Sciences, National Research Council. Publication 975. Washington, D.C.

TIMOTHY, David H., C. S. LEVINGS III, D. R. PRING, M. F. CONDE, and J. L. KERENIDE
1979. Organelle DNA variation and systematic relationships in the genus ZWA: Teosinte. *Proceedings of the National Academy of Sciences of the United States of America*, 76 (9): 4220–4224.

TIMOTHY, David H., Bertulfo PEÑA V., and Ricardo RAMÍREZ E., in collaboration with William L. BROWN and Edgar ANDERSON
1961. *Races of Maize in Chile*. National Academy of Sciences, National Research Council. Publication 847. Washington, D.C.

TING, Y. C.
1964. *Chromosomes of Maize-Teosinte Hybrids*. The Bussey Institution of Harvard University. Cambridge.

TOLEDO, Francisco de
1867. Memorial que D. Francisco de Toledo dio al Rey Nuestro Señor, del estado en que dejó las cosas del Perú, después de haber sido en él Virrey y Capitan General trece años que comenzaron en 1569. In: *Relaciones de los Vireyes y Audiencias que han Gobernado el Perú*. Publicadas de O.S. Tomo I. Memorial y Ordenanzas de D. Francisco de Toledo. Imprenta del Estado por J.E. del Campo. Lima.

TORRES, Barbara

1985. Las plantas útiles en el México antiguo según las fuentes del siglo XVI. In: *Historia de la Agricultura. Época prehispánica – Siglo XVI. 1*, edited by Teresa Rojas Rabiela and William T. Sanders. Colección Biblioteca del INAH. Instituto de Antropología e Historia. Mexico City, D.F. pp. 53–127.

TORRES UMAÑA, Calixto

1917. La nutrición en la altiplanicie de Bogotá. Influencia de la chicha sobre el metabolismo azoado. In: *Proceedings of the Second Pan-American Scientific Congress.* Vol. X. U.S. Government Office. Washington, D.C. pp. 52–111.

(Note: The reference Valdizán, 1990, makes to this paper is wrong.)

TOSI, Joseph A.

1960. *Zonas de vida natural en el Perú.* Vol. 5. Instituto Interamericano de Ciencias Agrícolas de la OEA. Zona Andina. Boletín Técnico. Lima. pp. 1–271.

TOWLE, Margaret Ashley

1952a. Descriptions and identifications of the Virú plants remains. Appendix 2. In: *Cultural Stratigraphy in the Virú Valley Northern Peru: The Formative and Florescent Epochs*, William Duncan Strong and Clifford Evans Jr. Columbia University Press. New York. pp. 352–356.

1952b. Plant remains from a Peruvian mummy bundle. *Botanical Museum Leaflets*, 15 (9): 223–246.

1954. Plant remains. In: *Early Ancón and Early Supe Culture*, Gordon R. Willey and John M. Corbett. Columbia Studies in Archaeology and Ethnology, Vol. III. Columbia University Press. New York. pp. 130–138.

1961. *The Ethnobotany of Pre-Columbian Peru.* Viking Fund Publication in Anthropology, Number Thirty. Wenner-Gren Foundation for Anthropological Research, Incorporated. New York.

TRUCCO, Giovanni (editor)

1935. *Grande Dizionario Enciclopédico.* Vol. V. Unione Tipografico-Editrice Torinese. Torino.

TRUE, D. L., Lautaro NÚÑEZ A., and Patricio NÚÑEZ H.

1970. Archaeological investigations in northern Chile: Project Tarapacá – Preceramic resources. *American Antiquity*, 35 (2): 170–184.

TSCHAUNER, Hartmut

1998. El Maíz y la Desigualdad Social en la América Precolombina. In: *50 Años de Estudios Americanistas en la Universidad de Bonn. 50 Years of Americanist Studies at the University of Bonn. I. Contribuciones arqueologícas. Contribuciones etno-históricas. Archaeological Contributions. Ethnohistorical Contributions*, edited by Sabine Dedenbach-Salazar Saénz, Carmen Arellanos Hoffmann, Eva König, and Heiko Prümers. Bonner Amerikanistische Studien, 30. Verlag Anton Saurwein. Markt Schwaben. pp. 321–342.

TSCHUDI, Juan Jacobo von

(1891) 1918. *Contribuciones a la historia, civilización y lingüística del Perú antiguo.* Tomo I. Colección de Libros y Documentos referentes a la Historia del Perú. Tomo IX. Imprenta y Librería Sanmarti y Cía. Lima.

TSUKADA, Matsuo, and John R. ROWLEY

1964. Identification of modern and fossil maize pollen. *Grana Palynologica*, 5: 406–412.

TURNER, Christy G., II

1978. Dental caries and early Ecuadorian agriculture. *American Antiquity*, 43 (4): 694–697.

TYKOT, Robert H.

2003. Comments. *Current Anthropology*, 44 (5): 695–696.

2004. Stable isotopes and diet: You are what you eat. In: *Proceedings of the International School of Physics "Enrico Fermi."* Course CLIV, edited by M. Martini, M. Milazzo, and M. Piacentini. IOS Press. Amsterdam. pp. 433–444.

TYKOT, Robert H., Richard L. BURGER, and Nikolaas J. van der MERWE

2006. The importance of maize in Initial period and Early Horizon Peru. In: *Histories of Maize: Multidisciplinary Approaches to the Prehistory, Linguistics, Biogeography, Domestication and Evolution of Maize*, edited by John Staller, Robert Tykot, and Bruce Benz. Academic Press, Elsevier. San Diego, London. pp. 187–197.

TYKOT, Robert H., and John E. STALLER

2002. The importance of early maize agriculture in coastal Ecuador: New data from La Emerenciana. *Current Anthropology*, 43 (4): 666–677.

UCEDA CASTILLO, Santiago Evaristo

1986. Le Paijanien de la région de Casma (Pérou): industrie lithique et relations avec les autres industries précéramiques. Thesis presented at the Université de Bordeaux to obtain the title of doctor. Bordeaux.

1987. Los primeros pobladores del área andina central. Revisión crítica de los principales sitios. *Yunga*, 1: 14–32.

1992. Industrias líticas precerámicas de Casma. In: *Estudios de Arqueología Peruana*, edited by Duccio Bonavia. FOMCIENCIAS. Lima. pp. 45–67.

UGENT, Donald S., Shelia G. POZORSKI, and Thomas POZORSKI

1982. Archaeological potato tuber remains from the Casma Valley of Peru. *Economic Botany*, 36 (2): 182–192.

UPHAM, Steadman, Richard S. MacNEISH, Walton C. GALINAT, and Christopher M. STEVENSON

1987. Evidence concerning the origin of Maíz de Ocho. *American Anthropologist*, 89 (2): 410–419.

VALCÁRCEL, Luis E.

1959. *Etnohistoria del Perú Antiguo. Historia del Perú (Incas)*. Universidad Nacional Mayor de San Marcos. Patronato del Libro Universitario. Lima.

VALDIZÁN, Hermilio

1990. La chicha, bebida de los primitivos peruanos. In: *Paleopsiquiatria del Antiguo Perú*, edited by Hermilio Valdizán. Introduction, compilation, and annotations by Javier Mariátegui. Universidad Peruana Cayetano Heredia. Fondo Editorial. Lima. pp. 135–153.

VALDIZÁN, Hermilio, and Ángel MALDONADO

1922. *La medicina popular peruana*. 3 vols. Imprenta Torres Aguirre. Lima.

Van der MERWE, Nikolaas J., J. A. LEE-THORP, and J. Scott RAYMOND

1993. Light, stable isotopes and the subsistence of Formative cultures at Valdivia, Ecuador. In: *Prehistoric Human Bone: Archaeology and the Molecular Level*, edited by Joseph Lambert and Gisela Gruppe. Springer-Verlag. Berlin. pp. 63–97.

Van der MERWE, Nikolaas J., and E. MEDINA

1991. The canopy effect, carbon isotope ratios, and foodwebs in Amazonia. *Journal of Archaeological Science*, 18: 249–260.

Van der MERWE, Nikolaas J., Anna Curtenius ROOSEVELT, and J. C. VOGEL

1981. Isotopic evidence for prehistoric subsistence change in Parmana, Venezuela. *Nature*, 292 (5823): 536–538.

Van der MERWE, Nikolaas J., and Hartmut TSCHAUNER

1999. C4 plants and development of human societies. In: *C4 Plant Biology*, edited by Rovan Friederich Sage and Russell K. Monson. Academic Press. San Diego. pp. 509–550.

Van der MERWE, Nikolaas J., and J. C. VOGEL

1978. 13C content of human collagen as a measure of prehistoric diet in woodland North America. *Nature*, 276 (5690): 815–816.

VARELA, Consuelo

1984. *Cristóbal Colón. Textos y documentos.* Preface and annotation by Consuelo Varela. Alianza Universidad (Editorial). Madrid.

VARGAS C., César

1962. Phytomorphic representations of the ancient Peruvians. *Economic Botany*, 16 (2): 49–142.

VÁSQUEZ, Mario C.

1967. La chicha en los países andinos. *América Indígena*, 27: 264–268.

VAUGHN, H. H., E. S. DEEVEY, and S. E. GARRETT-JONES

1985. Pollen stratigraphy of two cores from the Petén Lake district, with an appendix of two deep-water cores. In: *Prehistoric Lowland Maya Environment and Subsistence Economy*, edited by M. Pohl. Peabody Museum of Archaeology and Ethnology. Harvard University Press. Cambridge. pp. 73–89.

VAVILOV, Nikolai Ivanovich

1931. Mexico and Central America as the principal centre of origin of cultivated plants of the New World. *Bulletin Applied Botany Genetics and Plant Breeding (Trudy Po Prikladnoi Botanike, Genetike I Selektsii)* (Leningrad), 26: 135–199.

1949–1950. *The Origin, Variation, Immunity and Breeding of Cultivated Plants: Selected Writings.* Translated by K. Starr Chester. (Crónica Botánica, 13: 1–6). Waltham.

VÁZQUEZ DE ESPINOSA, Antonio

(1612–1621) 1948. *Compendio y descripción de las Indias Occidentales.* Smithsonian Miscellaneous Collections. Vol. 108. Smithsonian Institution. Washington, D.C.

VEGA CENTENO SARA-LAFOSSE, Rafael

2007. Construction, labor organization, and feasting during the Late Archaic period in the central Andes. *Journal of Anthropological Archaeology*, 26: 150–171.

VESCELIUS, Gary S.

1981a. Early and/or not-so-early man in Peru: The case of Guitarrero Cave. Part 1. *Quarterly Review of Archaeology*, 2 (1): 11–15.

1981b. Early and/or not-so-early man in Peru: Guitarrero Revisited. *Quarterly Review of Archaeology*, 2 (2): 8–13, 19, 20.

VIERRA, Bradley J., and Richard I. FORD

2006. Early maize agriculture in northern Rio Grande Valley, New Mexico. In: *Histories of Maize: Multidisciplinary Approaches to the Prehistory, Linguistics, Biogeography, Domestication and Evolution of Maize*, edited by John Staller, Robert Tykot, and Bruce Benz. Academic Press, Elsevier. San Diego, London. pp. 497–510.

VIERRA, Robert K.

1981. Big Tambillo Cave, Ac 244. In: *Prehistory of the Ayacucho Basin, Peru. Vol. II. Excavations and Chronology*, Richard S. MacNeish, Ángel García Cook, Luis G. Lumbreras, Robert K. Vierra, and Antoinette Nelken-Terner. University of Michigan Press. Ann Arbor. pp. 133–138.

WAGNER, G. E.

1994. Corn in eastern Woodlands late prehistory. In: *Corn and Culture in the Prehistoric New World*, edited by Sissel Johannessen and Christine A. Hastorf. Westview Press. Boulder. pp. 335–346.

WALLACE, Dwight T.

1962. Cerrillos, an Early Paracas site in Ica, Peru. *American Antiquity*, 27 (3): 303–314.

WANG, Huai, Tina NUSSBAURN-WAGNER, Bailin LI, Qiong ZHAO, Yves VIGOROUX, Marianne FALLER, Kirsten BOMBLIES, Lewis LUKENS, and John F. DOEBLEY

2005. The origin of the naked grains of maize. *Nature*, 436 (7051): 714–719.

WANG, Ron-Lin, Adrian STEC, Jody HEY, Lewis LUKENS, and John F. DOEBLEY

1999. The limits of selection during maize domestication. *Nature*, 398 (6724): 236–239.

WATSON, P. J.

1995. Explaining the transition to agriculture. In: *Last Hunters, First Farmers*, edited by T. D. Price and A. B. Gebauer. SAR Press. Santa Fe. pp. 21–37.

WATSON, S. A., and P. E. RAMSTAD (editors)

1987. *Corn: Chemistry and Technology*. American Association of Cereal Chemists. Saint Paul.

WATT, George

1889–1899. *A Dictionary of the Economic Products of India*. 7 vols. Superintendent of Government Printing, India. Calcutta.

WEATHERWAX, Paul

1918. The evolution of maize. *Bulletin of the Torrey Botanical Club*, 45: 309–342.

1919. The ancestry of maize: A reply to criticism. *Bulletin of the Torrey Botanical Club*, 46: 275–278.

1935. The phylogeny of *Zea mays*. *The American Middland Naturalist*, 16: 1–71.

1936. The origin of the maize plant and maize agriculture in ancient America. *University of New Mexico Bulletin*, 296: 11–18.

1942. The Indian as a corn breeder. *Proceedings of the Indiana Academy of Sciences*, 51: 13–21.

1945. Early contacts of European science with the Indian corn plant. Reprinted from *Proceedings of the Indiana Academy of Sciences*, 54: 169–178.

1950. The history of corn. *The Scientific Monthly*, 71 (1): 50–60.

1954. *Indian Corn in Old America*. Macmillan. New York.

1955. History and origin of corn. I. Early history of corn and theories as to its origin. In: *Corn and Corn Improvement*, edited by G. F. Sprague. Academic Press. New York. pp. 1–16.

WEBSTER, David, David RUE, and Alfred TRAVERSE

2005. Early *Zea* cultivations in Honduras: Implications for the Iltis hypothesis. *Economic Botany*, 59 (2): 101–111.

WEIR, Glendon H., Robert A. BENFER, and John G. JONES

1988. Preceramic to Early Formative subsistence on the Central Coast. In: *Economic Prehistory of the Central Andes*, edited by Elizabeth S. Wing and Jane C. Wheeler. British International Series (BAR) 427. Oxford. pp. 56–94.

WEIR, Glendon H., and Duccio BONAVIA

1985. Coprolitos y dieta del precerámico tardío de la costa peruana. *Bulletin de l'Institut Français d'Études Andines*, XIV (1–2): 85–140.

WELLHAUSEN, E. J., O. Alejandro FUENTES, and Antonio HERNÁNDEZ CORZO, in collaboration with Paul C. MANGELSDORF

1957. *Races of Maize in Central America*. Publication 511. National Academy of Sciences. National Research Council. Washington, D.C.

WELLHAUSEN, E. J., and C. PRYWER

1954. Relationship between chromosome knob number and yield in corn. *Agronomy Journal*, 46: 507–511.

WELLHAUSEN, E. J., L. M. ROBERTS, and X. E. HERNÁNDEZ, in collaboration with Paul C. MANGELSDORF

1951. *Razas de maíz en México*. Oficina de Estudios Especiales, Secretaría de Agricultura y Ganadería. Folleto Técnico 5. Mexico City.

1952. *Races of Maize in Mexico: Their Origin, Characteristics and Distribution*. The Bussey Institution. Harvard University Press. Cambridge.

WHITE, Shawn, and John F. DOEBLEY

1998. Of genes and genomes and the origin of maize. *TIG (Trends in Genetics)*, 14 (8): 327–332.

WHITT, S. R., L. M. WILSON, M. I. TENAILLON, B. S. GAUT, and E. S. BUCKLER

2002. Genetic diversity and selection in the maize starch pathway. *Proceedings of the National Academy of Sciences of the United States of America*, 99 (20): 12959–12962.

WILCOX, George

2004. Measuring grain size and identifying Near Eastern cereal from the Euphrates Valley. *Journal of Archaeological Science*, 31: 145–150.

WILKES, H. Garrison

1967. *Teosinte: The Closest Relative of Maize*. The Bussey Institute, Harvard University. Cambridge.

(Note: This is the doctoral dissertation Wilkes presented in 1966 at Harvard University, Cambridge, Massachusetts.)

1970. Teosinte introgression in the maize of the Nobogame Valley. *Botanical Museum Leaflets*, 22 (9): 297–311.
1972. Maize and its wild relatives. *Science*, 177 (4054): 1071–1077.
1977. Hybridization of maize and teosinte in Mexico and Guatemala and the improvement of maize. *Economic Botany*, 31: 254–293.
1979. Mexico and Central America as a centre for the origin of agriculture and the evolution of maize. *Crop Improvement* (India), 6 (1): 1–18.
1989. Maize: Domestication, racial evolution, and spread. In: *Foraging and Farming: The Evolution of Plant Exploitation*, edited by D. R. Harris and G. C. Hillman. Unwin Hyman. London. pp. 441–455.
2004. Corn, strange and marvellous: But is a definitive origin known? In: *Corn: Origin, History, Technology, and Production*, edited by C. Wayne Smith, Javier Betrán, and E. C. A. Runge. Willey Series in Crop Science. C. Wayne Smith, series editor. Hoboken. pp. 3–63.

WILLEY, Gordon R.
1966. *An Introduction to American Archaeology. Vol. 1. North and Middle America.* Prentice-Hall. Englewood Cliffs.
1971. *An Introduction to American Archaeology. Vol. 2. South America.* Prentice-Hall. Englewood Cliffs.

WILLEY, Gordon R., and John M. CORBETT
1954. *Early Ancón and Early Supe Culture.* Columbia Studies in Archaeology and Ethnology. Vol. III. Columbia University Press. New York.

WILLIAMS, L. R.
1980. Analysis of coprolites recovered from six sites in northern Chile. In: *Prehistoric Trails of Atacama: Archaeology of Northern Chile*, edited by C. Meighan and D. L. True. Monumenta Archaeologica 7. Institute of Archaeology, University of California. Los Angeles. pp. 195–204.

WILLS, W. H.
1988. *Early Prehistoric Agriculture in American Southwest.* School of American Research Press. Santa Fe.
1995. Archaic foraging and the beginning of food production in the American Southwest. In: *Last Hunters – First Farmers*, edited by T. D. Price and A. B. Gebauer. School of American Research. Santa Fe. pp. 215–243.

WILSON, David J.
1981. Of maize and man: A critique of the maritime hypothesis of state origins on the coast of Peru. *American Anthropologist*, 83 (1): 93–120.

WING, Elizabeth S., and Antoinette B. BROWN
1979. *Paleonutrition: Method and Theory in Prehistoric Foodway.* Academic Press. New York.

WISSINGER, A. K., D. H. TIMOTHY, C. S. LEVINGS III, and Major M. GOODMAN
1983. Pattern of mitochondrial DNA variation in indigenous maize races of Latin America. *Genetics*, 104: 365–379.

WISSLER, Clark
1945. Corn and early American civilization. *Natural History*, 54: 56–65.

WITTMACK, Ludwig

1880–1887. Plant and fruits. In: *The Necropolis of Ancon in Peru: A Contribution to Our Knowledge of the Culture and Industries of the Empire of the Incas, Being the Results of the Excavations Made on the Spot by W.(ilhelm) Reiss and A.(lphons) Stübel*, A. Nehring, L. Wittmack, and R. Virchow. Translated by A. H. Keane A.B.A.F.G.S. with the aid of the general administration of the Royal Museum of Berlin. Vol. 3. Cap. 13. A. ASHER & Co. Sole Agents for America Dodd, Mead and Company. New York.

(Note: The pages are unnumbered. The three plates illustrating the work done by Wittmack are 105, 106, and 107; plate 106 shows the maize remains.)

1888. Die Nutzpflanzen der alten Peruanes. *7th Congrès International des Américanistes*. Berlin. pp. 325–348.

WRIGHT, Stephen I., Irie VROH BI, Steve G. SCHROEDER, Masanori YAMASAKI, John F. DOEBLEY, Michael D. McMULLEN, and Brandon S. GAUT

2005. The effects of artificial selection on the maize genome. *Science*, 308 (5726): 1310–1314.

YACOVLEFF, Eugenio, and Fortunato L. HERRERA

1934. El mundo vegetal de los antiguos peruanos. *Revista del Museo Nacional*, III (3): 241–322.

(Note: For maize, see pp. 256–262.)

YOUNG, Arthur

1793. *Travels during the years 1787, 1788 and 1789*. 2 vols. Dublin.

ZÁRATE, Agustín de

(1555) 1968. *Historia del Descubrimiento y Conquista del Perú*. Biblioteca Peruana. Primera Serie. Tomo II. Editores Técnicos Asociados, S.A. Lima. pp. 105–413.

ZARRILLO, Sonia, Deborah M. PEARSALL, J. Scott RAYMOND, Mary Ann TISDALE, and Dugane J. QUON

2008. Directly dated starch residues document early formative maize (*Zea mays* L.) in tropical Ecuador. *Proceedings of the National Academy of Sciences of the United States of America*, 105 (13): 5006–5011.

ZEIDLER, James A.

1991. Maritime exchange in the Early Formative period of coastal Ecuador: Geopolitical origins of uneven development. *Research in Economic Anthropology*, 13: 247–268.

ZEVALLOS MENÉNDEZ, Carlos

1966–1971. *La agricultura en el Formativo Temprano del Ecuador (Cultura Valdivia)*. Talleres Gráficos de la Casa de la Cultura Ecuatoriana. Núcleo del Guayas. Guayaquil.

ZEVALLOS MENÉNDEZ, Carlos, Walton C. GALINAT, Donald W. LATHRAP, Earl R. LENG, Jorge G. MARCOS, and Kathleen M. KLUMPP

1977. The San Pablo corn kernel and its friends. *Science*, 196 (4288): 385–389.

ZIÓLKOWSKI, Mariusz S., Pazdur MIECZYSLAW F., Andrzej KRZANOWSKI, and Adam MICHCZYNSKI

1994. *Andes: Radiocarbon Database for Bolivia, Ecuador and Peru*. Joint Publication Andean Archaeological Mission of the Institute of Archaeology, Warsaw University & Gliwice Radiocarbon Laboratory of the Institute of Physics, Silesian Technical University. Warsaw-Gliwice.

Index